Praise for *Empire Express*

"*Empire Express* is more than a study of the building of a railroad. It encompasses the range of nineteenth-century American life as it swept up Native Americans, women, settlers, con men and speculators in one of man's greatest accomplishments."
—*The Denver Post*

"A big story, authoritatively told . . . thoroughly masterful."
—*The Boston Globe*

"*Empire Express* is one of those books that anyone with any interest in railroad history or the American West must acquire and keep close at hand. It is gargantuan, truly towering, and thanks to David Haward Bain's lengthy and painstaking research it is as complete as this subject can ever be. Bain uses the voices of the transcontinental railroad's builders to tell much of this epic tale. Furthermore, to enliven his narrative, he brings in contemporary events relative to this great American endeavor."
—Dee Brown, author of *Bury My Heart At Wounded Knee* and *Hear That Lonesome Whistle Blow*

"A breathtaking tale enthusiastically told—of vision, greed, adventure, courage, betrayal, accomplishment. . . . *Empire Express* is a spirited telling of a complicated tale."
—*Chicago Sun Times*

"Stunningly researched, prismatically written mix of Robert Caro, David McCullough, Shelby Foote and Connie Bruck."
—*Salon*

"One of the greatest of all American stories has finally found a chronicler up to the task of telling it. David Haward Bain has managed to encompass it all—genuine heroism and brutal dispossession, utopian vision and rampant corruption, technological wonders and war with the elements—in a vivid narrative that no one interested in the American character will want to miss."
—Geoffrey C. Ward, author of *The West, An Illustrated History* and co-author of *The Civil War*

"Monumental history . . . is exhaustively researched, even-handed in judgement and lucidly written. . . . Bain's work will be the definitive account of the transcontinental railroad for many years."
—*The Hartford Courant*

"This is truly a monumental work, equal to the monumental era it portrays."
—*The Florida Times-Union*

"A vast panorama, meticulously researched. Bain never forgets that two strenuously competitive companies were doing the building, one headed east, the other west. Every internal trouble the builders faced—grimly inhospitable terrain, avalanches, Indian battles, keeping track of supplies, and money, always money—was played out against this imperative need to hurry, hurry, hurry. You couldn't even take out time to hate your neighbor, and what a contentious bunch they were, in Bain's definitive telling of the tale."
—David Lavender, author of *The Way to the Western Sea* and *The Great Persuader*

"Displaying energetic research and enthusiasm for the subject matter, Bain brings the linking of the Atlantic and Pacific coasts, and the era that produced it, back to life."
—*Publishers Weekly*

"*Empire Express* is a brilliant work, a stunning fusion of splendid scholarship and graceful writing."
—*Kirkus* (starred review)

PENGUIN BOOKS

EMPIRE EXPRESS

David Haward Bain is the author of three previous works of nonfiction, including *Sitting in Darkness*, which received a Robert F. Kennedy Memorial Book Award. His articles and essays have been published in *Smithsonian*, *American Heritage*, *Kenyon Review*, and *Prairie Schooner*, and he has reviewed for *The Washington Post*, *Los Angeles Times*, *Newsday*, *The Philadelphia Inquirer*, and regularly for *The New York Times Book Review*. Bain teaches at Middlebury College, works with the Bread Loaf Writers' Conference, and lives in Orwell, Vermont, with his wife and two children.

Empire Express

Building the First Transcontinental Railroad

David Haward Bain

PENGUIN BOOKS

PENGUIN BOOKS
Published by the Penguin Group
Penguin Putnam Inc., 375 Hudson Street,
New York, New York 10014, U.S.A.
Penguin Books Ltd, 27 Wrights Lane, London W8 5TZ, England
Penguin Books Australia Ltd, Ringwood, Victoria, Australia
Penguin Books Canada Ltd, 10 Alcorn Avenue,
Toronto, Ontario, Canada M4V 3B2
Penguin Books (N.Z.) Ltd, 182–190 Wairau Road,
Auckland 10, New Zealand

Penguin Books Ltd, Registered Offices:
Harmondsworth, Middlesex, England

First published in the United States of America by Viking Penguin,
a member of Penguin Putnam Inc. 1999
Published in Penguin Books 2000

1 3 5 7 9 10 8 6 4 2

Maps by Northern Cartographic, South Burlington, Vermont

THE LIBRARY OF CONGRESS HAS CATALOGED THE HARDCOVER EDITION AS FOLLOWS:
Bain, David Haward.
Empire Express: building the first transcontinental railroad/David Haward Bain.
p. cm.
Includes index.
ISBN 0-670-80889-X (hc.)
ISBN 0 14 00.8499 1 (pbk.)
1. Railroads—United States—History. 2. West (U.S.)—History.
I. Title.
HE2751.B24 1999
385'.0973—dc21 99-33375

Printed in the United States of America
Set in Photina

For Mimi Aitken Duffy Bain
David Montrose Duffy Bain

"The book ends here, for we are not dealing with Western history. That history exists, one may remember, and its spectacle might be touched upon almost anywhere. Already in 1847 Asa Whitney, the dreamer of railroads, was by no means the figure of cloud-cuckoo land which he had been a year before—precisely as the abolitionists had, in that year, somehow ceased to be madmen. The spectacle of Western history might begin with the railroads, or with the stage-coaches that preceded them, or the pony-express riders—or with tall masts coming into the Bay of San Francisco, taller masts than any seen there before. . . . Or it might begin with spectacle's curiosa: the airship that was to cross to California in three days but somehow didn't, or a nester waking at midnight to see against the copper circle of the Arizona moon the silhouettes of Lieutenant Beale's camels. Or with the wagons that kept on coming year after year till Asa Whitney's dream took flesh, and very little difference between any of them and those we have followed here. Or agony giving a name to Death Valley. Or the mines in the canyons where the Forlorn Hope starved, or the mines anywhere else in the ranges of the West. Or the Long Trail and its herds, its ballads, and its too much advertised gunfire. Or the vigilantes, the Sioux and the Cheyenne rising, the army on the march. Or anything else from an abundance of spectacle."

—BERNARD DE VOTO,
The Year of Decision: 1846

Preface

I have always lived within the sound of a train whistle, whether it was the Pennsylvania (upon whose tracks countless pennies were flattened), the Baltimore and Ohio, the Long Island, the New Haven, Conrail, the Boston–Charles River freight yard, the IND, or the IRT. And for twelve years it's been Amtrak on the Delaware and Hudson tracks, six miles away across rolling farmlands and Lake Champlain. Train stories, train lore, train movies, and train songs chugged through my childhood—of course I had a Lionel set—and as an adult I'd rather take Amtrak than my car or a plane. No contest.

The first book I read on the first transcontinental railroad was one in the Landmark Books series. I was eleven. Little did I know that twenty-five years later, after I had begun writing books about politics and history, that an editor would see one reviewed, read the book, call me up, and issue a challenge: Would I consider writing a book on the first transcontinental railroad? Deep in the tunnels of my brain I heard a whistle calling. While it's perfectly possible that it was only the IND slowing for the Seventh Avenue subway stop in Brooklyn, where I then lived, I like to think it was an echo of old *Jupiter* or *Number 119* at the Golden Spike ceremony at Promontory. Nevertheless, I'm grateful to the challenger, Amanda Vaill—then of Viking and now a respected writer herself. With that challenge she allowed me to rejoin a lifelong love for trains, and a similarly long and deep fascination for the nineteenth century, particularly the Civil War and the old West.

Ellen Levine, my literary agent and dear friend of many adventures, was instrumental in making the dream a possibility. Our adventures were not over, and I trust will not be for many years to come.

At Viking, as the project took shape, I was so fortunate to have the faith and patience of Kathryn Court and Barbara Grossman, friends from my long-ago publishing days. Their encouragement across the years helped make this fourteen-year project what it is. And I'm grateful for all the good work and support of Stephanie Curci, and also Tory Klose, Janet Renard, Beena Kamlani, Gail Belenson, Jaye Zimet, Ivan Held, Paul Slovak, and Giovanni Favretti.

Soon after I began, I realized how many had previously attempted to tell the epic tale of the first transcontinental, and how much they relied on previously published work; a cycle of stories (some of them myths) thus became endlessly recycled and repeated. Aside from two very good biographies of Collis P. Huntington (by David Lavender) and Grenville M. Dodge (by Stanley Hirshson), there had been three books published in the midsixties, right before the Golden Spike centennial. But that was of no matter, for beckoning me were all the original sources, the handwritten letters, journals, business records, telegraph forms, official reports, and eyewitness journalism. Particularly helpful were the elaborately detailed inventories at the University of Iowa and the Nebraska State Historical Society (major Union Pacific repositories), at the University of California, Berkeley (the Central Pacific, where H. H. Bancroft's staff tirelessly followed his collection fever), at the California State Railroad Museum, and at the California State Library, where I also found, and copied with a blunt pencil stub, score upon score of handwritten subject cards for California newspapers. I also count myself lucky to have done two years of research at the extraordinary New York Public Library while the peerless card catalogues were still in existence.

As I forged ahead, two other books caught up with me in the late eighties. One, the authoritative official corporate history of the Union Pacific by Maury Klein, might have slowed my own research for a year or two as he had priority access to Union Pacific microfilm records, but it was worth the wait. His telling of one side of the transcontinental story (and the years beyond, to the end of the century) was well done, and his superior business sense helped me wade through high corporate financial records and dealings. My debt to him is real.

As I plowed through the mostly handwritten material, one particular collection became supremely important: the contents of Collis P. Huntington's New York office of the Central Pacific, with its voluminous letters from all the principals. Given to Syracuse University, and later microfilmed and widely distributed to libraries around the country, I found that earlier researchers had barely touched the surface of its thousands of handwritten pages, and perhaps to no wonder. The handwriting of the five principals often defeated a cursory reading, complicated by the fact that most of the six-, eight-, or ten-page letters had been microfilmed out of page order. I became familiar with their scrawls and pieced together the letters. What I found there proved that previous writers had been mostly defeated at getting into the collection—refutations of myths passed down for generations, and exciting, extraordinary voices unheard for 130 years, quotations any historian or journalist would behave shamelessly to obtain. The Union Pacific side yielded many new surprises, too; by going even to a much-quoted document and reading the original, one often finds quotations left out, inconvenient facts glossed over, and puzzle pieces useful somewhere else.

All along the way I was conscious of what I felt were the subject's marching orders: Put the story into a larger national context and take advantage of the exciting research done in recent decades on Native Americans, women of the plains and high country, immigrants, and other people below the radar scope of traditional historians' "great men" narratives. Placing these perspectives within the context of the transcontinental race, and the railroads' story into focus with larger national political and cultural events, was an important part of the mission.

I am most grateful to the following for helping me with research: Donald Snoddy, William Kratville, Deirdre Routt, Union Pacific Railroad Library; Annagret Ogden, Mary L. Morganti, Bonnie Hardwick, The Bancroft Library, University of California, Berkeley; Emily Levine, Chad Wall, James A. Hanson, Nebraska State Historical Society; Robert A. McCown, University of Iowa Library; Richard Terry, Kenneth Pettitt, California State Library; Carol A. Rudisell, Michael T. Ryan, Robin Chandler, Roberto G. Trujillo, Special Collections, Stanford University Libraries; Jennifer King, Lori Olson, Maxine Trost, Emmett Chisum, American Heritage Center, University of Wyoming; Mildred K. Smock, Council Bluffs Public Library; Peter Blodgett, Harriet McLoone, Daniel Woodward, Huntington Library; Christie Brandau, Ellen Sulser, Matthew Schaefer, State Historical Society and Library of Iowa; Anthony R. Crawford, Evan Williams, Kansas State University Library; Rick Stattler, Rhode Island Historical Society; Karon Tomlinson, Lake County Historical Society, Mentor, Ohio; Mae Bolton, Sacramento Public Library; Shirley Sun, Chinese Culture Foundation of San Francisco; Rene Morales, Gladys Hansen, San Francisco Public Library; Raymond Hillman, Stockton Pioneer Museum; Frank Gibson, Omaha Public Library; Gary K. Roberts, Nevada Historical Society; Deanna LaBonge, Oscar W. Ford, Joyce C. Lee, Millie L. Syring, Nevada State Library; Giaconda Capitolo, Salt Lake City Public Library; Linda Thatcher, Utah State Historical Society Library; Everitt Cooley, University of Utah Library; Gwen Rice, Wyoming State Library; Eric Leuschner, Oklahoma State University Library; Saundra Taylor, Indiana University Library; Margaret N. Haines, Oregon Historical Society; Hilary Cummings, University of Oregon Library; Thomas Heenan, Omaha Public Library; Benjamin Trask, The Mariners' Museum Library; Diane Skvarla, U.S. Senate Curator's Office; Carol A. Turley, University of Southern California Library. Staff too numerous to mention at the National Archives, the Library of Congress, the New York Public Library, the Mid-Manhattan Library, and the Brooklyn Public Library helped immeasurably in the early years of this project. Surely I have neglected to list other names, for which I apologize.

At Middlebury College, where I have taught since 1987, I was fortunate to have the excellent facilities of the Egbert Starr Library, not only because it is a federal repository but also because of its Interlibrary Loan department, which I used to the utmost. Deep thanks, then, to the indefatigable Fleur Laslocky, the tireless Lexa deCourval, and many student workers. Elin Waagen and Hans Raum helped many times. A number of college administrators aided my project, whether through verbal encouragement, adjusting my schedule, or support through the college's faculty development fund (which, late in this project, dispensed several research, travel, or materials grants). Thanks, then, to Bethany Ladimer, Alison Byerly, Robert Schine, John McCardell, Ron Liebowitz, Eric Davis, John McWilliams, Theodora Anastaplo, Susan Perkins, Susan Coburn, David Price, and Edward Martin. Carol Knauss, Stanley Bates, Devon Jersild, and Michael Collier of the Bread Loaf Writers' Conference were supportive over many years. Many thanks, also, to the alert Cates Baldridge.

I am grateful to Northern Cartographic, of South Burlington, Vermont, to Eileen Powers, Edward Antczak, and Robert Gagliuso, for the breathtaking GIS-based, terrain-shaded maps. Collectors of old books may note that I was inspired by the inkwork of one H. Scott, whose fine maps appeared in James McCague's *Moguls and Iron Men* (New York: Harper & Row, 1964), but I sought to provide more realistic depictions. Thanks to my Middlebury colleague Stephen Trombulak, I discovered Northern Cartographic.

This has been a long project. Expressions of encouragement from some people over the years, sometimes brief, undoubtedly lessened the toil, as did their example. Thanks to Jane Garrett, Charles Elliott, William H. Goetzmann, Justin Kaplan, Arthur M. Schlesinger, Jr., Geoffrey Ward, Richard M. Ketchum, Alvin M. Josephy, Jr., Shelby Foote, David McCullough, Richard Kluger, Martha Sanger, Shere Hite, Ron Powers, Paul Mariani, Robert Pack, William Kittredge, Terry Tempest Williams, Whitney Balliett, Tim O'Brien, Michael Arlen, Naomi Bliven, Mally and Jim Cox-Chapman, Don Mitchell, Alan Weisman, Jim Dumont and Karen Lueders, Brett Millier and Karl Lindholm, Nancy Rome and Allen Moore, and to all my supportive friends. Julia Alvarez—a sister in art if there ever was one—never failed to buoy me up. Thanks to her and Bill Eichner.

Jim and Trish Smyth, Sarah Duffy, Sheila Cohen, Bob Reiss, and Stacey Chase gave shelter during particular research trips, for which I was and am grateful.

Rosemary Haward Bain, John L. Duffy, and Katherine S. Duffy were my most faithful and critically helpful readers; they know that my gratitude for this, and a lifetime of gifts, buries the needle on the Richter scale. My father, David Bain, died eight months before this project began, but his enthusiasm for stories and personalities in history informed *Empire Express.* Christopher Bain, Terry Bain, and Lisa Bain never failed in their support, nor did their respective spouses, Andrea Bain, Marc Santiago, and William Schwarz.

Finally, it is dificult to put into words the gratitude I owe to my family for their faith and support, and for, too often, giving up comforts and companionship. Mary Smyth Duffy, my unfailing and unflagging partner, and Mimi Bain and David Bain, my wonderful children, endured much so that this book might be completed. Having finally done that, and being freed to the extent that one is freed from any consuming project, I hope I can begin to make it up to them.

July 4, 1999
Orwell, Vermont

Contents

Part I
1845–57: A Procession of Dreamers

1: "For All the Human Family" | 3
2: "Who Can Oppose Such a Work?" | 16
3: "I Must Walk Toward Oregon" | 26
4: "The Great Object for Which We Were Created" | 37
5: "An Uninhabited and Dreary Waste" | 47

Part II
1860–61: Union, Disunion, Incorporation

6: "Raise the Money and I Will Build Your Road" | 57
7: "There Comes Crazy Judah" | 67
8: "The Marks Left by the Donner Party" | 78
9: "The Most Difficult Country Ever Conceived" | 85
10: "We Have Drawn the Elephant" | 104

Part III
1863: Last of the Dreamers

11: "Speculation Is as Fatal to It as Secession" | 122
12: "I Have Had a Big Row and Fight" | 131

Part IV
1864: Struggle for Momentum

13: "First Dictator of the Railroad World" | 151
14: "Dancing with a Whirlwind" | 165
15: "Trustees of the Bounty of Congress" | 181

Part V

1865: The Losses Mount

16: "The Great Cloud Darkening the Land" | 205
17: "If We Can Save Our Scalps" | 219
18: "I Hardly Expect to Live to See It Completed" | 234

Part VI

1866: Eyeing the Main Chance

19: "Vexation, Trouble, and Continual Hindrance" | 252
20: "The Napoleon of Railways" | 261
21: "We Swarmed the Mountains with Men" | 281
22: "Until They Are Severely Punished" | 301

Part VII

1867: Hell on Wheels

23: "Nitroglycerine Tells" | 315
24: "Our Future Power and Influence" | 330
25: "They All Died in Their Boots" | 360
26: "There Are Only Five of Us" | 391

Part VIII

1868: Going for Broke

27: "More Hungry Men in Congress" | 435
28: "Bring On Your Eight Thousand Men" | 469
29: "We Are in a Terrible Sweat" | 506
30: "A Man for Breakfast Every Morning" | 550

Part IX

1869: Battleground and Meeting Ground

31: "A Resistless Power" | 594
32: "We Have Got Done Praying" | 645

Part X

1872–73: Scandals, Scapegoats, and Dodgers

Epilogue: "Trial of the Innocents" | 675
Notes | 713
Bibliography | 759
Index | 779

List of Maps

Page
25 Asa Whitney versus Senator Benton, 1845–50
Including Dates of Territorial Organization or Admittance to the Union
121 Central Pacific Survey, 1863
251 Union Pacific Railroad, 1866
257 Central Pacific Drives for the Summit, 1866
325 Union Pacific Railroad, January 1867
449 Central Pacific Route Across Northern Nevada, 1868
477 Union Pacific Construction Across Wyoming, 1868
593 Battleground in Utah, 1869

Part I

1845–57
A Procession of Dreamers

1

"For All the Human Family"

Nine weeks out of New York and bound for Macao, the leaky and overburdened merchant barque *Oscar* struggled to round the Cape of Good Hope and was becalmed. Its captain cursed and swore at their slackened sails and abused his crew, while the vessel's sole passenger, a pious and sensitive man, tried to ignore a tirade made worse because it was the Lord's Day.

As was common on Sundays, he tried to pass the time in prayer and meditation. But he found himself brooding over his lot in life. It was at such a time that Asa Whitney, staring over the ship's rail at the bright green sea and brooding, most strikingly resembled Bonaparte. The similarity in physiognomy had caused him no end of trouble during business trips to France, when strangers would stop and annoy him on the street. Nonetheless his admiration for Napoleon knew no bounds; his empathy over losing one's entire world had never been stronger than on this poky, Asia-bound barque—there was nothing but loss behind him, little promise ahead. By forty-five years of age Whitney had buried two wives and a child and lost all his considerable worldly possessions, and now he had started life anew. Even if we could have told him that in a few years he would not only have regained his wealth, and discovered a cause worth making his life's work, but also would come to be considered a prophet of a new age, it is unlikely that such a prediction would have allayed his bitterness. Still, he struggled to retain his perspective. "It certainly is a great tryal at my time of life," he had written in his diary, "to recommence the work, too in a strange foreign Land. Yet I hope it is Gods providence that guides me & I feel that I shall succeed. I hope above all things that I may yet be enabled to do some good to mankind & in some small degree make amends for the abuse of all Gods providences to me."[1]

Within days of doubling the cape, Asa Whitney was afflicted with boils.

He was the eldest child born, on March 14, 1797, to Sarah Mitchell and Shubael Whitney, near Lantern Hill in North Groton, Connecticut. For five generations in New England, Whitneys had engaged in farming or manufacturing. (Asa's fifth cousin, when not occupied by patent suits over his invention of the cotton gin, fabricated arms in New Haven.) Shubael, the son of an iron manufacturer, chose to coax corn from the rock-choked soil east of the Thames, raising nine children to help him in this effort. Meanwhile, he and his neighbors hired Indians from the Pequot reservation to hoe, paying them with rum. Apparently this arrangement pleased everybody, but as Asa grew into his teens he showed no interest in agriculture. He was likewise uninterested in going to sea on the countless whalers and sealers operating from the Connecticut coast. Before he was twenty he was in New York, engaged in that other great Yankee occupation, trade.

Beginning as a clerk with one of the city's largest importers of French goods, Whitney was promoted, and spent most of the decade between 1824 and 1834 in Europe as a purchasing agent. By 1832 he was a well-rewarded merchant who was about to be made a partner in the firm; that year he also acquired a wife abroad. Little is known about Herminie Antoinette Pillet Whitney except that she was French and that she died in New York City shortly after their marriage, on March 31, 1833, and that she was buried in the Trinity Churchyard in New Rochelle, close to where Asa had purchased seventy acres of land for their new home.

Whitney kept a lock of her hair for some years thereafter, but it was not long before he married again. Sarah Jay Munro was the daughter of a wealthy landowner and grandniece of the former chief justice and governor of New York, John Jay. The early years of the Whitneys' marriage were comfortable ones, for there was a great demand for French goods and Asa prospered. In 1835 he purchased two more tracts of New Rochelle land, and in 1836 he not only left his firm to begin a new importing partnership but bought a large commercial plot in Lower Manhattan, upon which he erected five wholesalers' buildings. Soon he had begun to build an imposing brick house in the Greek Revival style, in New Rochelle, completed in 1837.

It was around this time that Asa Whitney's upward-tending graph became a downward spiral, for as the Whitneys settled into their new home and the merchant saw to his expanded commercial interests, banks were closing, businesses were failing, crowds were rioting in the streets of New York and breaking into food warehouses—the beginning of that tribute to rotten banking and frenetic speculation, the seven-year misery known as the Panic of 1837—not a propitious time to find oneself overextended. Although he was not immediately affected, Whitney found it increasingly difficult to get by. His import business naturally required hard capital; moreover, he owed some $80,000 and interest on his Manhattan real estate, though tenants' rent came nowhere near his mortgage. Like a juggler whose arms grow weak from effort despite his skill, Whitney refinanced his commercial mortgage in March 1838 and took out a loan on the New Rochelle property in the December following but was still unable to make timely payments. In September 1840, he was faced with foreclosure.

An even greater tragedy struck while the merchant's case was before the magistrates. On November 12, 1840, Sarah Whitney died—"after a few days illness," as her obituary notices reported—which, one may surmise from family papers, occurred either after a miscarriage or following an unsuccessful childbirth.

Asa Whitney buried her beside his first wife and in his grief turned to face the courts. After foreclosure, his New York property went to auction, was bought by his mortgage holder for $80,000 (the amount of principal owed), and left Whitney holding a bill for over $10,000 in unpaid interest. He sold his house and remaining land, beginning to be, certainly, in an antipodal frame of mind, with but one word in his head: China, a place of dawning commercial promise where one could start anew.

Thus, as the *Oscar* was towed from the Pike Slip Wharf out to Sandy Hook and cast loose on Saturday, June 18, 1842, its heavily disappointed passenger could not be blamed for keeping his sight on the horizon, not back toward home. Whitney had put pride away, secured himself an appointment as purchasing agent for several firms trading in the Orient, and hoped to do a little business of his own on the side. For the long voyage he had packed a trunk full of books—including George Tradescant Lay's new guide, *The Chinese As They Are;* the life and writings of John Jay; a biography of Napoleon; a French grammar; the good Reverend William Wilberforce's *Family Prayers*—and several cases of wines.

But such would provide only limited diversion and small solace on a voyage whose misery would become memorable in that closing era of snail-paced, square-bowed, wooden-hulled sailing vessels, for the *Oscar* was loaded down like a coal barge—worse, even, for its cargo consisted mostly of lead ingots—and it was afflicted by the most adverse weather. Gales, calms, rough seas, and contrary winds followed the barque across the Atlantic in dreary procession. Whitney, a seasoned voyager, suffered from seasickness for the first time in his life, to which was added sleeplessness, rheumatism, his plague of boils, and growing dismay at the behavior of his only social companion, Captain Eyre.

The captain filled their quarters with cigar smoke. ("I cannot in anyway escape it," Whitney confided to his diary. "In consequence I have much of the head ache. What a vile practice, so useless, yes worse, so injurious to health & habits, for I have always found it creates a disposition to drink, if not to drunkenness, & so disagreeable to those who dislike it: that I sometimes think no real

Gentleman can smoak.") The captain was prone to tearing, profane rages aided by the seaman's astonishing vocabulary. ("Very disagreeable, presumptuous & wicked.") The captain seemed to take satisfaction from flogging transgressors in his crew, particularly the Chinese steward who appeared to be drunk one breakfast. "I did not see it & could not & I cannot bring my mind to believe in the necessity of such a discipline anywhere," Whitney wrote. "It is too humiliating, too degrading, too beastly, poor fellow I do feel for him. . . . these poor Chinese seem to be considered but dogs only fit to be kicked and flogged; this our Americans have learned from the English." Besides, Whitney added in afterthought, a steward punished thus could wreak revenge by poisoning all who dined in the captain's cabin.[2]

Still, being an affable man, Whitney took comfort from his books, from good weather, from sightings of other ships and of various inhabitants of the deep. "Thus far on our long voyage," he admitted with relief, "we are without an accident & all in good health."[3]

From New York to the Cape of Good Hope the sailing distance is eight thousand miles; from the cape to Dutch Anjier (Java Head), the gate between the Indian Ocean and the China Sea, another six thousand miles; from Anjier to the Portuguese colony of Macao on the Chinese coast, some two thousand more. When Asa Whitney was but a boy, the voyage between New York and China normally took six months, with runs of 125 days considered good, and by the 1830s this had been shortened to 100 days or less; in the year Whitney sailed, fast new China packets made the trip as short as 79 days. On the drawing tables of naval architects were plans for even speedier clippers, for a new era dawned.[4]

Nevertheless, a leaky relic with a tyrant for a master and a hopeless case for a passenger, the barque *Oscar* plodded eastward across the Indian Ocean, making land at Java for provisions after all of 107 days at sea. In port, Whitney could not bear to watch the Dutch subjugation of Javanese, and it moved him to philosophize. "Oh how long must the mighty oppress & brutalize the weaker," he wrote. "When I see human beings in such oppressive ignorance & servitude, I cannot but feel that they were created for a more noble & exalted purpose & that the purposes of a wise Creator are turned by the ambition & lust of Man or preparation of Nations perhaps for their eternal destruction, look at Spain, look at Portugal, & even Look at England too her time is allmost come. Her starving millions will not be willing to starve much longer, her wailing day must come & awful must be that day."[5]

He was relieved when the *Oscar* weighed anchor and proceeded northward into the torrid Java and China Seas, "full of fish and snakes," and "little wind & excessive heat." Passing Borneo, Palawan, and Luzon, evading reefs and Malay pirates, sustaining some damage and more delays due to typhoons, the *Oscar* cast anchor at Macao. It was Sunday, November 20, 1842. They had been at sea for 153 days—perhaps a record for slowness that year.[6]

Asa Whitney's business in China was to last for a year and four months. He arrived amid that Sino-British dispute recalled as the Opium Wars—a dispute characterized for three years by Chinese riots against the barbarians who had insisted on their imperial right to free trafficking in all commodities, especially opium, and by retributory British naval attacks upon heathen ports. A treaty had been signed in Nanking three months earlier, in August. A typically lopsided document it was—granting the British the island of Hong Kong, a cash indemnity, access to five ports, and license to profitably addict as many Chinese as they could manage. Whitney had little sympathy for the British and their imperial ways. ("Oh England," he wrote in one typical diary entry, "thine arrogance cannot be endured & thy pride must have a fall.") His sentiments were hardly improved when, immediately after he arrived in Canton, angry mobs plundered and torched some British businesses and cornered many Westerners (including Whitney and a group of fellow Americans) in their establishments. A tense night passed as the merchants could do little but peer out at the massed Chinese and at the firestorm raging toward their factory, but in the morning the Americans (and British posing as Americans, an irony not lost on Whitney) were allowed to evacuate.[7]

Affairs in China would settle down. As other foreign nations began to press for similar commercial access, Whitney found himself among a select few Americans arranging exportation of teas, spices, and other Chinese goods. There is no evidence that he trafficked in opium, as did many others; that would seem to have gone against his grain. He was a good businessman, though, dividing his time between Canton and Macao, and his profits mounted. Indeed, on April 2, 1844, when he rejoined Captain Eyre on the deck of the *Oscar* for its return voyage, he had assured himself of enough money to make further labors unnecessary for the rest of his life.

Any sort of idleness was not in his nature, however. Sometime during the grief-ridden year when Whitney had lost his family, home, business, and wealth, he had sworn to devote the remainder of his life to a higher purpose. "I hope above all things," he had confided to his diary, "that I may yet be enabled to do some good to mankind & in some small degree make amends for the abuse of all Gods providences to me."[8]

His return trip was tediously long, marred further by his cabin-mate's "segars" and rages and fondness for flogging seamen, but it gave Whitney all the time he needed to consider an idea that had been growing inside him for some time, perhaps encouraged by events at home reported in months-old newspapers. What began to take shape was a plan he thought would consign such long and uncomfortable voyages to history, put an end to the sort of un-Christian, colonialist abuses he had witnessed in the Orient, and place his little nation on a more equal footing with the great powers.

Perhaps fittingly for such world-shaking aspirations, the *Oscar* put in for a few days at the island of St. Helena, where Whitney was outraged to discover that the British had allowed the quarters of his departed, "misunderstood," illustrious doppelganger to be used as a stable. It is not recorded whether Bonaparte's living double excited any comment on St. Helena when he strode about the island, probably muttering under his breath at "English pride English Tyranny & oppression" as he committed those fulminations to his diary—"the settling day must, will come," he added, "& awful must be that day."[9] Staring from the heights beyond Jamestown to the sea, he thought that "the imagination almost pictures a Napoleon on every ridge, on every peak, a kind of awful supernatural sensation . . . different from any thing before experienced, like the child in the dark expecting any moment to meet a Spectre." But any ghosts Whitney may have encountered belonged, instead, to his own time and his own world—a world to which he was returning with a steadily developing agenda.[10]

Five months and nine days after departing Canton, Whitney stepped ashore at New York with joy and purpose. He tarried in the city for some weeks—probably disposing of a shipful of Chinese imports and counting his money—before moving onward to another Canton, some fifteen miles from the St. Lawrence River in upstate New York. It was there, as winter began its descent, that Asa Whitney put away his diary in favor of another document. It was a memorial addressed to the United States Congress. As he began working on it, his travel-stained little account book, consigned to history in his trunk, offered a hint and a caution about "the great purpose" to which he would devote himself. He had taken a steamboat to Albany, his diary recorded, from where he boarded a westward-bound railroad. "I was anxious to see the towns & villages through which we passed," he continued,

> *but, alas, in vain, time & space are annihilated by steam, we pass through a City a town, yea a country, like an arrow from Jupiters Bow. Schenectady, I can only say I passed through it because it is on the rout. . . . At Utica we stopped to dine, had only time to pass from the Cars to the Hotel & dined on the high pressure plan, they told me it was Utica but I have no memorial, I know nothing of it. . . .*
>
> *Oh, this constant locomotion, my body & everything in motion, Steam Boats, Cars, & hotels all cramed & crowded full the whole population seems in motion & in fact as I pass along with Lightning speed & cast my eye on the distant objects, they all seem in a whirl nothing appearing permanent even the trees are waltzing, the mind too goes with all this, it speculates, theorizes, & measures all things by locomotive speed, where will it end.*[11]

"Can it be happy," that diary entry had concluded in late 1844, "I fear not." Fatigued and out of sorts the merchant might have been, and not in step with the American pace after two years abroad. But if Whitney was truly fearful of an unhappy end he showed no other evidence of it—only industry and the most intense single-mindedness—in setting forth to harness the very contrivance that had set his head to spinning, in a plan he hoped would at once bring the world down to manageable size and make it a better place to inhabit. Asa Whitney, with no previous experience and having nothing but his faith and self-assurance to tell him he was not pursuing a chimera, began to outline how he would get a railroad across the vast, uninhabited middle of the American continent to the Pacific shores, where the lure of Asia beckoned, within reach. He would annihilate distance, yes—and with it, ignorance, want, and barbarism—through the ineffably promising devices of American trade and American Christianity.

Whitney's attention was first called to the importance of railroads as a means for the transportation of commerce as well as of passengers as early as 1830, he recalled later. It was only a year after British crowds had beheld the world's first steam-powered locomotive, George Stephenson's *Rocket,* draw a train of cars faster than a horse could haul a carriage. The *Rocket* trials in 1829 attained a top speed of twenty-nine miles an hour; a year later, when Asa Whitney paused during a buying trip to ride the newly formed Liverpool and Manchester Railroad, the locomotive sped them over a distance of thirty-four miles of solid English roadbed in forty-two minutes—a little over forty-eight miles an hour, he thought, though he may have been exaggerating. The merchant saw clearly, he said later, "their present importance and predicted their future importance to us as a means of communication with the Pacific." During his sojourn in the Orient, as the British secured commercial rights in China and it seemed that America would soon follow (as it did in July 1844), he foresaw "the importance to us if we could have a more ready, frequent, and cheap communication than the present long and dangerous voyage around either of the capes."[12]

In China Whitney had gathered much commercial information on "that ancient, numerous, and most extraordinary people," he would write. The principal object of inquiry was how to increase Sino-American commerce, for Whitney had chafed at the time and expense involved and at how limited the return was in comparison to the "almost boundless" possibilities. He also considered "the vast commerce of all India, of all Asia, which has been the source and foundation of all commerce from the earliest ages to the present day, possessed and controlled by one nation after the other, each fattening upon its golden crop, till proud England at last holds it in her iron grasp." This did not have to continue, Whitney noted. "She holds on, and will hold on until our turn comes, which will be different, and produce different results from all. We do not seek conquest, or desire to subjugate. Ours is and will be a commerce of reciprocity—an exchange of commodities."[13]

Whitney had much more in mind than mercantile matters—his plan fairly shone with global promise. His argument would grow fervent, a near-religious preoccupation for him as time passed, for the Pacific railroad idea, he would write, would not merely hold benefits for its projectors but for every American and a multitude beyond:

> *for the destitute overpopulation of Europe, without food and without homes—for the heathen, the barbarian, and the savage, on whom the blessings and lights of civilization and Christianity have never shone—for the Chinese, who, for want of food, must destroy their offspring—for the aged and infirm, who deliberately go out and die, because custom, education, and duty, will not permit them to consume the food required to sustain the more youthful, vigorous, and useful—and for all the human family.*[14]

For a merchant with no engineering ability, no political contacts, no experience in mounting any campaigns, especially of such national scope, Whitney had embarked on a project that seemed ambitious, quixotic, chimerical. However, in his absence from the United States—even before, when Whitney's whole energy had been directed at salvaging his business from creditors—the nation had begun a monumental transformation. "The mind too goes with all this," he had written, addressing not only his project but also the strange new American pace to which he had returned, "it speculates, theorizes, & measures all things by locomotive speed."

As he set forth to make his congressional memorial for a railroad to the Pacific as comprehensive as possible, a nation stirred.

It was a nation which, in 1844, some mossbacks believed had grown as far as nature and man's treaties would allow, and beyond which lay a dangerous overextension that threatened dissolution of the Union itself.

The stage of North America: Thirteen free and thirteen slave states extended westward from the Atlantic seaboard to the Missouri River—the sum of the United States. The two free territories of Iowa and Wisconsin waited in the wings for admittance, as did Florida. Another great chunk of the continent—Mexico, her medieval promise long faded—stretched improbably from the Gulf of Tehuantepec and Guatemala away to the Oregon border, all scattered, desultory rancheros and huddled mission settlements. There were the disputatious Oregon and Texas. The former extended from the Pacific to the Rocky Mountains and from Mexico to well north of Vancouver Island, being sparsely settled and occupied jointly by Britain and America; the latter had been for some eight years the independent Republic of Texas. If both were understudies, their parts awaited them.

The United States, British Canada, Oregon Country, Mexico, the Texas Republic—all encircled a vast and mysterious land, the subject of much speculation and not much careful thought. Call it Indian Territory for now, for it contained survivors of the displaced, decimated eastern tribes and the great unmolested, unsuspecting Plains Indians. As limitations to American growth, man's treaties had already proved to be the expedient instruments that they were intended to be by the enforcing party. But nature, the other great limitation, was not as malleable to national destiny, or so at least it seemed in 1844 as America stood on the eastern bank of the Missouri River and looked across to a hallucination known for thirty years as the Great American Desert.

Thomas Jefferson, who knew much, was ignorant about most of the territory he purchased unseen in 1803 at a bargain-basement price. The few settlements of the Louisiana Territory, he reported to Congress, "were separated from each other by immense and trackless deserts." Three years after this hearsay, William Clark and Meriwether Lewis returned from their examination of the country; they had found the plains to be simply dry and barren, though not desertlike. In that year, however, young Lieutenant Zeb Pike traveled the Far West and returned with the most fanciful impressions. "This area in time might become as celebrated as the African deserts," he wrote of the territory sitting between the meridian of the great bend of the Missouri and the Rockies. "In various places [there were] tracts of many leagues, where the wind had thrown up the sand in all the fanciful forms of the ocean's rolling wave, and on which not a spear of vegetable matter existed." Pike's visions of sand dunes, pathless wastes, and sterile soils were reported, widely read, and faithfully believed by geographers. The myth became innocently embellished by subsequent visitors, especially those in the party of Major Stephen H. Long, who traversed the whole area in 1820. It was reported to be "an unfit residence for any but a nomad population . . . forever [to] remain the unmolested haunt of the native hunter, the bison, and the jackall."[15]

Twenty-four years later the Santa Fe trader Josiah Gregg issued his *Commerce on the Prairies,* a book based on extensive experience on the plains. "These steppes," he wrote, "seem only fitted for the haunts of the mustang, the buffalo, the antelope, and their migratory lord, the Prairie Indian." Soon young Francis Parkman would see sand dunes along the Platte River, in his imagination extending this "bare, trackless waste" for hundreds of miles. Thus the future states of Kansas, Nebraska, Oklahoma, the Dakotas, Montana, Wyoming, and Colorado existed in the American minds of 1844 in hopelessness and sterility—fitting continental leavings for the aborigines.[16]

Little as the Great American Desert interested politicians and pioneers alike, temptations lay on its western and southern frontiers. As Asa Whitney composed his Pacific railroad memorial in the closing months of 1844, the upstate New York countryside rumbled with political activity, as was true all over the nation, with much attention being paid to the issue of expansion. Six months earlier in Baltimore the convened national Democratic Party plodded through seven deadlocked ballots before finally rejecting its obvious choice, Martin Van Buren. As former senator, governor, secretary of state, vice president, and president, Van Buren had, by 1844, served his country perhaps too well, but his failure this time around had less to do with his shopworn self than with his disinclination to invite war by annexing new territory—a position that was then at distinct variance with prevailing sentiments. Two canvasses later the Democrats acclaimed a dark horse, James Knox Polk. He had twice failed to be re-elected governor of Tennessee, but when he appeared in Baltimore, his proprietary urges toward hitherto disputed lands plainly in sight, Polk prevailed. In the ensuing presidential contest the opposing Whig party could muster little more than the slogan "Who is James K. Polk?" for their own candidate, Henry Clay, who was otherwise silent on the great issue of the day. That issue lay at the heart of the Democratic platform, but more important, it had already been accepted as a fait accompli by most Americans: the annexation of those title-clouded expanses known as Texas and Oregon—Mexico and England be damned.

Exactly a decade had passed since our neighbor to the south had reopened its Texas lands to American immigration after some years of nervous border restriction. Likewise, it had been ten years since the first Methodist missionaries had drifted to the bank of the Willamette River in Oregon Country, seeking Flatheads with a hankering for the Good Book (there were none). The latter territory had an agreeable climate and an excess of lush farmlands, and though it was jointly occupied with Britain there were relatively few British.

Oregon's emptiness beckoned. So did the equally virginal lands of Texas. By 1836, the number of American settlers in Texas had grown to nearly thirty thousand—ten times the resident Mexican population and more than enough to enforce a nascent Republic of Texas only weeks after the tragedies at the Alamo and at Goliad. President Andrew Jackson, in formally recognizing Texas sovereignty in March 1837, had less influence on encouraging further settlement than did the other great event of that season, which overshadowed it. The Panic of 1837 sent thousands of bankrupt and debt-ridden farmers of the Mississippi River valley flooding into Texas to join those who had preceded them. Others, their hopes dashed no less by the deepening depression, began to weigh the odds of the longer, more hazardous route to Oregon, across the Great American Desert. By 1839 some five hundred Americans had sunk their plow blades in the Willamette bottomland and a new destination had entered the dreams of would-be migrants: Mexican California.

Texas fever! Oregon fever! California fever! Rare was the American newspaper or magazine that did not carry a rhapsodic letter from a newly arrived settler in those and subsequent years. Farmers seemed to be spending as much time urging their fellow Americans to join them in paradise as they did in raising crops—that is, when they were not deluging Washington with petitions urging annexation.

If for many the lure of a new purchase on life was balanced by the numerous threats to life during the overland journey, news from those who had survived the ordeal was persuasive. Especially so were reports of the Bidwell-Bartleson party, which in the summer of 1841 followed the West's lure from Missouri, eventually splitting into two groups which attained Oregon and California after much hardship. Then, in 1843, young Lieutenant John Charles Frémont issued a report on his army exploration of the Oregon Trail from the Mississippi River through the South Pass and into the Wind River range of the Rockies. Published obligingly by the government, the path-follower's book was an instant success, with its descriptions and maps both a Bible and a Baedeker for thousands of potential migrants. And when the Democrats rallied behind James Knox Polk, with Texas and Oregon (and—who knows—California) at the forefront of their minds, the expansionist party prevailed, albeit narrowly, in the electoral college. Those faraway settlements seemed at once closer and more alluring. Meanwhile, an obscure merchant, recently returned from China, signed his name to a document which was handed to an upstate New York legislator, the Honorable Zadock Pratt of Prattsville, who packed it away for his trip to Washington and the second session of the Twenty-eighth Congress.

The subject of railroads seemed remote in the opening weeks of the congressional session. Only the prospect of admitting the Republic of Texas to the Union held any interest. But three days after the House passed a joint annexation resolution and a month before it joined the Senate in approving an amended measure, Zadock Pratt rose in the chamber. The title of the document he presented for consideration was *Railroad From Lake Michigan to the Pacific: Memorial of Asa Whitney, of New York City, relative to The construction of a railroad from lake Michigan to the Pacific ocean.*

All of the states east and north of the Potomac River, Whitney had written, were or soon would be connected with the waters of the Great Lakes by rivers, railroads, and canals. At that moment a chain of railroads was projected—in some places, already under construction—along the 840-mile route between New York and the southern shores of Lake Michigan. It was entirely practicable to extend the railroad from there across the unsettled lands of the West, through the Rocky Mountains, to the Pacific Ocean, some 2,160 miles. "To the interior of our vast and widely-spread country," he said, "it would be as the heart to the human body; it would, when all completed, cross all the mighty rivers and streams which wend their way to the ocean through our vast and rich valleys from Oregon to Maine, a distance of more than three thousand miles."[17]

The importance of such a route was incalculable, he said. Military forces could be concentrated at any point east or west in eight days or less. A naval station near the mouth of the Columbia River in Oregon, "with a comparatively small navy, would command the Pacific, the South Atlantic, and Indian oceans, and the China seas." Using a combined rail and steamship route between New York and China, which would require only thirty days, the products of American factories could be exchanged for Asia's rarities. Compare this to the round-trip sailing distance between New York and China (nearly thirty-four thousand miles, requiring up to three hundred days). World commerce would be revolutionized, with Whitney's Pacific route its channel. Each state and every town "would receive its just proportion of influence and benefits," he wrote, "compared with its vicinity to, or facility to communicate with, any of the rivers, canals, or railroads crossed by this great road."

Such easy and rapid communication, he argued, "would bring all our immensely wide-spread population together as one vast city; the moral and social effects of which must harmonize all together as one family, with but one interest—the general good of all." Moreover, because the destitute overflow population of Europe was beginning to clog the cities of eastern America, the railroad would attract throngs of hopeful farmers and workers to settle along its route,

> *where they will escape the tempting vices of our cities, and where they will have a home with their associates, and where their labor from their own soil will not only produce their daily bread, but, in time, an affluence of which they could never have dreamed in their native land. . . . Their energies will kindle into a flame of ambition and desire, and we shall be enabled to educate them to our system, to industry, prosperity, and virtue.*

All that was required to set this in motion, Whitney reasoned, was an elementary exchange. He asked that the United States set aside out of its public lands a strip of land some sixty miles wide and the length of his proposed route. Beginning at Lake Michigan, Whitney would sell this land—which would be settled and the proceeds of which would finance construction of his railroad. Section by section, the rails and their supporting population would leapfrog westward "so far as the lands may be found suited to cultivation." The cost of planting the railroad he estimated at $50 million, with a further $15 million for maintenance of the road until completion. The cost of building across uninhabitable terrain would be offset by this maintenance fund and by sale of the public lands—all proceeds to be "strictly and faithfully" applied to railroad construction, subject to whatever checks and guarantees Congress required. To determine the route he asked the legislators to order a survey between the forty-second and forty-fifth degree of north latitude from lake to ocean.

Only when the route was finished, when the travelers and commerce of the world crossed the nation in comfort and security, would the New York merchant collect his compensation. Whatever unsold land remained in that sixty-mile-wide belt would be deeded to Asa Whitney. It was that simple. Finally, tolls along the road should be kept low, to a level just above what was required for maintenance. The excess would "make a handsome distribution," he reasoned, "for public education."

Thus set forth before the House of Representatives, Asa Whitney's remarkable railroad proposal was referred to the Committee on Roads and Canals.

2

"Who Can Oppose Such a Work?"

Another figure had taken his place in a procession of dreamers that stretched back to the advent of steam railroading. Asa Whitney's predecessors, each of them urging that the Atlantic and Pacific coasts of North America be connected by twin ribbons of iron, were a fascinating and varied lot.[1]

It has been said that in 1819—six years after primitive steam engines were first used in Britain to haul coal, and ten years before the first real Iron Horse to reach America was given a trial run near Honesdale, Pennsylvania—the South Carolinian engineer Robert Mills, architect of the Washington Monument and other federal structures, proposed a "steam carriage" to run from the head of the Mississippi Valley to the Valley of the Columbia. Mills was a protégé of Thomas Jefferson and a student of Benjamin Latrobe, but despite his connections, his steam carriage idea was studiously ignored in 1819.

The merest suggestion of the idea of a transcontinental railroad surfaced in 1830, around the time that the first American steam locomotive, the *Tom Thumb,* was designed and built by Peter Cooper and employed on a thirteen-mile track between Baltimore and Ellicott's Mills, Maryland. A pamphlet was published by William C. Redfield, primarily urging that a "Great Railway" be constructed to connect the Atlantic States with the valley of the Mississippi. Redfield looked back in disgust at the vast expenditures of the War of 1812 and stated with confidence that an amount so freely and wastefully spent then could be used, in 1830, to build a rail route not only to the Mississippi but even all the way to the Pacific.[2] But two years later, after Redfield and his plan sank out of sight, the year that the former Philadelphia jeweler and bookbinder Matthias Baldwin began constructing *Old Ironsides*—the first of a long line of Baldwin locomotives—residents of Dunkirk, New York, convened a meeting at the Dunkirk Hotel on January 10, 1832. These citizens of Chautauqua County on the shore of Lake Erie resolved that "among the many reasons why a Railroad from the Hudson to Lake Erie should pass through the southern tier of counties [is that] it would be a strong and powerful link in a Railway to the valley of the Mississippi, and finally to the Pacific Ocean."[3]

It is likely that in gamely envisioning the commerce of the world passing through tiny Dunkirk, these burghers might claim credit for the first outright suggestion of a transcontinental railroad. But their place in our procession of dreamers is theirs only by dint of a matter of weeks. On February 6, 1832, a Michigan territorial newspaper, the *Western Emigrant* of Ann Arbor, published the first lengthy statement on the subject. Samuel Dexter, the editor, wrote,

> *It is in our power . . . to open an immense interior country to market, to unite our Eastern and Western shores firmly together, to embrace the whole of the fur trade, to pour those furs into India and in return to enrich our interior with the spices and silks and muslins, and teas and coffee and sugar of that country. It is in our power to build up an immense city at the mouth of the Oregon, to make it the depot for our East India trade and perhaps for that of Europe—in fact to unite New-York and the Oregon by a railway by which the traveller leaving the city of New-York shall, at the moderate rate of ten miles an hour, place himself in a port right on the shores of the Pacific.*

Leaving aside the question of what utility furs would have on the Indian subcontinent, Dexter's plan was quite detailed for its time. "The expense of it would not surpass one year of war," he supposed—$30 million—and the route could be completed in six years. If the government would not build it, he hoped it would allow a private company to do it with a grant of three million acres of public land. "It is one of those great projects," Dexter pronounced, "which none but a great nation could effect—but peculiarly adapted to the enterprising character of the people of the United States."[4]

Railways had assuredly taken hold in the minds of Americans. The country's first passenger service was under way on the Baltimore and Ohio, to be joined by the Philadelphia and Columbia, the Memphis and Charleston, the Boston and Worcester, the Lexington and Ohio, the New York and Harlem. The central patent office began to register a multitude of technological breakthroughs, the most far-reaching of which was Robert Livingston Stevens's system. In a burst of brilliance he conceived a flat-topped, T-shaped iron rail and a flanged wheel. These would replace the English-style, L-shaped rails and plain wheels which severely limited the stability, speed, and weight of trains. (Earlier attempts at employing rails made of wood, to which were fastened iron straps, were proving themselves to be extremely dangerous when the wooden rails warped off their beds, sometimes impaling moving railroad cars. Stevens spiked his T-rails to hardwood ties placed perpendicular to the track. Previously stone blocks had been used to support the rails, which reared up out of the ground at the first frost. Upon this system of T-rails, flanged wheels, and wooden cross-ties all further innovations were based—the counterbalancing cowcatcher and swiveling wheel trucks, improved boiler systems, and safety measures such as headlamps and steam whistles.[5]

Despite the Great American Desert and the lack of any firm Pacific Coast foothold, observers of the burgeoning railroad scene watched the developments and the chartering of more eastern enterprises—the Boston and Albany, the Philadelphia and Reading, the Central of Georgia, the Louisville, Cincinnati and Charleston—and some took their place in the procession of transcontinental dreamers. Dr. Samuel Barlow took up the cause in 1833, in the *Intelligencer.* Three years later arrived a summer number of the literary monthly *The Knickerbocker Magazine* in which Lewis Gaylord Clark (the editor of Bryant, Irving, Longfellow, and Lowell, and soon to become the bitter literary antagonist of Poe) asserted fathership of the idea. In the next year the Panic of 1837, variously called "the State Public Improvements Follies" and "the Great Land Speculation Frenzy," bankrupted a number of states and severely trimmed the number of hopeful railroad companies, but others continued to hope: the Erie and Kalamazoo, the Richmond and Petersburg, the Michigan Central and Southern. In 1838 the Reverend Samuel Parker, a pioneer Protestant missionary working the Oregon Trail to the Columbia, published his *Tour Beyond the Rocky Mountains,* which stated that a transcontinental railroad was entirely practicable.

Then there was the Welshman John Plumbe. As a boy he had emigrated to America and worked on a railway survey party in the Pennsylvania Alleghenies before settling in Dubuque, Wisconsin Territory, in 1836. Plumbe contributed to a number of eastern and local journals under the nom de plume of "Iowaian." In the *Iowa News* of March 24, 1838, "Iowaian" called for a rail connection between Lake Michigan and the Mississippi River, via Dubuque. This unmodest proposal grew during subsequent town meetings and petition-writing sessions. Finally the good citizens of Dubuque, with their eyes on gaining access to unsettled agricultural land, the Mississippi Valley fur trade, and north-central lead mines, composed a congressional memorial announcing their intention to complete an "Atlantic and Pacific Railway." Some weeks later congressional delegate George Wallace Jones submitted to the House of Representatives a petition "praying for the survey of a route for a railroad from the Mississippi river, at Du Buque, to Milwaukie, Wisconsin Territory."

"I was amazed at the temerity of my constituents," Jones recalled many years later, "in seriously sending me such an unheard-of prayer. Nevertheless, I felt in duty bound to present the petition, and did so, when it produced a great laugh and hurrah in the house, members singing out to me that it would not be long before my constituents would ask Congress to build a railroad to the moon."[6]

Though the Iowans' motion in 1838 resulted in some $2,000 being appropriated for a railroad survey between Lake Michigan and the Mississippi, their part in an "Atlantic and Pacific Railway" was not to be, although John Plumbe did not know this when he produced, the following year, *Sketches of Iowa and Wisconsin* ("one of the finest domains that nature ever offered to man"), in which he predicted a "free Railway, unparalleled in extent; and forming, when completed, the greatest thoroughfare in the world!"[7]

John Plumbe took his place in the growing procession—although, as will be seen, he would later attempt to push his way to the front of the line. Meanwhile, in 1840, the total length of American railways grew to 3,328 miles, nearly twice that of all of Europe's. If Mr. Samuel Dexter of the *Ann Arbor Western Emigrant* was still at work, he would surely have called that a testament "to the enterprising character of the people of the United States."

Enter Asa Whitney, fresh from China and full of ideas, to join our group, but no sooner had his memorial been registered as Document No. 72 of the Twenty-eighth Congress than another figure—a slightly shady figure—appeared. This was Mr. George Wilkes, a freebooting New York editor who had worked for a number of scandal sheets. His literary works included *The Mysteries of the Tombs: A Journal of Thirty Days Imprisonment in the New York City Prison for Libel* (1844) and *Life of Babe, the Pirate,* about a miscreant whom Wilkes met in jail. Having heard about Whitney's proposal when he emerged from prison, Wilkes hastily drafted his own memorial "to meet the demand of the public."[8] It emerged in *The History of Oregon, Geographical and Political,* a catchall of hastily digested material on the northwest quadrant. In it, he claimed that the project of a national railroad, "though generally denounced as visionary and impracticable, has long been the author's favorite idea."[9]

On the day before James K. Polk was to be inaugurated, March 3, 1845, Whitney's plan was tabled in the House Committee on Roads and Canals. Robert Dale Owen, congressman from New York, said he thought the subject was important—but it "ought to obtain the careful and deliberate attention of Congress" at a more convenient time. Owen added that the railroad memorial should have included a specific deadline for construction, that more guarantees were needed to limit speculation on the western lands in question, and that he assumed the military would be required to guard work crews in the wilderness. Finally, Owen said he was worried that building a railroad to the disputed Oregon Country would antagonize the British. One day later such concern was moot, as the newly inaugurated President Polk delivered his bellicose address in which England was notified that soon the Stars and Stripes would wave over all of Oregon, from sea to shining Rocky Mountains.[10]

Seeing that the political climate, belligerent as it might have appeared, favored discussion of the Pacific railroad, Whitney would not be daunted by the mere setback of having his memorial tabled. Instead, he stepped up his publicity campaign. He sent a letter on April 23 to the *National Intelligencer*—a letter widely reprinted in other newspapers with favorable editorial comment. "It is my intention," Whitney announced, "to pass over, examine, and partially survey seven or eight hundred miles of the proposed route for the Railroad from Lake Michigan to the Pacific." He planned to set out in late May from New York for Green Bay, Wisconsin Territory, proceed down Lake Michigan's shore to Milwaukee and thence westward to the Missouri River. Whitney, now forty-eight, invited "young gentlemen of high respectability and education," especially representatives of the South, to join him in the exploration. "The excursion will be pleasant, beneficial to health and useful in the knowledge to be gained of that vast country," Whitney wrote, "and, should the project for the railroad succeed, those who now accompany me can be usefully and advantageously employed in the great work."

Whitney was under way on June 2 with a number of positive newspaper clippings about his trip already in hand; he lost no opportunity to either write home himself or persuade others to do it. The *New York Tribune* soon published its Wisconsin correspondent's report of Whitney's arrival. "Our little Town has been honored with a visit from the 'Whitney Exploring Expedition,'" he said.

> *The party, consisting of the "Projector" and eight young men lodged with us last night and left here this morning en route to Prairie du Chien. They go today to Fond du Lac, where I learn they spend tomorrow (the Sabbath) with governor Dodge. They have chartered for the transportation of themselves and baggage, from Milwaukee to Prairie du Chien, two wagons, one with steel springs and box like your City furniture wagons, the other a common lumber wagon. Several others are expected to join them at Galena and other points, and at Prairie du Chien arrangements are to be made to supply them all with Indian Ponies, and outfits for the prosecution of their journey Westward. The party seem to be in the finest spirits—have each of them a coarse broad-brimmed straw hat, and with the exception of the Hero, guns of various calibre, with all the implements of sportsmen, and long-legged boots, India Rubber coats, blankets, etc.—according to their various tastes. The beards of the various "Boys" are now of a little more than a week's growth, and they are bound not to shave until they get back.*[11]

By July 1, Whitney's little group was in Prairie du Chien. At first they were slowed because only one "laboring man" accompanied them to do the heavy work; then they were daunted for lack of a guide across the increasingly unpopulated prairies ahead. Fifty miles later, at Fort Atkinson, they were still without a guide. Whitney was reluctant to lead the young gentlemen into "probable dangers and sure hardships and fatigues," but they insisted they could continue. Unused to any labor before, "they never flinched," Whitney wrote, but "were ready to wade through mud, water and grass to their necks, with their provisions on their heads, to swim rivers, to fell trees for bridges, and other fatigues necessary for the accomplishment of our object."[12] Apparently all this and more was necessary. "We have traveled only 10, 13, and 17 miles a day, since we left the Mississippi, on account of the heat," one young college graduate wrote. "In crossing the streams we sometimes had to unload the wagon two or three times a day, and lift it over by hand. Some of us have enjoyed the luxury of sleeping over night in wet clothes, from standing in the mud and water waist deep to lift out the wagon. This, however, we regard as only one of the varieties of our prairie life."[13]

In one perhaps apocryphal encounter with a large contingent of Sioux, the group was treated to a feast of baked dog. One elder offered his daughter in marriage to Whitney, who chivalrously refused on the grounds that the whites' tribe frowned on hurried marriages.[14]

From Fort Atkinson on the Turkey River, Whitney led his men, with compass and sketchy map in hand, across the various branches of the Cedar—where the forests diminished to almost nothing—to Clear Lake, to a branch of the St. Peter, to the Des Moines, by the headwaters of the Little Sioux and the branches of the Calumet, the Vermillion, the Jaques, until finally they arrived at "the great Missouri," some fifteen miles below the big bend. "A great part of the country over which I passed," Whitney said later, "had never before been traversed except by savages." From the Mississippi to the Missouri they had walked, ridden, and swam some five hundred miles of rolling prairie—"the finest country upon the globe," Whitney rhapsodized, "capable of sustaining more than three times the population of the same space in any other part of the globe . . . and undoubtedly the most healthy country in the world."[15] He had never found an atmosphere so pure, he wrote in a circular distributed and immediately printed by many newspapers in New York and in most of the country's large population centers. The soil was as rich as it could be; he had not seen even one-half acre of bad land. All a farmer had to do, he promised, was to plow, plant, and gather his crops. "I have found all I desired," he exulted,

> *and far more than I expected when I set out. . . . If Congress will give me the lands, in a very few months the work shall be commenced, and far sooner than I have dared to hope, it will be completed—when we shall have the whole world tributary to us—when the whole commerce of the vast world will be tumbled into our lap—when this vast and now useless waste and wilderness (and it ever must be so, without this road) shall become, not only the thoroughfare of the vast world, but its garden, feeding, clothing,*

comforting and enlightening millions, who are now starving, homeless, naked, ignorant and oppressed; and who can oppose such a work?[16]

While Whitney and his intrepid young assistants had traversed through an oppressive summer heat across the "useless waste and wilderness" of present-day central Iowa, meeting only encampments of Sioux and little to remind them of the East, it might have cheered the former merchant-turned-railroad-campaigner to know that he had already attained a national stature. His dispatches no sooner arrived in a newspaper office than they were typeset and placed on the next edition's first page. They attracted much attention and comment—the lands he described sounded as bountiful as those in Oregon and California. A line more than eight hundred miles long had once again been drawn toward the Missouri and the emptiness beyond, at a time when tens of thousands had felt their restlessness stirring. And Whitney was not the only wilderness traveler to be abroad during that spring and summer; by August, John C. Frémont and his armed contingent were four months west of St. Louis and bound for the Pacific slope, thinking less about how the approaching snows might block mountain passes and more about the ripe plum of California. With Mexican-American relations deteriorating rapidly—Mexico had broken diplomatic ties as its neighbor prepared to annex Texas—President Polk had sent General Zachary Taylor into the Southwest to defend a line "on or near the Rio Grande." And that August—as Whitney toiled across the prairies, as Frémont struggled over the mountains, as Taylor inspected his pickets and spat into the Rio Grande, as some five thousand American settlers in Oregon clamored for attention and eyed their few British neighbors with arrogance, a new term was coined in the East and spread with amazing rapidity. "Our manifest destiny," wrote John Louis O'Sullivan in *The United States Magazine and Democratic Review,* "is to overspread the continent allotted by Providence for the free development of our yearly multiplying millions."[17]

A steamboat from Fort Leavenworth carried Asa Whitney to St. Louis on September 19, 1845. He had been kneeling for twenty-six days in a canoe down the Missouri to the fort, and he arrived in St. Louis eager to make new converts. "Who can oppose such a work?" he asked all who would listen. "I have not exaggerated the results to flow from it. If you or any other intelligent man will calmly and deliberately look over this subject, you will, you must be convinced that it is not chimerical." He could begin ticking off the benefits on his fingers. "No man's rights to be abridged," he would say, "no man's taxes increased, and not even one cent asked for from any man. All I ask is, that which is now a great part useless, and ever must be without this road." It was not a sectional question, it had nothing to do with politics, he would say. All would benefit. "It is no stock jobbing or gambling scheme," he would reassure those burned in the Panic of 1837, "there being no company to manage, or stock to speculate upon, no one can be deceived or frauded."[18] Who could oppose such a work? Whitney could find none in St. Louis. "I find all here in favor of my project," he wrote home. Yet he fretted; the Congress would begin a new session in just over two months and it was imperative that the question be settled soon. "The lands are fast being taken up, from the Lake to the Mississippi," he wrote, "and will soon be so much so as to defeat the project," for his enterprise depended on a supply of public lands along his proposed route. Iowa would soon be made a state, and Wisconsin. Oregon, too, was "entirely dependent" on the road.[19]

It was time for action.

He had planned to go from St. Louis through Illinois, Indiana, Ohio, and Pennsylvania, to New York—all the way cornering newspaper editors, buttonholing politicians, persuading ministers, and holding forth in gatherings large and small. But a national convention was to be held in Tennessee in November, in which would gather railroad fanatics from all over: monomaniacs in Memphis. Whitney changed his plans and, towing a correspondent from the *New York Tribune*, he took a southbound steamboat. "It is but an act of justice to Mr. Whitney," the reporter wrote, "to state that his proposition meets with decided favor from the western people, and his amiable and communicative manner, his moderate and well-directed enthusiasm, make him immensely interesting to all on board."[20] What held the Westerners' interest was the fact that a local politician by the name of Stephen A. Douglas had leveled his sights on the railroad projector. As chairman of the House Committee on Territories and as a vigorous expansionist, Douglas published a pamphlet in which he objected to Whitney's proposed route. The pamphlet found its way on board before the steamer left St. Louis. What was needed was a proper survey, not a schoolboy's jaunt o'er the plains, Douglas intimated. And what about California? Surely with growing numbers of American farmers in that Mexican territory its manifest destiny lay in the Union. Finally, Whitney's projected path was too far north, the Illinois congressman complained; it should not begin at Milwaukee—it should originate in, well, Chicago.[21]

Asa Whitney's replies to these objections were tested on his fellow passengers—including, possibly, a tragedian named Booth who was on board (it would have been Junius Brutus—father of John Wilkes and Edwin Thomas Booth). The survey would come later under competent hands, Whitney answered; his expedition had intended to assess the public lands for agriculture, to form a picture of where building materials such as stone and wood were to be found, and see how many bridges would be necessary. As to California—why, by the time his rail crews were through a pass in the Rockies, both California and Oregon would probably be part of the Union. "The road can then extend to either point on the Pacific," Whitney would have exclaimed, "the distance from the pass being the same." As to an eastern terminus, the open lands decided where the route would begin, not the sectional interests of one politician.[22]

As the excitement of friendly debate over the Pacific railroad rose on the steamboat, the waters of the Ohio River receded, detaining Whitney until the Memphis Railroad Convention was well under way. Although he was able to persuade many dignitaries from the various states that his plan had merit, he was too late to maneuver it before the delegates. The convention adjourned without much being decided, and Whitney returned to New York and began to draft a new memorial to take to Washington.

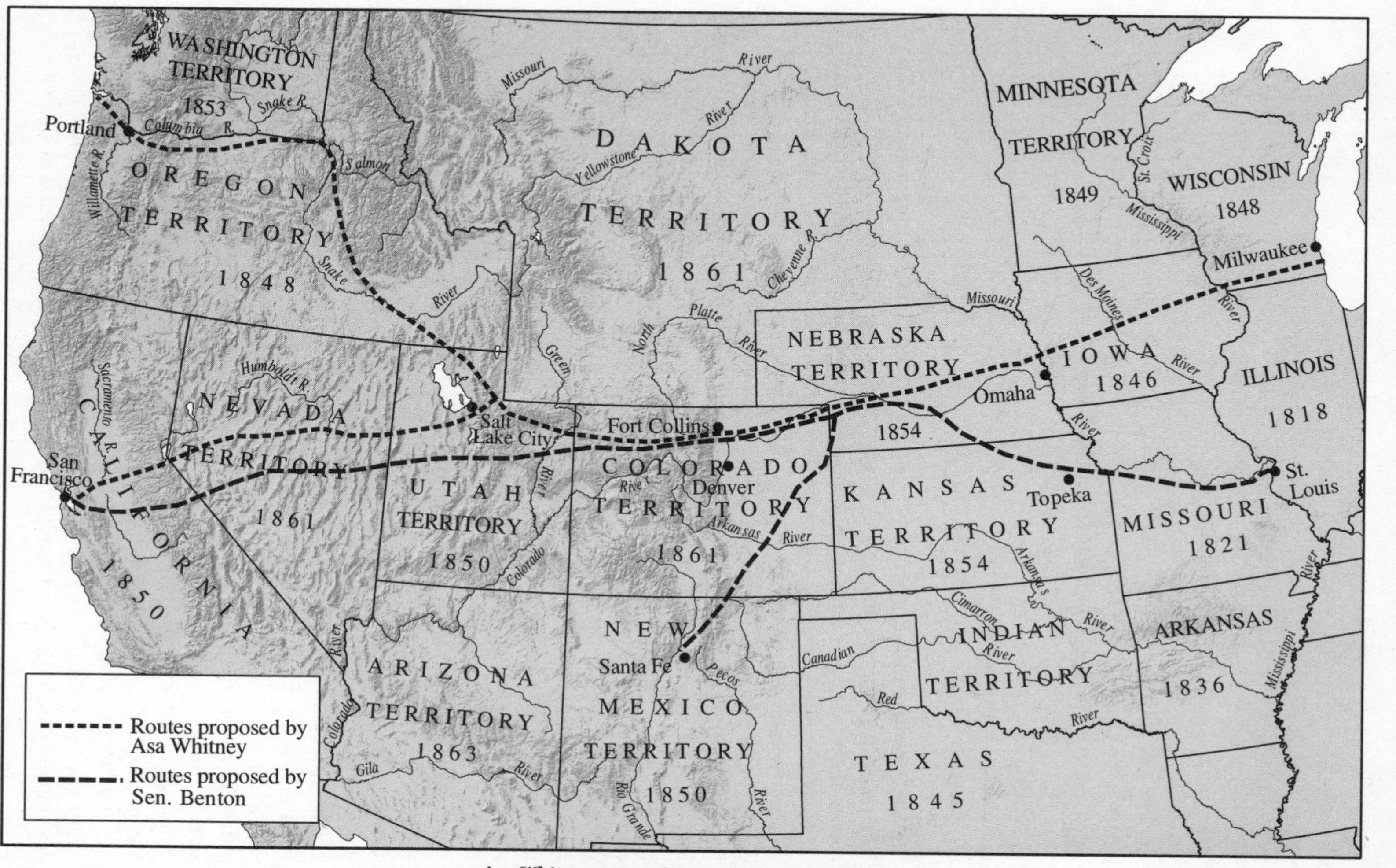

Asa Whitney versus Senator Benton, 1845–50
Including Dates of Territorial Organization or Admittance to the Union

3

"I Must Walk Toward Oregon"

The Twenty-ninth Congress convened. President Polk told the assembled delegates that henceforth the Monroe Doctrine contained a codicil: "the people of this continent alone have the right to decide their own destiny," he said. "No future European colony or dominion shall with our consent be planted or established on any part of the North American continent."[1] He ended the year by signing up Texas as the flag's twenty-eighth star. There were few in Washington who observed New Year's Day 1846 without some speculation about the imminence of war: probably with Mexico, probably with England. The Year of Decision—in Bernard DeVoto's phrase, defining not only choice but determination, the flowering of a national will—commenced. "Let me live where I will," wrote Henry David Thoreau, "on this side is the city, on that the wilderness, and ever I am leaving the city more and more and withdrawing into the wilderness. I should not lay so much stress on this fact if I did not believe that something like this is the prevailing tendency of my countrymen. I must walk toward Oregon, and not toward Europe."[2]

In late February, as DeVoto has juxtaposed the events, Frémont and his men were glowering in the direction of the Mexican post at Monterey and nearing the point of daring the "greasers" to do their worst; General Zachary Taylor's men continued to strut on the Rio Grande shore; President Polk briefly entertained the notion of plunking down gold instead of his countrymen's blood for the Southwest and the Far West; and thousands of persecuted Mormons emptied their Illinois town, Nauvoo, in favor of a westward-tending uncertainty on the trans-Mississippi frontier. The boundaries of the Great American Desert shimmered, expanded, contracted. In Washington, the distinguished senator from Illinois, Sidney Breese, presented Asa Whitney's new memorial, and it was referred to the Committee on Public Lands.

The document was more than twice the size of Whitney's effort of the previous year. In it Whitney elaborated his earlier arguments with a detailed description of his findings in the West, the fertile soils, the placement of timber and potential quarries, the probable bridge sites, the necessity of linking Oregon with the States before it decided to go its own way. Additionally, he had now had contact with the "numerous, powerful, and entirely savage" Indians and thought that his railroad would serve as a buffer zone to keep warring tribes from exterminating each other, preserving their race "until mixed and blended with ours." Thus had Whitney become the first railroad promoter to consider in any way the continent's first denizens. The Sioux were "ready and willing to sell all that may be desirable" for his project, he reported to Congress, "and for a very small sum."[3]

The promoter added one other significant element to his argument: he renounced all profits for the first twenty years of his road's operation, such money and titles of land which were not converted to railroad construction or maintenance to be held in trust by the government as security for the fulfilment of his contract. Whitney was approaching his forty-ninth birthday. "In all human probability," he said, he would in twenty years "be past the wants of this life." If he could be the instrument to accomplish "this great work," he said, "it will be enough"—he asked no more.[4]

To his satisfaction, the Senate Committee on Public Lands reported a bill drawn from his memorial on July 30. Much had occurred outside of the committee room in the interim: the Senate had ratified an Anglo-American treaty setting the border between British Canada and American Oregon at the 49th parallel; the much-anticipated war with Mexico had commenced, giving junior officers Grant, Sherman, Meade, Hooker, McClellan, Pope, Lee, Jackson, Beauregard, et al. the sort of military experience they required; Captain Frémont, after kicking sand from a breastworks in the face of the Mexican authorities in California, had retired, then returned to join revolting American settlers under the Flag of the Bear, closely followed by the U.S. Navy's capture of Monterey, Sonoma, and San Francisco; Colonel Stephen Watts Kearny's "Army of the West" had paused to peremptorily knock at the New Mexico border, announcing it would enter to ameliorate the conditions of its inhabitants.

Despite these distractions, the Senate ordered the public lands committee report on Whitney's Oregon railroad to be printed. This was done over the objections of the powerful Senator Thomas Hart Benton from Missouri, who thought Whitney's memorial "absurd," "ridiculous," and "impudent"—and, once he had time to consider the matter, would begin to hold that any Pacific railroad should have its eastern terminus in, well, St. Louis.[5]

But then, in July 1846, the senator had not yet considered the matter. Nor had he in December 1844, when Whitney first toured the Capitol to proclaim his work. "Old Bullion" Benton, before whose lance even windmills fell, whose foghorn voice in session caused all but a few of his fellow senators to wince and duck their heads, would typically have cut short his visitor's speech. "Impossible, sir," he is remembered to have scolded Asa Whitney and his projected Pacific railroad, "You are one hundred years before the time." Benton commanded his visitor to return when he had canvassed others in the Senate, which Whitney set out to do.[6] After this was done, however—when more often than not the merchant found his proposal falling on receptive ears—he had no desire to subject himself to more Bentonian bombast, and stayed away from that office. Habitually he remained in Washington while Congress was in session, so it is more than likely that some nineteen months later, after Senator Sidney Breese rose to advance the memorial, Whitney was sitting above in the visitors' gallery as the screech of Benton's chair echoed off the chamber walls, to be followed by a blast. The proposal was "an imposture, a humbug; it could have emanated only from a madman . . . science was unequal to overcome the Allegheny Mountains—and now Whitney proposed to scale the Rocky Mountains, four or five times as high! Why sir, 'tis madness!"[7]

Benton notwithstanding, the Senate recessed with the favorable report beginning to circulate around the country. To Breese's satisfaction, a former colleague from Mississippi, John Henderson, wrote him on September 19. "We shall expend at least $100,000,000 in our contest with Mexico," Henderson exclaimed. (Indeed, General Taylor's six thousand troops had already swept expensively southward into Mexico along the San Juan Valley, and the next day would begin their victorious engagement against General Amadia at Monterrey.) Such a sum, the Mississippian said, "would build this road and purchase so much of Upper California, as would fix its terminus in the bay of San Francisco. Who can doubt which would most promote our interest—our glory? I should have been glad to have you meet with Mr. Benton's support in this measure, but I trust you will not cower under his opposition, but push it with increasing zeal."[8]

Of course Whitney had the letter reprinted in the *Washington Daily Union*. His scrapbook of clippings was growing new leaves. "The project has been ridiculed as visionary, by those classes of persons, unfortunately too numerous, who always denounce as humbug what they do not comprehend," wrote the editor of the *Philadelphia Public Ledger*.

> *But ridicule from such sources is the common lot of all who are tall enough to see over the heads of the deriders. . . . Rumsey and Fitch and Fulton were ridiculed for dreaming of navigating by steam. . . . Clinton was ridiculed for undertaking to "cut a big ditch," an undertaking which has made the State and city of New York what they are. And now Mr. Whitney is ridiculed for undertaking a project, which twenty years hence, will be a "fixed fact," and thirty years hence, will be the avenue of all the trade between all Western Europe and all China and India. . . . And thus ever goes the world; fools laughing at what they cannot see, and wise men foreseeing what fools cannot comprehend.*[9]

Now no longer a shy man but an active publicist, Whitney began in October to tour "the western states," moving from Pittsburgh to Cincinnati, Louisville, St. Louis, Terre Haute, Indianapolis, Dayton, Columbus, Wheeling, Nashville, and back to Philadelphia. He garnered attention wherever he went, especially when he first appeared before an audience. As one newspaper observed, "Mr. Whitney is probably the most correct and striking representation of Napoleon Bonapart in personal appearance, that anyone has ever beheld, since the time of the great original; and has often, when in Paris, been annoyed by spectators on account of it."[10]

With hand not buried within his cloak but emphatically waving and grasping a pointing stick, his presentation seldom varied. "All the maps we have heretofore studied," he would say to the assemblies as he stood before two maps of the world, "have been made with particular regard to the position of Europe—Europe, Asia and Africa placed together, and our continent [to] one side of all, as if of no importance."

A stirring in those audiences of Americans. Their pride had recently swollen as American territory swelled. Whitney could pause significantly as they took in this cartographic insult.

Meanwhile, Whitney would say after a time, "here you see by this map, that we are in the center of all."

More stirring. A murmuring.

"Europe on the one side, with 250 million of population," he would continue, "and all Asia on the other side of us, with 700 million of souls. The Atlantic separating us from Europe, the calm Pacific between us and Asia, and you will see that the population and the commerce of all the world is on this belt of the globe—which makes a straight line across our continent." Those Ohioans, those western Virginians, Kentuckians, and Missourians, those Hoosiers and Keystoners, but most of all those Americans, would for the first time visualize the world in a new way with themselves at its heart. They would absorb Whitney's eloquent argument for a railroad, and, regardless of their particular remove from the line he drew between Lake Michigan and the Pacific, they would perceive the certain benefits. Perhaps they imagined themselves riding in comfort and security to a new beginning in Oregon or California.[11]

Whitney was "said to enlist warm advocates of his project wherever he speaks," reported the *Nashville Triweekly Union,* while the *American Railroad Journal* allowed that "the western community seem well disposed towards this stupendous enterprise."[12] Even the doubters maintained a modicum of respect, such as an editor of the *Cincinnati Daily Commercial,* who confined himself to a satirical account of Whitney's program:

Mr. Gullible, the President, took the chair, and after a few remarks, introduced Mr. Monchausen to the meeting. His statement was brief, plain, and business-like, in character with the subject, which, like the creations of genius, needs not the glitter of rhetoric for its embellishment, but shows best in the native, simple, yet awful grandeur and harmony of its proportions. He spoke of the relation of his project to other works, yet unlike any of them, inasmuch as it was to "unite neighboring planets in our solar system, and make them better acquainted with each other"—no capital was required, no stock was to be issued, no dividends to be made—the fruit of nature, it would create settlements, commerce and wealth, and stimulate production.[13]

Only in New York would Whitney encounter any trouble. There, his public meeting ended in anarchy. A group of National Reformers invaded the assembly and, without realizing how close their position was to Whitney's (they had neither listened to his speech nor read his proposal) they denounced the merchant as anti-Labor and took over the hall. Much later Whitney received an apology; the outburst, however, did gain him some adherents.[14]

To expansionists beyond the inner circle in Washington, the Year of Decision closed with a semblance of progress with new territory gained. Kearny's conquest, New Mexico, was annexed. Monterrey was Zachary Taylor's, while at year's end his men camped under a temporary armistice near Buena Vista. Another trans-Mississipi state, Iowa, the flag's twenty-ninth star, had been admitted—a symbolic advance westward. Before snow blocked the passes some 300 more settlers had arrived in California, and some 1,350 in Oregon, numbers which could be read as disappointing or promising depending on one's persuasion.

James Knox Polk derived little comfort from these gains—in fact, the president had passed beyond seeing the glass as half empty or half full to not seeing the glass at all. Rebelliousness was abroad; enemies lurked behind every bush. A Pennsylvania representative once thought to be a Polk man had been stricken with Van Burenitis in the Twenty-ninth Congress—by inserting a proviso into legislation, the annoying Mr. Wilmot seemed likely to bar slavery from any lands gained from the war. Of course the Whigs, who were gaining ground by calling Polk's war an unconstitutional grab for territorial spoils, had rallied behind the measure. The president could not even trust his own commanders of the Mexican campaign, especially Zachary Taylor—who, every time he ignored Polk's battle directions, emerged victorious over the Mexicans. Worse, Taylor would certainly be the Whigs' next presidential candidate. At his nomination Polk had pledged himself to only one term, which was now at midpoint; he had little time to simultaneously prosecute a successful war and defeat his own general. Moreover, state and congressional elections in the autumn had given Whigs and antislavery Democrats a troublesome margin. Consequently there was little for Polk to look forward to during the next session.

Members at the second session of the Twenty-ninth Congress, convened in December, were greeted as usual by a battalion of lobbyists, among them the genially persuasive Mr. Asa Whitney, "the Railroad Projector"; his was as dependable a presence as that of any of the legislators. Whitney's route now publicly divided on the western slope of the Rockies. One line veered to our new Northwest. And one extended to California, where, deep in the Sierra Nevada, the Donner party gazed into a sky that was aswirl with snowflakes and darkened by the horror to come.

Whitney's was not the only transcontinental memorial laid before that congressional session. The time was becoming more ripe. Dr. Hartwell Carver, of St. Louis, submitted his proposal for a charter to build along Whitney's route. He wrote that after a trip up the Missouri in 1837 "the propriety and the practicability of running a railroad across the Rocky mountains burst upon my mind with perfect conviction, and has ever since been the idol of my heart." He claimed to have published numerous articles that year, both in territorial newspapers (none have survived) and in the *New York Courier and Inquirer* of August 11, 1837 (it is not there), but he received neither "applause or encouragement" then nor any subsequent recognition. Were it not, he said, for the fact that he had been in Europe in 1844 when Asa Whitney began approaching editors and legislators, he likely would be receiving proper credit as the originator of the idea.

"The religious point of view alone, aside from any other, is enough to warrant this undertaking," Dr. Carver told Congress,

> *for it will open the way for the conversion of millions and millions to the true faith, and bring them fully under Christian discipline. Instead of the devoted devotees occupying two or three months going to Mecca to worship false gods, the true Christian missionary can go from this country and convert thousands to a saving faith in Christ, the true savior of mankind, in that time.*

This enterprise, he promised, "will bring about a kind of earthly millenium, and be the means of uniting the whole world in one great church, a part of whose worship will be to praise God and bless the Oregon Railroad." And think of the commerce! "The quality of teas would be much better coming to us in a few weeks of gathering," he said. "I suppose the flavor and quality of our teas now bear no comparison to what it would if brought directly to us, without crossing the equator twice. Methinks I can look forward, through the vista of time, and see countless thousands of our fair country women sitting of an afternoon leisurely sipping and drinking their tea, until they become intoxicated with the sweet flavored aroma of this delicious beverage and cry out, in sweet and musical accents, blessed be God, and the projectors and builders of the Oregon railroad, now and forever, amen."

Beyond the sweet and musical accents, Dr. Carver proposed to be deeded a transcontinental slice some forty miles wide (as opposed to Whitney's sixty) to be paid for in railroad stock certificates which Carver would have printed. He thought it would take him no more than six years to build the road at a cost of $60 million. While Mr. Whitney's proposal was evolving toward a nonprofit enterprise, the doctor would emerge from his deal owning all of the excess land and its timber and minerals, a remuneration estimated at eight million acres.[15]

It was the prospect of such a reward going to Dr. Carver that agitated a third projector of the Pacific road. Enter that relentless opportunist, Mr. George Wilkes of New York, looking rather more presentable than when he was last encountered—when the whiff of the Tombs still clung to him. Now Wilkes was reestablished; his Oregon-boostering book was selling well to Easterners, and he now presided over a new publication of astounding success, having issued for a year and a half the first numbers of *The National Police Gazette.* "Our city, and indeed the whole country," Wilkes had informed the readers of volume 1, number 1, "swarms with hordes of English and other thieves, burglars, pick-pockets and swindlers." Wilkes proposed to strip them of the advantages of "a personal incognito," by publishing a minute description of their names, aliases, and persons, together with a succinct history of their criminal acts. Wilkes's column, "Lives of the Felons," was particularly popular—except in the eyes of one Robert Sutton ("Bob the Wheeler"), a notorious burglar who so disliked the publicity that when he was released from prison he gathered his cohorts and attacked the *Gazette* offices.

Wilkes survived that assault and the scores of subsequent ones, both legal and physical, that dogged his *National Police Gazette.* His railroad memorial was written no less floridly than any of his "lives of the felons," and promised to place "the trident of the seas" in the grasp of the United States and bring "the proudest nations of the earth as suppliants to our gates." Wilkes warned the legislators to beware of Dr. Hartwell Carver and Mr. Asa Whitney and their plans. Carver was ingenious, he said, to apply for such an enterprise on credit—and his profit would be considerable even if he failed and paid a million-dollar forfeiture. Whitney, though, was "less moderate." Wilkes portrayed him as a supreme speculator. This was easily done, first by overlooking his character and second by omitting to mention that Whitney had bid to forgo all profits for twenty years when, "in all probability," he had said, he "would be past the wants of this life." Whitney, in Wilkes's view, was corrupt; he pictured "Poverty leaning on its spade before this Imperial Contractor" who would cynically dispense patronage. "We do not care to dwell upon the picture," wrote the editor of *The National Police Gazette,* "but we are free to say that a sordid spirit, possessed of such powers and advantages as these, could calmly reject the exchange of his position for the Presidency of the United States, as an unworthy compensation, and the mere temptation of weak ambition."[16]

After such accusations it would have been reflexive for any self-respecting congressman to suspect the memorialist's own motives—curiously, Wilkes did not specify how his connection with the railroad should be arranged or compensated. A man always with his eye on the main chance, Mr. Wilkes seldom put pen to paper without the expectation of reward. However, even more damning, Wilkes proposed the road to be built from the national treasury "without reference to any special mode of raising means," and that transportation on it should be free for all. Reading that, a legislator of any era would fling the proposal into the fireplace—where it would at least do some good.

The West and the East waited to be joined, and untold millions languished. China and India wallowed in paganism. Pioneers like the Donners were reduced to savagery in the snows when they could be rolling through mountain passes at ten miles per hour and sipping tea and praising the railroad in sweet and musical accents. Meanwhile, Mr. Whitney suffered in dignified silence while Mr. Wilkes called him a crook. Dr. Carver retreated to the position that he was the "Originator of the Pacific Rail-Road." Whitney had never said he was first in the procession of dreamers—but the doctor's claim to primacy spurred a raft of complaints from "Iowaian," Mr. John Plumbe, the Welshman from Dubuque, whose memorial in 1838 had proposed a railroad from Lake Michigan to the Mississippi, which had been derisively likened to "a railroad to the moon." Plumbe had become proprietor of a chain of photography studios in at least eight major cities—his "Plumbeotype" gallery had drawn an editorial rhapsody from Mr. Walter Whitman of the *Brooklyn Daily Eagle.*[17] Plumbe's "Iowaian" letters (and those of supporters he prevailed upon to write on his behalf) angrily asserted fathership of the transcontinental idea, a claim both unwarranted and in any case meaningless. To his satisfaction, "a large number" of Dubuquers gathered on March 26, 1847, to pass the following resolution: "Resolved, Unanimously, that the meeting regards John Plumbe, Esq., as the original projector of the great Oregon Railroad." It proved to be inadequate recognition. Ten years later, the embittered "Iowaian" committed suicide.[18]

With three memorials already before Congress advocating what would be deemed "the northern route," it was only natural in those increasingly bitter sectional times that Southerners would come forward proposing their own. Colonel James Gadsden of South Carolina had first made a general call late in 1845, to be answered by one Professor C. J. Forshey of Louisiana, who urged a route running from his home state across Texas and a strip of northern Mexico to the Gulf of California at Mazatlan.[19] General Samuel Houston would respond with his own idea later in the Thirtieth Congress, a line from Galveston to San Diego using the Gila Valley. Southerners pointed out that their routes would be a thousand miles shorter than Whitney's and would traverse countryside less severe in weather and terrain. In reply Whitney maintained that he cared not for sectionalism; the unsettled public lands of the North lying at the base of his scheme could not be moved to suit politicians. Moreover, the Southerners typically overstated the problems posed by snows. In any event, he would add with terse Yankee acumen, with a torrid southern route all perishables being sent to market would, well, perish. And the trade route between Europe and China, from which all America stood to benefit, would be substantially lengthened if the southerly line were plotted, as anyone with a globe and a string could determine for himself.[20]

Meanwhile, reports trickled back to the States telling that Vera Cruz was secured and that, at a heavy price, troops ascended Cerro Gordo's blood-slicked trails. Simultaneously, as the snow-blocked high mountain passes leading to California and Oregon's Elysian Fields began to soften, as John Charles Frémont journeyed eastward across the continent toward a court martial for unlawfully assuming the governorship of California, as more eastern farmers read their Frémont and their Hastings and their Wilkes and looked rebelliously at their stone-choked fields, so did agitation begin to rise for easier transcontinental travel. In April 1846, Brigham Young and his Pioneer Band, 148 in all, descended from their winter quarters on a prominence overlooking Council Bluffs. Putting the Missouri between them and fifteen thousand fellow Mormons who would wait behind for their summons, the pilgrims commenced their long journey to a haven in the valley of the Great Salt Lake. Some miles later at the Platte River, Young was anxious to avoid Gentiles, who habitually moved eastward or westward on the southern bank. So the Pioneer Band blazed its trail on the north bank—and thus left their wagon ruts for the transcontinental railroad to follow.

That year, order books of the Pennsylvania iron mills fattened satisfactorily and the nascent industry of the Iron Horse picked up. In 1847 Irish work gangs began scraping a grade along the Hudson's eastern bank between New York and Albany. As they labored, perhaps laughing to themselves that their line was marring views and cutting off river access for the patrician estates on the overlooking slopes, elsewhere attorneys drew up charters of incorporation for the Chicago and Rock Island, the Mobile and Ohio, the Pennsylvania, and the Hannibal and St. Joseph. The directors of the latter enterprise included a Hannibalian justice, Mr. John Clemens, whose young son Sam had just become a printer's apprentice. Justice Clemens and colleagues had lofty aspirations—that the Hannibal and St. Joseph would become the eastern terminus for Mr. Whitney's Pacific railroad.

For his part, Asa Whitney kept busy. He addressed whoever would gather to hear him out, and his speeches increasingly took on an urgent tone: the empty lands were rapidly filling up with settlers—soon he would be making adjustments in his proposal calling for an eastern terminus not on Lake Michigan but at Prairie du Chien on the Mississippi. To delay much further might delay the railroad indefinitely. His congressional allies continued to prod the memorial along, their cause buoyed as a succession of endorsements arrived in the Capitol. During 1847 and 1848 the state legislatures of New York, Connecticut, Rhode Island, Maine, Vermont, New Jersey, Pennsylvania, Maryland, Ohio, Illinois, Indiana, and Michigan, and even of Georgia, Tennessee, Alabama, and Kentucky—the South was not united against Whitney's northern route—passed favorable resolutions. Often they were unanimous; his was "the only feasible plan for the accomplishment of the work," they said.[21] "Already," writes DeVoto of 1847, "Asa Whitney, the dreamer of railroads, was by no means the figure of cloud-cuckoo land which he had been a year before—precisely as the abolitionists had, in that year, somehow ceased to be madmen."[22]

There was, however, a quality of madness abroad in the land. Not out of diplomacy and compromise, but out of an "easy" war with Mexico, and out of bellicose saber-rattling against Britain, the nation's "manifest destiny" had prevailed over barriers both natural and political. With the gain of Texas and New Mexico and California and Oregon, a great semicircle of potential was now bound to the eastern states. The vast emptiness in that embrace—the unknown represented by the Great American Desert, by the mountains and the plains, and by the network of raw, rutted narrow trails traversing it all—would somehow, someday be filled. But what more was possible in that time when, to many, nearly anything seemed possible? In the states, the acquisitive frenzy which followed General Winfield Scott's appearance in the Valley of Mexico was stupendous: troops swarmed over the mingled bodies of Mexicans and Americans at Contreras and Churubusco to make their slippery assault on the heights of Chapultepec, and, on the night of September 13, they broke through the old Spanish walls of Mexico City. At home, in response to news spread of the capture of the halls of Montezuma, the popular sentiment was "Take California, take the Southwest, yes—and, by gum, take Mexico!" The sentiments which would, in a few years, unleash freelancers like William Walker upon Baja California and Nicaragua were prevalent in the waning months of 1847—until President Peña y Peña and Polk's clerk Nicholas Trist settled on more moderate terms early the following year. Mexico retained itself below the Rio Grande and the Gila and Colorado Rivers. Like a victorious poker player, the United States raked in the present states of California, Nevada, Utah, and Arizona, and parts of Colorado, Wyoming, and New Mexico. But beyond the token purchase price of $15 million (and assumption of another $3 million in claims against Mexico), beyond the nearly $100 million spent on the army and navy, lay the grimmer cost. Nearly thirteen thousand Americans were not to return from the war—casualties of combat and, the greater part of them, of cholera and typhoid and sepsis—and another four thousand would return wounded. And of course, because the war had mostly been fought south of the Rio Grande, Mexican casualties were considerably higher.

Arguably, it had been a high price to pay.

4

"The Great Object for Which We Were Created"

Throughout the battles of August and September in 1847, throughout the hoopla over territory, throughout the negotiations and the peace, Mr. Asa Whitney journeyed across the East to soberly exhibit himself before state legislatures. To those politicians he talked with, each dispatch seemed to underscore the importance of the work, though the memorialist himself appears to have refrained, in a typically single-minded fashion, from publicly commenting on the war and its settlement. "His language is plain, pointed, unadorned, and business-like," commented one observer, "and he sticks to his text, the 2,400 mile Railroad."[1]

Sticking to his text, Mr. Whitney was in Washington at year's end as Senator Alpheus Felch of Michigan presented his memorial to the Thirtieth Congress.[2] Though it was referred to the Committee on Public Lands and printed, Whitney's proposal thereupon stalled. It became buried in committee—as deeply as the Rockies' defiles were clogged with snow, where, it happened, one John Charles Frémont struggled to put the rebuke of his trial, and his petulant return to civilian life, behind him. He hoped, in fact, to regain his hero's aura by discovering a route for the transcontinental railroad across the Continental Divide. Thus it is probable that by this time, in the early winter of 1848, Frémont's father-in-law, "Old Bullion" Benton, no longer believed that such a feat was "impossible," or "one hundred years before its time." No. The Thirtieth Congress was under way; with territorial organization bills for Oregon the legislators would begin their battles over the control of slavery in the states; Mr. Thomas Hart Benton was second to none in his love for the Union—but, in one of his greatest failings of vision, he considered such an issue to be overblown. Even as it affected the presidential campaign nominations and debates, split the Democratic Party, and overhung the congressional first session, the issue was of "no consequence" to the Missourian. In July, the docket was clogged despite the lateness of the session; Benton was still steaming over his failed one-man filibuster against the brevet promotion of General Kearny, under whose authority Frémont's court martial had been instituted; his impatience finally broke—and it broke upon Asa Whitney.

On Saturday, July 29, the Senate dispensed with various miscellaneous business, including discussion of a federal judgeship in Louisiana, when John Niles of Connecticut saw his chance. Niles had chafed as the memorial of his good friend Whitney had first languished in the Committee on Public Lands before referral to a Select Committee of Five, where it stood every chance of staying until the session's close. Therefore he rose and moved that his colleagues allow him to set a date for considering his own bill—granting Asa Whitney the land required for a Pacific railroad. It was important, he said, "to have a decision concerning this measure at the present session, because the public lands would probably be so disposed of before another session."[3]

John P. Hale of New Hampshire was recognized. "I hope this will not be taken up," he said, "as it is a most alarming measure. It gives away one hundred millions of acres of the public lands at one swoop! I am sorry to differ from the Senator from Connecticut—but I am opposed to throwing such an immense mass of land into the hands of speculators." Taking up this bill, he said, would "alarm the public mind." Niles responded that Whitney's company could not hold the lands; it only intended to make a public road through them. Anyway, the lands were currently worthless and would remain so until the railroad gave them value. "If it could ever be completed so as to make a common highway," Niles said, "it will bind the Union together."

Senator Benton obtained the floor. His tirades of earlier in the month—against Kearny, against Polk, against his colleagues—had hoarsened his voice. Because of a childhood bout with tuberculosis, which had killed five siblings, Benton hemorrhaged from his throat in the heat of peroration. But the wrath came upon him again as he rose, handkerchief ready to be blood-spotted once more, and expressed his "astonishment" that such a measure would be brought up out of turn. "At the very moment that I was looking into the records of the past," he exclaimed, "to find out some mode of settling the difficulty of governing the Territories of California and New Mexico, my ear is struck by this motion! I have studied the history of California long before Mr. Whitney thought of it. . . . I would never vote for giving a hundred millions of acres to any man. . . . We must have surveys, examination, and exploration made, and not go blindfold, haphazard, into such a scheme!

"I should not be astonished," Benton finished, his hoarse voice heavy with sarcasm, "if Mr. Whitney should come here with a large bill for damages against Congress—damages for going to all the States of the Union for recommendations!" With milder voices dissenting—including that of Senator John Bell of Tennessee, who praised Whitney's character, intelligence, and care in preparing the bill, and who lamented that the opportunity for a Pacific railroad might be "passing away forever"—the bill was laid on the table.

Whitney was not swift in answering Benton's attack—but then, life moved at a more leisurely pace in those days. A few months later his reply appeared in *Hunt's Merchants' Magazine:*

> *The unprecedented, and I may say, outrageous attack from the Hon. Mr. Benton, in the United States Senate, on the 29th July last, which, it appears, was caused in part by his fears that I may make a claim on the government for having remained at Washington during four or five sessions, having walked upon the carpets of the capitol and annoyed members of Congress, renders it not improper (even without his permission) to simply state my position, so obnoxious to that gentleman.*

He outlined his campaign from 1845 to 1848. "I have done all this," he said,

> *at my own expense, and have never asked congress to appropriate one dollar for me. Even the printing of maps to accompany reports of committees, has been objected to by Mr. Benton, and were furnished at my own expense; and in no instance has Congress paid for any extra printing. In addition to all my time, I have expended a very handsome sum of money, and have never made any claim upon Congress. . . . And now Mr. Benton appears to be horrified, from the fear of a claim . . . and that his mind may not be "disturbed," but be at rest on my account, I do hereby forever renounce any, all, and every claim upon Congress or the people, for my efforts to get a railroad to Oregon.*[4]

Some months later, still stung by the personal attack, Whitney must have had Benton on his mind when he wrote that "though the demagogue may rave and rage, it is against a destiny he cannot change, a power controlling all, and he must fall to the ground."[5]

The merchant was perhaps unaware that Frémont was searching the Rockies for a practicable route; he may have believed for a time that Senator Benton stood squarely against any notion of a Pacific railroad. But certain forces were unleashed on December 5, 1848—forces which would increase by several decibels the din for easier transcontinental travel, forces which could convince almost any doubting Thomas.

First of all, the nightmare that had troubled James K. Polk had come to pass: General Zachary Taylor had seemingly ridden from his triumph at Buena Vista in an unerring course over the presidential aspirations of his brother-in-arms and gadfly, General Winfield Scott, next trampling another fellow Whig, Henry Clay. Wheeling about to face the party of Polk, "Old Rough and Ready" beheld the Democrats split over slavery, with Martin Van Buren leading his dissident Free-Soilers to their own convention—an assembly of signally noble and sadly premature intentions. With his foes thus divided, Taylor had galloped over both Van Buren and the soft-on-slavery Michigan Democrat Lewis Cass—and the general had dismounted as president-elect.

Sourly, the outgoing Polk sent Congress his farewell message. "Peace, plenty, and contentment reign throughout our borders," he said, "and our beloved country presents a sublime moral spectacle to the world. . . . While enlightened nations of Europe are convulsed and distracted by civil war or intestine strife, we settle all our political controversies by the peaceful exercise of the rights of freemen at the ballot box."[6] (Somewhere, as his words were read, the clock of disunion was ticking away toward midnight.)

The late Mexican war had disabused the world of any notions that the United States was weak or lacked resolve, Polk continued. "Our country stands higher in the respect of the world than at any former period." He moved to look with satisfaction at the "immense additions to our territorial possessions," the value of which was impossible to calculate. And then he let loose the bombshell.

There is a fearful symmetry in Polk's first and last annual messages to Congress. In December 1845, he had notified the world that "the people of this continent alone have the right to decide their own destiny," that the age of European influence over North America was closed. And three years later, the territorial limits increased by more than a third, Polk did more for American continentalism and for the westward migration—by making the first official announcement that gold had been discovered earlier in the year in California. "The accounts of the abundance of gold in that territory are of such an extraordinary character," he told Congress, "as would scarcely command belief were they not corroborated" by army officers who had gone into the Sierra Nevada east of Sacramento and found four thousand prospectors already up to their waists in the gold-rich streambeds. Congressional eyebrows were raised even higher when their outgoing president advocated establishing a new branch of the U.S. Mint in San Francisco.

The news, of course, spread wildly. Within weeks the demographic profile of the United States had changed unalterably. The lure of easy riches blinded tens of thousands to the hazards of the long sea voyage, along which storms and tropical microorganisms lurked, or the equally dangerous transcontinental land route where deserts and mountains and aborigines waited. At the aspect of that convulsive population shift, even Congress in its dawning wisdom, took notice. But mere attention would not solve the problem. As the rail historian Edwin Sabin has noted, the "crowding events of the past span [were combining] to minimize the once splendid vision of the Emigrant contributor, whose railroad should make his United States 'the first nation of the world'; and of Asa Whitney, who would 'civilize and Christianize mankind,' and 'compel Europe on the one side and Asia and Africa on the other to pass through us.' They began to color the florid word-pictures of Benton, and place politics above patriotism, prejudice above prescience, spoils above principles."[7]

The first ship bearing Forty-niners was creeping up the California coast toward San Francisco (it arrived February 18, and even its crew jumped ship for the goldfields) when Old Bullion Benton told his colleagues that he presently would introduce his own plan for a transcontinental railroad. His would link San Francisco with St. Louis. His would set aside a hundred-mile-wide "central national highway" from western Missouri to San Francisco Bay, with fifty-mile-wide branches to Santa Fe, to the tidewater region of the Columbia, and to the burgeoning, flowering city of Brigham Young at the Great Salt Lake. The Benton route would be festooned with telegraph wires and initially carry a government-subsidized wagon course and a railroad whose tolls and taxes would be limited to costs of maintenance. Along the route each settler would receive land grants of 160 acres; parenthetically, the plan indicated preemption or extinction of Indian land titles, such as they were.

And Benton's plan was as short on engineering figures as any other before Congress, including Whitney's, since no formal surveys had yet been authorized. But, Benton said, "there is a class of topographical engineers older than the schools and more unerring than the mathematics. They are the wild animals—buffalo, elk, deer, antelope, bears—which traverse the forests not by compass but by an instinct that leads them always the right way to the lowest passes in the mountains, the shallowest fords in the rivers . . . and the shortest practicable lines between remote points. A buffalo road becomes a warpath . . . and finally the macadamized or railroad of the scientific man."[8]

Now Congress was confronted with four transcontinental plans: those of Asa Whitney, George Wilkes, and Hartwell Carver (all of which would begin at Lake Michigan, or the Mississippi, and run west to the Platte Valley, the South Pass, and beyond to California and Oregon), and that of Thomas Hart Benton (by way of the Kansas and Platte Rivers, following bear scat and elk prints through South Pass). Then, ten days after Old Bullion's announcement (his formal presentation came later in that second session), the South was heard from. First, Senator Solon Borland of Arkansas introduced a measure for a route to run from Memphis through Borland's home state and thus to the Pacific; then, Senator Samuel Houston launched legislation for a railway between Galveston and the Pacific.[9]

Congress was left with five railroad resolutions when Borland's measure was tabled. But five resolutions was four too many for any ready agreement; the clamor of the ensuing debates held in the parlors of the capital and in general stores from Maine to California and Michigan to Texas was a Babel of special interests: Whitney's plan, the arguers complained, was the brainchild of a diabolical speculator who was constitutionally unable to comprehend snow; Benton's, the reply swiftly came from the Whitney supporters, ran through timberless prairies until it collided head-on with Indian Territory, which treaties had made inviolable; Houston's, the arguers continued, traversed even more barren country until it fetched up at our Mexican border—and how could an American railway be built on foreign soil?

Now in his fifth year of agitation—his funds dwindling but not yet exhausted, his faith severely tried but not bested—Asa Whitney published in May of 1849 a pamphlet of over a hundred pages, entitled *A Project for a Railroad to the Pacific.* "My plan has become the foundation for others to attempt to build upon," he wrote in its preface, "but all [their] supposed improvements, yes, and more too, have been examined by me, and discussed with others, long ago, and thought to be not feasible.

"I have but one motive, or object, and that is to see this great work successfully accomplished, which would be a sufficient reward for my labors; and if there can be found a better plan, or a man whom the nation may think better qualified, then I am ready to support that plan, or sustain that man with my efforts, and all the information which my seven years' labors have gathered together."[10]

The pamphlet was not only encyclopedic but up to the minute. It included Whitney's exhaustive discussions on all the various railway plans, the routes' topographies, the potential world commerce, his head-turning theory placing North America at the center of the world instead of at its periphery—in short, the arguments he had been repeating and perfecting for five years. But then it moved beyond—to the developing gold-rush economy of the West, and the blooming shortage of food and supplies; to the problems inherent in oceanic transit, including the unhealthful isthmian shortcuts at Panama and Nicaragua; to the coming shift in population when the surpluses of Europe would flock to the American West. In a cursory flip through that extraordinary document a reader would behold the names of Senator Benton (his change of heart on the railroads was documented) and Colonel Frémont (his expeditions proved, Whitney claimed, that the snows of the North would not hinder his railroad), and those of eminent persons attesting to Whitney's plan. The reader would find memorials and maps; bills and charts; congressional committee reports; tables of tonnage of export and import for China, Java, the Philippines, and India; and resolutions of nineteen state legislatures and fourteen municipal public gatherings, all of which (of course, for why else would they be included) endorsed Mr. Whitney in glowing terms. No one person had ever devoted so much work to the Pacific railroad. The sheer concentration of that one mild-mannered and pious merchant, bereft of the sustainment of family but apparently buoyed by the enthusiasm of friends, towered over the massed effort of our entire procession of dreamers; few would ever equal it.

Asa Whitney's little book represented the peak of his effort; he was by no means finished, but, as Congress struggled with its sectionalism and as time passed, westward expansion—the filling in of this nation's empty spaces, especially those of the fertile, nearly unpopulated plains Whitney had explored with his party of collegiate adventurers only a few years earlier—was rendering his equation obsolete. With Minnesota now territorially organized, with Wisconsin now joining its neighbor Iowa in statehood, the pressures of population were inevitably seeing to it that unsettled lands along Whitney's route were becoming unavailable for use in financing a Pacific railroad by their very disposition. Already, something actually called the Pacific Railroad had been chartered in Missouri—it would, when completed in 1856 as the Missouri Pacific, extend only from St. Louis to Kansas City, but its significance would lie in its being the first railway built west of the Mississippi. In another year the Thirty-first Congress would embark on the great land-grant program that would ultimately transfer twenty-one million public acres toward the support of rail transportation in the settled states; a year more, and the new Illinois Central would become the first land-grant railroad—thus proving that Whitney's equation worked. But business in the Senate and the House of Representatives came to a close in the summer of 1849 with no significant movement of bills supporting Whitney—or, for that matter, any other railroad promoter. The first session of the Thirty-first Congress, scheduled to begin in December 1849, stood apparently as the merchant's last chance.

No frenzy of migration, no amount of pressure from rail enthusiasts, would be enough to deviate the Thirty-first Congress's attention from its thrall: the immensity of our new territories and the political task of carving them up according to sectional demands. There would be little time except for what the textbooks now compactly refer to as the Compromise of 1850—those ten months of exhausting, furious debates and arm-twisting—the resultant five bills which admitted California as a free state and abolished slavery in the District of Columbia but which, in a morally insane balancing act, organized the territories of New Mexico and Utah with the question of slavery open and, furthermore, streamlined fugitive slave laws throughout the Union.

"I fear the slavery agitation will prevent anything being done this session," ex-Senator John Niles of Connecticut wrote his friend Whitney. "The Missouri scheme, also, will be earnestly pressed by Col. Benton and others, and by dividing the opinions and views of members, will tend to lead to postponement, or non-action."[11] Niles had overestimated Benton's threat. By then Old Bullion, for too long inattentive to the slavery issue, had lost power among Democrats at home and in the Capitol; and he could look forward to passage of no further pet projects, especially his National Road to the Pacific. His last redemptive senatorial act would be to vote against dividing California; consequently he would lose his next election, and following a single term in the House would be cast back into private life forever.

But Benton aside, Niles was correct to temper his hope. Early action in the House had given him and Whitney some optimism—for some attention was spared for railroad legislation even in the face of the slavery question. At least two bills emerged favoring Whitney; one proposed in March in the House would have assigned the amount of $4,000 per year to the merchant for supervision of the project, the first compensation he would have received for his campaign.[12] Though this outraged Whitney's detractors, to his supporters it seemed eminently reasonable. One newspaper in New York praised his "Napoleonic perseverence" in the project, which had "forced universal attention to its grandeur and importance."[13]

Too, the House Committee on Roads and Canals issued a report in which it concluded that "the instincts of the American people seem to have leaped in one bound to a sense of the importance of a railroad connexion between the Atlantic and Pacific, and there can no longer be a doubt that this project is decreed in the public mind. Congress may lag behind, but cannot get before, the increasing eagerness of the people, in putting forward this stupendous plan." Furthermore, it noted that representatives could not "find any other route or plan besides Mr. Whitney's which promises to furnish any means adequate to the enterprise, or which will do anything considerable towards it; nothing, indeed, worth reckoning, except by dependence on the public treasury, or public credit."[14]

However, a motion to have the report printed was defeated and the work went no further.

Whitney's last opportunity to plead his case presented itself at the commencement of the second session. Thanks to the efforts of two congressmen from North Carolina and Georgia, the rules of the House of Representatives were suspended on January 13, 1851, and the body passed a resolution allowing the merchant to appear before them. One week later, for once not restricted to the Capitol's periphery but standing near its center, at the House rostrum, Whitney took courage at a hall packed with legislators and other spectators, and launched into an impassioned, two-hour address—on the importance and feasibility of his Pacific railroad, the settling of the public lands, the urgency for action then or never. Every steamer brought him letters from England urging him to present them his proposition, he told his listeners. Would it not be ironic, would it not be a tragedy, if Parliament should act where Congress had failed, if it would be on British Canadian soil that the world's commerce would travel, if the United States should not take up this supreme opportunity? His words to that attentive crowd were thought by observers to be eloquent and persuasive. But in the end, they were wasted. The campaign which had begun nearly a decade before on the deck of a poky, benighted barque as it sailed between a vanished America and a vanished China was, for all intents and purposes, over.[15]

The rest was dispiriting denouement.

In defeat, Asa Whitney sailed on the Cunard steamship *Asia*, on March 26, 1851, bound for Liverpool. It would have been merciful if only Whitney, and not any of his friends or the public, saw this pilgrimage in search of British support for his plan in all its tragic aspects. A man who had once fulminated in his diary against "English pride English Tyranny & oppression" could not have easily abandoned those beliefs or turned from his homeland, no matter how momentarily.

While he was away, American newspapers continued their coverage of his campaign whether they supported him or not. His most dependable critic, the *New York Herald*, ran a series of vituperative editorials, including this one of June 15:

> *We notice Mr. Whitney is seeking to get the mother country to aid him in making a Canada and Pacific railroad, and we care not if he succeeds, as, in such case, Congress will escape being bored any more about his humbug. He will never get the aid of the United States, in any way, to his scheme, by a donation of lands or otherwise. The road, if made, will be made by piecemeal, and be connected from time to time, as the settlements progress westward from the Atlantic, and eastward from the Pacific coast. A road, to extend upwards of two thousand miles, through a desert, through rugged and snow covered mountains, and a country peopled by warlike savages, is a work of a quarter of a century. Mr. Whitney's scheme is the most chimerical of all that have been proposed. Great Britain has too much sense to embark on it. If she is disposed, we shall profit by her folly.*

And, as dependably, newspapers, like the *New York Tribune*, that supported Whitney lauded the "singular clearness, enthusiasm and tact of the advocate" and the "grandeur and feasibility" of his program. With one voice they expressed hope that the British offers would be refused—that it would be America that profited by the work.[16]

A certain momentum carried Asa Whitney to England. But it was a momentum out of synchronization with British events. His addresses before various august bodies, including the Royal Geographic Society, received respectful hearings; interviews with parliamentarians, whose political support was needed, went well; a number of capitalists approached him to offer finances should he obtain Canadian land grants; and the press seemed to like his program and manner—but Whitney's campaign fizzled out. Twice defeated, his reserves of faith now as reduced as his purse, he returned home on the packet *Isaac Webb* on August 18. No newspaper remarked on his return.

Asa Whitney disappeared so completely from the public view that a decade later, with Congress still debating the Pacific railroad, a representative impatient with the legislative pace exclaimed that "the shade of Whitney—for I believe he is dead—ought to teach the gentleman from Vermont that this road will be the means of carrying the fruits of our abundant harvests to feed the famishing millions of the Old World."[17]

But Whitney was not dead. "I have isolated myself from the world," he would later write to a newspaper editor, "and have not written one word on the subject of a railroad to the Pacific during the past fifteen years."[18] Indeed, Whitney's return to private life was no more abrupt than he preferred; he was surfeited with bitter disappointment and had no more appetite for the public spotlight. In October 1852, nearly twelve years after his second wife had died, he was married to the widowed Mrs. Catherine Campbell of the District of Columbia, who was the daughter of a wealthy planter of Wilmington, North Carolina.[19] Together with his wife and sisters-in-law, Whitney bought a large farm outside the District in Maryland, which they called Locust Hill. Living well but quietly, a gentleman farmer, he managed a small dairy on his farm and became active in the local Rock Creek Church. In 1865 he addressed a letter to his sister, Lucy:

> *My health has been very bad through the summer and fall, I am now better, but Lucy I am getting to be an old man, I believe with March 68 years. My time here cannot be long, everyday I feel it more. You once said 'you have seen much,' yes too much. Much of it, as life, seems but a dream, worse far worse than a dream unless we fulfill the great object for which we were Created, to purpose to die. . . . I for myself have very little to do with the world. My intercourse with it has dwindled to almost nothing.*[20]

Typhoid fever took him on September 17,1872.[21]

"Mr. Whitney himself is forgotten," wrote one of his partisans in 1883, "and there is not even a station named for him on any of the great lines of continental railroads."[22]

5

"An Uninhabited and Dreary Waste"

At the risk of unduly coloring our parting image of Asa Whitney, certain words he wrote in his bitter denouement bear examining—they are instructive in understanding the hidden political forces against which Whitney struggled and ultimately lost. "But for Mr. Benton and others," he had written the editor of the *New York News* in June 1868, his only public utterance on the subject, "the Pacific Railroad would have been completed and now in successful operation." The failed promoter wrote that such figures—Benton and Douglas included—"never labored to bring forth a railroad to the Pacific, till it was made a gambling, stock-jobbing Wall Street and Threadneedle street concern, backed by some hundred millions of Government bonds, with an annual subsidy from the Treasury to meet the expenses for its management, to be provided for by an enormous and burdensome tax upon the labor of the country to pay the European stockholder his rich dividends."[1]

Whitney was understandably bitter that Congress would fail to carry a self-financing, altruistic railroad enterprise such as his, one that put the whole nation over sectional or private interests. But some of the heat of his complaint must be subtracted, for it is not accurate to assume, as he did at the time of his congressional defeat, that politicians were not fighting their private fights for a Pacific railroad—whatever their public pronouncements. Indeed, as will be seen, it would be the gamblers and stock-jobbers who were almost predestined to prevail.

It should be obvious that as the railroads were seen in all their potential, beginning in the early 1840s, so developed the political struggle for spoils. It had fanned inward from the Atlantic Coast at a slightly higher pace than the rails themselves, for it was fueled not by manpower or steampower but by the prospect of money. First, the coastal cities such as Boston, New York, Philadelphia, Baltimore, Charleston, and Savannah had vied for railroad termini and the business and prestige they would bring. As the routes lengthened, inland cities such as Atlanta, Chattanooga, Nashville, Chicago, Detroit, Buffalo,

Cincinnati, Wheeling, and Pittsburgh clamored for their share; but this level of competition became as nothing when the cities of the Mississippi River valley entered the picture—the westward tilt of population was under way, and with it vast opportunities for commerce. The gold rush, and California statehood (in 1850) opened what eyes were still closed in politicians' heads. By the time of Whitney's defeat—certainly within a year—it would be the rare legislator who was not involved in the scramble to promote his state or district's cause. Moreover, soon enough some might say that it was the even rarer public servant who had not somehow provided for his declining years by buying stock in a rail enterprise, or by purchasing land at a possible terminus or along a proposed route.

The extent, for instance, of Senator Benton's connections with wealthy businessmen and rail promoters in St. Louis will never be known; his tenacity in fighting for their interests might have only been politically, and not financially, motivated. But Benton would not have objected if his son-in-law had moved from recklessly slogging through snow-blocked Rocky Mountain trails in search of a rail route (an expedition financed by two St. Louis businessmen) to the boardroom of a railroad corporation; he certainly knew that Frémont had anticipated the transcontinental by purchasing California land from which he and Jessie Benton Frémont would profit.[2] For his part, Senator Stephen Douglas had allied himself politically and financially with Illinois and Iowa promoters—and he had hedged his bet by purchasing a large tract on the shore of Lake Superior (it would become Duluth), an obvious terminus for a far-northern rail route.[3]

The list of legislators personally interested in the transcontinental route goes from these towering figures on down the line to the midgets. But the antics of those in the Thirty-first Congress, when Whitney and plan bit the dust, paled before those in the following term. Most of the second session of the Thirty-second Congress (1852–53) was devoted to furious debates over a Pacific rail route. The clamor went beyond the struggle of North against South; it was an anarchy of state against state, county against county, city against city, township against township. By then no one was uninvolved in the question—and, because everyone was involved, the obvious result was stalemate.

But it was a stalemate in which the dim light of an exit glimmered slightly. If there could be no agreement on a single route, then at least all parties perceived that hard evidence was needed. Even Senator Benton and Mr. Whitney had been in accord in calling for official surveys. So, on March 2, 1853, Congress transferred its decision-making power to the secretary of war, Jefferson Davis of Mississippi, who was authorized to send his Army Topographical Corps to survey five possible transcontinental routes. Davis was given only ten months to complete the surveys and prepare a detailed report. It was a daunting, if not impossible, task.[4]

Prior to this, there had been only a few western surveys completed with the idea of locating a railroad. In the winter of 1848–49, John C. Frémont had ignored warnings of experienced mountain men and entered the Colorado Rockies with a particularly hard winter already descending. Though many in his party starved or froze to death, and others appeared to have survived by cannibalism, Frémont emerged alive and announced that he was convinced a railroad could bridge the mountains.[5] In stumping for a route he himself had found impassable, Frémont could offer only bravado, and not science, as his evidence. This could not be said, however, for subsequent efforts performed by the army. Beginning in 1849 and extending into 1852, the Army Topographical Engineers sent several expeditions across the Southwest, which located possible routes. Also successful was Captain Howard Stansbury of the corps, who in 1849–50 reconnoitered the Emigrants' Trail along the Platte River, determined a likely path through the Wasatch Range, and mapped the valley of the Salt Lake. Practicable these routes across the Southwest and over the Rockies may have been, but Congress had taken no action.[6]

The government surveys of 1853 were under way by early in the summer. They cost a fortune. Although they were supported at first by an appropriation of $150,000, it was later necessary to add another $20,000—and this vast sum was not to be expended solely on topographical engineers searching for railway gradients. Secretary Davis and his topographical chief, Major W. H. Emory, assembled a small army of engineers and explorers, yes, but also of anthropologists, botanists, cartographers, geographers and geologists, meteorologists, paleontologists, zoologists, and even water colorists. Railway gradients they would seek, but along the way they were to collect mammals, reptiles, amphibians, plants both endogenous and exogenous, fossilized woods, land shells, soils and incrustations, minerals, mosses, and liverworts—all for scrutiny back east—and they were to gather data on various Indian tribes. "Not since Napoleon had taken his company of savants into Egypt," writes the historian of the West, William H. Goetzmann, "had the world seen such an assemblage of scientists and technicians marshalled under one banner. And like Napoleon's own learned corps, these scientists, too, were an implement of conquest, with the enemy in this case being the unknown reaches of the western continent. The immense quantity of data collected by these government scientists constituted a plateau from which it was possible at last to view the intricacies of western geography."[7]

Instruments, notebooks, and collecting bags at the ready, the savants swarmed into the West. The first party, under Governor Isaac I. Stevens of the newly created Washington Territory, was ordered to traverse westward from St. Paul and the Great Bend of the Missouri, between the 47th and 49th parallels, in the hope of discovering a rail route between the watersheds of the Missouri and the Columbia Rivers. His reconnaissance completed, Governor Stevens not surprisingly called the route to his territory "eminently practicable."

The second party, keeping between the 38th and 39th parallels (the "great central route" touted by Senator Benton), was commanded by Lieutenant John W. Gunnison. It was directed to find a connection between Fort Leavenworth, the Arkansas headwaters, and the Great Basin, using the Cochetopa Pass or any alternates. Tragically, Gunnison and seven others were massacred in October by Utah Indians along the Sevier River, but not before he had reported that the 38th parallel route was impracticable due to the vast number of tunnels and bridges required. Later, Gunnison's assistant, Lieutenant E. O. Beckwith, took over the central route's exploration and reported success along the 41st parallel—from the Laramie Plains through the Wasatch or Timpanagos Canyons to the Great Basin, and thence across the basin to the Humboldt River and two practicable Sierra passes.

While Gunnison's and Beckwith's engineers surveyed these routes, Senator Benton schemed to keep his "grand national highway" along the 38th parallel under his control. Rebuffed in his attempts to have his son-in-law head the official expedition, he arranged a bankroll for Frémont's own party, which followed with great clamor on Gunnison's heels. But the freelancer again met disaster in the mountains, this time losing a man to the snows. Another unofficial party instigated by Old Bullion, that of Edward Fitzgerald Beale, carried its own press agent; in the end, however, their publicity would not offset Gunnison's negative report.

The third official party set out along a 35th parallel route under Lieutenant Amiel W. Whipple, from Fort Smith, Arkansas, west to Albuquerque and Yuma and on toward Los Angeles. When finished, Whipple confessed that his route was no doubt best.

Commencing somewhat later than the above expeditions, two parties under Lieutenant John G. Parke and Captain John Pope explored a route along the 32nd parallel; Parke worked eastward from Fort Yuma on the Gila to Tucson and El Paso; Pope continued across Texas in an arid line between the Rio Grande and the Red River. Following this, Parke and Pope were in accord that their route was unsurpassed.

Meanwhile, survey teams under Lieutenant R. S. Williamson explored California for suitable routes through the Coast Range and Sierra Nevada. In the north they deemed only two passes leading to the Sacramento Valley as practicable for rail use, discarding three others. Out of the possible routes linking the deserts with the Southern California interior, the 35th parallel route seemed far better than the 32nd. Later, other expeditions searched for rail routes between Los Angeles and San Francisco (a coastal road was found), and between San Francisco, Oregon, and Washington (two routes were found to exist on either flank of the Cascade Range).

Secretary Davis did not meet his ten-month deadline; his three-volume preliminary report appeared a year later, in 1855, and some months after that the full data began to be issued in a magnificent thirteen-volume, quarto-sized report. It contained hundreds of illustrations of geological formations, of plants and animals and birds, and of Indians, and it contained maps and profiles and charts and tables. And it contained the reports of engineers seeking the elusive transcontinental rail route: Governor Stevens touted his northern road from St. Paul to Washington Territory; Lieutenant Beckwith promoted his 41st parallel line; Lieutenant Whipple spoke for the 35th parallel; and officers Parke and Pope urged adoption of the 32nd parallel, a route unabashedly favored by Secretary Jeff Davis, Southerner, who in his zeal fudged data and budgetary calculations before Congress.[8]

Having asked to be delivered from stalemate and indecision, Congress found itself even more confused.

As the volumes made their expensive way from the government bindery to the office shelves of senators and representatives, one wonders how many congressmen troubled to read very far into that monumental collection of data; probably few political minds attuned to special interests were changed after the reports appeared, their scientific value notwithstanding. In the January 1856 issue of the *North American Review of Reviews*, after examining reports of the war secretary and his various reconnaissances, the editors admitted that the nation was still confronted with a number of choices, none of which was without its special problems. But the editors derived a cautionary overview of America's new western territories. "Before the accession of California," they wrote, "the western possessions of the United States were looked upon as a sort of fairy land basking under the influences of a most delightful climate, and enriched by the choicest gifts of Nature." But, they despaired,

> *our rich possessions west of the 99th meridian have turned out to be worthless, so far as agriculture is concerned. They never can entice a rural population to inhabit them, nor sustain one if so enticed. We may as well acknowledge this—and act upon it—legislate upon it. We may as well admit that Kansas and Nebraska, with the exception of the small strip of land upon their eastern borders, are perfect deserts, with a soil whose constituents are of such a nature as for ever to unfit them for the purposes of agriculture, and are not an expenditure of angry feeling as to who shall or who shall not inhabit them. We may as well admit that Washington Territory, and Oregon, and Utah, and New Mexico, are, with the exception of a few limited areas, composed of mountain chains and unfruitful plains; and that, whatever route is selected for a railroad to the Pacific, it must wind the greater part of its length through a country destined to remain for ever an uninhabited and dreary waste.*

Any real engineer would have admitted that the government's transcontinental reconnaissances were not practicable for a railroad project. Called to address a purely political question, the surveys had proved that tracks could be laid along a number of transcontinental routes—and indeed, in the fullness of time they would be. The political question had been addressed; it had not been answered. However, in such a stupendous project there was also the need of facing the economic question. A number of voices raised this objection as those expensive quarto volumes gathered dust on congressional shelves, but we shall listen to just one. He was a New Englander with a clear mind and a passion for numbers and calculations of any kind, but unlike some other figures who had played significant roles in coming up with a viable transcontinental railroad plan, he had failed to secure an appointment to a national military college for lack of political influence; otherwise, he likely would have taken part in the government surveys and our history would be different. But as it was, his engineering career was impressive and had been recently capped with a singular attainment: he had built the first railroad in the trans-Missouri West. His name was Theodore Dehone Judah.

"When a Boston capitalist is invited to invest in a Railroad project," young, brash, pragmatic Mr. Judah responded to the government reports,

> *it is not considered sufficient to tell him that somebody has rode over the ground on horseback and pronounced it practicable. He does not care to be informed that there are 999 different variety and species of plants and herbs, or that grass is abundant at this point, or Buffalo scarce at that; that the latitude or longitude of various points are calculated, to a surprising degree of accuracy, and the temperature of the atmosphere carefully noted for each day in the year.*
>
> *His inquiries are somewhat more to the point. He wishes to know the length of your road. He says, let me see your map and profile, that I may judge of its alignment and grades. How many cubic yards of the various kinds of excavation and embankment have you, and upon what sections? Have you any tunnels, and what are their circumstances? How much masonry, and where are your stone? How many bridges, river crossings, culverts, and what kind of foundations? How about timber and fuel? Where is the estimate of the cost of your road, and let me see its details? What will be its effect upon travel and trade? What its business and revenue? All this I require to know, in order to judge if my investment is likely to prove a profitable one.*

"When the friends of the Pacific Railroad can approach a capitalist and answer all these questions," Mr. Judah pronounced, "they may begin to hope for a realization of their wishes." He saw plainly that, for the present, the political question of the transcontinental railroad was a conundrum. "The advocates of any [route] do not believe that more than one road will be built," he said, and therefore they were all afraid "to give all a fair opportunity, lest their neighbor might get the advantage; they will probably manage to do as they have done before—defeat the measure."[9]

Probably Judah did not realize that he was nearing a solution—a simple, backward-entering solution that owed more to the ingenuity of the new West than the time-tested ways of the East. We can imagine, in fact, a shock of recognition among our Procession of Dreamers—Robert Mills looking up from his architectural drawings, Samuel Dexter pausing over tomorrow's editorial, John Plumbe lingering thoughtfully over a tray of developing solution, Asa Whitney turning from contemplating his green pastures. We assume they did not, but we know they should have, for their ranks had just grown again.

With all of that procession, it had been Asa Whitney who had succeeded in putting his dream on the national agenda. It would fall to Theodore Judah to make that dream a reality. But the two had something else in common: both—ironically, tragically—would be denied the fruits of their prodigious labors, despite the nation's incalculable enrichment.

Part II

1860–61
Union, Disunion, Incorporation

6

"Raise the Money and I Will Build Your Road"

He could be seen early any weekday morning, erupting from his boarding house on Fourteenth Street and charging down Pennsylvania Avenue—a teeming thoroughfare midway between mud and dust because it was springtime and it was Washington City, 1860. It is likely he would have skipped breakfast because he was absolutely, unswayingly possessed by his work.

The city in springtime would have been a noisome place, with all the winter's refuse defrosting in ugly piles along curbs and hulking in vacant lots, surfacing in drainage ditches and the canal. There would have been a flurry of speculative building—the noise of hammers and saws fashioning gimcrack structures from cheap pine would play a counterpoint to all life for the next twelve months. For most of Washington, the practice of politics was the practice of scrounging a living out of life's margins, picking at the leavings of the high-hatted individuals on the eastern and western ends of Pennsylvania Avenue. Each changing administration always set off a convulsion of appointment-seeking, a turmoil of advantage-taking, a spasm of leverage-bidding. There was no one in the city who was unaware that but one lame-duck year remained in President James Buchanan's term, and that a mighty change was on the horizon. Above the wafting aromas of wastes, pine pitch, and tar there was overlaid the scent of politics—and the certainty of a feeding frenzy beyond.

He would scarcely have noticed the flimsy structures (some brand new) lining Pennsylvania Avenue, the trash in its margins, or the shabby office clerks hurrying to their warrens, and he would have not been distracted by the din of carts and carriages and their cursing drivers—he would have swept by it all untouched, for he was anointed with purpose. Beggars accosting the handsome, bearded figure would have come away satisfied, for though he was inclined to watch his pennies he was a sensitive man. Looming before him, growing larger as he approached, was his goal, the unfinished Capitol building—where it was difficult to tell where the ruins left off and the improvements began. Its iron dome was exposed, waiting for the confection of stone that would rise above it; the structure's new wings were standing but still lacked adornment; a village of wooden shacks and workshops squatted in the mud at its foundations; the ground was littered with piles of materials and the odd statuary destined to preside over the finished building's sandstone and marble ramparts.

His engineer's eye would take note of any progress from the day before. Then Theodore Dehone Judah would pass inside—dashing up the Capitol staircase while clutching a sheaf of newspapers and pamphlets, ascending into the blue pall of cigar smoke on the second floor, unlocking a grandly paneled door of a committee room handed over for his personal use: the Pacific Railroad Museum. Presiding over this resort, he would welcome all visitors with their varying importance, while they would be dazed by his proficiency, his knowledge, his persuasion—the paragon of lobbyists, a beneficent zealot, not just let loose in the Capitol but assigned a lair.[1]

He was born thirty-four years earlier, on March 4, 1826, in the rectory of St. John's Church, in Bridgeport, Connecticut. His father, the Reverend Henry R. Judah, an Episcopal pastor, christened his second son for the good Bishop of Charleston, Theodore Dehone, a friend of the parents and the author of a number of Episcopal charges and sermons. The family lived in Bridgeport until 1833, when the Reverend Henry Judah was called to St. John's Church in Troy, New York.

It was there, at the junction of the Mohawk and Hudson Rivers, that young Judah spent the latter part of his boyhood. One gets an impression of sunlight filtered through both the smoky industrial skies over Troy and the stained-glass windows of an old stone church—because naturally his life revolved around St. John's; he would spend hours in the church at the organ, playing mostly by ear, and he drew most of his friends from the congregation. Considered a precocious, if moody, student, Ted was permitted to enroll three years early in the Classical Course of the nearby Troy School of Technology, an institution renowned for its training in civil engineering and later renamed Rensselaer Polytechnic Institute. While it is doubtful that the younger boys at the Troy School would have absorbed much of the engineering principles being taught to the advanced classes, there seems to have been a certain flow that affected young Master Judah. Senior students were instructed on the newly emerging technology of railroading, still in its infancy—the flanged wheels and T-rails of Robert Livingston Stevens, for instance—and presumably they would have studied their application in the yards of the Schenectady and Troy, and also they would have investigated other facets of railroading in that industrial crossroads.[2]

Some of it reached young Ted Judah in the year and a half he spent at the school. When his father died, aged only forty-two, and his mother moved her family down to New York City, the thirteen-year-old Ted had to put aside his studies. Moreover, any thoughts he might have had of following his older brother into a national military academy were put to rest. (Henry Moses Judah was enrolled at West Point, and Theodore had expressed interest in Annapolis.) Instead he took a job as a chainman on the Schenectady and Troy Railroad—an apprenticeship on a pioneer New York State line under the esteemed engineer S. W. Hall.

By the time he was eighteen, Judah was a surveyor on the New Haven, Hartford and Springfield, a line that would follow the easy grades of the Connecticut River from Long Island Sound up toward Vermont. First lugging his equipment along a Massachusetts section between Springfield and Brattleboro, Vermont, he was then sent down the river and promoted to location engineer on the Hartford section. Then he became assistant to the chief engineer, with an office in Greenfield, Massachusetts. It was while he was attending Episcopal services in Greenfield that he met Anna Ferona Pierce, the only daughter of St. James's senior warden. Her parents may have frowned upon a match between the two—Ted had the right pedigree and prospects, but certainly no family money, being a clergyman's son, while Anna's father was a prosperous merchant and cousin of the New Hampshireman Franklin S. Pierce, at the time a retired U.S. senator and some years shy of the presidency. However, the young couple prevailed. Ted Judah and Anna Pierce were married in St. James's on May 10,1847.

At twenty-one, despite modest circumstances, the newly married Judah had assembled an impressive dossier. He had been called to his profession when the railroad was still in its infancy, and, having proved himself at a time when civil engineers were at a premium and the northeastern states were becoming densely spiderwebbed with lines, he seemed to jump from job to job, besting hard tasks and taking his instruments and drafting tools on to the next assignment, sometimes even before his blueprints were transformed into working enterprises.[3]

Around this time we may trace him to Vergennes, in Vermont's Champlain Valley. There, for the Rutland & Burlington Railroad, he constructed a railroad span over an abrupt, deep chasm through which the Little Otter Creek tumbles on its way from the slopes of the Green Mountains toward Lake Champlain.[4]

The bridge was a difficult feat for the era, but a harder assignment awaited him. In western New York, a group of capitalists conceived the idea of controlling commerce between Lake Ontario and Lake Erie—a route meant to grasp some of the business passing through Mr. DeWitt Clinton's canal and through the Great Lakes route from the new western states. Their enterprise was incorporated in 1852 as the Niagara Falls and Lake Ontario Railroad

Company. Theodore Judah was named the associate engineer. He was given the difficult task of running a line down along the Niagara Gorge. There had been much doubtful talk in upstate New York about the practicability of such a route, but in the presence of the company's directors—all influential men in state affairs—the brash young engineer simply declared, "Gentlemen, raise the money and I will build your road," which they and he proceeded to do. In a remarkable fashion, and showing no restraint at the large and expensive amount of rock-cutting required, Judah ran his line down along a narrow shelf in the cliff between the Devil's Hole and Lewiston, with the river boiling far below.[5]

Writing many years after the fact, Anna Judah recalled, "He was the only engineer who believed in it, who had gone over the ground. Did it not show what was in the man then, to grasp the gigantic and the daring? The railroad down the gorge was a success and stands today a monument to his young power. Our cottage on the bank of the river between the falls and [the village of] Suspension Bridge is still there with the beautiful view, of both falls and whirlpool rapids below the suspension bridge. He selected the site, built the cottage, and there had his railroad office, and did his work for that—in those days—wonderful piece of railroad engineering and work." For years thereafter the line operated and in its later incarnation became part of the New York Central. It provided commercial shippers with an easy link between Lake Erie and Lake Ontario, and it afforded tourists, especially those seated on the western side of the train, with quite a nerve-racking ride down the canyon.[6]

However, the Judahs were not destined to take that ride, for Ted had moved on even before the grading was completed—thus probably just missing the opportunity to meet the future designer of the Brooklyn Bridge, John A. Roebling, whose task at the falls would be the awe-inspiring suspension bridge. By the time of the 1854 opening of Roebling's span and of Judah's line along the Niagara Gorge, Ted and Anna would be three thousand miles away. For a short while Judah was resident engineer on the Erie Canal's western half between Gordon and Seneca Falls—a job that could not have been very demanding, since the "Big Ditch" had been completed twenty-five years earlier. Soon he became chief engineer of the Buffalo and New York Railroad as it nosed across the state toward a link with the Erie road. It was 1854; legions of deadline-conscious government clerks were toiling away in the District of Columbia, where they hoped to make sense of the voluminous Pacific railroad surveys—still a year away from publication. Asa Whitney occupied a rocking chair in Maryland. And an entrepreneur sailed into New York Bay from San Francisco with a daring plan and a position to fill.

Originally from a farm in Maine, Colonel Charles Lincoln Wilson had seen action in the Mexican War before following the lure of gold to California. Like most of the Argonauts who amassed fortunes that endured, Wilson disdained standing waist-deep in frigid mountain streams, choosing instead to make his money from supplying the hordes of prospectors with their necessities. Charging as much as the traffic would bear—which was considerable—he operated a trading steamer that plied the Sacramento River. Later he built a plank toll road connecting San Francisco's North Beach with the Mission Dolores, and next a toll bridge over a slough at Eighth and Mission in the center of San Francisco. By 1852, Wilson met with a number of influential businessmen of Sacramento and San Francisco—including a banker, William Tecumseh Sherman, then in temporary retirement from the army—and they incorporated the first railroad west of the Missouri, the pioneer California line. Projected with $1.5 million of capital at a $100 a share, their Sacramento Valley Railroad was intended to control all the interior traffic of California. By running north from Sacramento and skirting the foothills east of the American River for some sixty miles to the commercial distribution center of Marysville, the line would pick up business from all the foothill communities and tie in with the scores of wagon roads that snaked down from the placer mining district in the Sierra Nevada. Thus the railroad company would dominate the supply lanes to eight counties—counties whose 239,000 citizens imported 162,700 tons of beans, picks and shovels, dynamite, and other freight each year. Wilson would be a sure winner from such an enterprise, since his steamships would form the link between San Francisco and Sacramento. It was an alluring but daunting plan; after all, California was eighteen thousand sea-miles from any source of supply in the East. Colonel Wilson and Captain Sherman, though organizing directors, cautiously committed none of their own money. Nonetheless, Wilson embarked for the East to hunt up investors and to hire a smart young engineer for the preliminary surveys.[7]

In New York City, Wilson called on the Democratic governor Horatio Seymour.[8] A Van Burenite, Governor Seymour was intensely interested in transportation matters—particularly the Erie Canal and the state's bustling railroad network, most of which was controlled by friends and associates. When asked to nominate someone to lay out Colonel Wilson's railroad, Seymour had a ready reply.

"We know a young civil engineer, just the man for you," the governor said, "if you can get him." He sent a telegram off to Judah in Buffalo, instructing him to come at once to New York City "on business."

"I am going to New York tonight," Judah told his surprised wife that afternoon. "I must respond to the call of such men, though I don't know what's up."

Three days later, Anna Judah received a telegram: "BE HOME TONIGHT. WE SAIL FOR CALIFORNIA, APRIL SECOND."

"You can imagine my consternation on his arrival that night," Anna recalled many years later. "It was all laid out in these words: 'Anna, I am going to California to be the pioneer railroad engineer of the Pacific coast.'" And her husband had plans beyond that singularity: he would somehow connect himself with the Pacific railroad. "He had always talked, read and studied the problem of a continental railway," Anna wrote, "and he would say, 'It will be built, and I am going to have something to do with it.'"[9]

Anna Judah's parents and brothers escorted the pair to New York, not knowing if they would ever see their only daughter again. "It was a sad parting," she recalled of their last moments on the wharf, "but we were full of youth and hope; they, of prayerful trust." And then they were on the water and heading for Greytown, Nicaragua; a mosquito-ridden isthmian transit followed, and then another steamer bound for California.[10]

Sacramento City stood at the looping confluence of the Sacramento and American Rivers, and despite its youth—it had been incorporated only five and a half years before—it was already beginning to resemble a southern river town. One of the most populous cities in the state, Sacramento had gamely kept up its appearances in spite of the march of catastrophes that had repeatedly trampled it flat—floods and more floods, cholera plagues, firestorms, street riots, outbursts of lawlessness and vigilantism. With each difficulty Sacramento had buried the bodies, cleaned up the mess, and erected new commercial and residential blocks. By the end of the year some five hundred brick and two thousand frame buildings would comprise the city. So promising was Sacramento's future in 1854 that it had won out over Monterey, San Jose, Vallejo, and Benicia in attracting the Democratic-controlled legislature. (That governing body also accepted a million-dollar contribution to locate there.) When the legislature convened, it was greeted by the requisite parades and windy oratory.[11]

The growing city certainly had aspirations of respectability. But there was no denying that, having been born during a boom, Sacramento still had an explosive side. Every commercial block seemed to have its taverns, gambling dens, and houses of prostitution, all hoping to fleece the hundred thousand or more transients who had passed through in those five years. And Sacramento would not soon lose its raffish aura. Young Sam Clemens was then occupied with Mississippi snags and currents, but in twelve years he would call the California capital the "City of Saloons." There were a good many of them, he would tell readers of the *Territorial Enterprise.* "You can shut your eyes and march into the first door you come to and call for a drink, and the chances are that you will get it."[12] The most elite establishment was the Tremont Hotel on J Street, a three-story brick structure housing one of the largest gambling saloons in the West, in which the minimum poker ante was $1,000, gold.

Much had changed in a short time. Only fifteen years before, the Swiss emigré Johannes Sutter had settled on the riverbank, later obtaining a nine-thousand-acre grant from the Mexican authorities. Under his guiding vision, the empty plain had grown into a veritable principality, which he named Nueva Helvetia—wheatfields, cattle pastures, vegetable gardens, and vineyards, all presided over by an imposing fort constructed of timber and adobe, and defended by twelve guns. Despite the high walls and the firepower, the enterprise was more Edenlike than medieval. Sutter was a generous employer of Indians. And he provided a haven for all passing emigrants, encouraging many to stay on, circulating the word that Nueva Helvetia (as well as all California) was a paradisical destination.

Sutter's generosity, as has been well documented, was his downfall. After his employee, the former New Jersey wagon builder James W. Marshall, bent over Sutter's millrace up at Coloma to investigate what could have been only iron pyrite but in fact was pure gold, the idyllic little world changed forever. With pride and enterprise Sutter and son staked out the future city of Sacramento on farmland later in that year of 1848, but before long Sutter watched as his workers deserted his fields and tanning vats in favor of the big bonanza in the hills, watched as his crops and cattle were openly plundered by hordes of goldseekers, watched as squatters began a long pattern of defiance by perching on his property and daring him to come closer. In disgust he sold the entire lot and bitterly retreated into obscurity on a farm along the Feather River.

The year 1852 would prove to be the peak year of the bonanza: prospectors plucked gold worth $81,294,700 from the Sierra gravel, but after that the easy pickings were over. By the spring of 1854, when our young engineer and his wife arrived, only fortune hunters with plenty of capital were succeeding. Only those who formed corporations could afford to buy not only canvas hoses and mechanized troughs to sluice and comb the banks from their streams, but also gargantuan machinery—ore-pounders and quartz-crushers. Whole rivers were being diverted into canals and flumes, ravaging the landscape, while the piney canyons resounded with the roar of infernal machines. Those prospectors who did not move farther afield wandered the streets of Sacramento and San Francisco, dispirited and disgruntled, looking for work. Land values declined, prices receded, and fortunes dematerialized, while over eighteen thousand more hopeful emigrants streamed across Panama, ten thousand across Nicaragua, and nearly an equal amount across the continent.

At least two of the trans-Nicaragua emigrants arrived at San Francisco Bay anticipating no immediate bonanza, only work with a modest reward. The Judahs stopped in San Francisco—Ted's brother Charles had gone west as a Forty-niner only to set up a law practice in that city—but the reunion was brief, for the young engineer could not be restrained from pushing on to Sacramento and uncrating his instruments. He hired laborers to stand at various road junctions and tally the number of freight wagons and passengers going by, as a way of anticipating the Sacramento Valley's rail business. He noted the nearly level grade of the proposed route, the absence of any irregularities requiring expensive rock-cutting or embankment-filling. And he pronounced that "with such a Road and such a business, it is difficult to conceive of a more profitable undertaking." He estimated that the entire forty-mile stretch from Sacramento to important wagon roads supplying Auburn, a conservative but practical beginning until Colonel Wilson could enlist more Boston and New York capitalists, would cost an average of $43,500 per mile.[13]

One of the crowd watching him bend and squint into his transit soon became a close and valuable friend: Lauren Upson, editor of the *Sacramento Union.* Upson was immediately affected by the engineer's enthusiasm. "Everything promises favorably," he told his readers on June 5, "and the engineers are going forward immediately to determine the absolute locality of the line." Shortly, he added, the company would be advertising for contractors.[14] Two weeks later, Judah's crew was planting stakes at Alder Creek, eighteen miles out from the city, while their chief was mentally removing granite blocks from an outcropping and imposing them across the American River near the town of Negro Bar. That settlement had been named for a group of black prospectors who in 1848 worked the sandy deposits there in the American River. The survey veered north at that point, finally halting eighteen miles farther, at the Auburn road. Then, while Judah rode out along the line to obtain the remaining right of way, the company directors signed a contract with the New York firm of Robinson, Seymour and Company. For the sum of $1.8 million, they would supply the Sacramento with its rails, rolling stock, and buildings.[15]

After committing themselves to such a sum, probably some company directors raised their eyebrows when Ted Judah issued his year-end report. The chief engineer enthusiastically predicted that his road could at anytime be extended up the American River to Coloma (where James Marshall waited near the foundations of Sutter's sawmill for the next inevitable claim jumper), and northward up the Sacramento Valley to Shasta (where endless mule teams still trudged up toward mountain mines), and south to the cities of Stockton and San Francisco. But why stop there? The Sacramento, he postulated, could even be run over the Sierra! Perhaps it was Colonel Wilson or Captain Sherman who first began to question Theodore Judah's sanity by nicknaming him "Crazy Judah."

Graders began scraping dirt at the end of R Street in February 1855, soon after construction supervisors John P. and Lester L. Robinson arrived from the East. In March heavy rains slowed the company's hundred workers, who hardly could have been blamed for pocketing whatever nuggets their shovels struck. Nor did the weather dampen Judah's enthusiasm. He was presented with a handsome ring on his twenty-ninth birthday; in characters almost too tiny to read, it bore the inscription "Sacramento Valley Railroad, March 4th, 1855, First gold ever taken from earth used in making Railroad bank." And the work on that pioneer line continued.[16]

Even more tangible than a muddy grade extending toward Negro Bar was the sight, in June, of a freighter tying up to the Sacramento levee with the company's first consignment of iron: not only rails but rolling stock, all brought around Cape Horn to San Francisco by the clipper ship *Winged Racer.* On August 9, the first rail, fabricated in Britain, was ceremoniously spiked to a California laurel tie. And two days later Judah and three officials took off their coats, spit on their hands, and lugged a handcar over onto the new rails, thereupon pumping themselves along the first railroad ride in California—a distance of four hundred feet. Ceremony followed ceremony: days later cranemen swung a locomotive, the *Sacramento,* off its freighter, and by August 19 citizens were cheering as dignitaries from San Francisco climbed aboard the train for an exciting ride out to the dusty railhead.[17]

Company directors beamed all the way—but there were a few flinty, calculating glances at the visitors. Many of them were seen as prospective investors—because the Sacramento Valley Railroad was already in financial trouble. Everyone felt the pinch in 1855—a number of banks had failed all over the nation, and people were feeling a suggestion of panic. The railroad was no exception. As was common in those days, only $300,000 in cash was actually paid to the contractors. Robinson, Seymour and Company had accepted the rest in the form of IOUs—$800,000 in capital stock, at par, and $700,000 in 10 percent twenty-year bonds. In July, consumed by interest payments to Robinson, Seymour, the overstrained railway company forfeited a large block of its remaining stock when investors who had bought on margins as slim as 10 percent defaulted on a call for additional payments. Other assessments failed, and thereafter the railroad was slow enough in payments to Robinson, Seymour to encourage the contractors to file suit and attach the company's rolling stock and rails. Most of the directors, including Colonel Wilson, lost their seats on the board and most of their initial investments in the road. The new directors—notably the brothers John and Lester Robinson, and a San Francisco businessman, J. Mora Moss—would, in time, profit handsomely from their acquisition.[18]

Despite these troubles, work continued apace. On New Year's Day 1856 the line was fully operative along the eighteen-mile stretch to Alder Creek, with the bridge over the creek nearing completion. Four miles ahead, on the southern bank of the American River, Ted Judah had completed laying out a plan for a new town whose name would be changed from Negro Bar to Folsom, after a New York army captain, Joseph L. Folsom, who owned much of the adjoining territory. The chief engineer bought a few town lots as an investment.[19]

This was a good idea. With the company's financial problems, and with a change in management, Judah was told that the railroad would build no farther than Folsom—and that his services were no longer required. Probably

Ted and Anna had places of honor on the grand-opening excursion (over a thousand went along) and probably they took a few turns around the dance floor during the celebratory ball held on Washington's birthday.[20] And just as probably the taste of ashes was in their mouths at the thought of all the California lines Ted Judah hoped to build from his position as the pioneer railroad engineer of the West, and of all of the towns he expected to found. And—paramount to all of these dreams—there was the chimera of the transcontinental. With all the oompahs and oratory, Judah's mind would already have been calculating at the next step necessary, for, as he had often said, "It will be built, and I am going to have something to do with it."[21]

7

"There Comes Crazy Judah"

He lost no time. As his wife recalled, everything he did from the time he went to California "was for the great Continental Pacific railway. Time, money, brains, strength, body and soul were absorbed. It was the burden of his thought day and night, largely of his conversation, till it used to be said, 'Judah is Pacific Railroad crazy.'"[1] So it was when he secured his second job in the West. Businessmen hoping to make something out of soured property investments in the swampy town of Benicia, on the northern edge of San Francisco Bay, enlisted Judah in a preliminary survey of a railroad up to a point opposite Sacramento.[2]

His work for the paper San Francisco and Sacramento Railroad took only a few weeks. But clearly his monomania was developing. "There is still another light in which your Road may be viewed," he told the directors and any prospective stockholders in his report of February 7, 1856. "It is its connection with the great Pacific Railroad."[3] Connection with that distant goal, he thought, might make the San Francisco and Sacramento Railroad eligible for federal land grants. It was a tantalizing notion; the recent land-grant program in the Mississippi Valley had encouraged settlers and railroad companies alike, to the region's everlasting benefit. Very shortly the Illinois Central Railroad would link Lake Michigan with the Ohio and Mississippi Rivers, an enterprise made possible by a grant of two and a half million acres to the railroad promoters. Judah was proposing the same sort of arrangement for California. However feasible the idea seemed, he saw it would remain simply an idea until he acquired the requisite political savvy to put it to use. And what better place to do his homework than Washington, D.C.? In April 1856—two years to the month after their departure from New York—Judahs disembarked on a Hudson River wharf, Anna to visit her family in Greenfield, Ted to take up the study of politics and lobbying in Washington and to observe Congress in action.

He would see plenty of it from the first time he edged his way into the spectators' gallery, for the walls veritably rang with furious debate. For two years since its passage, repercussions had rumbled from the bill introduced by Senator Stephen A. Douglas, which organized the territories of Kansas and Nebraska with the slavery issue left, as it was said, open. Douglas's unhidden lust for the presidency was a powerful reason for the Illinois senator to have promoted what amounted to a proslavery measure. But Douglas was a complicated man with complicated interests, one of the strongest being his desire to fill his region of "the West" with settlers and to locate the Pacific railroad across Nebraska with an eastern terminus, naturally, in his beloved Chicago. Thus, with his railroad investments, Douglas stood to profit from the Kansas-Nebraska Act, which perhaps allowed him to ignore the growing echoes of screams and conflagrations that by the spring of 1856 could nearly be heard all the way from lawless, divided, bleeding Kansas.

Legislators still felt contentious over the issue—how could they not be with Kansas's open struggle between slavers and abolitionists breaking out almost daily?—and every national and international issue seemed similarly tinged. The southern adventurer Colonel William Walker had entered the eruptive little nation of Nicaragua with mercenaries, had joined a successful rebellion, and now had proclaimed himself president. New England abolitionists would find another annoyance when President Franklin Pierce, their soft-on-slavery New Hampshireman, extended diplomatic recognition to Walker's regime—provoking many complaints in Congress about "the grey-eyed man of destiny," as Walker's adherents styled their pint-size Tennesseean hero. Perhaps, it was said, with Walker presiding over Nicaragua a trans-isthmian railroad would be constructed with imported and domestic slaves, to rival the line recently completed by northeastern American capitalists across Panama.

Thus was the North-South struggle extended to every facet of life and government—a statement with which the Honorable Charles Sumner of Massachusetts would have readily agreed had he been interrupted at his writing desk on the Senate floor, on May 22, by a friendly interrogator instead of Representative Preston Brooks. Without warning and from behind, the enraged young knight of the South began raining blows with his cane upon the head and shoulders of Senator Sumner, who collapsed in a bloody heap. The attack came in retaliation for a heated exchange over Kansas and slavery; Sumner had blasted a colleague, Senator Andrew P. Butler, and his state, South Carolina, which together had been too much for Congressman Brooks to ignore—an insult to his uncle, to his home, and to the beloved South all in one intemperate, abolitionist breath. Though prone to dueling, Brooks instead bushwhacked Sumner at his desk, putting him out of commission for four years.

It is not recorded whether Theodore Judah was sitting in the Senate gallery on the day of the attack—if he had been present he doubtlessly would have seen to the distressed ladies as they fled their seats or fainted, for his gallantry is well documented. But possibly he was watching committee hearings on legislation for interstate wagon roads. The issue was receiving more attention in the Thirty-third Congress than railroads; nonetheless it was very pertinent to Judah's quest: as Lauren Upson had declared in the *Sacramento Union*, wagon roads were "harbingers of the . . . great and magnificent scheme, a railroad to the Pacific," since the army-protected, federally built highways would encourage western emigration and hasten the necessity of a transcontinental railroad.[4]

There was much congressional talk on the wagon roads during that session—Judah scribbled reports and sent them back home, where they were picked up in the *Union*—but this was overshadowed by presidential preoccupations, for the three-way campaign was under way. In February, former president Millard Fillmore of New York was nominated to lead the American Party (known popularly as the Know-Nothing Party) in its relentless pursuit of Catholics and foreigners, who were ruining America. The Democrats picked the amiable old Pennsylvanian James Buchanan, ambassador to the Court of St. James, in June; shortly the embryonic Republican Party, a national entity only since earlier in the year, chose as its candidate "the Pathfinder," John C. Frémont, whose career in politics consisted of several days as self-proclaimed president of California and, later, an appointive half-year stint as the state's senator. Frémont actively promoted the Republican planks against slavery and in favor of a Pacific railroad in the subsequent campaign, but he did not prevail in the election on November 4, which consigned the republic to Buchanan's hands. Meanwhile, Theodore Judah—who registered and voted as a Republican as much for his abolitionist views as for his party's position on the railroad—had begun to make use of his first political lessons.

The months of note taking in committee rooms, of watching the endless debates (and probably nodding off once in a while), of standing at the margins of crowds in the Capitol lobby and listening to the colloquy between legislators and lobbyists, of assessing the steps necessary to create a national railroad, he finally poured into the composition of a thirteen-thousand-word pamphlet. He entitled it *A Practical Plan for Building The Pacific Railroad*, contracted a Washington printer, and published it at his own expense in January 1857.[5]

The Pacific railroad would never be constructed, he said, so long as it remained a political—and thus a sectional—question. Judah had watched legislators in action. "Any attempt at legislation is, for our project, the signal of defeat," he wrote. "We start with this assumed as an incontrovertible fact: we must build the road without legislation." Besides, he said, if the railroad was built by the government it would be "built by a political party [as] a stepping stone to power" and the control of all legislation.[6]

No, Judah wrote, when capitalists decided to finance it, the road would be done, and they would not be interested in doing so until a thorough and reliable engineering survey was completed. No casual government expedition such as those performed by the army would persuade investors to part with a penny. Supplied with a mere $200,000, four "efficient, practical" engineering parties could complete an eighteen-hundred-mile survey in eight months; provided with proper estimates, investors could build the road in ten years, and the best way for them to finance it was by private subscription—not by federal or state bonds, and not by relying on "moonshine speculators" who would suck the enterprise dry while allowing interest charges to take whatever was left. What was preferable, he continued, was a national subscription campaign.

"Let a great simultaneous effort be made throughout the whole country at one and the same time," he urged. "Let popular speakers be employed in behalf of the enterprise. Let it be impressed upon the public that it is a people's railroad—that it is not a stupendous speculation for a few to enrich themselves with. Show them that it is entirely in their hands and under their control; that its officers and managers are to be appointed by them, and hold office only at their pleasure. Explain that every agent in each county is under their eye; that the money which he receives from them is to be deposited as they direct." What farmer or laborer could not afford one $100 share, paid in ten equal yearly installments—a mere thirty-three cents per day?[7]

There was more to Judah's *Practical Plan:* He advocated the Platte Valley route across the Great Plains and a western terminus in the Sacramento Valley, saying that snows could be kept cleared, and hostile Indians could be kept at bay; he urged that a federal wagon road be built on that line to aid the railroad. In January 1857, the engineer mailed his pamphlet to all members of Congress and heads of all pertinent departments, and to others in high places. It is likely that more than one recipient's eyes glazed over at the tiny print and the closely spaced lines as he recalled a similar document sent some eight years before by an earlier dreamer, Mr. Asa Whitney.

But as readers pored over Ted Judah's *Practical Plan,* more than one must have emitted a *harrumph* of disbelief near the end. In his most visionary excess, Judah predicted that trains could speed across the continent at one hundred miles an hour—a forty-hour trip from coast to coast—by employing an enormous engine with driving wheels fourteen feet in diameter, which would pull the ninety-ton locomotive and double-width cars along a system of four railroad tracks. Why, the man may have been well intentioned, with some persuasive arguments, but Mr. T. D. Judah, Civil Engineer, San Francisco, was obviously, patently, a lunatic.[8]

But Judah was an ambitious lunatic. "Let the Company look about for a Practical Railroad Engineer," he wrote, "who has distinguished himself for the celerity of his movements and the accuracy of his work as Locating Engineer, and invest him with the entire charge of the enterprise."[9] Certainly, he was thinking, not many could approach his record, though he was not about to let his pamphlet be his only self-recommendation. As the *Practical Plan* was circulating around Washington in March 1857, James Buchanan's trunks were being unpacked at the White House. The president had occupied his office for only a few days when he received a letter from an earnest young man named Theodore Dehone Judah, who offered himself as engineer for the California and Utah sections of the newly legislated federal wagon road. Despite all of his efforts, though, Judah's letter and pamphlet were of about equal value in promoting himself and the Pacific railroad.

Undaunted, he returned to press his case in California, where he supported himself and his wife by picking up surveying work in the foothills of the Sierra for various interests, the first of which was the California Central Railroad—creation of his former employer, Colonel Charles Lincoln Wilson. The California Central was aimed to extend the line of Wilson's lost pet project, the Sacramento Valley Railroad, from its railhead at Folsom northward to the confluence of the Feather and Yuba Rivers and the bustling town of Marysville, seat of Yuba County.

Judah completed his forty-mile survey with alacrity. He also spotted an opportunity for himself. East of where the California Central line reached the town of Lincoln, the engineer found an indentation in the foothills about seven miles long. It reached up toward the Bear River and into the center of the gold-bearing quartz district of Nevada County, and it could handily support a short branch-line of the California Central. He began to envision a town he would call Centralia at its terminus; it would be a hub from which a number of turnpikes would radiate to other towns and stage roads. Finally, he hoped, he would have a chance to get in on the ground floor of an enterprise and make a big strike. Quietly, seeking to attract no attention, he began to lay plans for this short line he would call the Eastern Extension: of course his own surveying was free, but there were options to be taken out, claims to be filed, imaginary lots in his imaginary town to be bought, graders to be employed, iron and equipment to be ordered.[10]

Certainly Judah could not have begun the scheme alone. But he was firm that it would be his company, Theodore D. Judah and Company, that would build the little seven-mile line. So probably he enlisted a primary backer in his sponsor and friend Colonel Wilson. The colonel, having visions of his own transportation empire, had approved Judah's survey for his California Central, and he had ordered work begun on its magnificent bridge over the American River at Folsom—the magnificent, expensive bridge that, down the line, would help dash Wilson's hopes for the California Central, and Judah's for the Eastern Extension and his Nevada County empire.

Much would transpire before their hopes would be dashed, and by that time, at least, Ted Judah would be sighting an even grander scheme. Meanwhile, Colonel Wilson put Judah to work studying inclines northeast of Folsom, along the eastern bank of the American River, for a feeder line to run to the Placer County seat of Auburn. It was not long, however, before Judah saw that the route was impracticable, its projected costs much too high.[11]

Always ahead of him, as he squinted into his instruments and compiled his endless lists of figures, was the looming Sierra and the chimera that lurked somewhere across its summit—the railroad to the East. Most of the time there was room in his mind for little else. "Judah's Pacific Railroad crazy" was an assessment increasingly heard uttered by those who encountered him. Anna prudently tried to rein in his enthusiasm. "I would say, 'Theodore, those people don't care,' or 'you give your thunder away,'" she recalled. "He'd laugh and say, 'But we must keep the ball rolling.'"[12] And so he became known as "Crazy Judah." In one telling incident, which occurred in Sacramento, a young man named John McIntire was standing on Seventh Street with a wholesale grocer, Newton Booth, when the latter suddenly exclaimed, "There comes crazy Judah." McIntire was very interested, having never seen an insane man, and he carefully scrutinized the fastidious stranger with a Van Dyke and friendly eyes. But during the ensuing conversation he heard nothing out of the ordinary—in fact, Judah struck him as particularly sane and lucid. When the engineer had passed by, and McIntire asked his friend to explain, Booth replied that Theodore Judah was a gifted engineer who had become monomaniacal about the Pacific railroad. Booth said he had forbidden Judah from approaching him on the subject again.[13]

Of course there were many other Californians who were as fascinated with the project as Judah was—including a small group of citizens who joined him in yet another enterprise, the Sacramento Valley Central Railroad. No records of its incorporation have survived, but in the minds of its founders, including the attorney Amos P. Catlin, who had been a stockholder in the pioneer California railroad and who became Judah's close friend, the company was capable of extension from the state capitol all the way to the eastern states; in reality, Judah seems to have pushed his survey only to Roseville, perhaps ten or fifteen miles up the American River valley, before more pressing matters came before him.[14] By early 1859, he and his friend Lauren Upson of the *Sacramento Union* had decided that the only way Congress could be convinced to act on the railroad would be if it were deafened by the din of Westerners wielding their new economic and political power and demanding action. Through a series of speeches and editorials Judah and Upson succeeded in goading the state legislature to call for an official railroad convention. It was to be held in San Francisco in September 1859, with delegates from most California counties, from the new state of Oregon, and from the territories of Washington and Arizona.[15]

So it was that on September 20, several hundred impatient and loquacious Westerners, all men of substance who were committed to the notion of expediently lassoing East to West with twin bands of iron, filed into the San Francisco Assembly Hall to convene the Pacific Railroad Convention. It was not long before they saw that it would be rough going. "The prejudices of localities," commented the *Sacramento Union* on September 24, "have already and strongly broken out, and today there is no harmony of interest."[16] San Diegans, Los Angelenos, and Arizonans called, of course, for a southern route; just as predictably, northern Californians favored the so-called Central Route, the Platte Valley to Salt Lake to Sacramento line, but the choice of a Sierra pass—which would determine which communities derived direct benefits of the railroad—splintered them into bickering factions. The main concern of delegates from Oregon and Washington was to support the Central Route with the understanding that a northwest rail route would branch from Utah or the South Pass along the Oregon Trail. Throughout the clamor of debates, speeches, resolutions, counterresolutions, amendments, and parliamentary folderol, however, one raised voice attracted attention—that of the earnest Mr. Judah, delegate from Sacramento, listed as chief engineer of his paper entity, the Sacramento Central. He became a prominent force in the convention. Word quickly spread that he had conducted many surveys of the Sierra Nevada's western slopes, more than anyone else on the Pacific Rim; his command of facts and figures impressed even those Westerners who distrusted him for his identification with Sacramento interests and for the bias he must certainly have had toward a route beginning there.[17]

"Let us adopt as our motto," the engineer declared, "'the Pacific Railroad wherever it can be soonest and best built.'" Only a proper survey would determine the best route—in congressional galleries and hearing rooms he had seen delivered to the West a surfeit of impassioned addresses and little else, and now he was seeing the national stalemate in microcosm. As he had written in his pamphlet, Judah called for a federally supported survey and for federal land grants to the railroad builders, and he added a call for twenty-year bonds—yielding the usual 5 percent, to be federally guaranteed—to safeguard the company from being devoured by interest payments in the manner that had crippled the Sacramento Valley Railroad. Let the convention have one voice on this, he urged. Let selection of the route be left to the railroad company, he said, for after all, it would be the directors' investments that would see the project through. And my expertise, and my survey, added the fervent voice within him.[18]

Crazy Judah's years of monomania were paying off. In the face of his persuasion, delegates forgot their differences and rallied behind him. They adopted his three-point plan and added resolutions urging the California and Oregon legislatures to award $15 million and $5 million, respectively, to the railroads conquering the Sierra and the Cascades. They voted to print some twenty-five thousand copies of a report on their deliberations. They decided to reconvene in Sacramento, with the state legislature as invited guests, early the following year. And in the meantime, they appointed T. D. Judah, Civil Engineer, to be the convention's accredited lobbyist in Congress. Only nine days after their appointment, on October 20, 1859, Ted and Anna were ascending the gangplank of the steamer *Sonora*.[19]

Introductions among the passengers bound for New York began while the *Sonora* was still anchored in San Francisco Bay, and Judah discovered he had influential company: the newly elected Senator Joseph Lane of Oregon, and a freshman Democratic congressman from California, John C. Burch. "I lost no time," Judah reported, "in endeavoring to impress upon these gentlemen the views of the Convention."[20] Burch had no more than taken in Judah's name and pumped his hand before the engineer launched into his first pitch as an official lobbyist. And he found an avid listener. "We became immediate and intimate friends," Burch recalled many years later, "bound together by sympathy in a common purpose. . . . No day passed on our voyage to New York that we did not discuss the subject, lay plans for its success, and indulge in pleasant anticipation of those wonderful benefits so certain to follow that success."[21] As the ship conveyed them to the isthmus, as they made the sweaty crossing, as they steamed up the Atlantic coast, Judah dazzled the freshman congressman with his mastery of financial questions and his clear sense of how his legislation should be moved through the hazardous political swamp of Washington. Whether he could similarly dazzle Burch's more experienced colleagues, who had much on their minds, was yet to be seen.

Upon their arrival in the capital, on December 3, 1859, Judah may have wondered if the government would endure long enough for him to launch his campaign on its elected stewards. The men of Congress had begun trickling into town a few weeks earlier, and they had brought their pistols. If the vortex of disunion had a center, it had been located until the previous day in a federal cell in Charles Town, West Virginia. There prayed the old, half-crazed, rectitudinous John Brown until he was escorted to the gallows and his reward. Of course, the convulsion after his failed slaves' uprising did not end with his death, especially in the South. And Washington was a Southern town. Its sixty thousand permanent residents grumbled as the outsiders from points north and west climbed down from their carriages, moved their belongings inside, and prepared for the opening of the Thirty-sixth Congress.

On December 6, White House doormen admitted a small party of Californians, consisting of Judah, Senator William Gwin, and Congressmen Burch and Scott, for a presidential audience. They found the sixty-nine-year-old James Buchanan sagging and squinty from some old undisclosed neural accident and plainly paralyzed with worry over whether the Union would hold through the end of his term. Nonetheless a gracious host, he accepted a copy of the Pacific Railroad Convention's proceedings and Judah's memorial, and he "expressed himself generally in favor of a Pacific Railroad," as Judah recollected. Such a stance cost him no effort, and it would gain his petitioners nothing, but no matter: they could report that they had seen the president, even if Buchanan was but a powerless, preoccupied caretaker in desperate need of a successor.[22]

With Congress organizing and with the House bickering over the choice of its Speaker, Ted Judah spent a month touring "the West," "to endeavor to awaken as much interest as possible in our efforts."[23] He visited New York, Albany, Syracuse, Buffalo, Cleveland, Columbus, Chicago, and Cincinnati, finally returning to Washington on January 14, 1860. He soon became a familiar figure in government buildings and on city streets, and it was then that the engineer was accorded a privilege almost unknown for a lobbyist: with the aid of Burch and the approval of the Democratic congressman from Illinois, John A. Logan, chairman of the House Committee on Contingent Expenses, Judah was given nothing less than his own office in the Capitol building.

Known as the Old Vice President's Room, that barrel-vaulted chamber with its cavernous fireplace and sumptuous chandelier was across the hall from the Supreme Court and on the same floor as the halls of the House and the Senate. "This was considered a sort of general headquarters for the Pacific Railroad," Judah reported to his counterparts in California, betraying no surprise at the privilege and no question of its propriety. "I procured all the maps, reports, surveys and papers of every kind to be found on the subject, and it being so convenient to the Halls, many of the Senators and members were accustomed to drop in almost daily." He assured the railroad promoters back home that "no opportunity was lost to further our views and impress on them the importance in which the subject is held in California."[24]

With congressional minds focused almost exclusively on the topic of secession, Judah's railroad room offered a sort of haven from the cares of the day, where a troubled legislator could lose his thoughts staring into the emptiness between contour lines on a map of a western mountain range, and where an energetic, personable young man—a walking encyclopedia of transportation—stood ready to answer any questions. "His knowledge of his subject was so thorough," recalled Burch, "his manners so gentle and insinuating, his conversation on the subject so entertaining, that few resisted his appeals."[25]

Many of the visitors to the Pacific Railroad Museum, however, were already active converts—they could detect the whiff of dollars in the pall of cigar smoke, and in such an atmosphere spittoons rang like coins tossed on a table. Their association in Judah's headquarters would have been natural; no other railroad lobbyist could claim a more central location than the main corridor of the Capitol. And what naturally sprang forth during that congressional session in the Pacific Railroad Museum and in the halls and the committee rooms was a loosely organized brotherhood of legislators and lobbyists. All represented local financial and political interests in a broad northernly swath of states and territories from the Mississippi to the Pacific. They easily formed an alliance with those in the eastern states who stood to gain if their lines were joined with those projected in the West. Of course the national railroad gained much lip service from these men, but their more immediate concerns lay back home with voters and with investors. It had been established in the Mississippi Valley land-grant program that they would profit whenever federal funds or lands were turned over to aid railroads and development. Many of those earlier recipients had now consolidated much political power at home, and even within Congress. Every one of them grew more impatient by the day with the Southern obstructionists who were slowing commerce and development in deserving Illinois and Iowa and Nebraska and Missouri and Kansas, by rolling legislative boulders onto the tracks of the transcontinental and, by extension, on any feeder lines with a chance for free land and free money and the power such things bought. For the most part they were Republicans—their party would meet in Chicago in May to choose a presidential candidate and a platform, and, symmetrically, their Republican Party would nominate a former railroad lawyer named Abraham Lincoln and adopt a plank strenuously advocating the Pacific Railroad.

Months, though, before the convention, this particularly intense group of railroad buffs welcomed Mr. Judah into its midst. One of its acknowledged leaders was Samuel R. Curtis of the Iowa congressional delegation, who knew much about railroads from his earlier career with the predecessor of the Chicago and Northwestern, the old Lyons and Iowa Central company. Now he presided over the House Select Committee on the Pacific Railroad. Judah's memorial had followed a circuitous route through the building—from the House chamber, where it was introduced by Burch on February 16; to the Committee on Post Offices and Post Roads, where it was referred; and to Curtis's Select Committee, where it and all other pertinent matters were transferred on March 5. Judah became a familiar presence in the offices and meeting rooms of committee members as the Curtis bill was drawn. Most of nearly three months of committee time was lost in wrangling over legislative language; what emerged was a bill calling for a government loan of $60 million to the builders through the sale of thirty-year bonds yielding 5 percent. Further construction cash would come from private investors. And there would be land grants. And the loans and grants would also be given to those linking feeder railroads to the main line.

The Curtis bill also left the question of routes and surveys open for later solutions, omissions seen as much to give northern capitalists room to maneuver as to lull the House Democratic minority who doggedly supported the southern route. But on the latter point the Curtis bill stuck fast. For one more year the Pacific railroad had failed before creation, victim of the same sectional contentiousness that killed another effort of Ted Judah's: a separate bill to support the railroad with federal land grants to California. Deliberation on that and the Curtis bill was postponed until December and the next session.

All he had to take home to California was approval of an eastern boundary survey—something which, in less promising times, Judah might have pressed to direct. In the railroad bill's defeat, however, there was some hope for the future.[26]

And so one day legislators or lobbyists with time on their hands ambled over to chat with the intense young engineer and found he had cleaned out his Pacific Railroad Museum. If they had inquired further they would have found his rooms on Fourteenth Street also empty, for he and his wife were already sailing back to California. There would be twelve or perhaps fifteen weeks before snows would begin to obscure passes in the Sierra Nevada, and Ted Judah intended to do some tramping in the high altitudes. He had seen what was required for success in Washington. People were scrambling to prepare for the inevitable transcontinental legislation—the developing presidential campaign had too many contenders, but the Democrats were more divided than Republicans and a balance of power seemed ready to tilt, with one likely result being seen in the Republicans' sixteenth party plank supporting the Pacific line. Judah did not plan to return to the city until he had his own route and his own railroad company. Not even the southern obstructionists would hold him back then.

8

"The Marks Left by the Donner Party"

Since he had traveled west to seek his fortune, Daniel W. Strong had shown himself to be as civic-minded as the next man—perhaps more so. He had settled in the bustling town of Dutch Flat, a muddy scrape on the forested slope of the Sierra Nevada, some fifty-five miles east of Sacramento. Founded in the spring of 1851 by Joseph and Charles Dornbach, for whom it was named, Dutch Flat thrived thanks to the rich hydraulic mining available wherever a prospector chose to register a claim and direct his high-pressure hose. By 1860 the surrounding landscape had been so eroded and denuded by eager goldseekers that the town itself seemed poised to begin sliding down the slope. Nonetheless, its residents were putting down the kinds of roots that would keep the settlement anchored. Dutch Flat boasted of the largest voting population in Placer County, as well as a post office, several hotels, churches, and banks, a firehouse, a newspaper, an opera house, and any number of saloons. Most residents—even a parson—owned interests in local mining enterprises, and Daniel W. Strong was no exception, though for some years he had divided his time between creating gullies in the quest for gold and running a small drugstore. Thus he was known as Doc Strong to the people of Dutch Flat.

By the early summer of 1860, hordes of prospectors were making their way eastward over the mountains toward the newly discovered gold and silver strikes in the Washoe District (in what would soon become the Territory of Nevada). For some years the bulk of trans-Sierra traffic had been along a route which benefited settlements along another valley corridor, primarily the town of Placerville. Hoping to capture some of that business, Dutch Flat promoted construction of a wagon road that would run nearby and across the summit. A number of its citizens, led by Doc Strong, made a reconnaissance for a possible route. While ascending the long, high ridge that divided the American River's North Fork from the Bear and the Yuba Rivers, Strong saw that not only would the ridge support wagon traffic but also it offered the sort of easy, continuous rise needed for a railroad.[1]

Strong hastened home, and on June 26, 1860, he and forty-six other citizens began a subscription campaign to hire a preliminary surveyor. Pathetically, they raised less than half of their $500 goal. Apparently, though, that was enough to pay expenses of the Placer County surveyor, a rather lackluster individual named Samuel G. Elliott, who set out from the office of the *Dutch Flat Enquirer* on July 19 and disappeared from the public consciousness.[2] No further activity ensued—until a month or so later, when Doc Strong heard that the smart young railroad engineer named Judah was back in California after a long stay in the East, and that he was investigating various mountain passes for a new line, the line. Judah had reportedly crossed and recrossed the divide at Beckwourth, Henness, and Tehachapie, finding fault with all, and he was now back in Sacramento for resupply. Strong sat down and wrote him a letter.

On the voyage back home, Ted Judah had sworn he would not return to Washington until he had gone into the Sierra Nevada and had come down with maps, profiles, and estimates. "With facts and figures," Ted kept telling his wife, "they cannot gainsay my honest convictions—as now." In a gesture that shocked her, he presented the officials of the Pacific Railroad Convention with a bill for only $40 for printing circulars, overlooking their personal expenses of over $2,500 for the lobbying trip that had kept them in the East for nearly a year. Perhaps he meant such munificence to demonstrate his honest convictions; perhaps he believed the convention lacked the funds to reimburse him; or perhaps, it seems most likely, he anticipated appearing hat in hand before those men of wealth and importance with a new investment proposal before too long.[3]

Still, they were back in Sacramento and he had to support them both and finance his reconnaissances. It is possible the Judahs received some income from real estate investments in Sacramento or up in the foothills, and at the time nearly everyone in California seemed to own stock in one mining enterprise or another. But whatever they had, it was not enough, for upon their arrival in Sacramento, Ted Judah embarked on such a flurry to raise living expenses that it is hard to track him exactly.

First, there were his interests in the foothill railroads. During his absence, Colonel Wilson's California Central line, for which Ted still served as chief engineer and company officer, had graded the eight-mile stretch between Folsom and Lincoln. But there in March it had abruptly halted and the laborers had dispersed. The books of the California Central did not look good; in a first burst of show-offy enthusiasm the company had built a magnificent bridge over the American River at Folsom. The span stood useless, for it had emptied the company treasury. Since there was not enough money to continue grading up the Sacramento Valley toward the goal of Marysville, Colonel Wilson had gone east looking for more investors, casting an anxious eye toward the date that his initial shipment of iron and rolling stock was expected in San Francisco.

Anxiety is contagious. Ted Judah's hopes for his own enterprise—his seven-mile short line from the California Central's depot at Lincoln up into the Placer County foothills, his paper city in the quartz district that he planned to call Centralia, his network of short turnpikes—all that rode on the shoulders of his friend Colonel Wilson. Five miles of his line was graded—but even that would come to nothing without Charles Lincoln Wilson's success. So Ted did what an optimist should do: he wrote the *Sacramento Union* a glowing letter about the imminent success of Centralia, the Eastern Extension, the toll roads, and the California Central (in that order), and he predicted that all would be running and prospering by the first of the year.[4]

Then he went about his business as if nothing was amiss. "Three weeks from the time we arrived in California," recalled Anna Judah, "'the engineer' was in the mountains. I remained in San Francisco or Sacramento among friends."[5] By the time snow blocked the passes he would have crisscrossed the summit some twenty-three times on foot, on horseback, or in a little one-horse wagon, equipped with a compass, an aneroid barometer, and an odometer—examining all the previously touted emigrant routes, noting their inclines and the depths of their canyons and the number of bridges and tunnels required. All the while he stopped in each community to openly broadcast his intention to find a way to stretch a railroad over the Sierra Nevada, to solicit information, and to try to drum up a little business on the side.

The engineer in him saw opportunities for toll roads along every mountain indentation, for turnpikes between every little hill town; the opportunist in him lost no chance in pointing out these potential bonanzas to anyone who might pay for his services; the publicist in him made use of every editor in every little job-printing, newspaper-publishing office from valley to summit. We have just received a visit from the brilliant, genial young engineer of the first railroad in the West, the editor would obligingly write, and while searching for his Pacific Railroad route he has spotted something of vital economic importance to our region which we have somehow overlooked. Whether Judah's bait snagged any customers is not known—but if not, it was not for lack of trying.

His onetime employers, the Robinson brothers—who had yanked control of the Sacramento Valley Railroad out of Wilson's hands some years before by loading him down with carrying charges—saw an opportunity of their own in Judah's feverish activity and relentless publicity seeking. Of course they were aware of Wilson's current financial troubles, and they had designs on his California Central and even Judah's short line, but for the time being they kept their eyes on the engineer by making him a roving, commissioned freight agent, to solicit business for their Sacramento Valley Railroad while he was up in the mountains. He could not have made much money at it, but he made sure that every editor in the hills mentioned the sideline in any story about his explorations.[6]

During one of his breaks in Sacramento, Judah received a rather momentous letter. The next morning he was on the first stagecoach to Dutch Flat. The stage driver took him to Doc Strong's store. Strong recalled, "The next morning I went with him to the livery stable, got some horses, and we started up over the trail. There was no wagon road, nothing except the marks left by the Donner party, but there was a regular trail by which we went up to the summit." When they reached the top of that long rise, they stood while Judah took in the expanse of peaks marching north and south, and far below, in their midst, the picturesque, sparkling little lake that memorialized the emigrants who, under Jacob Donner, had blundered to their deaths. Doc Strong perhaps interrupted a reverie when he reminded Ted that just beyond Donner Lake was the Truckee River canyon, draining Lake Tahoe, which might form an easy corridor through the mountains ahead as it threaded downward toward the northeast and the Humboldt Plains of Nevada.

Judah's thoughts soared out across the great clear emptiness below as he interposed a thousand-foot iron line down the cliffsides past the lake and toward the green and foaming Truckee. He knew that any other route across the Sierra Nevada would have to contend with two parallel ridge lines separated by a deep trough, meaning two expensive ascents for any locomotive. The Truckee would be his natural passageway through the mountains' eastern heights. He knew then that he had found the answer to an engineering problem that had troubled him for years—and that the transcontinental would go through Donner Pass.

"We camped there over night in an abandoned herder's or stockman's house," Strong remembered, "and came back the next day. It rained all the way down, and we were pretty wet and did not feel very much like doing anything." Tired as he was, Ted Judah could not sleep or rest, and soon enough he was banging on Doc's shop door and, when Strong was roused, Judah called for pen and paper. He sat down and drew up what he called "Articles of Association." Judah shoved them across the counter and said, "Sign for what you want."

"You should sign first," Strong protested.

"No," replied Judah. "You are the party, and you sign for just the shares you want." He explained that state law required that there should be $1,000 subscribed for every mile of the line, and that 10 percent of the stock subscriptions had to be paid in before they could organize as a company. Strong and Judah had estimated the distance from Sacramento to the state line as 115 miles, meaning they had to raise a down payment of $11,500 against the total of $115,000. Doc Strong signed for fifty shares of stock valued at $100 per share and slid the document back across the counter.

"Well," said the engineer, "that is about right."[7] And he knew what their next task would be. Ted Judah still had retained much of his enthusiasm about a "people's railroad," as he had outlined in his *Practical Plan* of three and a half years earlier, when he had still believed in building a road without legislation. Let it be shown, he had written, "that there is not a man in the whole community who has hands to labor with who cannot afford to take one share of $100 and pay $10 per year, or three cents per day, in upon it; that there is no retail merchant, doing even an ordinary business, who cannot afford to take ten shares and pay in his $100 per year, or thirty-three cents per day. . . . There is no money to be borrowed at enormous rates of interest; no loans to be negotiated in Europe; no first, second or third mortgage bonds to be issued and sacrificed at one-half their value; there are no commissions to be paid to negotiators. There is no mortgage on the road; it is built and paid for as built in ready money. It is a clean thing, built and owned by the people, for its actual cost and no more."[8] Three and a half years of working toward his dream with little or no reward, much of it passed in cynical Washington City, had not totally dimmed his idealism. "Go around town," he told Doc Strong, "and see what other subscriptions you can get in the town from merchants, from the Ditch and Water Companies, and so forth."[9]

Two muscle-sore but elated stock solicitors were a far cry from the national publicity campaign and the army of exhorters that Judah had envisioned, and they found their results to be about as humble. They walked around Dutch Flat raising small stockholders here and there; none took more than twenty-five shares. In the nearby town of Auburn the merchants were distrustful of the enterprise unless they were put in control—how many drummers had appeared in town with similar wild ideas only to disappear with their money. Only the stockholders themselves would control the Pacific railroad, came a heated reply from Strong. His friend's idealism was catching.

Barely daunted, they split up. Strong headed into adjacent Nevada County "and got considerable stock taken." Next, he rode down to Sacramento and canvassed the streets, collaring merchants, saloon men, and draymen. Most of the citizens he stopped took from five to ten to fifteen shares of stock "simply to encourage it. They said that if anything came of it it was bound to make business for Sacramento, and they simply took hold of it to show their interest in the matter." Encouraged, he then took a steamer down to San Francisco. Apparently there he was not on as solid ground, not being well known—"the people there laughed at the idea." The only two with vision that he found in San Francisco were actually men from Grass Valley, up in Nevada County, and they both subscribed for twenty-five shares. Doc Strong headed back inland.[10]

Meanwhile, Ted Judah was similarly occupied, though on a larger scale. After all, he was the pioneer railroad engineer on the Pacific Slope and had hobnobbed with many wealthy and far-thinking Californians in boardrooms and in banks; he had won the trust of all those enthusiasts of the Pacific Railroad Convention; idealistic though he may have been, Judah had begun to recognize that his railroad could not be created solely on the subscriptions of small storekeepers.

Characteristically, he sat down and wrote a long appeal to be printed and circulated (first, an eighteen-page pamphlet dated November 1, 1860, followed nine days later by an abridged version for the newspapers) in which he announced that the long-sought passage over the Sierra Nevada had been discovered. According to the notes he had taken on that first reconnaissance, and the profile he had drawn in Dutch Flat, the railroad would climb at a steady, practicable rate over the 102-mile distance from Sacramento to Donner Summit with its elevation of 6,690 feet—and then easily descend the remaining 13 miles to the state line—in no place requiring a gradient of more than 100 feet per mile. This was well within the bounds of current technology, as Judah knew, for while in Washington he had paid close attention to what had been attained over the Alleghenies, the most prominent obstruction bested in the East; he had also corresponded with the Matthew W. Baldwin Company, locomotive builders in Philadelphia, from which he had the latest information on the operation of engines on heavy grades. Moreover, the Dutch Flat–Donner Pass route crossed the state at its narrowest point, saving many miles over other proposed routes; it also avoided a number of expensive geographical obstacles posed by those competitors.[11]

Four days after he composed his announcement, copies were on their way to capitalists in San Francisco, and Judah excitedly wrote a letter dated November 14 to his partner in Dutch Flat. "I have struck a lucky streak," he told Strong, "and shall fill up the list without further trouble. I have got one of the richest concerns in California into it, and will post you fully in a day or two."[12]

"He left me in high hopes," recalled Anna, who stayed at their hotel room in San Francisco. "A meeting was called at the office of a leading law firm and Mr. Judah left . . . firm in the faith [that] the gentlemen he was to meet that evening would give him the aid he required to make his survey the following spring."[13]

We can imagine the engineer's delivery and his listeners' calculating silence. The Pacific railroad bill was to be called up in Congress in a few weeks' time, he would have said. It was important to organize a company under California law before then so that they could win an official designation. In the mountains he had collected subscription down payments totaling $7,000, and therefore needed only a minimum of $4,500 more to meet the state's required 10 percent capitalization—a mere pittance to the wealthy gentlemen he addressed, but still crucial in making the railroad possible. The new company would use the money to finance a proper survey of the Dutch Flat–Donner Pass route, which in turn would bring in more investors and increase the company's standing in the eyes of Congress and the California legislature.

For, gentlemen, he would have continued, there was money to be made here. The federal bill would appropriate $13,000 per mile from the navigable waters of the Sacramento River to the base of the Sierra Nevada, and $24,000 per mile to the summit, and an additional $3,000 per mile for each degree of longitude crossed until the line reached the 109th degree. Support would also come from the state, and certainly the counties of Placer and Nevada would subscribe up to a half million toward construction of the lower portions of the road. And the iron and rolling stock could be purchased easily if they issued first-mortgage bonds of, say, $12,000 per mile. There would also be land grants: Mr. Burch would press a bill in the next Congress that would appropriate forty-four hundred acres of government land for each mile the railroad constructed. Not only would the gentlemen be connected with the most stupendous enterprise of the century, but they stood to realize a handsome profit once the railroad was completed, a matter of seven years at the most.[14]

To Ted Judah's utter chagrin, his potential benefactors refused him. There were quicker and surer investments to be had, they said. What they did not say, but what Judah realized only later, was that he was not bound to receive a friendly hearing anywhere in San Francisco; too many people there held interests in enterprises that were threatened by his railroad—the steamship lines, the stagecoach lines, the wagon and express companies, all of which could not hope to compete with a national railroad.

Ted was still fuming when he returned to his hotel. "Anna, if you want to see your friends in the morning," he exclaimed, "you must pack your bag and trot around and see them, for I am going up to Sacramento on the boat tomorrow afternoon.

"Remember what I say to you tonight so you can tell me sometime," he continued with unabated heat. "Not two years will go over the heads of these gentlemen . . . but they would give all they hope for in their present enterprises . . . to have what they put away tonight. I shall never talk nor labor any more with them."

No. Perhaps there were more investors to be wrung out of the mountain towns. But first, surely, in Sacramento he would find men of greater vision.[15]

9

"The Most Difficult Country Ever Conceived"

The notice appeared in the *Sacramento Union* late in that month of November 1860, thanks to Ted Judah's friend Lauren Upson. Citizens interested in hearing a presentation on the newly discovered railroad route over the Sierra Nevada were invited to the St. Charles Hotel on K Street. At least thirty men were waiting when Judah appeared, distributed his pamphlets, unrolled his charts, and sized up the audience.

Almost everyone was known to him, but in that assembly of blank, curious faces he was glad to see a few friendly ones. Doc Strong had come down from Dutch Flat. Near him was Benjamin Franklin Leete, a local engineer and bridge specialist with whom Ted had worked back in New York State and also later in 1858 on his first private railroad scheme, the Sacramento Valley Central, which had managed to run only a preliminary survey some ten or fifteen miles up the American River valley to Roseville before Judah's attention shifted to the Pacific Railroad Convention and his Washington trip. Other friends present were Amos P. Catlin, his attorney; James Bailey, a successful jeweler; Lucius A. Booth, a wholesale grocer who had been a delegate at the Pacific Railroad Convention; and Cornelius Cole, the district attorney for Sacramento County. The latter was one of the founders of the Republican Party in California. Cole had worked, as it happened, with many men in that assembly in the St. Charles Hotel to win the state for the president-elect, Abraham Lincoln, who had edged by in California as well as in the rest of the nation thanks to a crowded field. With his power and connections, Cornelius Cole was a good man to have present, but Judah may have not been so sure about two others who are reported to have been there: the brothers Lester and John Robinson of the Sacramento Valley Railroad.[1]

How much bad blood existed between Judah and the Robinsons will never be known. Some years later, a very bitter Lester Robinson looked at his dwindling resources and attempted to raise a serious scandal in retaliation against those he blamed. Ted never discussed the breach, so all historians have had is the word of Lester Robinson, who waited until Judah was in his grave (and beyond seeking a libel judgment) to call him a liar, a cheat, and a thief who discovered the Central Pacific route while working as an agent of the Sacramento Valley Railroad, to which should have gone the transcontinental plum. According to Robinson, his firm had hired the engineer to locate a toll road on the divide between the American River's North Fork and the Bear River, in the area around Dutch Flat.

This contention, however, has never been verified beyond the mouth of Lester Robinson. In all his visits to California newspaper offices, Judah—the relentless publicist who could never resist telling all in earshot what he was up to—was quite candid about his own rail reconnaissances and equally open about his work as a freelance freight agent for the Sacramento Valley, and there was no connection implied or any taken; moreover, all talk of practicable turnpikes in those scattered towns was likewise reported with the implication that he was seeking more work for himself, not as someone's employee. (This is moot, though; credit for discovery of the Dutch Flat–Donner Pass route went to someone who had never been connected with any firm, much less the Robinsons': Daniel W. Strong.) By this time in late 1860, it should also be noted, the interests of the Sacramento Valley Railroad had diverged from that of Sacramento itself, with the company's directors having increasing financial ties with competing communities—not the least of which was San Francisco, which had spurned Ted Judah's proposals. Finally, of all his contemporaries, Robinson was the only person to question Judah's character; within his charges were contained a number of statements later proven utterly false.[2]

This, however, is getting ahead of the story. It is enough to say that by the time the *Sacramento Union* announced the meeting at the St. Charles, the Robinsons had read a number of newspaper accounts of Judah's explorations and could not have been pleased, for what he proposed would not only compete with their enterprise but bury it. Also, the Robinsons were maneuvering to control a paper entity, the Sacramento, Placer and Nevada Railroad, which was projected to strike east from their own railhead at Folsom to the town of Auburn—in direct competition with the projected Central Pacific's. And so if they sat in the assembly (as Judah's friend, Ben Leete, recalled over sixty years later) and if they had already severed their connection (as Lester Robinson later claimed) with the young engineer—they would have sat there radiating a heat that did not warm, only alarm.

Under such circumstances Judah could only launch into the presentation he had delivered in San Francisco—about the wondrous long ridge ascent to Donner Pass which provided the cheapest and least-complicated rail route, about the portal through the eastern barrier along the Truckee River, about the need to get the enterprise under way in order to be officially recognized by Washington when the landmark legislation was inevitably passed, about all the financial support they could draw from the federal and state governments, about the riches and honors which would surely come their way. All that was needed to get the ball rolling was for Judah to conduct his proper survey over the magic route, and that could be financed out of the 10 percent down payments on stock subscriptions which would also satisfy the state's minimum requirement for incorporation: only $10 down per share, friends, and a new world would be theirs.[3]

When Judah finished talking, he was hardly mobbed by eager investors. Accounts vary on his audience's response. At the very least it was polite—his reputation as Crazy Judah seemed to finally be shed. A few men stepped over to sign up, among them some who offered bulk goods such as flour or potatoes in lieu of cash, which Judah could not afford to rebuff. But a pocket of them held back—seeming to defer to a prosperous hardware merchant, the immense, black-bearded Collis Potter Huntington, who seemed disinclined to get involved. "Huntington," one man said, "you are the man to give to this enterprise." After all, he had subscribed before to support wagon roads and the overland telegraph. But he shook his head doubtfully, and the responding shrugs from the others should have been visible to Ted Judah. The thing was just too big. However, as he stepped out toward the hotel lobby, Huntington paused. "If you want to come to my office some evening," he said to Judah, "I will talk with you about this railroad."[4]

Ted Judah could not have suspected it, but his salvation from failure and obscurity had just been delivered in the person of a brawny, no-nonsense, thirty-nine-year-old merchant with a cash register for a heart. And, like Judah, like most Californians, especially successful ones, Collis P. Huntington was a remade man. He was born near Hartford, Connecticut, into an impoverished farm family with nine children, and he received little schooling before leaving home—although in what he received, he did well in mathematics, history, and geography. "I could whip any boy in school, old or young; I excelled in geography," he recalled fondly many years later. It is interesting that in that one sentence he links the personal qualities of strength and aggressiveness to geography. Huntington had a gift for numbers and an inclination toward buying and selling. At sixteen he became an itinerant peddler, dispensing clocks and inexpensive tableware from his wagon as he roamed across many states, doggedly spartanlike in his living, saving much, spending little. At twenty-one he joined his older brother, Solon, in Oneonta, New York, where Solon had established a successful store. Two years later he became a full partner; the flinty tradesman even felt flush enough to marry a distant cousin from Cornwall, Connecticut, Elizabeth Stoddard. The Huntington brothers continued to prosper and made good reputations with farmers, factory owners, and wholesalers even all the way to New York City.[5]

Then came word about the California goldfields in December 1848, and Collis Huntington answered the call. He did not intend, like most others, to give up what he had at home and go crouch in frigid Sierra streambeds. No, he would find a place to open shop—the West Coast branch of S. & C. P. Huntington, Oneonta—and sell miners their pans and picks and shovels and socks. Bankrolled by both brothers' savings and mortgages, Collis left New York in March 1849, settling for, typically, steerage passage on the paddle-wheel steamer *Crescent City.* In the hold were his crates of woolen socks, patent medicines, rifles, and probably whiskey, for trade on the Pacific slope.

At the Panama isthmus he spurned a stagecoach, which he thought cost too much, and instead walked the twenty-four miles across. There he joined the miserable, heat-addled and fever-stricken legions of Forty-niners waiting for space on a Pacific boat to San Francisco. Instead of idly sitting by, drinking (he was abstemious) and gambling (he gambled only in business, even then with reasonable odds) or sinking into desperation (he was prone to self-pity though he would not indulge in it for very long), Huntington began to tramp back and forth across the isthmus, buying supplies from the locals and selling them at a high premium to his fellow fortune hunters. After six weeks, he was to boast later, he had put away another $3,000 for the California enterprise.

Finally, he found a place on the coal barque *Alexander von Humboldt.* Truly it was a ship for sufferers; jammed into makeshift accommodations, they were becalmed for days, and provisions grew so short that the passengers mutinied, set the profiteer owners ashore at Acapulco, Mexico, restocked, and sailed up to San Francisco. They finally entered the Golden Gate in late August and sold off the ship to recover the money they had been forced to spend on provisions. In those freewheeling times, the passengers easily got away with it, with no legal repercussions.

All during the autumn of 1849, Collis Huntington crisscrossed the foothills above Sacramento, establishing several little trading posts in tents and trying to figure what demands he could fill. It was exhausting and barely profitable work, as his letters home admit. But by early 1850 Huntington had found partners to open a hardware store on K Street in Sacramento.

Life would not immediately improve—and hard times would harden the man into a tenacious, frugal calculator who maintained a hale exterior to conceal a ruthlessness inside. He never tired of the combat of the marketplace. Like most other town merchants, his business kept getting knocked down by floods, fires, pestilence, and alarming fluctuations in customer traffic. But Huntington's mercantile savvy was uncanny. He developed a sixth sense of what would be in demand in the next season or even the next year. And often competitors would rush down to the Long Wharf in San Francisco to meet an incoming ship only to find that Huntington had spotted it with his field-glasses, gone out in a little sailboat, hailed the ship with a halloo and a wave of his trademark droopy, wide-brimmed straw hat, and bought the best goods for transshipment to Sacramento. He dispatched a flurry of letters to his brother Solon back in Oneonta: send more shovels, picks, plows, horse tack, wagons, wagon parts, wool socks, blasting powder, cooking utensils, butter, oil, dried fish, canned ham, whiskey, here's a chance to corner the market in whatever—there was almost nothing Huntington could not sell. But the real trick he pulled off was knowing when to order, how long goods would take on the long sea voyage from the East, and what conditions would prevail when they arrived.

As an employer Collis Huntington took a stern and paternalistic interest in the young men who clerked and toted for him. He insisted that they spend their evenings at home with lights out at nine, that not one drop of alcohol should ever pass their lips, that never—absolutely never—should they ever enter a gambling den or house of prostitution. He may have had no compunction about selling such places (ubiquitous as they were in Sacramento) their whiskey or provisions. But there were limits. Once he agreed to sell a shipment of preserved peaches to a Mrs. Caswell, proprietor of a whorehouse, if she would pay top dollar. When she asked that they be delivered, he refused. "You tell Mrs. Caswell that if she pays ninety-six dollars and takes the peaches along I have no objection," he intoned. "But my boys cannot go into her house." Somehow, though, his high moral character accommodated possibly the largest collection of dirty jokes on the Pacific slope, the sharing of which he was careful to limit to his own gender.[6]

But he demanded no more from his employees than from himself: work hard, live a quiet and dignified private life, scrimp and save, put down roots. And Huntington did put down roots over the years. He sent for his wife. He parted amicably from his first partners, went out on his own, took on another partner for a couple more years, bought a lot at 54 K Street and erected a new store, then rebuilt it after a fire, moved himself and his wife to an upstairs apartment, and finally, in 1856, joined forces with his next-door neighbor, an abstemious, penny-wise former upstate New Yorker by the name of Mark Hopkins. For the good of the business probably more than for the community's benefit, Huntington subscribed to civic causes and improvements. A Whig, he drifted into Know-Nothingism before lodging in the new Republican Party, becoming one of the earliest members of the California Republicans, largely because it seemed the only banner worth following but also because of its antislavery reputation, which he wholeheartedly supported.[7]

With all of this, he would certainly have been aware of how important a Republican Party plank was the call for a transcontinental railroad; a condition, after all, of being Californian was vociferous support of the railroad regardless of what party one belonged to. When Collis Potter Huntington followed his curiosity to a meeting on the subject at the St. Charles Hotel in late November 1860, and met Theodore Dehone Judah, something tugged at him—the merchant's savvy saw promise in so much for so little initial investment. Years and years later, a San Francisco journalist would call him "a hard and cheery old man, with no more soul than a shark." Well, there was something sharklike in his consideration of the matter, and probably, then, of the idealistic young railway engineer. But, Huntington would say, Judah was going at it backward, panhandling for small subscriptions. Efficiency was needed, and during their next encounter, upstairs at 54 K Street, Huntington made the engineer go over it all again.[8]

Judah did, and then he elaborated: Washington talk, insider talk. He told a little of what he had witnessed back in the Capitol's corridors, about conversations in his Pacific Railroad Museum, about the way legislators and lobbyists performed the alchemy of turning local special interests into law. He explained what lay behind the pending transcontinental railroad bill of Iowa representative Samuel R. Curtis, a powerful Republican whose connection to railroads went back at least eight years: that a durable political coalition of Midwesterners, New Yorkers, and New Englanders had arisen and was ready to face off against the Southern obstructionists to push through the Pacific railroad. The movement was hardly altruistic; it was spurred by the certainty of federal subsidies, land grants, and opportunities for enriching existing eastern and midwestern lines. Once the bill was passed and these plums began to be plucked, it hardly mattered if the national road was built at all. Judah knew these lawmakers intimately; his engineering expertise had encouraged and even sometimes guided them. Moreover, Judah would have added, this recent railroad coalition was Republican. And a Republican who felt great affection toward the railroad companies of Iowa and Illinois (and the political clout they carried) had just been delivered his ticket to the White House.[9]

Now two of Collis Huntington's primal instincts had been tantalized: his fondness for making money and his fascination for politics, at least the backroom variety. The California Republicans had allies in the East, too. He instantly realized that regardless of how high the Sierra loomed on the horizon out his windows, the Pacific railroad—and Huntington's share of its money and power—was hardly the chimera that outsiders still claimed.

Judah, eager to press his advantage, then delivered a coup de grâce. After he completed his survey, he said, and the actual preliminary work on grading and constructing had begun, they could run a wagon road up over the pass from Dutch Flat to the Nevada Territory. Tolls from freight and emigrants would bring in cash almost from the beginning.[10]

How could one lose? Huntington was convinced. And then he took the floor away from Judah and proceeded to lecture him: this going around knocking on tradesmen's doors, this advertising and this calling of open meetings, was no way to run a railroad. How long would it take to find all of these small subscribers, and what guarantee did Judah have that they would not disappear or default on paying anything beyond the meager down payments? No, Huntington would find four other Sacramento businessmen for a board of directors. Each would subscribe $1,500—the 10 percent down payment on 750 shares of $100 stock; together, this board of directors would pay for Judah's survey and whatever lobbying would be necessary in Washington. And the first person he would touch would be his own partner in the hardware business at 54 K Street: Mark Hopkins.[11]

If Huntington's managerial style in his store was to loudly humiliate one of his clerks who was careless over a dropped six-penny nail, it was Hopkins's to quietly rummage through an office wastebasket for the still-usable blotting paper. Despite their great outward differences, the two men were made for each other. Where Huntington was in every way a large man, Hopkins, though tall, was rail-thin; at age forty-seven he was much older than the average Californian immigrant, and because of a slight stoop and a quiet, lisping voice (he even stroked his long, wispy beard when thinking), he seemed a decade older. Thus he was known to all as Uncle Mark.[12]

But his whole life pointed toward an alliance with someone like Collis Huntington. Hopkins had been born in 1813 in Henderson, New York, on the eastern shore of Lake Ontario. One of eight children, as a boy he lived for some six years in St. Clair, Michigan, where his father ran a store. A vegetarian who was known to eat sparingly when at all, he was disinclined toward hard physical labor and so would always manage to work indoors. At fifteen, when his father died, Hopkins left school to help run the family store, later going into partnership in Lockport, New York, near Niagara Falls. An older brother tutored him in law for a while but Hopkins did not pursue it, instead going to work selling plows in New York State and Ohio, then settling in New York City as a bookkeeper in a commission house.

When word of the California gold strikes came, Mark Hopkins had risen to comptroller in the firm, but he lost no time in leaving, getting himself on a ship, the *Pacific*, which sailed the miserable, long route around the horn of South America. Under horrible conditions and a tyrannical captain, the passengers nearly mutinied. Hopkins, it was recalled, was the one who argued them out of such lawlessness and convinced them to complain to the American consul at their next port of call, Rio de Janeiro. Luckily for the passengers, the captain was removed and the rest of the journey lightened. In California, Hopkins and several comrades rowed a lifeboat up the Sacramento River, intending to ascend to the headwaters and camp near the gold mines. But they heard so many unfavorable rumors about conditions that they turned around, exchanged their boat for an ox-team, and returned to Sacramento. Along the way they ran out of food and were forced to break open the bones of a dead ox on the roadside and eat the marrow.

From such a low point, the frugal and hardworking Hopkins prevailed, buying a wagonload of supplies in Sacramento and selling the goods dearly over the winter up in Placerville (which was then called Hangtown). By the spring of 1850, back in Sacramento, he encountered a friend from the voyage of the *Pacific*, E. H. Miller, and they formed a partnership in a wholesale grocery business. After six years Hopkins moved next door on K Street when he became a partner in the hardware store of Collis Huntington. It was a good match; Huntington was the outside man, the hard-driving bargainer who never came out second, while Hopkins took care of the inside details of accounts and collections. Uncle Mark could not abide leisure, working doggedly late almost every night. Indeed, he did not find time to marry until he was forty-one; even so, it was too late to change old work habits. "He liked to work," recalled a nephew, "more than the laziest man in the world liked to loaf."[13]

Huntington grew to depend on Hopkins's quiet and deliberative manner and on his penchant for listening attentively to others' views, restating the question at hand clearer and in much fewer words than it had been originally posed, and then rendering a precise, almost judicial opinion. "Hopkins had one of these sharp analytical minds that could master anything he took hold of," Huntington once admitted. "I never thought anything finished until Hopkins had seen it. He never had anything to do with trade and never would. He had general supervision of the books and the papers, contracts, &c. When he said they were right, I never cared to look at them." Once Uncle Mark's mind was made up he seldom strayed. One associate, Charles Crocker, would later say that "when Hopkins wanted to be, he was the stubbornest man alive." But, in late 1860, after he had listened to Huntington's proposal about a railroad to the east, and stroked his beard for a while, Uncle Mark agreed to climb aboard the enterprise.[14]

That left three more directors with $4,500 between them. Buoyed by his partner's approval, Huntington lost no time in looking for them. Jeweler James Bailey came on. So did, briefly, an engineer named James Peel—but then he backed out.[15] Then there was Leland Stanford, a vain, dark, stolid wholesale grocer with a direct gaze. Stanford actually was somewhat of a slow thinker who spoke only after ponderous, eternal-seeming silences. Having arrived in California in 1852, he had been a relatively latecoming Argonaut. Born in 1824 outside Albany, New York, the son of an innkeeper, Leland Stanford had attended both the Clinton Liberal Institute at Utica and Cazenovia Seminary near Syracuse, though he neither distinguished himself nor graduated. However, he was interested in politics and the law, and so went to work for an Albany law firm as a clerk. Three years later, at twenty-four, he passed the bar and began to look westward for a town in which he could establish himself and prosper. For nearly four years Stanford practiced law in Port Washington, Wisconsin, not far up the Lake Michigan shore from Milwaukee; he sought a foothold in the local political establishment by running for district attorney as a Whig. He was badly defeated by the Democrats who controlled the county.[16]

Business, however, was moderately successful though hardly stimulating. Stanford grew dissatisfied with his routine of writing wills, defending clients in picayune disputes, and collecting bad debts. He married an Albany girl,

Jane Lathrop, and chafed at practice for another two years before fortune intervened. In March 1852, after a fire destroyed his law office and library and, it has been said, the businesses of most of his debtors, he grasped the opportunity to get out. If there was any insurance money it would be used as a grubstake to get him to California—to join his brothers who had become well established in the wholesale grocery business in Sacramento and in the foothill communities. He intended, like many Californians, to make his fortune quickly and return home; in 1855, after putting aside some money and going back east to rejoin his wife, who had remained with her parents, he even attempted to settle down in Albany. But he and his wife found the pace too slow. They soon left for California.

Stanford was allowed to take over his brothers' firm. By the time in 1860 when Huntington wandered over to his grocery store on Front and L Streets, Leland Stanford had prospered from the business, from rental income, and from mining investments. And, like Huntington and Hopkins, he had been in on the ground floor of the California Republican Party. It took several days for Stanford to ponder this railway business, but in the end he agreed to gamble his $1,500.

The last investor Huntington found was yet another shopkeeper, the proprietor of a dry-goods emporium on J Street. Forty-one-year-old Charles Crocker was a tall and brawny native of Troy, New York, the son of a liquor wholesaler who had gone bankrupt after the Panic of 1837. Charles was then fourteen, and when the elder Crocker moved his family west to get a new start in Indiana, Charles went too. He did not prove to be very good at farming despite his bulk and great physical strength—at least not in the eyes of his father. They quarreled often and had a final break after Charles, then seventeen, demanded to know whether his father wanted him to leave home. "Yes and no," came the reply. "Yes, because you are no use here. No, because I am afraid you would starve among strangers."[17]

Crocker did leave home but he did not starve. He worked first as a farm hand, then at a sawmill, and then, supported by his elder brother, Edwin Bryant Crocker, a lawyer, he opened up a little iron forge. The income from this was so meager that in 1849 Charles decided to put blacksmithing and Indiana behind him in favor of the California goldfields. With a party of other adventurers he crossed overland, taking the pioneer Platte Valley route from Council Bluffs on the Missouri River and finally joining his brother Edwin in Sacramento in March 1850. Mining, however, neither suited nor rewarded him. So, in the spring of 1851 Crocker opened up several small stores adjacent to mining camps, doing well enough that with a partner the next year he built a large store in Sacramento.

It is amusing to picture big Charley Crocker with hands like a pair of anvils reaching up behind his dry-goods counter to pull down a bolt of imported fabric or to measure out a length of satin ribbon. But Crocker adapted to this genteel work and prospered. He briefly went back to Indiana to marry the daughter of a former employer. When one of Sacramento's periodic firestorms destroyed his firm, he built another in its place. A comfortable living, marriage, and an expanding sense of civic responsibility led Crocker to follow his brother Ed into politics. When he ran for a seat on the city council as a Know-Nothing candidate in 1855, he won—as did fellow shopkeeper Mark Hopkins.

Charles Crocker's seat would literally have groaned under his weight, for the merchant now weighed over 250 pounds. With his seemingly limitless energy and his hearty backslapping gregariousness, Crocker was a popular politician. However, he would soon come to know what it was like to be an outsider. After his term Crocker would leave the Know-Nothings in April 1856 (so would Hopkins) to join the founders of the Republican Party in California, suffering as they all did from the smears and open aggressiveness of proslavery Californians. Being branded as degraded, radical, "Negro-worshipping" fanatics in the press and sometimes even in the Sacramento streets helped bind these burghers together. They all deserve credit, in fact, for adhering to abolitionist principles despite the possible economic consequences. That year their presidential candidate, John C. Frémont, would of course not prevail—not even in California, where he was well known. When Crocker and Leland Stanford ran for office in 1857—Crocker as state assemblyman, Stanford as state treasurer—both were defeated. But Republicanism was on the ascent, locally and nationally. And people like Charles Crocker, Mark Hopkins, Leland Stanford, and Collis Huntington would ultimately benefit greatly from their alliance.

Even by the time Huntington went over to interest Charley Crocker in joining them in a certain railroad venture, Crocker had gotten himself elected to the state assembly, squeaking by his divided opponents. Knowing the other prospective railroad directors as well as he did, Crocker did not think twice.

Crocker's $1,500 down payment brought Huntington over the top. His small directorate of successful merchants—two hardwaremen, a grocer, a drygoodsman, a jeweler—now had hold of a still-informal entity which Theodore Judah named the Central Pacific Railway Company. And, Huntington smoothly assured Judah, he would not be overlooked; thus it was marked down that the cash-poor and idea-rich Judah owned 150 shares, and that he would occupy a seat on the board with the businessmen. There was no question but that he would be an equal.[18]

Ted Judah had much to look forward to and little, it seemed, to fear. The survey could begin—Huntington had gotten everyone to put in $50 to get it going—and as chief engineer Judah could look forward to receiving a monthly salary of $100. He was more than usually agitated when he rushed back to his rooms at the Vernon House to tell his wife the news. "Anna," he said, "if you want to see the first work done on the Pacific railroad look out of your bedroom window. I am going to work there this morning and I am going to have these men pay for it!" His wife, the eternally supportive spouse through all of his struggles, took this in and sniffed. "It's about time," she declared, "that somebody else helped!"[19]

That morning, while his wife gazed proudly from their hotel window and while passersby paused momentarily to watch—this was in late December 1860 or January 1861—Crazy Judah squinted through his instrument at the levee on Front Street, aiming it up along the Sacramento River. If his transit had been a telescope he could have brought into focus the distant and massive mountain range, and along with it the knowledge that he would have to stretch his line over the summit at a point nearly seven thousand feet higher than the Sacramento levee.

Some part of his mind held the knowledge that three thousand miles away in Washington, the politicians were working on—or against—the railroad bill that would grant them their government charter. On December 18, 1860, the House of Representatives did vote to pass the Curtis bill. But after the short holiday break a similar bill introduced by California's Democratic (and Southern-sympathizing) Senator William M. Gwin grew heavy with amendments—or, more likely, pork, particularly to appease the restless Southerners. Such a bill, making so many other regional interests happy, would certainly have little trouble passing in the Senate. But the sticking point—and it turned into a killing point—was reconciling Gwin's fattened legislation with the leaner and more practical bill of Iowa's Curtis.

Sure enough, it came to nothing. But the optimistic Judah would have said the stalemate was already broken. And no citizen would be ignorant enough to disagree, unhappy as the events that were transpiring. With Lincoln set to be inaugurated in March, the cries for the South's secession had grown shrill. South Carolina had become the first to leave, on December 20, following this by seizing two federal forts and the federal arsenal at Charleston. Now the state forces were menacing holdouts at Fort Sumter. And beginning in January 1861 the rest of the southern states began dropping off, too. Every evening, as Judah edged his way up along the bank of the American River toward Folsom, his newspaper would report the new secessions—South Carolina, Mississippi, Florida, Alabama, Georgia, Louisiana, and Texas, thus far—and he would have read of the creation of the Confederate States of America; on February 4, the choosing of Jefferson Davis as president; and on March 4, under a clear sky, the Washington inauguration of Abraham Lincoln. Judah would be planting survey stakes in a line northeastward from opposite Folsom up into the hills at a steady incline, scratching endless figures in his notebook, while the hungry, weary, defeated federal troops of Fort Sumter steamed out of Charleston harbor on a relief ship; he would have been similarly occupied on April 15 as the president called for seventy-five thousand volunteer troops to quell the "insurrection." Judah would be working on April 19 when, in Baltimore, as a federal regiment marched south to defend Washington, Confederate-sympathizing civilians stoned four Massachusetts soldiers to death—the first casualties of war, quickly augmented when their comrades retaliated by firing into crowds, killing twelve civilians. The horror was well on, and the second wave of southern state secessions had commenced.

Even an incurable case of railroad monomania would hardly have diminished a citizen's chagrin at the bloodshed that certainly must follow. But a hard reality was incontrovertible: while Washington was nearly deserted, boarded up and fortified and flinching from the threat of engulfment, the chambers of the Senate and the House of Representatives were also emptied of all the Southern obstructionists who had been stalling a northern transcontinental route since the time of Asa Whitney. When the new session opened, some new version of the Curtis railroad bill would certainly prevail. And in the meantime, the Californians had much to do—Judah with his survey, Huntington with his canvas for more stock subscribers. And Stanford with his politicking; he had gone east to confer with President-elect Lincoln (and with Republican railroad allies in Congress) on January 21. Now, by the approach of spring as politics heated the air, Huntington, Hopkins, and the Crockers would have decided that Stanford had a good chance for the governorship. After all, their cause could only be helped. The convention in Sacramento would be held on June 18; they urged Stanford to return west in time to accept his draft.[20]

By April 30, the co-organizers of the Central Pacific called a meeting to elect a board of directors. Only Stanford, who had gone east, was missing. Named to the board, it seemed only natural, were the six "ground-floor" men—Judah, Stanford, Huntington, Hopkins, Crocker, and Bailey—with the addition of Lucius Booth of Sacramento and Doc Strong of Dutch Flat and Charles Marsh of Nevada City; on the board of directors the two latter men would give representation to the upland communities whose political and financial support was so important.[21]

After the meeting Judah bade goodbye to Anna again and left for the hills to join his survey party. Ahead was the most difficult part of the entire transcontinental route. Above Auburn, where the terrain of the divide between the American and Bear Rivers became complex, the number in his party had swelled to ten or fifteen, including Judah's brother Edward and the ever-faithful Doc Strong. The previous winter had been more dry than usual. Thus they were able to make good progress in following the retreating snows upward on horseback along a succession of ravines and their tributaries, edging around grassy side-hills, progressing up gentle valleys populated by grazing cattle or (sometimes) orchards, crossing creeks, streams, and stage roads, passing through gaps and divides, never too far away from the scattered, picturesquely named little mountain settlements such as Barmore, Clipper Gap, Illinois Town, and Gold Run. (Their citizens would come over to see what Judah's party was doing and sometimes volunteer to accompany them for a while.) They stopped frequently—to note ascent levels and take barometrical readings, and to measure the radius of a curve, the depth of a ravine below a projected bridge or trestle, the height of a ridge overhead, the depth of a needed tunnel, the character of rocks and soils for excavation and fill. It was precisely the routine carried out numerous times with previous railroad lines, but this was *the* route, including engineering tasks more daunting than anything ever attempted on the continent.[22]

It became obvious that the railroad line could reach the top of the divide between the American and Bear Rivers at Clipper Gap, some forty-eight miles from Sacramento, and from there could ascend steadily along the ridgeline for some twenty-five miles. The crest, not being uniform, necessitated running the line from gap to gap (the lower points along the ridge), skirting the intervening hills on their side slopes. But there would be no need to cross any of the major canyons or to bridge a principal river.

They camped out in tents every night with the temperatures plunging into the thirties until their water buckets would freeze. Several weeks into this phase, on Monday, May 18, 1861, Judah responded to his wife's first letter from Sacramento. "It has been impossible for me to leave to go down," he apologized. "There are ten men in my parties. And I am running a line through the most difficult country ever conceived of for a Rail Road. And had no time to explore it in advance. I get my breakfast and am off by sunrise every morning, and in the saddle all day long over hills mountains ravines & etc. And come home tired out. Then in evening I have four men at work till after ten PM plotting up work—and it gives me no time to come down. . . . Some time we will come out on horseback over this country together and you shall see what I have been doing."[23]

By early June they had passed Dutch Flat and were working along a steep side-hill above the fifteen-hundred-foot-deep Bear River Gorge, still breaking the morning's ice in their water buckets. If Anna Judah could look in his tent, her husband wrote, she "would say bueno. I filled the tick with Hay and have got my india rubber bag with an overcoat on it for pillow. The whole floor of both tents we have nicely carpeted with pine sprigs. I have got a pole in the centre with prongs on it, on which is hung a lantern, my saddle bags, and other fixings. I have a drawing board table made and set up on boxes, with a box for a seat with blankets on it, and here I am writing, on my desk, a candle on each side and lantern in the centre. Doctor Strong reading the *Union*. Douglass and Silby lounging, and a big fire just outside the tent."[24]

Not all was pleasant. Some nights it was all that they could do to simply fall into unconsciousness on their bedrolls. The surveyors had constant, often dangerous trouble with their new horses. Once, on a very steep hillside,

Strong's saddle began to tend downhill and he dismounted from his nervous black horse to tighten the girth. "His horse struck and kicked at him," Judah recounted, "broke loose [and] ran against mine, [which] started on the run down hill kicking and plunging; my stirrup broke, and off I went head first, and the black horse came Thundering down kicking and right on to me. Doc said he supposed I was killed, for he saw me down and both horses seemed jumping on top of me, and the black horse acted as though he wanted to jump on me, but he was under such headway that his feet missed me. Off they went home, the black kicked his saddle off and we picked up the pieces strung along the road. When we got into camp there they were. We spent the rest of the afternoon in taming them." Then, after two days and nights of solid rain kept them in their tents, a rattlesnake slithered out of the cold and damp underbrush to warm itself on a log near their fire. "Five rattles hang over our blankets now during the day," Judah wrote.[25]

Picking their way across rough and trackless terrain was a constant hazard as Judah ran lines on hillsides that would need to be graded. On June 13, at an elevation of five thousand feet above tide, some seven miles above Dutch Flat, he wrote to describe their line. It was "seven hundred feet below on Bear River side hill—the worst place I ever saw; the river is 1200 feet below us, and the top of ridge 700 feet above, in places so steep that if you once slip it is all over. To day the boys carried out ropes, the only way for them to get along being to fasten one end of a long rope around their waists, with a man above holding on with a turn around a tree. . . . They make only a mile all day."[26]

Within days of continuing ascent, about fourteen miles above Dutch Flat, they saw that below them the deep Bear River Gorge widened out to the mile-wide Bear Valley, along which trickled the small creeklike headwaters of the Bear, whose source was disclosed to be a marshy lake. Soon they came into view of what Judah would note as "a singular freak of nature": the South Yuba River, tumbling out of an impassable rocky canyon with nearly perpendicular walls toward the head of Bear Valley, veered suddenly northward, piercing a divide just north of the Bear, and disappeared into a canyon from two to three thousand feet deep. It was a breathtaking sight—and it seemed obvious that at one time the South Yuba had flowed into the Bear. Indeed, a hydraulic mining company down in the Nevada district had tapped the South Yuba and was conveying water in a canal down along the Bear Valley's upper slope.[27]

Here, at the near-juncture of the Bear and the South Yuba, they were closing in on the summit. Business matters, however, would summon the engineer back down to Sacramento at the height of summer in late June. And for a brief time, having been adjusted to kinder mountain temperatures and a more carefree atmosphere, Judah and Doc Strong would find the lowland conditions not only hotter but thicker—with ambition, intrigue, and the sort of glowering envy that spurs one to keep glancing over one's shoulder if one has any instinct of self-preservation. It had come time to elect a slate of corporate officers for the Central Pacific, and with that political necessity the tiniest of rifts would open, like an imperceptible crack in the ground opened by a deep, unfelt tremor, at first seeming insignificant but portending trouble nonetheless. Collis Huntington wanted and expected to be elected chairman of the board; he was certainly equal to the task and felt it his natural right as the one who had begun to stir Theodore Judah's dream into reality. But against his ambition loomed the stolid impediment of Leland Stanford, now returned from the East and now the promising Republican candidate for governor. Stanford, a man whose pride most certainly equaled any other man's, had swelled after months hobnobbing in Washington and New York society, and, surely, after consulting with the new president of the United States. He, too, felt equal to any task and just as deserving.

For parliamentary reasons Stanford was named the temporary chairman of the Central Pacific in advance of the vote. In turn he named Judah and Strong as a nominating committee. The engineer and the druggist, now fast friends, sensed an ally in Leland Stanford. So they jumped over what might have been a more obvious choice, Huntington, and nominated Stanford as president. The rest of the slate came out with Huntington as vice president, James Bailey as secretary, Mark Hopkins as treasurer, and Theodore Judah as chief engineer. And Collis Huntington could hardly hide his dissatisfaction at not being given the top spot. He as much as said so before all of them. But the nominators did not defer to his pressure. Saying nothing when the slate was elected, Collis Huntington showed only a sour face; he was not the kind to forget a slight, and on that day Ted Judah may have guaranteed himself one enemy he could not afford to have.[28]

Soon incorporation documents were filed with the state, on June 27. The engineer would not have tarried long in returning to the mountains—they were, in fact, so close to the summit! But it would take several attempts—sorties up blind ravines, planting lines of flags only to have to pull them up again when it got too steep to continue—until they arrived at an acceptable grade. Up a canyon at maximum grade, then entering the gently sloping Summit Valley, Judah's survey finally reached the crest of the Sierra at an elevation of 7,027 feet above the top of the Sacramento levee.

The old emigrant trail threaded nearby. Its wagon ruts were all that remained to mark the passage of the thousands who had used it. Of course there was only memory to commemorate those survivors of the ill-fated Donner party who had crawled through the fearsome snows of the Sierra fifteen winters earlier. As Ted Judah took his measurements within sight of the trail he could not have been unmindful of what the Donner survivors had contended with, and what doubting Thomases had always said about a central route through the mountains: that the heavy snows would make it impassable throughout much of the year. Judah had always said that his line could take care of the problem of the snows. The engineer believed he saw the supportive evidence he needed on the trunks of the slope's many fir and pine trees: axe marks chopped by snowshoers at a level of ten to thirteen feet above the ground, and the complete absence of moss below; also, he saw that the tree limbs up each bole to thirteen feet high were nearly all in their natural positions, whereas those above the snow line had been bent or broken by the weight of the snow.

Thus, he optimistically recorded, the greatest depth of undisturbed snow was thirteen feet at the summit, though with drifts going much greater. "This may at first seem to be a serious obstacle to the passage of railroad trains," he would write. "But this depth of thirteen feet is not the result of a single storm, but the accumulation of a number of successive storms occurring during the winter." Each would fall to the depth, perhaps, of three or four feet, successors piling on their predecessors. And since the route was almost exclusively a sidehill line, engines with snow plows could easily begin at the summit in both directions and push the snows off and below the grade. If the snow was cleared as it fell, the track could be open the entire winter. Besides, Judah knew, "the great dread and real danger of a storm in the mountains does not arise from the depth of snow, but from the entire absence of shelter and relief in the mountains, there being no houses of accommodation excepting upon the wagon across to Washoe." The Placerville wagon road was kept open for travel each winter, with hotels built every few miles along the route. "With a railroad built upon this route," he believed, "this objection would also be entirely obviated.[29]

Of course, in late June of any year it would take a feat of imagination to see the mountaintops thus blanketed in white. Now, Judah could stand at the summit and look eastward, as he had done once before with Doc Strong, on a nearly perpendicular, thousand-foot-high granite wall, looking down over the green forests to the mirrorlike little Donner Lake with its outlet descending and emptying into the Truckee River. Above and behind there was the second summit of the Sierra, while far in the distance rose the Washoe Mountains of Nevada. By now Anna Judah had ridden up to join her husband in his mountain camp for a few weeks. Genteel as her upbringing was, Anna seemed to take pride in being able to follow her husband anywhere. "It may be I was just the wife for him," she would write many years later, "for I never held back, always as he used to say, 'Had the right pair of gaiters on.'" Wearing pantaloons that would have shocked proper Greenfield society, she would march out of the tent carrying a sketch pad for the breathtaking views and a book for pressing the wildflowers which grew in profusion in the meadows of Summit Valley. Her sketches of Donner Pass and the lake were to be engraved on the original stock certificates of the Central Pacific Railroad.[30]

Now it was time to turn their horses to the eastern slope. Picking their way down a steep and rocky mountainside ("quite heavy rock-cutting" would be necessary, noted Judah) the surveyors encountered two ravines (the first was named for Doc Strong, perhaps because he found it) which allowed a fairly straight descent with only two sharp turns down past Donner Lake and along Donner Creek until it spilled into the tumbling green Truckee River. This they accomplished in just over eleven miles of maximum grades. Then for five miles they ran the line down the Truckee Valley.

The last survey stake was pounded with relief and no little delight—one hopes with some ceremony—into the sandy, rocky soil of the Sierra base not far from the California state line, 128 miles from Sacramento. Before he would turn back, though, Judah rode down along the presumed extension of the line, following the Truckee for twenty-eight miles to the western base of the Washoe Mountains, and thence another twenty-three miles to the Big Bend of the Truckee, at the edge of the Humboldt desert. No obstacles presented themselves.[31]

And now it was time to take his maps and notebooks back to his office in Sacramento. Judah returned in late July—and then would have heard that during a special congressional session convened by President Lincoln in July, Representative Samuel R. Curtis had reintroduced his Pacific Railway bill; issues of war, however, would continue to stall its passage, so hopes were pinned on the next regular session. Judah would also have read in one of the mountain newspapers or in the *Sacramento Union* the distressing war news, reported not just by correspondents but by scores of civilians (many of them Washington politicians and their wives) who on July 21 had gone out to Manassas, Virginia, with their picnic baskets to see a quick Union victory. Instead they witnessed a rout of Federal troops. For the first time was heard the hair-raising ululation of Brigadier General Thomas Jonathan Jackson's charging, shooting rebel infantry, and it would be in this battle of Bull Run that he won the name Stonewall. Now the combined Union and Confederate casualties were nearly five thousand, including almost five hundred dead. And a former associate of Judah's on the Sacramento Valley Railroad, William Tecumseh Sherman, for now colonel of an Ohio brigade, would consider the poor quality of his soldiers at Bull Run and begin to think, "Nobody, no man, can save the country."[32]

Such news as the Union defeat at Bull Run would have been felt hard, even three thousand miles away from the eastern seaboard in Sacramento, where Theodore Judah would have shrugged off war concerns and turned to writing his report, while assistants bent over profiles, surveys, and tables. The result would signify a stupendous leap forward in transcontinental railroad data, far beyond the dreams of an Asa Whitney or of any of the multitude who worked on the government's Pacific surveys. Theodore Judah's report, the draft of which would be presented to Congress as a memorial for recognition of the Central Pacific company as builder of the western part of the national line, methodically presented his argument in terms that would allow an understanding of mountain conditions and engineering details—even to a politician who had never been out West or to a capitalist who could not draw a straight line.

Expecting to pack all of his information for placement before the U.S. Congress, Judah also supervised the drawing of a colossal map—some ninety feet long, showing curves and tangents from Dutch Flat to the Truckee River on a close scale of four hundred feet per inch—and, on the same scale, a set of five 20-foot-long maps which traced the entire length of the route through the Sierra. There would also be other maps of lesser detail, sets of location sheets, gradient tables, barometric readings, and profiles of the line showing its entire length and inclination from Sacramento to the Truckee.[33]

In his report the hard figures would have stared back at his fellow directors, especially the shrewd merchants Huntington, Hopkins, Crocker, and Stanford. To be sure, the required construction work would be ambitious. And it would be expensive. The length of the line had proved to be 140 miles, costing over $12 million, or about $88,000 per mile. Although no major canyons or rivers needed to be bridged, a number of bridges and trestles were needed over low points. Moreover, there would be no less than eighteen tunnels to gouge through the granite mountains—some only four to six hundred feet long, but twelve over a thousand feet in length. Mining of tunnels would cost, he believed, $870,500. It, like certain other features in the report, would be an underestimate though made in good faith with the available data and technology. What remained to be seen at that time was how the cost could be borne by further stock subscriptions and by government subsidies.

The latter, of course, would follow a national railway bill that was to their liking. And even the local political situation, with all the boisterous, end-of-campaign hoopla of the races for governor and for several congressional seats, seemed to be steps in the right direction. As Judah gathered his material for the journey to Washington, he wrote to his friend Doc Strong in Dutch Flat, on September 4, 1861. "Our office work is getting along rather slowly," he reported. "I hope to get everything finished so as to be able to go on by 1st October, if they still desire me to go." It seems as if the engineer was having trouble interesting their fellow railroad directors in his problems. "Election and politics so monopolize everything here now that our people have very little time to talk railroad matters," he wrote. "A good deal depends on the election of Stanford, for the prestige of electing a Republican ticket will go a good ways toward getting us what we want."[34]

Two days later Leland Stanford won his residency in the governor's mansion, and three other Republicans gained congressional seats, including one Aaron Augustus Sargent, the editor of the *Nevada City Morning Transcript.* Sargent, at thirty-four a year younger than Judah, was counted as one of the engineer's most enthusiastic supporters. Cheerfully the editor had boosted him in his newspaper back at least to the year before, when Judah had roamed the Sierra looking for wagon and rail gradients and trying to blow a bubble called Centralia. Now his friend's enthusiasm could be put to good use in Washington.

Only the formalities of booking passage and packing waited, that and a commission from the board of the Central Pacific. On October 9, it adopted the following resolution: "RESOLVED, That Mr. T. D. Judah, the Chief Engineer of this Company, proceed to Washington, on the steamer of the 11th Oct. inst., as the accredited agent of the Central Pacific Railroad Company of California, for the purpose of procuring appropriations of land and U.S. Bonds from Government, to aid in the construction of this road." Sure enough, on October 11, 1861, a particularly hopeful band of railway promoters—Ted and Anna Judah (she with "the right pair of gaiters on"); their friend and fellow board director, the jeweler James Bailey; and the freshman congressman Aaron Sargent—ascended the gangplank of their steamer for the long trip back east and the contest ahead of them.[35]

10

"We Have Drawn the Elephant"

Washington, D.C., in November 1861 was a bleak armed camp surrounded by muddy breastworks and dotted with fortifications, its population swelled now by some two hundred thousand undisciplined soldiers and a multitude of camp followers. Everywhere one looked, even on the field in which stood the unfinished obelisk memorializing General Washington, the troops were relentlessly drilled and drilled in a desperate attempt to make them into fighting men. On a clear day the flag of the enemy could be perceived on the Virginia horizon. Washingtonians had even more to worry about, though, than the possibility of rebel troops engulfing them. Britain had, infuriatingly, recognized the Confederacy as belligerents. In response, a Union warship halted and boarded a British steamer in the Bahamas Passage, arresting two Confederate emissaries to Britain and France—precisely the sort of international offense that brought about the War of 1812. Now relations between Britain and the United States had deteriorated until the two nations seemed to be on the very brink of war.

Having been appointed by the president to replace Winfield Scott—old "Fuss and Feathers," the ancient, obese hero of the Mexican War—the new commander-in-chief, thirty-four-year-old George B. McClellan, seemed to be spending most of his daylight hours galloping from one reviewing stand to another, and most of his evenings basking in the attention of the capital's elite in their salons. He was a strutting, banty West Pointer from Philadelphia, his energies balanced equally between his high, nearly messianic self-regard (privately he mused on whether he should "take the dictatorship" of the country, of course for its own good) and his deeply held aversion to ending the pomp by closing with the enemy he thought badly outnumbered them. Weeks passed after his adoring soldiers had held torchlit celebrations of his appointment and there was no movement. In a short while criticism would begin to mount at the army's inactivity.

That on November 7 there had been an easy Union naval victory at Port

Royal, on the South Carolina coast, did not by much diminish the public depression over the terrible defeat at Ball's Bluff in Virginia, on October 21. There, some seventeen hundred Federal troops had been trapped on a cliff overlooking the upper Potomac. Forty-nine were killed, including their colonel, the popular new Oregon senator Edward Dickinson Baker, a close friend of the president's from early Illinois days, and 158 were wounded. Some 714, driven into the river in conditions likened to a turkey shoot, were missing. When the Potomac's floodwaters rose in Washington in early November, they brought with them many bullet-riddled or drowned soldiers' corpses from the upriver battle.

War had engendered an elephantine support system across the city. Like most other public buildings, the Capitol had been pressed into wartime service, beginning in the summer after the congressional special session had ended. A portion of the basement in the old building was turned into an army bakery; barrels of flour were stored in the crypt below the Rotunda floor; for a time the Rotunda served as a barracks for a Massachusetts regiment. Some fifteen hundred beds were carried inside to transform all the grand pilastered and vaulted corridors, the sumptuous offices, and the stately committee rooms into a hospital for the wounded. After the commissioner of public buildings complained, President Lincoln ordered the army hospital to vacate the building. Workers were barely able to get the place clean before the corridors began to fill again with returned legislators readying themselves for the opening of the new session.[1]

Legislators had their inevitable attendants. Anticipation was in the air. Expectations swelled the rosters of every hotel and boarding house across the city until good rooms grew impossible to locate. Hope sprang as eternally in the hearts of the legion of influence peddlers, contractors, suppliers, speculators, middlemen, office seekers, and inventors who gathered in hotel lobbies and restaurants and in the public corridors of the Capitol, as it did in the hearts of the tens of thousands of whores, pickpockets, and confidence men who roamed the streets and alleys in search of soldiers and teamsters. Reductively speaking, their goals were the same. War was, as always, good business. On the higher social echelon, especially, government disarray and bureaucratic inefficiency made good partners with waste and fraud—and the word was out all over the northern states: come feed at the trough.

The greatest trough of them all was made of granite and limestone and marble, and it was called the Capitol. About its still-unfinished exterior scaffolding still stood, though because the nation was going broke little work was being done; a steam derrick rose high above the colonnade where one day the iron dome would be placed. Inside, above a staircase of the House of Representatives, a prominent German painter, red-bearded Emmanul Leutze, had begun a monumental fresco commissioned in July; entitled "Westward the Course of Empire Takes Its Way," it would depict settlers in an ox-drawn wagon train struggling heroically through a mountain pass toward their promised land. It was a fitting time to be working on a painting of such a subject. As much as the war had thrown life into turmoil, the course of empire would again be served before the session would close. And there would be a good deal of private empire building.

It was common knowledge that the Pacific Railway bill would be reintroduced. Moreover, to some minds war conditions gave it a better chance at succeeding. The transcontinental could now be presented as a military necessity for troop movements and for binding the isolated Far Western states to the Union. As in every preceding session in which a similar bill was discussed, the great question of locating the transcontinental route was still open to debate and machination; it would be a debate somewhat compressed but no less spirited; the machination would be on a level hitherto not reached in Washington.

Now the field of players seemed to have narrowed to interests backing three routes: a northern route westward from Minnesota, a middle route from the western Iowa border, and a lower route from the border of Missouri. As congressional offices began to bustle with new activity, their antechambers were soon filled with railroad lobbyists waiting for appointments—lobbyists whose professional callings were as lawyers, judges, businessmen, bankers, and even former members of Congress, all good protective colorations—from a host of private interests. There was, of course, the Central Pacific, menaced by a California contender, the San Francisco and San Jose, and joined by such lines as the Hannibal and St. Joseph, the Perham Peoples', the Sioux City and Pacific, the Mississippi and Missouri, and the Leavenworth, Pawnee and Western. Other contenders were present, but they did not figure outwardly in the ensuing drama.

Every lobbyist had an inside track or, as the common term put it, a wire to pull. An old friendship. A relative. A common close acquaintance. A former or even current business relationship whether hidden or acknowledged. There was also the considerable and certainly timeworn aid of many legislators' veniality, with a rationalization open to all: any bending of ethics was for the national good. Of all of them, only the Hannibal and St. Joseph and the Mississippi and Missouri had as much as a yard of track on the ground. Adding to the crush were representatives of the eastern entities along each prospective route's latitude, each of which naturally expected to link up with the winner. They all gazed warily at one another, sized each other up, considered alliances, and perhaps managed a thin smile as their eyes gauged the precise place between the shoulder blades where a stiletto might be inserted.[2]

Two of these lines bear further scrutiny. The Mississippi and Missouri (M&M) Railroad, chartered in 1852, was the child of the Rock Island Line. The route of the latter somewhat resembled the eastern leg of old Asa Whitney's dream—the tracks went westward across Illinois between Lake Michigan and the Mississippi River with termini in Chicago and Rock Island. The Mississippi and Missouri Railroad was to extend the line between the two mighty rivers, across southern Iowa between Davenport and Council Bluffs, both thriving towns. Nagging financial problems, though, had halted construction of the M&M after only the eastern third of its projected length had been completed, to some fifty miles west of Iowa City. Ten years old, the line now languished, its directors hoping for a federal bailout.

And the M&M's high hopes were certainly realistic. For nearly eight years it had been dispensing campaign contributions, land deeds, stock, gifts, free railroad passes, and even contracts to selected members of the Iowa and Illinois congressional delegations. It had a deep friend in the person of Iowa governor Samuel J. Kirkwood, an influential Republican, and other allies had served as mayors of cities such as Chicago and Davenport. The Iowa state chairman of the Republican Party, Herbert M. Hoxie, was so deep in the M&M pocket that he needed a periscope. And the line had boasted other exquisite ties to the Republican Party, having saved the third congressional term of Samuel R. Curtis (now gone to the army) and having helped deliver the state of Iowa to presidential candidate Abraham Lincoln, the onetime Rock Island attorney. Lobbyists for the M&M had been camping out in Washington reception rooms since before the inauguration, sniffing out war contracts and quietly building support for the Pacific Railroad bill; three of their directors—Henry Farnam, Peter R. Reed, and Dr. Thomas Clark Durant—were frequently in attendance, too. They will be heard of again.[3]

The Leavenworth, Pawnee, and Western (LP&W) Railroad (it would later be known as the Union Pacific, Eastern Division, and the Kansas Pacific) had been chartered in 1855 to run westward from the Missouri River at Leavenworth (then the largest settlement in Kansas Territory) to Fort Riley (the first territorial capital). The distance was only one hundred miles. There were vague intentions to continue toward the Rocky Mountains, but by 1860 the company had performed a survey between Leavenworth and Fort Riley and nothing else. That year it was taken over by four Kansas promoters: A. J. Isaacs, James C. Stone, James H. McDowell, and the brilliant, aggressive attorney Thomas Ewing Jr., who dominated the firm and shaped its fortunes.[4]

Ewing was born in 1829 into a powerful, politically connected Ohio family; that same year his father took in a neighbor's redheaded, nine-year-old fatherless boy, William Tecumseh Sherman, who would be the younger Ewing's foster brother and lifelong friend. At nineteen Ewing had become a private secretary to President Zachary Taylor, who a year later named Ewing's father as secretary of the interior. Young Ewing moved to Leavenworth in 1856 to practice law. For the past several years much of his legal time had been devoted to keeping slavery from being instituted in Kansas when it attained statehood in 1861; he had served until the war as the first supreme court chief justice of the state. Ewing was not only a dynamic attorney; he was also a relentless businessman who knew how to use the law to his great advantage, who would not stop short at any ethical shortcut. His idea for the LP&W was to prod the U.S. Senate into passing bills giving him the right to buy, at criminally low prices, several hundred thousand acres of land owned by the Delaware and Potawatomi tribes—deeded to the Indians to compensate them for being forcibly removed from their homelands. After a high markup this land was to be sold to white settlers, with profits being theoretically applied to building a railroad into interior Kansas—the Leavenworth, Pawnee and Western. The scheme was, in the words of the Indians' attorney John Palmer Usher, a "gross, heartless fraud."[5]

Usher soon would join the winning side. Named assistant secretary of the interior by Lincoln, he would begin to draft a Pacific Railroad bill for Missouri congressman James Rollins. He would work alongside an old friend with whom he had read law, Henry Bennett, who was a veteran lobbyist. And Bennett had just been hired by the Leavenworth line as its chief lobbyist. When he was not writing the legislation (with specific naming of the LP&W) Bennett was very busy making his rounds in Congress. Of all the railroads whose lobbyists passed fat envelopes beneath the table, the LP&W seems to have been willing to spend the most. In one of the only documents ever to have surfaced concerning the underside of this particular bill, it was recorded that the LP&W distributed stock certificates with a face value totaling $4,137,000 in 1861 and 1862. Among the recipients were newspaper editors and reporters, former legislators with clout, and four sitting congressmen and four sitting senators (at least—the LP&W list which fell into the wrong hands was replete with the names of middlemen close to many other legislators; in some cases annotations openly indicated "blackmail"). At their giving, the stock and land deeds were not worth much—unless the LP&W became one of the "pet" railroads. In time it would be alleged that some years later this LP&W stock was "extinguished or ignored" by a slippery maneuver of the board of directors. However, stock value in 1862 or even at repudiation does not camouflage givers' and recipients' intentions in the second session of the Thirty-seventh Congress: it was compensation—for services rendered. Small wonder that some select legislators were now being called Railroad Congressmen, so avidly did they pursue support in return for favors.[6]

It was not on such a grand millionaire's scale as the Leavenworth line, but when the Central Pacific's chief engineer and accredited representative climbed the House staircase—passing the painter Emmanuel Leutze on his scaffolding, one fancies—he did so carrying a portfolio of Central Pacific stock. Its face value was $100,000. Of course the shares were worth nothing at present. Given to him by the board, part of it was meant to underwrite his personal expenses beyond the $1,500 he had drawn for what might prove to be a six- or eight-month stay. Undoubtedly, though, given the climate of easy virtue in Washington, most was meant to help secure aid from other lobbyists, newspaper editors, and Railroad Congressmen. When questioned by Congress many years later, Leland Stanford admitted that they had released the stock to

Judah "for expenses and to secure aid," though he volunteered no details about aid or beneficiaries. (Significantly, he was not seriously pressed by his interrogators.) "I think, however," Stanford laconically testified, "that he brought back most of it." Actually, according to one source Judah may have dispensed stock with as much as $66,000 in nominal value.[7]

But no record can be found. And what one is left with is the impression that Judah presumably received during his last lobbying of Congress: the days of visionaries had passed. Pragmatism and self-interest were now the rule. Judah, who had spent his entire adult life dreaming of a part in the Pacific railroad, clearly recognized the only route toward gaining it. And part of that route entailed alliances. Many years later the former LP&W executive James C. Stone would admit that in that session one "F. Judah" (a transcription error) of California may have also accepted twelve hundred shares of LP&W stock with a face value of $60,000.[8] If so, pragmatism, Gilded Age–style, prevailed, and a black mark must be entered against him for accepting the certificates as well as for distributing his own.

Not that slipping Central Pacific stock or land grants seemed to be that necessary in Judah's case. From the day in October that his steamer had passed through the Golden Gate to now, ambition and political skill ruled him mightily. And he had advanced accordingly. He had booked passage with his upland editor friend Aaron Sargent, now a freshman congressman. They spent the entire voyage huddled over strategies. In New York City, after Anna went north to spend time with her family in Massachusetts, he polished his *Memorial of the Central Pacific Railroad Company of California* for Congress. He printed a thousand copies, and he immediately sent them off to each of the senators and representatives; to the heads of government departments, such as the Department of the Interior; to President Lincoln; to sympathetic railroad men; and to the railroad journals (which published his report). He met in New York for a long time with California's Democratic senator James A. McDougall, the chairman of the Senate Committee on Pacific Railroads. "I left fully satisfied," Judah reported, "that in him the [Central Pacific] had a firm friend."[9]

Later in November, in Washington, Senator McDougall asked Judah to prepare a railroad bill, elaborating on the moribund Curtis legislation. But then McDougall decided that the best hope lay in going through the House first. Luck pushed things along; Congressman Aaron Sargent reported for duty and was assigned by the Speaker to the House Pacific Railroad Committee and to no other. It seemed an act of hardship, Judah reported, "in assigning a new member to no standing committee, but placing him on a special committee which had for many years been unsuccessful in their labors." Indeed, as one commentator quipped, such service was one of the more dependable forms of punishments available at the Capitol. In reality, though, Judah wrote, it "proved an act of great benefit to our future interests, as well as to the State of California, for Mr. Sargent, having no other committee duties to occupy his attention, took hold of the Pacific Railroad and devoted his time and energies almost exclusively to that subject." Sargent's newly drafted railroad bill was introduced into the House and referred to its Pacific Railroad Committee. For nearly two months nothing was heard; the committee did not even meet.[10]

Then, on January 31, 1862, with the ill-attended House meeting as a Committee of the Whole on the State of the Union and discussing the slavery question, the freshman congressman was recognized and rose to speak. Instead of slavery, his topic was of course the Pacific railroad. Undoubtedly Sargent's more senior colleagues started at his presumptiveness at bringing up a subject out of turn. "Although both Houses are now filled with the avowed friends of the [railroad] measure," Sargent intoned, "it almost seems as if its fate would be to be betrayed with a kiss. I now conceive it my duty—representing, as I do, a people whose safety, perhaps their power to remain a part of this Union, depends upon it—to speak plainly upon it, to arouse this House from its inaction, and convince it, if I am able, that this railroad is a necessity of the times—a great war measure—to be inaugurated now, if regard is to be paid to the most vital interests of the country."[11]

By all reports his speech was well received. It seems to have spurred action. The railroad committee immediately called a meeting, created a subcommittee under Sargent to draft one railroad bill out of the various ones in contention, and—of all things—it appointed Mr. Theodore Judah (who knew more about railroads than anyone in Washington) as its clerk. Later in the session he would also be appointed secretary of the full House committee and of the Senate's railroad committee. All of this, though hardly proper then or today, firmly put Judah and the Central Pacific into the center of things.

Now with access unequaled by any other citizen, Judah reopened his Pacific Railroad Museum in the Old Vice President's Room—unwinding his profiles and long maps of the proposed Central Pacific route through the Sierra and fixing them around the walls, making it possible for careworn, war-weary congressmen to forget their troubles and trace their way northeast from Sacramento through the most formidable obstacle to truly joining East and West, while performing a simple circumnavigation of the room. At various places above the map were hung Anna Judah's Sierra sketches, watercolors, and flower pressings.

There, one day, down from Greenfield for a visit, Anna found herself chatting with a genial attorney from Leavenworth, Kansas. It was Thomas Ewing, the aggressive genius behind the Leavenworth, Pawnee, and Western Railroad. In a few months, his behind-the-scenes work in Washington done and a railroad fortune virtually assured to him courtesy of the federal government and the Indians of Kansas, Ewing would trade civilian life for that of a ruthless Union Army officer in Arkansas and Missouri, who would clear four Missouri counties of rebel sympathizers by ejecting under threat of death every man, woman, and child from their homes. (President Lincoln approved the war measure, as would have, presumably, General Ewing's brother-in-law, William Tecumseh Sherman.) Now Ewing stood there surrounded by railroad maps, his ruthlessness in money and military matters invisible, as he listened graciously to Anna's stories of her husband's adventures in the Sierra. Turning to Ted, he said, "Judah, I see your wife has learned her lesson well. She's as enthusiastic as yourself and believes in the Pacific railroad without a shadow of doubt." He looked over to Anna. "Let us hope, Mrs. Judah, we will all go on a grand overland picnic to the Sierras—to Donner Lake sometime."[12]

Another visitor to the Pacific Railroad Museum would be Collis Huntington, who had arrived with his wife by steamer in New York on December 24.[13] Huntington lost little time in joining Judah in Washington. He would remain for two months—and how the former itinerant peddler must have relished standing now in the heady company of legislators, generals, government officials, and powerful lobbyists, as empire-building commenced. Huntington would have filled Judah in on details not communicated by letter. In October, he said, after the Judahs had left California, Doc Strong had led Huntington, Stanford, and Crocker up on horseback along the proposed route to Donner Summit. There, gazing down and out into the great gulf, they had pondered Judah's certitude that he could grade a trackbed down the cliffs a thousand feet to Donner Lake. It looked like an impossibility. "I'll tell you what we'll do, Crocker," said Huntington after a thoughtful silence. "We will build an enormous elevator right here and run the trains up and down it." "Oh Lord," moaned Crocker, "it cannot be done."[14]

What could be done during that expedition was to ride down the emigrant trail to the Truckee Meadows and proceed by wagon across the sagebrush flats to Carson City, where they planned to lobby the Nevada territorial legislature for a franchise to build the Central Pacific across Nevada. It was not to be an uneventful trip; their wagon connection was a day late, forcing them to make a chilly camp overnight alongside Donner Lake; the next night, down on the Truckee Meadows, they spent in a haystack. Then, the rear axle of their wagon cracked in two—partly because of the rough road, partly the passengers' elephantine weight: Crocker, Huntington, and Stanford each weighed well over two hundred pounds. Stanford, being flabbiest, was allowed to drive the crippled wagon to the next stage stop. The others footed it in, and they exchanged the disabled wagon for a good one.

In Carson City the quartet more than made up for their discomfort on the rough journey. Living well, they spent nearly two weeks meeting with the territorial governor and other politicians, forging alliances and presumably distributing Central Pacific stock, until they virtually assured themselves that they would receive a franchise for a "Nevada Central Railroad" link with their line, and that they would have exclusive use of the Truckee Canyon and its easy gateway through the Sierra's eastern summit. Also, they obtained a franchise for continuing their Dutch Flat and Donner Lake wagon road into

Nevada Territory, assuring them access to businesses of the rich Washoe mining district. The four went home considerably satisfied.[15] On November 27, 1861, they officially incorporated the Dutch Flat and Donner Lake Wagon Road Company, hoping to raise enough money by spring to begin work.[16]

Some twelve weeks later, in Washington, Huntington was tickled to see an old friend from his storekeeping days in Oneonta, Richard Franchot, who, after Huntington went west, had gone into the railroad business in Schenectady. Success and influence had propelled Franchot into Congress as a Republican in 1861—and now fate had made him one of the three-man House subcommittee pondering the railroad bill. (Huntington would learn of more luck—a distant Ohio relative, the radical Republican senator Benjamin Franklin Wade, would be added later in the session, on May 7, to Senator McDougall's Committee on Pacific Railroads.)[17]

Those two forceful railroad men, Collis Huntington and Thomas Ewing, neither of whom had ever built a railroad, were pleased at the way their legislators began crafting the 1862 railroad bill, with the secretary Judah advising and participating. Introduced by Missouri congressman James Rollins and favored by the subcommittee's chair, Aaron Sargent, the bill satisfied the local interests of the most aggressive and politically influential railroad concerns. It straightly addressed the controversial points which had so often killed its predecessors: the bill was specific about who would build the Pacific railroad and where it should go. And, in contrast to earlier bills, including that of Samuel Curtis, it did not specify that only one company would build the entire route; none of the lobbyists wanted so unwieldy a prize. In the West, the Central Pacific would rule, and the Nevada Central; in the East, the Leavenworth, Pawnee, and Western would share its reign with the Hannibal and St. Joseph and several Iowa-Nebraska concerns, each of which would build feeder branches into the Leavenworth main line. A newly chartered entity, to be called the Union Pacific, would construct the great middle along a still-undetermined line, from the eastern boundary of Nevada to the western boundary of Kansas.

Even before the Rollins bill was amended and reported out of subcommittee, those interests who had failed to grasp a slice of the pie began to complain. "A determined and bitter warfare kept up," Judah would write, "not only against our company, but finally against the bill itself. Pamphlets were written and laid on the desks of members and Senators, absurd statements with regard to bribery, fraud, etc., were freely circulated, and every effort made to poison and prejudice the minds of members against the bill, but as it appears, all in vain."[18]

Deflecting most opposition from its enemies, the bill moved implacably on. Sargent made adjustments to the original Rollins document, most notably (at Judah's insistence) by striking out the right of other California companies to unite with the Central Pacific road and thereby gain access to the federal largesse. Also, Judah had Sargent adjust appropriations, allowing more federal assistance for the sections going through the more difficult (and hence expensive) terrain of the Sierra and the Rockies. For some reason these priorities had been reversed in the original language, paying more for less work—which made no sense to the engineer. Tinkering done, the subcommittee approved and reported the bill to the House Pacific Railroad Committee, received its sanction, and had the bill printed for distribution. Then, in the early days of March, the bill stood open to criticism and amendments.

It seems appropriate that while the railroad bill was confronted with a larger scale of lobbying and legislative maneuvering, the name of Iowa's Samuel R. Curtis should have been heard in the frontline dispatches. Curtis, who had done so much for the Pacific railroad—and Iowa's Mississippi and Missouri Railroad—in the last session, had resigned his House seat in August 1861 to return to his beloved army as a brigadier general. Now he was riding at the head of eleven thousand Federals from Missouri into Arkansas, driving eight thousand Confederate soldiers ahead of them. After the quarry joined a much larger force under Major General Earl Van Dorn, the rebels, now swelled to seventeen thousand, wheeled and turned on Curtis's men, who dug in at Pea Ridge, Arkansas. After a bitter two-day battle, the Union troops prevailed on March 8, repulsing and scattering the much larger enemy.

Curtis's decisive victory against such odds could have been an inspiration at least to some of the railroad promoters in Washington, especially those of the Iowa and Illinois congressional delegations. Resolutely, they began to chip away at the provisions which left their states wanting. The debate continued through most of March and into April. Inevitably, though, it was interrupted by consideration of war tax measures—a hiatus finally punctuated by the horrid news from Tennessee about the bloody, inconclusive battle of Shiloh, on April 6 and 7, in which more than thirty-four hundred combatants were killed and eighty-five hundred were wounded (casualties almost evenly divided between both sides). The tax bill prevailed and Mr. Lincoln had more funds to continue pressing the war. Over the next three weeks there would be Union victories on the lower Mississippi, including the capture of New Orleans, and in Virginia, where McClellan's siege of Yorktown led to its abandonment by the Confederates.

In Congress, attention turned back to the railroads. Legislators backed by the Mississippi and Missouri Railroad renewed their campaign for supremacy. It would later seem to Judah, then in accord with the Kansans, that the Iowa and Illinois foes of the LP&W would sacrifice the entire bill if it were the only way to defeat them. One representative from Iowa, James F. Wilson, obtained the floor to criticize the hidden pressures and incentives which had so skewed the bill toward Kansas. "This Leavenworth, Pawnee, and Western Railroad Company is the pivot upon which the whole system of the bill turns," he complained. So cleverly had the Kansas people written it that it would take a dozen changes "to root the contract of the Kansas Company out of the bill."[19]

Others stood to denounce the bill and its choice of companies to build the outer ends of the route; or to suggest that the middle section over arid or mountainous terrain would never be built; or to object that timber and mineral rights, or sectional land grants along the right-of-way, should be denied to the railroad builders; or to urge that the motion be postponed until the next session or until the end of the war. Members turned to focus attention on the section of the bill creating the Union Pacific, with a chartering convention to be held in Chicago in September. Filling a long list of commissioners, some congressmen waggishly added the names of avowed enemies of the Pacific railroad. And there was the constant congressional companion of self-interest. Most notably, old Thaddeus Stevens of Pennsylvania, partner in an iron works that made track rails, insisted on an amendment stipulating that only American iron be used on the road. No one could object to that, so Stevens began to look forward to new revenues for his mill.[20] Others offered more substitutions, mostly voted down.

But Wilson and his allies, representing a belt of interests from Iowa through Chicago through New York and Boston, nonetheless turned the tide in their direction. By the time, in early May, that the bill was presented for what most everyone hoped would be the last round of amendments, the eastern terminus of the national line had been moved northward—to an undetermined point on the Iowa border. The Kansans were reduced to providing only a feeder line, though even that promised to be lucrative. The Pacific Railroad bill passed the House of Representatives, 79–49, on May 6—but the contest was by no means won yet.

The struggle over the route and over government favors shifted to the Senate, where three actors dominated. There was the bill's sponsor, James McDougall of California, advised by committee clerk Theodore Judah. There was the gaunt, raw-faced, opportunistic freshman senator James H. Lane of Kansas, who had just raised a platoon of "Frontier Guards" (mostly politicians) to safeguard Washington—they were now camping out in the White House. Lane had the president's ear on many matters. The Leavenworth, Pawnee, and Western had Lane's. Opposing him there was another radical Republican and close friend of Lincoln's, Iowa's second-term senator James Harlan, who had looked after the interests of Dr. Thomas Durant and his Mississippi and Missouri Railroad while in and out of his seat. In three years Harlan would, serving as interior secretary, secure himself a tiny notoriety in literary history by firing a government clerk by the name of Walt Whitman because the clerk had written an "indecent" book of poems entitled *Leaves of Grass.* But in 1862, with the railroad bill under consideration, Harlan was merely the chairman of the Senate Committee on Public Lands—an entity with enormous clout in any present or future matter concerning the railroad or the American West. Together with his Republican colleague from Iowa,

Senator James W. Grimes, Harlan repeatedly exerted the will of the Iowa company upon McDougall's legislation.

Most colorfully, Senator Harlan stood up on May 20 and threatened to move that the bill be recommitted to his Committee on Public Lands, for all intents and purposes throwing the railroad bill into a swamp for the remainder of the session—and probably for the duration of the war. Panic reigned in the Kansas and California camps; the perceived value of the Leavenworth stock certificates plummeted; the Central Pacific's now-unquestioned supremacy in the West seemed about to be put on hold. Ted Judah hastened off with Harlan and Grimes to work out a peace accord with the Iowans. The result, forged after several weeks of negotiation, was an all-out victory over the main line for the Mississippi and Missouri.[21]

The Leavenworth, Pawnee, and Western's $4 million had not prevailed in Congress, though its lobbyists had apparently outspent their competitors. In the end, coalition building and, probably, a longer pattern of congressional giving—"older money"—won out for the M&M and its forebear, the Rock Island. Their latitude—the Council Bluffs, Chicago, and New York latitude—won over that of Kansas City, St. Louis, Pittsburgh, and Philadelphia. Senator Harlan was speedily admitted into McDougall's Pacific Railroad Committee; the Kansans' line was shortened, and the Union Pacific's lengthened, by moving the stated meeting point northeast from the 102nd meridian (the border between Kansas and Colorado) to the 100th meridian (in the Platte Valley of Nebraska); and the Union Pacific was freed to build all the way to the Missouri River. The Senate bill added a clause from the House giving the builders of the main line extra time to build their road until the M&M finished laying track across the western two-thirds of Iowa. There was no question but that the Central Pacific would be given the right to the entire line in California, with the freedom to begin either in Sacramento or San Francisco.

When these changes were incorporated and with Harlan and associates backing the bill, it came up in the Senate for final consideration on June 17. Three full days of intense polishing followed. On June 20, 1862—the day Congress prohibited slavery in the territories—the Senate railroad bill easily passed by a vote of 35 to 5. With the House concurring in all the Senate amendments on June 24, the bill was carried to the president's desk.[22]

Lincoln signed it on July 1. And tradition has it that Theodore Judah thereupon sent a victory telegram to his colleagues in the little Central Pacific office at 54 K Street in Sacramento: "WE HAVE DRAWN THE ELEPHANT. NOW LET US SEE IF WE CAN HARNESS HIM UP."[23]

It was quite an elephant—exciting, ferocious, possibly ungovernable—dubious in many respects to the public interest and formidable both in spelling out the burden on the nation and in the rights and responsibilities of the railroad builders. At a time when the resources of the federal government were taxed to the limit, with McClellan's Army of the Potomac retreating on the peninsula, with the president having desperately replaced a poseur with a paper-pusher by naming Henry W. Halleck as new general-in-chief, the people were now committed, with this act, to do what had eluded them for nearly twenty years. Some twenty million acres of public land, and a $60 million loan, at least, were to be handed over to groups of obscure businessmen, most of whom had yet to prove themselves.

The legislative controls were numerous, but they also were replete with vague language and loopholes large enough to drive a Baldwin locomotive through. One of the most blurry areas concerned repayment of federal loans. The railroads were ordered, once they were completed, to place 5 percent of their annual net earnings in escrow toward interest charges and retiring the loans. But legislators did not even structure when and how interest should begin to be paid—whether during actual construction, or after partial service had begun, or after completion, or even when the bonds came due decades later.

The companies' share of the elephant could carry or crush them. On the face of it, the federal subsidies seemed generous. They were to be dispensed in the form of land grants and first-mortgage bonds fixed at 6 percent for thirty years whose amounts would be scaled as to terrain and distance: 6,400 acres of land would be given per mile of track constructed, in ten-mile strips alternating on either side of the four-hundred-foot right-of-way; mineral lands were excepted in the bill, though the railroad companies had the right to timber and stone on public lands; $16,000 in bonds would be given for each mile constructed on the easy grades at either end of the line, up the Sacramento Valley and across the Nebraska prairies; $48,000 per mile would be given for mountain portions, beginning at the western base of the Sierra Nevada and the eastern base of the Rocky Mountains, and both sections continuing toward the center for 150 miles; $32,000 per mile in bonds would be given for track stretching across the high plains.

However, there were catches and limitations. Not all of this money was really available for construction: 15 percent of the mountain funds, and 25 percent of the bonds for the easier stretches, would be withheld until the full line was in working order. And nothing whatsoever could be drawn from the U.S. Treasury until the railroad companies had each raised millions from investors and built forty miles of track. This might not be as easy as an optimist like Ted Judah might have thought: there was a war on, with quicker and surer ways to invest money, and since the government would hold first interest in the road, capitalists would balk at buying second-mortgage bonds with their longer, riskier recoupment waiting periods.

And deadlines had their own rewards and penalties, as did clauses giving the companies the right to build farther than their charters indicated. Each line had been granted two years to build the first fifty miles; only fifty additional miles per year were required thereafter. If the Central Pacific reached Nevada before the eastern roads, it could keep on building until it met them. Moreover, if the Leavenworth line reached the 100th meridian before the Union Pacific, it was free to press on toward California. But if the railroad was not completed between the Missouri and Sacramento Rivers by January 1, 1874, then the road—and all company properties, including tracks, rolling stock, buildings, and real estate, would be forfeited. That meant that if one party defaulted, it was up to the others to build the uncompleted sections to avoid being swept down the same drain. This could be a blessing or a curse depending on how one looked at it.

To someone like Theodore Judah, of course, it was an enticing possibility—although presumably business partners such as Collis Huntington or Mark Hopkins would have seen it more realistically. But there were more immediate concerns on the engineer's mind. Before the ink was dry on the completed railroad bill—certainly a week before President Lincoln picked up his pen to affix his signature—Judah prepared a map with the Central Pacific route drawn across it, and filed it in the office of the secretary of the interior. This action, stipulated in the legislation, would cause the secretary to withdraw the lands for fifteen miles on each side of the route from private enterprise and sale, keeping speculators at a relatively safe distance.

His bustle continued. On June 24 he received a happy document, fairly remarkable in congressional history. It was a letter of thanks signed by two representatives intimately connected with the passage of the bill—Aaron Sargent and James H. Campbell, chairman of the Pacific Railroad Committee—and cosigned by forty-two congressmen and eighteen senators, representing a wide range of local interests. "Your explorations and surveys in the Sierra Nevada Mountains have settled the question of practicability of the line," they wrote, "and enabled many members to vote confidently on the great measure, while your indefatigable exertions and intelligent explanations of the practical features of the enterprise have gone very far to aid in its inauguration." Mr. Theodore Judah, the irrepressible inside man, could take pride that the Central Pacific's grasp on the western terminus was unbroken; somehow, probably through the force of his personality and intelligence, he had managed to remain on good terms with so many of the combative, competitive Easterners.[24]

The testimonial safely packed away, on June 27 the Judahs left Washington for New York, to make arrangements for equipment for the first fifty miles of the line. The Central Pacific would need to import everything but its railroad ties and ballast from the East. And Judah audaciously intended to fill his shopping list and pay for it with federal bonds which would not even be issued until all the fifty miles of iron rails were spiked across California soil and the rolling stock planted on top of them. Owing to the national emergency, eastern mills had already begun to refuse to take orders; most order books were filled up for many months ahead and prices were climbing steadily. Triumph having elevated his head into the clouds, he intended to get his first fifty miles of track laid by the fall. Accordingly, as he would declare in his report, Judah found a broker to arrange the best possible terms with several firms for five thousand tons of iron rails; eight locomotives; and eight passenger, four baggage, and sixty freight cars, deliverable in January 1863. All sellers, he said, agreed to accept the promise of the bonds. Contracts were subject to final approval of the Central Pacific's board of directors within thirty days of being drawn. And apparently as soon as Judah returned to Sacramento the appalled directors wired to cancel the contracts; first there was the matter of raising money to pay for it all.

During contract negotiations Judah had kept another front open: that concerning President Lincoln. Sitting in Washington, the chief executive was obliged by the railroad legislation to decide exactly where the mountain sections in California and Wyoming Territory commenced so that the government bond levels could be set. Lincoln was also to determine where on the Missouri River bordering Iowa would the Pacific line begin. Moreover, because the legislators had been unwilling or unable to decide on such a basic decision as the gauge to be used on this new national road, Lincoln was expected to set the gauge. It would fall to Aaron Sargent (still in the capital until the session ended on July 17) to advance the Central Pacific's case before the president. Off went letters which Judah obtained from influential rail executives, espousing the five-foot gauge as the best. Five feet had been the width favored by the Baltimore and Ohio, and five feet was the gauge the Central Pacific and the Leavenworth, Pawnee, and Western wanted, though neither had yet built any track.

But any choice would be controversial and angrily opposed by other companies committed to other gauges. Nearly as difficult would be a decision on where the distant mountain ranges began. Sargent's repeated calls on the president brought no action. Lincoln was too preoccupied; he was busy toiling on the first draft of a document he would call a "Preliminary Emancipation Proclamation," a war measure which would free all slaves in the rebellious South, which he would submit to his cabinet on July 22. He would be persuaded to keep it secret until there was better war news to report. Meanwhile, Congressman Sargent had decamped for New York, and on July 21 he met Ted and Anna Judah on the steamer *Champion* for a triumphal return to the West—and to the monumental enterprise awaiting them.[25]

Part III

1863
Last of the Dreamers

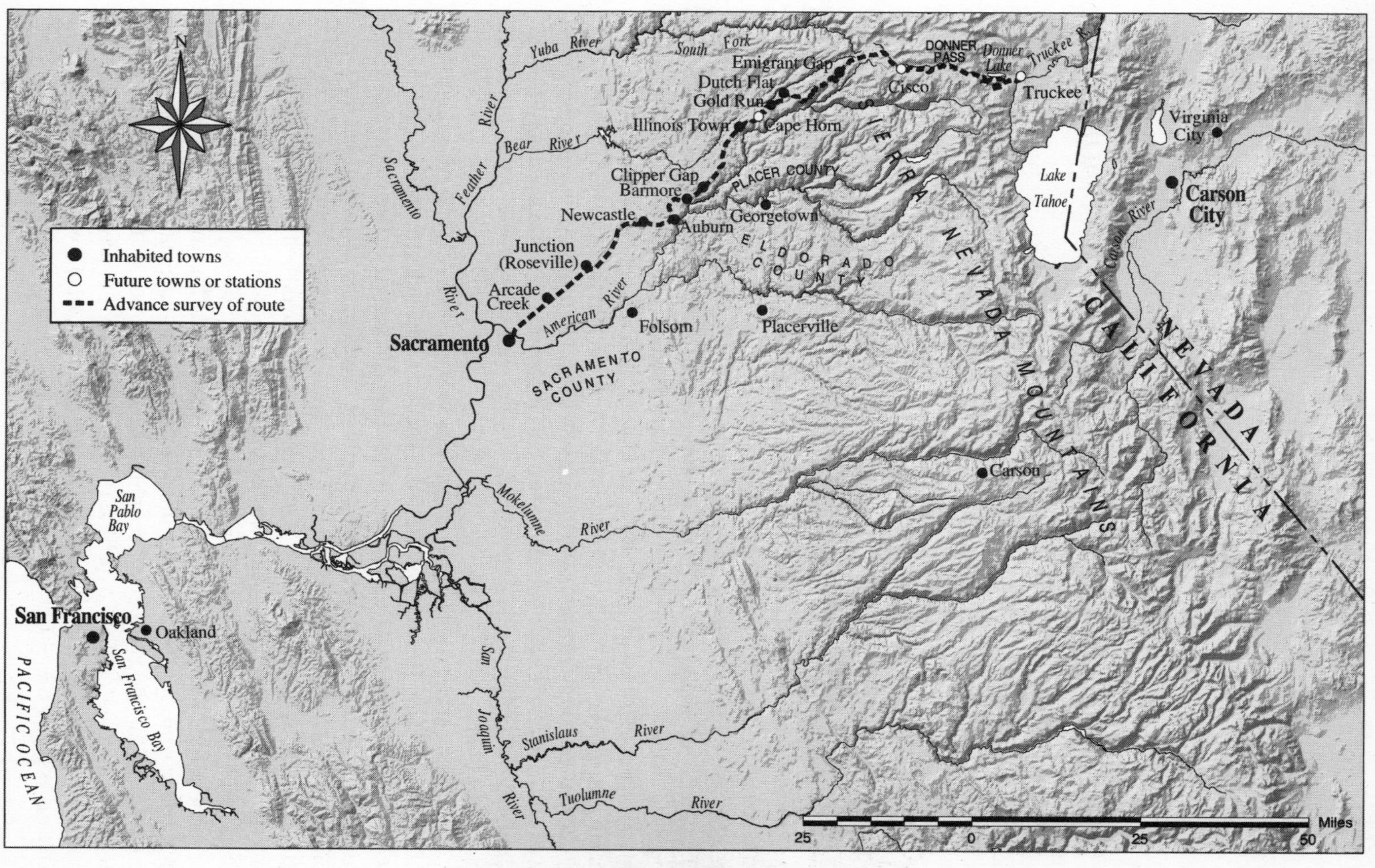

Central Pacific Survey, 1863

11

"Speculation Is as Fatal to It as Secession"

On the morning of January 8, 1863, the citizens of Sacramento looked apprehensively eastward to behold the sun rising above the blue Sierra Nevada into a clear sky. Relief triumphed; days of rain were over, leaving swampy Sacramento awash in mud. But the ceremony that noon would now go forward with ample crowds, oratory, and oompah-pah, under a smiling and propitious sky.

Down on Front Street near the levee, just above K Street, laborers swept the puddles off a simple, bare wooden platform and draped flags and patriotic bunting from its railings, while others distributed bundles of hay over the muddy assembly area for relatively drier footing. Two wagons filled with moist earth were driven up, stationed to flank the platform, and then decorated—including their teams of horses—with more flags and bunting. A large banner was unfurled and fixed to one of the wagons: it depicted hands clasped across the continent, the Atlantic joined to the Pacific, with the words MAY THE BOND BE ETERNAL.

By half-past eleven, as expectant crowds began to form, the ten members of the Sacramento Union Brass Band emerged onto the balcony of the American Exchange Hotel and began to serenade the onlookers with lively marches and popular tunes. One current song would be played to the great amusement of all—the ditty "Wait for the Wagon." The band would repeat the song over and over throughout the festivities to make sure no one missed the joke.

No one did. Nearly all of the population of the capital, and great numbers from outer districts, would soon turn out for the much-heralded groundbreaking of the Pacific railroad. For the most part, with the ground to be broken being so soupy, women ascended to the second-floor balconies of the commercial buildings for drier feet and a better view, or else they remained seated in carriages pulled up behind the crowd. "The great preponderance of pantaloons," grumbled a reporter for the *Sacramento Union*, "was a disagreeable necessity of the 'situation.'"

Shortly after noon, as Governor Leland Stanford appeared on the grandstand, burly, muttonchopped Charles Crocker lumbered forward as master of ceremonies to bring the expectant crowd to order and introduce the governor. In his twelve months in the statehouse Stanford had not improved much as a public speaker, but that day he did manage to rise to the occasion. "The work will go on," he assured his fellow citizens at one point, "from this side to completion, as rapidly as possible. We may now look forward with confidence to the day, not far distant, when the Pacific will be bound to the Atlantic by iron bonds, that shall consolidate and strengthen the ties of nationality, and advance with giant strides the prosperity of the State and of our country." Following Stanford, a cleric invoked God's blessing upon the enterprise and its directors and stockholders, and upon city, commonwealth, and nation.

Then Crocker stepped up again. "The Governor of the State of California," he bellowed, "will now shovel the first earth for the great Pacific Railroad!" Accompanied by cheers, the two wagons were positioned as Stanford seized a shovel, attacking a pile of dirt "with a zeal and athletic vigor that showed his heart was in the work," as the *Sacramento Union* observed, "and his muscle in the right place." The first clods of earth were deposited for the Central Pacific's railroad embankment as the beaming Crocker called for "nine cheers."

A raft of California senators and assemblymen followed Stanford on the grandstand, their speeches frequently interrupted by applause and cheers. One politician, A. M. Crane of Alameda, imagined a rail trip from San Francisco westerly toward the distant Missouri River: breakfast in the Golden Gate city, lunch in Sacramento, dinner in Carson City, and so on, speeding across the desert toward Salt Lake, the Green River in Wyoming, upward through the "magnificent scenery" of South Pass, over to the Platte Valley and the "emerald plains" of Nebraska, arriving finally on the Missouri riverbank some three and a half days after leaving the Pacific coast. Senator Crane did not have to remind his listeners that such a trip had taken him a hundred and more days on foot and by wagon some thirteen years before; nearly every adult man and woman in that Sacramento throng—nearly everyone all up and down the Pacific Slope—had spent up to a quarter of a year overland or at sea to get themselves to that place.

With their pride in their own forbearance and in their new golden land all wrapped up in the bunting of patriotism and piety, as the speeches went on and on (and on and on) they cheered the prospect of a shortened trip to their childhood homes; they clamored at every mention of Abraham Lincoln, the Union, and free soil; they applauded the notion of Nevada statehood, of a cluster of new states—"the happy homes of millions"—rising from the Dakota Territory, of precious metals, progress, and civilization riding the rails in a glorious procession. Dr. Hartwell Carver and Mr. Asa Whitney were at least in spirit in that crowd as one speaker proclaimed that over the road they dedicated that day "the skills of India, the rich tribute of China and Japan, the gold, the wine, and the wool of Australia, the treasures of California, and the spices of the East shall roll in a mighty tide of wealth such as mankind has never realized before. The stormy horrors of Cape Horn shall be forgotten; the sickly terrors of the Isthmus cease to be remembered."

At least fifteen hundred miles of trackless mountains, deserts, and prairies lay between those first few lumps of railroad embankment at the Sacramento levee and the nearest eastern railhead. In the triumph of the moment, in a collective act of will, that detail was—outwardly, at least, and only for the moment—expunged.[1]

In contrast to that happy-appearing tumult, four months before, in Chicago at the organizing convention of the Union Pacific Railroad, the mood had been as heavy as those first-flung wet clods of California earth. Organizers hired the capacious Bryan Hall for three days beginning on September 2, 1862, to welcome 163 commissioners whose task would be to elect officers and somehow get the new company moving. Fewer than half of that unwieldy bunch, many of them enemies of the railroad bill, showed up.[2]

This was probably a blessing. Of the 158 commissioners from twenty-five northern states named by Congress in the Pacific railroad Act (five more were appointed later, in August, by the interior secretary), a fair number were enemies of the bill and of the anointed railroads. Within the sixteen-member California delegation alone, half were, reasonably, allied with the Central Pacific, but the other half included sworn enemies such as John P. Robinson of the Sacramento Valley Railroad, and officials of a number of companies likely to lose business should the transcontinental line go through—such as the Pacific Postal Telegraph Company, the Pioneer Stage Lines, and the California Steam Navigation Company. The composition of other state delegations did not teeter so dramatically with opposing interests, but there were enough unfriendly names to make the prospect of the Chicago convention a potential headache.

Most enemies, however, stayed away from Chicago, finding it too much of a bother or too expensive, as commissioners were not paid. Many friends stayed away, too, including most of the Californians. (One who did attend, and who would continue to play a strong role, was California's Democratic senator James A. McDougall, chairman of the Senate Select Committee on the Pacific Railroad.) What little debate aired at Bryan Hall touched on issues yet unresolved—such as the eastern terminus on the Missouri River and the official track gauge (both in any case would be decided by President Lincoln). More unanimous was the feeling that the enterprise would never get off the ground—especially in wartime—without more assistance from Congress. Who indeed would be interested in buying second-mortgage bonds of a railroad that did not yet exist? Congress, they felt, would have to amend the 1862

Pacific Railroad Act, though the ink was hardly dry. Dutifully, amid speechifying plainly meant to bolster sagging morale, the delegates elected Mayor William B. Ogden of Chicago, a director of the Mississippi and Missouri Railroad, as president of the Union Pacific; Henry Varnum Poor, editor of the *Railway Journal,* was named secretary; Thomas Olcutt of Albany was elected treasurer; Peter Anthony Dey, an esteemed engineer associated with the Rock Island and the Mississippi and Missouri Railroads, was appointed chief engineer. President Ogden echoed the mood of the hall when he declared before all that the hard-won railroad act would have to be liberalized "before capitalists will be glad to take hold of it." He then continued: "This project must be carried through by even-handed wise consideration and a patriotic course of policy which shall inspire capitalists of the country with confidence. Speculation is as fatal to it as Secession is to the Union. Whoever speculates will damn this project."

Somewhere, the speculators guffawed, and sat back to bide their time. And after the convention closed upon resolving that Union Pacific subscription books be opened on November 1 in thirty-five cities across the United States, the portents of gloom in Chicago came true. There would be no rush of patriotic or evenhanded investors—only a polite knocking could be heard at any agent's door. Four months of earnest attempts came to this: the company signed up exactly eleven shareholders who spoke for thirty-one shares valued at $1,000 apiece. Only a single investor, one Brigham Young (who was anxious to get a railroad to his Salt Lake City), paid in full for his five shares; the rest paid only the required 10 percent. Of the Union Pacific commissioners—all prosperous men—only two reached for their checkbooks. There remained 1,969 shares to be sold before the company could be legally organized.[3]

Nearly as dismal was the experience of the Central Pacific, which opened its subscription books in Sacramento and San Francisco in late October, accompanied by a boostering editorial in the *Sacramento Union* and much energetic solicitation by the company board and its agents. On the first day investors in Sacramento signed for 3,642 shares, at $100 apiece and with 10 percent down. But beyond this promising start, there did not seem to be droves waiting.[4]

The directors even went out themselves to drum up business, stalking the plank sidewalks of the capital, hand-clasping and back-slapping other shopkeepers and sitting, hats in hand, before the city's most solid citizens. Charles Crocker recalled the typical response he received: "'Why, Crocker, are you crazy? You think of building a railroad across these mountains? Why, you have got a good business, what do you want more?' 'Well,' said I, 'I am going to build this railroad,' and they ridiculed me, and tried in every possible way to induce me to give it up. . . . They wanted to know what I expected the road would earn. I said I did not know, though it would earn good interest on the money invested, especially to those who went in at bed-rock. 'Well,' they said, 'do you think it will make two per cent a month?' 'No,' said I, 'I do not.' 'Well,' they answered, 'we can get two per cent a month for our money' [in the Nevada Comstock mines]."[5]

In San Francisco the story would be even more depressing: only three individuals bought all of fourteen Central Pacific shares. A distance yawned between that and their opening goal of capitalizing $3 million in stock. In fact, by year's end (as the annual report for 1862 would show) only $24,620 was actually paid into the treasury.

It was, simply put, no season in the Union for any speculation on the future that was not somehow connected to more killing. Twenty-five thousand combatants had been killed, wounded, or missing at Second Bull Run on August 29 and 30, which ended in a sobering defeat for General John Pope's Army of the Potomac, and resulted in Pope's replacement by the strutting George Brinton McClellan; two weeks later Confederate forces under the brilliant Robert E. Lee had crossed from Virginia into Maryland, with Stonewall Jackson taking the federal garrison at Harper's Ferry, before being stalled at Antietam Creek near Sharpsburg, Maryland, on September 17. It became the bloodiest day of the war, with nearly five thousand killed and eighteen thousand wounded on both sides. Antietam was merely a military draw, its only utility being that it drove Lee back off Federal soil and gave Washington a chance to declare a victory in an otherwise dark season. Seizing the political opportunity, President Lincoln released his Emancipation Proclamation on September 22, freeing all slaves in the rebellious states upon the first day of the coming new year.

But still there was little public confidence in the North, still reeling from its human losses. It was clear that from a military standpoint the Union was still adrift. McClellan had refused to press his advantage at Antietam just as he would decline to follow Lee and destroy him when such was possible. In due course he would be removed from command, sent to preen back home in New Jersey. Mistakenly, some mourned when they should have cheered. However, there was little time to cheer. McClellan's replacement, General Ambrose Burnside, a man long out of his depth, succeeded only in having his army pulverized as it attempted to cross the Rappahannock River at Fredericksburg, Virginia, on December 13. That Burnside would himself be swiftly replaced seemed a foregone conclusion.

Small wonder that patriotic and evenhanded capitalists were in no mood to snap up either stock certificates or mortgage bonds of the Union Pacific and Central Pacific companies, two railroads without a yard of track between them in late 1862. Meanwhile, the speculators so feared by William B. Ogden sat back, content to make their profits from safer industries more closely connected to war, while they waited for Congress to sweeten the railroad pot.

But work had to be started somehow, and for good reason the directors of the Central Pacific were worried. As of November 1 they had formally committed themselves to the government's terms as specified in the Railroad Act. Somehow forty miles of track, including an expensive bridge over the American River, would have to be constructed—as quickly as possible—before federal aid could commence. There was hope of raising cash in the East from mortgage bonds through intermediaries or through their own efforts. However, expectations of that yield ranged widely. Judah, no businessman, was characteristically optimistic; he put a high value on anticipated revenues from freight haulage and timber sales, which would surely attract capitalists. Huntington's prediction was, as usual, hard-nosed and conservative; if they unloaded any bonds at all in these tight times, they would be drastically discounted. And intermediaries might prove unreliable. Without any dissension it was decided that Huntington—easily the shrewdest among them—would as soon as possible go East. He would take a headsplitting task: unload their securities at the least humiliating discount, purchase equipment with Central Pacific bonds, and lobby as best he could in Washington and elsewhere to liberalize the 1862 Railroad Act.[6]

Until Huntington got rolling, what else could be done? Other possible outside funds—aid from the state or from interested counties—would require months of lobbying and snail-paced legislation, and probably public referendums. Even the carrot wielded by Judah to entice the associates into backing him some two years before was thus far producing nothing, although tolls from the Dutch Flat and Donner Pass Wagon Road might someday provide them with operating cash for the railroad. Late the previous May, Huntington had sent Judah's assistant engineer, Samuel Skerry Montague (formerly of the Sacramento Valley Railroad) up into the Sierra to locate the route, with work commencing later at a slow rate. But they would need to hire an army of laborers in the spring to finish it before anything tangible was realized.

The only immediate answer seemed to be for the directors to dig into their own pockets. Accordingly, Crocker, Huntington, Hopkins, and Stanford each bought 345 additional shares, while Judah, James Bailey, and Lucius Booth subscribed for more modest amounts. Next, hoping to improve upon their mediocre sales (and standing) in San Francisco, the directors moved to solidify an arrangement initiated the previous spring in Washington by Judah and Huntington in return for local political support of the railway legislation. The agreement, which ran outside the 1862 Railroad Act, called for the Central Pacific to assign its rights to land grants and government bonds between Sacramento and San Francisco to four Bay Area businessmen—Peter Donahue, Timothy Dane, Alexander Houston, and Charles McLaughlin—who had incorporated the San Francisco and San Jose Railroad. As the Western Pacific Railroad, they would continue building their road around San Francisco Bay to Sacramento by way of Stockton. The news was released on December

4; they prayed that the news that San Francisco would be the true western terminus of the Pacific railroad would pry open more pockets.[7]

In the meantime they had to address the question of getting construction started. It would be around this issue that the nearly imperceptible fissure that had opened within the board of directors during the election of company officers in mid-1861 would begin to steadily widen. The usual practice of other railroad companies was to hire outside contractors who would agree to be paid partly in cash but also in railroad securities, easing the cash burden on the corporate treasury. However, this arrangement included an obvious pitfall: all too often, before they finished, contractors had accepted enough stocks or bonds to allow them to wrest control of the company away from the orginators, reaping along the way cash from their inflated estimates. The experience of Colonel Charles Wilson and the Robinson brothers on the Sacramento Valley Railroad was only the closest example of such table-turning.

To Huntington, Hopkins, Stanford, and Crocker—who had taken to calling themselves "the Associates" to distinguish themselves from others on the board with whom they felt less affinity, who had each built himself up from hardscrabble poverty into a pillar of society, who each had a preternatural aversion to being hornswaggled in any deal, grand or petty—there was only one route open. To keep control of the corporation the Associates would have to maintain control of construction. The most obvious answer lay in the massive person of Charles Crocker, physically and temperamentally the best suited on the board to take on such work and the last of the Associates to find himself a useful niche in the Central Pacific: Stanford was clearly the outside political man, Huntington the wheeler-dealer, Hopkins the hawk-eyed keeper of accounts. True, Crocker had spent more than a decade behind a millinery counter. But at forty-one he still looked like a hod carrier; as he would be quick to rationalize later, "I had all the experience necessary. I knew how to manage men; I had worked them in the ore beds, in the coal pits, and worked them all sorts of ways, and had worked myself right along with them." After striking west from Indiana, he would continue to explain, he had led three combined teams of emigrants through all manners of difficulties all the way to California.[8]

There was another benefit: as the shrewdest shopkeepers in town, the Associates had always tried to turn, as Crocker often said, one dollar into a dollar and five cents. There were . . . possibilities inherent in construction contracts, as anyone who had even a nodding acquaintance with civic affairs knew. What public road, what bridge, what municipal building had ever been constructed free of padding? But that could be explored in time.

Almost from the inception of the Central Pacific, the Associates had been meeting separately from the rest of the board. Until December 1862, their fellow directors saw no reason to complain; after all, the four were friends with much in common.[9] But when they presented their plan to make Crocker the general superintendent of the line, with his taking on construction of the first eighteen miles of track, the rest of the board—Judah, Strong, Booth, and Marsh—objected strenuously. Beyond the fact that Crocker had no experience as a contractor (the engineer in Judah would not have wanted to ride across a bridge built by Crocker) lay a deeper suspicion which may or may not have been immediately voiced in a full directors' meeting: Judah and Strong, particularly, suspected that the other Associates stood behind Crocker—and that they intended to bleed the enterprise dry through willful mismanagement and inflated construction charges. By the following month, accusations of fraud would begin to surface in the press; fourteen stockholders (including, it must be noted, grocer Lucius Booth and a "B. R. Crocker," whose relationship to Charles Crocker is unknown) took it upon themselves to examine the contract, pronouncing it sound and even advantageous to the railroad; however, they did not look into the charge of any illegal interlocking directorship: "The high character of the members of the Board," they wrote the *Sacramento Union*, "renders it unnecessary to deny the charge." Suspicions, though, were hardly laid to rest by their "investigation."

Despite all objections, Crocker was awarded a $400,000 contract for the first eighteen miles, including the span over the American River; he was to be paid $250,000 in cash and would accept $50,000 in discounted Central Pacific stock certificates and $100,000 in discounted bonds.[10] Possibly at Judah and Strong's insistance, but more probably after consulting with his oldest brother, Edwin Bryant Crocker, who had been retained as the railroad's attorney, Charles Crocker was apprised of the legal penalties for conflict of interest. So he resigned his place on the board of directors, retaining all his stock, and his brother slipped into his seat. Two days later, on December 27, he put his signature on the construction contract between the railroad and Charles Crocker & Company. (He would cheerfully admit later that there was no "& Company.") To underscore his total commitment, Crocker began to sell off most of the contents of his store, allowing his clerks to pick up the rest at a good price. Then the directors scheduled their groundbreaking ceremony—with appropriate fanfare—for the eighth day in January.[11]

By the time Crocker stood puffed up and proud on the speaker's platform down by the levee, introducing a procession of dignitaries, the new construction boss with no prior experience had subcontracted his work to three builders who knew what they were doing, and who promptly signed on four additional firms to construct parts of their contracted mileage. From the foot of K Street up along eighteen miles of Judah's little flags to the little settlement of Junction (later Roseville), seven motley construction crews with their pickaxes and shovels and horsedrawn dirt carts would soon be attacking the Sacramento Valley soil, with Charley Crocker riding up and down the line learning what he could by watching. Crocker began to close the ceremony with his own choice words: "It is going right on gentlemen, I assure you! All that I have—all of my own strength, intellect and energy—are devoted to the building of this section which I have undertaken."[12] As he spoke, up the line a steam pile-driver was hammering thirty-foot-long redwood timbers from the Coastal Mountains into the silt at the bottom of the American River for the rail bridge.[13]

Had an enterprising reporter looked up at that moment at the assembled dignitaries, he might have noted two telling absences among the Central Pacific's board of directors: where was Theodore Judah? And indeed, where was Collis Huntington? The faithful railroad booster Lauren Upson, editor of the *Sacramento Union*, was present but made no mention of either in his article; the failure to give a nod in the newspaper columns to his intimate friend Ted Judah almost certainly signifies that the engineer was elsewhere. The only reason why Judah would have missed an opportunity to escort Anna to bask in the commencement of their dream is that he had already sensed that "his" railroad project was in danger of slipping away from him. In such a mood, he would have had no appetite to watch Charley Crocker strutting up and down the muddy platform, mugging for the crowd like a ham actor or a cheap politician and interjecting a comment or a huzzah at every lull in the program. ("It is going right on, gentlemen, I assure you!") Yes, there was trouble ahead for T. D. Judah.

And Huntington? He had left town two weeks before—for New York and Washington, by sea—head down, poised to charge rhinolike right into the financial nightmare confronting them all.[14]

12

"I Have Had a Big Row and Fight"

Washington in January 1863 was a chilly, depressing place, its military hospitals bursting with maimed survivors of Antietam and Fredericksburg, its streets teeming with men of military age passing through on their way northward from the Rappahannock trying not to attract attention—for the desertion rate of the freezing, disease-ridden, winter-camped Army of the Potomac had climbed to two hundred per day. Antiwar sentiments were more than palpable; legislators returned from home for the opening of Congress on December 1 reported a rising tide of anger (particularly in Illinois, Indiana, and Iowa, but also Michigan and Ohio) about casualties and emancipation of blacks in the South. Clearly, disturbingly, there was growing support for the rash of bluecoat desertions.

On January 20, a group of railroad congressmen and lobbyists—California's senators James McDougall and Milton Latham, and the Central Pacific's Collis Huntington among them—was ushered into Lincoln's own Cabinet Room. It was still a chamber of uneasy vibrations, Lincoln having weathered a bad outbreak of impatience among the Radical Republicans in the Senate and in his own cabinet only a few weeks earlier. Lincoln—the onetime Illinois railroad lawyer whom some in that room knew well from happier times, now careworn with a horrendous burden on his shoulders—listened as his visitors urged him to approve a five-foot distance between rails for the Pacific railroad. It was the gauge embraced by so many of his political supporters behind the Rock Island, the Mississippi and Missouri, and the Central Pacific.

And apparently Huntington had done some sly advance work before he found himself respectfully before the president. It involved Lincoln's longtime friend from Indiana, quick-stepping John Palmer Usher—who, a year before, had betrayed his Delaware and Potawatomi tribal clients by assisting in the land swindle that created the Leavenworth, Pawnee, and Western Railroad, before moving to the Interior Department to help draft the 1862 Pacific Railroad Act. Learning that Usher had just stepped into the post of secretary of the interior, Huntington located the boardinghouse where the forty-seven-year-old Usher lived and began to take his own meals there at Usher's own table. Casually, Huntington pumped information from the new interior secretary, finally learning that Usher had heard from directors of the Baltimore and Ohio and the New York Central, among others, and consequently was planning to urge friend Lincoln to adopt the 4-foot, 8½-inch gauge.

Hastily, Huntington found a New York engineer (from the Hudson River Railroad) who would be willing to amass documents and to supply an affidavit in favor of the five-foot gauge, and he also persuaded a noted railroad equipment supplier, Colonel George T. M. Davis, to accompany him to the cabinet meeting as their spokesman; Davis also happened to be a longtime friend of the president's. Lincoln recognized that no one in the room save himself and perhaps a few others was without a personal stake in the decision. After hearing presentations from both sides, Lincoln called for a secret ballot of his advisers. Usher and the peppery navy secretary, Gideon B. Welles, both felt that the cabinet had voted for the smaller gauge—but as Lincoln pocketed the ballots there was no way to know.[1] And the next day, after the president announced that he favored the five-foot gauge, Collis Huntington could stride through the blustery capital streets grinning in triumph, already composing victorious telegrams homeward.

There was more work to be done. Huntington set to work visiting iron factories and railroad equipment manufacturers, many of which Judah had visited in the summer; he found that prices had risen dramatically. Meanwhile, Senator McDougall was tireless in continuing his work on behalf of the Central Pacific. He introduced a sheaf of amendments to the 1862 Pacific Railroad Act, ones which Pacific railroad supporters universally felt would make the road more attractive to capitalists. The government support bonds, he urged, should be slid back to become second-mortgage instruments, allowing the company securities the less risky front spot; further, no government bonds should be held back until completion of the line. The rigid deadline should be relaxed and the threat of forfeiture lifted from the backs of the brave entrepreneurs willing to risk all for the line. McDougall guided these and other changes through the Senate until February 25, when the bill passed.[2]

Within a week, though, it became apparent that the companion House measure was to be sacrificed to a busy docket that understandably was weighted toward emergency measures such as the Conscription Act. Then disappointment piled on disappointment: a majority of senators rebelled with a bill to overrule Lincoln's decree on the railroad gauge; soon, on March 3, a House cabal led by New York's Erastus Corning hastily put over a bill also repudiating the president's selection in favor of the one lobbied by the Baltimore and Ohio and the New York Central—4 feet, 8½ inches. Representative Corning's other job was as president of the New York Central Railroad.[3]

News of the setbacks flashed ahead of Huntington, who was on his way to Boston to consult with the august brokerage house of Flint, Peabody & Company, with whom he had contracted to sell $1.5 million in Central Pacific securities. Without pausing, he wired a number of manufacturers to change his equipment orders, knowing that altering now to a narrower gauge would cost him money and time which he could ill afford. Then, when Huntington arrived at Flint, Peabody, he was thunderstruck to learn that the brokers were backing out of the arrangement. Though generally business was booming because of war, inflation and wild currency fluctuations had the investment community in turmoil—and with this bad news from Washington, surely Huntington would understand that at this time they could not lend their name or support to his enterprise.

The scrappy former itinerant trader soon found himself back down on the street, seething at the patrician moneymen who had so callously dashed his hopes. There was no point in taking his portfolio of bonds to another broker; word would travel fast, and all of those crooks worked together anyway. Instead, in a brilliant flash of insight, Huntington marched over to the Boston office of a company with which he had conducted a healthy long-distance trade for a number of years. The brothers Oliver and Oakes Ames, third-generation owners of Ames & Sons, ran the largest factory for shovels, picks, hoes, and hand tools in the nation out of Easton, Massachusetts; their famous "Old Colony" shovel had given them the nickname of "the Kings of Spades." All through the 1850s, Collis Huntington had sold tens of thousands of Old Colony shovels to gold rushers, and there was even one feverish period when he cornered the California market in them.[4]

In Huntington's portfolio sat letters of introduction from two leading California businessmen, Darius Ogden Mills of San Francisco (a banker who had incidentally been named as one of the 158 commissioners of the Union Pacific), and Orville Lambard, a Sacramento capitalist. Knowing that his record of credit with Ames & Sons was impeccable (he was always scrupulous about paying bills on time), and realizing that the Central Pacific's success would ensure the sale of tens of thousands of more Old Colony shovels, when he was ushered into the office of Oliver Ames, Huntington put forth an argument for a loan. Up to $200,000 was needed, Huntington said, so he could begin ordering railroad equipment, start tracklaying, and qualify the Central Pacific for government money. For collateral he offered his company bonds. After considering the proposition overnight, Ames agreed to the loan with one flint-edged Yankee stipulation: Huntington and his board would have to personally guarantee any semiannual interest payments on their company bonds. Huntington gulped and agreed.[5]

Along with the money, Oliver Ames handed Huntington letters of introduction to iron manufacturers with which Ames did business. Over the next several weeks, Huntington visited all of them, completing his shopping list for a fortune's worth of rails and rolling stock, as earlier, on the strength of modest down payments, then-worthless bonds, and personal assurances. In New York he made all the obvious rounds hawking Central Pacific securities, but had more success borrowing seed money from bankers like William E. Dodge—again on his personal guarantee, which he said he would enforce for at least ten years. "I will guarantee it," he would say, "because if the Central Pacific ever stops short of completion C. P. Huntington will be so badly broken you will never spend any time picking him up."[6]

Back in Sacramento, as news of Huntington's contracts with suppliers arrived, and as work gangs slowly progressed up the right-of-way, Leland Stanford was busy trying to find money to pay for it all. Taking advantage of his dual status as governor and Central Pacific president, Stanford applied what today seems like an unseemly amount of pressure to persuade the California legislature to permit the county governments of Sacramento and Placer to invest in the Central Pacific. San Joaquin and San Jose Counties were authorized to invest in the Western Pacific Railroad, with San Francisco free to back both companies. The exchange of railroad stock for twenty-year, 8 percent county bonds—which would pour up to $16 million into the Central Pacific treasury—would have to be approved in public referenda.

Immediately the opposition rushed in, attacking on many fronts. In the hillside communities bypassed by the proposed rail route there was the strafe of old complaints that the Dutch Flat route could never work, that success would be found only in overlooked, already-proven passes.[7] In the Sacramento Valley there was the dependable villainy of the Robinson brothers, who began whisper campaigns and threatened suits against both the route and the directors of the Central Pacific in hopes of ruining the struggling company's prospects and stepping in themselves. In San Francisco there were legions put up by all the transportation and freight interests for whom the Pacific Railroad was anathema; their voice was heard most dependably in the daily *Alta California*, which was unstinting in its criticism of the Central Pacific and anyone connected with it, and further in a relentless hail of broadsides and pamphlets decrying the "Dutch Flat Swindle," which seemed to keep any number of printing presses busy throughout the month of April. Stanford and his cronies, the rumors and gossip reported, had no intention of building the national railroad over the Sierra. Instead they would siphon off the public money, lay their rails only to Dutch Flat to meet their wagon road from the Nevada Territory mining district, and laugh as they counted their profits.

The Central Pacific and Lauren Upson's *Sacramento Union* determinedly returned the enemy fire all the way along the craterous route toward the polls in May. They gained morale from another Stanford "special" in the legislature in April: $15 million in state bonds would be donated to the Central Pacific. With each completed twenty-mile section they would receive $10,000 per mile. For its part the railroad agreed to transport the state militia for free during times of crisis, and it signed over title to a granite quarry to the state, promising to transport the quarried stone for nothing.[8]

And such success raised the controversy over county support to a fever pitch. On referendum day both sides seemed to deploy everything they had—including, not surprisingly, cash for votes. Of course each blamed the other for the identical sin. The *Sacramento Union* claimed that the opposition "had active strikers in every Ward, and paid them to the tune of six thousand dollars, with a promise that they should have a much larger sum provided the Pacific Railroad proposition was defeated. Another quiet operation was getting out thousands and thousands of circulars, full of misstatements and falsehoods, and distributing them, in the dark, broadcast over the city, even to thrusting them obtrusively into nearly every man's domicile."[9]

Meanwhile, the Central Pacific seemed to have been busy. "We have affidavits," reported rail historian Stuart Daggett in affidavitese nearly sixty years later, "that Philip Stanford [Leland's brother] went to the polls in San Francisco in a buggy, carrying a bag of money; that the said Stanford put his hand frequently in the bag of money and took money, some $20 pieces and some $5 pieces, to a considerable amount therefrom, and scattered the said money among the voters at the said polls, at the same time calling on them to vote in favor of the said subscription." Another witness alleged that he tried to count how much gold was being given away, but "he could not tell . . . as the crowd around the buggy of the said Stanford was so great." A third said that Stanford had even thrown handfuls of money among the voters "and thereupon they scrambled among themselves for the same."[10]

When the smoke had cleared and returns were counted, the Central Pacific had won dazzlingly: San Francisco, Sacramento, and Placer Counties all voted to exchange their bonds for the railroad's stock, the latter in less of an overwhelming fashion than in San Francisco and Sacramento, where the margin was nearly two to one. Sacramento voters also deeded thirty acres of city property, including more than a third of a mile of waterfront, to the railroad.

The victory, though, was blemished by writs and continued bad publicity; within hours of the referendum in Sacramento, the Robinson brothers obtained an injunction against the bond transfer, which held for two weeks.[11] In San Francisco another suit by other business interests blocked that county's aid for more than a year, and no sooner was it settled than city officials refused to release any bonds; the railroad did not receive a penny until it had expensively fought through a thicket of legalities which took it all the way to the California Supreme Court, in 1865.[12] With what little credibility the Central Pacific gained at the polls eroding, the situation was hardly helped on May 21 when Governor Stanford named his fellow board member and partner's brother, Edwin Bryant Crocker, to be chief justice of the California Supreme Court; true, it was a temporary seat (filling an unexpired term for seven months), and during that period no cases affecting the line would be submitted to that bench—but it looked bad, nonetheless. And such political misjudgments would soon, in June 1863, lose Stanford his own supporters and turn him out of the governor's mansion when his first term expired.[13]

All this must have seemed tawdry to Theodore Judah, who appropriately followed most of it from above, in the mountains, while taking more measurements for the line. Little of what he heard pleased him, especially one piece of business involving the "legal" boundary of the very mountain range he knew so well. Judah of course had helped guide the 1862 Pacific Railroad Act through Congress, and he had taken pains to have the scale of federal subsidies set with an emphasis on the steep climb over the Sierra Nevada: once construction began on the incline, the loan rate would rise from $16,000 to $48,000 per mile, for a total distance of 150 miles. And the decision as to where that line lay still fell to one overburdened chief executive some three thousand miles away in the White House, who in his whole life would never set eyes on the Sierra Nevada or the Sacramento Valley.

The trouble over this began with Stanford and Crocker, who had unrolled the profiles of the road and measured backward over the mountains from the green and level Truckee Meadows to the 150th mile point. This proved to be where Arcade Creek meandered along the Sacramento Valley floor into the American River: terrain as level as any tabletop, and only seven miles from where they sat in Sacramento. "I thought a good deal about it," recalled Crocker many years later. "We were very hard up, and we wanted to get the base down as near the river as we could." (Speaking to congressmen at the time, Crocker was quick to add "dealing justly with the government.") How to qualify for the higher rate? Stanford inventively found the solution when he recalled a whimsical little item of intellectual conversation which for some time had been making the rounds of social gatherings: where, a questioner would ask, was the base of the Rocky Mountains? The surprising answer, according to James Dwight Dana's famous *Manual of Geology*, was not somewhere in Colorado Territory, but on the western bank of the Mississippi River.

Certainly this logic—as assuredly academic as it might have been—could be applied on the Pacific Slope! The governor forthwith sent his state geologist, Josiah Whitney, off with Crocker to look over the line by buggy. At Arcade Creek, Crocker unrolled his road profiles, showed them to Professor Whitney, and then read him the provision in the railroad act: 150 mountain miles from Truckee Meadows was Arcade Creek, and although the nearest mountain they could see was some 15 miles away, the profile did show a barely perceptible rise beginning where their feet were planted on the flat, flat ground. "I did not ask him to do anything," recalled Crocker, "except that I wished him to decide where true justice would place the western base of the Sierra Nevadas. 'Well,' he says, 'the true base is the river, but,' said he, 'for the purpose of this bill, Arcade Creek is as fair a place as any.'" As Stanford later requested, Professor Whitney put his opinion in writing, on March 23. Two other experts—Edward Fitzgerald Beale, U.S. Surveyor General for California, and J. F. Houghton, the state's surveyor general—concurred with their own letters of support.[14]

Judah would have nothing of it. As an engineer he knew where the serious road cuts and fills would begin—some six miles east of Arcade Creek, near Roseville—and where the road would begin its true ascent—when it crossed an imaginary line drawn between Folsom and Lincoln; not until it drew near Newcastle, twenty-one miles from Arcade Creek, would the incline be appreciable. He had already filed his report and his profiles with the federal government which attested to this, and which plainly showed Arcade Creek to be on valley land. As a director of the line Judah was asked by Stanford or Crocker to sign an affidavit for the more liberal viewpoint. "We cannot sign it," he told his friend Daniel Strong during a number of consultations that spring, "because the foot-hills do not begin here according to our surveys." The difference, of course, was not merely academic; it meant nearly half a million dollars in government bonds. Neither Judah nor Strong, nor fellow director James Bailey, wanted to be parties to such "a fraud."[15]

But the others on the board did not trouble themselves with such scruples. They signed the affidavit and dispatched it to the Interior Department. For the rest of the year the final decision would be held in abeyance—for most of that time still a matter of contention within the Central Pacific board. Finally, Stanford would ask Aaron Sargent to go directly to Lincoln with it; upon that Railroad Congressman's intercession, Lincoln would sign the order deeming Arcade Creek the western base of the Sierra Nevada. Some time later, Sargent encountered his friend Noah Brooks, Washington correspondent for the *Sacramento Union,* and gleefully told him of the process. "Here," Sargent laughed, "you see my pertinacity and Abraham's faith moved mountains."[16]

It was not a joke to Theodore Judah—he believed it illuminated the vast moral differences between a visionary such as he regarded himself and the gang of corner-cutting merchants he saw the Associates to be. "I cannot make these men, some of them, appreciate the 'Elephant' they have on their shoulders," he exclaimed to his wife. "They won't do what I want, and must, do. We shall just as sure have trouble in Congress as the sun rises in the East if they go on in this way . . . something must be done. I will not be satisfied before Congress and the world!"[17]

Increasingly, he had a gnawing suspicion that resources raised for the railroad were being diverted to pay for construction on the Dutch Flat wagon road, an enterprise whose financial matters and interests were kept away from him. How else in that stretched-thin spring could the Associates be paying two hundred laborers up in the mountains?[18] In the months since the ground-breaking ceremony, grading work of the child's-play stretch from Sacramento to Roseville had sputtered along fitfully, free of any organizational plan until all seven subcontractors gave up, turning the jobs back to Charles Crocker & Company. Finally, approaching the end of April and after four months of work, some seven miles would be ready for ties and rails, with four additional miles having culverts installed.[19]

Having Crocker as general superintendent was still a sore point with Judah. But what other plans were being concocted? Lines of decision making were murky at best, and he felt a growing estrangement. Soon the stakes grew rapidly higher when Hopkins called him in. To Judah's astonishment Uncle Mark coolly asked for a payment of $1,500, representing 10 percent of Judah's original 150 shares. The nonplussed engineer replied that upon organization of the Central Pacific he had been recompensed for his initial Sierra survey work and promotion efforts by a paper entry of the first 10 percent payment for those shares; Hopkins should remember—he had been there. Hopkins shrugged. He knew of no such agreement, he said. Rising to what he now recognized as a challenge, Judah somehow put together the money from his limited resources.[20] "Some of those days," his wife recalled many years later, "were terrible to us!"[21] Determined to get his position clarified, at the board of directors meeting on April 1 he presented an accounting of his time and insisted upon a settlement of the question, How was he, the linchpin of the Pacific railroad, going to be treated? After some hemming and hawing the meeting adjourned without a decision or even a clear acknowledgment of his right to stock previously promised him.[22]

By the first of May, seeing that for some reason there were no plans to call the regular monthly board meeting, Judah stormed into the company office. "I had a blow-out," he reported to Doc Strong, "and freed my mind, so much so that I looked for instant decapitation. I called things by their right name and invited war; but counsels of peace prevailed and my head is still on. My hands are tied, however." He tried to remain philosophical about things, he told Strong with undisguised disappointment, "and devote myself with additional energy to my legitimate portion of the enterprise." He held to the thought of getting away from the noisome, scheme-filled climate in Sacramento by putting parties on location of the second fifty miles by summer.[23] The surveyors would be up in the mountains at the anticipated time. But the peace Judah hoped for could withstand neither the workings of the Associates nor his own thinning temper—especially after Collis Huntington returned to California in late June.

War in the boardroom had its mirror in the East, where more grievous stakes prevailed. In early May—as Judah unleashed his temper at Hopkins and Crocker—near the thickety settlement at Chancellorsville, Virginia, Union forces under Generals Joseph Hooker and Daniel Sickles reeled from a surprise assault from the Confederate armies of Robert E. Lee and Stonewall Jackson, costing both sides some 3,240 lives and 18,675 wounded; one of those killed was Jackson, by mistake mortally wounded by his own men; it precipitated mourning all over the South. All during the month of June—as Judah saw it—he stood on zealous, lonely guard over construction bids received for thirty additional miles of grading and bridging from Roseville to Clipper Gap, vigilant lest they all be discarded in favor of the "inside man."[24] Meanwhile, on the eastern outskirts of besieged Vicksburg, Mississippi, "Unconditional Surrender" Grant was steadily lobbing artillery shells across the city's defenses toward a capitulation on July 4; the Union and Confederate casualties of the entire eight-month campaign would reach nineteen thousand, with thirty thousand other rebels captured. Then, during the last three days of June—during which C. P. Huntington stepped off the gangplank in Sacramento as if he were a general assuming command to restore discipline and press for victory—Lee's gray column paused in its drive onto Pennsylvania soil; meanwhile, a onetime surveyor for the Long Island Railroad and the Alabama, Florida, & Georgia Railroad, "a damned old goggle-eyed snapping turtle"[25] by the name of George Gordon Meade, replaced Joseph Hooker as Army of the Potomac commander, at Frederick, Maryland. The peaceful crossroads town of Gettysburg waited, two days' march away across the Pennsylvania border.

News of that ghastly battle between July 1 and 3, in which some fifty-one thousand were casualties, arrived in fits and spurts over the telegraph in Sacramento, drawing anxious assemblies to the bulletin boards at the *Union* offices. While Californians strove to piece together a cohesive account of the bloodiest battle on the continent and discern what kind of victory it had been for Meade's Federals, there would be no indication either on the boards or in the newspaper's columns that an intense, though bloodless, struggle was escalating right at home. In the Central Pacific office the overproud, defiant Judah had put up with Crocker's rough and hale ways, had endured mandarinlike Hopkins's byzantine accountant's concerns, had tried to fathom Stanford's whimsical leanings from one side to another on many important issues. Huntington, though, was an irritant of a higher magnitude. For six months in the East the merchant had wheeled and dealed for loans and equipment, returning with order receipts for six railroad engines, forty boxcars or flatcars, six passenger cars, two each of baggage cars and handcars, and, with all its hardware, some five thousand tons of iron rails—$721,000 was the total cost, paid for with discounted company bonds with the directors' personal guarantee of the interest.[26] Quite rightly, he was feeling cocky at having negotiated the largest deal of his life. He had taken in the news of strife within the board and coolly prepared to ride himself over the bodies of those standing in his way.

Judah and Huntington had spent little time together since early the previous fall, when they had been anxious to still criticism over the railroad's choice of the Dutch Flat–Donner Pass route. In September 1862 a party consisting of the two, along with Congressman Aaron Sargent and fellow board member Charles Marsh, had traveled by wagon over the Yuba and Beckwourth Passes and rejected them as too expensive or long. Then, looking into the canyon of the Middle Fork of the Feather River, Judah and Huntington had left the others, hired a Chinese porter, and descended along the river in a chasm from two to three thousand feet deep.[27] It was a rough, dangerous, cliff-hanging experience for seven days—perhaps seventy miles in all—and one can only guess at the private assessments of character and strength made by the two as they stared through the flames of a campfire every night. Judah recorded no personal specifics about the merchant. Huntington's only telling utterance would be aired some years later: of the engineer's flaws, he thought, the most irritating was his "cheap dignity"—and a "low cunning" which, issuing from Huntington, was interesting indeed.[28]

"Huntington has returned," Judah sent word to Doc Strong in Dutch Flat, "and seems to possess more than usual influence. Stanford, who I told you was all right, is as much under their influence as ever." The fight had nearly as many fronts, indeed, as Gettysburg. Of course one revolved around construction contracts and Judah's suspicion that somehow the Associates secretly shared Crocker's interests—as negligible, to be sure, as they currently were. After sealed bids had been considered for the thirty miles between Roseville and Clipper Gap, Crocker had received only two miles due to the engineer's pressure.[29] Now Judah felt he had drawn a line. "I have had a big row and fight on the contract question," he informed Strong, "and although I had to fight alone, carried my point and prevented a certain gentleman from becoming a further contractor on the Central Pacific Railroad at present." Of course, the "certain gentleman" was Charley Crocker. On another front was the wagon road, that presumed drain on the railroad's resources, the "Dutch Flat Swindle." "They have been consulting and looking over the way every day," Judah wrote, "and do not hesitate to talk boldly, openly before me, but not to me, about it. They talk as though there was nobody in the world but themselves who could build a wagon road." Along with suspicion there was also something of injured pride in Judah's tone—perhaps because he was now being treated like a servant, or because he had been elbowed away from any involvement in the road—but it led him to speculate to Strong whether they should organize their own freight road company. Nothing, though, would come of it.[30]

Then there was the bedrock issue of how the true originators of this enterprise were to be treated. More than three months had elapsed since Hopkins had abruptly thrown open the question of Judah's being recompensed for his early work. Ever since, the engineer had been unable to get his position clarified. Finally, though, with Huntington back, the board seemed free to address it: it voted him some $91,000 in Central Pacific stock "for his services as agent in the Atlantic States, and prior to the organization of the company," stock which at present was worth less than half of face value.[31] Strong, however, who had discovered the route and then found Judah, and who had tagged along with him during the entire reconnaissance for day wages, did not fare so well with company stock. "I offered the resolution at $5,000, and tried hard to get it," the engineer reported, "but from the indications of old sores in a debate, was afraid of upsetting the whole thing." The druggist was to get $2,500 "for previous services," to be credited as five assessments on his interest. Doc Strong began to wonder if all this was worth the trouble. But as Judah requested, he prepared his stockholder's proxy and began to collect others' up in the mountain towns for the company's annual meeting, on July 14—when they hoped to elect supporters and tip the balance of the board of directors their way.[32]

On the appointed day the Huntington faction handily prevailed—when the board admitted Philip Stanford (the alleged Central Pacific bagman in San Francisco during the referendum) and Dr. John Morse (a valued friend of Huntington's since they had sailed to California together in 1849). Huntington moved quickly to consolidate his winnings and turn out all the dissenters on the board. Over Judah's heated objection, he halted laborers who were expensively filling in a swampy area along the levee (where the proposed line was vulnerable to flooding) with granite riprap. "It will cost $200,000 at least to put the road here," the merchant rasped. "It must go up I Street."[33] He vetoed a plan for a grand Sacramento terminal which Judah had designed as a visual testament to the Central Pacific's solidity, impractical (and inaccurate!) as it might have been. Instead, Huntington quickly sketched out a simple batten-and-board shed—one door, a few windows, sloping roof—which could and did double for tool storage. With the company desperate for cash, and with only sixteen months left until the federal deadline for building fifty miles of track (November 1864, according to the 1862 railroad act), something daring had to be done. Huntington's answer was to step up the assessments of stockholders, and in the meantime have all the board members pay in what they still owed for stock—even if it meant mortgaging their homes or businesses—to keep the crews working.[34] "Huntington and Hopkins can, out of their own means, pay five hundred men during a year," he declared. "How many can each of you keep on the line?"[35]

It was a bold challenge to Judah and Strong, with their meager resources, and to the jeweler James Bailey and the grocer Lucius Booth, who were nervous about going so far out on a financial limb. And Judah's counterproposal for fund-raising, to mortgage the rolling stock on order but not paid for, was hardly advisable. Bailey, who had just been elected secretary of the Central Pacific, refused outright to comply with Huntington's demand. And he refused to take up Huntington's retort that Bailey should either buy him out or sell his interests to Huntington. "Then there's only one alternative," Huntington replied. "The work must stop at once." By nine o'clock that night he had made a circuit of all their construction camps, ordering workers off. The next day he coolly gave Bailey and Judah two weeks to buy him and his associates out.

With the crews idle and beginning to drift back into town or up toward the mining districts, the engineer and the jeweler went off in search of capitalists. After several days of silence Huntington was handed a cable from Bailey, in San Francisco. He reported he had found a buyer—the San Francisco banker Charles McLaughlin, who was behind the Western Pacific Railroad. But Bailey had neglected to tell McLaughlin whose interest he was buying, and when the banker found out he hastily backed away. "If Old Huntington is going to sell out," he explained, "I am not going in. Just what sends him out will keep me out."[36]

No one else could be found in San Francisco. By early August, as Huntington's deadline came and went, the discouraged Judah and Bailey were back in Sacramento. Bailey resigned his board appointment and sold off all his Central Pacific stock. But Judah had no intention of withdrawing. He must have begun writing letters to potential sponsors in the East and received indications of interest—for in September he made plans for going to New York. He began to gather funds for travel and several months' stay. On September 28—a week after news of General William S. Rosecrans's disastrous defeat at Chickamauga, Georgia, hit the streets of Sacramento—Judah sold his interest in the Nevada Rail Road Company (the proposed extension of the Central Pacific line across Nevada Territory) for $10,000 to Charles Crocker. Probably the engineer thought of it as a sort of loan, for there is no doubt that he intended to get his stock back later.[37] On September 30, his last official act was to write his assistant, Samuel Skerry Montague, appointing him "Resident Engineer" to "assume charge of the Engineer Department and supervise the work during my absence."[38] Three days later on October 3, Ted and Anna Judah boarded the Panama-bound steamer *St. Louis.*

As the ship chugged down the coast, Judah, his mind full of plans and purpose, wrote a letter to Doc Strong. "I was so much hurried and distracted with events and the hurry of getting off," he said, "that I did not write you before leaving. There is not much that I can say now that is very interesting. I have a feeling of relief in being away from the scenes of contention and strife which it has been my lot to experience for the past year, and to know that the responsibilities of events, so far as regards [the] Pacific Railroad, do not rest on my shoulders." What now occupied him was more far-reaching; though much of this is still shrouded in mystery, Judah had apparently secured a verbal agreement with the Associates that, as chief engineer and board member, he had the right to buy out each of them for $100,000. If he failed to find backers, the Associates could purchase his interest for the same amount. Clearly, Huntington believed that Judah would fail. But the engineer predicted to Strong that the Associates would shortly be out of the picture. "If the parties who now manage to hold to the same opinion three months hence that they do now, there will be a radical change in the management of the Pacific Railroad, and it will pass into the hands of men of experience and capital. If they do not, they may hold the reins for awhile, but they will rue the day they ever embarked on the Pacific Railroad."[39]

No written record has survived to give any indication as to who his mysterious backers were. "Everything was arranged for a meeting in New York on his arrival," asserted Anna Judah some years later, "gentlemen from New York and Boston who were ready to take their places."[40] She believed that one of them was none other than sixty-nine-year-old Cornelius Vanderbilt, whose steamship interests had made him a multimillionaire, and who had begun to be interested in branching out into the railroad business.[41] Other speculation has mentioned the Boston-based Oakes and Oliver Ames, who had lent money to the Central Pacific.[42] But as the steamer *St. Louis* coursed southward toward Panama, Judah finished his letter to Doc Strong naming no names—though exhibiting a self-assured mood about prevailing over the Associates by the end of 1863. "If they treat me well," he said, "they may expect similar treatment at my hands. If not, I am able to play my hand. If I succeed in inducing the parties I expect to see to return with me to California, I shall likely return the latter part of December; if not, I shall stay later."[43]

Work on the railroad had resumed after the great boardroom struggle, and it would continue—partially thanks to the sudden, court-ordered release of support bonds from Sacramento and Placer Counties, worth a sum of $550,000 when exchanged for a hundredth of that face value in Central Pacific stock shares.[44] Throughout September, ships bearing iron rails from the East in hundred-ton lots, and ties of redwood harvested in the Coastal Mountains, had eased up alongside the Sacramento levee almost daily.[45] Up the American River, the final inspection of the iron and redwood bridge, all 192 feet of it, pronounced it and the long line of trestle on either approach to be ready for rails.[46] Then, on October 5, the Central Pacific's first locomotive arrived, for the moment under hawsers on the deck and in the hold of the schooner *Artful Dodger* and completely disassembled. It had already been christened the *Governor Stanford* by Collis Huntington, homage to a now-lame-duck political career.[47]

At Aspinwall, on the Caribbean side of the Panama isthmus, travelers came under a furious deluge of rain. Theodore Judah purchased a large umbrella and escorted his wife from the hotel to their ship's stateroom, remarking that it was a shame not to give it greater use. He then went to the aid of the many unescorted women and children who had begun arriving in coaches at dockside. "He could not see them exposed to the rain," Anna Judah remembered, "and not try to do his part and more for women and children who had no one to help them. I feared for him and remonstrated, for I knew he was doing too much, but he replied, 'Why, I must, even as I would have someone do for you—it's only humanity.'"

That night he had a terrible headache.[48]

The situation with Judah's friends and allies in the Central Pacific continued to erode. James Bailey was back in his jewelry shop, bitterly wishing never to have the word "railroad" uttered in his presence (and perhaps even then planning to abandon California entirely for Rochester, New York, where he would end his days).[49] Doc Strong tended his business up in Dutch Flat, not bothering to go to any more board meetings in Sacramento since the decisions were obviously being made elsewhere. On September 30 he wrote to Leland Stanford asking that he be named agent to look after the company's timber interests in the mountains. Stanford sent a reply on the third, smoothly promising to urge his appointment when such matters were taken up. "Indeed, doctor," Stanford assured him, "I shall always be glad to serve you whenever it may be in my power." Strong knew not to hold his breath.[50] Also in October, the strength of Collis Huntington in the company began to solidify like mortar when the absence on the Central Pacific's board of directors caused by the resignation of James Bailey was filled by E. H. Miller Jr. The former Forty-niner shipmate and onetime business partner with Mark Hopkins, Miller also replaced Bailey as secretary of the Central Pacific and brought to that capacity his friend's obsessive attention to detail. He, too, had been active in the formation of the California Republican Party. Soon Ed Miller began to liquidate his business holdings to devote all his time to the railroad enterprise.

The headaches grew intolerable to Theodore Judah as their ship steamed across the Caribbean. He was seized by a racking fever and took to his stateroom berth, his every need attended by Anna. By now she may have learned from the ship's bibulous surgeon that her husband's symptoms spoke of an illness much more alarming than one brought about by exertion or dampness; the curse of isthmian travel, "Panama fever," or yellow fever, had found yet another host. Once, lying so quietly in his berth that Anna thought him asleep, he roused up, feverish but also still warmed by the prospect of joining with the Vanderbilt people to turn out Huntington and the other rascals. "Anna, what cannot I do in New York now," he exclaimed. "I have always had to set my brains and will too much against other men's money. Now, with money, equal, what can I not do!" They were unaware that, back in California, rumors had begun to be spread (and published) that Judah—the most famous railroad engineer on the Pacific Slope—had sold out of the Central Pacific and was using the profit for a grand tour of Europe with his wife.[51]

Three weeks after the exciting arrival of the *Governor Stanford,* on October 26, the first rails of the Pacific railroad were laid across their ties and spiked—with no ceremony whatsoever, which puzzled Lauren Upson so that he remarked about it in the next day's issue of the *Sacramento Union.* Years later, Collis Huntington, who had avoided the groundbreaking ceremony also, would offer an explanation. "Now, if you want to jubilate over driving the first spike here," he claimed to have told his associates, "go ahead and do it. I don't. Those mountains over there look too ugly, and I see too much work ahead of us. We may fail, and if we do, I want to have as few people know it as we can. And if we get up a jubilation, everybody will remember it. Anybody can drive the first spike, but there are many months of labor and unrest between the first and last spike."[52]

Eight days after Judah fell ill, the ship came to a halt alongside a New York City wharf. One of the first off was a stateroom attendant who leaped ashore with a note summoning a prominent physician, Dr. F. N. Otis, with whom the Judahs had sailed several times when he was ship's doctor on the *Illinois,* and who loved Judah like a brother. They waited for the rest of the passengers to disembark, finally learning that Dr. Otis was away from his office. The ship's current staff physician had been drinking, and he seemed anxious to get on to more pressing business, especially after the weak and feverish Ted Judah was borne down the gangplank and placed in a carriage. "I kept him up," Anna recalled, "by dipping my finger in the brandy bottle and having him take it that way." Perhaps she should have offered some to their escort; finally, on the way to the Metropolitan Hotel, the Judahs dismissed their impatient physician and he thankfully stepped off at Wall Street. Fortunately for their spirits, Dr. Otis awaited them at the hotel.

There was little they could do but watch the exhausted thirty-seven-year-old Judah die, a process which took a week. Anna Judah never left his side. He died in her arms on November 2. Anna accompanied his body on its final train ride to her childhood home in Greenfield, Massachusetts, where she buried him in the quiet country cemetery of the St. James Episcopal Church under a granite stone suitably engraved "He rests from his labors." While Anna was still benumbed by grief, two unidentified businessmen with whom her husband had corresponded about taking over the Central Pacific, who had been barred from Judah's sickroom in New York, appeared at her parents' door. They sought to somehow continue the plans they had made with the engineer, but all they could do was gaze at the new sod over his grave and go home empty-handed.[53]

News of the engineer's death was published in California on November 5,

1863, prompting much mourning in places outside the boardroom at 54 K Street in Sacramento, and causing Daniel Strong to immediately resign his director's seat. (For posterity Stanford would write to urge him to stay—which Strong did not bother to answer—and at the next annual meeting Strong would be voted off the board anyway.)[54] For their part, the Associates called a directors' meeting to endorse ("with unfeigned sorrow," they felt necessary to reassure) a pious resolution of sympathy to the widow and Judah's mother and siblings.[55] Samuel Skerry Montague, the capable and honest engineer sponsored by Judah himself, inherited his job. Until his last day in New York, Theodore Judah remained on the payroll of the Central Pacific at $10,000 a year as chief engineer—despite rumors promulgated by antagonists such as the Robinson brothers, the editors of the *San Francisco Call*, and (probably) Collis Huntington himself, that he had walked away from the project.[56] Most virulently, the Robinsons used Judah's death (and the point of law that one cannot libel the dead) to contend that he had repented of his choice for the Dutch Flat–Donner Pass route, quit, and was bribed with $100,000 to keep his mouth shut.[57] It was a nasty work of revenge (and a long-lived one) to pay back Judah and the Central Pacific for declining to buy out the Robinsons' undercapitalized, deteriorated, wrongly gauged, foreign-manufactured trackage of the Sacramento Valley Railroad when the Robinsons attempted to bail out. Such calumnies sometimes outlive their inventors.

Affairs of the Central Pacific moved on, as they had to, with the federal deadline for forty completed miles of track now only twelve months away. Montague and a team of assistants were immediately dispatched to the Nevada territory, to survey another forty miles of the route past the California border to the Big Bend of the Truckee River, a task completed in December. Work began in Sacramento shops to fabricate passenger coaches. And within two days of Judah's death notices, technicians first fired up the steam in the *Governor Stanford*, readying it for an unveiling and trial run on November 9. Crowds assembling on that day beheld a glorious sight: the forty-six-ton, fifty-foot-long locomotive and wood tender gleaming as brightly and colorfully as circus wagons, painted predominantly in royal maroon and apple green with eye-catching accents of blue, red, and orange, with shiny brass fittings and gold stars painted on the driving-wheel hubs, with tall gold initials, C.P.R.R., emblazoned on the tender. Dignitaries who had been invited to sip champagne and take the inaugural ride along the Central Pacific's first two completed miles of track waited in vain for a place, shouldered away by a raffish crowd of boys and young men who scrambled aboard the wood tender only to have it wheeze to a halt: the engine valves could not build up enough pressure and had to be adjusted. That night the youthful hangers-on got another chance at a ride when the problem was fixed, as would the dignitaries the next afternoon when the flag- and bunting-festooned *Governor Stanford* chugged all the way to the railhead—at Twenty-first Street, where the champagne awaited. Among the many toasts was one to the memory of Theodore Dehone Judah.[58]

Collis Huntington was not there among his fellow directors and friends to raise his glass, having been called to jury duty.[59] But he would have been in no mood for celebrating, still eyeing the calendar on the wall and those "ugly mountains" out the window on the horizon, and ever mindful of their scarce assets and plentiful debts. How to keep grading past Roseville toward Newcastle in the foothills? The troubling figures would have tap-danced through his brain. Their balance sheet would show that they had paid for $118,000 in rolling stock and $947,000 in iron rails and labor on what really amounted to personal IOUs; Crocker, presently owed $48,000, accepted payment in company securities but grumbled that henceforth he would accept them only at a 50 percent discount, probably adding ominously that he would have to be raising wages to compete with the Nevada silver mines. And the black side of the Central Pacific's balance sheet could hardly compete with the red. According to Hopkins they had only $7,000 in the treasury; the eighth stock assessment of 10 percent would trickle in over the next month, hardly enough to do much good; there were, finally, those freshly printed county bonds, which could only be disposed of in the East at punishing discounts. Thinking gloomily, no doubt, of going back to endure winter in New York and Washington (more than one, most likely) while slaving to keep the enterprise alive, Huntington booked passage for himself and his wife and their adopted daughter. He would pass as much paper as he could manage. And he would nose around Washington to see how more government aid could be pried from the Lincoln administration; the congressional amendment to the 1862 Pacific Railroad Act, dead since early spring, would have to be somehow revived.

And certainly it was time to closet himself with the man he would soon be calling Mr. Union Pacific.

The Huntingtons were possibly in steamy transit across Panama—avoiding the infection which had carried away Theodore Judah only weeks before—on December 2, 1863, when, at the intersection of Seventh and Davenport streets in the raw settlement of Omaha in Nebraska Territory, with the muddy Missouri River flowing a pebble's throw away, the first frosty clods of soil were disturbed in a "gala groundbreaking ceremony and banquet," as promoters touted it, for the Union Pacific Railroad. A thousand Omahans attended to cheer as Nebraska governor Alvin Saunders gouged the Missouri bottomland with his shovel, accompanied by cannon fusillades and brass-band fanfares. After the usual excess of oratory, after congratulatory telegrams from President Lincoln, Governors Stanford and Yates (California and Illinois would avidly watch the Union Pacific's progress), from Brigham Young of Salt Lake

City and from the mayors of Denver and Chicago, dignitaries and celebrants alike followed the mayor of Omaha over to the Herndon House for their gala banquet and ball. "Mr. Union Pacific"—a lean and reckless engulf-and-devour shark of a speculator by the name of Thomas C. Durant—was half a continent away, in New York City. But on his behalf, his agents in Omaha enjoyed themselves. Exquisitely, extravagantly, the blowout would all but empty the corporate treasury.

Part IV

1864
Struggle for Momentum

13

"First Dictator of the Railroad World"

His air of mystery was purposeful, as carefully nurtured as a hothouse orchid. Every move was calculated to keep men of wealth and influence as much in the dark as the lowest unskilled laborer who toiled out on his right-of-way. Thomas Clark Durant was a smoke-filled-back-room man—brooding, sharp-eyed, retiring—an absolute genius of manipulation, whether it be of high finances or of his fellow man. Now a general waiting for his command, the tall and lanky, stoop-shouldered and scraggly-bearded businessman would soon earn a title as "first dictator of the railroad world."[1] He could be impetuous. Durant had been among the curious crowd of Washingtonian civilians who had gone out to view the battle at Bull Run. "When the stampede began," a contemporary recalled, "he was the angriest man I ever saw. He picked up a stake and sprang right in front of the running soldiers, and in spite of their muskets and fixed bayonets, hit right and left, shouting: 'Go back, you damned cowards—go back!' And a good many of them actually did turn back."[2] In business and personal matters, his unpredictability could drive associates mad, just as his energy could exhaust them. Possessing "a capacity in mental strength and energy which seemed to grow in proportion to the difficulties that he was compelled to encounter," as an associate once said, Durant had an intriguing background.[3] He had once actually been a doctor—one, it was said, of promise. It is hard to imagine, however, a greater gulf between the ethics implicit in the Hippocratic oath and the iron-bound allegiance of self-interest embodied in his every move.

Such an evolution must be studied. Durant was born in 1820 at Lee, in the center of the Berkshire Hills in Western Massachusetts. His father, also Thomas, a successful merchant and manufacturer, might have planted his factory on the bank of the Housatonic, which rose not far away. A grandfather had been an officer in the Revolution and a member of the Boston Committee of Safety. While still very young, Thomas Durant aimed himself toward medicine and surgery. He enrolled at the Albany Medical College, emerging with his license, heaped with honors, at the age of twenty. His specialty was ophthalmology.

But within three years Durant had his eye on a more remunerative frontier than that of practice and teaching, exchanging a procession of corneal ulcers and cataracts for the export trade in flour and grain. An uncle made him a partner in Durant, Lathrop & Company, an Albany shipping firm with branch offices in Boston, Chicago, and New York City, and Durant was placed in charge of the latter office. Primarily he oversaw a brisk trade in shipping cereals to the European ports aboard a fleet of ships owned or controlled by the firm; it was precisely the time in which another merchant, Mr. Asa Whitney, newly returned from China with a fortune as well as a burning mission, might have been found standing in the same mercantile exchange as young Dr. Thomas Durant. The juxtaposition is nearly bizarre.

In the realm of business Durant had gifts. He steadily expanded the holdings as well as the scope of Durant, Lathrop, especially into transporting supplies to the "West"—Illinois, Michigan, Minnesota—by way of the Great Lakes; increasingly, it moved construction material and other freight for the emergent railroad companies there, accepting securities in payment. While negotiating the bonds on Wall Street, Durant realized he had other gifts. His personal wealth grew accordingly. He married in 1847, siring a son and a daughter.[4]

Durant moved into the railroad world in 1851, becoming an associate of the skilled civil engineer Henry Farnam, a man of shrewd business sense and integrity. Their brief association would enrich them; it would also illustrate how easily young Durant could abuse trust. Farnam had been born forty-eight years earlier in Cayuga County, New York. He had started as a "ditch man," surveying for the Erie Canal and acting as chief engineer for the Farmington Canal, linking New Haven, Connecticut and Northampton, Massachusetts. The latter enterprise ultimately failed, though Farnam recovered and even prospered by running a railroad line along the canal right-of-way. He was called to Chicago in 1850 to begin work on one of the first rail lines out of that growing city, the Michigan Southern—projected to shoot east across the Michigan peninsula toward Lake Erie, ending in the Michigan town of Hillsdale. Durant aided him in the project. It was completed in just over a year, closing the link between Chicago and the East.

In 1853, Durant, then only thirty-three, joined Henry Farnam as a full partner in the contracting firm of Farnam and Durant. At first things went well. Their first project was the Chicago and Rock Island Railroad, which would stretch westward from Lake Michigan, reaching the Mississippi River in early 1854. The Chicago and Rock Island was nominally headed by John Adams Dix, a New York lawyer with powerful connections in and out of the Democratic Party; a former U.S. senator, Dix would within time occupy several influential positions—as New York City postmaster, as treasury secretary under President Buchanan, as major general of the Union Army, as minister to France. Moreover, for a number of years he would prove to be a useful figurehead for Durant, having no objection to a good salary and generous stock benefits for a no-show job which in turn lent Durant and his enterprises the political connections and prestige they desperately needed.

Of course finances between the two firms began to entwine as the contractors accepted more securities in lieu of cash; soon Farnam would succeed Dix as president of the Chicago and Rock Island. And after Farnam designed and constructed the first wooden railroad bridge across the Mississippi, between Rock Island and Davenport, he, Durant, and the other promoters of the Chicago and Rock Island cast their gaze farther westward across Iowa, chartering the Mississippi and Missouri Railroad, with Dix acting as president. The M&M aimed to traverse Iowa from Davenport to the Missouri, with the rail-emptiness of the West beckoning on the river's far side. Thus did Thomas Durant, M.D., enter the Pacific railroad sweepstakes in 1854.

Iowa, however, would prove more stubborn as a site for a railroad than Illinois. What happened there would also help to drive a wedge between Farnam and Durant. At a maddeningly slow pace the line moved out of Davenport across the gentle, nearly obstacle-free prairie. It laboriously reached the state capital of Iowa City, only forty-three miles away, barely in time to qualify for a $50,000 bonus, though to earn it employees were forced to hand-push a disabled engine into town.[5] Periodically, as the M&M ran short of cash, work beyond on the right-of-way sputtered to a halt; laborers dispersed only to be called up again. At one time a squabble over where the line would be stretched beyond Des Moines froze work and all but paralyzed the board of directors, which had split into factions according to where members had speculated in land or businesses. Riverbank villages in both Iowa and Nebraska began a bitter competition for the prize of being named the M&M's western terminus, the strongest competitors being Council Bluffs, Iowa, and its neighbor on the Missouri's western bank, Omaha (the two settlements could muster only a shaky alliance at best), and Florence, Nebraska, six miles above Omaha. Offers of large land grants from the state of Iowa sweetened the pot considerably.

Presiding over this chaos of self-interest were Durant and Farnam—who entered the Missouri River terminus controversy with their hip boots on. Previously, the company had promised to grade across mid-state Polk County and through Des Moines if citizens paid it $300,000 in bonds. Adjacent Jasper County seemed ready to approve $200,000. With Durant remaining in the shadows (though the scheme bore his stamp), Henry Farnam went out to persuade citizens of Council Bluffs and Pottawattamie County to part with $300,000 in bonds to induce them to begin grading eastward from Council Bluffs. That Farnam and Durant had already invested considerably along the anticipated right-of-way in Council Bluffs seems not to have been considered germane to Farnam's sales pitch. Then, across the river, Omaha and Douglas County subsequently parted with $200,000 in bonds under the same inducement.[6]

Then the Panic of 1857 washed like a tidal wave all the way from New York in August, ending the rosy expectations of progress. Land prices plummeted; innumerable businesses failed. And bitterness bloomed forth within the partnership of Farnam and Durant. To Farnam's horror he discovered that the Doctor had been pledging the securities of their construction firm to underwrite Durant's personal dabbling in the stock market. For more than four years, in his association with Durant, Farnam had been, as always (as his son later recalled), singularly consistent: "he did not have one standard of conduct for the counting-room and another for the home. He did not stoop to practices in public life which he would have scorned in his relations towards his friends. He was always straightforward, open, true," and always—young Farnam emphasized—"charitable towards the shortcomings of others." But this latest act was a serious betrayal. Farnam's own finances were affected, tied up in litigation. He tottered near bankruptcy.[7]

"I thought a week ago that I was a rich man," he wrote to a confidant on August 29, 1857. "I now find the concern so involved that we cannot possibly go on, and the firm must make an assignment tonight or Monday. The loss of property is nothing, if I was only sure that I had enough for the support of my dear wife and family; to lose everything now is rather more than I can bear." Durant himself was relatively unscathed from the panic. Infuriated, Farnam denounced his friend and partner—Durant (he would later haughtily say) had neither a "high sense of honor, sound judgment, [nor] financial skill." Only with, as Farnam's son later recalled, "prompt and vigorous measures," was he able to save himself—and the experience took much of the allure of business away from Henry Farnam.[8] Of course the days of his association with Thomas Clark Durant were numbered, though it would take some time to fully extricate himself.[9]

The M&M plummeted into receivership. But the citizens of Iowa had much more to regret than that. Polk County and Des Moines received nothing for the $300,000 given to Farnam and Durant. For an identical figure, Pottawattamie County had an empty, useless railroad grade trailing four miles out of Council Bluffs, which had cost the contractors $4,000. Somehow the other $296,000 was never accounted for.[10]

But perhaps some of the missing money took a roundabout route to the pockets of various Iowa and Illinois officials, state and federal, over the following five years. To be sure, almost nothing more was built of the M&M line, it having been stalled at Grinnell, 120 miles from Davenport; indeed, the connection across the state would wait until 1869. Durant, though, now had his eye on the Pacific railroad, of which the Iowa road could be a part. Accordingly, the Doctor built lines of a different sort—lines of influence. Assiduously, through the spring of 1862 when Congress readied to pass the Pacific Railroad Act, he dispensed campaign contributions, company securities, land deeds, and other gifts to where it counted.

In Washington, he and Farnam made many calls. It would be among the last instances of cooperation between the two, for in June 1863 Farnam, tired of business and weary of keeping his eye on Durant, decided to resign his railroad posts and stalk out of their partnership. He would live out his days surrounded by family members in New Haven, Connecticut, becoming a generous benefactor to Yale and dying five weeks shy of his eightieth birthday in 1883. "His name was never tainted," commented Yale president Noah Porter in a glowing eulogy whose sole bitter observation could have referred to only one man, "with the suspicion of being a manipulating director, who 'slaughtered his own railway in Wall Street,' or a 'receiver,' who 'wrecked' it for his personal profit. Nothing was more noticeable than the detestation in which he held practices of this sort, and the mixture of simplicity and scorn with which he looked upon manipulations of every kind which deviated from the obvious rules of business sagacity and forecast."[11]

With a different kind of sagacity, Durant's Capitol campaign of 1862 helped inspire the ferocious behind-the-scenes congressional maneuvering which resulted in the comatose M&M's prevailing over its bitter rival in Kansas, the Leavenworth, Pawnee, and Western. Durant would be present at the organizing convention of the Union Pacific Railroad in September 1862. And he would bide his time until the time was ready to strike. A year later—as Theodore Judah left California to retrieve the Central Pacific even as Collis Huntington solidified his own strength on the Central Pacific's board of directors—Dr. Durant prepared to seize the infant Union Pacific.

He had no rivals. In a year few subscribers had appeared. In August, Durant himself put down 10 percent of a subscription of $50,000; then he began looking for others, offering at least a dozen friends and associates a risk-free deal: Durant paid their subscription deposits of 10 percent himself with stock remaining registered in their names, and pledged to protect them from any loss, if need be, by taking the stock off their hands. "I finally got my friends to make up subscriptions to the amount of $2,180,000," he would later boast to congressional investigators, "by furnishing three-fourths of the money to make the subscriptions myself."[12] That the maneuver was illegal—individuals could own no more than two hundred shares—was beside the point.

With the required minimum of two thousand shares finally attained, the Union Pacific was now entitled to be free of government commissioners and to run itself. On October 30, the stockholders assembled to elect a board of directors of thirty, and a slate of officers. Durant arranged the company presidency for John Adams Dix—his figurehead on the Rock Island and the M&M—who, a year and a half earlier, had accepted a major general's commission in the Union Army and was now commanding the Department of the East; three months before, he had led the army's suppression of the New York City draft riots. Of course Dix would have no time for administration, but the company would benefit from his prominence. Durant, elected by his friends to the post of vice president and general manager, given a place on the board of directors as well as on the executive and finance committees, would be in complete control.

To his compliant associates Durant revealed just how long before organization he had been at the helm. Three months before, and at his own expense, he had commissioned field examinations of the prospective route between Omaha and Salt Lake City, under the direction of Peter Anthony Dey, chief engineer of the M&M, who had now been given the same post with the Union Pacific. Dey, overseeing four parties of engineers and a geologist dispatched by Durant to search for coal and iron-ore deposits, was now somewhere out at the far end of the Platte Valley—the old emigrant trail—looking for the most advisable railroad grade over the Rockies. All the while, Dey grumbled of his undying distrust of Dr. Thomas Clark Durant.

"I fear," he had written an associate a year earlier, "that Durant has eternally damned the Mississippi and Missouri." Nothing in the intervening twelve months had persuaded him to change his opinion—and although the prospect of engineering west of the Missouri was an exciting one, Dey had bad feelings about any further venture under the Doctor.[13]

Born in 1825 in the hamlet of Romulus, New York, in the Finger Lakes region, Peter Anthony Dey had originally intended to be a lawyer. A short period in the legal department of the New York and Erie Railroad persuaded him to take a different direction. He taught himself engineering, eventually moving westward to the Rock Island as a division engineer, then winning a promotion to chief engineer of the M&M. During the latter project he put down roots in Iowa City, eventually to the point of being elected mayor. Dey was an honest man of considerable technical talent—and he, like Farnam before him, did not approve of the "high-handed" business tactics and decisions emanating from Durant's office. He objected to his employer's constant interferences, taking offense at the frequency with which Durant declared a dividend for the line's stockholders (and, principally, himself) and thus diverted money away from the engineering and construction departments, running the unfinished railroad into the ground.

There seemed to be no way that their personalities could ever mesh. "Mr. Durant has got the whole thing in his hands," he would write with characteristic bitterness a few months after being given his Union Pacific post, "but he is managing it as he does everything else—a good deal spread and a good deal do-nothing. He considers it a big thing—the big thing of the age—and himself the father of it. Durant needs common sense more than anything else. . . . If the geography was a little larger I think he would order a survey round by the moon and a few of the fixed stars, to see if he could not get some more depot grounds or wild lands or something else, that he don't want, and he does not know what to do with when he gets it. I have been so disgusted with his wild ideas that I have been disposed repeatedly to abandon the whole thing."[14]

Dey's confidant, to whom these complaints were addressed, had his own experience with Thomas Clark Durant, though his ambition and acquisitiveness, far deeper than Dey's could ever be, would give him more patience in dealing with the Doctor's viccissitudes—at least for a while. Grenville Mellen Dodge was born north of Boston in Danvers, Massachusetts, in 1831. He was educated at Durham Academy in New Hampshire and Norwich University in Vermont, emerging at age twenty from that private military college with a degree in engineering. Of average height and slight build, wild, nervous, and aggressive in temperament, constitutionally able to play the bully upon occasion, Dodge went west with his engineer's papers to Illinois, securing work as a surveyor on the Illinois Central Railroad around the town of Peru.[15]

Within a year, in 1852, he jumped over to the Chicago and Rock Island, where he worked as a rodman under Peter Dey, coming to have nothing but respect for the engineer some six years his senior. "He was a man of great ability, probity, and integrity," Dodge would recall.[16] Likewise, Dodge impressed Dey enough so that in early 1853, when the latter was named chief engineer of the Mississippi and Missouri Railroad, he picked Dodge to assist him, the first of many acts of encouragement. Before winter that year young Dodge had traversed the only partially settled Iowa terrain, becoming particularly enthralled by Council Bluffs, and had penetrated several dozen miles into Nebraska Territory to determine the general line between Omaha and the Platte River. His faith in that latitude as thc one destined for the Pacific railroad would thenceforth be as unshakable as his hope that being connected to it would be his path to wealth and power.

With the exception of some eight months spent homesteading with his wife, father, and brother on the bank of the Elkhorn River in the newly organized Nebraska Territory, in 1855, Grenville Dodge devoted all of his energies toward the railroad's connection to his own destiny. Still surveying under Peter Dey for the M&M, Dodge opened a real estate business in Council Bluffs that benefited from his advance knowledge of where the line would be stretched; in such transactions he frequently acted as a front for the railroad's officials, including both Henry Farnam and Thomas Durant, in their own speculations. With Dodge's influence in Council Bluffs, Farnam and Durant secured the county bonds that would prolong the life of the M&M while giving the river town its own four-mile-long railroad grade eastward to a dead end on the prairie. Moreover, young Dodge evidenced a keen understanding of politics, lobbying local and federal officials for the M&M, and working on behalf of Republican Party interests in Iowa; in 1860 he would have a good deal to do with re-electing Iowa congressman Samuel R. Curtis, chairman of the House Committee on Pacific Railroads, and at least a visible role among the local party faithful in helping to sway the state to Abraham Lincoln.

By then Grenville Dodge had already made the acquaintance of the Illinois attorney. Lincoln had previously done much work for the region's railroads (indeed, most of his income had derived from them), especially the Rock Island—which he successfully defended in a celebrated suit over Henry Farnam's wooden bridge over the Mississippi.[17] In August 1859, Lincoln rode to Council Bluffs to inspect seventeen building lots offered to him as security in a $3,000 loan to an associate; he attended a reception in his honor and delivered a speech on slavery. He also stood atop the yellow bluffs and looked down at the Missouri River and out toward Nebraska, the topic of railroads as naturally on his mind as on that of any Westerner in 1859. He heard one of his hosts predict that one day a railroad line would pass by where they stood and thread off toward the western horizon. "Not one," Lincoln answered, "but many railroads will some day center here." Stopping at the Pacific House, Lincoln heard that another guest, a young engineer by the name of Dodge, had just returned from an exploring trip in the Platte Valley, and that he knew more about railroads than any "two men in the country."

"After dinner," Dodge recalled, "while I was resting on the stoop of the Pacific House, Mr. Lincoln sat down beside me." Lincoln was a skilled and mannerly interrogator; he had no trouble learning about Dodge's fervor for the Platte route and for the part Council Bluffs and Omaha waited to play in the Pacific railroad—nor did the attorney find it difficult to extract the young man's detailed findings about the way overland from Omaha to the Platte, and the possible routes over the Continental Divide toward Utah. "By his kindly ways," Dodge would confess ruefully, Lincoln "soon drew from me all I knew of the country west, and the results of my reconnaissances. As the saying is, he completely 'shelled my woods,' getting all the secrets that were later to go to my employers."[18] In words perhaps similar to those uttered by him forty years later, Dodge made an impression. "The Lord had so constructed the country," Dodge would say, "that any engineer who failed to take advantage of the great open road from here west to Salt Lake would not have been fit to belong to the profession; 600 miles of it up a single valley without a grade to exceed fifteen feet; the natural pass over the Rocky mountains, the lowest in all the range; and the divide of the continent, instead of being a mountain summit, has a basin 500 feet below the general level."[19]

Lincoln did not forget his conversation with the young engineer; indeed, a few weeks after his inauguration, they renewed their acquaintance when the two met again at the White House, the topic of a Pacific railroad naturally taking second place to the gathering crisis in the South. "I shall bring the country out all safe," the president assured Dodge.[20] Although Dodge had gone to the capital with the dependable brigades of office seekers and patronage peddlers which always descend on a new federal administration, after Fort Sumter he stepped out of the pack. Despite the call of his Iowa associates to jump into war profiteering, with moderate difficulty he arranged for a political appointment as a colonel of an Iowa regiment. "I go into the field in twenty days," he wrote to his mother on June 21, "with as fine a body of men as ever drew a sword or shouldered a musket. I go into this war on principle. Pecuniarily it will ruin me."[21]

For seven months Colonel Dodge and his infantry regiment, the Fourth Iowa, did little but drill for war, although Dodge did sustain a minor injury when a small pistol he normally carried in his greatcoat pocket discharged, wounding him in the thigh. Nevertheless, from the time of his commission his many political friends in Iowa and Washington lobbied the War Department and the president for Dodge's promotion to brigadier general,[22] part of the incessant rain of similar entreaties which Lincoln and War Secretary Simon Cameron were forced to endure. Cameron's successor, Edwin M. Stanton, punctuated the issue by requiring that only officers who proved themselves in battle could win promotions.

For Colonel Dodge, his opportunity came during the campaign around Pea Ridge, Arkansas, in February and March 1862, when he acquitted himself well as an officer. Additionally, three horses were shot out from under him in battle, his clothes were riddled with bullet holes, the knuckles on his left hand were scraped by passing grapeshot, and he was bruised when an overhead tree limb was dislodged by gunfire and swept him from the saddle. Apparently, though, this was not sufficiently heroic. For the rest of his life, official records to the contrary, Dodge would claim that he was severely wounded at Pea Ridge—lying hurt on the battlefield for ten days, then being "hauled 250 miles over rough roads in an ambulance," the general prognosis being "that I could not live."[23] Within time, even the injury inflicted two months before his battlefield baptism, by his own pistol, would be transmogrified into a Confederate-inflicted wound; these were only a part of a much wider tendency on Dodge's part to revise the facts of life—which, judging from the frequency with which they were repeated by journalists and historians, he did with a great amount of success. But insofar as Pea Ridge was concerned, the colonel's assertion that he led his soldiers with skill and valor is upheld by the official record. This news, aided no doubt by his legion of political supporters, led to Dodge's promotion to brigadier general in April 1862.

This was the spring, of course, in which the Pacific Railroad Act of 1862 was thrashed into form by Congress, assisted by such players as Theodore Dehone Judah, Collis Potter Huntington, Thomas Ewing, Henry Farnam, and Dr. Thomas Clark Durant. Thus it was a fitting season for the new Brigadier General Dodge, aged thirty-one and with one military campaign notched onto his saddle, to be assigned to command the Union forces' rebuilding of a 150-mile section of the Mobile and Ohio Railroad, between Corinth, Mississippi, and

Columbus, Kentucky. Despite harassment from Confederates under the tenacious cavalry leader Nathan Bedford Forrest, who had destroyed all the bridges, burned depots, and pried up and twisted rails, Dodge performed commendably. He cut down continued depredations through the inventive use of a system of blockhouses along the route; the stratagem was subsequently adopted by the Federals to safeguard other contested stretches of southern rails, and indeed began to contribute toward the gathering of Union strength against Vicksburg.

This did not go unnoticed. As early as October 1862, Dodge was called to Jackson, Tennessee, to the headquarters of the newly installed commander of the western armies, Ulysses Simpson Grant—so quickly, in fact, that Dodge did not have time to change out of his muddy work clothes. Joined on the railroad platform by the chief of staff, General John Aaron Rawlins, Dodge began to apologize for being so poorly dressed. "Never mind about the clothes," he was told. "We know all about you."[24] Further assurance came with Grant's praise for his work, and a promotion to a divisional command in the Army of the Tennessee. Thus his duties expanded—beyond directing patrols and rebuilding railroads to also include, during that buildup of Grant's Vicksburg campaign, the gathering of intelligence. In previous campaigns in Missouri and Arkansas, Dodge had already proved to have talent as a spymaster; now, in late 1862 and 1863 his line of undercover operatives eventually reached some one hundred men—and the intelligence they passed on proved to be invaluable to the Union effort. Grant's friendly feelings toward Dodge were shared by his staff, including Generals John Rawlins and William Tecumseh Sherman, and their support would do much to aid Dodge's subsequent career, even in civilian life when he had returned to engineering and that towering feature of his postwar life, the Union Pacific Railroad.

A portent unexpectedly appeared in June 1863 when General Dodge received a summons to the adjutant general's office in Washington. "I was somewhat alarmed," Dodge recalled, for on his own authority he had stirred up angry protests from fellow Union officers and civilians alike by equipping a group of black men at Corinth, Mississippi, with rifles to safeguard a camp of escaped slaves, then called "contrabands." Hastening to Washington, he was directed to the White House. Sure that he was to be scolded by the commander in chief, Dodge instead found Lincoln cordially recalling their conversation on the stoop of the Pacific House in Council Bluffs nearly four years earlier, and now pointedly in the mood to discuss railroads. Where on the Missouri River, he asked, did Dodge think he should locate the eastern terminus of the Union Pacific Railroad? Politicians, capitalists, and their lobbyists had been parading their competing opinions before him, and Lincoln was eager to hear the honest opinion of an expert and an officer. As they bent over a map, naturally the former employee of Peter Dey and Thomas Durant told Lincoln that Council Bluffs, his adopted home, was the place.[25]

And how, Lincoln wondered, did Dodge think he could get the enterprise moving? "I told him very decidedly," Dodge recounted, "that in my opinion private capital could not be obtained to build the road," and that the government should do it. Lincoln responded that this was impossible in wartime—but that "the Government would give the project all possible aid and support" and "make any change in the law or give any reasonable aid to insure the building of the road by private enterprise."[26] Then, the officer replied, recalling a prescription written to him earlier by Dr. Durant, the 1862 Pacific Railroad Act would have to be amended; no investor, in these times or better, would buy second-mortgage bonds; instead, the government's subsidy bonds would have to take a second seat. The president said he would see what he could do.[27]

Leaving the White House in a high state of excitement, Dodge lost no time in proceeding to New York, where he reported to Durant and his associates: "Lincoln will help us."

Then it was back to Mississippi and the closing scenes of the Vicksburg campaign, during which Dodge's cavalry and infantry protected Grant's left flank, and his spy network collected intelligence on Confederate movements and even on conditions endured by the civilians of besieged Vicksburg as they huddled in caves and trenches and dreamed of food and clear water. After the surrender on July 3, Dodge was rewarded with a promotion to command in the Army of the Tennessee's Sixteenth Corps, and again he began to dream of a major generalship; his supporters were now pressing hard, buoyed by word that General Grant had put Dodge at the top of his candidates' list.

Grenville Dodge soon began to register interest in him from another quarter, however. "[Durant] asked me this afternoon if you could be induced to leave the army and take hold of the Union Pacific Railroad," wrote Peter R. Reed, an Illinois businessman and lobbyist working for Durant, on August 20.[28] The Doctor urgently needed Dodge's help, he said; President Lincoln was still not ready to declare a winner, but suddenly it seemed that Durant's enterprise (and with it, Council Bluffs, Omaha, the M&M, and the Rock Island) was in danger of losing its primacy. The ghost of Thomas Hart Benton had risen again, and with it, St. Louis. This was because weeks before, in a strange, ultimately untenable alliance, John C. Frémont, famous explorer and onetime presidential candidate, and Samuel Hallett, a shady new York investment broker who had previously raised money in Europe for the Atlantic and Great Western Railroad, had picked up the charter of the Leavenworth, Pawnee, and Western; the LP&W had been a moribund concern since its loss to Iowa and Chicago interests during the behind-the-scenes maneuvering over the Pacific Railroad Act a year earlier. Frémont may have finally shown himself as the incompetent officer he always was, in disastrous wartime posts in St. Louis and in the Shenandoah Mountains, but though now thrust back into civilian life he still had political support. And cunning: the LP&W was thereupon renamed the Union Pacific Railway, Eastern Division, spawning much confusion in the minds of the public, especially potential investors, who might be diverted from purchasing stock in the Union Pacific Railroad; for some mysterious reason no suit for infringement was filed. The Missourians' intention was to push their line westward to the 100th meridian which, according to the Railroad Act, would make the Missouri-Kansas latitude the main stem of the Pacific railroad. Thus the Iowa-Nebraska trackage would be reduced to a humiliating branch line.

Dodge went on leave. Ostensibly he did so for medical reasons; physically slight, he was undoubtedly worn out, but he wasted little time either in bed in Council Bluffs or even resting on the dais of a hometown reception, ball, and oyster supper held in his honor.[29] Instead he hurried east to New York to huddle with Durant.

By then it had been announced that the LP&W had actually broken ground, in Kansas City on September 7. Now Durant fully recognized the immense value in Dodge's twofold talent as a lobbyist as well as an engineer. In grumpy Peter Dey, his current chief engineer of the M&M and obvious candidate for the Union Pacific, Durant may have had talent and dependability, but could Dey (out that fall inspecting the Platte Valley for Durant) boast of private conferences with the president of the United States, of the affection and gratitude of high military leaders who might one day move into politics, of a legion of political cronies across the nation's capital?

The answer, of course, was no. Several times in those autumn months, before and after the October 30 stockholders' meeting when Durant had himself appointed vice president and general manager of the Union Pacific, and secured seats on the executive and finance committees, Durant urged Dodge to resign his army commission, offering him the position of chief engineer—his salary would be a respectable $5,000 per year. Durant went so far as to offer to leave the position open until the spring, to give him a chance to make up his mind.

Dodge could not be persuaded.[30] Moreover, he refused to go to see Lincoln once more, as Durant urged. That, he reasoned, would strain the association. On his way back to his command, he wrote the Doctor from Cairo, Illinois, to reassure him. "I probably can do you more good in my present position while matters are being settled by Congress and others." Already, he said, he had made sure that they could get almost anything from the Iowa legislature, and he promised to keep his eye on the state's congressional delegation. And Durant would need a full-time lobbyist in Washington—Dodge recommended his crony, Herbert M. Hoxie of Des Moines. After the war, above all, Dodge assured the doctor, "[I wanted] to identify myself with the project in some active capacity." In the meantime, though, there was the issue of money, a continuing trouble to the engineer. When the president named the Union Pacific terminus, "telegraph or write my Brother at Bluffs, so that he can invest a little money for me or else give me a chance with you if you make any . . . *bear this in mind.*"[31]

But everything depended on fixing the location of the starting point. With General Dodge returned to rebuilding southern railroads, this time between Nashville and Decatur, Durant summoned Peter Dey to New York with his preliminary maps of the Nebraska terrain west of Omaha. As soon as could be arranged, the doctor sped to Washington for an audience with the president. Therein a curious controversy was born in a strange little waltz figuratively danced over the Missouri River. Along with Interior Secretary John P. Usher they pored over Dey's maps. Durant pointed to the western bank of the Missouri; having secretly appointed an agent to begin buying land, he pleaded for Omaha.[32]

In reply, Lincoln pointed across the river to Council Bluffs. He was cognizant that he possessed some seventeen Council Bluffs building lots held in collateral for a personal loan to a friend—the inspection of which, more than four years before, had led to his chance first meeting with Grenville Dodge. "I've got a quarter-section of land right across there," Lincoln said, in Usher's recollection, "and if I fix it there they will say that I have done it to benefit my land." A small pause for deliberation may have been registered as Usher and Durant waited for the pronouncement. "But I will fix it there anyhow."[33]

Lincoln wrote out the executive order. In two days he was scheduled to deliver one of the most eloquent speeches of all time on a bloodstained battleground in Pennsylvania, but for some reason, as Durant stood by, the same hand that had penned the Gettysburg Address wrote out the following:

> *I, Abraham Lincoln, President of the United States, do hereby fix so much of the western boundary of Iowa as lies between the North and South boundaries of the United States Township, within which the City of Omaha is situated, as the point from which the line of railroad and telegraph in that section mentioned shall be constructed.*

He handed the note to Dr. Durant, neglecting to make a copy for himself, and the Doctor thrust it in his pocket. It would remain in his personal possession until 1874, when he allowed it to be photographed as evidence in the complicated legal wrangle over the location of the Union Pacific's eastern terminus. In those parts, in 1863 as now, the boundary between Iowa and Nebraska was quite wide and rather wet, with only a ferryboat franchise connecting the two states; practically, the western boundary of Iowa ran along the shore at Council Bluffs, and the eastern boundary of Nebraska ran through Omaha. Despite the out-of-character obfuscation, Lincoln plainly meant the eastern terminus to be in Council Bluffs, probably—with his kind of lawyerly thoroughness—making sure that it should be Durant's responsibility to soon build and maintain a railroad bridge over the Missouri River. But Lincoln's language had enough space for interpretation in it to drive a ten-car train through it sideways. So Durant thrust the note into his pocket, took his leave of the president, and went out to announce to the world that the Union Pacific terminus would be in Omaha.[34]

Bridging the Missouri could wait.

14

"Dancing with a Whirlwind"

For Peter Anthony Dey, the westering terrain of Nebraska had grown as familiar to him as his dooryard in Council Bluffs, the line beyond—through the Laramie Mountains and the Medicine Bow Range, across the Great Divide toward the Wasatch and Great Salt Lake—as comprehensible as the right-of-way he had staked across Illinois and Iowa. Beginning in September 1862, sent into the field by Durant and Farnam, the engineer had personally scrutinized the length of the rivers Platte, North and South, had ridden up Lodge Pole Creek and the Cache la Poudre, had forded Green River and followed the Bear, had edged through both Timpanagos and Weber Canyons, and had ascended the mountain passes of Cheyenne, Bridgers, and Berthude (the last in the company of its discoverer). When he made reference to gradients, curvatures, road cuts, tunnels, and bridges, to depth of snow, width of valleys, height of overshadowing cliffs, to timber (cottonwood, pine, cedar) and to minerals (coal, iron ore, lead, borax, sandstone, hornblende), Peter Anthony Dey knew from whence he spoke. Nine-hundred-sixty miles from Omaha to the Great Salt Lake valley, he opined, and he would not be far off the mark. "I am satisfied," he would conclude in a report shelved by Farnam for a year but finally delivered when it was needed to Dr. Durant, "the cost of the road will be less, and its business far greater, than its most sanguine friends anticipate."[1]

And Peter Dey badly wanted to be part of it. "All the professional ambition that I ever had," he had written to Durant from Omaha on January 4, 1863, "pointed to no position to me as desirable as that of Engineer of this road."[2] Their great differences in temperament and ethical outlook continued, though, to weigh heavily on the engineer, whose sensibility seemed to register another bruise with the arrival of every new imperious, peremptory telegram from Durant in New York. Immediately after President Lincoln had chosen the initial point, the Doctor scheduled December 2, 1862, for the lavish groundbreaking ceremony and speechifying that would send an amplitude of hot air into the atmosphere above the Omaha bottomland (and temporarily bankrupt the company coffers), but with Dey in charge he could not resist pushing. "You are behind time for so important an enterprise," Durant scolded in a wire received on Monday, November 30. "Break ground on Wednesday."[3]

After the ceremony, Dey began to locate the line, and was pleased when the city of Omaha granted the Union Pacific its riverfront and much of its bottomland; too, Douglas County awarded some six thousand acres. Dey nonetheless watched with horror as land offices were besieged by speculators grabbing up land around Omaha and up along the Platte, thinking, as his friend Dodge would voice, that the company would have to buy its way "clear through to the Pacific."[4] In the flurry it was hard to tell friend from foe, or even to trust his employer to do the right thing for the company. Several times during the rest of the month he urged Durant to get the Omaha land office closed. He heard nothing in return. "Don't communicate anything to parties from or in Nebraska," Dey pleaded to Durant in an extraordinary telegram sent on December 4. "Give me by resolution of the Executive Committee full power to fix location as a finality where I please."[5]

But even Dey was kept in the dark. He had given the Doctor a choice of running the line south of Omaha through the town of Bellevue, an old Indian mission and fur-trading center, or around the north end of Omaha through the village of Florence. In the engineer's opinion Bellevue was best; though nearly six miles longer than the Florence route, it featured less expensive, lighter, and more uniform grades.[6] However, on December 4, Dey received an order to begin at Florence. This, he soon learned, would be nearly impossible, as the company did not and could not control the land over which the line would run; moreover, construction would simply be too expensive.

Anxious for a solution, wanting to take advantage of their local momentum, Dey began planting stakes in a different line altogether around the southern limits of Omaha, wiring the Doctor to assure that he could save upward of $8,000 per mile. Within a week he could report that he had run the line across the eighteen miles of rolling prairie, easily cresting the intervening valley of the Elkhorn River to the Platte, at Fremont Village. In reply, Durant sent only criticism. On Christmas Eve, his patience thin, Dey offered to "cheerfully" resign his place as of January 15, 1863. Durant's response arrived on January 3: he had arranged for the company's executive committee to finally give Dey the official appointment of chief engineer, and for his own good, the engineer should start taking Durant's acerbic wires as the "friendly advice" they were meant to be.[7]

It was a weak assurance, ensuring further frustration down the line—not very far. In mid-January 1864, with no forewarning, Dey was ordered to survey a completely new line from the Missouri River at DeSoto, a steamboat town some twenty miles north of Omaha, and the seat of Washington County. Dey was flabbergasted—the new route would obviously bypass both Council Bluffs and Omaha, against the president's directive, and it made even the undesirable line from Florence look good. No, Dey's route west from Omaha to Fremont was best—but, as he complained in a letter to Grenville Dodge, "I cannot make him see it . . . and if I attempt to put a little common sense in his head, he flies off in a fit of excitement." Dey felt "so completely disgusted with his various wild ideas," he said, "that I have been disposed repeatedly to abandon the whole thing. I hate to do it as there is a great future in this thing, if judiciously and prudently managed."[8] Frustrated with the knowledge that if the road was run rationally they would have already completed a gradient a fifth of the way across the state to the Loup River fork, nearly ninety miles away, the engineer sent a plea to Durant in New York. "It is difficult to make surveys," he wired, "without forming some idea of what you are doing and what it is for."[9]

Dodge, rebuilding railroads in Tennessee, tried to piece together the puzzle of Durant's business decisions. A dispatch from the Omaha banker and Union Pacific agent Augustus Kountze reported from Washington that Durant was rumored to be planning to merge his M&M with the Cedar Rapids and Missouri line, an extension of the Galena and Chicago Union Railroad, with the intent to build westward from the M&M's current railhead at Des Moines over a single trans-Iowa line that would bypass Council Bluffs and Omaha by at least twenty miles. Durant was doing everything he could, Kountze charged, "to slaughter us."[10]

Dodge lost no time in wiring the doctor. "My plan will be carried out or the work abandoned," replied Durant. "Iron is being shipped from St. Joseph to Bellevue. This is too important an enterprise to be controlled by local interest. The road can be built by the Kansas line if in no other way. No road through Iowa will terminate at Omaha." Dodge's letter streaked back to Durant in New York: he was inviting the wrath of the president and two state legislatures, Dodge warned, and all for a handful of engineering headaches; Peter Dey's surveyed gradients out of Omaha were the best available—and Durant would even have to ultimately contend with the mighty Missouri: it was only three miles wide at Council Bluffs, but eight miles wide at De Soto. These rumors of Durant's plans were not simply conveyed to Dodge in Pulaski; they were zooming to legislative halls from Nebraska to New York, and flying unerringly to Wall Street, and finally ricocheting into the White House. Pressed to clarify his decision favoring a Union Pacific terminus at Council Bluffs, Lincoln scrawled a message which he read to the Senate on March 7, 1864: it should be "on the western boundary of Iowa," opposite Omaha.[11]

A few weeks after news of Lincoln's proclamation would have reached General Dodge, a letter from Dodge's brother, Nathan, who was watching the family interests in Council Bluffs and Omaha, went a long way in explaining the alarming puzzle of the shifting rail line. It was, Nathan pronounced with rueful envy, "a stock operation entirely." First, he said, the doctor had announced that the Union Pacific would connect to his Mississippi and Missouri line, and M&M stock began to rise to $149 per share. Durant then discreetly sold his M&M stock, reinvested in the Galena and Chicago Union, and announced to the world that he would build the Union Pacific to connect with the Galena's subsidiary, the Cedar Rapids and Missouri. Immediately the M&M stock plummeted to $111, the Galena shot up, and Durant sold his Galena and rebought his M&M. "In other words," Nathan reported, "he gets back home and makes in the round trip for him and his friends $5,000,000. It is the smartest operation ever done in stocks and could never be done again."[12]

It was no way to run a railroad—but it certainly seemed like a sure path to an immense financial empire for Dr. Thomas Clark Durant. One can only imagine their welter of emotions, only one of which would have been envy, if the relative annual salaries of a brigadier general or a chief engineer were stacked against the income of that stoop-shouldered, inscrutable corsair.

By then the Thirty-eighth Congress had been in session more than three months and, amazingly, Durant had also been active on that front. Just before its Christmas recess, the Senate had created a new standing committee on the Pacific railroad, which had the effect of granting permanent status to the railroad senators' coalition and beefing it up in anticipation of improving the 1862 Pacific Railroad Act to the greater benefit of investors. As soon as Congress reconvened in the early part of January 1864, Dr. Durant appeared, satchel bulging, at the head of the pack of railroad promoters. Collis P. Huntington of the Central Pacific was there. And John C. Frémont and Samuel Hallett, of the Kansas entity now known as the Union Pacific, Eastern Division, were also present, though no longer speaking except through their attorneys. Frémont and Hallett had broken in a vehement disagreement over whether Hallett should have awarded himself the company's construction contract, which he had done without authorization from Frémont in December, at the same time and in the same manner that he had also raised construction funds by bringing in a group of outside investors—including none other than Thomas C. Durant of the Union Pacific, who now possessed one-third interest in the rival line's building contract. Durant, unsurprisingly, was planning to put his support behind Sam Hallett, whose pocket reportedly contained the interior secretary, John Palmer Usher; their moves in the capital would be coordinated as pressure to write the new Railroad Act was refined and directed, but it would be Dr. Durant's satchel that would begin to disgorge the wherewithal.[13]

Details would not emerge for many years. But when Durant moved among congressmen, their agents, associates, and relatives, he did so with funds totaling some $435,754.21 from the treasury of the Union Pacific, a feat of fiscal magic in view of its depleted resources, and when the campaign was over and the new legislation was written to his satisfaction, the money had gone elsewhere. So had about a quarter-million dollars in bonds handed over to a lobbyist and attorney named Joseph P. Stewart, who disbursed them to recipients he would later refuse to identify when hauled before a congressional investigating committee. Stewart had been paid $38,000 by Durant and Hallett, and $2,000 by Huntington, for his work, which had kept the legislation moving when controversy over the Kansas line's directorship threatened to stall it, and all Stewart would say to questioners was that he had not handed anything directly to an elected official. For his reticence he would be found in contempt of Congress and imprisoned. Durant, when invited in for questioning, would shed no light on the $435,754.21 beyond the admittance that he had lived and entertained lavishly while in the capital, and that the money had been charged to a line item on the Union Pacific books called a "suspense account"—construction funds, supposedly, spread out over the life of the entire building project. There was, he assured the committee, "no necessity" for bribery at that time.[14]

Nonetheless, no one could deny that the Doctor had been busy, though not too occupied to lay plans in yet another direction. Serious students of Durant's activities have frequently suggested that it was of little matter to the vice president of the Union Pacific whether the Union Pacific actually wended its way across the prairies and plains and mountain chains to its presumed meeting point with the Central Pacific, so long as he was able to turn a dime into fifteen cents and one million into five or ten. Thus far, by early 1864, it had all been expenditure and side bets. No guarantee sat in anyone's strongbox, and there was a war on. The way to rake it in—as everyone knew who had looked at a public works project since Roman times—was during the construction phase. It had become ready, with Nebraska sod finally turned over and enabling legislation being negotiated toward its close, to set up a system for constant revenue into the private coffers of Thomas Clark Durant and a small group of associates. To make this possible, the Doctor had a consultation with one of the most improbable characters to take a breath in the nineteenth century, a merchant and railroad promoter by the most appropriate of surnames: Train.

He would not shake a man's hand. But George Francis Train called himself Citizen (and he considered equal any other human being, man or woman, of any sort of complexion); in seven years, after failing to win the American presidency as an independent candidate, he would then be one of the prime movers of the Paris commune. "Been in fifteen jails," he would chortle in his memoirs, "without a crime." Among the charges would be obscenity—in the celebrated aftermath of Henry Ward Beecher's adulteries, he came to Victoria Woodhull's defense by printing in his own newspaper the earthier verses from the Holy Bible. He would cheerfully deem himself the "Champion Crank," and tersely record in his autobiography, "Now lunatic by law, through six courts." He would be an enthusiast of psychic telepathy, through which he proposed to double his age. He had seen something of everything, and by 1870, when he would travel around the world in eighty days, he would give a French writer by the name of Jules Verne a subject worthy of a novel. And, thank goodness, George Francis Train knew something about finance, and he knew something about trains.

He was born on March 24, 1829, in Boston, the son of a successful merchant who had the misfortune of moving himself and his family to New Orleans in time for a yellow fever epidemic. After losing his wife and three daughters, Oliver Train dispatched his three-year-old son back to Boston alone, by sea, and then himself expired. Train grew up on his grandmother's farm in Waltham, Massachusetts, and ran away at fourteen to avoid apprenticeship. After a time working for a grocer in Cambridge, Train was united with an uncle, Enoch Train, a shipping merchant, who gave him a clerkship. At twenty-one the manager of an English branch office in Liverpool, he later was sent to Australia to establish another shipping firm where his first year's commission was some $95,000. In 1856 he returned to New York and began to write prolifically, first in a number of columns for James Gordon Bennett the elder in the *New York Herald,* then in a series of discursive, opinionated travel books. Two years later, Train may be traced to Britain, where he finally began to be interested in railroads; he floated the initial mortgage bonds of the Atlantic and Great Western Railroad, and devoted a great deal of energy to a campaign for street railways in London and Birkenhead, to no avail.[15]

Train was forced to abandon England in 1862, the culmination of too many caustic comments about British society and too many intemperate speeches supporting President Lincoln and the survival of the Union. It was said, however, that before he left, learning that the Union Pacific Railroad had been organized, Train directed parting words to Prime Minister Benjamin Disraeli. "You go to India by your Suez canal," he said. "I'll go home, build a railway across the continent, and beat you to the goal."

He lost little time. In wartime New York there were a number of sure ways to make money, one being in the illegal traffic in contraband cotton, and one of his partners in this venture was Dr. Thomas Clark Durant. And when Durant approached Train in the summer of 1863 with his no-fault subscription scheme to get the Union Pacific moving, George Francis Train took twenty shares. He became an active and effective lobbyist in Washington. As it became apparent that fall that Omaha and Council Bluffs would get the president's nod, Train made haste to Nebraska and began buying up land, eventually holding title to some five hundred acres of Omaha real estate bought at rock-bottom prices. And on December 2, 1863, when Omaha soil was first disturbed for the railroad, George Francis Train, attired characteristically in a splendid white suit, was the keynote speaker. "America," he bellowed out over the assembled heads, "possesses one-half the common sense, three-fourths the enterprise, and seven-eighths the beauty of the world." With the

Pacific railroad in place, he predicted, America would control the world's commerce.[16] The speech was, in the opinion of the *Omaha Nebraskan*, "the raciest, liveliest, best natured, and most tip-top speech ever delivered west of the Missouri." And even gloomy Peter Dey had to concede that Train was an "attractive talker" at the ceremony: it was "the finest address that I ever heard on such an occasion."[17]

Train was tall and dark, wore a mustache and a goatee, and had winning ways. Most politicians grew to like him; journalists, whether they liked him or not, could not help but be fascinated by his eccentricity and his wealth. Albert D. Richardson of the *New York Herald* was, in 1864, doing time in a Confederate military prison for defying the batteries at Vicksburg, but he later came to know Train well. "Curiously combining keen sagacity with wild enthusiasm," he once wrote after encountering the land baron in Omaha, Train was "a man who might have built the pyramids, or been confined in a strait jacket for eccentricities, according to the age he lived in. . . . Since he began to make money, people no longer pronounce him crazy." Train, he said,

> *Drinks no spirits, uses no tobacco, talks on the stump like an embodied Niagara, composes songs to order by the hour as fast as he can sing them, like an Italian improvisatore, remembers every droll story from Joe Miller to Artemus Ward, is a born actor, is intensely in earnest, and has the most absolute and outspoken faith in himself and his future.*[18]

A reporter for the *Louisville Courier-Journal* waxed even more eloquent when he said that Train was

> *a locomotive that has run off the track, turned upside down with its cowcatcher buried in a stump and the wheels making a thousand revolutions a minute—a kite in the air which has lost its tail—a human novel without a hero—a man who climbs a tree for a bird's nest out on a limb, and in order to get it, saws the limb off between himself and the tree—a ship without a rudder—a sermon that is all text—a pantomime of words—an arrow shot into the air—the apotheosis of talk—the incarnation of gab. Handsome, vivacious, versatile, muscular, as neat as a cat, clean to the marrow, a judge of the effects of clothes, frugal in food and regular only in habits. A noonday mystery, a solved conundrum—a practical joke in earnest—a cipher wanting a figure to pass for something; with the brains of twenty men in his head all pulling in different ways; not bad at heart, but a man who has shaken hands with reverence.*[19]

Early in 1864, the noonday mystery met with Durant and another director of the Union Pacific, the Connecticut railroad builder Cornelius S. Bushnell, to lay plans for a separate construction company that could generate profits all along the right-of-way. To do so, and to obscure their movements, they looked to buy a preexisting company charter that would be flexible for their needs. In Connecticut, Bushnell found one, but then a wire from the "solved conundrum," Mr. Train, directed Durant's attention to Philadelphia. His idea, Train recalled many years later, "cleared the sky. It made the construction of the great line a certainty," and, in the same manner, profits to its makers. Train had remembered that a few years before, while in Paris, he had been attracted to the Perrère brothers' inventive new method of finance, which allowed individuals to obtain credit based on personal property and real estate, and which the Perrères had used to construct many French civic improvements, including railroads and the Paris Gasworks. The construction corporation had been called the Crédit Mobilier de France. In the charter of an unnoted, moribund little Pennsylvania corporation called the Pennsylvania Fiscal Agency, Train found an immense flexibility that could accommodate not only the new credit principles but also the idea of limited liability for the company stockholders; in other words, investors would not be liable for corporate debts, as with the current Union Pacific charter, to the full extent of the individual's personal worth—instead, their liability would go only as far as each had invested. The Pennsylvania Fiscal Agency had been chartered in 1859 by the state legislature; its incorporators had simply put up 5 percent of its stock subscriptions, turned to other pursuits, and waited for the company's utility to show itself.

This came nearly five years later, on March 2, 1864, when George Francis Train and his psychic telepathy presented himself at their offices and offered them $25,000 for the charter—$5,000 in cash and the rest in bonds of the new and improved entity. The following day the transaction was made, with Durant named as president and board member; Train, and later Bushnell, was also a director. "In order to make it suitable for our uses," Train remembered, "I wanted its title changed." In homage to its model the entity was henceforth to be known as the Crédit Mobilier of America. He went to the state legislature and was told the name change would have to be "engineered"—"I paid $500 for having this done." Not long after, Train was talking with William Henry Harding of the *Philadelphia Inquirer* and told him of the transaction. Harding had a good laugh. "I could have had it done for $50," he said. Nevertheless, the change in name became effective on March 26, 1864.[20]

Durant would hold the Crédit Mobilier of America in abeyance for some months—probably directing more of his attention to influencing the developing Pacific Railroad legislation in the capital, or to smuggling contraband rebel cotton, or even to riding poor Peter Dey, out there on the cold and windy Nebraska prairie, who was having his own share of problems. Dey had returned from a few late-winter weeks in New York, where he had tried to arrange for surveying and construction parties for the spring and summer, but he was hampered by a shortage of labor and no end of exorbitant prices, and he could find no one willing to take a construction contract under such conditions. Still feeling reverberations from the Doctor's route-switching stock scheme earlier in the year, Dey's consultations with Durant in New York had given him no more confidence than before. "Durant is vacillating and changeable," he wrote to Grenville Dodge in Tennessee, "and to my mind utterly unfit to head such an enterprise. He has the position and money to run it, but it is like dancing with a whirlwind to have anything to do with him. To-day matters run smooth and tomorrow they don't. . . . If there were parties managing in New York that would be governed by what I write them and furnish the money without desiring to meddle with the details I could build the work for less money and more rapidly than can be done the way they propose to do it."[21]

Struggling to get any Nebraska grading done whatsoever, unaware of what the dancing whirlwind and the noonday mystery had concocted to bedevil his future dreams, Dey would have only been more agitated had he known that in faraway California, the Central Pacific builders, despite their own nagging problems, were grading and laying track even then at a rate of half a mile a day.

On February 18, 1864, the *Governor Stanford* stood murmuring in its maroon-and-gold proudness at railhead, sixteen miles outside of Sacramento. Thirty-two distinguished Californians disembarked stiffly from the open-air platform car on which they had gamely ridden, seated on improvised benches; enclosed passenger cars were still being constructed in Sacramento. Leland Stanford and Mark Hopkins, no less sooty than the assortment of state politicians and railroad directors and other dignitaries, led the excursionists to carriages for an examination of the work in progress. Two miles up the slight grade lay the settlement of Junction (where the short-line California Central paused on its way from Folsom to Lincoln) although Junction had recently been renamed Roseville (a more apt name would have been Figville or Grapeville). Beyond Roseville, the land rose appreciably into oak-forested foothills, traversed by the Central Pacific's gradient pointing toward the pine-clad slopes of the Sierra. Newcastle, a mining camp perched atop a hill overlooking the highly auriferous Secret Ravine, some twelve miles nearer to Omaha, was the day's goal for the railroad excursionists.

Among them was Lauren Upson, editor of the *Sacramento Union* and dear friend of the departed Theodore Judah. Despite any lingering loyalties to the outmaneuvered chief engineer, Upson could not contain his avidity for what had been accomplished and his appreciation of what lay ahead, as the party disembarked frequently to peruse various sites. They saw where the Central Pacific depots and turntable were to be constructed, at Newcastle, which at 970 feet was some 800 feet higher in altitude than Roseville and more than

900 feet higher than Sacramento. Truly, Upson wrote, "the labor of ascending the mountains is fairly begun." His eye took in the steep grade being smoothed out by gangs of laborers, and, just past Newcastle, looked on as workers attempted to scratch a sixty-foot-deep trough, the exact shape of a modest wedge of pie balanced on its apex, through the solid, cement-hard indurated gravel of the Bloomer Divide. "Bloomer Cut" would soon extend to more than eight hundred feet long, requiring some five hundred kegs of black blasting powder each day. It was backbreaking work for the laborers hauling away rock debris in wheelbarrows and one-horse carts, of course, and figuratively as difficult to the Central Pacific treasury. Yet the editor was convinced, he told his readers, that the completed passenger ascent would be as smooth and comfortable as that he had found on the initial sixteen miles: "a person could read almost as comfortably," he said, "as in a rocking chair at home."[22]

Would that construction were actually going as smoothly. As Upson composed his rhapsody on the return trip to Sacramento, Stanford and Hopkins were likely less cheerful, despite the high spirits of their excursionists. At mile 18 at Roseville, the railroad's initial construction contract—with hale Charley Crocker—was up. Owing to the hue and cry against Crocker's "inside spot" orchestrated by Theodore Judah in the summer of 1863, the line from Roseville past Newcastle, miles 19 through 31, had been let to a handful of outside contractors: two mile-long sections to Cyrus Collins & Brothers, four sections to Messrs. Turton, Knox & Ryan; three to C. D. Bates; two to S. D. Smith, and the final two miles to Crocker. Of these, Collins's sections 19 and 20 had to be taken over (by Crocker) when the contractor went bankrupt; results on the other sections, moreover, neared chaos. The other independents continually squabbled with one another and vied in poaching their competitors' work gangs, bidding the price of labor ever upward. "The men on the upper section would go to Sacramento to get men for the work," Huntington complained some years later, "but it was pleasanter on the lower section and contractors there would scoop those men in as they were going on to the work on the other section." Added to this was the problem of overall timing and allocation of resources. "We wanted the work done as nearly as possible in consecutive order," recalled Stanford. "We often found, however, that the contractor the farthest off would finish his work first. The others would be slow, and there was constant trouble from this source."[23]

Even so, progress was measurable. Exactly four weeks later, on March 19, rails were spiked by Crocker's gangs some distance past the station in Roseville, and once more Lauren Upson swung aboard the train in Sacramento for an excursion among the bigwigs. Almost two-thirds of the state legislature and their families squeezed into two just-completed passenger cars, each with a capacity for sixty and both painted a merry hue of yellow, or onto seven platform cars outfitted with benches. Banners and patriotic bunting were draped on the *Governor Stanford,* and as it chugged out of Sacramento a ten-piece band seated just behind the locomotive on a platform car struck up the dependable old chestnut "Wait for the Wagon," and throngs of excursionists waved handkerchiefs and small flags at the unlucky ones left behind on the platform.

"Everyone was pleased with the smooth and steady motion of the train," wrote Upson, "which showed how well the new track had been laid." Nearing the end of the twenty-two-mile journey, he continued, "the excursionists soon penetrated among beautifully rounded foothills, gemmed with groups of green trees—oaks, scrubby pines and buckeye clumps—with rocks picturesquely arranged in small cliffs, chasms, and grottoes, while through the little valley brooklets meandered, albeit turbid with the inevitable reddish yellow soil of California." The railhead proved to be alongside some new granite quarries, and as soon as the train stopped the cars emptied. "The children," Upson noted, "of whom there were enough for a small regiment, scampered and tumbled about, wild with delight, gathering bouquets of wild flowers, only restrained by constant warnings against poison oak; matrons and maidens wandered off amongst trees and rocky knolls, according to their 'own sweet will'; while grave legislators and solid men generally gathered around the quarry in groups, conversing learnedly and geologically." Then, the editor reported, "it was discovered that a large stock of baskets had been piled upon the turf, each of which contained a dozen bottles with something in them. Also, that there was a bountiful bread and cheese accompaniment." Band music accompanied the victuals, but serious discussion of an impromptu "dance on the green" was ended by a warning locomotive whistle and the sight of squall clouds bearing down on them from the Sierra.

The expedition "triumphantly ran away" from the showers, and the *Governor Stanford* pulled them all back into Sacramento in an excess of merriment and self-congratulation.[24]

Four days after the excursion, the *Sacramento Union* reported that two new tank locomotives, the *C. P. Huntington* and the *T. D. Judah*, had arrived, unassembled, in San Francisco on the ship *Mary Robinson*, products of the firm of Danforth, Cooke of Patterson, New Jersey; also, that the ship *Success* had moored with a supply of rails, and fifteen boxcars, thirty-five platform cars, and two passenger cars, all knocked down but ready for assembly.[25] The day after this report, on March 25, the Central Pacific unveiled its second operating locomotive, *Pacific*, manufactured in Taunton, Massachusetts, by the William Mason Company, and at 47½ tons slightly larger and stronger than the *Governor Stanford*. Joining them in operation during the second and third weeks of April were the two little tank engines, each when fully loaded with water and wood some eighteen tons.

Crowning this flurry of activity were the first revenue trains of the Central Pacific. Three carloads of light blue granite from the trackside digs at Rocklin had rolled down the twenty-two miles to Sacramento on March 25.

The high quality of the stone and its accessibility promised to defy competition; within months, state commissioners would contract for all the granite required to complete the state capitol building. And not only freight began to roll. On April 25, regular passenger service to Roseville commenced—thrice daily in each direction, with connections to the California Central. At the end of the first week the Central Pacific had carried 298 passengers in those bright yellow cars, earning all of $354.25. It was not much to crow about, but it was minted money, and there was more where that came from.[26]

And it was direly needed. Against those three carloads of granite and those promised to come there towered the costs of the imported equipment—nearly $14,000 for the *Governor Stanford,* for instance, and more than $15,000 for the *Pacific.* Against three hundred passengers clutching tickets there was a veritable mountain of wooden kegs—containing the black powder necessary for blasting the way through Bloomer Cut—five hundred kegs per day at up to $6 per keg. Every piece of rolling stock, every rail and tie, every sweating laborer's pickax stroke diminished the Central Pacific's treasury, increased its obligations, and focused the eyes of the Associates on their partner Collis Huntington and, more important, on what was occurring that spring, not very far away from him, in Washington City.

His first teacher had been Theodore Judah. Two years earlier, the engineer had showed him the rudiments of legislation and introduced him around. Now Huntington's instructor was entirely more formidable—Mr. Union Pacific himself, his tools a squad of influence peddlers, his lavish presence at the Willard Hotel, and his "suspense account" of cash and securities. For a part of the late winter and spring—up to six weeks—Dr. Durant was confined to his hotel room with an undisclosed illness, while his agents Samuel Hallett and Joseph Stewart carried out his bidding. Durant asked for no vouchers or accounting of specific amounts, telling them to get the job done and not bother him with petty details. Huntington paid close attention to what followed.[27]

Two bills amending the 1862 Pacific Railroad Act inched along that spring. In the House, a freshman congressman from Iowa, Hiram Price—a Lincoln stalwart who was pledged to advance the cause of Iowa railroads for Henry Farnam and Thomas Durant, and who had joined the small chorus of midwestern politicians calling for a military promotion for Grenville Dodge—introduced a bill on March 16. It was handed to the House Select Railroad Committee, where resided a number of close friends. Thaddeus Stevens, the craggy old radical Republican from Pennsylvania, whose iron foundry would not be participating in the Pacific railroad owing to its recusal by Confederate guerrillas who had torched it, was not only chair of the Select Railroad Committee, but also of the powerful Ways and Means Committee; he fervently believed in the urgency of speeding the project along as a war measure.

By Stevens's side within the committee, and at a slightly lower end of the political spectrum in terms of experience and influence, were two freshman congressmen, well-meaning and energetic men: Cornelius Cole of California, and Oakes Ames of Massachusetts. The former district attorney of Sacramento County, the forty-one-year-old Cornelius Cole had been in on the creation of the Republican Party in California and also of the Central Pacific Railroad; when Theodore Judah had called his open meeting for potential investors at the St. Charles Hotel on K Street in November 1860, Cole had been there and bought a few shares. Now, obliged by commendable (and rare) ethics to sell his modest investment, Cole could be a dependable and outspoken ally. He also held the confidence of the secretary of state, having clerked in William Seward's New York law office twenty years earlier, when partners still called that imperious reformer governor but not yet senator.

Oakes Ames, "King of Spades," the broad-shouldered, prosperous sixty-year-old manufacturer of Old Colony shovels—purveyor to the California gold rush and the Washoe and most past, present, and future construction on the Pacific Rim—likewise had acquired an interest in the Central Pacific when old Huntington sought him out for a loan for nearly $200,000 in 1862. Ames, a founder of the Massachusetts Republican Party, had first become actively interested in railroads in 1855, when he joined several other New England investors in a land speculation in Iowa, anticipating the arrival of the Chicago, Iowa and Nebraska Railroad from its starting point on the Mississippi River. Ames became a principal stockholder and director of the line in 1856, as it continued westerly across the middle of the state toward the Missouri River; succeeded by the Cedar Rapids and Missouri, it was of course the major competitor against Farnam's and Durant's Mississippi and Missouri in the trans-Iowa sweepstakes with the looming prize of the Pacific railroad beckoning them on. Earlier in 1864, Ames had briefly found himself in the curious position of owning Cedar Rapids stock alongside his fast-trading nemesis, Dr. Durant, who then attempted, in vain, to merge the two rival companies at Des Moines. The Doctor's sleight-of-hand and business audacity would be only a taste of what was in store for Oakes Ames—although, as a successful businessman at the dawn of the Gilded Age, Ames understood how diversification and swift-footedness were the keys to immense wealth. And, like politics, business made strange bedfellows. In 1862, answering the legislative call for domestically made materials for the Pacific railroad, the Spade King had bought a third-interest in the Lackawanna Steel Corporation, whose premier product would be rails.[28] But now, assuredly, pure civic duty beckoned. Having answered the will of the people in his Massachusetts district, Railroad Congressman Ames took his seat in the exact House committee that would affect the future of the Union Pacific and the Central Pacific—and with them the trans-Iowa railroads, the Old Colony shovels, the Lackawanna rails, and, as it happened, the fate and the reputation of Oakes Ames himself.

A month after the select committee bent over the bill it emerged, not surprisingly, as a railroad man's dream. It allowed the Union Pacific and Central Pacific companies to issue their own first-mortgage bonds equal in value and interest as well as in length of term—6 percent, for thirty years—to what was procured from the federal government, which now was relegated to the less desirable position of second lien-holder, with no prospect of payment for those three decades. The grant of public lands to the companies was doubled, made even more attractive by the grant of coal and iron rights from those 12,800 acres per mile. As the companies graded each twenty-mile section, they could apply for two-thirds of the government bonds before even laying a tie or spiking a rail, regardless of whether the sections were continuous; this would speed their work enormously, allowing them to employ multiple labor gangs at once. Dr. Durant, especially, would have been pleased to see that the limitations on individual stock ownership were rescinded; both Durant and Huntington were gratified to see that the government would no longer be allowed to withhold a portion of its due bonds until the line was completed.

The House version of the railroad bill sat, pending discussion. Elsewhere in the city, military hospitals began to swell with the latest wave of casualties from the swampy, blood-soaked thickets below the Rapidan River in Virginia. The Army of the Potomac, under the command of the newly promoted Lieutenant General Ulysses S. Grant, had fought there over two days, May 5 and May 6, to a strategic draw against General Robert E. Lee's Army of Northern Virginia. Lee's rebels lost about 7,500 men in the Battle of the Wilderness, while the Union forces suffered some 17,500 casualties. Instead of retreating, however, Grant turned south toward Richmond, doggedly attempting to pound his way past the Northern Virginians' flank. Almost immediately the forces clashed once again at the little crossroads village of Spotsylvania Court House. Over the ensuing twelve days of vicious trench warfare, some twelve thousand Federals fell. "We have met a man, this time," remarked a Southerner of Grant, "who either does not know when he is whipped, or who cares not if he loses his whole army." Cold Harbor and Petersburg awaited the survivors.

Meanwhile, back in Washington, while the battle of Spotsylvania still raged, the Senate's standing committee on the Pacific railroad reported its bill to the Senate floor on May 18. The document was less generous than its companion in the House, although still an improvement over the 1862 act. The railroad companies would be permitted to float their own 6 percent, thirty-year bonds at three different levels, depending on what terrain they happened to encounter: on the low fields and prairies, $24,000 per mile; on the high plains, $48,000 per mile; in the mountains, $96,000 per mile. Though in this form of the legislation, the government would not issue its own bonds, it would guarantee interest on the company paper for twenty years, actually paying the first year's interest outright as a gift. And the federal government would in return be given a deed of trust to the full length of the Pacific railroad, in case of default—although the power of federal seizure, should the system fail to be completed by July 1, 1876, was deleted from the bill.

Wording of another particular in the Senate bill, however, prompted a short, intense, behind-the-scenes struggle between Collis Huntington and Thomas Durant: the Central Pacific was to build only to the boundary line between California and Nevada, hardly out of the shadow of the Sierra Nevada, with only a line over the worst geography in North America to show for it. Nearly apoplectic, Huntington confronted the Doctor—and one can imagine the scene whether it occurred in a corridor outside the Senate chamber or back at the Willard Hotel: brawny, gesticulating Huntington dashing his hat to the floor; calm, smirking, mandarinlike Durant rubbing his hands; Huntington shouting, "How dare you try to hog all the continent?" and Durant smoothly replying, "Well, how much do *you* want?" And Huntington, probably caught up short by the Doctor's parry as well as his self-assurance, with hardly any money or maneuvering room of his own, could only think about Nevada mines and all the trade they would generate, and so instead of throwing Durant's outrageousness back in his face by demanding rights all the way into Wyoming, he replied, "Give me Nevada," and settled for 150 miles east of the California border, barely halfway across the state in the Reese River valley.[29] Twenty-five years later, as an immensely wealthy and powerful public man, Huntington could afford to boast that he had scoffed, even, at the 150-mile limitation in the law: "I said to Mr. Union Pacific, when I saw it," he would claim, "I would take that out as soon as I wanted it out."[30] Nonetheless, even if he actually said these words to Durant, his bravado carried no force.

Shortly after this little backroom deal, on May 23, the Democratic senator from San Francisco, James A. McDougall, rose to inform his colleagues of the alteration. "The representatives of the Union Pacific as well as of the Central Pacific Company concur in this amendment," he said, and with very little further discussion, the bill was passed, 23 to 5.[31]

Not for another three weeks did the House version of the amendment come up for discussion. By the time it did, on June 16, General Grant had been badly repulsed during a horrifying frontal assault at the crossroads town of Cold Harbor, and he had already begun to lose again at the Confederate railway hub at Petersburg. The warlike verbal atmosphere in the Capitol building in some ways mirrored conditions southward. Representative E. B. Washburne of Galena, Illinois, a fifty-seven-year-old radical Republican who worked closely with Thaddeus Stevens and sometimes had the ear of the president—remembered especially on that day as the congressional sponsor of his hometown associate, Ulysses S. Grant, but also widely feared for his wasplike protection of the public treasury and any legislative raids upon it—led a stinging assault on the railroad bill. The House measure was, he would say later, "the greatest legislative crime in history."[32]

William Holman, Indiana congressman, echoed this theme: the legislation may have been advertised as a war measure, he said, with the heartfelt support of President Lincoln, but the corporations were only out for themselves—"the patriotism of this thing," he intoned, "does not weigh a feather in the estimation of these people." Others spoke. Debate surged for five days until the evening of June 21, when Washburne glared out at the galleries "packed with people interested in the measure, by lobbyists, male and female, and by shysters and adventurers hoping for something to 'turn up.'" The bill, he said, was "the most monstrous and flagrant attempt to overreach the government and the people that can be found in all the legislative annals of the country." Perhaps Washburne's lamplike eyes rested on the smooth Dr. Durant and his lieutenants, up in the House gallery, as he thundered that the entire Pacific railroad had been snatched up by "Wall Street stock jobbers who are using this great engine for their own private means."[33] These objections, answered by Railroad Congressman Hiram Price of Iowa (there was no more of an immediate burden upon the Treasury than with the original bill of 1862, he claimed) and by Washburne's erstwhile ally, Thaddeus Stevens (who called Washburne "my friend, the detective," zealous finder of "hobgoblins" in the bill), began to lose their force.[34]

Though Washburne threatened to move for postponement until the next session, his support steadily eroded over the next week of debate. Durant and his squad of agents—John B. Stewart and George Francis Train among them—kept themselves busy, fanning out bonds and cash to the elected and their relatives and stand-ins. Collis Huntington, too, did what he could with his limited means.[35] And on June 29, the matter was temporarily transferred off the House floor when, in an effort to reconcile the wildly different bills of the House and Senate, a joint committee was formed. Thaddeus Stevens, James Harlan, and California's John Conness and Cornelius Cole, railroad experts all, dominated the committee, tipping the joint bill overwhelmingly toward the more generous House version. Two amendments slid over from the Senate: the government's presence within the Union Pacific's board of directors would be increased from two to five, and it was further ruled that no significant management decision could be made without at least one federal director present. On July 1, 1864, the amended bill passed in the Senate and was quickly paraded past weary House members. Despite the jack-in-the-box objections of Congressman Washburne, Stevens pushed it out of the chamber with a vote of 80 to 13. President Lincoln—renominated four weeks earlier by the Republican Party and facing a wartime election in the fall—signed the bill the next day, and it was law.

15

"Trustees of the Bounty of Congress"

As news of the bills' passage was telegraphed to Omaha, a worried and distracted Peter Dey could hardly pay it heed. His four-year-old son, Anthony, had contracted scarlet fever during one of the city's epidemics, and for days the engineer and his wife, Kate, kept a constant vigil in the boy's room. On the afternoon of July 4, 1864, their doctor gave him up, and he died that evening. The Deys buried the boy in Buffalo, New York, on July 11. Upon their return, Peter Dey attempted to focus on Union Pacific matters.

It was a frustration, pure and simple. Even while Dey had been huddled over his feverish little boy, none other than George Francis Train, the apotheosis of talk, the incarnation of gab, had appeared on his doorstep during a whirlwind junket to raise interest across Nebraska in his candidacy for senator. Train strutted around Omaha, taking credit for events in which everyone knew he had no part, intimating that Nebraska's sole hope of seeing a completed railroad track rested on his political fortunes, boasting of the half-million dollars he'd made already on the Kansas line, and even offering a bribe of twenty shares of stock in the Crédit Mobilier to a powerful Republican newspaper editor. The editor spurned it and vowed to ruin Train's prospects. He was not alone. "I could do nothing for Mr. Train," Dey wrote to his backer, Dr. Durant. "A moderate amount of discretion would have made him very popular in this territory. . . . [W]hen he asks his friends to exert themselves for him he should not furnish others clubs to fight them with."[1]

Even more difficult were Dey's prospects along the proposed railroad line, the first forty miles of which he needed to grade. His chief headache was the scarcity of labor. On April 2, 1864, he wrote to Durant with the warning that there was trouble ahead and he had better send some reliable contractors. This produced nothing in six weeks. "It is impossible," he wrote Durant on May 14, "to do anything in the way of letting this work now without some provision for furnishing men. I had on my file of letters in February quite a number of applications for contracting. I have telegraphed and written every responsible contractor that we were prepared to receive proposals for this work. The uniform response has been that they were with the present scarcity of men afraid to take the work at any price to be finished in a specified time. I do not know what to do—the question with me has been whether it would not be policy to import a thousand men or more. We can work them to advantage for a long while."[2] In response, Durant and his agents in Washington attempted to persuade the War Department to ship brigades of freed slaves from their southern contraband camps to the Nebraska plains, but to no avail. One contractor wheedled by Dey into service struggled for a month over two difficult mile-long sections just outside the city limits, and begged out of the contract, remaining on the job only after Dey pledged to advance his expenses himself. Another contractor by the name of Carmichael put some 250 laborers to work, including fifteen female members of the Omaha tribe, but their progress was dismayingly slow in the hilly plateaus west of Omaha. As the summer's heat set in it became apparent that even paying the expenses of workers out to Nebraska from the East was doing little good. "Every [wagon] train for Bannack," from which Montana settlement news had come of gold strikes, "takes from the work from ten to twenty men to drive the teams," Dey wrote Durant on June 20. He proposed buying "a lot of single and double wheeled scrapers and run them with horses and cattle—one hundred of these scrapers will do the work of five hundred men."[3] Even if such a solution had been practical under the circumstances, it was not acted on.

Equally difficult was the search for adequate hardwood timber for railroad ties. Dey purchased nearly three hundred acres of hardwood far up the Missouri River, from which the Union Pacific's newly designated superintendent of ties and timber, M. B. Sprague, succeeded in cutting nearly twenty thousand ties and twelve hundred cords of wood with a workforce imported from Illinois. There weren't many other stands available, and it soon looked to Dey as if they would be relying, at least for the moment, on soft cottonwood. Nonetheless, as men drifted away to other regions, Sprague was hampered by his own worker shortages. Durant telegraphed the suggestion of importing loggers from Canada, but Dey replied on June 20 that Canadians were "the hardest set to manage that I have encountered."

Remarkably, throughout much of the spring, the chief engineer had yet to receive a paycheck from Dr. Durant despite several exchanges about the salaries of his division engineers ($1,800 annually); first and second assistants ($1,200 and $900, respectively); rodmen, chainmen, and axemen ($50, $45, and $40 per month), said amounts being part of the problem in securing employees. "I have had no information as to what my own compensation is to be as yet," he wrote on April 9, "and feel some solicitude to know. As my expenses are large it becomes a matter of some moment with me."[4] After nearly three weeks went by with no answer, Dey wrote again. "For the last four years my income from my services has been considerably less than my living expenses," he pleaded, "and I can not feel satisfied to stay here without earning enough to pay my way and something in addition."[5] Finally, the board of directors set Dey's salary at $7,500.

As Dey fretted over personal finances, timber, labor, and the cares of the Union Pacific along the right-of-way leading out of Omaha, he at least knew he could rely on the four location engineers he had hired to determine the final line all the way into the valley of the Great Salt Lake, men with whom he had worked on the Iowa and Illinois lines. Ogden Edwards was instructed to locate the line west of Omaha for the first hundred miles, joining and following the sluggish Platte. Francis M. Case was sent out with his small party to Colorado Territory, where he was to examine various passes in the Front Range between the Cache la Poudre and the Arkansas Rivers in the slim hope that the Union Pacific might run through Denver. James A. Evans's group was dispatched into the high country of southern Wyoming Territory, to find a line through the Rockies to the Green River. And Samuel Benedict Reed was put in charge of the terrain from the Green River of southwestern Wyoming across the Uinta Mountains and the Wasatch Range in Utah Territory to the Great Salt Lake.

Reed was a Vermont native who had grown up in western New York, where as a young man he labored briefly as a rodman on the Erie Canal enlargement before moving west to work on the Detroit and Pontiac Railroad and the Michigan Central Railroad, followed by the Michigan Southern, the Rock Island, and the Burlington and Missouri; thus, he had a role in building the first railroad into Chicago, the first from Chicago to the Mississippi, and what was deemed the first important bridge over the Mississippi. He was forty-five when he joined the Union Pacific as division engineer with the awesome task of spanning the western high country.[6] Owing to the number of people heading west, Reed was stuck in Omaha for nearly all of April while he waited for a seat on a stagecoach. With Dey worrying about the waste of clear surveying weather, Reed and an assistant went down to Atchison by boat and found a coach on May 7. The dusty, bone-jarring ride out to Salt Lake City consumed two weeks, but when they arrived in that Mormon settlement there was the friendly reception of Brigham Young himself, shareholder in the Union Pacific and booster of his verdant paradise-on-earth as a railroad destination, who furnished Reed with fifteen men, three wagon teams, equipment, and provisions. When Reed was short on cash on payday, with the wired funds delayed in transit, Young induced the men to remain at work until the draft arrived. Then they set off. "All told," Reed wrote to his wife, Jenny, back on their farm in Joliet, "teamsters, cooks, engineering party and night guard, we are seventeen. The big dog makes eighteen."[7]

After a few days of training, Reed turned his party up the Great Salt Lake shore for forty miles to the place where, in a short time, the city of Ogden would be staked out; just to the east was the mouth of the Weber River—its canyon a deep gouge into the wild, towering Wasatch Mountains. Reed carried with him a letter from Brigham Young ordering any Mormon settlers they encountered to sell them provisions, and, as they began their ascent along Weber Canyon, he wrote home that "we are in no danger of famine as long as we are in the settlements." They were well supplied with fresh mutton, eggs, milk, and butter, and hooked many large trout in the clear mountain streams. Reed exulted in the outdoor life, sleeping on buffalo and beaver robes. "The scenery is magnificent," he wrote, "mountains composed of granite and gneiss towering four to five thousand feet almost perpendicular above us. The deep narrow gorge in which the river runs is only about 300 feet wide and is the wildest place you can imagine."[8]

Although they were well into June, as they pushed higher up Weber Canyon, past Devil's Gate, nights grew bitterly cold—on June 18, Reed woke up stiffly to find a heavy frost over everything, "ice as thick as a window glass in the dishes standing out last night."[9] Intersecting streams swollen by snowmelt hampered their progress, but after three weeks they neared the Mormon settlement of Echo, once a Pony Express stop, where Echo Creek tumbled down through its own canyon from the northeast, marking the continuation of the old Emigrant Trail. Here they turned up the tributary canyon and thereby crossed over the spine of the Wasatch to Sulphur Creek, over the Wyoming border. As a railroad line, his notes revealed, it seemed almost too perfect to believe. To make sure it was the best, Reed backtracked to try the next eastward pass from Weber—along Chalk Creek to the Bear River—but that only proved Echo Canyon's superiority. Before returning to Salt Lake City, Reed laid down lines in the dry, mountainous, sagebrushy southwestern corner of Wyoming Territory, in the Green River region, near the close of James Evans's tangent; he presumed Evans was progressing westward across the Continental Divide.

James Evans, in fact, was making progress, though with no little trepidation, even with the proximity of a thickly armed cavalry escort. There was trouble afoot—Indian trouble—that would continue to shadow all of their efforts for years to come. Ironically, his worries and those of untold thousands could, in that pivotal year of 1864, be traced to another Evans (no relation)—John Evans of Evanston, Illinois, an incorporator and director of the Union Pacific, currently serving as appointive governor of Colorado Territory.

At age fifty, Governor John Evans had amassed his wealth in land and railroad speculation in Illinois and Iowa before being sent by President Lincoln to Denver in May 1862 as the second territorial governor, to replace an incompetent. Governor Evans viewed the post as a way to increase his fortune—Colorado was on its way up, with hardly tapped mineral riches and a spectacular, barely settled terrain—and his political power, once the territory, created by Congress in 1861, achieved statehood. When that occurred, Evans had his eye on the U.S. Senate, and as a prominent Republican, he would likely attain a seat. Denver, too, had a future in Evans's view, as a part of the Pacific railroad and as the principal transportation hub of the West. At the Chicago organizing convention of the Union Pacific, new Governor Evans looked down on all his fellow incorporators—knowing, probably, all of them by first name—and touted Denver and the Rockies' Berthoud Pass as the only sensible route across the Continental Divide.[10] By early 1864, Evans assessed the political climate in the West and the political fortunes of Abraham Lincoln, about to seek his second term, and saw that statehood could be an issue in the coming election; indeed, the president's supporters in Congress might fight for admittance of a potentially wealthy territory before November. (Nevada, with its immense silver strikes, was also a candidate.) But the governor needed to attract the attention of Lincoln and Congress, and since their focus was the war raging far from the Territory of Colorado, Evans set out to manufacture a serious threat to his peaceable kingdom—indeed, a war—and, in his quest for more wealth and more political power, he got his war.

The Cheyenne and the Arapaho inhabited the high plains and foothills between the Platte and the Arkansas Rivers, through which all emigrant trails crossed. Generations before, both tribes had been displaced by other tribes and by whites east of the Missouri River and both had become mounted hunters of the prolific buffalo herds in the high country. In 1835, the Cheyenne and Arapaho, who coexisted amicably, had each split into two tribes, some locating in the general region of the North Platte River (present-day eastern Wyoming and western Nebraska) and others, being more inclined to coexistence with the whites, in eastern Colorado Territory near Bent's Fort, on the Arkansas River. A treaty council at Fort Laramie in 1851 reinforced the boundaries of these hunting grounds. Few whites had ventured into the large oval between the Rockies and the 101st meridian, and the North Platte and the Arkansas Rivers—until the Colorado gold rush in 1858 magnetized the region, planted white settlements, often by force, scratched more emigrant trails across hunting grounds, drove the buffalo eastward, and prompted innumerable (and inevitable) small clashes. Trouble was particularly bitter in Colorado—most white settlers viewed Indians as subhuman candidates for extinction, and the feeling was mutual among the young firebrand tribesmen. But peace prevailed for the most part, and by 1861, what had been taken by force was ratified by treaty, at Fort Wise, on the Arkansas River. The Southern Cheyenne and Arapaho—or, at least, the few chiefs who could be persuaded to sign—agreed to exchange their extensive hunting lands for an arid and desolate little triangle of sand between the Arkansas River and Big Sandy Creek some two hundred miles southeast of Denver. They were allowed to leave their reservation to hunt, and indeed did not spend much time there. Remarkably, despite continuing provocations by settlers, the fiercely proud warriors refrained from striking back.[11] When there was serious trouble in the West, it was elsewhere, inspired by broken treaties—in Minnesota and the eastern part of South Dakota with the Sioux uprising of 1862, and, farther west, in Utah, Nevada, and Idaho Territories, with the Ute, Shoshoni, and Paiute.

Nevertheless, Governor Evans's plan was to force not only the recalcitrant Southern Cheyenne and Arapaho but their fiercely militant brothers, the Northern Cheyenne, into the 1861 treaty reservation. In the long run they could be eliminated. Thus, white emigration to Colorado could rise, increasing Evans's own political base. The Northern Cheyenne refused to even discuss the move, and for a time it seemed as if the Commissioner of Indian Affairs, William P. Dole, would support them. Colorado settlers sent up a hue and cry, and pressure applied against Dole's superior, Interior Secretary John Palmer Usher (betrayer of his former legal clients, the Delaware and Potawatomi, on behalf of the Leavenworth, Pawnee, and Western Railroad),[12] not surprisingly reversed Dole's position. With none of the chiefs interested in meeting with him, Evans felt encouraged to use force. He was hampered, however, by a lack of troops. The War Department refused to return any of the Colorado regiments sent east to fight Confederates, ignoring both Evans's constant entreaties. They escalated in intensity, finally predicting an Indian war in the spring; and there were like-minded calls of the newly installed commander of the Colorado Military District, Colonel John M. Chivington, a fire-breathing giant and former Methodist preacher.[13] Chivington's hope, come statehood, was in the House of Representatives.

Hunger helped to precipitate Evans's and Chivington's war. In the winter of 1863 the Cheyenne and Arapaho were badly racked by malnutrition and disease. No buffalo could be found within two hundred miles. By early 1864, with many Indians dying, several wagon trains were attacked and stripped of provisions, the whites being left for the most part unharmed. Local press reports inflated the incidents shamelessly, but the defining moment came in late March, when Evans was told by the army that all troops formerly at his disposal would be shifted to Kansas to fortify against an anticipated Confederate onslaught.

Immediately, the governor detected a pattern of Indian uprising east of Denver. The reports had to do with scattered, small-scale instances of rustling horses and cattle, with actual Indian complicity never substantiated (though it was certainly possible, given their desperation). Uprising detected, war commenced. In April, Colonel Chivington dispatched cavalry units with orders to "kill Cheyennes wherever and whenever found."[14] One detachment of the First Colorado galloped into a Cheyenne encampment in a canyon of the Cedar Creek, a feeder stream of the South Platte. "We commenced shooting," reported Major Jacob Downing. "I ordered the men to commence killing them. . . . They lost . . . some twenty-six killed and thirty wounded. . . . I burnt up their lodges and everything I could get hold of. . . . We captured about one hundred head of stock, which was distributed among the boys."[15] Loot thereupon became an added inducement to the troops. Another unit under the command of Lieutenant George S. Eayre drove eastward across the state line into Kansas and was reinforced by an artillery battery. Encountering several deserted Indian villages, the soldiers plundered and then destroyed everything left.

Another Kansas expedition under Lieutenant Eayre, on May 16, northwest of the Arkansas River, did not find the Cheyenne village empty. Chief Lean Bear, a peaceable man, put on the medal given to him by President Lincoln the previous year in Washington and, as he rode out with his son to greet the approaching soldiers, brandished papers attesting to his cooperativeness. When the chief and his son got close enough, the cavalrymen blew them off their saddles. In an instant the village men were swarming out against the Coloradans, killing four and wounding three. Another chief, Black Kettle—an important, peaceable leader—succeeded in halting the warriors as Lieutenant Eayre's men retreated to Fort Larned, on the Pawnee River, where they telegraphed news of their victory.[16]

"We are at war with a powerful combination of Indian tribes pledged to sustain each other and drive white people from the country," reported Governor Evans in an appeal to General Samuel R. Curtis, commander of the Department of Kansas at Fort Leavenworth. He made no mention that it was the whites who had started the war. An investigating officer sent by Curtis to Colorado found no general uprising, but added that "great caution" would have to be exerted to avoid widespread bloodshed. "It should be our policy to . . . stop these scouting parties that are roaming over the country," the officer telegraphed General Curtis, "who do not know one tribe from another and who will kill anything in the shape of an Indian."[17]

His words went unheeded. Meanwhile, a trader named William Bent, who had great and long-standing sympathies for the Cheyenne people—he had taken a Cheyenne wife and sired two sons—appeared before Colonel Chivington at Fort Lyon with a peace message from Black Kettle, who said he represented the majority of Cheyenne. But Chivington curtly refused to meet the chief. War, he said, would continue. Bent reminded him that many white settlers and emigrants would perish if the conflict widened. The settlers, Chivington replied, would have to protect themselves. Things quickly escalated. On June 11, a personal dispute between four young Arapaho warriors and a rancher named Hungate at Box Elder Creek, only twenty miles east of Denver, erupted in murder; the Arapaho killed Hungate, his wife, and two small children, scalping and mutilating them. Their corpses were put on display in Denver, raising a hysterical public outcry for vengeance against all Indians. "Indian hostilities on our settlements commenced. One settlement devastated 25 miles east of here," Governor Evans wired Washington, "murdered and scalped bodies brought in today. Our troops near all gone."

Ten days later after this report, Evans issued a proclamation. All Plains Indians were ordered to register at the nearest fort, encamping there, lest they be "killed through mistake": the northern Cheyenne and Arapaho were ordered to northeast Colorado, to Camp Collins; the southern Cheyenne and Arapaho were assigned to Fort Lyon; the Sioux, to Fort Laramie; and the Comanche and Kiowa, whose hunting grounds were south of the Arkansas River, were ordered to proceed to central Kansas and Fort Larned. Freedom and treaty territories were to be exchanged for "food and protection" and, it was underscored, their lives.

Trader Bent continued to press for peace but it was too late, for the governor's war was finally fact as a series of monumentally stupid and callous blunders ignited Indian rage. Outside Fort Larned, the Kiowa chief Satana, formerly a friendly man, was threatened by a sentry and shot him; Satana then made off with all of the post's grazing horses. A Southern Arapaho chief, Left Hand, still desiring peace with the whites, took a small party to Fort Larned with the offer of helping to recover the horses, but a drunken officer fired an artillery shell over their heads before they got near enough to be heard. Directly north, at Fort Kearney on the Platte River in Nebraska, the commander, Brigadier General Robert B. Mitchell, bereft of judgment (his few soldiers were posted along the thin line of the emigrant trail some eight hundred miles long between Kearny and South Pass in the Rockies—giving him no strength to boast of) and overloaded with racial pride, called in a council of various Sioux leaders, including the cooperative chiefs Spotted Tail and Bad Wound of the Brulé Sioux. General Mitchell lectured them about behaving themselves—they were to stay away from the Platte Valley all summer—or he would "station a soldier to every blade of grass from the Missouri River to the Rocky Mountains."[18] Everyone present knew it was an empty threat.

Not ready to play the part of chastised children, or to give up their treaty rights, or to die, the newly converted enemies of the whites joined their militant brothers and began to strike back. General Chivington's words—"the settlers will have to protect themselves"—could have been emblazoned in the sky above the Platte and Arkansas Valleys and the Continental Divide in July and August of 1864. Some two hundred whites—the majority of them civilians—would die that summer. Along the Platte Valley, Sioux and Northern Cheyenne and Arapaho picked off stagecoaches and wagon trains and raided isolated coach stations, settlements, and ranches, killing, scalping, mutilating, looting, and burning, attacking with skill and stealth and vanishing before troops could respond. The spread-thin, undermanned army could not keep up with it all. Southward, not just along the Arkansas River but in many places across Kansas, war parties of Kiowa, Comanche, Southern Cheyenne, and Arapaho, struck similarly and vanished. General Curtis opened new posts and fanned his regiments outward from Fort Larned, finding no Indians but finaly reopening the Santa Fe Trail and the branch road to Denver up the Arkansas. Denverites, threatened by food shortages but not by hostile parties, tightened their belts and responded hysterically to every atrocity report.[19]

Governor Evans fanned the hysteria, doggedly seeking more attention from Washington while mounting his own statehood campaign at home. "The conflict is upon us," he proclaimed to Coloradans. He authorized "all citizens of Colorado, either individually or in such parties as they may organize, to go in pursuit of all hostile Indians on the plains, scrupulously avoiding those who have responded to my call to rendezvous at the points indicated; also to kill and destroy as enemies of the country wherever they may be found, all such hostile Indians."[20] His messages to Washington went out every few days. "The alliance of Indians . . . is now undoubted," he telegraphed War Secretary Edwin M. Stanton on August 10. "A large force, say 10,000 troops, will be necessary to defend the lines and put down hostilities. Unless they can be sent at once we will be cut off and destroyed." He fired off another wire the following week: "We are in danger of destruction both from attacks by Indians and starvation. It is impossible to exaggerate our danger."[21] Meanwhile, in southeastern Nebraska, there was uninflated peril as Cheyenne raids in the vicinity of Little Blue River left several dozen whites dead and two white women and five children taken hostage. A full-scale search failed to find them.

During the summer Chief Black Kettle of the Southern Cheyenne, and other elders, had attempted to halt the younger warriors' raids, to no avail. When news almost simultaneously arrived of Governor Evans's edict against loosely defined "hostiles" and of raiders among his tribe who had brought in the white hostages to his camp, Black Kettle resolved to restore control. Ransoming the four youngest hostages and securing safety for the others, he held a council in which the chiefs authorized contacting the whites for a prisoner exchange and peace. On September 12 an extremely tentative and uneasy talk was held between Black Kettle's group and the commander of Fort Lyon, Major Edward W. Wynkoop, at the Southern Cheyenne camp on the Smoky Hill River in Kansas. Wynkoop, an honorable man, was persuaded of their good intentions. He promised to escort them under the safety of the American flag to Denver.

Seven chiefs, including Black Kettle, were finally ushered into the Colorado capital on September 28, with their white flags and concessions to confront Governor Evans with the ugly prospect of peace. The governor's statehood referendum had gone down in defeat only weeks before. It had been killed as much for the local campaign of accusations that he had fomented this Indian war for his own political purposes as it had to keep Coloradans out of the draft—but the referendum's defeat could not have helped the cause of peace in his mind. On his desk was a telegram from General Samuel R. Curtis from Fort Leavenworth: "I want no peace till the Indians suffer more."[22] Colonel Chivington, his military adviser and political crony, sat beside him. Major Wynkoop, young and inexperienced, had brought these chiefs in on his own, and he would be censored for it. What should he do if he made peace, the governor rhetorically asked Wynkoop in private, before the Indians had been admitted? He had raised a new regiment—the Third Colorado—for the express purpose of killing Indians. The soldiers, signed on for a mere hundred days, were thirsty for quick and cheap retribution. He listened to the chiefs' explanations—that they, at least, had made no alliances with the warlike tribes, that they were pledged to peace—but he made no promises. Colonel Chivington would only tell the Indians that if they wanted peace and were willing to disarm, they had to move their camp to the walls of Fort Lyon. "Go to Major Wynkoop," he said.[23]

With the exquisite clarity of hindsight one perceives that young Wynkoop's days as a commander were at that moment numbered—as much, on a more grievous scale, as many lives of the would-be allies. Within weeks more than six hundred peaceable Arapaho were encamped outside the walls of Fort Lyon, accepting rations from Major Wynkoop because they had nothing else to eat. Out on the plains, the Southern Cheyenne were gathering in preparation to also turn themselves in at the fort. Then, abruptly on November 4, Wynkoop was relieved of his command—for "letting the Indians run things at Fort Lyon."[24] His replacement, Major Scott J. Anthony of Chivington's Colorado Volunteers, stopped their rations, temporarily disarmed them, and ordered them to move some distance off to Sand Creek. Informed that the Southern Cheyenne had finally assembled, Anthony instructed them to wait at Sand Creek also. When a small group of unarmed Arapaho appeared, grasping buffalo hides which they hoped to trade for food, Major Anthony had his pickets fire over their heads. Then, on November 28, Chivington himself arrived at Fort Lyon at the head of a large column of cavalry—the Third Colorado, raw recruits who had chafed under taunts in Denver that their brief enlistment would be over before they saw any action. Intentions could be read in a sudden change of sentries with orders to let no one leave the stockade. After nightfall seven hundred cavalrymen—Chivington's, ranks swelled by Anthony's available men, accompanied by four light mountain howitzers—rode out in the direction of Sand Creek.

At dawn atop a bluff above Sand Creek, the soldiers looked down on the village of Southern Cheyenne and Arapaho, some 550 men, women, and children altogether, including Chiefs Black Kettle and Left Hand. "Kill and scalp all, big and little," Chivington ordered; "nits make lice."[25] As the bluecoats prepared to move in, a few early risers among the tipis sounded an alarm. Black Kettle, seeing the whites, quickly raised a white flag and the Stars and Stripes and shouted to his people not to be alarmed. Almost simultaneously, Chivington thundered out in his angriest fire-and-brimstone, vengeance-is-mine voice. "Remember the murdered women and children on the Platte," he screamed. "Take no prisoners." The valiant Coloradans charged.[26]

Such carnage is difficult to relate. The heart rebels. And this narrative is a chronicle of the coming of the railroad to the West, the knitting together of the continent, the consolidation of empire, the realization of good people's dreams and of venal men's ambitions. But it is the Native Americans' story, too, for their territory was breached and their way of life was an impediment and they were brutally, calculatedly shunted aside. They will be present on every subsequent page, for it was here, in Colorado and Kansas and Nebraska and Wyoming in the summer and fall of 1864, as some railroad men in civilian garb tried to go about their business along the Platte and Republican, the Green and the Cache la Poudre and the Echo, that others set out to sign the Indians' death warrants and thus stepped into infamy. Some, like General Curtis and Governor Evans, linked as they are to the railroad's story, transferred their taint onto the enterprise; other railroad men waiting in the wings would soon enough, like Grenville Dodge, take up participation in both sagas; as it was, careers in the military and in engineering went, often enough, side-by-side in the years leading into the Gilded Age. In the aftermath of Sand Creek in those bitter waning days of 1864, with the Shenandoah Valley campaign over, with haggard Abraham Lincoln resting for a few moments after defeating the strutting blowhard George McClellan at the polls, with the state of Georgia overrun by exultant Union forces under Tecumseh Sherman, initial news of the engagement in Colorado was of a starry victory in equal contest against the enmassed savages ready to help pull down the Union.

Then complications emerged—reports of fallen wounded executed on the ground, babies and children shot down and bayoneted, women killed and their private parts cut away for trophies, skin peeled and fingers and ears taken from the few Indian males who were downed, skulls smashed, scalps pared. One hundred and fifty dead were left behind at Sand Creek—two-thirds of them women and children—as the "Bloody Thirdsters" rode off looking for more easy victims in a looping course back toward Denver, where cheering civilian throngs and the proud, beaming Governor Evans awaited them. Within six weeks, though, the atrocity reports had circulated enough throughout the nation that the army was forced to launch its own inquiry, followed by congressional probes, but a call for Chivington's court-martial fizzled away to nothing when he resigned from the army and escaped jurisdiction. For his part, Governor John Evans, whose ambition had touched off the conflagration, would hold office only for another half-year, squirming in the revealing congressional committees' limelight, until his removal. He would remain connected to various railroad enterprises, especially the Union Pacific, for some years to come, while the retaliatory war continued to threaten settlers and railroaders all the way across the plains.[27]

Confronted by unceasing fears and sporadic Indian raids on outposts and emigrant wagon trains, Peter Dey's location engineers went about their field surveys with varying results. "Indian depredations in the latter part of the year were so serious," reported Dey to Durant and the Union Pacific board, "that all that had been marked out was not accomplished by the parties in the field." James Evans inched gingerly westward through the Black Hills toward

Green River, hoping that his small military escort looked sufficiently forbidding. Even at the beginning of his long reconnaissance, following Lodge Pole Creek westward across panhandle Nebraska, Evans had suffered premonitions of disaster, especially after encountering a Sioux party led by a gray-haired warrior who said he had ventured east and been impressed by the white man's power and numbers. "I do not like the idea of your coming here," the chief said through an interpreter. "This is the Indian's country. Besides, over the first range and beyond you will find many deserts which have no crossings nor water—and how can the iron horse succeed in passing over the country when it is impossible to go through the mountain range and the desert beyond?" Later, at the lower end of Bitter Creek, seeing the alkaline deserts for themselves, the surveyors were menaced by a large number of Arapaho, and then shaken down for food and presents by other parties; then they were forced into a skirmish with Cheyenne during the last leg of their trip, after their military escort took a suspicious-looking Indian captive and then shot him as he attempted to escape.[28]

But despite the constant nervousness, and the further inconvenience of snow blindness which plagued everyone, Evans recorded a profile that was, in Peter Dey's words, "remarkably free from difficulties of construction" but for the crossing of the Black Hills, which was, simply, "forbidding." In only fourteen miles, the ascent from the east was no less than fifteen hundred feet, with a similar rapid descent to the Laramie Plains. South of Evans's line, below the usual crossing of Cheyenne Pass, engineer Francis Case noted a better way across the Black Hills: Antelope Pass, reachable from either Lodge Pole Creek or the South Platte, with a saving of some five hundred feet in altitude but at what Dey saw as "a considerable sacrifice of distance." All of Case's other explorations of the Front Range, notably at Hoosier and Berthude Passes, showed a southern route to Denver to be, as far as he could tell, at least somewhat impracticable. But as reports came in, the great unknown was closest to home, in the western half of Nebraska's Platte Valley beyond stolid Fort Kearney. "We have three times attempted to [connect the Platte with the Republican Valley]," Dey recorded, "but the large bodies of hostile Indians that were congregated on the Republican, near this point, have prevented."[29] Likewise, for this wholly understandable reason Evans was deterred from surveying South Pass. Nonetheless, from his opulent surroundings in New York, the Doctor continued to issue orders that the territory be surveyed before winter set in. Not without an escort of one hundred men, countered Dey finally; "the part of the Republican Valley that you want explored is the camping ground of several Indian tribes, very hostile."[30]

Dr. Durant, too, was feeling hostile, confronted by irritating blank spaces on the Union Pacific route map overdue at the Interior Department (it would finally be filed in October) and by various disturbances on his periphery. First, there was the annoying Kansas line—the Union Pacific, Eastern Division—

still enmired in legal trouble with the struggle for corporate control between Pathfinder Frémont and New York financier Hallett. Under the latter's temporary control, the Kansas firm had seemed to draw momentum from the 1864 railroad act's passage in July, spurring construction of its first forty miles. Mandated by law and qualifying Hallett and backers for the first handsome federal subsidy, the opening stretch between Wyandotte, on the Kansas River opposite Kansas City, and the town of Lawrence had been widely touted in July to be completed almost immediately. Issuing rosy press releases from the railroad office in Wyandotte, Samuel Hallett set a ribbon-cutting ceremony for August 18 and invited a reviewing-stand's worth of government dignitaries and prominent businessmen, among them Thomas Durant. Hallett even offered free passage out to Kansas City to all grandees. The Doctor still had his outstanding loan to Hallett to think about, giving him his one-third interest in the construction contract. The investment, though, seemed questionable due to rumors that the Kansas trackage was shoddy and downright dangerous. Indeed, the line's chief engineer, Orlando A. Talcott, had disavowed what Hallett had hurriedly constructed, earning himself instant dismissal. Talcott immediately wrote to President Lincoln and warned that the Kansas road was horribly substandard and could never qualify for federal subsidies. Bridges would topple, culverts collapse, embankments settle and wash away, trains derail. Hallett was masterminding, he complained, "the Bigest swindle yet."

The president naturally had referred Talcott's letter to his Interior Secretary, John Palmer Usher, and Usher had just as quickly sent a copy to his associate Samuel Hallett. Furiously, Hallett wired his brother in Kansas; he may have been a financier but Samuel Hallett was a two-fisted one who more than once had threatened the lives of those who stood in his way, and if anything, his brother Tom was less refined. Ordered to find Talcott and thrash him, Tom Hallett did so. It was not difficult, since the engineer was slightly built and partially crippled by a stroke. Talcott vowed revenge. When Samuel Hallett returned to the company's headquarters in Wyandotte, on July 27, Talcott limped out from cover in an alleyway and shot him in the back.[31]

Hallett lived only a few minutes, and Talcott escaped to the Colorado camps (where he evaded capture for some fifteen years—when he was finally arrested, extradited, and tried, the jury had no heart to convict him), but of course there was no ribbon-cutting ceremony in August, no opening of service until the line was improved, and, of passing irritation to Dr. Durant, no hint of return on his investment, much less a quick profit on the $450,000 Kansas line bonds he held.[32] A prominent St. Louis banker, John D. Perry, whose larger investment in Hallett had given him about half of the line's first construction contract, stepped in to take control of the Kansas line. Frémont was bought out, the company began to be stabilized, and its threat of winning the main-line sweepstakes of the Pacific railroad would begin to diminish. For months all concerned would be tied up in a struggle over the company's board and the veracity of the first construction contract. Nevertheless, the Union Pacific, Eastern Division, had trackage, no matter how substandard. Durant's Union Pacific had a gradient out of Omaha to a dead end on a prairie, where grasses bent under the hot, late-summer wind, and where, not far away though just beyond hearing, the yips and whoops of war cries could almost be discerned.

Dr. Durant's correspondents kept him aware of his great competitor in the Far West. Central Pacific tracks had been extended past Roseville in early summer to the hilltop mining camp of Newcastle, mile 31; daily passenger service opened on June 6. Soon after, the Associates announced the opening of their wagon toll pike, the Dutch Flat and Donner Lake Road, a vast improvement on the one competing freight route (via Folsom and Placerville) to the Carson Valley of Nevada. Teamsters could save three days by traveling between the Central Pacific's railhead at Newcastle and the rich mining camps around Virginia City, according to the Associates' claims, and take advantage of lighter grades. Passengers, too, could make the crossing of the Sierra in record time: by train from Sacramento to Newcastle, and by the coaches of the California Stage Company on to Virginia City, it took about seventeen hours.[33] Hotels had been opened along the route.

Within a short time the Dutch Flat road would earn the company a million dollars annually, but that happy prospect was still on the horizon. Bills had to be paid: twelve hundred laborers, many of them Irish immigrants scooped off the quays in Baltimore, New York, and Boston, were on the company payroll in 1864, toiling away at the Bloomer Cut and on the established grade for some twelve miles above Newcastle; two engineering parties—one just ahead of construction, one pressing upward toward the Sierra summit, pursuing the goal of the continuously ascending grade—struggled against the rugged terrain. Prior to the passage of the railroad act in July, no federal subsidy could be earned until continuous track had been spiked for forty miles; the new version credited the company for its first twenty miles, but Washington was slow to send the money. The state-level subsidies were similarly unforthcoming. Central Pacific bonds, their 7 percent interest guaranteed by the state after an uncomfortable deliberation in the legislature, began to move—but then were halted when a suit of unconstitutionality was filed.[34] The company would not prevail in that fight until early 1865. There were moments in the summer and fall of 1864 when not even the personal guarantees of Huntington, Stanford, Crocker, and Hopkins seemed enough to keep the Central Pacific afloat. Seventeen days of that summer went by when the treasury was utterly bare.

Matters were certainly not helped by a plague of negative publicity masterminded by the Central Pacific's mortal enemies in San Francisco—the bankers, coach-line operators, and so on—and by its direct railroad competitors, notably Lester L. Robinson, who still hoped to run a line over the mountains via Placerville and the Johnson Pass. The *Alta California* was the most prominent of local newspapers bitterly critical of every move by the Central Pacific, "full of abuse and of ridicule," recalled Charley Crocker, "every day."[35] Robinson, though, whose animus had thwarted the Associates time and time again, who would not pause at slandering the dead Theodore Judah, was, with his associate Charles E. McLane, the most dangerous foe at this time. Robinson bombarded his supporters in the California and Nevada legislatures with complaints and innuendoes and charges of chicanery and corruption on the part of the Central Pacific directors. The great Dutch Flat Swindle controversy—that the Associates had no intention of building higher than the wagon road terminus—was resurrected yet again. The campaign had at least partial effect. The alarmed officials of Placer County sent a delegation to scrutinize the books of the Central Pacific, looking for all that corruption; they found none, and indeed announced that there was no hint of evil doings.[36] The Nevada legislature, in the middle of drafting a constitution for statehood scheduled to take place in November, was prodded by Robinson to write in a clause awarding some $3 million to the first railroad to enter Nevada territory. It seemed innocuous, but it set off alarm bells in Sacramento. Leland Stanford hurried to address the legislators, pleading for the Central Pacific's primacy and warning that Nevada's "reward" would create a disastrous contest that could destroy the good-faith efforts of his railroad. Better that the state just say and do nothing, Stanford argued. Solely because of his personal plea, the clause was dropped.

These were indeed difficult times, with progress often seeming at a standstill, and had the Associates not agreed to be personally responsible for ten years' worth of interest on company bonds, they would likely have walked away from the enterprise after midsummer in 1864. "I would have been glad when we had thirty miles of road built," recalled Crocker, "to have got a clean shirt and absolution from my debts."[37]

Perhaps it was on one of those summer days when the Central Pacific treasury was empty, but nonetheless, three thousand miles away in Washington, and only three days after the freebooting Samuel Hallett was drilled from behind and fell to dust, a sharp-eyed government engineer by the name of Silas Morrow Seymour, aged forty-seven, dispatched a telegram to the Union Pacific headquarters in New York. "If I can be of service under new state of things in Kansas," he wrote, "command me."[38] Always the opportunist, Seymour currently worked for the Interior Department as chief engineer on the Washington Aqueduct, and had supervised construction of a railroad bridge over the Potomac. He had been introduced to Dr. Durant by a mutual acquaintance,

John F. Tracy, president of the Chicago and Rock Island line, in Washington while Congress was still in session and the railroad bill still being debated. Durant enlisted Seymour in his lobbying efforts—literally installing him in a parlor room just off the lobby of the Willard Hotel, where maps and profiles of the Union Pacific route (so far as it was then known, given the troubles west of Fort Kearney) had been unfolded over most horizontal surfaces. Seymour's job was to study the maps as hastily as possible and discourse learnedly on them as legislators and lobbyists passed through on their way to the dining rooms and bar. That Durant was gaining a useful ally deep inside the Interior Department, who knew most people of consequence in the capital, was not lost on either.

After passage of the railroad bill, Durant told Seymour that he would be appointed a consulting engineer to the Union Pacific. Not for two months would the Washington engineer obtain his appointment in writing or begin drawing his annual salary (of $5,000), but Seymour's letters to Durant that summer and fall speak of an undisguised eagerness to make himself not only useful but indispensable to the Doctor, just as the Doctor's responses (and occasional nonresponses) speak to his fondness for keeping his people dangling on a string without clues to the larger picture. "I will lose no time in responding to your call," wrote Seymour, willing to meet Durant in Washington, New York, or even the Saratoga spa where Durant took his family in the hot months. "I would like if consistent with your views to spend a few days in New York in looking over your maps and engineers' reports," Seymour urged, "and, after that to go West with you for the purpose of obtaining further information as I can, by a personal examination of some of the proposed routes as far west at least as the 100th parallel; and also such details as may be practicable of the survey now being made beyond that point—all of which I feel will be necessary before I can be prepared to render you such advisory aid as I would like to be able to do."[39]

Somehow, though, the prospect of enduring the Nebraskan heat and dust in Seymour's company, coming, no doubt, in proximity to roving bands of Cheyenne and Arapaho, did not entice Dr. Durant away from genteel Saratoga Springs or from his periodic late-summer duties in the opulent Union Pacific offices at 13 William Street in New York. Two and a half months and a handful of letters later, having supportively purchased Union Pacific stock and gotten involved in Durant's "southern matters," or the smuggling of contraband cotton into the North, and having journeyed out to Chicago, Wyandotte, and Omaha himself, Seymour, growing anxious, wrote again. "I would like you to make up your mind with reference to the duties you would like to have me assume," he said, repeating his interest in examining the company's maps and profiles and learning something of the company's history. By then, Seymour had it in mind that he would be assuming the post of chief engineer, and had a high enough opinion of himself to come away from an interview with Peter

Dey with that impression reinforced; Seymour thought that Dey would acquiesce and be demoted to his assistant (there is nothing among Dey's papers to indicate that this was so, much less discussed). Seymour told the Doctor that he was confident that Interior Secretary Usher would release him from his Washington job. "If, in addition to this," the hopeful engineer continued, "I could be appointed Consulting Engineer of the Kansas Division and of the Mississippi & Missouri R.R.—at a small salary upon each—I would arrange in the spring, or sooner if necessary, to leave here entirely, and move my family to New York."[40]

Durant had about as much intention to put Seymour in the top spot in Omaha as Dey had knowledge of it. As a matter of fact, he had not got loose of the desire to convince Brigadier General Grenville Dodge, attached to the Army of the Tennessee during the summer's drive to Atlanta, to leave the service and become chief engineer. Dodge's martial spirit had lately begun to waver again; he was not advancing in rank fast enough, and as a political appointee had become sensitive to the scorn of regular army officers—he had even gotten into a brawl with a subordinate West Pointer over his tactics, earning a bloody nose and a challenge to a duel. Then, in late August, major newspapers in New York carried reports that Dodge had been killed in action just outside Atlanta. A few days later his relatives and friends, and Dr. Durant, learned that Dodge's death had been greatly exaggerated—at the front lines, he had stooped to peer through a peephole at the Confederates' position when an enemy bullet tore through the barricade, fractured Dodge's skull, and knocked him unconscious. He was sent north to recuperate, and though his health and spirits were never lower than when he stopped in New York to meet with Durant, in late September, he continued to decline the civilian post. Lengthy interviews a few weeks later with Grant in Virginia and with Lincoln at the White House, which reinforced his friendly ties with both men, led to a promotion: Dodge was appointed commander of the District of Missouri, headquartered at St. Louis.[41] To Durant, it was a mixed blessing. The chief engineer post was becoming a vexation: Dey wanted the job he now held, Seymour begged and maneuvered for it, and Dodge flat-out refused it. Though General Dodge might be of aid to the Doctor in Missouri, through which his smuggled cotton passed ("Dodge can help us now," Durant crowed upon learning of the assignment),[42] he was becoming irritated with Peter Anthony Dey. Seymour had fit nicely into his plans, especially in terms of keeping him aware of the Interior Secretary's every thought, but Dey . . . Dey was another story.

If he was a religious man, Peter Dey might have considered Job in the declining months of 1864: his little boy dead in July, and the plague of reversals within the engineering department of the Union Pacific—the Indian troubles, the lack of momentum in the locating surveys, the critical absence of laborers, the difficulty in obtaining ties and other materials, the mercurial Dr. Durant.

Then there was the matter of having his experience and professional judgment called into question. Building on the surveyors' field notes in the summer, Dey had considered the construction cost per mile over the first one hundred miles west of Omaha, including bridges and track sidings, and estimated the cost at being somewhere between $20,000 and $30,000 per mile; to be on the safe side, allowing for the scarcity of labor and the wartime costs of materials, he went with the higher figure of $30,000 when he sent his estimate on to Durant in New York. Soon Dey found himself receiving Joseph E. Henry, who had worked for Durant as construction supervisor on the M&M, who had recently moved to oversee the shambles of the Kansas line down in Wyandotte and who served Durant on the Union Pacific board. Henry relayed a directive from Durant: the construction figures had to be much higher, $50,000 per mile. Dey could see no reason for such an increase other than profiteering, not even treating seriously Henry's vague reference to wider road embankments to avoid washouts. But he acquiesced at the time.

It was not until November 1, when Henry reappeared in Omaha with new orders, that astounding new pieces of the puzzle surfaced. Henry bore orders from Durant appointing him construction superintendent on behalf of a contractor whose name Dey had never heard in any railroad context save his occasional ride on one. Moreover, Henry gave Dey—relegated now to simply chief engineer—a copy of the construction contract let to one Herbert Melville Hoxie of Des Moines, Iowa.

Dey knew Hoxie well. Born in New York and raised in Des Moines, the son of a tavern keeper, thirty-six-year-old "Hub" Hoxie was a close friend of Dey's friend and former assistant on the M&M, Grenville Dodge; he was a political fixer and perennial office seeker who had risen in influence along with the Republican Party in Iowa, ascending to chair the state Republican committee during Lincoln's triumph at the polls in 1860. His spoil for those services was an appointment as a federal marshal, in 1861 (which, based on his public character record, had caused some controversy); it was his first real job in some years. From the moment Fort Sumter was fired on until the present, Hoxie had schemed for a way to make fast money on the war.[43] As a promoter and errand runner, he had done favors in the state legislature for Durant's railroad interests, but he had never shown any skills beyond native cunning and smooth politicking. His appointment as contractor for the Union Pacific (initially for the first 100 miles but soon extended, though Dey was not officially informed, to 247 miles)—at $50,000 per mile—was absolutely confounding. To Dey, there could be only one explanation: Hub Hoxie was acting as a front for Thomas Durant. "No man," he wrote to General Dodge down in St. Louis, "can call $50,000 per mile for a road up the Platte Valley anything else but a big swindle—and thus it must stand forever."[44] Over the next five weeks he pondered the state of affairs as his blood pressure rose.

Dey's suspicion was correct: during a meeting in Saratoga in late July or early August, the Doctor had offered Hub Hoxie a $5,000 fee and $10,000 in Union Pacific stock, along with the promise of a job, to take the construction contract and immediately reassign it to Durant and his associates. Hoxie submitted "his" bid—it was the only bid, and it was drawn up by Durant's attorney, Henry C. Crane—on August 8, covering the first one hundred miles. Total expenditures for all stations, machine shops, and other structures would not exceed $500,000. Bridges would each cost no more than $85,000. The contract claimed liability for iron only to $130 per ton unloaded at Omaha, with the railroad company paying any excess. In return the contractor would be paid $5,000 cash per mile, and would be issued 6 percent stock certificates against completion of the first twenty miles of track—when, as detailed in the new railroad act, first-mortgage and federal bonds, and land grants, could then be awarded.

A special committee of the Union Pacific's board of directors consisting of John Adams Dix, George T. M. Davis, and Cornelius S. Bushnell reviewed Hoxie's proposal and formally accepted it on September 23. When, on October 4, the bid was increased to cover 247 miles, for $12,350,000, the committee accepted Hub Hoxie's proposal with no debate. In return for this increase, the impecunious Hoxie pledged to "subscribe, or cause to be subscribed for, five hundred thousand dollars of the stock of your company."[45] Four days later he signed the contract over to Dr. Durant, joined by four other men of means, Union Pacific stockholders Cornelius Bushnell, Charles A. Lambard, H. W. Gray, and H. S. McComb. To actually fulfill the construction, the five paid in some $312,000 in cash against a total subscription of $1.6 million. Durant's part of the pledge was $600,000, with one quarter fronted in cash. All the complex transaction required now was a formal transfer to the former Pennsylvania Fiscal Agency, renamed Crédit Mobilier of America by the noonday mystery, George Francis Train, to limit the investors' personal liability should the railroad somehow not get built.

Much of this would remain far over Peter Dey's head for quite some time. All he had to fuel his righteous indignation was the gap between a generous $30,000 per mile and an outrageous $50,000 per mile, the difference spelling doom—of that he was convinced—for the Pacific railroad. It was the M&M debacle all over again, Durant sucking the orange dry before the entity even had a chance.

Then came the crowning blow.[46] From Washington came a letter from the "consulting engineer," Silas Seymour. In it Seymour proposed changing Dey's direct route already surveyed and graded between Omaha and the Elkhorn River—at a cost to the company of $100,000, said route approved by President Lincoln only a month before, on November 4—in favor of a circuitous route. This "oxbow" route veered southward from Omaha along the valley of the Mud Creek to connect with the West Papillion Creek valley, striking west and finally reaching the Elkhorn River. Seymour claimed that his way would save the company money and take advantage of slightly better grades. It seemed like balderdash; on reflection Dey would see that it would actually be more difficult and expensive. But riding on Dey's chagrin over the Hoxie contract was this new illumination: Dey immediately saw that the difference between the located line—of which Seymour proposed to abandon some fourteen miles—and Seymour's new oxbow route of twenty-three miles was a gain of nine miles. That would be an extra cost to the company and a gain to the purported contractor of some $450,000. One could envision at that moment in Dey's suspicious mind the route of the Union Pacific as supervised by Seymour and Durant, winding in expensive curlicues off toward the Far West, running the railroad utterly into the ground.

Peter Dey began composing letters. "This part of the road was located with great care by me," he wrote to Durant.

> *You even animadverted on my going into the neld personally to examine the proposed lines; you also promised to have the lines scrutinized by a committee of engineers nearly a year ago. The line as located by me has been approved, and the location has been acted upon for a year. It is too late, after spending so much time and money on the construction, to go back and consider relative merits of this and other lines. The present location is right, unless it is desirable for the company and government to make a longer road, more bridges, heavier excavations, and spend on twenty miles the money which should be expended on one hundred miles of road. Your views favored the economical policy, which was certainly the true policy of the company. I acted upon it deliberately and, as I still think, wisely.*
>
> *In view of the decided advantages of this route and the expenditures already made, it is in my opinion altogether out of the question to modify the location to meet the undigested views of Mr. Seymour, who can not know the relative advantages of one route over another, because he has not been over the country, and, from the tenor of his letter, not even examined the profiles in the New York office.*[47]

Seymour later admitted that his inspiration for the oxbow had come not while looking over the maps, which he had not seen, but while he stood on the pilot deck of a steamboat heading downriver from Omaha toward St. Joseph. From that prospect on the Missouri River he saw the cottonwoody indentations of the Mud and Papillion Valleys. "I became satisfied that it could be done," he later fervently told a government investigator. "The opinion was then formed that a very palpable engineering mistake had been made."[48] The equally palpable extra profit for Durant that the ambitious, job-hunting Seymour had delivered was not mentioned.

For Peter Dey there was no question, however, that arguing with Dr. Durant, or even with Silas Seymour, would get unsatisfactory results. The future of the whole enterprise lay at stake, and to Dey it was a question of honor. A year before, when pushed up into a smaller corner, he had exercised his only option by flatly resigning his job. And Durant had backed down. This time the engineer applied his tactic to the only honorable man he perceived in a management position with the Union Pacific: Major General John Adams Dix, commander of the Department of the East in New York City and nominally president of the Union Pacific.

Back during his boyhood in upstate New York, Dey had looked up to Dix, a state official and Jacksonian Democrat whose home was a hundred miles away in Cooperstown and whose antislavery sympathies had allied him with the "Barnburner" Democrats during the furious 1844 electoral vote struggle that had denied Kentuckian Henry Clay the presidency in favor of James Polk. Dix, and his New York allies in that epic struggle, he considered full of "integrity, purity, and singleness of purpose," he would tell Dix now. He had witnessed nothing to change his mind when working for the Rock Island and the Mississippi and Missouri, when Dix, as titular president, had occasionally come west from his executive haunts in New York. It was to Dix, then, that he would present his case.

"I hereby tender you my resignation of the position of Chief Engineer of The Union Pacific Rail Road," Dey wrote on December 7, 1864, "to take effect on the thirtieth of December, one year from the date of my appointment. I am induced to delay until that time, that I may combine the results of the Surveys of this year, and present them to the Company, and to myself, in a satisfactory manner. My reasons for this step, are simply that I do not approve of the contract, made with Mr. Hoxie for building the first hundred miles from Omaha west, and I do not care to have my name so connected with the Rail Road, that I shall appear to indorse this contract."[49]

"You know the history of the M. & M. road," he wrote in a separate, more personal note to Dix, "a road that to-day could be running to this point if its stock and bonds only represented the amount of cash that actually went into it. My views of the Pacific Railroad are perhaps peculiar. I look upon its managers as trustees of the bounty of Congress." A trustee such as Dix should be warned, he thought, that before the rails ran up into the mountains, the company might be dead. And he appealed as much to their shared upstate New York roots and political sympathies as he did to what he called "integrity, purity, and singleness of purpose"—Dix was marked, he said, with such qualities, and Dey's boyhood faith was still unshaken. "You are doubtless uninformed," he went on, "how disproportionate the amount to be paid is to the work contracted for. I need not expatiate upon the sincerity of my course when you reflect upon the fact that I have resigned the best position in my profession this country has ever offered to any man."[50]

He sent off the letter and waited for a reply, for some sort of action that would vindicate his boyhood faith and his professional reputation. If Peter Dey was hoping that the general would refuse to accept his resignation and reach down from his lofty place in New York and pluck the crooks out of the Union Pacific Railroad Company, then he would wait for an eternity and he would still not be satisfied.[51]

Part V

1865

The Losses Mount

16

"The Great Cloud Darkening the Land"

Beset by the devils of flood, famine, and overpopulation, neglected by the corrupt imperial government, threatened by encroaching outsiders from the north, militias, secret societies, and finally an all-out civil war, the peasants of the harsh coastal mountains of the Sinning district of Kwangtung Province, in southeastern China, had listened in 1849 with no little joy as market-goers returning from the nearby city of Canton related news from America: about the *Gum Shan*—"Mountain of Gold"—discovered only months before, with fortunes ripe for the plucking (or panning) to any sturdy and enterprising man willing to mortgage his family's future for the price of a sailing-ship ticket. And so, as the farmboys and failed merchants of Vermont, New York, North Carolina, and Ohio had struck out for California in 1849, so did peasants of Kwangtung in 1850 and 1851—traveling by junk to Hong Kong, signing on for passage to San Francisco on credit against future earnings in the mines, jamming like livestock in the holds of sailing vessels for the two-month voyage—close to three thousand immigrants by the end of 1851, rising to a high of twenty thousand in 1852, showing only occasional abatement thereafter.[1]

The peasants of Kwangtung were indentured in California to locally run Chinese district companies, signed on for up to five years of labor at comparatively low wages until their tickets were paid; they then filtered out into the streams and rivers of the Sierra slope in search of gold. Enforcement thugs and old-world penalties awaited slackers and deserters, but the life, hard as it was, promised rewards, particularly the hope that two or three hundred dollars could be amassed—enough by Kwangtung standards for a return home and a luxury retirement. Until that happy time, which rarely attended any of the immigrants, there were only the diversions of gambling, prostitution, and opium, establishments for which sprang up in the Chinatowns of San Francisco, Sacramento, Stockton, and Marysville, and in the smaller mountainside encampments beneath roofs of canvas.

Even so, life was hardly pleasant. The Chinese worked in severely regimented gangs of thirty or forty, panning and sluicing and hoping that rival Chinese associations would not move in on them and that white miners would not jump their claims. Often they would take over sites abandoned by whites as unpromising, lessening the chances of being jumped. Their willingness to work hard for low wages, and to take on any kind of labor, sparked an inevitable jealousy among white Californians. And predictably for a state in which racial attitudes were primitive at best, where competition was fierce, where rule of law was sometimes spotty, the Chinese immigrants endured a backlash of florid variations. These went beyond the harassments by individual ruffians (though such were frequent and often boiled into mob actions), and took root in California officialdom. First, local jurisdictions stepped in, such as Marysville, where Chinese were restricted from filing mining claims. Then there was the series of restraints from the state legislature: the Foreign Miners' License Tax Law in 1850 (subsequently raised in 1852); the Act to Provide for the Protection of Foreigners, and to Define Their Liabilities and Privileges (more taxes) in 1853; the Act to Discourage the Immigration to This State of Persons Who Cannot Become Citizens Thereof (1855), another tax; the Act to Prevent the Further Immigration of Chinese or Mongolians to This State (1858), setting an absolute deadline for entry, twenty-seven years ahead, in 1885; and the Act to Protect Free White Labor against Competition with Chinese Coolie Labor, and to Discourage the Immigration of the Chinese into the State of California (1862), establishing a monthly capitation tax (called the Chinese Police Tax) of $2.50 gold, with penalties of imprisonment and property seizure for nonpayment. Other California statutes forbid public schooling for "Mongolians, Indians, and Negroes" (1860) and excluded "Mongolians and Chinese," and Indians "or persons having one half or more Indian blood," from giving evidence in court against or on behalf of a white man (1863), an affirmation of rulings by the California Supreme Court going back to 1854.[2]

Local politicians quickly discovered in that nativist climate that promises to somehow address the "yellow peril" paid off at polling places. Notably, Leland Stanford had found it profitable when running for governor; then, at his inauguration he renewed his promise to aid the legislature in whatever measure it took to halt immigration of "the dregs of Asia," the degraded Chinese, and for the rest of his short term gave active support to the tax and restriction efforts.[3]

Nonetheless, the lure of *Gum Shan* was too strong, and despite the taxes and levies, Chinese continued to cross the Pacific. Not all worked at the sluices and panning boxes, for they became a cheap and convenient, if despised, labor pool as houseboys, launderers, gardeners, cooks, and fishermen. "They are a harmless race," commented Mr. Samuel Clemens after he arrived in Nevada and California and had a look around, "when white men either let them alone or treat them no worse than dogs; in fact, they are almost entirely harmless anyhow, for they seldom think of resenting the vilest insults or the cruelest injuries. They are quiet, peaceable, tractable, free from drunkenness, and they are as industrious as the day is long."[4] It would have seemed quite natural to find them along railroad rights-of-way, but this seems largely not the case—with the exception of the San Francisco and Marysville Railroad, where, in 1859, a commerce journal noted that nearly a third of the five hundred graders were Chinese, and also the San Francisco and San Jose Railroad, which employed some segregated Chinese work gangs in 1863 as strikebreakers.[5] But this was soon to change, in early 1865.

On January 7, days after the California Supreme Court upheld the constitutionality of the state railroad aid bill, releasing $1.5 million in state-guaranteed 7 percent interest bonds, the *Sacramento Union* posted a notice that the Central Pacific required some five thousand laborers immediately "for constant and permanent work." At present, Charley Crocker had about six hundred laboring on the twelve-mile stretch between Newcastle and Clipper Gap; the new infusion of cash from bond sales, in addition to wagon-road revenues, made the Associates' optimism soar. Unfortunately, despite the newspaper advertisement and similar handbills distributed at every post office in the state, fewer than two hundred new faces turned up in the construction chief's office. As was often the case in those days, most were Irish. The man they would work under, as a matter of fact, was Irish, too, and newly hired by Crocker that same month, but if they expected any special treatment because of their shared heritage, their hopes would soon be exploded.[6]

His name was James Harvey Strobridge and he was born in 1827 in Albany, Vermont, in the state's remote Northeast Kingdom. While a teenager he had worked on the building of the Vermont Central Railway, the middle link between Boston and Montreal, and he was promoted to foreman. Then Strobridge had gone west with the gold rush. Drifting back into construction when prospecting and other endeavors failed, he served as a foreman on the Placer County canal and on the San Francisco and San Jose Railroad before taking subcontracts with Charley Crocker—first grading along the Sacramento waterfront and then tracklaying up the line above Roseville. It was there, the previous summer, while clearing the way through the deep pie-wedge of Bloomer Cut, that Strobridge had led the way back into a still smoky blasting area only to have a delayed charge shower him with granite; a piece took care of his right eye.[7]

Now wearing a black eye patch, Strobridge was a commanding presence at an inch over six feet and made of solid iron—but not just because of appearance. A pure fire-breather with a demon's temper, asbestos lungs, and the sharpest, most profane tongue in the state, he used physical fear as his prime managerial tool. Like Crocker, Strobridge was a vehement teetotaler, and in all ways he was strictly the sort that Crocker would like. Crocker knew himself as a "mad bull" whom Mrs. Crocker found to be "overbearing and gruff" in the infrequent times he could be found in their Sacramento parlor; he had learned how to get the job done in his extraordinarily hard young days in Marshall County, Indiana. "I knew how to manage men," Crocker would recall. "I had worked them in the ore-beds, in the coal-pits, and worked them all sorts of ways, and had worked myself right along with them." As a labor boss Strobridge was his ideal, although at the beginning of their long and close relationship even Crocker found him excessive. "I used to quarrel with Strobridge when I first went in," he would tell a biographer. "Said I, 'Don't talk so to the men—they are human creatures—don't talk so roughly to them.' Said he, 'You have *got* to do it, and *you* will come to it; you cannot talk to them as though you were talking to gentlemen, because they are not gentlemen. They are about as near brutes as they can get.'" Charley Crocker was the first to admit that he agreed.[8]

Strobridge took over managing the labor force at a salary of $125 per month. With work for four thousand going unfulfilled, Crocker suggested they hire Chinese to wield picks and shovels. Crocker considered himself prejudiced, but Strobridge was even more so. "I will not boss Chinese," Strobridge retorted heatedly. "I will not be responsible for the work done on the road by Chinese labor." They were strange and unsavory, he thought, not hardy enough, unskilled, and they ate disgusting things. There the matter stalled. "Our force, I think, never went much above 800 white laborers with the shovel and the pick," Crocker would say some years later, "and after payday it would run down to six or seven hundred, then before the next payday it would get up to 800 men again, but we could not increase beyond that amount." The white workers were, Strobridge would add, "unsteady men, unreliable. Some of them would stay a few days, and some would not go to work at all. Some would stay until payday, get a little money, get drunk, and clear out." As usual, there was also the lure of the mines.

The unsatisfactory state of affairs cried out for a solution, until the end of January, on payday. As Crocker and Strobridge were supervising, Crocker noticed a little knot of Irish laborers huddled together and looking dissatisfied. "There is some little trouble ahead," Crocker muttered. As a delegation approached them, Crocker spoke up. "Go over to Auburn," he told Strobridge in a carrying voice, "and get some Chinamen and put them to work." The Irishmen overheard, forgot their wage dispute, and begged them not to hire Chinese.

But Strobridge finally agreed to try fifty Chinese as fillers of dump carts—the least skilled work available, serving under white foremen—and hired them through a labor contractor at a dollar a day, or $26 per month; with this, they would board themselves. (The white workers received $30 per month, including board.) "I was very much prejudiced against Chinese labor," admitted Strobridge years later. "I did not believe we could make a success of it." Within a short time he was proved wrong, and he hired another fifty, followed by another fifty. He contracted with several labor firms, some white-owned, some Chinese, which combed the coast and then sent recruiters to Kwangtung. The total number of Chinese would eventually exceed twelve thousand. Strobridge would also learn they could do considerably more than merely shovel dirt and rocks.[9]

Four hundred miles west of Omaha, near where Lodgepole Creek flowed into the South Fork of the Platte River, lay the rickety, rawboard settlement of Julesburg, Colorado Territory. "The strangest, quaintest, funniest frontier town that our untraveled eyes had ever stared at and been astonished with," wrote Sam Clemens as he passed through in 1861 on a stagecoach bound for Nevada.[10] The junction of the Overland and Oregon Trails ran right through town, and because no emigrant wagon trains passed in wintertime, the three or four dozen residents had battened down for a quiet hibernation—relieved that Julesburg had somehow been spared during Indian raids of the summer and fall. There was little likelihood that trouble would occur in January 1865, with plains grazing for Indian horses so severely limited; raids, if they came, would wait until spring. But on January 7, only thirty-nine days after the massacre at Sand Creek, a combined force of one thousand Northern Cheyenne, Arapaho, and Sioux, under the leadership of the Brulé Sioux Spotted Tail, abruptly thundered into Julesburg. Some survivors from Sand Creek were among the raiders. They killed a number of civilians and soldiers from nearby Fort Sedgwick and scattered remnants of the townspeople out among the cottonwoods and riverbank thickets. The raiders swept on against Platte Valley stage stations and wagon trains carrying supplies for Denver, causing food shortages and panics. They methodically tore down miles of telegraph wire; all communication between the East and not only Denver but Salt Lake City and San Francisco was thereby halted.[11] "At night the whole valley was lighted up with the flames of burning ranches and stage stations," remembered a survivor of Sand Creek who had joined the northern warriors.[12] Little Julesburg had hardly returned to normal on February 2, when revenge parties returned. This time they plundered the town and burned it to the ground.

Eastward, at Fort Kearney on the Platte River, the newly installed commander of the Department of the Missouri, which had been enlarged with the Indian war to encompass Nebraska, Colorado, Wyoming, Utah, Kansas, and Montana, launched his campaign to pacify or eradicate the Plains Indians. General Grenville Dodge had passed the month of January in St. Louis, his men pursuing confederate guerrillas and bringing prisoners in for hasty military trials and hangings. This, and harsh measures against civilians, earned him a reprimand from the more lenient President Lincoln. Dodge, however, was unrepentant. His attention then diverted by Indians, he moved westward with several Kansas regiments. He dispatched units to restore telegraph service and posted cavalrymen in the subzero weather all along the line between Omaha and Denver. Employing the intelligence methods that had served him so well in Tennessee and Arkansas, Dodge sent spies (in this case friendly Indians, half-breeds, and whites with Indian wives) out to keep him informed of developments. They were also instructed to spread the news among the tribes that the new commander was invincible, even supernatural. As a matter of fact, some of the older leaders had either met or heard of Dodge a decade before, when the engineer was busy surveying around the rivers Platte, Loup, and Elkhorn for Farnam and Durant. He had been nicknamed "Long Eye" for being able to squint through a surveyor's glass and see objects two or three miles away. Some thought he could shoot that far. Now, Long Eye Dodge focused his men on lightning tactics—holding to ridges, wasting no shots unless the enemy was in range, wearing the warriors down through "continual following, pounding, and attacking," in Dodge's words. "Their life and methods are not accustomed to it."[13] Dispatches went out to all district commanders on the plains: "Attack all bodies of hostile Indians large or small," he ordered. "Stay with them and pound them until they move north of the Platte or south of the Arkansas."[14]

The tactics of perseverance—themselves an echo of General Grant's in the South—had their effect. Revenge for Sand Creek now too costly, the northern war parties dispersed in mid-February, most heading northeast to hunt on the Powder River in Wyoming Territory. Under Dodge's command, a swift and savage campaign against the southern tribes in the vicinity of the Arkansas River valley sent the Indians off toward the Wichita Mountains, and they sued for peace. Communication and unescorted wagon traffic along the major trails were completely restored—but Dodge had no faith that peace could hold come spring.

Despite the many matters holding his attention, Dodge kept himself informed of affairs with the Union Pacific, and its own labor problems, and he even evolved a solution more inventive—and, certainly, more ironic—than his earlier proposal to use contrabands (freed slaves from the Deep South) in work gangs on the right-of-way. He would soon write Herbert Hoxie to offer the Union Pacific "from one to two thousand captured Indians for work upon the grade, the company feeding and clothing them." Hoxie, in relaying the proposal to Durant, endorsed it. "The clothes would consist of a little calico & a few blankets," Hoxie said. "The rations would be beef & corn. They would grind the corn—or roast it to suit themselves. The General would furnish a guard to make the Indians work, & to keep them from running away." Even as slavery in the southern states was on its deathbed in those early months of 1865, Dodge was suggesting to resuscitate it, in Nebraska, no less.[15] Durant took no action.

Hub Hoxie, by then, had ceased drawing salary as U.S. Marshal in favor of a full-time job with the Union Pacific, as shipping agent for all construction materials and equipment passing through the riverside towns of the Missouri. His construction contract, upon which he had of course done nothing before it left his hands, was reassigned in early March to the Crédit Mobilier, which, although required by Pennsylvania law to keep an office in Philadelphia, actually moved in next door to the Union Pacific office in New York. Henry C. Crane—Durant's private secretary—was listed as assistant treasurer. The five Union Pacific stockholders who together had subscribed $1.6 million when buying the Hoxie contract—Cornelius Bushnell, H. W. Gray, Charles A. Lambard, H. S. McComb, and Thomas Clark Durant—took equal stock in Crédit Mobilier. They lost little time in hooking additional wealthy investors, although reeling them in involved delicacy and patience over some months. However, the new limit on personal liability provided by the Pennsylvania incorporation proved irresistible to a handful of well-to-do New Englanders, a veritable cross-section of Yankee entrepreneurship.

Most significantly, there were the Massachusetts Ames brothers, the "Kings of Spades," Oliver and Oakes Ames—the latter with helpful influence as a U.S. congressman and continuing interests in Iowa railroads, primarily the Cedar Rapids.[16] As it turned out, Oakes Ames brought in nearly all the other primary investors in Crédit Mobilier, such as several Bostonians: contractor Oliver S. Chapman, friend and associate of the Ameses; merchant and congressman Samuel Hooper; stockbroker and railroad capitalist John Duff; the clipper ship magnates William T. Glidden and John M. S. Williams; and the Bank of Commerce president Benjamin E. Bates. From Connecticut came the prominent railroad contractor Sidney Dillon, who had built a number of lines in New England, New York, and New Jersey. From Rhode Island there were the woolen-mill owners Isaac and Rowland G. Hazard.[17] Most of these men joined Oakes Ames because they had profitably pooled their investments before, in the Chicago, Iowa and Nebraska Railroad, the Cedar Rapids and Missouri Railroad, or in the Iowa Land Company speculation, all in the late 1850s; additionally, some had recently joined Ames in yet another line, the Sioux City and Pacific, another feeder line to the Union Pacific.[18]

The Crédit Mobilier seemed like the best and most likely investment of all, certainly the safest, even if the Union Pacific took years to be built. And it is even reasonable that Congressman Ames appealed to their patriotism. Weeks before, probably on the night of January 20, President Lincoln is said to have summoned Oakes Ames to the White House to discuss the Union Pacific—the lack of progress, the shortage of interested capitalists, the national need. In a latter year and in an unfriendlier political climate, Ames would recall the president's words. "Ames, you take hold of this," Lincoln said. "If the subsidies provided are not enough to build the road ask double and you shall have it. The road must be built, and you are the man to do it. Take hold of it yourself. By building the Union Pacific, you will be the remembered man of your generation."[19]

With his checkbook, at least for the present, Ames was complying, while, humble to the last, the true remembered man of his generation was readying himself for his second inauguration, on March 4. With the federal swath across Georgia under Tecumseh Sherman, and his subsequent fiery drive through the Carolinas, with the sealing of the last Confederate port of Wilmington and the spreading food shortages and even riots across all of the South, the end seemed in sight; the cement of Confederate resolve seemed to exist only in the ego of Jefferson Davis at Richmond and in the sheer tenacity of the outnumbered General Robert E. Lee at Petersburg. Thus Lincoln's words on March 4, 1865, extraordinary in their eloquence and quiet, dignified morality, were especially poignant to the silent audience gathered in Washington before the newly completed Capitol dome: "With malice toward none, with charity for all, with firmness in the right, as God give us to see the right, let us strive on to finish the work we are in, to bind up the nation's wounds, to care for him who shall have borne the battle, and for his widow and orphans; to do all which may achieve and cherish a just and lasting peace among ourselves and with all nations."

Many in that solemn assembly had tears in their eyes, though it is not recorded whether any glistened from several of the president's fervent Republican supporters present in Washington—Congressman Oakes Ames, Acting Chief Engineer Silas Seymour of the Union Pacific, and, representing the Central Pacific, Collis Huntington and the alliterative California congressman, Cornelius Cole. These had more than one reason to be present in town that week. The day before being sworn in, Lincoln had, as one of the last acts of his first term, signed an amendment to the 1864 railroad legislation—rushed over to the White House immediately after its passage in Congress, reportedly after herculean efforts on the part of Representative Cole. It allowed the Union Pacific, the Central Pacific, and the Kansas road to borrow money in advance of their unbuilt lines, by issuing first-mortgage bonds at the mileage rates pegged by law and actual terrain, up to one hundred miles ahead of the continuous, completed track. It was a timely windfall for all, but especially for the Californians, inching their way upward into the serious part of the Sierra Nevada, needing to commence work on tunnels, trestles, and bridges—at best, simultaneously—if the Central Pacific was to meet its deadline of completing one hundred miles.

Collis Huntington had been in Washington for weeks, lobbying hard for the bill's passage. He also had other tasks. Even with newly freed bonds, and growing squads of Chinese laborers, the California work could not advance without more blasting powder—which it was consuming at horrendous rates—but wartime had of course made powder scarce for civilian applications. In that late winter, then, Crocker had fired off a wire to Huntington in Washington calling for five thousand kegs of black powder. Huntington, seeing the legions of War Department clerks between him and his goal, went to the top—Secretary of War Edwin M. Stanton—and found the cause of the bottleneck. Stanton turned him down, citing necessities of war, but the Californian thought he saw a more venal side, as he usually did: there could be only one reason for noncompliance, and that was that the secretary required a bribe. "He looked up like a hog," Huntington remembered, "as he was." Stanton, honest and a hidebound patriot, though decidedly prickly, asked him what he intended to do with the powder and was told that it was for the Central Pacific and for Huntington's and Hopkins's store, for the miners—he had, Huntington added, never previously had trouble procuring black powder in the three or four years of wartime. "What you have been doing for the last three or four years," retorted Stanton, hardly bothering to look up, "is nothing to me. I shall not give you any permit." "There is certainly one thing you can't do for me," replied Huntington haughtily, "and that is treat me like a gentleman, for you are incapable of it." The up-by-his-bootstraps onetime handcart peddler strode out of the former bookstore clerk's office after this withering remark and went directly to the White House. The president, of course, admitted him, being as usual interested in seeing to details his assistants wished he would delegate to others. Huntington told Lincoln nothing of his interchange with Stanton—only that the Pacific Railroad was in dire need of powder. "'Well,' Lincoln replied [according to Huntington, some years later], 'that seems proper to me. Yes, that is very proper.'" Then, the Californian noted, "he wrote on a card, which I have preserved, 'Mr. Stanton: Mr. Huntington requests permission to send 5,000 kegs of powder from Boston to California. Unless you know of some good reason why this should not be done, you will please give him the necessary permit.'" Huntington pocketed the card "and said nothing—although I have since regretted not telling Lincoln that his Secretary was a hog. I did not return to Stanton, but telegraphed an order to Newhall of Boston: 'Ship me the powder, and for any damage you may hold me responsible.' The powder was sold, shipped, and paid for, without Mr. Stanton's permit and, as far as I know, without his knowledge."[20]

As he sat listening to Lincoln's address on March 4, Huntington probably had every right to feel smug.

Events in the ensuing six weeks moved rapidly. Collis Huntington and Dr. Durant moved among bankers and captains of industry with their company bonds; Peter Dey settled back into domestic life in Iowa City and looked about for prospects; General Dodge returned to St. Louis to consult with his superiors about Indian policy; Cheyenne, Arapaho, and Sioux concerned themselves with hunting and gathering a distance from the whites; Irish and Chinese laborers scratched away at the earth's crust on the prairie or on the Sierra slope as weather lightened toward springtime; an actor by the name of John Wilkes

Booth, who had attended the inauguration, conspired with a small, ragtag group of Wasingtonians to kidnap President Lincoln and ransom him for Confederate prisoners of war, only to have the plot fail. In Virginia, General Phil Sheridan's victory at Five Forks took five thousand Confederate soldiers prisoner; with the nine-month deadlock at Petersburg crumbling, Robert E. Lee, his lines broken in three places, advised President Davis to evacuate Richmond, a task accomplished when Davis and his entire cabinet hopped a freight train heading south—"government," said an observer, "on wheels."[21] Lee, his Army of Northern Virginia leaking men, his hopes of resupply by rail dashed by bluecoat cavalry, moved westward across countryside bereft of forage in the general direction of Lynchburg, closely pursued by Grant's Federals. President Lincoln went by military railroad—an engine and one ordinary passenger car clattering across old and new battlegrounds—to Petersburg to view the ruins, the day before he would visit Richmond and sit in Jeff Davis's office chair. "The lines upon his face seemed far deeper than I had ever seen them before," recalled a correspondent of Lincoln as the train whistled into Petersburg. "There was no sign of exultation in his demeanor."[22] Six days later Lee and Grant sat down over terms in the parlor of a farmer named Wilmer McLean, in the village of Appomattox Court House, and on April 12 the tattered remnants of the Confederate Army surrendered their arms and their flags.

On the morning of Good Friday, April 14, 1865, John T. Ford, manager and proprietor of theaters in Philadelphia and Baltimore and Washington, sent out bundles of handbills advertising the evening's benefit performance at his place on Tenth Street in the capital—"Tom Taylor's Celebrated Eccentric Comedy," *Our American Cousin,* starring "The Distinguished Manageress, Authoress and Actress," Miss Laura Keene. "This Evening," the headline read, "The Performance will be honored by the presence of PRESIDENT LINCOLN." It would be Miss Keene's last night, the poster added. That morning, before he could see the announcement of Miss Keene's last night, the actor Booth stopped by Ford's Theatre to pick up his mail, as usual, and was told by a stagehand—perhaps it was the same stagehand who would receive an eight-year sentence for holding the reins of Booth's horse later that night—that President and Mrs. Lincoln would attend.[23]

That night, "God's instrument" limped out into the darkened theater alleyway and four, then six soldiers bore the dying president out the front exit to wander up and down the street for five minutes. Mrs. Lincoln led the lamentations, until soldiers found their way to a boardinghouse across Tenth Street and a spare room with a too-small bed where the man could die. The newly reborn nation went into the most profound shock, from which it would never, never recover.

Stanton handled all the funeral arrangements; a year and a half earlier he had supervised the relief of Chattanooga by transferring some twenty thousand soldiers and equipment by rail, in just over a week; now he constructed a memorial on iron rails stretching from Washington to Springfield, Illinois, which was a masterpiece of administration. The route closely followed President-elect Lincoln's triumphant path from his home to the capital in 1861. Thanks to the efforts of John William Starr, a Lincoln enthusiast and rail buff, author of *Lincoln and the Railroads* (1927), the itinerary has been preserved; it is at once a fascinating record of the state of American transportation as it existed just before the Gilded Age and an indication of how utterly chimerical, how revolutionary it still was that merely two corporations proposed to span a combined distance of 1,660 miles from the Missouri to the Sacramento.

Following a service in the East Room of the White House on April 19, and after the president's mortal remains lay in state in the Capitol rotunda, they were driven to the Baltimore and Ohio Railroad station where they were placed aboard a special nine-car train. "Coffin that passes through lanes and streets," Walter Whitman of Brooklyn and Washington was writing, even then, "Through day and night with the great cloud darkening the land."[24] Everything was draped in mourning. The funeral coach itself, commissioned originally for the president for his personal travel, had been built inside the seized stockade at Alexandria by the U.S. Military Railroad. It had never been used. "Just when the fact of its being built came to [Lincoln's] knowledge I do not know," recalled Colonel D. C. McCallum, who was present in the shops, "but as I recollect it, some of the New York newspapers opposed to his administration took up the matter and presented it in a very unfavorable light." Lincoln "utterly refused" to accept the luxurious car, he said, "or ride in it during his lifetime." The upholstery was dark green plush, the woodwork was black walnut, and the curtains were light green silk. "The ceiling was paneled with crimson silk, gathered into a rosette in the center of each panel. The American eagle with the national colors appeared in a large medallion on each side of the exterior."[25] The car, which, contrary to popular belief, was not armored, stood next to last in the procession—the other coaches, and the engine, were all new—as the train eased from the B&O depot and headed north. A pilot engine preceded it.

"From Washington to Springfield the train entered scarcely a town that the bells were not tolling," recorded Ida Tarbell in her passionate, two-volume biography of the president, in 1900, the work preceding her muckraker classic on Standard Oil. "The minute guns firing, the stations draped, and all the spaces beside the tracks crowded with people with uncovered heads."[26] "With the waiting depot, the arriving coffin, and the sombre faces," Whitman wrote, "With dirges through the night, with the thousand voices rising strong and solemn."

Mary Todd Lincoln had asked that the remains of their eleven-year-old son Willie, who had died in the White House of "bilious fever" in 1862, be disinterred to accompany his father ("he was too good for this earth," the president had mourned). Tad and Robert Lincoln were aboard, along with their mother, according to John Starr, although every other historian says Mrs. Lincoln was too prostrate to make the journey. Lincoln's secretaries, John Nicolay and John Hay, were there. Numerous military and civilian dignitaries, and a special Guard of Honor, and men from the Invalid Corps (who attended the body) crowded the coaches. There were no meals served on trains. There were no sleeping accommodations.

There were also no direct railroad lines. At Baltimore, the president's coffin was conveyed from the B&O to the rotunda of the Exchange Building, where it was opened and exhibited to many thousands of mourners. Then, the train was put onto the Northern Central Railroad tracks to Harrisburg, and after a three-hour viewing in the Pennsylvania state capitol, and a night of rest, the party left on the Pennsylvania Railway tracks, the engine bell tolling mournfully. "The school children along the line had strewn flowers over the tracks in such profusion at different stations," wrote Starr, who quoted the train engineer, "that it was with difficulty [that] the engine passed through, as the wheels in crushing the flowers became so slippery that the train almost stalled more than once." Independence Hall in Philadelphia sheltered Lincoln's remains over the weekend. "The sidewalks in front of the old Hall were littered with crushed hoop-skirts and bustles, then in vogue, that had become disarranged and broken in the congestion," the Pennsylvania engineer recalled. He was at the head of the train which left Philadelphia before dawn on Monday, April 24, and stopped it briefly at Trenton. "At many points arches were erected over the tracks," wrote Tarbell. "At others the bridges were wreathed from end to end in crape and evergreens and flags. And this was not in the towns alone; every farmhouse by which the train passed became for the time a funeral house; the plow was left in the furrow, crape was on the door, the neighbors were gathered and those who watched from the train as it flew by could see groups of weeping women, of men with uncovered heads, sometimes a minister among them, his arms raised in prayer."[27]

The Pennsylvania line terminated at the ferry slips of Jersey City. A draped ferry took the coffin and all attendants across the Hudson to the Debrosses Street slip, while another boat was readied for the funeral coach and the honor guard car, which were taken to the Hudson River Railroad depot. Half a million mourners shuffled by the open casket at city hall in a day. "The train went up the Hudson River by night," wrote the secretaries Nicolay and Hay in their ten-volume biography (1890), "and at every town and village on the way vast crowds were revealed in waiting by the fitful glare of torches; dirges and hymns were sung as the train moved by." At the end of the line, in East Albany, train and cargo parted company temporarily as the casket and attendants crossed the Hudson by boat and the train took the long way around to the New York Central's Albany depot by way of the tracks of the Troy Union, the Rensselaer and Saratoga, and the Albany and Vermont, really the matter of a short distance across an important crossroad twained by the Hudson. Meanwhile, at the New York state capitol, many tens of thousands from upstate New York, Massachusetts, and Vermont filed by. "Every train, boat and omnibus leading into the city was crowded with people wishing to pay their last tribute," Starr wrote. Twelve hours after they had set their burden down, Lincoln's pallbearers began the process again—to the New York Central for the Buffalo run. "As we sped over the rails at night," recalled Chauncey M. Depew of that line, "the scene was the most pathetic ever witnessed. At every cross-roads the glare of innumerable torches illuminated the whole population from age to infancy, kneeling on the ground, and their clergymen leading in prayers and hymns."[28]

Booth was cornered in a barn near Port Royal, Virginia, and shot.

Buffalo to Cleveland over the Erie and North East, and the Cleveland, Painesville and Ashtabula: "Despite a heavy rainfall," Starr found—rain bled the dye from draped buildings, staining them black—"the well-loved features were viewed during the entire day at the estimated rate of one hundred and eighty persons per minute." Cleveland, Columbus, and Cincinnati line overnight to Columbus and a day's viewing in the Ohio state capitol rotunda. "Night did not hinder them," wrote Tarbell. "Great bonfires were built in lonely countrysides, around which the farmers waited patiently to salute their dead. At the towns the length of the train was lit by blazing torches. Storm as well as darkness was unheeded." Columbus and Indianapolis Central Railway, also overnight, and a day in the statehouse at Indianapolis: "as Indiana had been his home from the time he was seven years old," recalled Starr, "until he attained his majority, the affection displayed here was, if anything, more pronounced than before, the falling rain undeterring the multitude from turning out en masse as a mark of respect to the martyred President. To this point also came a delegation from Kentucky, the state of his birth."[29]

"The dim-lit churches and the shuddering organs—where amid these you journey," wrote Whitman, "With the tolling tolling bells' perpetual clang, / Here, coffin that slowly passes, I give you my sprig of lilac." By then it was midnight on April 30 and it was time to move on to Chicago, and the funeral train chugged over the Lafayette and Indianapolis, the Louisville, New Albany and Chicago, and then the Michigan Central, a journey of some eleven hours. "Whenever we halted," recalled the Washington correspondent Horace Gobright, "flowers were brought into the funeral car, and placed upon the coffin by the delicate hands which had culled them for this purpose." Lincoln rested in the rotunda of the Chicago courthouse—where in life he had frequently practiced—for wholly two days. "The eastern members of the cortege could not repress surprise when they saw how Chicago and the Northwest came," writes Starr, "with one accord, with tears and with offerings."[30] Finally, then, the train was hooked up, and all the family members and dignitaries and honor guard and correspondents climbed aboard for one last journey down the Chicago and Alton Railroad tracks to Lincoln's hometown at Springfield.

General Dodge was waiting there, sent from the War Department with his staff and some of his command to attend the services. "I took my place at the head of the military procession," he wrote many years later, "and it was the saddest sight of my life. The streets were lined with thousands and thousands of people, evidently in great sorrow and distress and at every step, we could hear the sobs of the sorrowing crowd and every little while a negro would come out and drop down on his knees and offer a prayer. There was hardly a person who was not in tears, and when I looked around at my troops, I saw many of them in tears. As we paid the last rites to this great man, the sorrow was universal, for it was one of the greatest calamities of this or any other nation."[31] Thus came the end, on May 3, of that continuous funeral—which had stretched from Washington to Oak Ridge Cemetery, seventeen hundred miles over fourteen days, along the tracks of fourteen railroads—with its many hundreds of thousands of participants: "It would have been impossible," recalls Gobright, "to render greater honors to any mortal remains."[32]

The greatest friend of the Pacific Railroad was gone—Lincoln standing atop Council Bluffs looking westward across the Missouri and predicting "not one but many railroads will center here," whittling on the porch of the Pacific House with Grenville Dodge and "shelling his woods" about the route across the Plains and the Divide, leaning over a profile map with Dr. Durant, exhorting Dodge and Ames and countless others in Congress and business to do their parts. "The sweetest, wisest soul of all my days and lands," said Whitman. Taking his place in the White House was the fifty-six-year-old Tennessee tailor and Unionist politician Andrew Johnson, who at best was indifferent to the whole enterprise, feeling no regional tie to the route and enjoying no personal relationship with anyone connected with it. If anything, Johnson's sentiment would have been as it was in 1849, during his third term as Democratic congressman, when Southerners agitated for a transcontinental route from Raleigh, North Carolina, across Tennessee and Arkansas, and westward from Texas, complicating the mission of Asa Whitney and irritating the competing Thomas Hart Benton. For the rest of Lincoln's hard-won second term—all forty-six months of it—Andrew Johnson, who would certainly invite trouble upon himself when he attempted to continue Lincoln's federalist policies for the vanquished South—would barely lift a finger in the Pacific railroad's support.

17

"If We Can Save Our Scalps"

Mourning cast a pall and slowed government workings to a near standstill in Washington. "The President's death," wrote Silas Seymour to Dr. Durant, "seems to paralyze everything here." In a nearby office at the Interior Department, Lincoln's old friend from their circuit-lawyer days in Indiana and Illinois, John Palmer Usher, was so bereft that Seymour made terse note of it: "Secretary Usher takes it very much to heart."[1] In a month the interior secretary would break with President Johnson over his first efforts at Southern Reconstruction, and trade government service for a position as chief counsel for the Union Pacific Railroad. But now, in late April 1865, the secretary would have affixed his signature to a plan that would subsequently cause him some corporate discomfort; it was a route map for the Central Pacific, slipped in by the devilish Collis Huntington, and it drew a line straight from Sacramento to the briny shore of the Great Salt Lake in Utah, in defiance of the restriction of the 1864 act limiting the Central Pacific to only 150 miles of Nevada right-of-way. Audacious as ever ("I said to Mr. Union Pacific, when I saw it, I would take that out as soon as I wanted it out"), Huntington maintained exquisite timing during the month that the war ended and the president was murdered. No one would notice the switch for some time.

News arrived from California—in a letter from the Central Pacific's attorney, Judge Edwin Bryant Crocker, to Congressman Cornelius Cole—regarding the progress of work gangs toiling up past the black oak and manzanita-covered Sierra foothills. "We have now about 2,000 men at work with about 300 wagons and carts," the judge reported, "and I can assure you they are moving the earth and rock rapidly. We are now on some of the heaviest work in the mountains, but so far we have been very fortunate in meeting very little hard rock. You will be astonished when you come back and see the amount of work we have done. A large part of our force are Chinese and they prove nearly equal to white men in the amount of labor they perform, and are far more reliable. No danger of strikes among them. We are training them to all kinds of labor, blasting, driving horses, handling rock, as well as the pick and shovel. We want a body of 2,500 trained laborers and keep them steadily at work until the road is built clear across the continent, or until we meet them coming from the other side."[2]

In a few weeks, with the gentling weather on the Sierra slope, the labor force would recede to the lure of prospecting higher in the mountains, or in Nevada, or in the new fields of Idaho. "They come and go, much like 'tramping' Journeymen Mechanics," complained saturnine Mark Hopkins in one of his regular letters to Huntington. "There are today not above 1,600 men on the work. Two thirds of them are Chinamen—and without them it would be impossible to go on with the work. But China laborers are coming in slowly, so that Charley thinks the force will steadily increase from this time on."[3]

Charley Crocker now possessed a contract to provide construction to the California state line, with competitive bidding abandoned—there were no more sops to criticism about the "inside man" after the chaos of contractors below Roseville. He received, as had been the practice since September 1863, only five-eighths of his pay in gold, with the remainder in Central Pacific stock valued at fifty cents on the dollar. At least some of that money to cover the legions of workers and carloads of equipment appeared fortuitously in early April after the California Supreme Court ruled that San Francisco County bonds worth some $400,00 and owed to the Central Pacific for nearly two years should be paid; the bonds had been offered by county supervisors who balked at the voters' decision to purchase $600,000 in Central Pacific stock, but even this arrangement had been tied up in the courts for years. Moreover, the necessary legal fees had eaten away $100,000.[4] Nonetheless, the infusion was cause for relief in the boardroom, and more was due. Four weeks later, on May 12, Huntington accepted federal bonds with a face value of $1,258,000, representing reimbursement for thirty-one miles of track completed by the end of 1864. Of course, despite the fact that currency was finally rising in value, the bonds' value in gold was considerably less than their face: $953,030.[5]

The money was spoken for as it arrived—but at least they could measure progress in the continuation of trackage; work trains chugged up into the knoll-and-hollow mining camp of Auburn—founded in 1848 as Wood's Dry Diggings but renamed a year later by gold-seekers from Auburn, New York. Auburn sprawled at mile 36, some 1,360 feet in altitude, and its residents cheered with the arrival of passenger service on May 17. The added receipts would swell Central Pacific's revenue for May 1865 to $22,939.36, with a much higher figure expected for June—especially as the line lengthened another seven miles to the ponderosa lumber camp at Clipper Gap, altitude 1,751 feet, on June 10. Indeed, June receipts for freight, passengers, mails, express, and the company's Sacramento wharf rose to $32,429.07—"business," Hopkins wrote to Huntington with great satisfaction, "is constantly increasing."[6]

So, too, was their payroll. By midsummer, shiploads of seasick farmboys from Kwangtung, arranged by several labor contractors in San Francisco, had begun to dazedly descend gangplanks into the new world of *Gum Shan,* their destinies being not gold like their older brothers and fathers but the rough tumbled Sierra granite itself. As one they were attired in rough, baggy homespun, scratching from lice acquired during their crowded months-long passage, their pockets emptied by gambling, their shaved heads and long braided queues announcing their raw naiveté long before it would be seen that they spoke only Cantonese. By the chattering carload they took their first train ride up to ear-popping Clipper Gap and dispersed into tent cities on the hillsides, looking to the more seasoned California Chinese for translations and cultural touchstones. Already the enculturated Chinese had begun to take on new responsibilities. "We tried them on the light work, thinking they would not do for heavy work," recalled Crocker. "Gradually we found that they worked well there, and as our forces spread out and we began to occupy more ground and felt more in a hurry, we put them into the softer cuts, and finally into the rock cuts. Wherever we put them we found them good, and they worked themselves into our favor to such an extent that if we found we were in a hurry for a job of work, it was better to put Chinese on at once." Labor troubles among the whites, a constant concern, continued to fan the expansion of duties. "At one time when we had a strike among our Irish brothers on masonry," Crocker reported, "we made masons out of the Chinamen." Supervisor Strobridge retorted, "Make masons out of Chinamen!" Crocker's reply brought the Irishmen back to reality: "Did they not build the Chinese Wall, the biggest piece of masonry in the world?"[7]

The new workers were organized into gangs of twelve to twenty men, including a foreman and a cook. The foreman kept records, collected the gang's pay (it had risen by midsummer to $27 per month and would soon climb to $30), deducted for food and for any outstanding amount owed to the labor contractor, and dropped the remainder of gold coins—perhaps $20—into the hands of his charges. The cook obtained food, usually dried, from Chinese merchants in Sacramento and San Francisco, and it was decidedly unlike anything the whites had ever seen before. The cookfires sent up fragrant clouds of peanut oil and garlic, of simmering white rice or clear noodles, of stir-fried cuttlefish, abalone, shrimp, and oysters, of mushrooms, bamboo shoots, bok choy, mung beans, snow peas, and kelp, and, most often on Sundays, of pork or chicken slaughtered on the spot. Seeing this alien cornucopia, the Irishmen shuddered—and turned smugly to their unvarying menus of boiled beef, boiled potatoes, boiled beans, boiled coffee, and bread and butter. And the following day the mountainside was waiting for all. Toiling in the sun, everyone turned frequently to the circulating waterboys, but there again stood the cultural difference: the Chinese drank lukewarm tea, boiled at dawn in great pots and poured into glass jugs for distribution to the gangs, then dispensed from powder kegs throughout the day; the whites merely drank dippers of unboiled water, more often than not bringing on the dysentery which plagued their ranks and caused absenteeism to soar.[8]

The quality of the Cantonese—they were often called "Celestials" for their belief in a celestial kingdom—was exemplary. "They are ready to begin work the moment they hear the signal," observed the Reverend Augustus W. Loomis, a Presbyterian missionary who took a longtime interest in the California Chinese, "and labor steadily and honestly on 'till admonished that the working hours are ended. They have no storytelling; they have no sentinel set to watch while his companions enjoy their pipes, and to pass the word when the 'boss' comes in sight. Not having acquired a taste for whiskey, they have few fights, and no 'blue Mondays.'"[9] They further astonished their white co-workers by their daily practice of bathing and changing clothes before dinner, by their frequent laundering and mending of apparel. Their only vices, in fact, were gambling and perhaps an occasional quiet pipe of opium on a Saturday night before dreamland, contrasted to the clamorous whiskey-soaked revels and brawls on the next hill in the Irish camps.

It was around this time that someone related a story to Charley Crocker, which he subsequently told many years later to a congressional committee seemingly bent on halting Asian immigration. "We were working in the heavy work just [beyond Auburn], and the stage road from Virginia City passed alongside of the railroad at that point," he recalled.

> *There was a very heavy cut there, and we had employed white laborers at that time to work on it. We, at that time, thought that only heavy things could be done by white labor; but there were some Chinamen near by. This gentleman spoke to one of these laborers, asked him what wages they were receiving. I think we were paying $35 a month and board to white laborers, and $30 a month to Chinamen and they boarded themselves. Said the workman, $35. The gentleman remarked, "That is pretty good wages." "Yes," says he, "but begad if it wasn't for them damned nagurs we would get $50 and not do half the work."*[10]

More than one observer called the Chinese "antlike" in their industrious swarming over the Sierra slope, but in fact all people—whites and Asians, workers and foremen—looked like ants, which is to say, tiny and indistinguishable to anyone standing on opposing heights looking out over the yawning gulfs of empty space to the work crews and their busy aggregate movement, an isolated clang of a pickax somehow carrying all that distance, an inexplicable, silent puff of smoke suddenly joined by the disembodied thud of a placed charge, the entire enterprise dwarfed by those russet palisades, the tumbling cliffs, the dark green cedar and pine forests, the immensity of what lay ahead. Now they were crawling up the long, high ridge that divided the American River's North Fork from the Bear and the Yuba Rivers, as first appraised by Doc Strong and the stalwarts of Dutch Flat, and as certified by Theodore Judah as offering the sort of easy, continuous rise needed for a railroad. At Clipper Gap the line reached the top of the divide, and the ridgeline beckoned for some twenty-five miles ahead, but, as Ted Judah had predicted four years earlier, the broken crest made it necessary to run from gap to gap, skirting intervening hills on their side slopes. One rail historian, Wesley S. Griswold, has delightfully likened the ridge to "an enormous fossil spine," with track having to "wind around successive vertebrae and cross gaps between them."[11] Judah had hoped that in the time-honored engineering tradition, gaps could be filled with earthen embankments, but experience showed that the soil was too thin to meet their needs. Until earth fill could be transported from elsewhere, Crocker's builders relied on temporary trestles. These were constructed of extraordinary pines hauled down from forests around the Puget Sound—sunk into masonry at the bottom of the ravines, soaring in pairs forty, fifty, sixty, seventy, even one hundred feet high, tied together with horizontal timbers and crowned by ties and iron rails, swaying slightly in the winds and creaking in a way to terrify the workers who trudged across them, trying not to look down to the broken, tumbling riverbeds far below.[12]

Higher up the ridge, Judah's former assistant and successor, Samuel Skerry Montague, directed the survey parties as they sought to justify or improve on the four-year-old survey. Montague's new observations below Dutch Flat resulted in a savings of a half mile, some track curvature, and $50,000 in construction estimates, but for the most part the earlier line held. Above Dutch Flat, however, Montague's assistant engineer, Lewis M. Clement, saw that Judah's misgivings in the fall of 1863 were well placed. For speed's sake he had originally adopted a very heavy line along the northern slope of the ridge overlooking the Bear River, involving a sinuous and steep grade for nine miles, which would have been within congressional guidelines for engineering but would have expensively provided a slow, uncomfortable transit. Judah intended to send out parties to resurvey later, but then, of course, he died. Now, the engineer parties edged out along the ridge's southern slope, along the American River's North Fork. During numerous experimental forays, Clement's party laid down line after line along ridges, across creeks, canyons, spurs, and ravines, making plans for tunnels and then abandoning them, thinking ahead under that hot July sun to the winter conditions that would build up thirty-foot snowdrifts and the thaws that would precipitate avalanches down toward the imagined trackbed. Finally the engineers attained a good southern slope line, saving money and needing lighter maximum grades than originally projected: 116 feet per mile—the maximum allowed—was the rule for ten and a half miles above Dutch Flat, but for the next twenty-four miles, the rise was only 81 feet per mile on average, all the way to the summit. While all this was going on, other parties toiled above, running more experimental surveys across the summit and down to the state line in a coordinated campaign that would continue in the Sierra until winter closed its door. Meanwhile, the burgeoning workforce, now four thousand strong, inched its way upward.[13]

General Dodge was back in St. Louis by the first of May, having ridden with his soldiers in the Lincoln funeral procession in Springfield. Although many matters crossed his desk concerning the Indian threat to settlers and travelers within the Department of the Missouri, Dodge found himself with railroad business on his mind. Durant was still dangling the chief engineer's post, with a handsome salary, before Dodge's eyes, but it seemed strategic to the general to continue to resist Durant's advances, although he indicated enough interest to keep the Doctor clamoring. He wrote to Dr. Durant on May 1 to say that he had consulted with different members of the next Congress and that they were growing impatient. "There is no doubt but that you will have trouble in next Congress," he warned, "unless you get to running at least forty miles of the Union Pacific Rail Road before December 1st. It should be one hundred miles to make all things safe." Dodge proposed an additional flurry of action: first, "the combination of all of the connecting lines east to New York and Boston so that we can depend upon their influence," much as, he noted, the directors of the Cedar Rapids company (Oakes Ames, primarily) had forged strong, enviable political links with the eastern railroads and suppliers; second, get Congress to pay for completing the moribund Mississippi and Missouri Railroad across Iowa, giving the Union Pacific an unbroken supply line to its origin on the Missouri mudflats in Omaha; third, to establish a relationship with Brigham Young or some of his fold in Utah, "to work on the road through the mountains, say through the difficult passes"; fourth, to organize "a pleasure party across the Plains in connection with Ben Holiday. Select members of Congress and Senators and make the trip a pleasant one. Have along some able, discreet man who can point out the country and its advantages to them, and keep before them the construction of the Union Pacific R.R. This will do more good than all the money you can spend in Washington and will bring you support from a class of men whose hearts will be with you on the merits of the case alone." Perhaps Dodge shifted uncomfortably in his hard army chair for a moment before writing, more than a little wistfully, "I know if I were footloose it could be accomplished."

"The war is about to close," he wrote, for even then the remnants of the Confederate Army were tramping homeward, "and the great project before the people will be the Union Pacific R.R. Company. You should have this in view and allow no outside projects" (such as cotton smuggling, war profiteering, real estate scheming, or byzantine stock manipulating, he might have wanted to add) "to attract your attention." Then the general laid the butter on: "It is *big* enough," he told the doctor, "for the *biggest* man in America."[14]

Thomas C. Durant responded well to fawning, having a high opinion of himself, although by May he had embarked on a new recklessness which paled at least some of his earlier endeavors. Silas Seymour's proposed oxbow route—abandoning the original grade out of Omaha in favor of a circuitous, nine-mile route south almost to the riverbank town of Bellevue and then westward, gaining Durant and the other contractors nearly half a million extra in cash and thousands of acres in land grants—was now the reality feared by Peter Dey.[15] Having convinced the five government directors of the Union Pacific to approve the new and longer route under the false impression that it would obtain radically lighter grades, but without the mandated permission of the president of the United States, Durant ordered construction to begin. It ignited a conflagration of complaint in Omaha, which began to spread eastward to the larger towns and their newspapers and ultimately to Washington.

Durant's swift retort was telegraphed to construction supervisor Jacob House: do nothing permanent in Omaha, and survey the short distance—merely three miles—between Bellevue's riverbank and the oxbow line; to shipping agent Hoxie went the peremptory order: land all railroad supplies at Bellevue, not Omaha. The alarmed Hoxie, with his alliances and few investments in Council Bluffs and Omaha, dashed off an alarm to General Dodge: "The Doctor is pursuing a suicidal course." On the letter's heels came one from the general's brother, Nathan Dodge, who reported that the *Chicago Tribune* was calling the Union Pacific a "rotten" tool of crooked Wall Streeters. "I don't want you mixed up in it in any shape whatever," Nathan Dodge cautioned; "keep clear of it until after the investigation."[16]

Seeing the city's bright commercial future in serious danger, an Omaha citizens' committee fired off a protest to Durant in New York. He sent his reply on June 9: "Shall make no promises as to crossing the river. We had made our arrangements to build at Omaha. We have had enough interference. We shall consult the interests of the road whether the citizens aid us or not." Durant threatened the committee not to oppose the oxbow route down Mud Creek—otherwise he would see to it that the Cedar Rapids road would cross the Missouri at De Soto, and his M&M line would veer north to unite with that. "Your people and papers will destroy the last chance you have," he warned, "for the terminus of our road at your place will not help you if there is no road to connect east. If any more obstacles are thrown in the way, we shall make application to the President to change the terminus."

The reply was telegraphed the next day, bluster not disguising the capitulation nor the town's essential powerlessness: "The people here will be satisfied with the Mud Creek route, if Bellevue movement is abandoned and permanent buildings be erected here at once. Omaha must be the only point of connection with the Missouri river; without this there will be trouble."[17] Within a matter of days the triumphant Durant sent word to resume work at Omaha. "We trust," sniffed the *Omaha Republican* on June 16, "we have seen an end of the game of 'fast and loose.'" Work down the Mud Creek could commence—though there was still the legal matter of the president's approval. At Andrew Johnson's behest, the new interior secretary—it was Iowa senator James Harlan, a booster of Durant's causes in Congress, appointed by his longtime friend Lincoln just before the assassination—named a government commission to assess the oxbow. The impartial commission consisted of Colonel James H. Simpson, who served in the Army Corps of Engineers; Springer Harbaugh, the government director of the road; and Silas Morrow Seymour, late of the Interior Department and now a full-time consulting engineer with the Union Pacific. The new impartial commissioners headed west in July to make their judgment.[18]

The matter was not settled as far as a number of newspapers in the region were concerned; the *Omaha Republican* was still incensed, but its furor paled against that of the *Chicago Tribune*. "Why should the Union Pacific Railroad be so anxious to put an extra million of dollars into the pocket of the contractor," complained the *Tribune*, "after having arranged to give him a profit of three millions on his job of one hundred miles originally? There can be only one answer. The Directors are the real contractors and are building the road at the largest possible expense to the Government and the least possible expense to themselves." Nothing les than "a set of unprincipled swindlers" was running the Union Pacific—"that such a contract could be made at all is monstrous. That Congress can allow it to be consummated is incredible." The *Tribune* urged Secretary Harlan to "put his foot down upon this contract" and "protect the people from such an outrage."[19]

Not surprisingly, Harlan stamped his approval upon the commission's findings.

Would that another of Dr. Durant's plans was going as well. With the new federal law allowing the sale of Union Pacific first-mortgage bonds for one hundred miles of anticipated railroad, Durant had not even waited for certificates to be printed and swiftly put them up to sale. Alarmingly, almost none sold. The Doctor appealed to his friends in Washington. One name sprang to the minds of Secretary Harlan and Supreme Court Chief Justice Salmon Chase: a prominent New York broker, John Pondir. When called on, Pondir declined to buy any bonds, but he did agree to arrange a Bank of Commerce loan to the Union Pacific, for $1 million, with the bonds as security; it was hardly largesse, marking them down to 90 percent and charging a heart-stopping 19 percent interest. At this critical stage there was little choice but to forge ahead. "We were deeply in debt," remembered Oakes Ames of the Union Pacific's financial climate at the war's end, "and very much embarrassed."[20]

Ames did what he could at rounding up support, both for the Union Pacific and for Crédit Mobilier through the spring and summer, calling on a succession of bankers and businessmen and wheedling large sums for them. He then found a promising source for Crédit Mobilier investors right in the Capitol building: Congressman Samuel Hooper and John B. Alley subscribed, without apparent qualms that they, like Ames and other legislators in the future, were buying into a corporation under their influence as lawmakers. Oakes Ames's longtime co-investor, the Maine-born Charles A. Lambard, was similarly busy at rounding up support. However, despite (or perhaps because of) his years of association with Thomas Durant as an original Union Pacific director, as a co-smuggler of contraband southern cotton, and as an adopter of the Hoxie Contract, Lambard was now allowing himself a glimmer of doubt in the Doctor. Money poured out of Durant's office—much of it, to be sure, for iron, equipment, ties, and payroll—and Lambard somewhat cautiously affected a jocular tone. "You do spend an *awful* pile of money," he wrote Durant on July 11, "& I doubt if we can raise enough to keep you going."[21]

The day before, with a complete lack of fanfare, a small work gang standing on the Missouri mudflats above Omaha watched as several of their number unceremoniously hoisted the first rails of the Union Pacific onto the trackbed and pounded the spikes home.

The rails pointed west, to trouble, the main object of attention for General Dodge. Given some ten thousand new soldiers, a number of them reconstructed Confederates, the Missouri Department commander itched to use them. Along the Arkansas River, bands of Southern Cheyenne, Arapaho, Kiowa, Comanche, and Apache had come to life in the spring, moving furiously against stagecoach lines and scattered settlements, killing many whites. Dodge would have unleashed his troops but for public opinion. While the army, under Commanding General Grant, was now poised to strike, the nation was weary of warfare and politicians were quick to respond. Weeks following the Sand Creek massacre, Wisconsin senator James R. Doolittle, head of the Senate Committee on Indian Affairs, opened an investigation into the Coloradan volunteers, Colonel Chivington, Governor Evans, and indeed into the overall federal policy regarding Indians. By spring Senator Doolittle had led a peace commission into the Arkansas basin, carrying with him an endorsement from President Johnson, stilling Dodge from action with the judgment that all the southern tribes desired peace with the exception of the Cheyenne, who were rightly upset over Sand Creek. The senator put his blessing on negotiations.

Dodge called for a harsh military solution in the south. Overruled on that front by civilians, with a cavalry column under Brigadier General James H. Ford placed on standby at Fort Larned on the Arkansas, Dodge shifted his attention to the northern tribes, finding welcome support from Grant and Grant's new divisional commander, Major General John Pope, a West Point engineer and veteran of the Mexican War, whose defeat at the second battle of Bull Run had exiled him west to try his skill against Indians; Pope oversaw Dodge's Department of the Missouri, and the Departments of Kansas and the Northwest (the latter encompassing Wisconsin, Minnesota, and the eastern Dakotas). War parties of Sioux, Arapaho, and Northern Cheyenne had begun riding against white settlements and travelers, including, on May 19, against a small detachment of Dodge's soldiers caught on the trail between Fort Kearney and Leavenworth, with two killed and the injured left to die, and on May 20, against a stage rushing for Atchison.

Dodge proposed a three-pronged attack on the northern plains. One column five hundred strong, under Colonel Samuel Walker, was to move north from Fort Laramie, Wyoming, toward Montana. Another detachment of one thousand, led by Colonel Nelson Cole, would follow the Missouri past Omaha to the Loup Fork, and thereafter into Montana on the Powder and Tongue Rivers, linking up with Walker. And a third column of five hundred was to leave Denver for the North Platte River and Wyoming Territory, pushing farther to Powder River; these soldiers were under the command of an aggressive, red-headed officer decidedly in the good graces of Grenville Dodge, Brigadier General Patrick Edward Connor.

Originally from County Kerry, Ireland, Connor had grown up on the streets of New York City; uneducated, he had escaped a life of ditchdigging by enlisting in the army. He fought in the Seminole War and after a short return to civilian life went from New York to Texas, where, in 1846, he joined the Texas Volunteers to fight in the Mexican War, finishing as a captain in 1847. He had gone to California during the gold rush and later settled in Stockton, participating in militias. Upon the outbreak of the Civil War, Leland Stanford's predecessor in the statehouse appointed him a colonel of the Volunteers, and Connor had led his regiment along the emigrant trails in Nevada and Utah Territories to a new command of the District of Utah. He arrived at Salt Lake City in September 1862, hating everything about the district but especially its Mormon colonists. "It will be impossible for me to describe what I saw and heard in Salt Lake," he wrote a superior, "so as to make you realize the enormity of Mormonism; suffice it, that I found them a community of traitors, murderers, fanatics, and whores."[22] (A long life as a civilian in the City of the Saints would only harden his views.) As a military commander in 1862, echoing the opinions of much of the nation at the time (and indeed his later commander, Dodge), Connor saw the Mormons as a threat to the Union and to Christendom in general, with their polygamy and theistic society. He lost no time in building a fort for his headquarters, locating it on a prominence overlooking Salt Lake City and training a cannon on Brigham Young's house and church. For his part, enduring a barrage of harassment for a year, Young forbade his followers from having anything to do with the Yankees, leaving it to a small committee to trade with Connor's quartermaster.

Fortunately for the Latter-day Saints, the army had other enemies to pursue. With emigrants constantly passing over the trails to Oregon and California, Shoshoni tribal parties had waylaid travelers, killing a number. Connor dispatched troops northward to the desert country along the Humboldt River, where they were to capture and execute the guilty parties; failing that, they simply killed any male Indian they encountered. A greater clash occurred north of Salt Lake City along the Cache and Bear Rivers. There, Mormon settlers in greater numbers had been seizing lands belonging to the Northwestern Shosoni, forcing a great alteration in tribal hunting and gathering and, as winter settled in, bringing on widespread starvation. The Shoshoni retaliated by stealing cattle, food, and property from the white settlers. Two small expeditions sent north in November 1862 by Colonel Connor first attacked a large Shoshoni encampment in Cache Valley, and then returned a week later, murdering four hostages when some stolen cattle were not returned.

It was kerosene flung on fire. Revenge and retaliation dogged white settlers throughout the winter. Finally, in late January 1863, Colonel Connor himself led a two-pronged column of Californians northward through a howling blizzard to the Shoshoni winter camp on Bear River. The Indians were well entrenched, but combined infantry and cavalry tactics soon had them surrounded. The ensuing rout quickly turned into a massacre, with warriors, women, and children shot down indiscriminately, the survivors tortured and butchered, and female captives brutally raped—"some," reported a few Mormons along as guides, "were used in the act of dying from their wounds."[23] The Shoshoni chief, Bear Hunter, was captured, tortured, and killed with a red-hot bayonet thrust through his ear into his brain. Some 250 others were killed, with army losses being 23 killed and 14 wounded. Colonel Connor was promoted to brigadier general for his command of what is now known as the Bear Creek Massacre. Over the ensuing months, frightened tribes of Shoshoni, Gosiute, and Bannock in the area of Utah, southern Idaho, and southwestern Dakota Territory (or Wyoming) who had learned of Connor and Bear Creek lined up to sign treaties. Connor's name endured in those parts along with his actions. Some years later, by the time the Central Pacific ventured across Nevada and into western Utah, their laborers would encounter no troublesome Indians, "one reason being," construction boss James Strobridge would recollect, "that General P. E. Connor was sent out with a thousand soldiers a few years before and he cleaned up the country, destroying men, women and children indiscriminately."[24]

All this was history when Grenville Dodge addressed the situation in the spring and summer of 1865. General Connor had tried to widen the war to include the Mormons but was muzzled by new administration appointees. However, in 1865, Connor had been promoted as a brevet major general "for gallantry and meritorious service," and transferred to oversee the newly created District of the Plains, covering a vast and contentious area between Salt Lake City and Fort Kearney in Nebraska, including the high-traffic Overland and Oregon Trails. In March, Pope and Dodge sent him north into the largely unknown buffalo hunting grounds of present-day Wyoming and Montana; Dodge ordered him to chase and eliminate as many Indians as possible, and to run off any survivors into Canada. Connor told his soldiers that the Indians north of the Platte "must be hunted like wolves," and that he would stand for no parley. "Attack and kill," he ordered, "every male Indian over twelve years of age."[25]

It was a bad time to be caught on the plains, regardless of who one was. With the Civil War over, with thousands demobilized from the armies and reunited with families, the westward impulse was impossible to ignore. Goldfields in Montana and silver mines in Utah and Nevada lured many, while the City of the Saints and the blossoming farmlands in all of that unclaimed Indian Territory beckoned especially the Mormon families finally coming from the East. To the Sioux, Northern Cheyenne, and Arapaho north of the Platte, among them survivors of Sand Creek, the stream of outsiders crossing their lands on the whites' established trails had swelled to a mighty river, and worse, many were taking shortcuts across lands never before trespassed; bluecoats patrolled the emigrant routes and established new forts on their hunting grounds. They could not be everywhere in force, however.

Brutality ruled with an even hand. An escorted wagon train of Mormons on the North Platte bank was halted by a war party of a thousand, who, in a two-day battle, scalped and mutilated twenty-six soldiers. A stagecoach station in the Laramie Mountains of Wyoming was attacked by a force of two hundred Cheyenne, who killed five civilians and took two young girls as captives. A wagon train of Montana-bound gold-seekers foolishly traversing the Powder River hunting grounds of the Western Sioux, thinking themselves the vanguard for a new emigrant trail, were confronted by nearly the entire encampment of Sioux and Cheyenne under the leadership of Chiefs Red Cloud and Dull Knife. Told that the Indians would fight all white men "until the government hanged Colonel Chivington," the whites placated them with a large peace offering and were allowed to depart—although separate bands, caring nothing about the parley with Red Cloud and Dull Knife, continued to harass and pick off stragglers all the way to the goldfields. General Connor, meanwhile, was equally active. He constructed and garrisoned a fort in the center of the Powder River hunting grounds, modestly naming it after himself. Riding out ceaselessly in search of the enemy, somehow missing the larger forces, Connor's troops and their mercenary scouts—Pawnee Indians, mortal antagonists of the Sioux and Cheyenne—encountered small bands and wiped them out: twenty-four Cheyenne in mid-August (they were carrying scalps and booty) and a similar number, captured nearby after a raid against the bluecoats, executed and scalped.

Then, at the summer's end, Connor's forces saw the smoke from a large Arapaho camp on the Tongue River, west of the Powder; it was the peaceable tribe of Chief Black Bear, some 250 lodges, tending a large corral of three thousand ponies, sitting complacently far away from the strife and the white interlopers. When the cavalry charged into the village, the Arapaho were chased outward so quickly that many children and women were cut off and cut down by the soldiers. All the lodges were torn down, heaped with clothing, possessions, and the tribe's entire store of winter food, and torched. A thousand ponies were rounded up by the Pawnee. When they left, sixty-three scalped Araphao, among them the women and children stragglers, were left behind. No, it was no time to be abroad.[26]

At some point in the late summer or early fall of 1865, it finally began to dawn on the administration that the plains expeditions of Generals Dodge, Connor, Walker, and Cole were serving only to inflame the situation. Militarily, tactics that worked on paper, or at least against the Rebs, were proving ineffective. In unwieldy numbers—regiments and brigades rather than companies and battalions—the bluecoats fanned out against an evanescent enemy, requiring a level of logistical report which simply could not be provided. Over and over, particularly under Connor's command, troops found themselves chasing after only thirst and starvation, sometimes coming very close to wiping themselves out when supplies collapsed. Moreover, the Indians had little trouble evading the huge, slow-moving columns with their pack trains, their extraordinary clouds of dust and din.[27]

The entire campaign would ultimately be viewed as an expensive failure, a repudiation (if one chose to see it that way) of a military solution on the plains. Insight, however, was at a premium. And to the rough-riding commanders facing what they believed to be a lower form of human life, nothing racked their faith. In the field, there were the satisfactions of scalps, booty, and their own loud momentum. And as the departmental commander, Grenville Dodge was pleased, particularly with Patrick Edward Connor. He recommended that the Irishman be promoted—because of his "energy, capacity and fidelity to duty."[28] Above Dodge's head, however, Connor was not so satisfactory; less than because of the massacres than because of his virtual abandonment of two-thirds of his troops to awful deprivation, Connor earned a large share of blame for the Powder River failure and was relieved of his command. Angered, he returned to Utah without even bothering to write a report. Connor would be honorably mustered out in the spring of 1866, later entering the newspaper and mining businesses. A lifelong friend of Grenville Dodge, he would receive a Union Pacific contract for supplying ties in Utah, in 1868.[29]

Surprisingly, work on the Union Pacific continued out in those killing fields. A thousand laborers graded and laid rails in Nebraska, almost too worried about raiding parties to notice that they were not being paid; after hanging on for two months on contractors' grub and no cash, they finally were satisfied. Beginning in May, engineering parties had moved out tentatively into the landscape. Ogden Edwards's group, armed to the teeth and escorted by bluecoats, took measurements along the Platte Valley out past the Loup River. Samuel Reed's party consisted of himself and two assistants, backed up by Mormon porters furnished by Brigham Young in Salt Lake City and, when needed, by cavalrymen. James Evans occupied himself in the Black Hills.

Reed covered a good deal of terrain. Seeking to find a way from Salt Lake City to Denver, he crossed the Wasatch Mountains from Weber Canyon southeastward to Green River, finding nothing practical; next, he tried the old emigrant route through South Pass, but saw that it was too long and too prone to blockage from snow. Thus it looked as if Denver would be ruled out. Back in Salt Lake City, Reed reprovisioned and set out westward through the blinding alkali desert toward the Humboldt Valley. He found virtually no obstructions to a railroad line in nearly three hundred miles—"without a cut or embankment exceeding 15 feet or grades exceeding 75 feet per mile," he would write to Durant—but the team suffered mightily in that late-summer desert from thirst and the glaring sun, finally turning back seventy-five miles from the farthest edge when their water gave out. The "high rugged range" of the Humboldt Mountains, he reported, disclosed three possible traverses, only one, some twenty miles north of the famous Humboldt Pass, seeming practicable to Sam Reed. A telegram from Durant ordering him to push farther west to the Truckee Valley and the California line arrived too late in the season to provision such an extension. Disappointed, Reed rode back to Salt Lake City to settle his accounts with Brigham Young, before going home to Joliet to write his report.[30]

James Evans's engineering party operated in more hazardous territory. In the obstructive Black Hills again, he addressed the oversteep, rapid ascent and descent which, a year before, had seemed too much to meet government limits. Only by climbing along the South Platte and Cache la Poudre Rivers in northern Colorado Territory could he mark a barely acceptable route; only by clambering through Antelope Pass could he draw a line down to the Laramie Plains; another line, from Lodgepole Creek to Crow Creek, and thence across what would be named Evans Pass, offered another alternative. He emerged with good, solid data but many misgivings, and not just about the steep, impractical terrain. "The hostility of the Indians (everywhere) made explorations extremely difficult and dangerous," he wrote in his report. "Until they are exterminated, or so far reduced in numbers as to make their power contemptible, no safety will be found in that vast district extending from Fort Kearney to the mountains, and beyond."[31]

Before too long General Dodge himself was addressing the riddle of the Black Hills. Returning with his brigade in mid-September from the Powder River expedition, Dodge rode south from Fort Laramie; habitually, he took a few men and, leaving the main body of troops and mule teams plodding below, rode up all the approaches and passes into the Black Hills from Fort Laramie south, scanning the range overhead through powerful binoculars in search of an undiscovered pass for the railroad. Throughout, he followed the lead of a civilian mountain guide, Leon Pallady, whom he had hired in the nearly deserted Julesburg a month before. On the day his army reached Lodge Pole Creek, the general took perhaps half a dozen men and Leon Pallady and followed the old emigrant trail up the creek to the summit of Cheyenne Pass. Still in the forenoon, they struck south along the crest of the mountains "to obtain a good view of the country," Dodge reported, "the troops and the (wagon) trains." They had entered the valley of a tributary of Crow Creek when, about noon, "we discovered Indians, who, at the same time, discovered us." The mounted war party—as many as three hundred—was between Dodge's little group and their main column. "I saw our danger," the general wrote, "and took means immediately to reach the ridge and try to head them off, and follow it to where the cavalry could see our signals. We dismounted and started down the ridge, holding the Indians at bay, when they came too near, with our Winchesters." Pausing during the descent as his men ignited signal fires and shot toward the encircling warriors, Dodge's trained eyes noted a crucial feature about the ridge they were following: "it led down to the plains," he said, "without a break." He turned to his guide with unconcealed excitement. "If we save our scalps," he exclaimed, "I believe we've found a pass through which the Union Pacific can go."[32]

"It was nearly night when the troops saw our smoke signals of danger and came to our relief," Dodge recalled. The Indians disappeared into the hills. With the escort, Dodge picked his way down the unbroken ridge all the way to the plains, estimating the descent at something under one hundred feet per mile. It would measure out at ninety. Rising to a pass some 8,236 feet in altitude, Dodge's discovery, when finally the Union Pacific reached it, would stand for years as the highest point reached by any railroad in the United States. He would give it the name Sherman for his "great chief," William Tecumseh. At its base down on the Laramie Plains, as Dodge rode off to rejoin the dusty caravan, he noted a lone pine tree marking the entrance to the Rockies.[33]

18

"I Hardly Expect to Live to See It Completed"

In the early evening of Thursday, August 17, 1865, at the Hang Heong Restaurant on Dupont Street in San Francisco, some thirty-five prominent American businessmen and thirty leading Chinese gentlemen assembled to honor four dignitaries newly arrived from the East. At heart all four visitors were journalists, though two had prospered in national politics, and all four were in a position to aid the cause of enterprise in the West. In a matter of days they would be minutely examining the Central Pacific, but here they would be confronted by a more challenging cultural experience, quite unlike anything they had previously encountered. Most important, there was the Honorable Schuyler Colfax, the forty-two-year-old Speaker of the U.S. House of Representatives. Son of a Revolutionary War general, Colfax had grown up in Indiana and gravitated toward journalism, writing for, editing, and finally buying the *Register* in South Bend. A frequent political correspondent for Horace Greeley's *New York Tribune,* Colfax had entered Whig and Republican politics early. His supporters called him "Little Greeley of the West," while his enemies named him "Smiler" Colfax, and as a radical Republican and architect of Reconstruction, Colfax would earn many enemies. But there were assuredly none abroad that night, and no ill-will.

Accompanying Speaker Colfax was William Bross, the fifty-year-old lieutenant governor of Illinois and editor of the *Chicago Tribune,* who had been dogging the heels of Dr. Durant, and who had been singularly unimpressed with the Union Pacific's lack of progress that summer; Samuel Bowles of Massachusetts, the waspish but talented forty-year-old editor of the *Springfield Republican;* and Albert D. Richardson, thirty-two, Greeley's favorite, the daring young war correspondent of the *New York Tribune* who had spent nineteen months in a Confederate prison. "I have sat at good men's feasts," Richardson would confess to his readers after this San Francisco repast, "both to the stalled ox without hatred, and the dinner of herbs where love was. I have enjoyed the hospitalities of Mexican haciendas, Arapahoe lodges, choctaw cabins, negro huts and rebel prisons; but this was a new gastronomic and social experience."

With today's cultural sophistication it is hard to appreciate how alien was their experience as they were conducted to the second-floor banquet hall, handed ivory chopsticks, and confronted with no less than 325 separate dishes. Richardson's partial catalog listed "bamboo soup, bird's nest soup, stewed sea-weed, stewed mushrooms, fried fungus, banana fritters, shark fins, shark sinews, reindeer sinews, dried Chinese oysters, pigeons, ducks, chickens, scorpion's eggs, watermelon seeds, fish in scores of varieties, many kinds of cake, and fruits ad infinitum." The courses took three and a half hours. "Mr. Bowles partook from about a dozen" dishes, recalled Richardson, "Mr. Colfax from forty; I suspended somewhere in the seventies; but Gov. Bross religiously tasted every one."[1] Richardson was an adventurous type; less so was Samuel Bowles, who was uncomfortable being in such close proximity to Chinese. Looking around at his hosts, comparing them to other Chinese merchants he would meet in the city and to the laborers he would glance at out on the trackbed, Bowles had already made up his mind about the Chinese. "There are a few men of great intelligence and wealth and ability among them," he would say. "These are of larger stature and finer presence than the rest, whom although not the poorest and most debased classes of the Chinese—not the Coolies proper—are yet of a low type, mentally and physically, and show little capacity for improvement. Most of them can read and write, but all their education lies in a simple, narrow range, and here, as in their work, they all show a certain sure and uniform attainment, beyond which it seems impossible for them to go. They can beat a raw Irishman in a hundred ways; but while he is constantly improving and advancing, they stand still in the old ruts." The toasts and mutual compliments, the outpourings of champagne and black tea affected him only a little. He noted with a hygienic and racial fastidiousness that his hosts, in urging delicacies upon him, were sharing the very chopsticks they placed in their own mouths. "I went to the table weak and hungry," he would write, "but I found the one universal odor and flavor soon destroyed all appetite; and I fell back resignedly on a constitutional incapacity to use the chopsticks." Before the end of the evening, Bowles slipped out with a sympathetic banker to an American restaurant down the street—"the lost appetite came back," he celebrated, "and mutton-chops, squabs, fried potatoes and a bottle of champagne soon restored us."[2]

The Colfax Excursion, underwritten by the companies with federal contracts to deliver the mails, had begun on May 22 at Atchison, Kansas, where the Speaker and his three invited companions had arranged to rendezvous. At Fort Leavenworth the coach they would board arrived bristling with arrows, its badly shaken passengers recounting an Indian attack some 150 miles out. Colfax's company was forced to delay until an escort arrived with the newly installed General Patrick Edward Connor, who accompanied the stage for its first two hundred miles to Fort Kearny. They would see no hostile parties during this part of the journey, only many long wagon trains on the green prairie, "the wagons . . . covered with white cloth, each drawn by four to six pairs of mules or oxen, and the trains of them stretched frequently from one-quarter to one-third of a mile each," reported Samuel Bowles. "As they moved along in the distance, they reminded me of the caravans described in the Bible and other ancient Eastern books." They were well armed, "alert for an Indian attack, but none appeared; and not even buffalo or antelope gave us opportunity to test our vaunted prowess with firearms."[3] For his readers, Bowles noted some Mexican terms in currency out west—"an inclosure for animals was called a 'corral,'" he wrote; "a house of turf and mud is of 'adobe;' and a farmhouse or farm a 'ranch.'" As the plains became more barren, with a profusion of prickly pear and sagebrush, they frequently passed the bleaching bones of oxen and horses and, sometimes, the raised, open-air burial structures of the Plains Indians.[4]

All the soldiers who accompanied them and guarded the stagecoach stations were former Confederates—"they styled themselves 'galvanized' Yankees," wrote Albert Richardson, and they "were faithful prompt and well-disciplined." Four and a half days out from Atchison, their coach rolled into Denver. Festivities and tours of the mining district behind them, they moved westward. "Beyond Denver," wrote Richardson, "the road had been practically closed for several weeks by Indian hostilities." Two hundred miles out, war parties had driven off the livestock from three coach stations; Colfax was told that "two emigrants were found dead upon the road—one scalped, the other with throat cut from ear to ear, and thirteen arrows in his body." It "obliged us to lie by, in dreary cabins, in the wilderness of the Bitter Creek desert country," wrote Bowles, "till other horses and soldierly protection could be procured."[5] At the North Platte ford, awaiting a ferryboat, the excursionists saw an attack beyond the opposite bank on an eastbound Mormon wagon train. "The Indians came over a hill," said Richardson, "in a sharp dash upon the animals, hoping to stampede and secure them." The emigrants were able to drive off the attackers before Colfax's escort could reach them.[6] Despite these excitements, the Colfax Excursion passed through the mountains to the Great Basin, and Salt Lake City, and across the broiling desert to the mountains of Nevada and California, without further incident. "Seven weeks of steady journeying," exulted Samuel Bowles, and they reached San Francisco on the eve of the Fourth of July.[7]

Between their arrival on the Pacific Slope and their inspection of the Central Pacific, the Colfax excursionists visited Oregon and Washington, and combed northern California; they also toured Yosemite Valley. At the end of August, they addressed nearly their last task of the long trip when they left San Francisco by steamer for Sacramento, where a smiling Leland Stanford awaited to put them aboard the railroad. The Associates planned to spare no effort for the important cause of public relations—they were no doubt aware that Colfax's expenses were largely being borne by a nemesis, Ben Holladay, the "Stage Coach King," and they had been wearily fighting the dependable rumors about the impossibility of their enterprise. Indeed, Mark Hopkins would soon write Huntington in New York to report that "Touting Colfax & his party over the route & removing the erroneous impressions created by the Holliday & McLane crowd [Joseph McLane, owner of the rival Placerville Railroad] together with our daily routine business, has given us the busiest month just past that ever I have seen."[8] With Stanford and the Easterners aboard, the special train chugged up to the railhead, fifty-five miles out, at the little mining camp called Illinoistown, altitude 2,422 feet. Settled in 1849 when it was the head of "wagon navigation" to the gold mines, where supplies were loaded on mules for the final stretch to the isolated camps of Red Dog, Little Rock, and You Bet, Illinoistown had been known for its first few years as Alden Grove; it was about to weather its second change of identity, as Stanford ceremoniously told his visitors over iced champagne: henceforth, Illinoistown would be known as "Colfax."[9] The Speaker of the House was pleased, flattered, humble.

The dignitaries transferred to horseback and climbed some twelve miles upward through cuts of vivid reddish rock and soil along the trackbed, passing numerous deep streambeds, the sites of frenzied hydraulic mining operations which had yet to cease. Stanford told them that the Central Pacific had four thousand laborers at work on this section of the railroad, the proportion of Chinese having risen to 90 percent. "They were a great army laying siege to Nature in her strongest citadel," wrote the enthusiastic Albert Richardson. "The rugged mountains looked like stupendous ant-hills. They swarmed with Celestials, shoveling, wheeling, carting, drilling and blasting rocks and earth, while their dull, moony eyes stared out from under immense basket-hats, like umbrellas. At several dining-camps we saw hundreds sitting on the ground, eating soft boiled rice with chop-sticks as fast as terrestrials could with soup-ladles." Elsewhere, Richardson would continue riding his insect simile: the Chinese were, he wrote, "swarming among the Sierras like flies upon a honey-comb."[10] Perhaps "bees" would have been more apt—the Cantonese were currently working against the toughest terrain yet encountered in the entire project.

At Gold Run, twelve miles farther than Colfax and eight hundred feet higher in elevation, the riders changed to a six-horse stagecoach, and they continued up the Dutch Flat Wagon Road, "commanding," noted Richardson, "an endless sweep of dense forest and grand mountain, among graceful tamaracks, gigantic pines and pyramidal firs. Immense barns beside the mountain houses attest the length and severity of the winters. At many points we found the surveyors awaiting our coach to receive their letters and newspapers." Two hours after sunset, the stagecoach pulled over the summit, "when its wild gloomy grandeur is far more impressive than by day. It is boundless mountain piled on mountain—unbroken granite, bare, verdureless, cold and gray. Through the biting night air we were whirled down the eastern slope for three miles to Donner Lake, blue, shining, and sprinkled with stars, while from the wooded hill beyond glared an Indian fire like a great fiendish eyeball."[11]

Later Richardson would learn that the "Indian fire" high up on the mountainside was actually the campfire of Samuel Montague's railroad surveyors (presumably under the direction of the engineer L. M. Clement) shivering in the late-summer high-altitude cold while Speaker Colfax and company were welcomed into the fire-warm and pleasantly appointed hostel called the Lake House. The next day they spent with the surveyors "among the precipitous granite ledges." Already, he noted with foreboding, "snow lined the higher ravines and icicles hung from the water-tanks on the stage road." He learned that two laborers on the Dutch Flat Wagon Road had perished in a snowslide nearby during the previous winter: "it buried them fifty feet deep," he wrote. "In spring their bodies were found standing upright, with shovels in their hands." That final excursionary night, as the Easterners and their host Stanford joined the engineers in the Lake House's candlelit tavern parlor for a "long earnest conference, to determine upon the route near the summit," they pored over "maps, profiles and diagrams, held down at the edges by candle-sticks to keep them from rolling up." Richardson could not help but be amused at the democratic sight as he helped, with his words, to seal the Associates' public relations victory. "On their knees," he wrote, "were president, directors and surveyors, creeping from one map to another, and earnestly discussing the plans of their magnificent enterprise. The ladies of our excursion were grouped around them, silent and intent, assuming liveliest interest in the dry details about tunnels, grades, excavations, 'making height' and 'getting down.' Outside the night-wind moaned and shrieked, as if the Mountain Spirit resented this invasion of his ancient domain."[12]

If the "Mountain Spirit" resented the engineers, then what would it have thought of the toilers on the west slope of the Sierra, especially those addressing the forbidding, obstructive cliff face above Colfax? Named Cape Horn by Californians who well remembered their seagoing difficulties rounding another geological protrusion at the root of South America, this sheer mountainside towered over the American River, two thousand feet below, rising in the blue mist. With an incline of seventy-five degrees, not even the sure-footed surveyors had broached it; they had merely sighted on the far end and drawn a steadily climbing, wholly theoretical line across the cliff. The Cantonese were to take care of it, producing a seven-foot-wide shelf from which a wider platform could be scraped and blasted—and they were aided by picks, shovels, hand drills, wheelbarrows and one-horse dump carts, and the unlikeliest of materials, the humble, bendable reed.

Daily progress was measured by inches, kegs of black powder consumed by the hundreds, as slowly a narrow shelf lengthened alongside the yawning gulf and its roiling green river far beneath the laborers. A month passed. Cape Horn resisted. The ingenious breakthrough has been attributed to Samuel Montague, who was certainly resourceful, but it is far more likely that inspiration came from the Chinese themselves. Back at home, longer than anyone could remember, in places like the deep canyons of the Yangtze workers had been lowered down the cliff face by rope in large woven baskets, which enabled them to work in more efficient lines across the heights. If Strobridge were to send down to San Francisco for reeds, the Cantonese would weave such baskets and submit themselves to this new, highly dangerous work. It was a safe bet for Strobridge—what could he lose but a few Celestials?—so he agreed. With materials in hand, the Chinese worked through the evenings fashioning waist-high baskets, with four woven grommets to hold the rope, and decorated with symbols to ward off both mishap and evil spirits. Two men were required to lower a swaying basket down the cliff, its occupant sometimes losing his hat to the gusts of wind, until the man reached his appointed place and began hammering and drilling at the granite face. Then there was the cautious tamping of powder into the hole, the difficulty of striking a match to a fuse in the wind, the wild shout of success, and the worker scrambling out of his basket, up the ropes, the alerted men above hauling him up as fast as they could, the dull crash of the explosion just below him, the spraying granite dust and rock fragments arcing out over the gulf and down to the distant American River, containing—this time but not always—no unfortunate immigrant whose unrecoverable bones would never return to the homeland.[13]

As Charley Crocker had proclaimed to a *Sacramento Union* reporter in the summer, "the work goes bravely on." The *Union* agreed: it was "the largest construction enterprise in the world, not excepting the Suez Canal." The Central Pacific, its forces swollen to 6,000 (the majority Chinese), running 6 locomotives and 134 freight and passenger cars (with more on order or being locally constructed), had by October 1865 laid some 135,000 redwood ties, and 6,000 tons of iron rails, and had constructed 215 stone or brick culverts. The company was consuming, when Cape Horn loomed, some five hundred of the seventy-five-pound kegs of black powder every day, six days per week. As Leland Stanford reported to President Andrew Johnson and Interior Secretary James Harlan, the twenty-five-mile section above Colfax, which would stretch beyond Dutch Flat, was "being rapidly prepared for the track."[14] Rapidity did not denote ease. On the heavily timbered mountainsides, all trees were removed in a swath from sixty to two hundred feet wide. "Where there were any light banks," James Strobridge would recall, "the stumps had to be grubbed out until you could get everything cleared within 2 or 3 feet below the roadbed; all stumps and everything, of course, had to be removed." The higher they climbed above Colfax, he said, "it became much heavier."[15] Moreover, for ten and a half miles of this distance they imposed the maximum legal grade of 116 feet per mile, and in five instances along the new section rather tight curves were necessary. In addition there were now tunnels to contend with.[16]

No less than ten tunnel projects were begun between the summer's end and the fall. All of course had to be high enough to accommodate the tall engine smokestacks, and all would be wide enough to eventually fit two tracks when traffic demanded. The greatest burrowed beneath the Sierra summit, and it was to be sixteen hundred feet long; soon it would be addressed from opposite sides of the summit, with workers chipping and "shooting" (the California word for blasting) toward a central meeting point. Seven miles away to the east lay another long tunnel site, at nine hundred feet. The five others, at crucial intervals between Dutch Flat and the Truckee River, though shorter, proved to be no less difficult. Against the uncooperative granite, the average progress won with hand tools and black-powder charges was only seven inches per day.[17]

It was to be billed as a "Grand Excursion." It could have been not unlike the gala Colfax Excursion—bunting, brass bands, and all—when the Central Pacific's Leland Stanford popped champagne corks for dignitaries and reporters. And Thomas C. Durant journeyed all the way from New York to Omaha to preside. His illustrious guest of honor for the initial ceremonial jaunt on the Union Pacific's tracks, at that point in November all of sixteen miles, would be one of the great heroes of the Union, Major General William Tecumseh Sherman, since summertime the commander of the military division of the Missouri. With headquarters in St. Louis, Sherman now oversaw the military departments of the Ohio, Missouri, and Arkansas—everything north of Texas and north of the Ohio River, out across the plains, across the Rockies, and including Utah and Montana. His power, prestige, and influence would be of immense value to Dr. Durant and the Union Pacific. His brother, Charles Sherman, served as a federal director of the line. His friendship with Grenville Dodge would be strategically important. And of course Sherman's interest in railroad enterprises was long-standing, going back to his brief period as a civilian in San Francisco where, as a banker, he became a vice president of the Sacramento Valley Railroad, the penultimate engineering triumph of Crazy Ted Judah. Upon reaching St. Louis, Sherman recalled, "I put myself in communication with the parties engaged in the work, visiting them in person, and assured them that I would afford them all possible assistance and encouragement." Durant, he thought, "seemed to me a person of ardent nature, of great ability and energy, enthusiastic in his undertaking, and determined" to get the road built.

In his welcome at the Missouri ferryboat dock, in his appointments at Omaha's best hotel, in his escort down to the start-of-track, Sherman was spared no courtesy. Standing steaming if not gleaming proudly there behind a small crowd of railroad workers and curiosity seekers was Union Pacific Number One, the corporation's first engine, conveyed in July over the Hannibal and St. Joe on a flatcar and pulled up the Missouri on a barge, and named in his honor: the *General Sherman.* And then the war hero was invited to hoist himself up behind the *General Sherman*—not into a passenger car, for the Union Pacific did not yet possess one, but onto one of two humble flatcars. Nail kegs had been fixed down as seats. The scourge of the South found himself a perch, and Mr. Union Pacific sat down beside him, with the attending army officers and Citizen Train and the engineer-at-large Silas Seymour and at most a couple of dozen others, as the *General Sherman* lurched forward and steamed them from the town down the aptly named Mud Creek and around Seymour's oxbow and out toward the Papillon River, smoke and cinders sometimes drifting through the swaying celebrants. Where the track ended, and where the prepared gradient picked up to run another hundred miles out onto the prairie, they were all subjected to speeches. The general's thoughts naturally went back to Sacramento (Ted and Anna Judah somewhere there in the crowd, although Sherman left no acknowledgment of the engineer who had made all that was subsequent possible). He reflected on the earlier promise and jubilation. And between then and now lay the great gulf of the war and all the dead and the waste. "When the orators spoke so confidently," Sherman would write in his memoirs, "of the determination to build two thousand miles of railway across the plains, mountains, and desert, devoid of timber, with no population, but on the contrary raided by the bold and bloody Sioux and Cheyennes, who had almost successfully defied our power for half a century, I was disposed," he would confide primly, "to treat it jocularly."[18]

The ironic humor—of nail-keg seats, of the high-flown rhetoric, of the great hostile void awaiting the enterprise—was certainly in keeping with the situation. With months of productivity squandered by the oxbow controversy, with shortages of laborers and then of cash to pay them, it had gotten to the point that Hub Hoxie could boast to Dodge that they were making a half mile per day, with all of eighteen miles "in running order"[19]—this in a land with no mountains, no Cape Horn, no tunnels, no towering trestles. It was a time when humility might have been called for, but humility would have buried them. Hoxie, that vain office seeker, trying to grow into his job, saw his best efforts running aground as with the low waters of the Missouri, which hampered heavy boat transportation and the delivery of their thousands of wrought-iron rails and bridge timbers and ties, or disappearing altogether, as with boat captains who demanded pay up front and who could usually do better elsewhere.

The problem with ties was a serious vexation. The nearest hardwood tracts lay up the Missouri, and Durant frowned on the expense of tapping the relatively plentitude in Minnesota and Wisconsin. By summer's end four steam sawmills on the river north of Omaha were cutting and shaping ties—forty thousand had been sighted in July waiting to be rafted downstream. Apparently, Paul Bunyan could not have done a better job at clearing them. Indeed, as J. Sterling Morton has complained in his history of Nebraska, "the destruction of our finest forests—and especially of our precious hardwood trees—in the Missouri valley will always be resented as an act of vandalism which no exigency such as they might plead could excuse or palliate."[20] But ecological considerations aside (for of course there were none of those at the time), with engineers calling for twenty-five hundred ties to the mile, the paltry local forests would not stretch far.

The only solution—a bad one—was to make the hardwood stretch by interspersing ties of cottonwood, five soft to one hard, which could be found not only higher in the Missouri Valley but also growing along the Platte and its tributary valleys. Cottonwood had the staying power of pasteboard and a spike-holding power somewhat better than hardtack; cottonwood began to rot, local wags could claim, before the last axe-stroke brought it down. On a trackbed it might last a year or two. This could be extended at least past government inspection and some traffic, Durant had found, by employing a new technology called Burnettizing. A tenth of a mile's worth of ties would be stacked inside a gigantic steel cylinder at the Omaha shops, and pumps would create a vacuum inside; a zinc solution, forcefully injected into the cylinder and by capillary action into the outer pores of the lumber, would produce a burnettized tie. It fairly reflected light, like metal, and weighed nearly as much as metal, and it resisted decay a little better than untreated cottonwood. It was cheap. It would have to do. It would have to do as well as the lightweight iron rails ordered by Durant, each some six pounds lighter per yard than the federally mandated standard of fifty pounds per yard, and as well as all the attendant ironwork—the train sidings, even lighter, and the rail fasteners, called "chairs," which tended to rock and vibrate and were certainly less secure than the most up-to-date technology of the tightly bolted "fishplate"; if there were cut-rate spikes available they certainly would have showed up on the Doctor's order pad.

Economies notwithstanding, the money frittered away distressingly. Congressman Ames's great stock-selling efforts of the summer, along the lengths of Pennsylvania Avenue and Wall Street and Tremont Street, continued into the fall until it became necessary for the Crédit Mobilier to issue more stock to advance greater capital. Nearly an extra million could come available after the announcement in September. To Dr. Durant, new stockholders meant money but also the inconvenient sloppiness of accountability. Ames and Lambard had each objected to his profligate spending. Now the "Clipper Ship King," John M. S. Williams, wrote to complain. He had gone down to New York to look at the books but had been stolidly rebuffed by a Durant subordinate, and when he sought an audience with the Doctor himself he could only look meaningfully at him through a crowd. It was not fitting acknowledgment for his $50,000 cash investment. "We are associated with you," he wrote,

> *in the management of the Credit Mobilier & thus become in fact (with you) the actual managing Builders of the U Pac R Road (*& will be held responsible as such*)—We have an Organization ahead of us—the* Rail Road Company*—in whose management you lead—but in which we are not recognized, thus you have power and authority which we do not.*
>
> *But for the purposes of the Credit Mobilier, it is necessary that we should know* all *that is going on, & that I should have free access to all Books and papers.*[21]

Williams added pointedly that John Duff, the Boston stockbroker whose own involvement was $100,000, had read and approved of his letter. To relieve himself of this headache, Durant agreed to admit John Duff and Sidney Dillon, the contractor, into the Executive Committee of the Crédit Mobilier's Railway Bureau. It was an unpalatable but necessary move. The Doctor could probably count on the unswerving support of five members of the committee; four, including the newcomers, were wild cards. Even with Durant himself as chairman, he could no longer assume an unobstacled management. Truly, it would be a problem which would not go away.

And in that season of fits and starts with the Union Pacific and the Crédit Mobilier, one final irony must be recorded. When Ames and Durant moved to increase the capital stock of Crédit Mobilier to avoid complete paralysis, it has been obscurely claimed that one of the first new investors to come on board was a most familiar name—not a large investor, and not even a name on which Durant would choose to capitalize for whatever public relations value it would hold for those in the know—but if the assertion is so, the Doctor would probably not suppress a small smile of triumph when he made out a stock certificate in the name of his former chief engineer, Peter Anthony Dey.[22] Having been asked in July by Nebraska governor Alvin Saunders to consult with the Interior Department's engineers regarding the oxbow controversy, Dey had begun to see the future. "The company will have it all their own way," he wrote to George Francis Train, "probably little local opposition will be made. The belief prevails [in Omaha] that the influence of the company is, and is likely always to be overshaddowing, that Mr. Durant is supported by a large political power which can control the questions in issue."[23]

"After Appomattox," a number of chroniclers have noted, "the nation faced west." Thousands that summer had made the crossing already, toward Utah, Nevada, Montana, Oregon, and California, to the mines and fertile fields and forests of El Dorado wherever for them it happened to manifest itself—and tens of thousands hunkered down to one more winter in the East or the Midwest, managing resources, against joining a great thrumming pulsebeat when the sap rose again. More than ever, the Pacific railroad was part of the national enterprise, and now, with the battlefields already obscured by a season's verdant growth, fertilized by blood, more than ever the Union Pacific and the Central Pacific would both expand, extend, and clarify that enterprise and all the energy behind it, implying a crushing force. With every eastward turn of the Central Pacific's driving wheels, with every westward lurch of the Union Pacific work train, the invisible line between them would grow more palpable.

Signals clashed over the heads of the great indigenous people in the intervening wilderness: the obvious messages of the white travelers and their billowing wagon trains, and the thunderous bluecoat cavalries; the awkward, confusing overtures of peace missions enacted by Congress in the spring, which sent negotiators to the upper Missouri and to the Little Arkansas to investigate matters and work toward a temporary settlement. A few peaceable tribes signed for peace by the autumn, giving an illusory surcease to those most concerned for the welfare of the railroads, the Homestead Act, and the mining interests at the far end of the still-faint Bozeman Trail. To the northern Plains Indians, however, there was only the baleful memory of General Connor, General Dodge, General Pope, and General Sherman. A bad winter awaited them, one of unending cold, famine, and disease. In Washington, Interior Secretary Harlan submitted his annual report to the president, who passed it approvingly to Congress. "In flagrant violation of treaties which had been observed by us with scrupulous good faith," he wrote, "and in the absence of any just ground of complaint, these confederated Indians entered into an alliance with the rebel authorities and raised regiments in support of their cause. . . . The perfidious conduct of the Indians in making unprovoked war upon us has been visited with the severest retribution."

After having erased Bear River, Sand Creek, Tongue River, and all the more ambiguous clashes, leaving only perfidy and barbarism on the other side, Harlan reported the half-hearted treaties in place by the end of the year before taking up the academic question of "total destruction of the Indians . . . openly advocated by gentlemen of high position, intelligence, and personal character." But he rejected genocide first on civilized grounds, though basing his argument chiefly on financial considerations. "The attempted destruction of three hundred thousand of these people, accustomed to a nomadic life, subsisting upon the spontaneous productions of the earth, and familiar with the fastnesses of the mountains and the swamps of the plains, would involve an appalling sacrifice of the lives of our soldiers and frontier settlers, and the expenditure of untold treasure. All the military operations of last summer have not occasioned immediate destruction of more than a few hundred Indian warriors. Such a policy is manifestly as impracticable as it is in violation of every dictate of humanity and Christian duty."[24] Treaties and patience were his stated hopes, especially with the Pacific railroads showing such promising progress in the year, with great hopes for the future. But with Tecumseh Sherman in charge of the military district, they at least could look forward to his brand of military efficiency.

"This is a great enterprise," General Sherman is said to have remarked to others who had balanced on nail kegs during the "Great Excursion" of the Union Pacific as the seasons prepared to shift, "but I hardly expect to live to see it completed."[25]

James Strobrige bossed his legions of Chinese and Irish laborers above Colfax, in the late autumn grubbing and blasting and excavating up that colossal ridge at altitudes ranging from twenty-five to thirty-five hundred feet, just below the snow line. He remembered the change of seasons coming early, commencing with pelting rains and ushering in "the three worst winters that ever were known in the Sierra Nevada Mountains." First there would be mud to rival even what Strobridge himself had encountered as a boy in northeastern Vermont, where the mud is so infamous as to be given its own season. "The winter of 1865–66 was a very wet winter," he would say, "it was terribly muddy. It was impossible to haul goods to the amount required for our use, and we were obliged to pack it over the mountain trails, and we had to pack bales of hay, even, a distance of 25 miles on animals, and pretty nearly all of our supplies for four or five months were furnished them in that way." Even the indomitable stagecoach which ran over the Dutch Flat Wagon Road from the railhead in Colfax to Virginia City was "stuck in the mud for six weeks in the streets of Gold Run," barely ten miles above Colfax.[26] "The soil is of a peculiar character," Leland Stanford would explain to congressional investigators some years later. "It is a red, spongy, thick soil, and takes up and holds water." When the stage was abandoned, when the railroad contractors' wagons were paralyzed, when for two months everyone and everything ascended the slopes on the backs of horses "until they could get up on the snow," they still "pushed the work up there as fast as possible."[27]

Snow had begun steadily falling on the summit in late September (after a warning fall in August); it arrived in quantity only a few weeks later. What followed was a steady pummeling of snowstorms. Six hundred laborers, working against the rock above Dutch Flat, were soon inundated nearly as badly as were the tunnel camps at Grizzly Hill and on either end of the Summit Tunnel, and at the lesser tunnel projects on the eastern slope. Down in relatively balmy Sacramento, Stanford and Hopkins (particularly Uncle Mark, who ventured out on the line perhaps once a year) could not initially understand why Crocker did not push his men faster. "I would say, 'Here is this snow in the way,'" Crocker recalled, "and I would ask, 'Who will pay for removing it?' And they would say, 'We will pay for it.' Of course I was not going to pay for it. I was not going to get any pay except for the removal of the rock. I know there was a 60-foot ravine that we cleared of snow and pitched it over and over and over before we could get to the rock. The engineers would not allow the track to go on snow or ice."[28]

After Stanford had seen it, he was a believer—the snow was "somewhere from 30 to 40 feet deep, as near as we could measure," he would remember. "The snow would fall sometimes five to six feet in the night; I believe in one case that nine feet fell in a single night. It obstructed all the roads, and made it almost impossible to get over the mountains." To do so they used not only horses but ox teams, "moving constantly trying to keep the road clear and the snow packed so that when the storm was over we could pass along with our material and could transport our iron." They "oftentimes had long tunnels under the snow to reach the mouth of the tunnel in the rock where we were doing our work."[29] Thus, for weeks at a time, the workers lived a troglodyte's existence, entirely underground, as the tunnel work continued; not only were crews going at the Summit Tunnel from both ends but the work expanded to run three shifts per day.

It was quite a contrast to a sixteen-mile Grand Excursion with nail-keg appointments, to lightweight rails and cottonwood ties: regular passenger trains of the Central Pacific rolled on schedule up and down fifty-five miles of line into the foothills at twenty-five miles per hour (freight ran at about half that speed) over heavy-duty iron and redwood; of the rolling stock, the Central Pacific owned six locomotives, six first-class passenger cars, two cabooses, one baggage car, thirty-nine boxcars, sixty-five platform cars. Twenty new dump cars were being bolted together down in Sacramento; a new locally fabricated locomotive was to arrive in that month of December; one passenger and seven heavy freight locomotives were somewhere at sea. "We have gone far enough already to convince the most incredulous, not only of the entire feasibility of the route, but that the work can and will be accomplished within the time stated," Stanford reported to President Johnson and Secretary Harlan. "The whole cost of the work done and materials purchased amounts to the sum of $5,596,476.89."

Impressive indeed; however, as the annual report would divulge, the Central Pacific was indebted by more than $3.2 million. It was no wonder that, with some polite urgency, Stanford urged the federals to issue bonds for the work now completed—the company had completed forty-three miles before seeing a penny of that federal aid, and the heavy work subsequently performed had dug it into a deep hole. With the exceptionally heavy and expensive work ahead for them in 1866, he wrote, "it will be seen that any great delay in receiving the means provided by Congress, will necessarily cause a serious derangement of our plans, and hinder the progress of the work."[30]

Huntington had by then succeeded in persuading a New York investment firm to sell the company's pitifully discounted securities. But not until they held the federal bonds in their hands could they relax and face the intimidating promises made to the president and Congress for progress and expansion in the coming year.[31]

Similarly at the end of the construction season, with the ground freezing and storms howling down upon the Nebraska plains, the Union Pacific had managed to complete forty miles of road after a furious regrouping of energy and resources. Now the line reached to the intersection of the Elkhorn Valley and the Platte, a few miles short of the stagecoach station town of Fremont, a Republican settlement incorporated during the 1856 presidential campaign of the Pathfinder. Dr. Durant's annual report was waiting to be finished when he journeyed out from New York City to inspect everything with Silas Seymour in late December, and to fire his younger brother, William Frank Durant, to whom he had briefly entrusted the reins out in Omaha. In his report he would disclose the purchase and arrival of four first-class locomotives, two passenger cars, twenty-five baggage cars, thirty-four platform cars, and nine handcars. Wood and iron for another twenty-two miles stood waiting for a thaw; beyond stretched an additional sixty miles of completed grading, and, amazingly, grading teams had gotten well into the next stretch of one hundred miles. The claims looked good on paper, and of course the Doctor would breathe easier when the railroad's share of federal bonds were in his hands. But Durant knew he had a large job ahead: to maintain control of the Crédit Mobilier in the face of his partners' uneasiness; procure a mammoth workforce for the spring, hoping to take advantage of the legions of newly demobilized veterans, mostly Irish, who would be desperate to resume productivity; bend the U.S. Cavalry and the Interior Department to the needs of the Union Pacific Railroad; and somehow transform the hodgepodge of little men running the construction crews and conveying supplies into a more efficient, even militarylike organization. That last task would be the key: to bring order out of chaos.

Part VI

1866
Eyeing the Main Chance

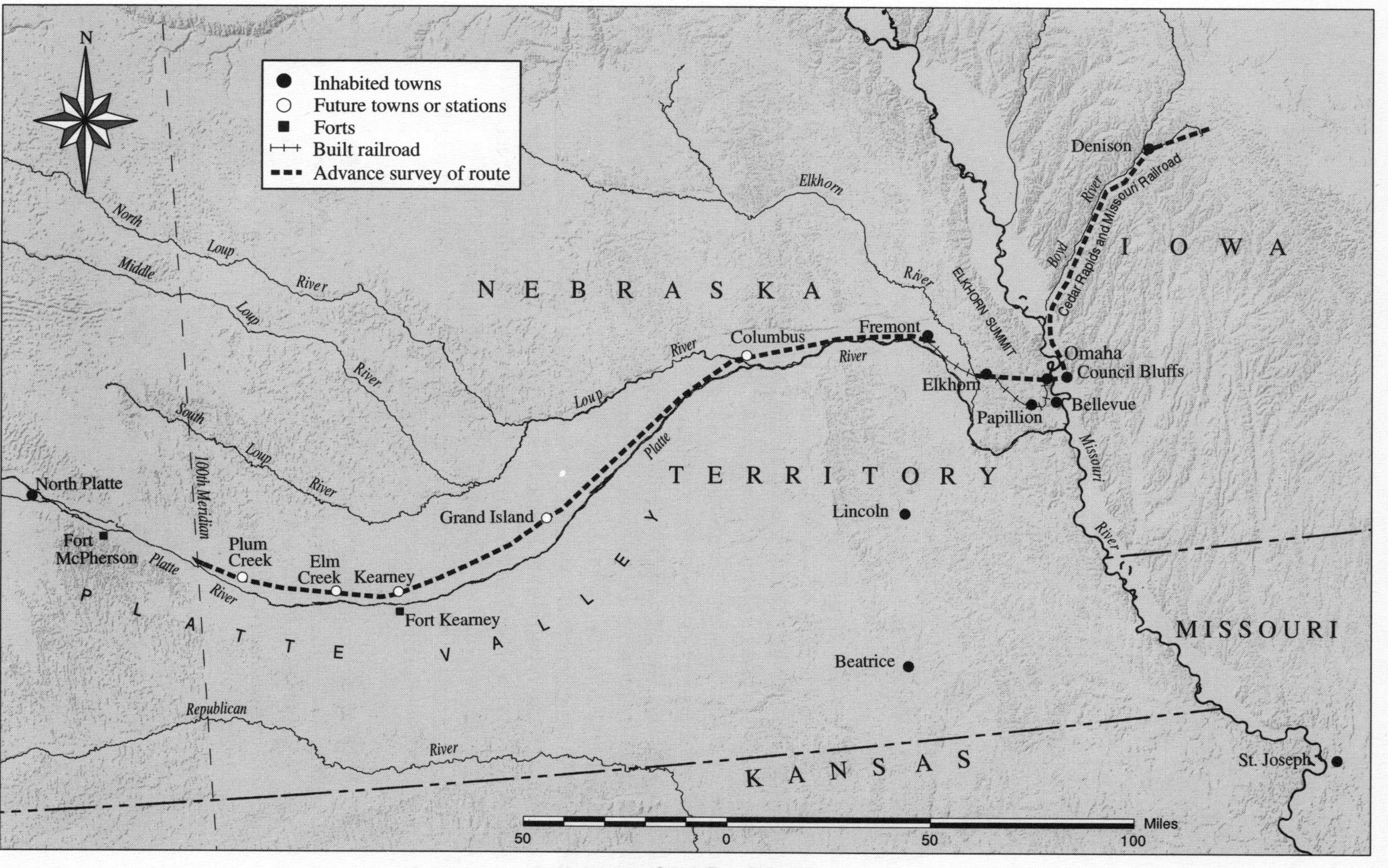

Union Pacific Railroad, 1866

19

"Vexation, Trouble, and Continual Hindrance"

Bending over his desk in the Omaha engineers' office of the Union Pacific, with the Missouri River not far away, all in ice floes, Samuel Benedict Reed was surrounded by stacks of his field notes and measurements from the previous year of 1865. It was now January 14, 1866, and work showed no sign of abating. He pushed aside his cost estimates and his report in progress to write his wife, "Dearest Jennie," with whom he had spent too little time at year's end in Illinois before heading west again—summoned by Durant to Omaha and a promotion to engineer of construction. His duties were to include supervising the grading, bridging, tunneling, and tracklaying, keeping abreast of all materials used and funds expended. "This does not look much like going home very soon," he apologized. "Still, I'd rather remain here than go to the mountains next season." Dr. Durant will still nearby, bustling about, he reported, though his brother Frank Durant had slunk out of town after his discharge—"the work was too much in his rattle brain to manage successfully." But more immediately it was time to share his pleasure. "You may be surprised some time to see my name in congressional proceedings in relation to the Pacific Rail Road," Reed wrote.

> *Col. Simpson of the Topographical Engineers paid me a very handsome compliment at the lunch on the day of the inspection of the first 40 miles of road. He gave us a toast to which I was expected to respond: "S. B. Reed the able, energetic pioneer Engineer of the Union Pacific Rail Road Company, who crosses Mountains, Streams, and deserts which have baffled the best Engineers of the Topographical department of the Government." As speech making is not my fort I called on George Francis Train to respond for me, which he did in a very handsome speech which was very loudly cheered.*

Within a few days, Durant, Train, Seymour, "and other notables" returned east, leaving him virtually alone. "It will seem lonely here after so much excitement," he wrote on January 17, and four days later he was thinking of Jenny and his two little daughters, who must be "sadly disappointed that I am not to spend the winter at home as we fondly anticipated." But career matters were intervening in his family life: "Mr Durant," he wrote, "if I can see correctly thinks well of me and if I remain in the only business I seem to be fitted for perhaps it is best for me and for you pecuniarially that I should make an effort to command the confidence and respect of those in authority on this great National highway acrost the continent." He still had his two reports to complete, one for the company and one for Congress, but he planned to ask for "a few weeks furlow to rest my brain and muscle before entering upon the active duties of another year's work. One thing, if I remain here I shall not stand in fear of loosing my scalp by the Indians."[1]

Back in New York, Durant with some relief claimed the first released government bonds for the forty miles of work approved by commissioners earlier in the month—$640,000 in thirty-year, 6 percent currency bonds, the receipt of which triggered the issuance of Union Pacific first-mortgage bonds of the same value of $640,000. The latter Durant was to use as new collateral for outstanding loans.[2]

With funds coming in from Congressman Ames's drive for the Crédit Mobilier, it was time for the Crédit Mobilier to let a new construction contract—and this time Durant's choice was brilliant from both practical and political views.

For Jack Casement, who, with his younger brother Dan, would be given that contract, a defining moment had occurred less than five years before, in August 1861. Freshly elected major of a battalion of the Seventh Ohio Volunteer Regiment, Casement found himself part of the advance guard of Federal General W. S. Rosecrans's command in the narrow Kanawha Valley of Western Virginia. A large Confederate Army swooped down on the Union men as they were eating breakfast near Cross Lanes on August 26, killing twenty and capturing one hundred; the officer in charge, Colonel Erastus Tyler, fled with some two hundred men, leaving another four hundred in panic under heavy enemy fire. Major Casement took command of the chaos and safely rallied the scattered soldiers into an orderly retreat across the heavily forested mountains to Elk River and thence to Charleston. Honored and later promoted to colonel, then commanding the 103rd Ohio Volunteer Infantry Regiment, which was posted under General Sherman during the Georgia campaign and the capture of Atlanta, Casement had made his mark by seizing initiative and leading his force away from defeat if not annihilation; subsequently, his military command never advanced at the expense of his men. In selecting him for the construction, Dr. Durant was banking on Casement's doing the same for the Union Pacific. Forty miles completed in the previous year's time was, regardless of one's point of view, pretty miserable. Moreover, government deadlines did not waver.

John Stephen Casement's parents had emigrated from the Isle of Man in 1828, settling near Geneva, Ontario County, New York. Casement was their sixth child and the first born in America, in 1829. When he was about fifteen the family moved to Michigan, and Casement's formal schooling ended soon thereafter when his father died. His first work was as a water boy for the Michigan Central (Sam Reed was another veteran). The kind of kid who constantly pestered more seasoned workers on higher levels with questions, Casement advanced steadily in the railroad business. Barely five feet tall, he was nonetheless as strong and tough as a buffalo, and was not averse to using his fists to settle difficulties. In 1850, moving to Ohio to work on the Cleveland, Columbus, and Cincinnati, he rose from common tracklayer to foreman, and following completion he bossed a gang on the Lake Shore Railroad, later becoming an independent contractor over a number of that road's sections; after graduating from the Cleveland Commercial College, his brother Dan, three years younger, joined him as bookkeeper and cashier for the brawny little firm. Working northeast of Cleveland on the Lake Erie shore and boarding in Painesville, Jack met and married a local farmer's sixteen-year-old daughter, Frances Jennings, in 1857; the right-of-way ran across her father's farm some three hundred yards from their house. Dan would follow suit with another Painesville girl a few years later, and the town would become their home. When the war broke out, the Casements were at work on the Philadelphia and Erie and the Erie and Pittsburgh. Dan stayed at home and ran the contracting business. Jack went to war.[3]

Weeks before Appomattox, Colonel Casement was brevetted brigadier general. For the rest of his life he went by the honorific—General Jack to friends, family, employees and competitors. And it was as General Jack, in the months following his return to civilian life, that Casement paralleled Dr. Durant in his cotton schemes in North Carolina.[4] Relying on military and political ties (he secured a handful of business recommendations from former comrades-in-arms, notably a fellow Ohioan, Major General Jacob Cox), Casement took out a $105,000 loan from the New York firm of Fisk & Hatch and went to Charlotte, North Carolina, across the territory his unit had scourged, in search of exportable cotton. The torrid months of July and August were excruciating as much for the weather as for his utter failure to find cotton and get rich quickly. "I got here too late," he wrote his wife, "and it is now a little too early, but things begin to look a little better. . . . I am disgusted with this kind of life and believe I will know enough in future to stay nearer home."[5] Up in the Platte Valley, as Casement worked on his life's largest enterprise, those associations—with the Sherman brothers, with Phil Sheridan, with Jacob Cox (who would become governor of Ohio and then Secretary of the Interior), and of course with Grenville Dodge—joined with his military approach toward organization and discipline, would soon begin to get results.

How General Jack made the connection to Dr. Durant is obscured, whether it was in cotton dealings, through associations at Fisk & Hatch (where Durant was frequently to be found), or through their prime mutual interest in railroads. After his debacle in the South, Casement swore to his wife that he would go back to the enterprise he knew the best—although his promise to "stay nearer home" would go unkept. Correspondence places him on the stage road between Des Moines and Council Bluffs at Christmastime, heading to Omaha presumably for a conference with Durant, Train, Hoxie, and Reed.[6] Appointed in February, the Casement brothers drew up a contract in Durant's New York office on February 6: they proposed compensation of $800 per mile but Durant talked them down to $750; they would lay the track "as fast as required" but not to exceed one mile per day—unless the Union Pacific (for its own or for Durant's own reasons) instructed them to slow to a half mile per day. The company would furnish locomotives, wood and water, and, interestingly, the undercarriages, or trucks, for special cars the Casements proposed to build themselves. "We will commence," wrote the Casements, "as soon as the Company requires, & will lay track during the whole of the present season," and adding, "and longer if the Company requires." Three weeks later, on February 27, General Jack had finished up his personal affairs in Painesville, and he wired the Doctor that he was leaving for Omaha the next day. "Casement has contracted for tracklaying," Durant telegraphed Sam Reed in Omaha. "Will probably want our men. Aid him in building boarding cars. Make arrangements to furnish 3,500 ties per day after river opens." The enterprise was ready to begin—in earnest.[7]

Meanwhile, at Fort Leavenworth, General Dodge was sick and tired of army life. From a distance he had gotten involved in Iowa politics during the autumn's senatorial race, resisting friends' calls to run himself but exerting as much influence on the outcome as he could. If politics and certain postwar spoils did not, for the moment at least, tempt him, he was also cool on the prospect of joining the Union Pacific despite all the Doctor's blandishments. "I am afraid I might have trouble with Durant," he wrote his brother in January, "though that doesn't enter into my decision." Although there were salary and stock benefits, "the vexation, trouble, and continual hindrance one continually meets with in building railroads, especially when we are subject of the whims of the money market to get along, is terrible, and I cannot always last with my brain and nerves strained to their utmost tension all the time. However, I have got to decide in a short time, for as soon as the river opens in the spring, I am going to leave the Army." Durant continued to woo him. On March 1 he wired from New York that he wanted to confer with the general in a few weeks when Durant would go west. His renewed offer of the chief engineer post—"Would like you to superintend survey"—was terse (and carelessly worded enough to confuse Dodge and a number of historians, who thought

Durant was proposing merely the post of field surveyor). The offer drew a long response from Dodge. Characteristically, the Doctor was running his organization with only himself knowing the big picture, he said. Each department chief reported directly to Durant in New York and otherwise operated without consultation in the field, "jealous of his power and rights." It was not conducive "to harmony, energy, economy or celerity."[8]

His chance to expand his theme came on April 24 in St. Joseph, when he conferred with Durant. By then he was feeling the certitude of his own momentum—he had sent his resignation to Grant on March 10—and he minced no words with the Doctor. He had seen in the army what disasters followed a divided command. "I would not accept and be responsible for the road unless given absolute control," he said. With Indians about, with a scattershot cavalry approach to dealing with the tribes, with hardly any civilian authority between the stagecoach stop at Fremont and the first Mormon settlement in Utah, Dodge would not stand for indiscipline in the Union Pacific ranks or for interference from New York. Dodge would "obey orders and insist on everyone under me doing the same." Durant swore he would give him control. Dodge would receive $10,000 per year in salary and stock benefits in the Crédit Mobilier. And so they shook hands, even if a little warily. A few days later, on May 1, Dodge received permission from Tecumseh Sherman to leave his command. As soon as he could wind up his affairs, he was on the next coach to Omaha.[9]

Now in their eleventh year of partnership, Collis Huntington and Mark Hopkins almost never saw each other, separated as they were for most of the year by the width of the continent. Huntington lived mostly in New York City, though he frequently made raids on Washington or forays into the lairs of Boston capitalists or dashes into Pennsylvania iron mills and New Jersey locomotive factories. He maintained a faithful correspondence with "Uncle Mark," who kept to his abstemious accountant's life in Sacramento. Hopkins's interior life was stormy, however, thanks to the cruelly long hours, the crushing pile of details, the cliff-hanging excitement of the Central Pacific's balance sheets, the teetering levels of cash on hand. Huntington became concerned for his old friend, hearing from Senator John Conness that the strain was seriously eroding Hopkins's health. "I don't think I shall break down under it," Hopkins reassured him in a letter on February 20, "though for a time last year I was not as strong as usual and I had neuralgy or rheumatism in my neck & head which prevented reading & writing with comfort—and for a time I did not lie with comfort the latter part of the night. That gradually wore off & I think I am usually well. A change would be agreeable to me but as our affairs are situated I don't desire it. . . . Until we can scream a locomotive's whistle on the other side of the summit, so as to feel ourselves well *out of the worst,* I don't intend to ask or suggest any change."[10]

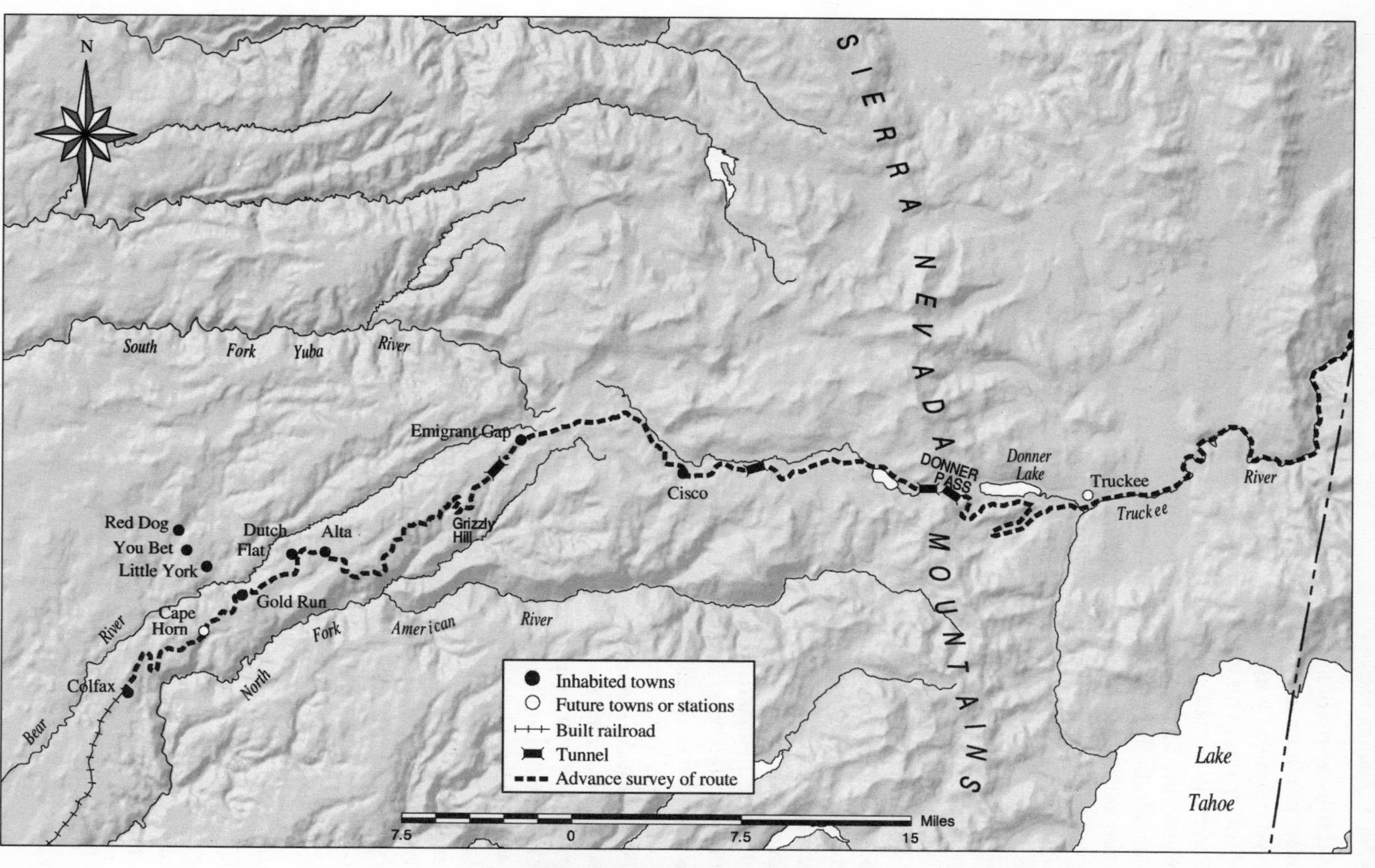

Central Pacific Drives for the Summit, 1866

If he were thinking of a palpable date on a calendar for his relief, at the very minimum it would be two years—at least that is what the eminent railroad engineer George M. Gray had estimated at year's end in a highly complimentary report on the Central Pacific sent to President Johnson and Interior Secretary Harlan. "In fact," Gray wrote, "it is quite a remarkable feature of your route that so elevated a mountain range can be surmounted with such comparatively light grades and curves, and at a cost which will favorably compare with other important railroads, long in successful operation."[11] Such good publicity—it could have been penned by the dead hand of Ted Judah, and it was quoted in the *Sacramento Union* and circulated past all the naysayers on the Pacific Slope—might have slightly lessened the weight on Mark Hopkins's frail shoulders. But there were certainly other concerns, both financial and political.

First there was the worry about the Dutch Flat Wagon Road. Granted, it had generated good tolls of $24,000 by year's end, leaving a balance debt of $239,000 for construction and upkeep. There had been labor troubles in mid-1865 when teamsters combined into a protective association and fixed their wagon freight prices, deigning to touch Central Pacific freight only upon payment of a surcharge of $1.25 per ton. They quickly became "masters of the situation," Hopkins noted, with no teamster daring to go against the association lest they "find nuts and linchpins gone, their harnesses cut or their oxen stampeded or otherwise disabled & injured." Nevertheless, it was widely acknowledged that the Associates now controlled all the freight and passenger traffic over the mountains—and would prevail over labor. And late in the year, when the teamsters had all but left the wintry slopes, the railroad directors stepped in and formed a freight concern called the Central Transportation Company, which had broken the monopoly and thus far been able to operate with modest profits with the roads opened. Anticipated tolls on the Dutch Flat Wagon Road itself, moreover, would be "considerable," he thought, but in the end "wagon transport will be superceded by the Rail Road long before the Wagon Road can pay for itself." If it had been politically expedient—and it had not been, with echoes of the "Dutch Flat Swindle" still reverberating off canyon walls—they should have just sold their turnpike to their railroad, and paid themselves off.[12]

Surpassing the wagon road, though, was a whole welter of problems rooted in the bedrock issue of control over mountain resources. One former friend of the Central Pacific, ironically, seemed at the heart of it. Nursing a grudge against the Central Pacific for not contributing to his failed senatorial bid, Aaron Sargent, the onetime Railroad Congressman during the Judah campaigns in the Capitol, was seething, having joined forces with a dependable band of enemies of the railroad, including certain miners, state legislators, and competitors from Placer County. It was certainly ironic that the Central Pacific had christened its first California-built locomotive the *A. A.*

Sargent, launching it at the Sacramento shops in ship fashion with a bottle of champagne during a riotous New Year's Day celebration.[13] Ignoring the gesture, Sargent and his minions were eyeing the proposed federal land grants to the Central Pacific (draped over the mountain range like a beaverskin cape, hiding who knew what riches beneath the muddy soil). The land grant was outlined in the 1862 Pacific Railroad Act. But the grants were certainly going to be cast as infringements on "the great mining interests of California" if Sargent had his way, Hopkins predicted. There would be "a howl." "The mining interests are sensitive," he told Huntington, "& demagogues with Sargent at the head are endeavoring by falsehood & misrepresentation to mislead and deceive." As far as Hopkins was concerned, it would be better to sacrifice future land value for currency or gold.

Lawsuits skulked like panthers in the shadows of every outcropping. Freebooting entrepreneurs, who earlier might have been occupied in jumping goldfield claims, had examined the railroad's intended route map and begun seizing title to timberlands and watercourses in the Sierra. "Timber grabbers" and "water sharks," Hopkins called them, complaining to Huntington that one individual had just blazed a property line around "a large timber tract, and a very valuable Hay Ranch, which the R. Road grant would perhaps interfere with." Water speculators eyed every spring and brook along the old Donner route, aiming to make the railroad pay for every quart gulped by a cart pony, every gallon poured into the boiler of a locomotive. Indignantly, Hopkins told his partner about having just received a demand for several hundred dollars from one speculator—charges for watering horses (four gallons per day), and workers' drinking, cooking, and washing-up water from an unoccupied and apparently unclaimed natural-running stream. The Central Pacific was granted the right by Congress to take "earth, stone, timber, & other materials' from federal lands during construction. Such a right had to be extended to water, Hopkins stressed, and Huntington had better busy himself making sure the California water companies did not build a legal dike around the railroad through legislation in Washington.[14]

Meanwhile, the overly wet winter continued to swell the Sierra streams, turning the wagon road into a morass, frequently halting traffic, and slowing the railroad construction crews with constant landslides. The worst occurred in March just above the settlement of Secrettown, where engineers had earlier intended to locate a tunnel through the hundred-foot hill. Instead, to save heavy, extensive masonry, the laborers produced an open cut. They were nearly finished when a section several hundred feet long caved in. It "slowly continued to slide in," Hopkins wrote, "imperceptible to the eye, yet continually moving. All the space that 500 men could make during the day with all the cars and wheel barrows . . . would be filled by the next morning, & sometimes bulged up higher than the original surface." Finally Strobridge suspended work there and sent the teams higher up the hill "to decrease the weight of the moving mass." Into that muddy cut they had poured nearly $50,000, Hopkins dolefully reported. He had already noted that freight and passenger receipts were depressed by the sodden winter, but this seemed a final insult.[15]

But by late March he was able to look forward with some good humor. Huntington had wired him on March 13, advising receipt of $640,000 in federal bonds for the unfinished twenty-mile section east of Colfax, which Hopkins estimated would be completed in June. At that same time, he told Huntington, they would be eligible for another $640,000 in bonds for the next unfinished section, where even in March tunnelers were at work. Some 730 dumpcarts with teams and "men in proportion" were toiling between Dutch Flat and Blue Canyon; with track laid across the spindly-looking Long Ravine Bridge in early April, nothing but lack of iron would slow them on the western slope. And the enterprise now expected fresh aid in Washington: Senator Cornelius Cole, their old political crony from Sacramento, victor over Aaron Sargent, was returning east for the next session. "We loaned him $5,000," Hopkins confided to Huntington, "which he will repay us in service as he may be able to render in any way not inconsistent with his official function. You will see him, of course, and you can rely on his cordial friendship."[16]

Back in the fall, it had seemed to the Associates on the West Coast—Hopkins, Stanford, and the Crocker brothers—that an all-out effort was necessary throughout the winter of 1865–66, Hopkins wrote to Huntington. "As *we* see it," he said, "it is either a six month job or an eighteen month job to reach a point where the road will earn *a heap* & where in construction we can make *a pile.*" In February it had seemed necessary to increase their labor force to six thousand men. To push ahead, to make up for lost time, that estimate would rise in the spring to ten thousand. A lot of newly released first-mortgage bonds had better be swiftly sold before the auburn mud of the Sierra began to dry.[17]

20

"The Napoleon of Railways"

I would give a good horse to be with you today," declared the romantic Jack Casement in a letter to his wife, Frances, on March 18, 1866. Since it was a Sunday, the one day of rest in his hectic and demanding week, the contractor was feeling "very uneasy and lonesome." The twenty-six-year-old Frances was similarly disposed, being alone in Painesville and still grieving the death of their four-year-old son in November 1865. Jack had arrived in Omaha on March 8, crossing the Missouri by dodging ice floes in a skiff, lodging in a crammed, poorly run hotel though making arrangements to rent a house in the summer when his wife would join him. He lost no time for work. All through the rest of March, carpenters were kept busy by the impatient, diminutive Irishman with the unforgiving military bearing. Upon the undercarriages supplied by the Union Pacific, workers were hammering away at four strange-looking new construction cars. They were each eighty-five feet long; two and a half of them were rolling bunkhouses, soon to be thick with three-tier bunks with a corridor down the middle and lit by skylights on the roof; one and a half served as dining cars, also containing a businesslike kitchen and the Casement brothers' office. Up near the ceilings, in racks, were a thousand army-issue rifles in case of Indian attacks. Completing the squadron were a number of boxcars outfitted for a bakery, a butcher's, a colossal pantry with sacks of flour and oats, and other supplies. Omaha was bursting with out-of-work laborers, some from the previous season's construction force but many others demobilized from the northern and southern armies and desperate for a place. "Could get plenty of men here if I was ready to go to work," General jack wrote his wife. "We will have lots of hard Driving to do if they can get the Iron fast enough." The Missouri would prove to be late in opening and then delayed in rising enough to accommodate freight boats laden with iron from St. Joseph. Not until Monday, April 6, when the first crafts were unloaded, was the laborers' army to commence tracklaying.[1]

Back in New York, Dr. Durant kept the telegraph clerks busy with a flurry of peremptory orders, trying by sheer force and bluster to affect affairs half a continent away. The questions flew out to Reed: When will you be ready for government commissioners? When will the second hundred miles be located and mapped? Where are the ties? Organize a surveying party at the 100th meridian and wait for my orders, Durant told James Evans; send more iron, send locomotives, send ties, send bridges, send sawmill, he wired factories and shipping agents in Pennsylvania, New York, Ohio, and Missouri; buy a freight boat, name it the *Elkhorn,* register it in my name for the Crédit Mobilier, he telegraphed Springer Harbaugh, his agent, board member, and government director. And build two barges! The wires must have hummed, with each dispatch a potential headache for its recipient. Hurry the iron! Can you raise a sunken cargo of iron? Where are the ties? "Why is not track laid faster," he shot to Reed on April 14 from Chicago, collect. "Why are the timbers for bridges not out. Why is the work so much behind."[2] Aiming west toward the Nebraska bustle, he tarried in Chicago for board and stockholders' meetings of the Rock Island and the Mississippi and Missouri, hearing the happy news of large dividends and plans to put the M&M through to Des Moines. He had made sure the government inspectors heading to look over the Union Pacific line would have transportation, furniture, and bedding, and on April 17 learned they had accepted sixty-five miles of track as "A-1." And on the same day he closed out the purchase of the elegant railroad coach used for the Lincoln funeral train; the President's Car, as it would always be called, with its gilt and plush and silk, was to be readied for the completion of Durant's trip to Omaha.[3]

Durant arrived in time to watch the Missouri leave its banks and rise to within five feet of the permanent buildings built for the Union Pacific on the Omaha floodplain, carrying away about four thousand new ties and cutting the track to the machine shop. However, he quickly wired Boston investor John M. S. Williams, who was breathing down his neck, "There has been no interruption in track laying."[4] On the same day, May 2, crews finished a temporary bridge across the Loup River fork, fully seventeen hundred feet long, so that the track crews could continue without delay; in the fall it would be replaced by an iron Howe truss bridge sitting on masonry piers.

In mid-May, Grenville Dodge was in Omaha in the little red-brick engineer's office with the Doctor.[5] Years and years later, when he began to publish his memoirs, Dodge would always imply that he had directed all phases of the project—construction, supplies, maintenance, even operation. His actual job, signified in the terse Durant telegram, was to "superintend survey"—supervising the engineering reconnaissance, managing the profiles and reports, and laying out rail yards, junction points, and new town grids. No one person on the spot could have done what he later claimed to have managed, but nonetheless the job facing Dodge in May 1866 was gargantuan, something which perhaps few others could have completed as well. Again and again, he found himself reaching outside his outlined duties, to good effect. But he did not build the railroad without help.[6]

And before he could so much as roll up his sleeves, the headquarters received a visit from Dodge's friend Tecumseh Sherman. The general had gone out on an inspection tour of his military division, first to note the progress of the Kansas Pacific line (still identified then as the Union Pacific, Eastern Division), which had progressed beyond Lawrence almost within rifleshot of Topeka; then, by wagon, Sherman had ridden north to the Union Pacific railhead where the Casements' men were hard at work. Riding past them, he had repeatedly heard cries of recognition: "Uncle Billy! Uncle Billy!" Throngs of his men, many attired in pieces of their old uniforms, were now among the laborers. "Seeing them," Lloyd Lewis has written, "made him feel like the godfather of the Union Pacific."[7] Sherman had caught a ride on a Union Pacific train back to Omaha, an improvement over his previous flatcar-and-nail-keg journey. Regarding the newly civilian Dodge in his dirty working clothes, Sherman laughed: "He looks like a muledriver." He conferred with Durant and noted the stacks of ties and iron in the Union Pacific yards—enough, he was told, for fifty miles. Sherman wrote Grant that the Doctor had contracted for another 150 miles' worth of materials, that the Casements' men would reach mile 100 in mid-June, and that by year's end the rails would reach the second hundred-mile mark, nearly to Fort Kearney. It would be a boon to the army, he told Grant, but the year might prove to be a busy one for his cavalry to help ensure it coming to pass.[8]

To accomplish all of that would take determination. After Sherman departed for Sioux City and St. Louis, leaving Dodge, Durant, and Reed to the intimidating matters at hand, Durant backed up his grant of total control by furnishing Dodge with a letter instructing all division engineers in surveys on the Republican and Platte Rivers, and also in Colorado, Utah, and Nevada, and all other Union Pacific engineers, and all right-of-way agents, to report only to Dodge.[9] It was to James Evans, waiting two hundred miles to the west in the vicinity of the Plum Creek trading post and not far from the 100th meridian, that Dodge's attention was most clearly focused; as division engineer, Evans would be responsible for the Laramie Hills/Black Hills crossing. "That really is our first objective point," he wrote Evans on May 29, "as it fixes our line east and west." It was paramount that they locate a line through the hills in the autumn so that the Casements' graders and suppliers could be moving in that direction before winter closed in. Evans was to explore all possibilities—but to pay special attention to the route up Lodge Pole Creek from the Platte to the Lone Tree Pass discovered by Dodge seven months earlier as he fled the Crow raiding party. It was Dodge's private view that no better transit could be found.[10] He organized an armed support party and sent it off to Evans.

Meanwhile, there was much terrain to survey. Percy Browne, a young assistant engineer trained at the Troy Institute who had begun only two years before as a rodman, was to head a party investigating the northern Colorado mountains, through Clear Creek and the Berthoud Pass, and also Boulder Pass, since the desire to run through bustling Denver was still strong, if not directly then at least with a connection to the main Union Pacific line. (Former Colorado governor John Evans, now that his opportunities were gone to use the Cheyenne, Sioux, and Arapaho as sacrifices to his own political aspirations, was ceaseless in his promotions for railroads into the territory.) From the Platte to the Lodge Pole, surveying would be done by L. L. Hills. Another party headed by Thomas H. Bates would locate lines through the eastern Utah mountains and westward toward California. J. E. House, who had assisted Dodge during the M&M surveys in the 1850s, would be headquartered at Omaha and would attend to the Missouri River soundings when not overseeing the engineers' office while Dodge was out in the field.

As the engineers packed their gear, shouldered their rifles, and headed westward, the tracklayers moved on at a walking pace, laying an average of 1¾ miles per day, though sometimes as many as 3 miles, thanks to the Casements' military planning and precision. Nebraska had not seen such determination since Brigham Young had led the Saints westward up the valley, leaving wagon ruts for the Union Pacific to follow. And there was the broad Platte: "a dirty and uninviting lagoon, only differing from a slough in having a current," one traveler wrote,

> *from half a mile to two miles wide, and with barely water enough to fill an average canal; six inches of fluid running over another stream of six feet or more of treacherous sand; too thin to walk on, too thick to drink, too shallow for navigation, too deep for safe fording, too yellow to wash in, too pale to paint with—the most disappointing and least useful stream in America.*[11]

At the absolute railhead—beyond the boarding cars and the flatcars piled with ties and rails, with a smooth grade stretching into the distance parallel to the sluggish Platte, and the ties bedded and waiting—small horsedrawn "lorry" cars heaped with iron advanced as gangs on either flank removed rails and trotted them forward to lay them parallel in their "chairs." Other gangs stood ready to leap in, gauge and spike them, with the lorry already moving another few feet while the process continued anew; when the lorry was emptied, the laborers simply unhitched it and tipped it off the tracks into the ditch, another having appeared in its place. "Thirty seconds to a rail for each gang," a visiting journalist would be moved to write, "and so four rails go down a minute. . . . Close behind the first gang come the gaugers, spikers and bolters, and a lively time they made of it. It is a grand Anvil Chorus that those sturdy sledges are playing across the plains. It is in triple time, three strokes to a spike. There are ten spikes to a rail, 400 rails to a mile."[12]

One of the laborers in that rolling assembly line was a teenager from Painesville, Ohio, Erastus Lockwood, whose older sister had married Dan Casement. His job was to walk just behind the lorries, dropping spikes for the hammerers who closely followed (a job he performed for some three hundred miles until he was injured and laid up for the rest of that season). In the evening, with the boarding and dining cars pulled to the front near where the next day's supplies were waiting, Lockwood retired to a berth not inside but on top of the cars. "To tell the truth," he recalled, "we were troubled by 'cooties,' and a companion and myself escaped them by making our bed between two skylights on the roof. Our tailor was interested enough to make, out of an old tent canvas, a cover for our roof bed. This cover was supported by a pole reaching from skylight to skylight, and about two feet above us as we lay on our blankets. In case of a storm we could button our cover to a row of brass buttons on the edge of the roof of the car." Their fare consisted unvaryingly of beef, bread, and coffee (a cowherd followed the workers, periodically replenished as bovines went to the knife), and the workers were charged $20 per month for board. The daily pay for graders and teamsters was $2.50; spikers received $3.00; ironmen, who had to be brawny but quick and light on their feet, got $3.50 to $4.00; and the foremen were paid $125 per month.[13]

At this point there was little for the men to do with their money but drink. Dr. Durant saw this as a problem which could only worsen, so he ordered Sam Reed to take measures. Out on the line in the stagecoach station town of Columbus, founded at the confluence of the Platte and the Loup Rivers ten years earlier by settlers from Ohio, Sam Reed may have sighed upon reading Durant's telegram: "Make a side track west of Loup Fork for second base of supplies where you deposit materials," he had written on June 1. "It will not answer to have men so near whiskey shops in Columbus."[14] To effectively keep them away, anyone knew, Reed would have had to post armed guards on the railroad bridge. But before too long, to the laborers' delight, the pleasure purveyors were going to follow their trade west—on iron wheels. The Casements' army would get its own horde of camp-followers. And even General Sherman had mixed feelings about what would ensue; there might be a military benefit to letting the men blow off steam. Nearly ten years before, he had semiseriously written his brother about this very subject. "So large a number of workmen distributed along the line," he said, "will introduce enough whiskey to kill off all the Indians within 300 miles of the road."[15]

Those train gangs moving along the Platte as well as the engineering parties advancing in the foothills and upper valleys of the Rockies had another thought than whiskey in their minds—the worry about raiding Indian parties in the warming spring days as they began to range from their winter camps. Predictably for the times, the whites had difficulties telling friend from foe, but for the friendly Pawnee along the Platte, the work trains became even more of a diversion than they had been the previous summer. W. B. Doddridge, who would run the telegraph key at Grand Island, recalled with the patronization of the era that the "railroad company, with the idea of retaining their friendship, gave them the privilege of the free use of their trains. Without stint the Indians availed themselves of the privilege. As a matter of fact, it spoiled them. They spent most of their time, weather permitting, on the trains coming and going hither and yon, like children with new toys. For sanitary reasons the only restriction imposed by the railroad company was that they should ride on car platforms or on top of cars. This suited them very well; they preferred the roof of a box car, where the air was fresh and exhilarating. Joy riding only was the purpose of their trips."[16]

But the Pawnee were one tribe out of many. For the Sioux, Cheyenne, and Arapaho it had been an unusually cold winter with little to eat, but emissaries from the U.S. government had begun assuring the tribes that peace was at hand. A large parley was to be held at Fort Laramie in June; there would be many peace offerings available, including food and valuables. Indians up north in the Powder River country under the influence of the Oglala Sioux warrior-leader Red Cloud were disposed to wait peacefully and see. So were the large bands of Brulé Sioux under the shrewd and valiant Spotted Tail, who had left their winter camp on the Republican River to begin hunting south of the Platte. Ten years before, Spotted Tail had spent almost a year in the whites' military stockades after a series of skirmishes with the cavalry; awed by superior numbers and technology, he had emerged convinced that in a war with the whites the Indians would be destroyed. Known from then on as a conciliator, he nevertheless had his limits—such as in the aftermath of Sand Creek when he had led the raid on Julesburg in January 1865. He also had no good feelings for the Pawnee, whom he had fought over hunting lands for most of his forty-odd years, and who had donned blueshirts in 1864 to fight for Uncle Sam. Spotted Tail, in fact, deeply hated the Pawnee. However, he was favorably disposed toward the coming peace negotiation in June 1866, and ordered his warriors to stay away from the whites, especially the bustling work crews north of the Platte.

It was, though, an impossible situation to control after the winter of deprivation; small raiding parties, young hotheads in the main, ignored the directive and slipped across the river to steal food and the large prizeworthy horses and mules tethered so temptingly under light guards.

The tracklayers had by then advanced nearly to the little riverside settlement of Grand Island. One day, Erastus Lockwood recalled, the workers were alerted to a band of horsemen advancing up the slope from the Platte toward the railroad bed—it was a band of Brulé Sioux, seventeen in all, under the leadership of Spotted Tail himself. A halfbreed interpreter, Pat Mullaley, was with them. "While the interpreter said they had come to see how we laid track," Lockwood wrote,

> *I have always thought they came on a spying expedition. However, our superintendent received them cordially and showed them the process of track laying in all its aspects. They were taken through the boarding cars at which time they discovered the . . . U.S. rifles stacked horizontally in the roofs. It was interesting to see the expressions on their faces when discovering these guns. I was following them and noticed one Indian put his hand out of a window and measure the thickness of the wall of the car. As he looked to another Indian, I could imagine hearing him say: "I wonder if a bullet could go through?"*

The visitors were also shown the butcher's car with its hanging sides of beef, and the baker's car with ovens and plentiful sacks of flour. After their winter of hunger, the Sioux were impressed. Back down on the tracks, someone suggested that the visitors show them how accurately they could use their bows and arrows. "I put a shovel upright in the ground about 60 feet away," remembered Lockwood. "Sixteen of the Indians put their arrows through the hole in the handle, while the seventeenth hit the handle at the hole, knocking the shovel over. He felt quite disgraced."

The railroaders proposed a new contest: a race between Indian ponies and their locomotive. The Sioux readily agreed, although Spotted Tail was initially reluctant to accept the supervisor's invitation to ride in the engine cab. Finally he climbed aboard with the interpreter. "The Indians lined up abreast for the word to go, which was presently given," recalled Lockwood.

> *and away they went. At first they outdistanced the locomotive, which so pleased them that they gave their Indian war whoop. Presently the engine gathering speed, overhauled them. The engineer as he passed opened his whistle, which so startled them, that all, as if by word of command, swung to the offside of their ponies, hanging with their arms around the ponies' necks, holding on with their left legs over the backs. Of course this incident ended the race, and the engine and Indians, the latter much crestfallen, returned to the boarding train.*

It was only good manners in the customs of the Plains Indians that food and presents be given visitors by their hosts, but the Sioux had to prompt the railroaders for a meal. "We set before them an abundance of food," said Lockwood, "but in fear of being poisoned, they would not eat anything until we had tasted it all—they then devoured everything in sight."

Then relations turned ugly—perhaps the railroad foreman insulted him with his manner; perhaps it boiled down to a shrewd test of will, more important to either side in that moment than marksmanship or speed. But according to Lockwood, Spotted Tail

> *demanded that their ponies be loaded with sacks of flour and quarters of beef. Our superintendent refused this demand, saying they could have all they wanted to eat, but could not carry anything away. In answer to this, Spotted Tail threatened to come over some night with three thousand warriors and clean us out. The superintendent told the interpreter to tell Spotted Tail what he was going to say. Doubling up his fist and placing it against Spotted Tail's nose, he let out a string of oaths such as I never heard before. Spotted Tail made no answer, but all mounted their ponies and rode away. That was the last we ever saw of them—they had seen the guns! However fearing they might return, we doubled our night patrol for some weeks.*[17]

This was the only band of hostile Indians that Erastus Lockwood was to see at the site of tracklaying operations, there near Grand Island in the late spring of 1866; in the course of events in 1866, other railroaders would have different tales to tell.

What was also occurring at Grand Island was symbolic of the dilemma at large. The Platte Valley, rich buffalo hunting territory for unknown generations of the Crow and the Pawnee, and then for the interloping Sioux, Arapaho, and Cheyenne, had been destined for this final struggle—with the Iron Horse as referee—since Coronado and his band of conquistadors had wandered in the dry plains north of the Arkansas River in search of golden Quivira and heard rumors of a large river valley farther to the north. Later, the French *voyageurs* and American fur traders made the Platte their highway—it was, in fact, emissaries from John Jacob Astor who noted and described the seventy-mile-long island in the Platte, called La Grande Isle by the gallic trappers, in 1812—and explorers such as Frémont popularized the natural westward route for the Conestoga hordes, saints and gentiles, fortune hunters and adventurers. Iowa promoters had exported some thirty Germans to Grand Island in 1857 to start a city that with the coming of the railroad would become—they were utterly convinced—the new national capital. (Of course, the promoters went bust but the Germans stayed.) Sioux such as Red Cloud, who had been born in a Platte encampment somewhere westward, paused in their hunts to see villages like Grand Island grow slowly over the next few years—slowly, that is, until the smoke on the eastern horizon that was the Union Pacific advanced closer in the spring and summer of 1866, stopping every twenty miles to plant a depot, erect yards, and survey for town lots. When the tracklayers had finished and moved on and the carpenters had raised a station house and the surveyors had drawn their plats and left their little fluttering flags, the little town of Grand Island compliantly began dismantling itself from its place on the riverbank and moving upslope a couple of miles to straddle the iron bands and claim its future.[18] If there was any time when the realization of and the fury at what was so implacably occurring to their hunting lands burned hotter in the minds of Red Cloud's and Spotted Tail's folk, it could not have surpassed what was undoubtedly in the mind of the conciliator Spotted Tail as he galloped away from the iron men's worksite at Grand Island, with the image of an angry Irishman's fist just a hair's breadth from his nose, with the scream of a locomotive whistle in his ears.

The congressional deadline loomed. The pressure was up. By July 1, 1866, the Union Pacific needed to have one hundred miles of finished track, complete with stations, watering and fuel facilities, and sidetracks, and it needed to have regular passenger and freight service running in and out of Omaha along that hundred miles. If it failed, the Union Pacific would forfeit its charter. Dr. Durant hectored Sam Reed by telegram incessantly until June 4, when the glad news arrived: Jack Casement's men had spiked the rails past the hundred-mile post. "Well done," the Doctor telegraphed Sam Reed, Jack Casement, and the railway operations superintendent Webster Snyder, no doubt causing them to blink. "You have given Kansas line a hard tack. Take a basket at our expense only don't stop the tracklaying." There would be little time to enjoy any celebratory champagne, even at Durant's expense. The pressure continued to get passenger and freight cars finished, to erect buildings along the way to give some semblance of permanency; in his expensive rush, Durant even ordered several ready-made houses from a firm in Chicago and had them shipped to Reed with instructions to plant them on the line, lending an empty stretch of prairie at least the façade of habitation—all for the government commissioners' benefit. "How much track laid Monday how much Tuesday and Wednesday," he demanded one day in June, and the next, "If you do not average one and a half miles per day telegraph the reason that it may be remedied." In the second week of the month, alerted that the government commissioners had left for Omaha, he fired off directives telling both Reed and Dodge to stay close to the inspectors when they arrived and to "offer only 40 miles" for scrutiny. "Provide them with a car and refreshments, but do not make it an excursion," he ordered. With the deadline galloping closer, it was better to save time, money, and energy for the July 1 final inspection. The line continued to push westward, arrow-straight, under the bright Nebraska sky.[19]

A young surveyor from James Evans's party, Andrew Rosewater, was in the entourage and recalled that a large body of Pawnee, friendly to the whites and camped not far away, came down to see the first train. "The locomotive wheels were painted a bright red, with lavish brass trimmings, there being wide brass bands around the boiler at intervals of about a foot," he recalled. "All this brass was burnished and the locomotive looked splendid. It attracted the Indians who came with their squaws and papooses and were mounted on ponies. Upon a given signal the engineer blew a few loud blasts of the whistle, and ejected long spurts of steam from the cylinders at either side of the iron horse. The din was deafening. The Indians took fright and ran off in every direction." In time the local Pawnee, at least, got used to the railroad. "It was a common thing to see races every day along the track," wrote Rosewater, "the Indians whipping their fleet-footed ponies to their utmost speed in trying to out-run the locomotive."[20]

With the day of government scrutiny come and gone, after the three commissioners rode out in the Lincoln Car from start to finish and the road was accepted, it seemed almost an afterthought, even extraneous, to the overall effort. But it caused a flurry of cheer in Omaha, where many merchants were already fattening from the railroad's business. A month earlier, the Doctor had brightened the city's commercial prospects by proving that freight barges (loaded with iron, no less) could ply the Missouri from Council Bluffs and St. Joseph, something no one had thought possible. It had literally opened floodgates of materials for the railroad, and local dealers and merchants had already begun to follow suit. Now Durant's main cheerleader, the *Omaha Herald* (the *Republican* would continue to distrust him and his motives), noted that the inspection car had been pulled from Omaha to the Elkhorn station in a speedy sixty-five minutes (it was some thirty-one miles), "notwithstanding that many cattle were found on the road, causing some delay." Durant had, in the *Herald*'s view, earned a crown: he was "the Railway King," the "Napoleon of railways."[21]

Napoleon could not control the U.S. Congress, though, as new legislation moved toward a vote in Washington and Durant found himself spread too thin. The respective bodies' railroad committees were at work on amendments to the Pacific Railroad Acts of 1862 and 1864 which would significantly change the climate. California's senator John Conness, acting on behalf of Collis Huntington and the Central Pacific, was active, opposed mainly by Senator John B. Alley of Massachusetts, the champion of the Ames brothers and the various New England directors and shareholders of the Union Pacific and Crédit Mobilier. Durant got busy, traveling to the capital in June while the debate raged, twisting arms, writing letters. To Cornelius Bushnell in New Haven he urged that the Senate bill be "looked after"—who could he send to pressure the Connecticut senator Lafayette Foster, on the committee? To John Adams Dix, his figurehead president, he entreated that he do the same with his contacts. In the House, Ames and his allies did what they could.[22]

But it was to no avail. Collis Huntington was in town, also, with fewer distractions and possibly—for this fight, at least—with more friends. The postwar Congress, members up to their nostrils in Reconstruction controversies, still befuddled by new priorities and emerging lines of power, was confused by the number of railroads covered under the amendment—the Union Pacific barreling west in Nebraska, the Central Pacific drilling east in California, the Union Pacific (Eastern Division) heading west in Kansas. Congress was also institutionally absent-minded, forgetting the definitions of main lines and branch lines—under the original act, the main line of the Union Pacific would commence only at the 100th meridian, with the rails leaving Omaha and Kansas City deemed mere competing branch lines. Congress was even lazy, not troubling to check old issues of the *Congressional Globe* for the original bill's language or the terms of the earlier debate. When Senator Conness rose during the debate and addressed the issue of the 1864 limitation of 150 miles upon the Central Pacific, he traded on his own bluster and his colleagues' laziness by brazenly claiming that the 150-mile limit had not even appeared in the bill's original language. "It was stolen in," he thundered, "through the corruption of some parties and the clerk who eventually made the report." It was a bold, barefaced, unequivocal lie. "What I say," he ended with a flourish, "cannot be contradicted."[23] And no one contradicted him. The resulting act, confirmed by the House and signed into law by President Johnson, immensely benefited the Central Pacific and the Kansas road. The former was authorized to build eastward from California past Nevada until it encountered the Union Pacific, wherever that might be; the latter was freed from its earlier requirement to link itself to the Union Pacific "main line" at the 100th meridian, instead being permitted to build through and past Denver—making it a strong competitor, although it would have to join the Union Pacific within fifty additional miles.

It equaled, mourned Oakes Ames, "two well-nigh fatal blows, from the effect of which complete recovery is impossible."[24] He did overstate the case, as it turned out, but the hour clearly belonged to the former pushcart peddler from sleepy Oneonta. "I told Mr. Union Pacific," Huntington had once crowed about the 150-mile limit, "that I would take that out as soon as I wanted it out." He did so, he claimed, "without the use of a dollar." When Senator John B. Alley, whose votes and support had been bought by the Union Pacific, who had purchased five hundred shares of the Crédit Mobilier from Oakes Ames, confronted Huntington and accused him of bribing his way to victory, Huntington coldly replied that he had indeed gone to Washington with a satchel of money to be spent if necessary. He had, though, sat in the Senate gallery with a little spyglass—perhaps the same one he'd once used to spot approaching ships in San Francisco Bay—and he had studied every face in the Senate. Only one man there was corrupt enough to sell his vote, said Huntington with great relish, "and you know devilish well I didn't try it on *you!*"[25]

The ramshackle port of Aspinwall sat on the miasmic Caribbean shore of the isthmus, awaiting a brief wakefulness from the shrill steam-whistle of an arriving ship or railway engine crossing from Panama City, or else awaiting the departure signal that would send the town back to its tropical sleep. "The dreariest, wretchedest, most repulsive city of fact or fiction," the roving correspondent Albert Richardson would call it.[26] By the time of his visit, however, the bodies had been found and buried, the ruins scraped away and dumped out in the jungle, and the wharf rebuilt, and Richardson's eye probably missed the upper part of a ship's smokestack being lapped by the harbor waves as his steamer purred out toward deep water.

On Tuesday, April 3, 1866, months before his visit, the malarial torpor of Aspinwall had been quite shattered. The little train to Panama City was late in leaving the depot. Alongside the wharf of the Panama Railroad Company, the steamship *European* was tied up and being unloaded, with crews of Jamaican and Indian stevedores busy at their tasks, supervised by the ship's officers and railway clerks with their pads and bills of lading. About 7:00 A.M., someone in the ship's nearly empty hold knocked, kicked, dropped, spindled, or crushed a small wooden crate, one of seventy about to be hoisted onto shore and shipped west for a California boat. A second later that man and all those with him disappeared in a bright yellow flash of noise and fire. The upper parts of the *European* catapulted into the air as concussion and the thick steel plates of the hull slammed outward into the wharf, disintegrating some four hundred feet of it nearly to the freight house, the heavy iron and slate roof of which was lifted upward only to crash down, crushing the walls and all inside; another freighter on the opposite flank of the wharf, its hull outwardly intact, sustained grave damage when the concussion snapped its internal twelve-inch iron girders like pipestems. All of the warehouses, stores, hotels, and saloons on the waterfront were badly damaged. Not a whole window of glass remained in the entire port. Minutes later, volunteers began crawling out onto the burning wharf and ship to aid the survivors and retrieve body parts from the ruins and the harbor. Perhaps fifty persons had been killed.[27]

The news would not reach California until two and a half weeks had gone by. A ship would take a report from Panama to New York, later telegraphed to San Francisco, and copies of the *Panama Star* would arrive soon thereafter. By then it would be too late.

At about the same time as the *European* exploded—or even a week earlier—Panama steamers landed two consignments on a San Francisco pier, forty small wooden cases from the *Sacramento* and a larger crate from another ship, both consignments originally sent from Hamburg by way of Liverpool, both billed as ordinary freight. Most of the small cases went to the construction suppliers of Bandmann, Nielson & Co., San Francisco, where they awaited use in a special warehouse; three, consigned to the Central Pacific Railroad, were transferred to the Steam Navigation Company and sent up to Sacramento, where they sat for ten hours in the crowded freight shed before being taken by the regular train on April 14 up to Dutch Flat for experimentation the next day. On April 16, the *Alta* printed a short paragraph relating to those tests at trackside, of the substance in question, called, variously, glonoine or Nobel's Patent Blasting oil or nitro-glycerine. Six separate blasts satisfactorily threw many tons of rock out of the way. "The nitro-glycerine can be handled with more safety, and can be used with less expense," read the report, "than powder, because of the smaller number of holes or blasting used." It would be a great benefit to California "where so much blasting is done, and where miners' wages are so high."[28]

On that same day of April 16, on the San Francisco wharf, a freight superintendent noticed that the remaining crate, the larger one, was leaking a rancid-smelling oil. It had been destined for a party in Los Angeles, but now was draining all over another box for Idaho City which was later found to have contained silverware; both crates were refused due to arriving in damaged condition and they were sent to the office of Wells, Fargo and Company on the corner of California and Montgomery in downtown San Francisco to be inspected. There, the freight clerks for Wells, Fargo and the steamship company left them in a courtyard and went upstairs to the Union Club to lunch, before returning with two porters and a hammer and chisel to pry open the crates. The Union Club rooms were still busy with diners past 1:00 P.M., with a large staff of waiters, busboys, and cooks in attendance; next door to busy Wells, Fargo, at Bell's Assay Office, technicians bent over their instruments; in one building beyond, clerks of the Spring Valley Water Works returned to their paperwork; a number of pedestrians walked on California and Montgomery streets; the proprietor of the assay office, G. W. Bell, was in the alley and had just been about to climb on his horse for a ride to Petaluma when one of the freight agents called him over, knowing he was a chemist, to look at the oil.

At that moment they were all blown to atoms. The freight office disintegrated. Clerks, waiters, cooks, and diners were blown to bits. Bell's office and the water works office, both new, tidy, two-story brick buildings, were thrown off their foundations. People down the street heard what they thought was an artillery explosion. Glass, mortar, stone, brick, timbers, and human tissue shot into the air, raining down on survivors and on the neighborhood streets and rooftops as a pale yellow cloud of acrid smoke rose several hundred feet, like the vapor from a volcano, and quickly dispersed. The dead not torn apart were neither burned nor scorched but looked as if they had been struck by lightning.[29]

The news flashed up to Sacramento, where the Central Pacific directors fretted that the railroad would be blamed even in part for the catastrophe, which killed up to fifteen, grievously wounded that many and more, and sent scores to the hospital. Darius Ogden Mills, the influential banker well known by all the Associates, whose letters of recommendation were still clearing the way for Huntington in the eastern credit houses, had been one of those injured, cut in the head by flying debris as he dined at the Union Club. But before any attention was focused on the Central Pacific, and more important, before anyone in charge could really absorb the implications of what had just occurred in San Francisco, calamity struck the railroad workers. A nitroglycerine sample being used near Camp 9 at Gold Run detonated in a crowd of laborers, killing three Chinese and three whites. The construction foreman was blown to pieces, "part of him," reported the *Sacramento Union*, "not found."[30]

Only on Friday, April 20, did people begin to realize that the railroad was experimenting with a dangerous unknown commodity—and begin to object; the *Sacramento Union* reported that day that the Central Pacific cases of nitroglycerine shipped through the week before had been labeled as such—though no one knew what the dangerous substance was—and sent, like the San Francisco cases, as ordinary freight. "Terrible destruction" on the river steamer or on the Sacramento wharf had been avoided only by luck. More criticism of the importers (including the Central Pacific) followed. "Had one of these boxes commenced leaking, or received a slight blow while on board the steamer, crowded with hundreds of human beings," the *Alta* declared on April 22, "the catastrophe which rent the English steamer at Aspinwall into fragments in an instant, and hurled every person on board into eternity without a moment's warning, would have been thrown into the shade, and belittled to such an extent as to be hardly worth mentioning, by what would have followed." Other dispatches reported a recent nitroglycerine explosion outside the Hotel Wyoming in New York City, and briefly there were rumors (soon disproved) that the British steamer *Granadian* and all aboard had been blown up in the Caribbean while transporting the oil. New York fire marshals had just seized hundreds of pounds of the solution, and a man who had shipped some to California was arrested. In the resulting public hysteria, all stores of nitroglycerine in San Francisco were seized by authorities and taken out to a barge in the middle of the bay; officials quickly passed laws forbidding its transportation through the jurisdiction of the bay city and the capital.[31]

But Charley Crocker, Sam Montague, many of the railroad's engineers and foremen, and a throng of curious workers, had excitedly watched a Dutch Flat boulder ten feet in diameter turn to powder, a steep bluff disintegrate into gravel, a cliff face slide away like cornmeal. The nitro salesman had poured a bit on a boulder and lit it, and it had not exploded, only burned with a barely perceptible flame. The consulting engineer, William B. Hyde, had declared it "free from all danger" if carefully handled. Better, though, it was about eight times as powerful as the currently used black powder, and progress in mountain work might be improved as much as one and a half times. Such economies demanded attention, the calamities in San Francisco and Gold Run and Aspinwall notwithstanding. After all, as Crocker would note, "We burned 500 kegs of powder in half-a-mile's distance every day, throwing off the rocks," and black powder was always in short supply on the coast. If banned it was from transport, nitroglycerine might be manufactured from its component chemicals on-site. With the public furor taking a long time to subside, and while the company directors decided how to proceed, the obdurate Sierra summit waited.[32]

With black powder, picks, shovels, and grub-hoes, the workforce toiled upward, most of them on the twenty-six-mile section between the Secrettown trestle and Yuba Gap. Stanford, the Crocker brothers, and even Mark Hopkins toured the line in early May as far as Emigrant Gap. The spring rains and mudslides had delayed progress toward Dutch Flat, Uncle Mark reported in his regular letter to Huntington in New York, but nevertheless with most of the heavy cuts and fills completed, with the grade mostly cleared and nearly all of the culverts in place, "we feel confident we can complete the track to Yuba Gap in all the month of August."

In order to feel really confident, though, it would be necessary for Huntington to shoulder another burden, virtually guaranteeing to send $250,000 in gold every month. Hopkins's plea was being sent on the heels of a cautionary letter from his partner to make the payrolls light until at least midsummer. But too much was a stake. "We have $250,000 invested in tools & appliances for carrying on the work," Hopkins declared. "After fifteen months of persistent effort we have now got a thorough organization of trained and tried Superintendents, Foremen, Wagon Masters, teamsters, Rock Blasters, Bridge builders, Black smiths, commissaries, Cooks, Butchers, Masons & laborers—every place filled with the right man & the right man in the right place as nearly as can be done in 15 months."

Despite the heavy work ahead, Hopkins felt they could lay track to the summit by autumn and have grading well along beyond the unfinished summit tunnels, to the Truckee River; they could lay temporary track *over* the summit so that they could be laying track on the eastern slope in the spring of 1867; with the tunnels completed later in that year, they could look forward to unabated winter work thereafter. It was ambitious, perhaps crazy, but worth "more than ordinary sacrifice." If Huntington could pledge to raise a quarter million each month on first-mortgage bonds sold in New York, Boston, and London (he was about to dispatch an agent to England), and with the railroad earning at least $50,000 per month from regular operations, then Hopkins was certain they could do it. Stanford had been trying to raise a million dollars from San Francisco bankers, though they all knew that was a market with limited prospects. Still, it was something else to hope for—and hope was something they had to keep in ready supply.[33]

Chief Engineer Samuel S. Montague was, like Hopkins, thinking ahead. In April he sent teams of surveyors over the mountains to Nevada to make their measurements across the state. Nevada was a high-altitude washboard, facing in the wrong direction for the eastward-bound railroad surveyors with their need of a steady grade, nothing but a succession of mountain ranges aligned north to south—the Humboldt Range, the Santa Rosa, Shoshone,

Monitor, Ruby, Schell Creek, and Pilot Ranges, just to mention the majors—with but one valley crossing it: the Humboldt River in the north, route of most of the emigrant wagon trains. It was high and dry land, forbidding in most aspects but most so across the grim Forty-Mile Desert, which stood between the meadows of the Truckee at the base of the Sierra Nevada and the farthest reach of the green and turgid Humboldt. It was a terrible crossing. As Samuel Clemens had noted just a few years earlier, travelers were "walled in by barren, snow-clad mountains. There was not a tree in sight. There was no vegetation but the endless sage-brush and greasewood. All nature was gray with it. We were plowing through great deeps of powdery alkali dust that rose in thick clouds and floated across the plain like smoke from a burning house." And travel through the dust was exceedingly unpleasant.

> *We were coated with it like millers; so were the coach, the mules, the mail-bags, the driver—we and the sage-brush and the other scenery were all one monotonous color. Long trains of freight-wagons in the distance enveloped in ascending masses of dust suggested pictures of prairies on fire. These teams and their masters were the only life we saw. Otherwise we moved in the midst of solitude, silence, and desolation. Every twenty steps we passed the skeleton of some dead beasts of burthen, with its dust-coated skin stretched tightly over its empty ribs. Frequently a solemn raven sat upon the skull or the hips and contemplated the passing coach with meditative serenity.*[34]

About to make the same trip, one of Montague's chief surveyors would echo Clemens's sentiments. Butler Ives, born in the Massachusetts Berkshires ten years after neighbor Thomas C. Durant, in 1830, had meandered west as a young man, becoming a deputy surveyor in Oregon Territory in 1852. Now, for the Central Pacific, his explorations that year would embrace an arid, corrugated rectangle some 450 miles long by 150 miles wide, for which he would be paid $200 per month and expenses. His job was that of a "reconnaissance, not a careful/complete RR survey," Montague ordered; two other parties would follow, making the necessary slow and careful measurements. "The country is very dry, barren, and mountainous," Ives wrote to his brother in April, "with numerous alkaline valleys and plains, sort of a purgatory on the outer rim of the happy land of Mormonism." Heading a party of eight men, all mounted and armed against "*uncivilized* Indians" whom Ives did not realize had mostly been killed or driven off by General Connor in the early part of the war, the engineer was charged to reach the Great Salt Lake, investigate routes around the southern as well as the unexplored northern shore, to resupply in Salt Lake City, and return by the safest, most practicable way to join the other parties in Nevada.

After a punishing trip across the Forty-Mile Desert, and a stumble across the mysterious Humboldt Sink, where the river coming down its canyon from the northeast flowed out onto a vast evaporative sandy plain and disappeared, and then a reconnaissance up the dusty, throat-searing Humboldt Trail between mountain ranges, Butler Ives ran into a setback when his assistants began to threaten to strike. They demanded a $5 salary increase, to $50 a month. Montague, back in Edenic Sacramento, was notified but could not understand why they should be paid more than his other parties; surveyors in the Sierra received $40, and others in Nevada got $45. Act according to your circumstances, he wrote to Ives, "but do not pay more than $50." With that problem solved, Ives pushed east to the source of the Humboldt, leaving rock cairns to mark his passage, and thence over the Pilot Range and down onto the burning salt flats of Utah, tending southeast toward the City of the Saints. That long summer expedition to the Great Salt Lake would convince him, and Montague, that the only route across Nevada would be the Humboldt, and that it would be no picnic, with certainly no dessert. By September Ives, having returned by way of the northern shore of the Great Salt Lake, would join the two other surveyors' parties—to narrow their concentration to the best route over the Pilots.[35]

Back on the western Sierra slope, Crocker and Strobridge were having their own labor trouble. Since May, feeling the pinch from both directions, Crocker had renegotiated his compensation from the railroad. Henceforth as contractor he was to be paid in Central Pacific stock at the rate of thirty cents on the dollar, down from fifty cents. More so than ever, then, his solution to the slightest hint of collective organization was to rule "with an iron hand." On one particular payday, when the two were out paying off, Crocker noticed a little knot of Irish workers talking together. "Strobridge, there is something breeding there," he said, and the labor boss agreed: "They are getting up a strike or something of the kind." Crocker said to let him handle them, and went on with his work. "They are coming," warned Strobridge a few moments later. "Get ready." Crocker waited a beat and then turned around. "Strobridge, I think you had better reduce wages on this cut," he said in a voice meant to be overheard. "We are paying a little more than we ought to—there is no reason why we should pay more on this cut and on that tunnel than on the other work. You better reduce them about 25 cents a day." The men huddled again, until finally one stepped forward. "We thought, sir, that we ought to have our wages raised a little on this tunnel," he said. "The tunnel is very wet and the cut is wet." "Well," replied Crocker, "I had just been telling Strobridge that you are getting too much and that he better reduce it." "Yes sir," said the worker. "I heard it, but your honor, I think you better not reduce it. We thought we ought to get an advance, but you ought not to reduce it certainly." Crocker looked to his foreman. "Well, Strobridge," he said. "What do you think? Can we afford to pay them that wages?" "Oh, I wouldn't make a fuss over it," came the reply. "We had better let them go on at the same figure." The men went back to work—their grievance deflected for at least a while.[36]

Strobridge, when dealing with his Chinese workers, was blunter. Whenever the men declared a strike, he waded into them brutally. "Strobridge was ever on the job," recalled J. O. Wilder, who, still a teenager, had just joined the railroad as a surveyor's assistant and was working in the steep slopes above Cisco. "They feared him in their hearts as much as they did the Chinese devil." Wilder's admiration of the labor boss's tactics was undiminished more than fifty years later. "He was a fine general," he recalled.

> *He had a mild but firm way, which was in the form of a pick handle, in dealing with these fellows. He had but one eye, yet he could spot the ring-leaders at one glance and would bring his persuader into action and was not particular where it landed, for he was a past master in this line. Inside of five minutes you could not find a Chinese in camp, and could hear them say "muckahigh" as they went to their work with Strobridge acting as escort, for many times he was called upon to settle discords and confusion.*[37]

Despite pervasive troubles with mudslides in cuts at Gold Run and Tunnel Hill, rails were laid into joyous Dutch Flat on July 4; the festivities extended up and down the slope throughout the day, beginning in Sacramento with a Fourth of July parade, to which the burghers of Dutch Flat were conveyed by train. The Central Pacific Railroad Company, with its armada of floats, commanded the greatest attention: Hopkins, Stanford, and the Crocker brothers rode in the head carriage, saluting the crowd and brandishing the shovel used to turn the first earth of the enterprise. They were followed by seven horse-drawn carts crowded with Chinese workers, and then by a locomotive, a flatcar, a passenger car, and a freight car—which, though rolling up the street on the Central Pacific tracks, were each pulled separately by mule teams, with cheering railroad workers aboard all. Following, Stanford led an excursion train up to Dutch Flat for a celebratory picnic.[38]

Far away in New York, Collis Huntington still exulted about his victory over Mr. Union Pacific, with the new legislation now signed by President Johnson. Soon he would have another laugh over Dr. Durant, or so he thought. Strangely, Durant had greeted the news of his legislative defeat by announcing to all that the Union Pacific would press on undaunted, that indeed its summer projections were so vigorous that he would be buying a million ties and at least sixty thousand tons of rails before wintertime. When this news of intended purchases reached the suites of the nation's ironmakers they naturally hiked their prices. Huntington was incensed; Durant's big mouth was going to not only cost him plenty, it was also going to soak the Central Pacific. By then Huntington would have heard Hopkins's demand for a steady supply of money and materials. Soon, he recalled, an iron manufacturer's agent, George T. M. Davis, appeared to ask how much the Central Pacific wanted during the rest of the year. "Don't talk to me, Davis," retorted Huntington, "for I am mad. I have 15,000 tons of rails on the way to California, and if they don't lay all those rails this year, I will have to get rid of those we have left on hand." Davis was puzzled—he thought he had kept tabs on the Central Pacific's requirements, and with Durant's bullish claims to lay sixty thousand tons by winter, well—but Huntington cut him off. "You know he will do nothing of the kind. The Kansas people are growling—they won't want a rail. Sioux City won't start to build until the Cedar road is secured. The Central branch is not organized and will not begin."

Davis reported the dreary projection to his superiors. Word got around, and the price hike was rescinded. Then Huntington stepped in. "I ordered 10,000 tons from the Cambria," he chuckled many years later, "10,000 from the Bethlehem, 10,000 from the Simonton, and scattered orders among various other makers till I had secured 66,000 tons. Not a boy in my office knew that I wanted rails. I telegraphed all my orders at once, and every one of them was accepted by telegram within an hour."[39]

Traditionally, this triumphal anecdote of Huntington's has been reported as if Dr. Durant had seriously blundered by announcing his intentions to buy iron, sending the prices upward. Huntington, of course, was a genius at portraying competitors badly, and he has often been persuasive. "Far more of a showman than a businessman" is the assessment of Durant by the previously most careful transcontinental chronicler, Wesley Griswold. But this was the same Dr. Durant who had used planted rumors to get his way in Omaha over the oxbow route down Mud Creek; this was the same Dr. Durant whose business leaks in 1864 regarding the Mississippi and Missouri and the Cedar Rapids roads, and whose behind-the scenes stock manipulations, had gained him a quick fortune and the queasy admiration of Grenville Dodge's brother Nathan. "He gets back home," Dodge had written at the time, "and makes in the round trip for him and his friends $5,000,000. It is the smartest operation ever done in stocks and could never be done again."[40] Durant was a showman, yes—in a behind-the-scenes way—but he was also a sharp businessman. Somewhere in this episode is buried a fast buck or at least the expectation of it; either personally or through associates, holding iron securities or some other connective stock, Durant must have expected his injudicious remark to cause iron prices to rise, and his personal portfolio to go up—the only acceptable direction.

Still, from Huntington's perspective the Californian could crow. To H. H. Bancroft's researchers, years later, he boasted how astute he had been in the indirect tussle with Durant and then with an agent of a large shipping line he wanted to use to ship all the new iron. "I want to get a good ship, one that will be steady and safe," he told the agent, E. B. Sutton. "You go out and give me a list of what you can find." When Sutton called on him with a few names and rate quotations, he found Huntington in a picky mood. "Too high, I can't take one of those," he said. "I am in no hurry, for ships are coming in all along." Time elapsed, and Sutton reappeared with a list of twenty-three ships. "I noted down the vessels as we were talking," Huntington said, "and suddenly said, 'I will take them.'" Take what? Sutton asked. "I will take those ships," said Huntington, "if they are A-1." Sutton demurred. "I thought you wanted only one," he responded. "I will have to have two or three of them myself." "Not of these, you won't!" Huntington shot back.

"I secured them, and these vessels took out some 45,000 tons of rails," Huntington said, finishing his anecdote with high satisfaction. "Mr. Sutton afterward said that these ships would have cost me at least $10 a ton more had he known that all of them were wanted—a clear gain of $450,000 for my company."[41]

21

"We Swarmed the Mountains with Men"

It was, as Huntington and Durant were well aware, a season of politics, that summer of 1866, with larger issues crowding out mere adjustments to existing railroad legislation. President Johnson had proclaimed that the insurrection was officially over on April 2 (in all the Confederacy save Texas, which would bow under control in four months). Closer to the White House, though, rebellion over suffrage and southern representation continued to glow like molten iron within his own party, markedly under the leadership of the radical Republican Thaddeus Stevens of Pennsylvania. The year-old Freedmen's Bureau, empowered to care for freed slaves and administer over abandoned southern lands, was enlarged by Stevens's congressional rebels to adjudicate civil rights cases through military tribunals, with the bill being passed over the president's veto in July. A bill granting citizenship to African Americans and for that matter to anyone born in the United States (except Indians) was also passed over Johnson's veto. A constitutional amendment (the fourteenth) granting citizenship in a like manner and extending federal protection of rights flew through Congress and was dispatched out to the states for ratification. With the radicals in Congress treating the South like a conquered province, with the lines from Johnson's dragging heels leaving gouges on Pennsylvania Avenue, with most southern states rejecting the Fourteenth Amendment, attention shifted that summer to the nearing congressional elections where Reconstruction policies would either be endorsed or repudiated. Andrew Johnson's pitiful bid to create a third, more moderate party in August would fail just as badly as his stammering campaign tour would in the autumn. He would be greeted by insults and jeers in every town he entered, and be shouted down by ruffians in some. There would be only two real bandwagons then: the tiny one of the Democrats, and the juggernaut of the Republicans.

In his own inimical way, even the noonday mystery, George Francis Train, was pinning his hopes on political success with a base in the state of Nebraska, stumping for a senator's seat, counting on the votes of all the settlers who had bought land from his syndicate, the Crédit Foncier. "This corporation had been clothed by the Nebraska legislature," Albert Richardson would soon report, "with nearly every power imaginable, save that of reconstructing the late rebel States." Train himself owned some five hundred acres in the city of Omaha, in Richardson's eye now "the liveliest city in the United States," and had been busy planning towns all along the railroad right-of-way. More so than the gratitude of settlers, more than the aid of local politicians who were always glad to mount a bandstand and bask in the reflection of Train's white suit and thundering good oratory, there was the required support of his various corporate partners, without which he knew he would not even get to stand before his voting public. When, after the majestic dust of the state convention had settled and Train found himself bereft of nomination, his fury could not be contained. "George Francis was *as he says* sacrificed by the Rail Road men and *cursed loud and long,*" Sam Reed wrote to his wife on July 18. "All the blame was attached to RR men, and he gave us all notice that none that are now in good places will be here in 30 days from the time of his starting for NY, including T.C. Durant. One stagecoach or steamboat could not hold two such men. . . . George Francis goes from Neb. disgusted with western politics and I think with a large flea in his ear."[1]

The apotheosis of talk (and his threats were just talk) was not the only railroad man campaigning that season. There was also the Union Pacific's new chief engineer, Grenville Dodge of Council Bluffs and of the Iowa Regency Republicans who had controlled the state for years. In his memoirs Dodge would claim that he was above politics and that he was almost surprised to be nominated for a congressional seat from Iowa in the summer of 1866, and that he was elected that fall without raising a finger on his own behalf.

This assertion, repeated as gospel by railroad historians, was finally refuted by Stanley Hirshson, Dodge's good biographer. The general had maneuvered in the Iowa senatorial campaign in the winter of 1865–66, and Engineer Dodge would take an active, albeit behind-the-scenes part in the Iowa Republican congressional canvass. Regarding this episode, Hirshson says, "Iowa contained no heroes, only villains." In a compact block of weeks away from the Union Pacific office, Dodge betrayed the Republican incumbent, John Kasson, who for five years had been an ardent, influential Dodge lobbyist for army promotion, pay, and recognition. He stood quietly by as his supporters, including Hub Hoxie and the publisher of the *Des Moines Register,* both inflated his war record and embarked on a vicious smear campaign of Kasson's name: starting by calling Kasson a copperhead and a traitor, the *Register* escalated it to "apostate whoremaster," taking advantage of the fact that Kasson had just gone through a messy divorce in which his wife had charged him with adultery. Correspondence with his wife was widely published.

Then, as Hirshson noted, the Dodge campaign received reinforcements from an unexpected source. Kasson's former brother-in-law was the Reverend Doctor William Greenleaf Eliot of St. Louis, the founder of Washington University, the head of the western branch of the U.S. Sanitary Commission, and the future grandfather of T. S. Eliot. The Reverend William Eliot published a pamphlet asserting that Kasson "was a diseased man from dissipation, so that it was not safe for any woman to live with him. He acknowledged to me having been so diseased, and that he had had criminal connection with three different women." After the convention to Des Moines, Grenville Dodge emerged as the Republican candidate; the humiliated Kasson, who had earlier written his former friend that "I could not believe you would wish to take bread away from me," got behind Dodge's candidacy and the principles of Reconstruction in the subsequent campaign. Nationally the Republicans would collect two-thirds of each house in November. And the fifth Iowa district would be sending a new representative—a shy, unambitious, apolitical war hero, as he was described by the Regency's *Des Moines Register.* "The myth," writes Stanley Hirshson, "of Dodge's indifference to politics was being firmly established."[2]

Even out on the plains, in Indian country, it was a political summer. Government administration had been delegated to underlings; policy was dangerously adrift. The interior secretary was nominally James Harlan of Iowa (this was a year after he had fired "the Good Gray Poet," Mr. Walter Whitman, from his position as first clerk in the Indian Affairs Office in Washington). Harlan was weary of the cabinet and displeased with Andrew Johnson. In Iowa's January special election Harlan had won back his old Senate seat, which he was to take in 1867 for a six-year term. By July, with the president stumping for his moderate third party, Harlan left the cabinet and returned to private life in Iowa, which for him meant seeing to local political matters. Much of his time he devoted to working for Grenville Dodge's congressional campaign.[3]

His reluctant replacement, named in summer and succeeding Harlan when he officially stepped down in September, would be Orville Hickman Browning, who acceded to the embattled President Johnson's plea for help. This did not bode well for the Union Pacific. Browning, a longtime associate of Lincoln's, was a balding, fastidious Illinois lawyer who had been appointed to fill out Stephen A. Douglas's Senate term after the latter's death. An old-fashioned, Whiggish sort of Republican who was given to wearing anachronistic ruffled shirts and grew queasy with radical "excesses" like emancipation and suffrage, Browning had worked in Washington since 1863 in an influence-peddling law firm with Thomas Ewing Jr. of the Kansas line. This was not Browning's only reason for looking askance at the Union Pacific. He was also closely tied professionally and financially to a Detroit railroad entrepreneur and lawyer named James F. Joy, whose every association—with the Chicago, Burlington, and Quincy, with the Hannibal and St. Joseph, with large tracts of Kansas City real estate—would square him off against the Union

Pacific. Browning also had little sympathy for the army, chiefly on political grounds. Therefore, his tenure as secretary would cover some of the most difficult times to face the Interior Department and the Bureau of Indian Affairs, a succession of troubles in which the Plains Indians confronted Browning's department on one flank with the U.S. Army on the other.[4]

But meanwhile, in June, with Harlan attending to the Iowa canvass, and with Browning working loyally for Johnson's interests in Washington and Philadelphia and elsewhere, there were the Indian peace talks to get through.

Held as previously announced at Fort Laramie, the parley was a disaster from the start, being two weeks late in convening when the Northern Sioux of the Powder River camps had to be found and coaxed in. Indian Affairs was represented by a well-meaning conciliator, E. B. Taylor, who knew little about Indians and whose advisers on the treaty commission had scant knowledge about the situation on the plains, but who was unshaken in his belief in a humanitarian policy. As soon as the whites' treaty was laid out, the Powder River leaders, especially Red Cloud, loudly complained: the treaty sanctioned a trail straight across the valuable hunting lands of Powder River, a trail vigorously contested by the Sioux since it was created three or four years earlier by a freebooter named John M. Bozeman, who had sought a fast, cheap route between the Oregon Trail and the new Montana goldfields around Virginia City. The Bozeman Trial, leaving the North Platte river about 70 miles west of Fort Laramie, skirting the Bighorn Mountains before joining the Yellowstone and then the Missouri headwaters, had in just a few years seen a great deal of action as a steady stream of gold-seekers and settlers' wagon trains thundered through, ignoring the dire warnings of both the cavalry and the Sioux, despoiling the terrain and beginning to ruin the Indians' hunting. General Patrick Edward Connor's troops, after erecting Fort Connor where the Bozeman struck the Powder River, had ranged all through the country in 1865, enlisting hostility among the peaceable Arapaho and the resistant Sioux. Now Fort Connor was to be called Fort Reno, and the Bozeman Trail was to be officially recognized.

Commissioner Taylor answered Red Cloud's angry protests with soothing tones. He knew little about the Bozeman Trail, but guaranteed that it was just a small matter. The Sioux seemed somewhat mollified at the close of the day's discussions. But then, later in the afternoon, a Sioux leader by the name of Standing Elk spied a large wagon train—two hundred vehicles, seven hundred soldiers, families and personal possessions, a regimental band, followers, and even a herd of a thousand beef cattle—approaching the fort, halting and camping for the night. Standing Elk went to investigate. He asked the commanding officer, one Colonel Henry B. Carrington, where they were going. The colonel answered, "Powder River," to fortify and protect the trail. His order was to refurbish Fort Reno on the Powder, establish a large new fort up that river's tributary, and build a third post on the Bighorn River. Within minutes all the Sioux camps knew.

The following day, Chief Red Cloud spoke angrily, calling Taylor a liar and calling the whole treaty process a sham. "Great Father sends us presents and wants new road," he said, "but White Chief goes with soldiers to steal road before Indian says yes or no." Issuing a stream of threats, he led his Oglala delegation out. The Northern Sioux disappeared across the North Platte, prepared for a last stand on the Powder River. Commissioner Taylor was left with the conciliatory Brulé Sioux under Spotted Tail, whose hunting territory was unaffected by the Bozeman controversy, and a handful of likewise disinterested Southern Oglala. They, convinced that open war would signal an end to their people, signed the now-worthless treaty. E. B. Taylor—rightly called by one commentator "a fanatical optimist" for his dogged refusal to see matters as they drearily stood—proclaimed that the Laramie talks had succeeded: "Most cordial feeling prevails." Red Cloud was an insignificant leader, he told Washington. The Northern Oglalas would cause no trouble. Within weeks, Spotted Tail was to report that many of his firebrand warriors were deserting his camps south of the Platte and heading toward Powder River. If this continued, the tribes that agreed to peace would begin to disappear, ceding power to Red Cloud and the others, who were ready for war.[5]

Dodge himself felt the inevitability of it all, writing to Grant that "our Rail Road will do more toward taming Indians than all else combined." And that first skirmish involving the Union Pacific, which may have involved Spotted Tail's deserters, would occur in August, where Plum Creek entered the Platte, and where a trading post and former Pony Express station stood on the south riverbank; already the townspeople had begun dismantling their homes and businesses and were moving across the river to get with the railroad. They would name their new metropolis Lexington, commemorating the Battle of Lexington. It was 230 miles west of Omaha, and the railroad had run on another ten miles, though only work trains were moving. Dodge was at the end of track watching the Casement brothers' teams at work when he received a frantic wire from the Plum Creek telegrapher, who had spied a war party of Indians ride up from the south, splash across the drought-reduced Platte, and fall upon a stalled freight train. After taking the crew hostage, the Indians began to torch the freight cars. Dodge hurriedly gathered twenty riflemen from the work crew, piled them into his office car ("a traveling arsenal," he recounted), and an engine quickly hauled them the ten miles to Plum Creek. The raiders were preparing to assault the whistling new arrival when it suddenly disgorged men with blazing rifles; all were army veterans and knew what to do when Dodge deployed them to either side of the embankment. "They went forward as steadily and in as good order as we had seen the old soldiers climb the face of Kenesaw under fire," the former general recalled proudly. The Indians abandoned all thought of their captives and quickly escaped on their ponies.[6]

All along the line, and out onto the prairie, graders and construction crews kept one eye on the horizon for any suspicious movement. They all became, wrote Reed to Durant, "very timid and at the first appearance of Indians would all leave the work." More often than not the Indians would turn out to be friendly Pawnee, who were hunting across the sand hills and creeks north of the Platte, but the anxious laborers were taking no chances.[7]

In mid-August, some would take solace from the presence of General Sherman, who appeared in Omaha with his small retinue and with his brother, Senator John Sherman, to make an inspection of the work to date. Before they headed west on a special train, the general wrote to Grant through his aide, General John A. Rawlins. After having conferred with Dodge he was persuaded that the Union Pacific's progress "surpasses everything in the way of rapid construction I have ever known." Passenger service now extended 150 miles, he said, and a new section was about to be opened to passengers which would reach Fort Kearney, some 194 miles from Omaha. Dodge had told him that they had enough iron and ties to build another hundred miles that season. He projected that the tracklayers would reach Julesburg, or Fort Sedgwick, by April of the next year. "I am perfectly satisfied that this road is in excellent hands," Sherman wrote, "and I propose to give them all the protection and encouragement we can."[8]

When Dodge had gotten Sherman's party out to Fort Kearney, and the general had looked over the situation, he was inclined to be cynical about the magnitude of the Indians' threat. Since assuming command in the West he had seen a good deal of opportunism in the settlers' alarmist complaints. He had been plainspoken about how scattered farmers and merchants profited whenever large columns of cavalry appeared. "As usual, I find the size of Indian stampedes and stories diminishes as I approach their location," he wrote to Rawlins on August 24. But "there is a general apprehension of danger, though no one seems to have a definite idea of whence it is to come. I have met a few straggling parties of Indians who seemed to be pure beggars, and, poor devils, more to be pitied than dreaded." The more Sherman traveled over the ensuing weeks, inspecting the Casements' progress and venturing out in an army ambulance-wagon or on horseback, the more he was convinced of this. "All the Sioux have been driven west from Minnesota and the Missouri River," he wrote a few days later, "and the mountain region of Montana, Colorado, and Utah is being settled up with gold miners and ranchers, so that poor Indian finds himself hemmed in. The Indian agents over on the Missouri tell him to come over here for hunting, and from here he is turned to some other quarter, and so the poor devil naturally wriggles against his doom." It was a trenchant analysis. The settlers wanted the army "to kill all the Indians," he wrote still later. "I will not permit them to be warred against," he vowed, "as long as they are not banded together in parties large enough to carry on war."[9]

More potential complainers, certain headaches for the army someday, he knew, were arriving by train and by wagon every day to stake out their farms. Often, as he had seen in Kansas in midsummer, they were locating themselves far from neighbors and offering themselves as tempting targets for livestock raids, shakedowns, or worse. Already there had been a flurry of interest in the next hoped-for metropolis, planted on a flat plain across the Platte from the Kearney post, whose supporters would in future years advance it as a candidate for the state or even nation's capital. A good-sized contingent of Swedish immigrants had begun congregating there in 1865, but in August Grenville Dodge laid out an official town grid, setting prices for town lots. The prices were tempting: corner lots went for $150 and inside lots for $100 near the station, with discounts made out on the theoretical edge of town, of $75 and $50. Buyers had to pay only one-third down with the balance paid in up to two years; an interesting codicil of the agreement was that the purchaser had to plant shade trees within twelve months.[10]

On one particular September night in Kearney, a special passenger train stopped for the night on a siding, its occupants improvising beds with boards, cushions, and blankets, upon the backs of seats. Aboard was the roving *Herald* correspondent, Albert Richardson. He had no sooner returned to New York on a Panama steamer following his "eight-month wanderings" from the Missouri to the Pacific and back again when he set out again to see the railroad's progress. It may have been during his brief rest in New York that he visited Dr. Durant in the Union Pacific offices. They were, he would write, "among the most elegant in New York. Brussels carpets, and black walnut and marble counters in the rooms of the managers, rare statuary and choice paintings surprised the eyes of visitors. Dr. Durant's horses were the envy of Central Park, and his yacht was the admiration of the New York Yacht Club. I have seen him entertain a party of ladies and gentlemen upon it, down the bay, through an entire forenoon, as if he had not a care in the world beyond the comfort of his guests; and at one o'clock say nonchalantly, 'Well, goodbye, I must go ashore; I have a million of dollars to pay before three o'clock. Have your sail out, and don't return till you get ready.'"

The correspondent had taken a palace car west from New York to Chicago to Atchison, and after a tour of Kansas he had taken a steamer from St. Joseph to Omaha. Impressed with the "wonderful vigor" and growth in Omaha, he had boarded the U.S. Commissioners' train for an inspection trip to the end of track, 240 miles west where another 20-mile section would be accepted. "Having traveled to Fort Kearny seven times by wagon and coach," Richardson noted, "I found accomplishing it by rail in a few hours decidedly agreeable." The next morning, resuming the westward journey, the passengers saw hundreds of antelope from their train. "Some came within two hundred yards, curious to scrutinize the iron monster screeching into their vast domain," Richardson said. "While in motion we aimed hundreds of rifle-shots at them from the car windows. A single one, from General Merrill, took effect, and sent its beautiful victim limping into the sand-hills." After such valiant sport, the party reached the end of track, seeing the long sleeping and eating cars of the workmen, "who press forward so fast that only portable dwellings will serve them." They labored, Richardson marveled, "with the regularity of machinery, dropping each rail in its place, spiking it down, and then seizing another. Behind them, the locomotive; before, the tie-layers; beyond these the graders; and still further, in mountain recesses, the engineers. It was civilization pressing westward—the Conquest of Nature moving toward the Pacific."[11]

The torrid drought of the summer had finally broken in early September, and with the seasonal turn toward autumn weather other travelers were at large or making plans in that direction. Dodge, in his official capacity, would be called upon to cater to two separate junkets. Both would be important to the fortunes of the Union Pacific. The first, however, though less of a logistical chore, offered more potential trouble to the chief engineer. Colonel Silas Seymour, that wild card of a consulting engineer whose murky job description and obsequious loyalty to Dr. Durant had bedeviled Dodge's mentor Peter Dey, who would become the burr in Dodge's blanket, appeared in Omaha for an inspection trip of the work and for a personal examination of passes in the Rockies and the Laramie/Black Hills. He was accompanied by a government director of the Union Pacific, Jesse L. Williams, a Lincoln appointee and a civil engineer from Fort Wayne, Indiana. For the moment there was one significance for Dodge to make wary note: Williams had sided with Seymour on the oxbow route against the strenuous counsel of Peter Dey. To have two competing engineers sniffing through his territory must have rankled, especially given his ego. Moreover, only a few weeks before Dodge had received his best surveyor's preliminary report on the Black Hills line. James Evans had endorsed Dodge's dramatic, Indian-harassed discovery of the previous year, the Lone Tree pass to Crow Creek. Dodge had quickly forwarded Evans's preliminary line to Durant, calling it the most superior.

And here was Seymour with his habit of finding fault with the diligent work of others. The consulting engineer had reached Omaha via Pittsburgh and Chicago on the luxurious Lincoln Car, from Iowa City taking the necessary stagecoach. Dodge received them cordially, of course, and received fulsome treatment from both. At the end of track, after a suitable inspection, Seymour and Williams had ridden up to Denver, enjoying everything about the Rockies, including an early snowstorm, and subsequently had examined areas where coal and iron might draw the interest of the Union Pacific or its investors. They rejoined Dodge, and James Evans, at Fort John Beauford (soon to be renamed Fort Sanders) on the Laramie River, for a close look at the competing routes through the Black Hills. The two city slickers were delighted. "We were fully mounted as cavalry men," Seymour wrote to Durant on September 27, "on U.S. horses with our carbines dangling at our sides and our revolvers belted around us (except Mr. Williams who unfortunately had no pistol)." Seymour proudly reported that they had carried "four different kinds of shooting apparatus, so that on the whole we presented quite a formidable & warlike appearance . . . it was exceedingly fortunate for the Indians that they kept out of our way." They had been warned that "Indians indulge in mule stealing (and sometimes in scalping their owners)," Williams noted in his own letter, "having recently taken seventy mules from a transportation train." The engineers were escorted by twenty infantry, half of whom were mounted. With that reassurance, the party took on more of an air of a holiday where fantasies of Wild West prowess could be indulged. Six miles beyond the fort, they sighted a majestic elk stag grazing at the edge of a lake. No thought was given, of course, to letting it survive. "Several shots were fired almost simultaneously," wrote Seymour, "and after staggering a few rods he fell. When we reached the noble animal, life was extinct." It weighed at least eight hundred pounds. While soldiers carved out twenty-eight pounds of steaks for the party, leaving the rest for the buzzards, the "splendid" antlers were removed. Seymour, they all decided, would take them to New York and present them, "with the united compliments of the party," to Dr. Durant. Soon they would grace the Union Pacific's main office suite.[12]

The boyish party atmosphere continued as they ascended the western slope of the Black Hills to the virginal Dale Creek valley. Halting there, Seymour waxed eloquent:

> *We had at last reached the realization of our hopes and dreams, and were actually "camping out" in the mountains. We could roll in the long grass, drink our fill from the sparkling stream, sing and halloo as loud as we pleased, without disturbing any one outside of our own little party. The Indians might be watching us from some of the surrounding crags, and coveting our scalps as trophies for the adornment of their wigwams; or might be planning an escapade for our stock; but what matter—we all felt that innate sense of security and reliance upon ourselves, which always accompanies a wild and roving mountain life; and which, we felt confident, would enable us to cope successfully with five times our number of these savage denizens of the forest.*

Living out his fantasy as they continued the reconnaissance, Seymour enthusiastically recorded attempts on the life of a lone antelope on the plain between Lone Tree and Crow Creeks, on a herd of thirty elk in Lone Tree Creek valley, all of whom escaped unscathed, and on a large antelope in Box-Elder

Valley, which they surprised while it was napping and which was killed with one shot. Dodge, "whose reputation for skill in bagging much larger game had become so well established during the late war," preened Seymour, "immediately pronounced me the *huntist* of the party, and awarded me the beautiful skin as an additional trophy."[13]

From Dodge's point of view, the trip, which ended with some smug self-congratulation soon thereafter, was purely wasted time. His efforts to keep the two visiting engineers entertained and his earnest argument for the Lone Tree–Crow Creek route were unavailing. Seymour and Williams would soon espouse the route up the Cache la Poudre line. This was partly to take advantage of coal deposits, but also because of its easier reach to Denver (which they had visited twice during the trip, listening to the political arguments of their prominent hosts). Dodge himself had already gone on record as being against catering to Denver interests. "The people in Colorado are hostile," he had written Durant in September, "from the fact that they look upon the other road as their road and ours as having no interest in common with them."[14] But inexplicably, according to the engineering considerations Dodge brought up, Seymour and Williams were urging the Union Pacific into more difficult and more expensive terrain. Both Evans and Dodge had gone over it: the Cache la Poudre line exceeded Lone Tree by nearly two thousand feet in ascent and descent, with higher maximum grades. It even required a tunnel. Dodge could not believe that anyone could seriously dispute him on this. Before too long he would have to journey to New York to make his case personally before the Union Pacific directors.[15]

Before that, though, he had to play the host one more time.

Thanks to Durant's incessant goading by telegraph, to an increase in the Casement brothers' workforce, and to an escalating system of bonuses and inducements, there had been a furious amount of new track work. Durant had ordered Reed to increase passenger ticket prices by 25 percent to all stations west of Fremont, telling him also to "run no trains except construction with passenger cars attached."[16] Reed predicted that iron would be spiked past the 100th meridian in early October. This was significant. The Pacific Railroad Act as amended in 1864 had stretched a finish line of sorts along the meridian; whichever railroad—Union Pacific, Kansas line, or whomever—ran across the prairie and puffed out its iron chest and broke through the line would win the right to keep racing toward a meeting point with the Central Pacific.

On October 6 with a few hammer blows the Union Pacific thereupon became the undisputed main line—finally. Dr. Durant, having decided to throw a blowout to end all blowouts, issued a new string of demands to his employees in Omaha and then sent out no less than three hundred invitations. They went everywhere, and if all responses had been positive the locus of political and economic power in the United States would have shifted, albeit temporarily, to the undistinguished Nebraskan prairie: President Johnson was invited along with his entire cabinet; all members of the House and Senate, most of the commanders of the army and navy, all foreign dignitaries resident in Washington, government directors and commissioners of the Union Pacific, board members, large stockholders, captains of industry, railroad moguls whether they competed or not—all were respectfully invited to join one of two special trains heading west toward Omaha, hop-skipping by stagecoach and river steamer to Omaha where the parties would combine for the ceremonial trip out to the meridian. Bring the wife and kids.

Not everyone on the Doctor's grandiose list accepted. But the roster of two hundred was certainly distinguished. John Sherman and peppery, powerful Ben Wade of Ohio and New Hampshire's J. W. Patterson appeared for the Senate, joined by Nebraska's first two senators-elect; a dozen representatives included the Crédit Mobilier's railroad congressmen John B. Alley and Rutherford B. Hayes of Ohio, who in a decade would survive the major railroad scandals and electorally squeak into the White House; the diplomatic community sent an earl from Scotland and a marquis from France; many giants of capital came, including George M. Pullman (who had lent Durant palace cars for the occasion) and Joseph Medill (owner of the *Chicago Tribune*), railroad tycoons, and of course the Union Pacific board members present included Dillon, Lambard, and Duff, all immensely wealthy men long before they had met Durant. And yes, George Francis Train, and his wife and maid, were there. Grenville Dodge, wife and daughter, were there, as was Hub Hoxie and wife, the brawny little Casement brothers, Dan with his wife, and numerous others—journalists, editors, two photographers. And there were no less than two full musical groups, the oompah of the Great Western Light Guard Band of Chicago and the celebrated Rosenblatt's Band of St. Joseph. ("Both wind and string instruments," Hoxie had explained to Durant. "A good one for St. Joseph [but] not number one for New York.")

They had been plied with rich fare even before they landed in Omaha, and fêted at the Governor's Ball, at every opportunity orating and congratulating one another. Colonel Seymour, who had met them at the steamship dock, noted that they had been "evidently delighted, and somewhat astonished to find themselves, after a week's journeying westward from New York, still among people of wealth, refinement, and enterprise." Poor Sam Reed, who had exhausted himself with preparations, would have been dazzled by all the pomp as well as by the expense. "I hope," he had written his wife, "that there will be but few in the party." But by the time they had arrived, he was in bed with typhoid fever. The list of excursionists, as published en route in their own newsletter, had so many honorables, governors, generals, colonels, reverends, doctors, and judges, and so many of their grande dames, that the train, as it rumbled importantly away from the yards and machine shops and round the oxbow and up the Platte Valley, could hardly contain itself.[17]

There were nine cars pulled by two flag-decorated locomotives: behind a baggage and supply car, there was a mail car "fitted up as a refreshment saloon," a cooking car constructed by the Pittsburgh, Fort Wayne, and Chicago Railroad for the Union Pacific, followed by four passenger coaches built by the Union Pacific in Omaha, followed by the Lincoln Car ("the private property of Mr. Durant," noted Seymour, "and . . . therefore devoted principally to his own personal friends and their families"), and finally by the "magnificent" directors' car, built also by the Pittsburgh line. Dodge, as general host, was everywhere. He "rendered every assistance in his power," Seymour noted, before modestly admitting his own contributions: "the Consulting Engineer, by his timely presence, was enabled to relieve the others from much, if not all the *heavy standing around.*" The train proceeded at a purposefully slow pace down the "fine valleys" of Mud Creek and the Papillon. Many aboard exclaimed in "wonder and admiration," Seymour added, as the Great Platte Valley opened to view. After several stops to examine the bridge structures over the Papillon and Elkhorn Rivers and to admire the "commodious" depot buildings and water stations at Fremont and North Bend, the weary excursionists reached Columbus just after dark.

Just past the station buildings they found "a brilliantly illuminated encampment, which covered several acres of beautiful ground . . . so arranged as to afford comfortable accommodations for all who wished to leave the cars and enjoy the novelty of a night's sleep in camp." Inside a large dining tent they found food to vie "with those found upon the table of our Eastern hotels." But after dinner the real entertainment began. At Durant's behest the railroaders had hired a large number of Pawnee from the nearby reservation, and the Indians proceeded to perform a war dance. "Of all the wild and hideous yells, grotesque shapes and contortions that have ever been witnessed by a civilized assemblage in the night-time upon the plains," Seymour exclaimed,

> *this was most certainly the climax. The light of the moon, aided slightly by that of a dim camp fire, was barely sufficient to enable the spectators to distinguish the features and grotesque costumes of the savage performers; and the congregation of lady and gentlemen spectators were only too glad to know that the Indians were entirely friendly, and catering only for the amusement of the company, instead of being enemies, dancing and gloating over their scalpless bodies.*

Tingling with the thrill of it all, the refined company turned in soon thereafter. Then, just before dawn, the excursionists were jolted awake by "the most unearthly whoops and yells." Durant and Dodge had arranged for a sham Indian raid by the faithful Pawnee as the morning wake-up call. Disordered heads of the excursionists popped out of tent flaps. Many screamed, wondering "whether they were to be immediately roasted alive, or allowed a short time in which to say their prayers, and write a few parting words to their distant friends," as Seymour recounted. After calm was restored and the joke explained, and after a sumptuous breakfast, the tents were struck and the travelers climbed aboard the train once again.

East of the Loup Fork bridge the train halted once again on a high embankment. The Pawnee reservation was not far away, and Dodge had arranged for a mock Indian battle between the Pawnee and some of their number dressed up as Sioux warriors. "The shock of meeting was grand and terrific," wrote Seymour.

> *Horses reared and plunged against each other. Indian grappled Indian, and both fell to the ground in deadly embrace. Rifles, revolvers and arrows were discharged apparently with deadly effect. Riderless horses, and horseless riders were to be seen roaming wildly over the plain. And all was confusion and intense excitement, until at length the victorious Pawnees brought their vanquished enemies into camp, amid the most tempestuous shouts of triumph and exultation.*

Afterward, Dr. Durant tossed several hundred dollars' worth of presents into the crowd of Pawnee. There was a terrific scramble. And then all visitors climbed smoothly aboard the train and it chugged off westward, passing the embryo towns of Silver Creek, Lone Tree, Grand Island, Wood River, Kearney, Elm Creek, Plum Creek, and Willow Island.

At mile 279, about 8:00 P.M., the train halted for the night opposite Fort McPherson, whose soldiers were charged with protecting the gentle party from real marauders. On the following day Dodge ran a train out to the end of track, some ten miles farther, where Jack and Dan Casement had their men work for the party's brief enjoyment, laying eight hundred feet of track in a half hour. "So we go," exulted the editor of the *Railway Pioneer,* "on our march to the Pacific!" And that night, back at the McPherson encampment and after a musical program, they were treated to a magnificent fireworks display, "much to the amazement, no doubt," said Seymour, "of the distant savages and wild beasts, who might happen to be the witnesses of this first exhibition of the kind in the great Platte Valley." Then, after more music, the celebrated phrenologists, Fowler and Wells, mounted the dais. "The professors . . . called the listeners one at a time to the dais," recalled Erastus Lockwood, "and made phrenological examinations, reporting their findings in witty remarks to the audience assembled. George Francis Train was one of the victims, and when his diagnosis was given as colossal conceit, the tent went into an uproar—all of which Mr. Train took in good part."

As if music and fireworks and the cranial bumps of Mr. Train were not enough, on the eastward return the tiring excursionists were treated to a celebration at the Hundredth Meridian monument (during which time many photographs were taken), and a close examination, near Kearney, of a vast prairie-dog colony, some twenty-five square miles in size "and by far the largest town through which they had passed since leaving Chicago," Seymour wrote. Curious animals who peered out of their burrows were answered with prodigious gunfire; it wounded a number but netted the huntists only one victim, which was surrendered to the cook. "It has come to a pretty pass," joked one dignitary, "if this grand excursion is reduced to such a strait that its guests are obliged to subsist on *prairie dog.*" Their humor was increased after dark, during their last hours on the plains, at the lower end of the Platte Valley: Durant had ordered that the prairie be set afire for the excursionists' delight. "The flames extended in an unbroken line a distance of from fifteen to twenty miles," Seymour recorded, "and one end of the belt of fire was so near, that we could feel the heat, and distinctly hear the roaring and crackling of the devouring element, as it swept over the plains with almost railroad velocity, and shot up its forked flames into the sombre smoky sky."

Make no mistake. This excursion was, in Seymour's estimation, "the most important and successful celebration of the kind, that has ever been attempted in the world." It also garnered an immense amount of positive publicity for the Union Pacific, coming at a good time when the board of directors' annual meeting was nearly upon them and when bond selling would have to be increased over the winter months to pay for the next impulsive push westward. Certainly a number of those bonds would have to pay for the prodigious cost of Dr. Durant's Hundredth Meridian Excursion.

But the last word on that extraordinary event, and on all it signified to the well-heeled celebrants and even more so to the people and the way of life the railroad would displace, belongs to that consulting engineer on the Empire Express, Mr. Silas Seymour, and it comes in the form of the notes he jotted down just after the mock Indian battle at the Loup Fork bridge and published in a popular book. Not many years later, many of those Pawnee, having first been moved aside by the Sioux and Arapaho and then by the government whose shelter they would then seek, would find their way into the proud employ of one William Frederick Cody ("Buffalo Bill") and his Wild West Show, the sham and the drama and the tingly romanticism of which must have found its inspiration in the Hundredth Meridian Excursion of October 1866. Then, as the Pawnee had fought over the spoils strewn by Dr. Durant, the occasion had moved the consulting engineer to new heights of fancy, a veritable rhapsody of racial superiority. "Perhaps no better illustration could have been given," he said,

> *of the extremes of civilized and savage life, standing face to face with each other, than the one now before us. On the one side was the track of the Union Pacific Railroad, upon which stood that great civilizer, the locomotive and train, looking westward over the Loup-Fork bridge, fifteen hundred*

feet in length; and in the foreground stood the group of excursionists, composed of beauty, intelligence and refinement; while, on the other hand, were grouped these uncouth savages, many of them almost in their normal state, except for the profuse display of feathers and trinkets which bedecked their persons; low and brutal in their habits, and mentally elevated but slightly, if at all, above the level of the beasts that inhabit this vast and beautiful country with them.

But the laws of civilization are such that it must press forward; and it is in vain that these poor ignorant creatures attempt to stay its progress by resisting inch by inch, and foot by foot, its onward march over these lovely plains, where but a few years since, they were "monarchs of all they surveyed."

The locomotive must go onward until it reaches the Rocky Mountains, the Laramie Plains, the Great Salt Lake, the Sierra Nevada, and the Pacific Ocean. Lateral roads must also be built, extending in all directions from the main line, as veins from an artery, and penetrating the hunting-grounds of these worse than useless Indian tribes, until they are either driven from the face of the earth; or forced to look for safety in the adoption of that very civilization and humanity, which they now so savagely ignore and despise.[18]

The Fourth of July parade in Sacramento had been a great blowout, affording the Central Pacific many pages of good publicity. The mule-driven railroad cars proceeding through cheering sidewalk throngs, the beaming Chinese on their floats, the proud dignitaries in their bunting-adorned carriages consumed vast amounts of newspaper ink. Indeed, a banner unfurled from the Crockers' carriage said it all: "The Pacific Railroad a Fixed Fact!" And the publicity was not over.[19]

Then, two days later on the western slope of the Sierra at Secrettown, Charley Crocker was just returning from an inspection trip when he encountered a reporter from the *San Francisco Alta California*, who had just stepped off the cars from Sacramento. The writer, identified in publication only as the Traveling Correspondent (T.C.), had just noted that Secrettown was now "a flourishing entrepot of trade, containing four shanties." Crocker regretted being unable to escort T.C. over the line but he placed his mule at his disposal and ordered that road superintendents should extend him every courtesy—provided that he stick to the facts in his report. This T.C. promised to do.

Mounting Crocker's steed, he began the ascent. He noted the thousand-foot-long Secrettown trestle, ninety feet high at one point, and the remarkable Tunnel Hill excavations which had so vexed the engineers (and Mark Hopkins) in the spring when mudslides kept filling in the road cut; they had solved the problem by sculpting six back-sloping benches above the line—some 450,000 cubic yards of slippery soapstone and goop were removed. T.C.

paused to remark on the numerous hydraulic mining ditches of Gold Run (about which the Central Pacific had already been forced to pay damages of $30,000 when the line intersected them). He proceeded to dusty Dutch Flat, which had continued to grow despite the disappointment of being located a half mile from the depot. The intervening land, T.C. noted, had grown up in newcomers—it was "odoriferous to the nose and beautiful to the sight with male and female celestials." Above Dutch Flat there was the Alta station, the highest point to which the Central Pacific had service at present, where he lodged for the night.

The physical beauty of the upper mountains struck T.C. as strongly as did the difficulties presented to the railroaders by terrain, all during the next day. He would note a profusion of culverts, built of granite, and a network of ditches leading to each one "for carrying off every stream of water that could possibly threaten embankments with washing away." He would be assured that "no water, except that falling directly from the heavens, can touch the embankments." Wonder piled upon wonder. At Blue Canyon, he saw "the heaviest filling on the road"—600 feet long, 85 feet high in the center, requiring 120,000 cubic yards of earth "to fill up its gaping mouth." All of it had been done by the little Chinese with mule-drawn dump carts. He inspected the Central Pacific's first tunnel at Grizzly Hill and the Horse Ravine—cut five hundred feet through rocky cement and boulders, now standing at twenty feet wide and high. But he reserved his unqualified admiration for the prospect at Green Valley—where Leland Stanford had begun his California existence as "an honest miner." The reporter marveled at a 150-foot-deep railroad cut, some 300 feet long, which had been gouged using water sluices and hoses. "The immense force displayed by it in carrying off earth and large boulders was astonishing," T.C. said. "The boulders and debris were vomited out on the steep side of Green Valley, where they bounded from rock to rock with reverberating thunder, carrying away trees and making dents of from ten to twenty feet in depth in the earth until the bottom was reached some 1,500 feet down."

Soon he was identifying with the targets of that monstrous talus through more than mere imagination. Now beyond a finished grade, threading through rough terrain where throngs of Chinese picked away at the mountainside, accompanied by a road superintendent by the name of Morris, T.C. followed an often indistinct pack trail used by engineers and inspectors along the sloping south side of Yuba Valley, about a half mile from its bottom and the Yuba River. Shortly they wandered off the trail, "and then ensued a stumbling over granite boulders, fallen trees, and water courses, through brush and manzanita," he noted,

> *sometimes on horseback, sometimes on foot, for about three miles, such as I never in my most daring dreams of riding dreamed of. We kept constantly*

looking for the trail or for some path which would take us back to the line some quarter of a mile above us, where drilling parties were at work all along blasting granite. We were constantly stumbling upon immense blocks or boulders of granite, which powder had hurled down, and at one time a small blast was fired almost immediately above us, which sent heavy rocks and stones down in our vicinity in a manner not at all reassuring to the safety of our persons or craniums.

Not long after, the pair found their way to safety. After a night's rest at a place called Polly's Station on the Donner Lake Road, he went out with Lewis M. Clement, location engineer from Dutch Flat, to the summit, higher and higher up the Yuba Canyon, "through a country of solid granite formation, where drilling and blasting are the weapons used for the industrial progressive warfare."

The work was, he pronounced when he was finished looking at it, "Titanic."[20]

As it was. And it picked up in speed. A week after T.C. published his dispatch in the *Alta California*, passenger and freight service extended to the rude little settlement of Alta, altitude 3,602 feet. Of course an excursion train of ten cars ran up the sixty-nine miles from Sacramento, filled with dignitaries. "The excursionists were supplied with cold water, lemonade and a stronger beverage which may be called Pacific Railroad Punch," reported the *Union*. "Among the men, the last named drink seemed to be the favorite. When Pacific Railroad Punch is freely circulated, speechmaking is almost sure to follow." There was no lack of speeches that day. In addition, said the *Union*, "three cheers were given for C. P. Huntington, absent at Washington, three for E. B. Crocker, absent at Washoe, three for Lauren Upson, an early editorial friend and advocate of the road, and three more for the *Sacramento Union*. The memory of T. D. Judah, the original projector of the road, was proposed and drunk in silence."[21]

Efforts intensified. In a few weeks the *Union* announced that the line had advanced far enough past Alta to ensure service to Cisco, twenty-three miles beyond, that season. "The work of railroad building in this lofty and rugged region—ascending from thirty-six hundred to nearly six thousand feet above the sea—is very heavy," reported the *Union*. It was "a succession of deep cuts and huge fills; but the host of Celestial laborers, under the energetic control of Saxon and Celt, is so numerous that the mountainous obstacles are 'here today and gone tomorrow.'"[22] By then Crocker's laborers were redeployed in three eight-hour shifts, making the work constant; those on the night shift, working on the mountainside and in the tunnels by torchlight or lanterns or even bonfires, were happiest and most productive; it was a torrid summer, "hot enough just about now," commented the *Dutch Flat Enquirer* on July 18, "to give a thick beefsteak a decent broil."[23]

But hot as it might have been, this was the Sierra, and it had its own climate. The young flagman J. O. Wilder, at work in the survey crew on the crags above Cisco, recorded seeing snow still on the north side of Castle Peak in early August. It might have been a harbinger. Just a year before, in August 1865, the summit work had been all but abandoned when snow arrived. Three shifts—and redoubled efforts—would avoid that. Work up to the announced goal for the year, tiny Cisco at mile 92, was complex. In addition to the recently finished Grizzly Hill tunnel there was the 271-foot Emigrant Gap tunnel (Tunnel 2), which was nearly completed, and the hundred-foot-high trestle bridge across Butte Canyon, in addition to all the rock cutting, grubbing, and grading. "As fast as the gangs of Chinamen were released" from each of these obstructions, recalled the engineer John R. Gilliss, "they were hurried to the summit to be distributed among the tunnels in its vicinity"—there would be seven tunnels crowded within a two-mile stretch. "As an illustration of the hurry," he recalled "walking two miles over the hills after dark, and staking out the east end of Number 12 by the light of a bonfire; at nine o'clock the men were at work."[24]

With their thoughts of the coming winter, the Summit Tunnel (Tunnel 6) naturally commanded not only a strategic but an emotional urgency, being upon completion the longest on the line (at 1,659 feet) and also the deepest (at most, 124 feet). It took many of the men, and when progress at its western and eastern ends was unsatisfactory, engineers decided to sink a vertical shaft down through that solid granite to strike the midpoint of the tunnel line. Then drilling could proceed on two new fronts. It was a good idea, and work on the shaft—roughly eight by twelve feet—began on August 27. "For the first thirty feet it was sunk at the rate of a foot a day," recalled Gilliss, "after which progress slackened, from delay in hoisting the material with a common hand derrick." He calculated that they had forty-two feet to go.[25]

Steam power was plainly the answer. Mechanical engineers in Sacramento quickly scoured the region and found the necessary equipment in the yard of the old Sacramento Valley Railroad, now called the Sacramento and Placerville. It was none other than the first railroad engine to run on California soil, the noble little *Sacramento,* carted round Cape Horn in the spring of 1855 and christened with champagne by Ted and Anna Judah. By California standards, eleven years was old—and the *Sacramento* was about to be sacrificed to drill Judah's line through the summit. Like eager cannibals, mechanics unbolted and pitched the proud smokestack and cab and tank, sending the remaining carcass up to Gold Run on a platform car. Here, at trackside, the engine was jacked up and its wheels removed. Using traveling jacks, it was moved fourteen inches at a time to a gargantuan logging truck fitted with wheels two feet wide. Once it was bolted and braced, a teamster named Missouri Bill hitched ten yoke of oxen to the enterprise. A squad of freighters watched anxiously along with their foreman, a Central Pacific wagonmaster named Pratt. Missouri Bill cracked his whip.

The young surveyor's assistant J. O. Wilder was helping to lay out the town of Cisco and its sidings, but sometime later the whole story of the engine's ascent was told to him by Missouri Bill. "All went well till a half-mile east of Dutch Flat," he recalled.

> *It was here the engine received its first cussing, for she had been the cause of stampeding a ten-mule freight team. Anyone familiar with mules knows that when they start something they usually finish it to the Queen's taste, with broken harnesses and tug-chains. This happened every day while on her way up the mountain; even the stagecoach horses would balk at the sight of her, and it finally became necessary to blindfold teams of mules and horses to get them to pass, for they would leave the road and take to the hills or the ravines, whichever looked best to them; and they weren't particular, either, about what they took along with them. They would endeavor to kick themselves loose of everything before starting for the bushes.*

As if this were not enough, the teamsters had to negotiate many soft places in the road, some mudholes being as much as twenty-five feet across. Small trees were cut and laid lengthwise and crosswise, and branches heavily scattered over the top. Often when the grade got too heavy for the oxen, the wagonmaster conscripted a passing freight team for a lift up the grade. Thus they finally reached the divide above Emigrant Gap. "From here to Crystal Lake," recalled Wilder, "it is down grade—the largest problem of all, for Pratt had to make assurance doubly sure, taking into consideration the heavy weight. This part of the locomotive's journey was fraught with fear, for one mishap meant loss of time and perhaps engine. With heavy logging chains and chain tackle made fast to big pine trees, they would let down as far as the tackle would permit, then block it. This was repeated time after time until the bottom of each grade was reached."[26]

While all of this was occurring, carpenters put up a large structure over the head of the shaft, fifty feet square, which housed the hoisting engine and cable drums (the shaft had been partitioned with timbers into two) along with a machine shop and forges. Moreover, there was generous room for fuel storage—always there was an eye toward winter. Eighty-five days after the shaft was commenced, they hit bottom. Two platoons of Chinese went down, at first sweatily working back to back, nosing east and west. Steadily the burrow enlarged, as did the workforce in those last anxious weeks before the winter storms began to hammer them. The Emigrant Gap tunnel (Tunnel 2) had been finished on September 21, with those crews, and still others below Cisco, redistributed on the mountaintops. As Leland Stanford would say, "We swarmed the mountains with men."[27] The *San Francisco Bulletin* could not help but be awed by the industry. "The road is entitled to be ranked among the most remarkable achievements of science and labor combined," it exclaimed. "The elevation which it surmounts exceeds that of all but one of the passes of the Alps, and is the greatest yet reached by any railroad in the world."[28]

22

"Until They Are Severely Punished"

Seated in his elegantly equipped office at 20 Nassau Street in Manhattan, his staff bustling away and attendant to every whim, Dr. Durant nonetheless found it difficult not to feel ill at ease, even beleaguered in the closing weeks of 1866. A letter from Samuel B. Reed, dated November 29, was at hand, assuring him that Reed's long bout with typhoid was now over; all the aggravating construction delays in late October and November might be partially blamed on Reed's indisposition (and inability to urgently supervise), most pointedly on the Loup River bridge, which had halted track construction for weeks until only on November 24. Reed, about to go home to Joliet for a visit to his wife and daughters, thought that the Casement brothers' tracklayers would reach the end of the third hundred-mile section in less than a week. However, the grading on the fourth hundred was "not as well advanced as it should be," he admitted. "The Indian scene and severe storms has drove most of the men off the line. Since I have been able to do anything I have used every effort to get as much grading done as possible."[1]

It was not much solace, Durant would have thought sourly. And more pressing and urgent reverses could be found anywhere he looked—on his desktop, in his office, in his life.

He was not entirely well. All those telegrams to Omaha—each tiny detail seen to during the spring, summer, and fall, each representing a small portion of his finite mental energy, a waking thought in the middle of the night, an idea scrawled on a scrap at a Manhattan curbside, each mounting upon the previous, exacting a toll upon his powers—sometimes it must have seemed as if he were balancing the entire railroad on his brow. Throwing his huge Hundredth Meridian blowout on the plains came close to finishing him. Then a worried George Francis Train had implored him to see a doctor in late September. "Do it or you will have a stroke of paralysis," he wrote. "You cant strike the Almighty in the face as you do without getting a lick back. If you get sick the Road will go to the devil."[2]

But foremost in the Doctor's own mind, before his own slouching health, there was the matter of his slipping control of the Union Pacific. On the board of directors there were now dissenters. On October 3, the annual election had deposed three board members he counted on in favor of three he could not, all originally brought into the ring by Oakes Ames. They were Sidney Dillon, the sharp-eyed railroad contractor from Connecticut, an early Crédit Mobilier stockholder and board member; the Bostonian John R. Duff Sr., also ensconced in the Crédit Mobilier for a year and a half and, like Dillon, a wary, experienced railroader; and Oliver Ames, stolid and retiring brother of Oakes Ames, in many ways the emotional counterpart of Mark Hopkins in his like for the shadows of the many boardrooms and office suites he inhabited and, in his family business, the true King of Spades while Oakes tended to national affairs in Washington. The first business of this new Union Pacific board was to let a new construction contract, and Durant naturally assumed that there would be a pro forma extension of the old one. When the Casements' tarriers had spiked rails to the 100th meridian, the Hoxie contract—overtaken, of course, by the Crédit Mobilier—had expired, even as the Casements had continued westward toward the Loup crossing and their graders had edged closer to the Sioux and Arapaho. Durant thereupon moved, with great self-assurance, that a Union Pacific contracts committee under Cornelius Bushnell arrange with the Crédit Mobilier for the new work.[3]

While the others on the board waited for the committee's report at the November meeting, Durant moved ahead—independently, as usual—and signed his own contract. It was with a Chicago bridge builder, L. B. Boomer, who had built bridges for the M&M, and it was for a distance of 153 miles beyond the meridian, which, judging by the Casements' speed, at least, was not a very great distance. The document was an oddity, cloaking somewhere deep in its language or else somewhere hidden in Durant's intentions an advantage to the Doctor which is not visible today. Either that or it was a ghastly miscalculation. Remarkably, Boomer proposed to build to the North Platte River for $19,500 per mile, and beyond for $20,000. It was far below Hoxie's $50,000 per mile for the smooth Nebraska tabletop, and beyond the North Platte the terrain would begin to get interesting. Where was the *profit?* Where *was* the profit? Durant had held Dodge's estimates close to his vest for weeks. Was Durant expecting to pile another layer of expense over Boomer's, and then take out a larger contract for the Crédit Mobilier (or himself), raking off the difference?

Boomer himself began an obscure maneuver, selling his contract to an intermediary named N. A. Gessner. Clearly things were still in motion. But whatever the plan, it fell flat. At the next meeting, on November 14, to Durant's utter surprise, Bushnell's committee moved a Crédit Mobilier contract from the clipper ship magnate John M. S. Williams (another early stockholder). Williams proposed to be paid $50,000 per mile on the plains and $85,000 in the mountains, and he would build 650 miles. Before any action could be taken, though, the Union Pacific board decided to table the question, for Engineer Dodge was on his way east to argue for his Lone Tree route against whatever Seymour espoused. The actual route through the Black Hills/Laramie Hills would crucially affect construction costs from the North Platte westward. The board prepared for a Dodge appearance on November 22. Durant prepared for the board.

From Dodge's point of view, his presentation represented the difference between Seymour—the dilettante who had paid more attention to shooting game and playing Wild West during his "reconnaissance" in the Rockies than he had to hard engineering facts—and himself, who stood for some of the best minds in the business. James Evans had run no less than five lines between the Cache la Poudre and Laramie Canyon, finding nothing that compared to the Lone Tree route. Percy Browne had scoured the mountains west of Denver and found nothing suitable, but he had found a decent line north from that city to connect with Evans's at Lodgepole Creek. Evans's party had already run a good line from Denver north to Crow Creek, so one way or the other Denver could be somewhat mollified with a hundred-mile branch line to the Union Pacific. In addition, the landscape north of the city was rich with iron and coal deposits, according to their geologist; it would be a profitable branch for the company. Another crew under Thomas Bates, sent all the way to scrutinize Utah and Nevada, reported that the railroad could run either north or south of the Great Salt Lake (he favored the north); if they wished away the Central Pacific they would take the obvious route through the Humboldt and Truckee Valleys. Intermediate passage between the Laramie and the Utah basin, Dodge told the board, would be solved in the spring and summer. He lost no opportunity to give his men credit. "Often threatened by Indian attacks, sometimes without escorts," he said, "they have all had narrow escapes, have had stock stolen, camps attacked, and been caught in heavy snowstorms, in extreme cold without fires, but, as yet, we have not lost any lives or any stock of great value."[4]

It was an impressive presentation. Moreover, the Lone Tree route with its gentle, uniform grades, its lack of sharp curves, its prospect of less blasting and cutting, would be surprisingly cheaper to construct. At the point on their route where government subsidies increased from $16,000 to $48,000 per mile to reimburse for more difficult terrain, the board was finally able to contemplate high profits. There was no point in looking any further; on November 23 Dodge's line was approved, along with his branch line to Denver.

This was good news to the Doctor, but he found defeat waiting for him on the following day when the board met to elect a temporary president of the company. John Adams Dix, that tireless supporter of President Johnson during the summer and fall campaign, had been rewarded with the minister's post to France. Durant, who for three years had been running things behind Dix's figurehead, was ready to ascend to the throne. It did not work out that way. Before the election began it was revealed that the Doctor had presumed to sign his construction contract with L. B. Boomer "on the company's behalf." The New Englanders were furious and quickly paid him back. When ballots were tallied Durant found himself both last and least, having received only his own vote; two votes had gone to the Union Pacific treasurer, John J. Cisco; thirteen went to Oliver Ames. After the congratulations were over the board added a flourish: in a resolution moved by Charles A. Lambard, it was forbidden for any individual to make contracts or indeed do anything on the board's behalf without its permission.

Durant retired to his lair to lick his wounds. And to plan—both on his own behalf and on the railroad's. At the beginning of November the North Platte had been bridged—a long trestle on cedar pilings—to the place where, at mile 291, Dodge had laid out the town of North Platte, a junction point. Regular service from Omaha commenced on December 3. By then, the track-layers had gone on to complete the 305th mile before quitting for the winter. Even with his short leash, as he saw to details in New York and Washington and Omaha and the end of track, Durant had maneuvering room. The new year would prove to be an exciting one. And as a minor but interesting footnote, one of the last details he attended to in that year of 1866 was to write to George E. Pullman in Chicago to recommend one "Eli Lafox (colored)," whom Durant wished "to become acquainted with the duties of attending inspectors' car on our road. Will you please have him employed in your cars for the present so that he may become familiar with the work?" Thus Eli Lafox may have become the first in a long line of Pullman porters of color.[5]

After Grenville Dodge concluded his presentation to the Union Pacific board he was off to Washington, not just for the railroad but for himself. On election day, the voters in his western Iowa home district had followed the lead of the state Republican machine and elected him congressman with a clear majority. It was a field day for the radical Republicans all over the nation, to Andrew Johnson's great chagrin, and on the day Dodge was chosen by the will of the people he had been high in the Rockies near Boulder, holed up in a cabin with a surveyors' party while a snowstorm raged outside. Years later he claimed with some believability that in the storm he lost track of time and was not aware that he was being delivered into Congress on that day. Six weeks later, he entered wintry Washington; his swearing-in was months off, but he nonetheless made some important rounds. He conferred with General Rawlins and was closeted with General Grant. He also paid a visit to Interior Secretary Browning, to give him details of the company's plans for 1867, particularly about the Lone Tree line. Oliver Ames, as new Union Pacific president, had also just written the secretary, bringing up the fact that the railroad was about to pass into more expensive mountain territory, earning itself a higher rate of government bonds. "The maps prepared by Gen Dodge . . . show that the ascent of the Mountains commence when the new line adopted by the Company crosses Crow Creek," Ames had written. "I would therefore respectfully suggest that the Forty-Eight Thousand dollars of Bonds of the United States . . . shall commence on the crossing of Crow Creek." Of course there was no mention of the fact that Dodge's magic route was going to cost the Union Pacific less than anyone had dreamed.[6]

The grouse would get on a log and hammer with their wings. On the Sierra slope in the late weeks of fall, this promised a bad dream, portending a rushing, severe winter, and J. O. Wilder in his surveyor's camp began to prepare himself. In late October in Sacramento, the Central Pacific had unveiled its answer to the mountain snowstorms, a gargantuan snowplow. An assemblage of iron over wood, looking awkwardly like a big black ship perched on a railway car, the plow was 11 feet high, 10 feet wide, and 30 feet long. Engineers swore it would scoop up drifts with its forward end, a wedge which hung down to track level, and fling the snow with its superstructure to either side of the track by as much as 60 feet.[7] Wherever there was track they were confident they could clear it.

All of this had been theoretical in Sacramento in mid- to late October, when the only precipitation seen on the lower slope was a steady rainfall. No reverses of weather could prevent Sacramento's Metropolitan Theater from filling to the bursting point on Thursday evening, October 11, when Mark Twain walked out into the lights to deliver his celebrated lecture on the Sandwich Islands, which had delighted an audience and all of the press in San Francisco some days before. "Without means and without employment," he would admit in *Roughing It,* he had returned to San Francisco from Hawaii with no prospects, and out of desperation wrote a humorous lecture, hired a hall, and plastered fliers and advertisements all over town. He had "showed it to several friends," he said, "but they all shook their heads. They said nobody would come to hear me, and I would make a humiliating failure of it." It was rather the opposite, and Twain moved his show on to Sacramento with "an abundance of money." Presumably he had been partially forgiven by the capital's leading citizens for calling Sacramento the "City of Saloons" only six months earlier in the *Territorial Enterprise.* Twain was soon to be permanently leaving for the East—by ship this time—and this mild, rainy Sacramento evening was one of the final appearances of the West's new literary hero. But on the upper slopes, the rain was making life difficult for the railroaders. "No water, except that falling directly from the heavens," the engineers had confidently claimed to Traveling Correspondent, standing over their ornate drainage systems some months before, "can touch the embankments." But then two miles west of Cisco, a heavy rain in October overfilled a lake upslope from a hundred-foot-high trestle bridge over Butte Canyon, and when the lake burst its banks it swept away four bents from the center of the bridge. "That bridge had to be replaced," recalled A. P. Partridge, "with the road blocked to Blue Canyon. Well, we went to the woods and hewed the timber, hauled it to the track by main force, then got some ox teams and hauled it to the bridge and repaired the break." Before the season would be through there would also be a number of serious mudslides, in the deep cuts and fills between Secrettown and Gold Run, and on an embankment between Emigrant Gap and Cisco; even closer to the city there was a washout of a trestle only nine miles from Sacramento at Dry Creek, causing the derailment of a passenger train running in the night at full speed. "No one was injured," E. B. Crocker reassured Huntington in a letter reporting the incident, "more than by bruises."[8]

The snowplow did not have to wait very long. There was no snow at the summit on October 30, when the tracks were laid to Emigrant Gap, but it began falling in earnest in November, though "just enough to stimulate without delaying the work," recalled Engineer John R. Gilliss. Snow filled in the cuts and collected overhead of the line, smoothing out the harsh sides of Donner Peak above them and soon cutting off a trail between Tunnels 8 and 9 on the eastern side—"it remained impassable until spring, and communication had to be kept up by the wagon-road, five or six hundred feet below," Gilliss noted. He could stand at the bottom on the Dutch Flat and Donner Lake Road and look way up to the work crews and never fail to be moved, day or night, but especially after dark. "The scene was strangely beautiful at night," he recalled. "The tall firs, though drooping under their heavy burdens, pointed to the mountains that overhung them, where the fires that lit seven tunnels shone like stars on their snowy sides. The only sound that came down to break the stillness of the winter night was the sharp ring of hammer on steel, or the heavy reports of the blasts."

Thanks to the redeployment of crews two months earlier, by the time winter really sat down on the Sierra the tunnel headings were all under ground. "The work was then independent of weather," Gilliss would say, "except as storms would block up tunnel entrances, or avalanches sweep over the shanties of the laborers." In the last week of November, despite the snows, Strobridge's tracklayers spiked their way into Cisco, some ninety-two miles from Sacramento and exactly 5,911 feet above sea level—"a higher altitude than is attained by any other railroad in America," the *Sacramento Union* would proudly exclaim. "If keeping the cars running to Cisco from this place can be done," said the *Dutch Flat Enquirer,* "it will be done. There may be slides, either above or below this place, during the rainy season, which will stop them for a time, but labor and perseverance will in the end succeed. The old 'iron horses' make day and night hideous with their shrieks and puffs and groans as they try to get up hill. A 116 feet grade, with a dozen heavily loaded freight cars, is no 'sardine.' It 'makes the fur fly,' but they manage to do it somehow."[9]

The summit now lay only twelve miles away, but Cisco was enough for the year. Passenger service there would commence late in December, and long before then Crocker would have sent two thousand workers over the top and down into the Truckee Canyon, below the snowline, to begin grading that section. They would be supplied by ox teams on the Dutch Flat road. Eight thousand others were reported at year's end to be employed in the Central Pacific's twelve tunnels. "It is expected," said the *Sacramento Union*, "that during the year 1867 the road will be completed and in operation to the eastern line of the state, from which point the work of construction toward Salt Lake is comparatively light and can be prosecuted at all seasons of the year at the rate, it is believed, of a mile a day, and to reach Salt Lake City of the 1st of January, 1870."[10]

It was a good sort of dream to have. "The snow has been no trouble at all so far," wrote E. B. Crocker to Huntington on December 28. "It is the least of our troubles—& we no longer fear it." That, however, was merely dreaming, as events would soon unfold.[11]

Out at the end of track at North Platte, there was no shortage of dreams—some visionary, some mercenary, and some merely earthy. What had been a brushy, unpopulated, low-lying delta between two prominent bluffs at the confluence of the North and South Platte Rivers, was now on its way to becoming a city of more than five thousand. Jack Casement had frantically worked day and night to complete the North Platte bridge and now was doing the same to ready his winter quarters. Relying partly on prefabricated buildings and on cheap, hastily erected structures, he put up a warehouse (boarding as a subcontractor about 150 laborers in it until the company bunkhouses were finished), a wash house, and a mess hall; he built an ice house, a blacksmith shop, and a slaughterhouse. As a personal side business he started a general store which he planned to leave in the care of a relative; he intended to take advantage of his free freight rate on the Union Pacific, shipping mercantiles and foodstuffs alongside rails and track equipment—and make a killing. Outside the so-called town limits, Casement established a ranch to supply the workers' beef requirements; after the first of the year he hoped to pick up three hundred head of cattle in Ohio for his ranch. The gangs were a hungry lot—and as these sidelines disclosed, so was little Jack Casement.[12]

Even so, with several thousand suddenly idle workers concentrated in one place, he would have ample competition to serve them. The first trader had arrived on November 9, almost literally on the heels of Grenville Dodge and his town surveyors; a storekeeper, Andrew J. Miller, moved a building from Coldwater, Nebraska, to North Platte, becoming the town's first citizen and businessman. Another man named John Burke contributed North Platte's second building, a log structure used as a hotel which he had moved from Cottonwood Springs. In a giddy month or two the town would bulge with more than three hundred permanent structures and a number of temporary ones, such as a mammoth tent erected next to Miller's trading post by a man named McDonald, who furnished it with a bar, billiard tables, and gambling devices. And it would be McDonald who would endure the most competition. Each arriving train brought more camp followers to North Platte, and it became easier to find a drink of whiskey than a drink of water. Saloons and more saloons, dice and roulette parlors, and houses (and tents) of prostitution proliferated. Day and night the streets were jammed with drunken railroaders, prostitutes, pimps, pickpockets, and cardsharks—but no lawmen. Winter snows arrived in December and if anything North Platte heated up. In the spring, when the end of track would be pushed farther west and the laborers moved on, the camp followers would simply fold up their tents, dismantle their gimcrack structures, crate their whiskey and corral their women, and ship them out to the new end of track to start up all over again. It was Samuel Bowles of upright Springfield, Massachusetts, who would look at the scene, appalled and outraged, and give it an immortal name: "Hell on Wheels."[13]

There was at least one family man in that wild and woolly town, and Frances Casement, back in Ohio with a new infant son, John Frank, pined away for her husband's reappearance in the household and fretted about his drinking; but she was blessedly ignorant of the moral climate in North Platte. Her husband would be stuck there until almost Christmas. "Dear Jack," she had written on November 25, "do get home as soon as possible—and darling be careful of your health—and for the sake of our little boy more for your own sake, beware of the tempter in the form of strong drink. There, I thought I was going through with this letter without a mention of that—but I can't help it now." Many of her letters, in fact, mentioned his drinking habits—in her later years she would become a temperance leader in the Carry Nation era in Ohio—and his replies were gallant and forbearing and of little reassurance to the lonely, insecure Frances. "We are getting quite a Ranch," he told her jocularly on December 23. "It is nice and pretty for so far west . . . no wine or liquor—we have the only temperance house in this country." By then he may have just received her most insistent letter of the waning year: "If you don't come home & stay with us some this winter," she warned, "you will never know any thing more of this baby than you did of your first." That mention of their deceased five-year-old, now a year gone, must have stung—but General Jack could not help having to go wherever his business took him. The railroad which was to forge a new union in the states was pretty hard on family life in that Christmas season of 1866, and the Casements' story was not a rare one.[14]

In the faraway stockade of Fort Philip Kearny, on the Bozeman Trail, some 450 miles northwest of the railroaders' noisome town, the condition of apartness from loved ones was identical for most—though some of the officers, at least, had the comfort of wives and children with them at the fort: twelve women and eleven children had accompanied them up the trail that summer and watched as the point-tipped log palisade had been erected around a cluster of raw buildings, their only relief from boredom and anxiety the succession of emigrant wagon trains approaching the fort, pausing, and then moving on. The post commander who had encouraged the dependents to come for morale was Colonel Henry Beebee Carrington. Connecticut-born, Yale-educated, a schoolteacher and lawyer whose political appointment in his adopted state of Ohio as adjutant general of militia had actually been a boon to the Union cause, Carrington had organized nine volunteer regiments for meritorious service in Western Virginia—parenthetically, Jack Casement had served in one. Later in the war, transferred to Tennessee to oversee military tribunals, Carrington would preside over the trials of the Confederate guerrillas so earnestly chased down by Grenville Dodge's spy network. After the war, of course, after Carrington had been ordered to establish forts in Powder River country, he had been the officer whose truthful answer about the Bozeman Trail had caused Standing Elk and Red Cloud and the rest of the Northern Sioux to stalk out of the Laramie peace talks. There is something rather poignant about Colonel Carrington—cautious, truthful, diffident, punctilious, cerebral (he is chiefly known today as a historian, with five good books to his credit), tragically ignorant of the ways of the Plains Indians; sometimes, as at Fort Laramie, at the wrong place at the wrong time; now, at Fort Philip Kearny, though credible and well-meaning, about to suffer unfairly for another's rash actions and really sheer idiocy. Soldiers and civilians along the dangerous Bozeman Trail would be affected. Jack Casement's and Grenville Dodge's men would be affected. A new row of dominoes was poised to fall. It would transpire just out of Colonel Carrington's eyesight but within, most definitely, earshot.

It would be seen thirty years later in Luzon and Mindanao and a hundred years later in Vietnam—establish a post in a quiet, unpopulated, uncontested place, and almost before one could turn around it would be teeming with enemies. The rolling terrain east of the Big Horn Mountains was the richest and the most beautiful of all of the hunting grounds in Powder River country, and the Crow and Sioux tribes had bitterly fought over it for many years. Now, with the bluecoats, the tribes were at truce—and the trespassers were in trouble. Frantically trying to put up enough cordwood for winter warmth and enough hay for their dwindling livestock, parties of soldiers sent out to forage for supplies were being attacked by the Sioux, who had managed to kill several whites and drive off many of their horses and mules. Carrington had failed to counter these attacks because he had been supplied with mounted infantrymen, not cavalrymen. The former were not very good at shooting astride a galloping horse, and the arms they had were old-fashioned muzzle-loading rifles. In November, however, the colonel had been allayed when a company of cavalry joined the nervous occupants of the fort.

Carrington's relief did not last; one of the new men was a loose cannon.

In joining Carrington's force, the 2nd Battalion of the 18th U.S. Infantry Regiment, Captain William J. Fetterman was returning home; he had been an active, decorated member during the war. "He was a type that occurred frequently in the Indian-fighting Army," writes Ralph K. Andrist in his eloquent saga of the Plains Indians War, *The Long Death*, "and often did not last very long at the trade. It was a type that was arrogant, contemptuous of Indians and their fighting ability, usually excessively brave, a braggart, and loud talker." Patrick Connor had been of the same type. So would be Custer. Fetterman, whose ties with many of the unit were stronger than Carrington's, ties forged in combat while their organizer had been bogged down by administration, immediately began to question and even deride his superior's caution and respect for the Sioux capabilities. "Give me eighty men," Fetterman would often exclaim in public, looking scornfully at his professorial commander, "and I would ride through the whole Sioux nation."[15]

Vanity and racism were deadly. None in the fort knew for sure, but there were some four thousand warriors camped up along the Tongue River for forty miles. Their leaders had watched the soldiers for months, probing for every weakness, and they had seen that the whites were so prideful that they could effortlessly be drawn into ambush. In particular the cavalry officers tended to ride far ahead of their men, gloriously, futilely waving their sabers, and were easily picked off. In December, after a few skirmishes served as rehearsals, Red Cloud and the other chiefs had decided to try for a large operation in which a large number of the soldiers could be cut down. For his part, Captain Fetterman volunteered to be the prey, proposing to take one hundred mounted men to destroy the tribal villages—but Colonel Carrington refused his permission for the foolhardy mission. On December 21, though, the Indians were ready for him. To begin it, one contingent attacked a wood party and its escort, which had gone about six miles away from the fort, behind a screening prominence called Lodge Trail Ridge; another war party of ten waited until the inevitable rescue force had charged out of the stockade. The ten were bait. Leading the rescuers was Captain Fetterman himself. Carrington had assigned another, more disciplined officer to lead the rescuers, but zealous, vainglorious Fetterman outranked the other and had demanded to be given the plum. Before Fetterman went out with his eighty men, the commander was explicit: "Support the wood train," ordered Carrington. "Relieve it and report to me. Do not engage or pursue the Indians at its expense. Under no circumstances pursue over Lodge Trail Ridge." He repeated the order twice.

Instead of going directly to the wood gatherers, Fetterman galloped with his men in a loop and disappeared over an intervening hill, attempting, it seemed, to cut off the attackers from the rear. But the soldiers came upon a small group of mounted Indians who began screaming insults, and the bluecoats chased them to the top of Lodge Trail Ridge. The warriors were joined on the far side of the ridge, below the soldiers, by a few more, amounting to perhaps fifty. Ignoring both his commander's orders and the object of his rescue mission, Fetterman could not resist the easy prey and ordered his men over the ridge, and they flew down the hillside with rifles blazing. They have been likened to "a crew of schoolboys racing to see who would be first in the water."[16] As many as two thousand Sioux, Cheyenne, and Arapaho warriors were waiting for them, concealed behind boulders and in the high grass; Red Cloud himself insisted he was there, too.

Colonel Carrington had realized soon after Fetterman's eighty had thundered over the hill and disappeared that they had no surgeon with them. He sent an assistant surgeon and four escorts. They returned in only a half hour, saying that they could not proceed because "hundreds" of Indians had been spotted on Lodge Trail Ridge and in the valley beyond. Carrington could hear distant rifle fire. By the time a large relief column was assembled, the shots had died out. And when the rescuers reached the scene they found that all were dead—the eighty-one bodies had been stripped, scalped, and mutilated. It was said later that Fetterman and another officer had stood at arm's length and shot one another through the temple to escape being left alive, but the autopsy did not support this romantic theory; a comrade of Red Cloud's named American Horse later asserted that he had clubbed Fetterman to the ground and slit his throat. Fifty or sixty Indians had also been killed, though their bodies had been removed when the warriors withdrew.[17]

The nation was stunned by what the army and the press would label "the Fetterman massacre." It had not been a massacre in the way that Sand Creek had been a massacre; it had been a battle, and without a doubt the most significant battle thus far to be fought on the plains. In every settlement from Omaha west to the Salt Lake—raw, new North Platte included—farmers, ranchers, miners, and railroad workers shivered a little more than they normally would have as the snows began to descend over their holidays, while everyone, white and red alike, waited to see what would happen next.

Thinking about the future, William Tecumseh Sherman lost little time putting pen to paper. The sympathy he had shown the "pure beggars and poor devils" just four months earlier, the forbearance that had him pooh-poohing settlers' calls for annihilation only three months before, was entirely run out. "We must act with vindictive earnestness against the Sioux," he wrote to Grant on December 28, "even to their extermination, men, women and children." Back in September, after he had gone out along the Union Pacific tracks and beyond, assessing the danger when the danger had ridden north to Powder River, he could not have predicted this. "God only knows when," he had told Grant back then, "and I don't see how, we can make a decent excuse for an Indian war."[18]

Apparently he had his excuse. And his friend Dodge vehemently agreed. Imploring Sherman to send another ten thousand soldiers to the West, he was free with advice, hoping that the Johnson administration would "not do by you as they did by me—get you well after the Indians, just ready to punish them, and then under the hue and cry of wrong, cruelty, &c., stop you." No, Sherman should go after the tribes "early with good officers who never give up but follow them day and night, until doomsday if necessary—until they are severely punished for their past wrongs and feel our power, so that they will in the future respect us. They look upon us now as a lot of old women, who do not know whether we are for war, or peace, or both." Dodge said he intended to make Fort Sanders in Dakota Territory (now Laramie, Wyoming) in a year's time—but as much as he and the Casements had been hard-pressed to find laborers in 1866, the coming year would be crucial. Especially after what had just happened at Fort Philip Kearny. There was now the worry that the surveyors and track gangs would "leave because of what you and I may know—it is hard to make a lot of Irishmen believe. They want to see occasionally a soldier to give them confidence, and that is all we need to get labor on the line."[19]

Paying attention to political as well as economic realities, Sherman would decline to send such a huge fighting force—even for the nation's project, the Pacific railroad. But he promised Dodge that together, using the resources already in hand, when the track gangs had begun edging into the Black Hills, "we can act so energetically that both the Sioux and Cheyennes must die, or submit to our dictation."[20]

Part VII

1867
Hell on Wheels

23

"Nitroglycerine Tells"

The Judge was in a deliberative frame of mind. Edwin Bryant Crocker, general agent and attorney for the Central Pacific, paused in the midst of one of his long dispatches to Collis Huntington in New York. He blotted the excessive ink in a downstroked number. He was thinking about granite, about time, and about nitroglycerine. Carnage of only eight months before on the mountainside, in the bustling center of San Francisco, and at quayside in Panama, was forgotten—it did not even enter his equation. The Summit Tunnel through the pinnacle of the Sierra Nevada was exactly 1,658 feet long. This is what engineers had calculated after assiduous measurement of the solid granite. On New Year's Day of 1867, hand drills, picks and shovels, and much black powder had increased the heading length to twenty-nine feet. This left merely 1,367 feet to go. "We are only averaging about one foot per day on each face," E. B. Crocker wrote to Huntington on January 7, "& we have come to the conclusion that something must be done to hasten it." Even if, at best, the Chinese drillers managed three feet per day at the western and eastern ends and at the bottom of the new central shaft, it would be more than fifteen months before they would be through. They could not afford fifteen months.

Crocker had a recent issue of *Scientific American* on his desk, in which experiments of the Boston and Maine Railroad were detailed. Drilling through the spine of the Berkshires toward North Adams in the ambitious Hoosac Tunnel, engineers reported dramatic results using nitroglycerine instead of black powder. If the Berkshires, then why not the Sierra? In the wake of the tragedies of the past April, only five hundred to six hundred pounds of nitroglycerine could be found in all of California. Emboldened by the Hoosac trials, Crocker planned to buy all left in the state and run tests at the summit. A British chemist named James Howden had recently appeared at the Central Pacific office, claiming that he could manufacture the dangerous compound on-site, eliminating the hazard of transport. His equipment would not exceed

$100. The company had decided to go ahead. Howden would be paid $300 per month—a lot, perhaps, for the small amount he would produce, but probably worth it. The Judge asked Huntington to research manufacturing patents—"we don't want to get involved in a suit about it," he explained.

Judge Crocker was troubled. The company owed about $600,000 locally, on top of the large debt Huntington was generating with his order book in New York. One financial panic, he wrote, and "we would be in a bad fix. Everybody seems to anticipate a crash soon, but probably that fact keeps it off, as it induces caution, preparation for it. If nobody was in debt there could be no crash."

Despite his worry Crocker was in favor of pushing ahead vigorously—on all fronts. Younger brother Charley had been increasing his labor force. Stanford had been busy with bonds. Hopkins looked for advantages in every column of sums. And the Judge took care of the angles. Like sharks, they had to keep on swimming. One way to help keep them afloat, he thought, was public relations—the Central Pacific had just arranged a discreet relationship with the Pacific correspondent of the *New York Herald* and *The World*. The reporter had promised "to weave into his correspondence matters relating to our R.R.," Crocker said. "He is a vigorous writer, & shows his letters to us so that we can correct any errors." Huntington himself could help with public relations: Crocker sent him a bundle of breathtaking stereographic views of the mountains, many duplicate sets totaling some four hundred shots, showing the railroad's works and progress. They could be handed out to investors and legislators.

Another way to make them look better to investors was to look at the lightly surveyed deserts of Utah as their natural right—and indeed, why not more? By then, Crocker wrote, Huntington should have received a new location map he had sent on December 18. It marked a route for the Central Pacific from the Big Bend of the Truckee to Humboldt Wells—Huntington should lose no time in getting approval from the secretary of the interior. And he should expect a new map very soon that would brashly claim a line from Humboldt Wells to curve around the north end of Great Salt Lake. In the late autumn their surveyors had even ventured up Weber Canyon from the salt valley, before being halted by the snows. The Judge felt that with the Union Pacific's line finally having reached high country, their progress would slow considerably—any claim to the contrary was "all bombast." With determination they could build to a meeting point with the Union Pacific—at Fort Bridger in Wyoming! Despite Huntington's recent entreaties to save money, reduce the winter workforce, and wait for the bond market to improve, this was no time for timidity.[1]

It was a dramatic contrast. Huntington, three thousand miles away in New York, working his way through lending houses and security firms, could afford to criticize them for not getting farther along by the end of the year. He was not looking at the Sierra granite every day, or toiling to make limited resources stretch as far as possible, or contending with the California labor market. It was easy in the New York suites to react to a market dip by urging his Pacific associates to lay off most of the laborers for a few months, but then the Irish foremen and the Chinese would disappear—and all the work creating an efficient organization would have gone for nothing. It would be fatal.

Beginning in these early months of 1867, surviving correspondence between Huntington and his partners quadrupled, quintupled—possibly because he began saving and filing them, but more probably because the tension as well as the stakes were growing monthly and they had to stay in continuous contact, the letters usually arriving in a packet of three or four on a steamship from New York or San Francisco, all imbued with urgency, seething with personality. Hopkins and Stanford wrote several times in a month, but Judge Crocker became Huntington's principal conduit for information, sending his four- or five- or eight- or sixteen-page letters every couple of days.

It is an extraordinary record. In the hundreds of pages of handwritten letters poured out by E. B. Crocker in the subsequent year and months before a paralytic stroke would still his voice, one sees a quite different picture of hale and hearty and aggressive Collis Potter Huntington—the public man, the man of his own memoirs—as he confides his fears of reverses and ruin to his stalwart friend in Sacramento. And one finds a growing appreciation of the sheer iron-bound will of Charley Crocker's older brother—sitting amid the clerks in the front office as head of the "Legal Department," but frequently hopping aboard a train to prowl through the work camps and tunnel headings to report progress firsthand—whose contribution to the overall effort at this crucial stage, often undervalued by historians, was to be so important in keeping Huntington on track, and maintaining their partners' confidence. Seldom would any three of the Associates feel bold at the same time. But usually the Judge would be there, pushing them on.

A push is what everyone needed, particularly when faced with the weather. Severe winter rainstorms had wreaked considerable damage on the line, interrupted communication to Colfax for a week and to Alta for two weeks, until January 2, and it did not seem likely that they could reopen service to Cisco for two or three weeks more. December and January would be "one continued storm" according to Charley Crocker. "Snow 3 or 4 days then rain one or two days then snow 2 days then Sunshine one or two days—then Storm in about same proportion & so on during the two months." Snow on the mountains piled up. The Judge rode over a twenty-foot-high drift in the Summit Valley in late January, and when he approached the Summit Tunnel he found a "curiosity." Both ends had been completely snowed over in the succession of storms, "but they have tunnelled under the snow, so that they haul out the rock through both rock & snow tunnels—& it works first rate." The snow tunnel on the west slope, he told Huntington, was at least three hundred feet long.[2]

An even more remarkable burrow beneath the drifts could be found connecting the two ends of Tunnel 8 on the high end of the eastern slope, working around a steep bluff some two hundred feet above the wagon road. Inside, every few yards, workers cut windows in the sides of the tunnel to admit light, at one point constructing a staircase down to the blacksmith's shop, which had been deeply buried beneath the snow. Not far away, in a deep ravine between Tunnels 7 and 8, completely filled with snow, they excavated a snow cavern above work abandoned at the first storms, building then a large stone culvert and the foundation of an immense retaining wall; smaller tunnels headed off to quarries, shops, and staging areas.

Even with many of the laborers living a troglodyte's existence, and with work continuing beneath the surface, it was a winter for the record books—life was miserable and often extremely dangerous. Engineer John R. Gilliss recorded no fewer than forty-four storms that winter. The worst began on the afternoon of February 18, continuing for four days and depositing six feet of snow on the mountains. For five days nothing fell, though heavy winds piled drifts so high that the snow tunnel to the eastern end of the Summit Tunnel had to be lengthened by fifty feet.

Then, for five days it snowed again, "making," said Gilliss, "ten feet snow and thirteen days storm." Nonetheless, the engineer thought the storms were "grand." He was living in quarters in the narrow east end of Donner Pass, which concentrated the summit winds to alarming speeds. "About thirty feet from our windows was a large warehouse," he recalled.

> *This was often hidden completely by the furious torrent of almost solid snow that swept through the gorge. On the cliff above, the cedar trees are deeply cut, many branches of the thickness of a man's wrist being taken off entirely by the drifting snow-flakes. No one can face these storms when they are in earnest. Three of our party came through the pass one evening, walking with the storm—two got in safely. After waiting a while, just as we were starting out to look up the third, he came in exhausted. In a short, straight path between two walls of rock, he had lost his way and thought his last hour had come.*

Avalanches were commonplace. Outside Tunnel 10, which was being drilled through the aptly named Cement Ridge only two miles east of the summit, one slide killed "fifteen or twenty Chinamen"—Gilliss could not be bothered to get the body count right. But he clearly recalled one episode near the close of a storm around the same time, in which an avalanche crushed and buried a large log house with a plank roof. Sixteen men were inside—a Chinese work gang and their subcontracting supervisors, three Scots brothers. This was at daybreak, and the storm finally ended at noon. "Towards evening," Gilliss said,

> *a man coming up the road missed the house and alarmed the camp, so that by six o'clock the men were dug out. The bulk of the slide had passed over and piled itself up beyond the house, so that it was only covered fifteen feet deep. Only three were killed; the bunks were close to the log walls and kept the rest from being crushed. The snow packed around the men so closely that only two could move about; they had almost dug their way out; over the heads of the rest little holes had been melted in the snow by their breath. Most of them were conscious, and, strange to say, the time had passed rapidly with them, although about fourteen hours under the snow.*

This alarmed the other workers, whose main camp was overshadowed by a cliff wreathed in heavy snow, a veritable iceberg against the winter sky which could descend at any moment. The next day someone climbed to the top—"to reach the overhanging snow required courage and determination," Engineer Lewis Clement remembered, "and the call for volunteers for this daring undertaking was always answered"—and he planted a powder keg behind the accumulation and set it off. "A white column shot up a hundred feet," wrote Gilliss, "and then the whole hill-side below was in motion; it came down a frozen cascade, covered with glittering snow-dust for spray. It was a rare sight, for snow-slides are so rapid and noiseless that comparatively few are seen." Seeing the snow level rising, being within a wind's whistle of Donner Pass, easily and poignantly reminded the engineer of the earlier tragedy twenty years gone.[3]

After each storm subsided, supply roads had to be cleared, for even sleighs could not pass. Road gangs with shovels and heavy ox sleds would emerge. Newly fallen snow, being powdery, could engulf a man to his waist or shoulders. "Into this," remembered Gilliss,

> *the oxen would flounder, and when they lay down, worn out, be roused by the summary process of twisting their tails. I saw three in one team so fortunate as to have had theirs twisted clear off, none left to be bothered with. The men were as regardless of themselves as of their animals. They took life easily in fine weather, but were out nearly all the time when it stormed. Late at night they could be seen shovelling on a bad drift at the corner of the warehouse, where the wind heaped in the snow faster than they could dig it out, and then a denser mass of flying snow would hide them altogether.*

The trail-clearers experimented with Canadian snowshoes—the paddle-shaped, rawhide-webbed contrivances strapped on one's feet—but soon they adopted "Norwegian snow shoes," as Gilliss called the odd, unfamiliar appendages, "each a strip of light wood ten to twelve feet long," with straps and a "balancing-pole to steady, push, and brake with. The latter will be seen all-important, as a speed of twenty-five to thirty miles an hour is often attained on a steep hill side." For several winters a Norwegian named Thompson carried the mails across the mountains on these "shoes," knowing that their proper Norwegian name was "skis," sometimes making forty or fifty miles a day.[4]

Much depended on the snow-shovelers and the teamsters. Across the mountains, deep in the Truckee Valley, thousands of graders sent over in early winter had to be kept busy and fed; that they were toiling within sight of the old ghostly Donner encampment was not lost on anyone. The Associates had hoped to send iron over the summit on ox-pulled wagons during the winter of 1866–67—enough rails to lay track over the new grades, and engines and cars to begin transporting materials—but given the severity of the season and the number of times that any travel whatsoever was halted, it was all the Central Pacific could do to keep food and light equipment in the hands of their stranded but still-busy legion.[5]

As for the track up to Cisco, the elephantine snow plow kept it mostly open, though men with shovels were constantly called in. The trains were "running through walls of hard crystalized snow," Hopkins informed Huntington in late March, "as high as the top of the cars from Emigrant Gap to Cisco."[6] Although they pushed as hard as they could, any peace with the snow felt tenuous. No one doubted that when the spring rains arrived, dislodged snow would wreak havoc with the rails. Desperate for a solution, Stanford had called a meeting in January with the Crockers and with the engineer in charge of bridge and trestle work, Arthur Brown. They decided that the best snow insurance would be to roof the entire line of exposed track over the mountains. The extraordinary cost and trouble was "appalling," said Brown later, "and unprecedented in railroad construction, yet there seemed to be no alternative." Scrawling out some preliminary shed designs, knowing what the massive timbers would be meant to withstand, they decided to have some five miles of experimental snowsheds built through cuts and ravines above and below Cisco, and in a longer stretch between the Summit Tunnel and Strong's Canyon—"about 2 miles," Judge Crocker reported to Huntington. "[I]t will be best to cover the road all the way, to keep it clear of drifts & slides." Construction could not start until summer. The other thirty-seven miles would be addressed later.[7]

As the flakes continued to fall, engineers far below the earth's surface concentrated their attention on the stubborn Sierra granite. In early February, the Judge reported that they had safely moved some nitroglycerine from San Francisco Bay to the Summit Tunnel and tried it out. "It works well," he told Huntington. Less oil was required than the standard black powder, meaning that the three-man drill teams could use smaller bits and shallower holes, saving both time and money. The difference between a two-and-a-half-inch-wide hole and one an inch and a quarter wide represented a lot of saved sweat. Moreover, even the smoke cleared quicker. Howden, the British chemist, was soon put to work in the explosive's manufacture, although he told Crocker that he would need a ready supply of components. Nitric acid was easily had in San Francisco, but they also needed an extraordinarily pure form of glycerine, which was more difficult to procure. Accordingly, Huntington was directed to contact Price of London, the best maker, and ship it hastily before they used up local supplies. Some weeks later, with Howden manufacturing the explosive every day, the Judge could exult that "the use of nitroglycerine expedites work fully 50 per cent if not more." In the summer they could move even faster, but now he could predict that they would finish the Summit Tunnel by the first of October. "I know everybody that passes by the work says we can't get through in 2 or 3 years but *we* know better."[8]

One particular pair who had passed by and predicted that they would be tied up inside the mountain for two years happened to be Union Pacific spies—they were the young engineers Thomas Bates and Fred Hodges, sent by Dodge and Durant to ride the cars and sniff over the works. "They were quite inquisitive," Crocker wrote to Huntington, "but they got very little definite information." Montague had given them only the exact length of the Summit Tunnel, and the gross length of all of their tunnels, but they were not permitted to watch any blasts; then, when Bates and Hodges went down to look at the work along the Truckee Canyon they just shook their heads at the vast amount of work ahead. Apparently they instantly wrote Durant in New York, who was soon taunting Huntington that the Union Pacific would be whistling across the California border before the Central Pacific's coolies had broken through the mountaintop. "If Mr. Durant thinks it will take 2 years to get through the Summit Tunnel," the Judge wrote Huntington, "you can easily undeceive him" with the figures he had been sending. In the past seven days they had extended the headings in the east and west ends by 11 and 14 feet and in the central shaft by 11 feet. Soon they would be far enough along so that they would not have to waste time hoisting the drillers out of the central shaft before touching off an explosion. And they continued to do better. Howden raised his chemical production to one hundred pounds per day, and in one of the stormiest times of the winter, in the first week of March, their combined advance was 50.5 feet. "You see," Crocker told Huntington, "we are getting up pretty near to 2 ft. per day per face. *Nitroglycerine tells.*"[9]

Despite the Judge's efforts to encourage Huntington in faraway New York, the Central Pacific's vice president was still doubt-ridden and depressed. "It costs a fearful amount of money to pay all the bills," he wrote to Mark Hopkins on March 12. "I sometimes think I would change my place for any other in this world." Worries about the paltry sales of bonds and the slowness of the mountain work, and persistent feelings of disconnection from his trusted partners were deepened that spring when Hopkins and Huntington concluded to sell their successful hardware business at 54 K Street, which had proved to be a distraction. Huntington at first waffled, but when Hopkins remained firm ("My judgment is unchanged," he had written his friend in February), Huntington agreed that it would be a good idea to sell it to their longtime employees—he would leave the details entirely up to Hopkins. Still it was hard to part with the business he had begun in 1850 and run with Hopkins for more than a decade. "While I would like very much to go to Sacramento and take you by the hand and talk over matters as in years gone by, but if I should sell all my interest at 54 K, I think that I had rather you would do it while I am here, than do it myself. I do not think I could sell all my interest there and walk out of the old store without dropping a tear on the threshold, for the old place is somehow dear to me with all my losses there by fire and flood."[10]

Not far away from Huntington's little one-room office in downtown New York City, at the palatial headquarters of his foe Dr. Durant another kind of sellout had begun. It would spark a fierce though outwardly genteel war in the boardroom of the Union Pacific, which would embarrassingly spill over into the courts. The Bostonians—new president Oliver Ames, brother Oakes, and their Yankee supporters—still smoldered at the Doctor's effrontery in unilaterally drawing the Boomer contract for 153 new miles west of the 100th meridian, which they had belatedly learned about, almost in afterthought, on November 23. For six weeks the question of who was to next contract for construction—and thus who was to control the only sure profit on the road at that point—was left unaddressed. During that period few company decisions were made on any matter in New York as well as in Nebraska, causing alarm. Something was afoot.

Then, on January 4 and 5, the first shots were fired: first, there were calls for an audit of the Doctor's business doings and for rigorous record keeping, both obvious slaps across Durant's face; next, the board took up the matter of the fifty-eight miles built in the fall from the 100th meridian to North Platte, which of course had not been covered by the original Hoxie contract. The Bostonians put forward a resolution extending the Hoxie–Crédit Mobilier contract by those completed fifty-eight miles, for which the Crédit Mobilier would be paid $50,000 per mile. Thus Durant's Boomer contract was figuratively crumpled and discarded, and with it Durant's obscure plan (if that is what it was) to solely rake in the new profits.

Strangely to the Bostonians, during this vote the doctor either stalked out or abstained (with three other Union Pacfic directors)—the minutes of the meeting report he did not vote. The measure carried in a tally of 8 to 4. (The dissenters were the government-appointed directors, apparently leery of controversy but generally inclined at this point to go with Durant.) If anyone thought that Durant would go quietly they were wrong—a mindreader present in those tense minutes would have caught the Doctor's unspoken intention: *I'll see you in court!*[11]

Meanwhile, as for weeks, Samuel Reed and the Casement brothers cooled their heels at the St. Nicholas Hotel, eating bad, overpriced food and venturing out but little into the jostling, indifferent Manhattan crowds. "The Casements . . . are about as tired of waiting as I have been," Reed wrote to his wife on January 16. "Dr. Durant says he cannot give any positive instructions until he knows how and by whom the work is to be done." Reed was bored and restless and plagued with a bad cold and he missed his wife, who had been able to join him in the city for only a few days before returning to Joliet, but Jack Casement's impatience and loneliness exceeded his own. "I want to get home," the little general wrote his wife; "this town has no charms for me. I have not experienced so much pleasure from any Source here as in reading your letter that told me you loved me." Thrust into a romantic mood, he continued, soaring to what was for General Jack a high plateau of self-expression: "Darling allow me to assure you that I most heartily reciprocate your love." The waiting stretched into another week, toward the next Union Pacific board meeting on January 23, while Casement and Reed—and, no doubt, the Bostonians—wondered what the Doctor would do. "Mr. Durant is full of suggestions," Reed confided to his wife on January 22, "and, if I mistake not, he will set some trap into which they will all fall and allow the work to go on as heretofore."[12]

At the next board meeting, Durant strode in with two pieces of ammunition. One was a written protest against the board's action. It would be shortly and smoothly expunged from the minutes by his enemies. But the second could not be so easily dispatched: it was a court injunction barring the Union Pacific from extending the Hoxie contract and from distributing any money covering the disputed fifty-eight miles, and in effect paralyzing the board from letting a contract on the line west of North Platte. It came with extraordinary timing. Casement, Dodge, Hoxie, and Reed had passed on stacks of unpaid bills from the close of the season. Old notes were about due. New York bankers—with their facility for sniffing out trouble and dissension and slamming shut their lending ledgers—were too close for comfort. With the Bostonians cornered for the moment, the board voted to rescind the Hoxie extension and substitute a revised version of the Crédit Mobilier plan of John M. S. Williams, offered in mid-November and tabled, which covered the 58 completed miles and extended the line by another 650 miles. This also went nowhere past Dr. Durant.

Moreover, now edged out of control of Crédit Mobilier, Durant maneuvered the Union Pacific board to appoint a new committee of directors to negotiate with the contractors; the hitch was that no one on the committee could own part of the Crédit Mobilier. Check again. "If the Doctor and his friends are thrown out," the disgusted Reed wrote, "he can, if disposed, throw such obstacles in the way as to prevent the rapid prosecution of the work, and I would not be disappointed if he does so, even to the ruining of himself and friends financially." The engineer chafed in his hotel room into the last week of January. Finally he reported a small breakthrough to his wife, though weeks later he would not be surprised when even this compromise would be doomed to unravel: "They have let the contract to Crédit Mobilier from the 100th Meridian to the base of the mountains," he said, "but will not let over the Mts. until the president decides where the mountains commence." Oliver Ames had proposed the crossing at Crow Creek to President Johnson, but the White House had been silent. Thinking of his own job, Reed thankfully added that "no changes have been made in the management at Omaha."

But a few days later, during a dinner party on January 30 at the home of George Francis Train, Durant took Reed aside for a long, private talk. Then Reed was worried that their work would not progress as quickly in 1867 as during the previous season, and that the new leadership of the Crédit Mobilier would have a succession of financial troubles. "Moneyed men are afraid of committees," he wrote, "and will be very careful how they put their money into the work when they have more than one man to deal with." The boardroom battles were anything but over; in fact, matters would continue to deteriorate. But he had his own work to do, and it was finally time to get to it. He left on that evening's train for Chicago and points west.[13]

Reed walked into his Omaha office on February 13, in the teeth of a severe winter storm. "The county is flooded with water and ice," he wrote his wife. The rail yards and shops near the river were entirely covered, "nearly as bad as during high water last season."[14] For once the weather could be used to their advantage, though. The Missouri River had frozen over, solid enough to support a constant traffic of freight wagons, sledges, and sleighs, opening up new possibilities of commerce between Omaha and Council Bluffs; test borings showed it to be sixteen inches thick—enough, even, to hold up a railroad train. Four weeks before, the Cedar Rapids/Chicago and North Western line, under the leadership of the New Jersey baron John Blair, finished the breakneck race across Iowa against the M&M—begun more than fourteen years earlier—and eased its first train into Council Bluffs, the welcoming orchestra's fanfare nearly drowned out by stiff off-river winds. With the prospect of an unbroken supply route from Chicago, Union Pacific crews frantically erected a temporary railway bridge on the ice, giving them nearly two months of easy transport before the breakup. Iron began flowing west to a massive supply yard at North Platte, awaiting the resumption of trackwork.

Grenville Dodge was at his engineer's desk when Reed arrived in Omaha; he had stopped only briefly to attend to family and constituents in Council Bluffs. Dodge had heard from Sherman by then and knew that he could not

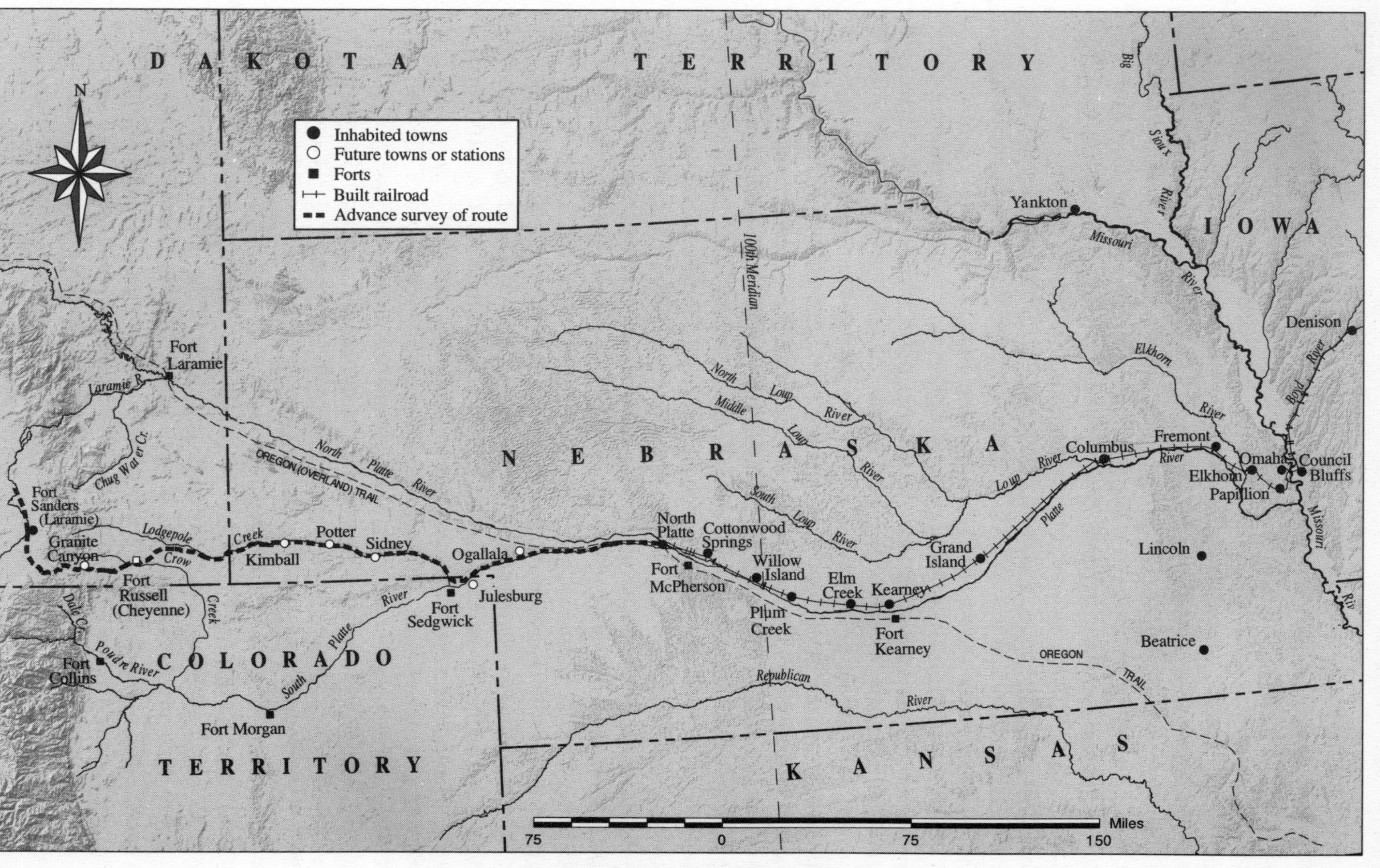

Union Pacific Railroad, January 1867

expect ten thousand bluecoats to defend their surveyors and tracklayers. Sherman had appointed a new commander of the Department of the Platte: General Christopher Columbus Augur, a muttonchopped forty-five-year-old former classmate of Grant's at West Point, later the academy's commandant of cadets, who had distinguished himself in war service in Virginia and Louisiana and who had been severely wounded by Stonewall Jackson's boys at Cedar Mountain. After the war, Augur had been detailed to investigate the Lincoln assassination conspiracy. His experience with Indians had been limited to the Bannock and Northern Paiute of Oregon, in the 1850s, but he had Sherman's confidence as a good commander and administrator, and Dodge had Sherman's assurance that the army would do what it could for the railroaders. But as Dodge wrote to Durant, there just weren't enough troops—and the demands on them were "so pressing that I fear *we* shall suffer." However, he amended, they had agreed to work closely with Dodge as to deployment of work gangs and troops. "Indians are *bad* threatening the Platte route but I think *we* and *our material* are safe."[15]

He had already told Sherman of his goal for 1867, and Sherman had thought it "almost a miracle." The proposed line stretched 286 miles all the way to Fort Sanders in Dakota (or, later, southeastern Wyoming Territory). It evenly and miraculously climbed from the altitude of 3,000 feet above sea level at North Platte to its high point in the Black Hills of 8,242 feet. It would follow the South Platte River through soapweed hills to the Colorado border, gaining about four hundred feet, dipping into Colorado only long enough to touch the stagecoach station of Julesburg before joining the tributary Lodgepole Creek, named because Plains Indians gathered mountain birch poles for their tepees on slopes above the headwaters in the Black Hills. The graders and tracklayers would then follow the Lodgepole Valley back into panhandle Nebraska and westward through cliff-dotted, high rolling plains, across the Wyoming border, eventually leaving Lodgepole after more than a hundred miles. Approaching the Black Hills, they would strike Crow Creek, cross it, and then climb Dodge's divide to Lone Tree Creek and the high point of the line. Casement's men would then cautiously descend to a strenuous crossing at the deep valley of Dale Creek and thence to the Laramie Plains, rolling northward up to the Laramie River and the welcome sight of the palisaded Fort Sanders. The proposed line would be a miracle, all right—and, scattering antelope and buffalo, it would go right across the hilly hunting grounds of the Northern Cheyenne and Arapaho.[16]

As eager as everyone was to begin the work, winter still had its heel ground down on them all, whether they were scrutinizing paperwork in an Omaha office or hunkered out in frigid North Platte, where it was impossible to get a good night's sleep due to the all-night revelry punctuated by gunshots. In mid-February a weeklong blizzard stopped all movement across Nebraska and Iowa; workers had barely dug out when another rolled in, this time paralyzing the states in drifts for ten days. "We are still laboring hard to clear the track," Sam Reed wrote his wife on March 27. "We work hard everyday and during the night . . . as much snow [will] blow in as the men have shoveled out during the day. Erastus has just sent me the following telegram from Grand Island. 'We are out of luck in this county. Wind blowing and snow drifting worse than ever. Half the men either blind or frozen, looks bad.'"[17]

Back in New York, winter conditions were equally severe: after a procession of heavy snowstorms in January, when the East River froze over and this rarity halted ferry service, thousands of Brooklynites and Manhattanites took to trudging across the river on foot. Dr. Thomas Durant of Brooklyn may have been among them—although, as Albert Richardson had observed, he always worked "like a galley slave," and "sometimes he was hardly in bed for a week," and thus may have just lodged in a Manhattan hotel if not simply in his office's elegant plush chaise longue when fatigue overtook him. And it was a tiring time for all as money pressures mounted, especially upon the Bostonians as they tried to keep up with the Doctor.

Several attempts in early February had been frozen when one side or the other refused to compromise. Durant began boasting he had raised a month's worth of expenses by negotiating a million-dollar bank loan, but the Bostonians considered its terms near-usurious; nonetheless, the money was slow to arrive. Desperate to meet contractors' obligations, the Ames group individually began making personal loans to the Union Pacific and also in February moved to increase Crédit Mobilier stock by $1.25 million, offering prospects a bonus thousand-dollar Union Pacific first-mortgage bond for every thousand dollars subscribed. Then the Crédit Mobilier board offered to lend the Union Pacific its new capital of $1.25 million—if the company (meaning, in this case, the obstructive Durant) would finally settle the Hoxie contract to the 100th meridian and extend it another one hundred miles at a cost of $42,000 per mile. Along with this, the Doctor was pleased and struck by the irony of a proposal to spread out his lofty "suspense account" over the entire line—as he had originally intended.

However, although this latest compromise package was approved by the Union Pacific executive committee, it smashed flat against the larger and immovable board of directors. Over two heated days of negotiations and votes, the board angrily rejected solutions to the Hoxie contract problem as well as to Durant's suspense account problem. It accepted the Crédit Mobilier's loan (as if there were a choice by this point). Impatient to let out new work, it approved a new proposal from John M. S. Williams; in this, some 100 prairie miles would be built at $42,000 per mile, and another 168 high-plains miles would be built at $45,000 per mile. At this time so perilous to the hopes of the Union Pacific company as well as the Crédit Mobilier, the irascible Durant demonstrated his financial independence as well as his rejection of the entity he had helped create, by selling out $100,000 of his personally held Crédit Mobilier stock. It was as if he had leaned out of his lifeboat to duck the head of the foundering Crédit Mobilier underwater. He exclaimed to his fellow stockholder, the Massachusetts congressman John B. Alley, that "the Crédit Mobilier should never have another contract while it was under the control of the men who were called the 'Ames party.'"[18]

Then, in late March, matters between the Doctor and the Bostonians entered a new phase of hostility during the scheduled back-to-back stockholders' and board meetings of the Crédit Mobilier and the Union Pacific. One resolution at the morning meeting (which Durant did not attend) boldly authorized an investigation of Durant's Crédit Mobilier stock account—to see if he had paid for it all properly, especially since he was now selling some out. Another resolution warranted the latest Williams contract and directed that it be sublet to the contractors at $25,000 per mile, which ensured a stockholders' profit of $17,000 to $20,000 per mile sweated out by the Irish work gangs. Later in the day, Durant stormed in to file another protest against the Union Pacific board for considering the Williams proposal; it was quickly expunged from the record. Continuing to ignore him, the board moved to end the Hoxie contract controversy by paying the Crédit Mobilier less than $2 million for the work to the 100th meridian. There was not enough money in the treasury, of course, so the matter was closed with an IOU. To all of this, the Doctor replied as he had two months earlier. He obtained another restraining order, which stopped the whole thing in its tracks.[19]

Out west, the situation was mirrored in swelling rivers and streams. "There is an immense quantity of snow on the plains and in the mountains," Samuel Reed had told Jenny on March 27. "I expect very high water and we may lose some bridges." Of course he was correct. By the time Jack Casement joined Reed in Omaha, two weeks later, regular service had been halted along with the special supply trains. "It is raining this evening and making the prospect as gloomy as possible," the little general wrote Frances Casement on April 13. "The Missouri is coming over its Banks and rising all the time. Miles of Road is washed away in the Platte Valley so that we cannot get over the Road for a few days, even if it stops Raining. We are all in a heap." There was, he mourned, "too much cold water for comfort or profit."

A good deal of the Casements' work was ripped out by roaring water and ice floes: at the confluence of the Elkhorn and the Platte, a mile was underwater; westward, about five miles of track was completely undermined when its embankment dissolved; near Grand Island, the surging Platte left its banks and diverted in a mad arc which severed the track in two wide gaps; at the Loup River bridge, the river tore away miles of approaches but left the span itself oddly intact; the last structure completed before the snows had closed in December, the North Platte bridge, was also spared along with its approaches. Reed glumly reported all of this to the home office, and as he began to make arrangements to repair all of the damage, he found he had more stormy weather approaching: Dr. Durant was coming west to inspect the work—and with him Oliver Ames. The two, still battling, were bound to strike sparks in Omaha. It passed across Reed's mind that he might be fired. "I do not feel any trembling in my boots," he wrote to Jenny, trying to keep her from worrying, "let what will come. I have a clear conscience because I know that I have done my whole duty for him and the company he represents."[20]

24

"Our Future Power and Influence"

The springtime punishment of storms and floods also hit northern California. By April 8, 1867, after a month in which the giant snowplow was in constant use, the *Sacramento Union* hazarded that the snow season, "one of the fiercest ever known in the Sierra Nevada," was over. It congratulated the Central Pacific: "The railroad men, though sometimes stopped at Colfax or Alta, persisted in grappling with the obstacle, determined to prove that even in this terrible season, without the instruction of experience and bothered by the settling of a new road-bed, they could run trains to Cisco." It was noted that for large stretches of the line snow could be simply pushed off the embankment and down into ravines—but that in roadcuts, where there was nowhere to push the snow, the company had been forced to have shovelers transfer it into freight cars returning empty from above. They rolled the snow-laden cars into Sacramento, "to the infinite delight of snow-balling youth." And the newspaper marveled at the changes in public consciousness that the railroad was making. "There is snow on the hills above Alta, within seventy miles of Sacramento, where the gardens are in bloom and the fields are arrayed in a refreshing green," the writer said. "And the swift transition from the reign of Spring to the Arctic realm of the Sierra is one of the most novel phases of experience in California."[1]

Almost within hours of this announcement that the snow season was over, the Sierra was slammed again. At Emigrant Gap and above some five feet of new snow fell, making the accumulation deeper than at any time in the season; a week later there was another five feet to contend with. With the lengthening days, Charley Crocker informed Huntington, the snow was "now going rapidly by sun heat—but in all probability we will not be able to commence work above Cisco before the 1st July unless we shovel a large amt. of snow away to get to the work." Inevitably, the extra snow meant extra run-off, with a resulting mayhem upon culverts, embankments, and bridge approaches all the way down to the capital. "Providence seems to be rather against us this winter," E. B. Crocker added in a separate letter.[2]

Still, the Associates felt as if they were holding their own. The Casements' Irishmen may have been spending their Nebraska winter heading only to perdition, but the Central Pacific's Chinese had moved ahead under the watchful eye of Charley Crocker and the brutish voice of Jim Strobridge. The Summit Tunnel, in late April, was considered to be halfway done thanks to Chinese prayers and nitroglycerine, and both Crocker brothers felt it would certainly be ready for track by October 1; happily, they had been within two days of running out of the explosive when a fresh shipment of five thousand pounds of glycerine arrived from the East. There had not been a single accident with it. Carpenters hammered busily on a sturdy prototype snowshed, equipped with a steep roof and massive, doubled-up rafters, for later tests. If they were unable to address the heights over the railhead at Cisco until the snows retreated, there was at least remarkable progress shown by the isolated gangs on the eastern Sierra slope. Charley Crocker rode over the top on April 18, finding eight feet of snow at Donner Lake but below, at the pine-shaded confluence of the Little and Big Truckee, only two feet thereabouts and on down.

All along the craggy canyons of the Truckee Crocker found much to be cheerful about. As the *Union* had reported earlier in the month, "Spring has fairly opened in the valley. The fields are green and grass is beginning to be of service to stock. The river is full of trout, making it a pleasant place to visit at this time."[3] One tunnel was through. Timber for six smaller bridges between Proper Creek and the state line was just being unloaded even as masons completed their foundations and pier footings. A number of sections had been graded and were ready for track. "We think we can follow the snow up & cover the work with men as fast as the snow disappears," Crocker told Huntington. "We will be compelled to use every exertion possible to connect across the summit this fall."[4]

Charley Crocker was busy managing his army, and frantically recruiting new ranks. He had hired a Chinese artist to translate their recruitment offers into Chinese and engrave them on a woodblock, printing some five thousand handbills to be distributed all over California and in China. ("The Chinamen all understand it," the Judge had noted in January, "but it is hard for them to translate it back into English.") On April 8, the *Grass Valley Union* noted that "The Chinese are swarming in the direction of the Central Pacific Railroad, and the company has commenced sending them in large numbers beyond the summit of the mountain preparatory to the work of the coming season." Agents of the company were scouring Nevada County and all of the mining districts, it said, and "within the next thirty or forty days there will be at least twenty thousand of these prospective unbleached American citizens scratching gravel on the great national highway." Alarmed nativists in the mountains would later be reassured that the number would be far more modest: Crocker hoped for a force of eleven thousand Chinese and twenty-five hundred Caucasians for the season.[5]

Briefly—very briefly—there was the hope that the former progress which had made them proud would soon be dwarfed by the newest emerging technology. At the Summit Tunnel, workers unpacked a newfangled labor-saving gadget which had been exciting the interest of those down in Sacramento since they had heard of it—a steam-powered, compressed-air-driven drill. The mechanic who had brought it up wanted it to be connected to the "Black Goose" hoisting engine, drawing off a little steam, but Strobridge appeared and refused to let them stop the regular work to make the adjustment. He was adamant—even when told the maneuver would take only two hours. His motivation seemed to be partly distrust of the modern contrivance—after all, he had been supervising human muscle over rock since his young Vermont days and that is what he knew. Also, though, there was a little logic, since the hoisting engine had its own limitations; as Hopkins commented, "our progress had reached a point at which it was equal to the capacity to raise." But he bought the engine and sent up what was necessary to connect it with the present boiler.

It never would be connected, however, at the summit or anywhere on the line. "Strobridge's refusal to let steam be taken from the boiler there now and Charles' apparent indorsement surprises me," the Judge confided to Hopkins on April 1. "It looks as though both were set against drilling machines." Stanford went up himself to try to persuade Charley Crocker to persuade Strobridge. The Irishman was unmoved. "It puts me out of all patience to see how that drilling machine matter is mismanaged," the Judge complained to Hopkins on April 15. "I sat right down and wrote the inclosed letter to Charles. Perhaps it is not just right, but I believe it is the truth, every word of it. If you think best send it up to him. The truth is things have got to such a pass that there can't be a thing done unless it suits Strobridge. Whenever a man gets Charles' confidence, he swears by him and all he says or does is right." Even Stanford gave up. "I fear the drilling machines will prove useless," he told Hopkins. "There does not appear a will that they should succeed, and usually where there is no will there is no way."[6]

All this new activity required a commensurate energy in raising money. The Judge investigated different bond schemes (such as an 8 percent gold bond pledged to proceeds from their land-grant sales) and became interested in contractors' methods, asking Huntington to find out about the Union Pacific approach—he had seen a newspaper report of Dr. Durant's restraining order against a Union Pacific contract and wanted to get to "the bottom of this—what ground or objection did he found it on?" Perhaps there was something they could learn from the Doctor by knowing how the Union Pacific paid for its road—and how the principals profited from it, even now. For his part, Stanford made the rounds of the larger California banking houses and representatives of the great eastern capitalists. It seemed expedient to let them see deeper into the company's affairs than they ever had before, unrolling maps and profiles and showing them at least some of their ledgers. "They all open their eyes in astonishment at the magnitude & value of our assets," Crocker reported. "In fact they see the big thing—they see millions of securities to negotiate, exchange, sell, & etc. They all want the business." In particular Stanford courted the former California senator Milton Latham, who now managed the London and San Francisco Bank, for the Associates wanted to get their securities moving in European markets. Latham's party of eight ladies and gentlemen were taken on a special capitalist's junket up to the summit to see the work above Cisco and then to ride by sleigh down to Donner Lake. They were even escorted into the Summit Tunnel, "and I can tell you," reported the Judge, "that impresses one."

Both he and Stanford urged Huntington to spend a few weeks in London, where they were certain he could "lay the bonds on the English market." They arranged more excursions for wealthy and influential San Franciscans, a number promising to spread the good word about the Central Pacific when they embarked on their summer tours of Europe. In that spring of mounting excitement and confidence, it seemed to Judge Crocker that their credit had risen so strongly that they could borrow whatever they needed. It made him a little afraid, confronting the eagerness of so many lenders—and here, very briefly, he had a glimmer of understanding what nervousness Hopkins constantly endured.[7]

Three thousand miles away, Huntington was equally busy raising money. The aggressive bond marketing company of Fisk & Hatch in New York had been handling Central Pacific bonds since 1865, to reasonable return, but until the tracks actually crossed the California border, it would decline to sell bonds for any mileage into Nevada, even if the bonds were legal. So Huntington took the bonds out himself, selling small numbers in New York City and Boston. He obtained the aid of the Central Pacific's former Railway Congressman, Richard Franchot of Ostego, who sold some in upstate New York. These did not spread very far; Huntington's greater contribution came through borrowing hundreds of thousands of dollars that year, and millions by the time the story was over—borrowing against the company's government bonds, its California county securities, against each of the Associates' personal credit. Always there was that ability to make a quarter into thirty cents, a dollar into a dollar and a half—now on what he would have considered a wondrous level a decade before. "He had an instinct for bargains in money," marvels his official biographer, David Lavender. "Whenever he sensed that interest rates were about to rise (they were much more volatile in those days than now), he borrowed heavily at the lower rates and loaned out the money himself until the railroad's creditors began dunning him. It was a dazzling performance—but risky."[8]

Fueled by equal parts of borrowed money and optimism, there were grandiose hopes afoot—and indeed, considering the stubbornness of the Sierra, the state of the Central Pacific's treasury, and the personal finances of all of the Associates, the aspirations given voice in the spring of 1867 might have seemed delusionary. It seemed, in fact, as if ambition had been rising at the same rate as the snowfall on the mountains, and it opened a rift in the Associates which might well have wrecked them all—ambition it was, and acquisitiveness.

Imagine it from the point of view of the cautious mandarin Mark Hopkins, that abstemious, indoorsy, quiet hardware-store accountant risen to Central Pacific treasurer, upon whose hand the proposed acquisitions could be numbered. First there was the Sacramento waterfront. Early in the year the company petitioned the city of Sacramento to vacate an overflooded area of the city, west of Sixth Street and north of I Street, which had earlier contained streets during a drier and more optimistic chapter of the city's history. Railroad workshops and other improvements would be placed on that swath once the company had moved enough fill and granite from its mountain quarries to raise it out of swamp and river surges. On April 1, when the city's board of trustees voted to approve the gift, the *Sacramento Union* voiced its support: "Certainly the municipal corporation of Sacramento cannot have for many years any purpose of redeeming from the waste of waters and otherwise improving the locality of these mythical streets," it said—let the work be done by others.[9] This enterprise, uncontroversial within the ring, had merit in Hopkins' eyes.

Second, there was the matter of a knobby little island in San Francisco Bay which in the Spanish era had been known as Yerba Buena Island, "but which," a waggish columnist for the *Sacramento Union* wrote in March, "latter-day authorities persist in calling 'Goat Island' and 'Wood Island,' for the reason, I suppose, that neither wood nor goats were ever known to exist there." At least half a dozen entities now wanted Goat Island—chiefly the Central Pacific. For some time the Associates had considered the prospect of extending their line from Sacramento down to San Francisco Bay. Judah's original reports had mulled over possibilities, and though they had ceded those rights five years ago, the matter could be reopened. While in earlier times it may have seemed natural to join or run a long line southward around the muddy edge of the bay and then up to San Francisco, what now seemed more alluring was to capture all the traffic of the Orient themselves—by establishing a Central Pacific deep-water port and terminal. Two possible sites, discussed for some time, shared the dubious distinction of once being designated as the state capital before being vacated in favor of Sacramento; there was the sleepy little army-post town of Benicia, on the northern shore of Carquinez Strait,

Theodore D. Judah, pioneer California railroad engineer, taken shortly before his untimely death in 1863. *(California State Library: 1428)*

Collis P. Huntington, the hardware store-keeper-turned-empire builder for the Central Pacific, late 1860s. *(The Huntington Library: HEH 61/1/2/13)*

Leland Stanford, one-term California governor and titular head of the Central Pacific, late 1860s. *(California State Library: 8527)*

Charles Crocker, brawny Central Pacific contractor, late 1860s: "I knew how to manage men." *(California State Library: 1732)*

Edwin Bryant Crocker, "the Judge," motive force in the Central Pacific who sacrificed his mind and health to the enterprise, taken around 1868. *(California State Library: 22,974)*

Mark Hopkins, Central Pacific treasurer "who liked to work better than the laziest man liked to loaf," late 1860s. *(California State Library: 23,461)*

Dr. Thomas C. Durant, Union Pacific vice president and "first genius for manipulation," 1869. *(Adirondack Museum)*

Gen. Grenville M. Dodge, Washington, 1865, before he became the Union Pacific chief engineer. Photo by Mathew Brady. *(Library of Congress: US262-15959-815738)*

George Francis Train, "the noonday mystery," Union Pacific stock promoter and land baron, 1860s.

Gen. John S. Casement, the Union Pacific's banty tracklaying contractor, 1865. *(Library of Congress: B812-9072)*

Rep. Oakes Ames, the "King of Spades" who distributed railroad stock around the Capitol. Photo by Mathew Brady. *(Library of Congress: BH83-2394)*

Samuel B. Reed, Union Pacific construction engineer, 1868. Photo by Andrew J. Russell. *(Union Pacific Museum: Russell S5)*

Gen. William T. Sherman, commander on the plains and friend of the Union Pacific. *(Library of Congress: B8172-6454-815738)*

Brigham Young, Mormon patriarch and champion of Salt Lake City, late 1860s. *(LDS Archives: P94/1-1)*

Red Cloud, courageous leader of the Oglala Sioux, late 1860s. *(Library of Congress: LCUSZ62-103232)*

Spotted Tail, the iron-willed conciliator of the Brulé Sioux. *(Library of Congress: LOT12337-2)*

Central Pacific locomotive in the 63-foot-deep Bloomer Cut. Photograph by Alfred Hart. *(Union Pacific Museum: Hart 12, x605)*

Prospect Hill Cut, looking down at the Central Pacific's Chinese workers and their one-horse dump carts. *(Union Pacific Museum: Hart 82, x167)*

Central Pacific's Secrettown trestle (mile 62.5), 115 feet long, 90 feet high, 1866. *(Union Pacific Museum: Hart 48, x179)*

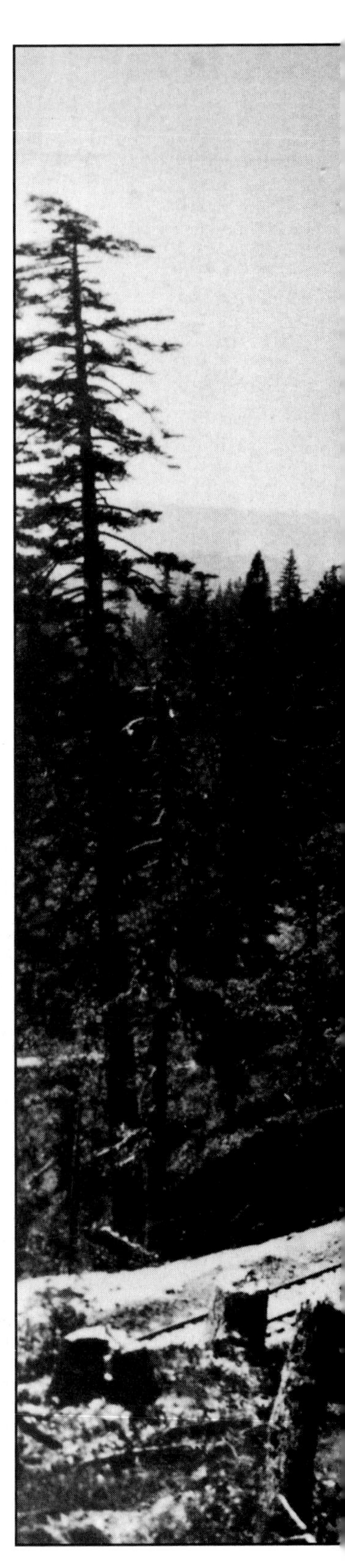

West of Fort Point, 76 miles from Sacramento, a Central Pacific train puffs uphill.
(Union Pacific Museum: Hart 151, x2994)

Long Ravine Bridge from below, 120 feet high, 56 miles from Sacramento, 1865.
(Union Pacific Museum: Hart 42, x6224)

Locomotive *C. P. Huntington* rounding Cape Horn. The American River is 2,000 feet below.
(Union Pacific Museum: Hart 44a, x781)

(Left) Central Pacific snowsheds under construction in the Sierra. *(Union Pacific Museum: Hart 247, x6351a)*

(Below) Locomotive *C. P. Huntington.* *(Union Pacific Museum: Hart 169, x600)*

(Above) Depot at Cisco. *(Union Pacific Museum: Hart 185, x101)*

(Right) Bank and cut, Sailor's Spur, showing effects of one-horse dump carts and enormous patience. *(Union Pacific Museum: Hart 90)*

View from Donner Summit, with Donner Lake below, fall 1866. *(Union Pacific Museum: Hart 125, x6302)*

Virginia Street, Reno, Nevada, shortly after its founding. *(Union Pacific Museum: Hart 286, x636)*

(Above) Central Pacific tracklaying, Humboldt plains, Nevada. *(Union Pacific Museum: Hart 317, x6807)* *(Below)* Curving rails with hammers, Ten Mile Canyon (mile 435), Nevada, 1868. *(Union Pacific Museum: Hart 333, x156)*

(Above) The Central Pacific construction train pauses, Nevada palisades, 1868. *(Union Pacific Museum: Hart 338)* *(Below)* Central Pacific construction boss James Strobridge and workers at "Camp Victory," Utah, after their ten-mile day, April 20, 1869. *(Union Pacific Museum: Hart 350)*

Thomas C. Durant at the 100th Meridian, Cozad, Nebraska, 1866, where ties await rails. Photograph by J. C. Carbutt. *(Union Pacific Museum: Carbutt JC208, 5-65)*

Union Pacific excursionists disembark at Omaha for the 100th Meridian gala, 1866. *(Union Pacific Museum: Carbutt JC200)*

(Above) Pawnee warriors stage a Wild West show for the 100th Meridian excursionists, 1866. *(Union Pacific Museum: Carbutt JC207)* *(Below)* At the 100th Meridian, Thomas C. Durant poses in front of the locomotive *Idaho* and the excursion train. *(Union Pacific Museum: Carbutt JC205, 504801)*

Jack Casement's grim and sweaty tracklayers, Nebraska, 1866. *(Union Pacific Museum: Carbutt JC209)*

Columbus, Nebraska: saloons and railroad hotels planted on hunting grounds. Photograph by A. J. Russell. *(Union Pacific Museum: Russell 7-28)*

Union Pacific construction officials at their headquarters, Laramie, Wyoming, 1867–68.
(Union Pacific Museum: Russell 5-1)

West approach, Dale Creek Bridge, Wyoming. *(Union Pacific Museum: Russell H2-23)*

Union Pacific officials (seated at table) Silas Seymour, Sidney Dillon, Thomas Durant, John Duff; (standing) H. M. Hoxie, Samuel Reed. *(Union Pacific Museum: Russell H5-3)*

Fort Sanders showdown, 1868, including Sidney Dillon (left), Generals Sheridan (with Dillon), Grant (straw hat, center), Dodge (doorway), and Sherman (to Dodge's left), and Dr. Durant (on fence). *(Union Pacific Museum: Russell 5-141)*

(Above) "General Jack" Casement, Union Pacific construction boss. *(Union Pacific Museum: Russell 5-135)* *(Below)* Carmichael's Cut, Bitter Creek, Wyoming, 1868. *(Union Pacific Museum: Russell 11-35)*

Citadel Rock and temporary and permanent bridges, Green River, Wyoming, 1868. *(Union Pacific Museum: Russell 2-39)*

Mormon graders west of narrows, Weber Canyon, Utah, 1868. *(Union Pacific Museum: Russell 11-26, R0372)*

(Above) The "Great Z" track, Wasatch rim, Utah, 1869. *(Union Pacific Museum: Russell 2-42)*
(Below) Newfangled steam shovel at Hanging Rock, Echo Canyon, Utah, 1869. *(Union Pacific Museum: Russell 11-55)*

(Above) Brooks Cut near Wasatch station, Utah, 1869. *(Union Pacific Museum: Russell 1-46)*

(Right) Cliffhanging Mormon surveyors passing equipment upward, 1868. *(Union Pacific Museum: Russell 6-35)*

Dan Casement (in doorway) and construction managers, Echo City, Utah, 1868. *(Union Pacific Museum: Russell 5-60)*

Tunnel 3, Weber Canyon, 1868–69. *(Union Pacific Museum: Russell 2-35)*

(Above) Excursion party of Sen. William Patterson pauses at Devil's Gate, Weber Canyon, Utah, 1869. *(Union Pacific Museum: Russell 2-33)* *(Below)* The last siding, Promontory, Utah. *(Union Pacific Museum: Russell 1-47)*

Celebrants pose before the Central Pacific's *Jupiter* and the Union Pacific's *No. 119*. *(Union Pacific Museum: Russell 1-47)*

More dignitaries and spouses arrive to pose. *(Union Pacific Museum: Russell 1-31)*

Union Pacific construction officials, including (seventh from right), G. M. Dodge and (to his right) Silas Seymour and S. B. Reed. *(Union Pacific Museum: Russell 5-41)*

(Right) Union Pacific dignitaries, including T. C. Durant (left of track with gloves) and (right of track) John Duff (standing on tie), Sidney Dillon and Silas Seymour to his left; G. M. Dodge (second to Seymour's left) squares his foot in embankment. *(Union Pacific Museum: Russell 1-24)*

(Below left) The crowd presses in as the last rail is placed. *(Union Pacific Museum: Russell 1040)*

(Below right) Gov. Leland Stanford and T. C. Durant brandish their hammers. G. M. Dodge is at Durant's left. *(Union Pacific Museum: Hart 356, 631)*

With engines *Jupiter* and *No. 119* touching pilots, Central Pacific engineer Samuel Montague shakes hands with Union Pacific engineer Grenville Dodge. *(Union Pacific Museum: Russell 1-2-3)*

Dignitaries and laborers mingle at the Golden Spike ceremony as telegraphers tap bulletins at trackside. *(Union Pacific Museum: Russell H-1-20)*

Cartoonists had a field day after the Golden Spike. Three years later they would have the Crédit Mobilier scandal. *(Frank Leslie's Newspaper)*

and there was Vallejo, ten miles farther up San Pablo Bay with an inviting offshore site called Mare Island (which would soon attract the attention of the U.S. Navy as a facility). But even more attractive to Huntington—because the choice offered fewer obstacles—was the three-hundred-acre government reservation at Goat Island, lying in the bay midway between bustling San Francisco and swampy Oakland.

Huntington spread the word that the Central Pacific had no interest in the island itself but for enough right-of-way to run tracks around its base to a depot and port, reached by a causeway out to the island. Privately, though, Huntington and Judge Crocker had more on their minds. Of the three hundred acres, the federal government would probably reserve between a fifth and a third of the land for military reasons; at present half a company of soldiers stared at their calendars there and longed for the fleshpots of San Francisco. In his loftier imagination Crocker began to fill in the remaining empty land—with "good substantial wharves," fireproof brick warehouses, grain elevators, flexible transfer facilities—and he saw opportunities in the isle's highland for rock fills and the creation of yet more territory on a big shoal off the island's western shore.[10]

Without tipping his hand as to such grandiose plans, Huntington's assignment was to soften Congress for any grants that were necessary. There was the question of leasing the federal land (for which Congress's permission was necessary), and also the closing off navigation between Goat Island and Oakland. So, treading carefully, Huntington began to exclaim in Washington (where he had gone to arrange for legislation) that Goat Island was "the most natural ocean terminus of the Pacific Railroad." Only ninety-one miles lay between this projected "chief seaport" and Sacramento. This, and the fact that in such an arrangement San Francisco would be left wholly out of the picture except for token ferry service—and would thus, optimistically, wither away—appealed to many in the state capital, as well as (of course) Oakland.[11] Huntington was confident enough that he could wrangle a right-of-way from Congress that Mark Hopkins made little objection, though he had no interest in extending the line down to the bay or in any other direction save eastward, over the mountains.

But it was the third sudden business opportunity which caused the mild-mannered treasurer to all but blow his top. It was complicated. It involved a spiderweb of mostly imaginary California branch lines, but chiefly the feeble Western Pacific Railroad, an entity chartered to connect Sacramento to the bay. Interestingly, the rights to a bay connection to Sacramento, originally belonging to the Central Pacific, had been assigned by Huntington to San Francisco interests during the furious maneuvering which had accompanied the 1862 Pacific Railroad Act in Washington. On paper it went by a roundabout route down the San Joaquin Valley to Stockton and then westward through foothills down to the shallow southern end of the bay and the marsh-edged town of San Jose. There it had a natural connection to the new proprietors' San Francisco and San Jose Railroad, which had been chartered way back in 1852 but had taken a decade to come to life. A single 175-mile line from San Francisco to Sacramento was therefore a possibility, but Western Pacific ran into a string of troubles. Only its first twenty miles out of San Jose had been constructed by the winter of 1866–67, and financial problems had not only halted the line at the little mill town of Niles, where Alameda Creek emptied into the bay, but whittled the Western Pacific down to one man, Charles McLaughlin, who was debt-ridden and desperate to get out.

For several years Judge Crocker had wanted them to buy the Western Pacific, and when he learned of McLaughlin's latest plight he knew they could "get it cheap." Only Governor Stanford seemed to pay him any attention. McLaughlin then added a winning sales argument to his moribund firm and few prospects: he had been offered rights and property for a rail line northward from Niles up the bayside and over extensive mudflats to Alameda and Oakland. McLaughlin allowed Stanford to assume that this would be part of the deal—and the Oakland businessman who controlled the waterfront, Horace W. Carpentier, chimed in with an offer of some muddy, tide-lapped Oakland real estate.[12] Earlier this may not have been as interesting a deal to Stanford, but now one could stand there at the water's edge and look across to placid little Goat Island, with the city of San Francisco behind it in the fog; one could feel the salt breezes coming in from the Orient. Still, Hopkins was unmoved—he thought the world of Judge Crocker and Stanford, he told Huntington, trusting them in all aspects but their financial judgment. Charley Crocker sided against his brother. And as to Huntington, who was silent enough on the subject of the Western Pacific to encourage both sides that he favored them, from faraway New York and Washington he had been developing his own enthusiasms: he flirted with the idea of running a simple line from the capital down the Sacramento River to the bay by way of Antioch, but even more fervently with the notion of having the Central Pacific acquire a handful of struggling little railroads *north* of Sacramento.

It was as if five men had gone in on the price of a lottery ticket and, in a surplus of reckless enthusiasm well before the contest would be decided, three had scattered to the winds, each brandishing an arm's-length shopping list. The bills continued to pile up, loans and mortgages matured, interest came due, armadas bearing rails and locomotives were expected regularly, brigades of hungry workingmen were deployed all over the Sierra, and nevertheless the shopping carts loaded with their shimmering prizes rolled to a stop before Mark Hopkins and his ledgers. Goat Island. Western Pacific. And here was Huntington with his desires plucked from the shelf of the northern Sacramento Valley: the California Northern, the California Central and Yuba, the California and Oregon. The nearly bankrupt California Central line left the Central Pacific station town of Roseville and went north up the valley to within seven miles of the Yuba River at Marysville. It had been chartered in 1853 by Ted Judah's friend, Colonel Charles Wilson (late of the pioneer Sacramento Valley Railroad), but Wilson had long ago been deposed by the Robinson brothers and was now trying to seize the California Central back from their bankrupt successors. Wilson owed more money to the Central Pacific than he could ever pay himself, but the Associates had kept stringing him along in case he and his plum became useful. The California Northern, a creation of the Sacramento speculator Sam Brannan, left Marysville and went up the Feather River to the faded foothill town of Oroville, of gold rush fame; Brannan had intended to finish the seven-mile gap to the California Central before losing all his money. Finally, the California and Oregon did not even exist—it was a paper entity consisting of a map (with a line drawn from the Sacramento Valley across the Klamath Mountains into Oregon) and a congressional land grant. The promoters had failed to build even a mile of track, and along the way had come to owe money to the Associates, and as security the Central Pacific had written a mortgage on the land grant. It is not surprising that Huntington—the merchant who loved to corner markets—saw all of these weaknesses, obligations, and dependencies, and began to think: *Empire!*[13]

They all were thinking big. The Judge and the Governor thought their scheme would control all business to the South, to San Francisco, and to Asia; Huntington had his eye on locking in San Francisco, northern California, and Oregon, and ultimately locking out the Union Pacific from the rest of the Pacific Northwest and Idaho. No wonder an eastern terminus for the Central Pacific at Green River in far-off Wyoming was so attractive: from Green River one could build northwest toward Oregon. The single tentacle of the Central Pacific, draped up the western slope of the Sierra and brought to a halt at the summit, was (in their fondest aspirations, at least) growing a whole octopus.

Here is where trouble broke out among them. The discussions went back and forth, around and around, for months. Then, in late March, the focus narrowed to the subject of the Western Pacific after Stanford talked with Charles McLaughlin (by now desperate) in San Francisco and they agreed on general terms for purchase. The Governor then went up to Sacramento to confer with Judge Crocker and Hopkins; as always the latter spoke disapprovingly of anything which would deter them from their main enterprise. But he did not say no. On the strength of this, Crocker and Stanford went back to arrange details and Hopkins returned to his books, not fully realizing what was transpiring. They stayed in San Francisco for two weeks to get an agreement in shape. "I would give a good deal if Huntington could only be here for a few days to consult," the Judge wrote to Hopkins at one point.[14] But he went ahead. On April 9, Hopkins received a telegraphed summons to get Charley Crocker and come down to sign the papers. Hopkins replied that he could not leave—but he said in his mandarin way that they should do what they had to, subject to consultation; he was still against it. He went back to his books. The next morning Judge Crocker energetically appeared in the office with a carpetbag full of papers, having come up on the first morning's steamer. He spread the papers out. His brother came in. The Judge told them to hurry, read, and sign, so he could return on the next boat. And then Hopkins—in his most emphatic manner—planted both his heels and refused. Charley Crocker joined him.

The Judge looked at their stony, irritable faces and was "astonished," as he would complain to Huntington ("surprised beyond measure," is how Hopkins saw him). How could they turn down such a bargain? But almost as important, how could they have let him waste two weeks in negotiation? This was a terrible embarrassment—how could he ever show himself to the business community again? For a long moment the Association hung there between them. Finally, Hopkins sighed, locked his office door, and began a condensed study of the documents, the Judge hovering nearby to point out finer details and explain. Stanford had vowed he could raise the purchase money on Western Pacific bonds, and to this the treasurer nodded—Stanford probably could. But of course it was more complicated than that. Finally Hopkins grimly looked up: within twelve months they would have to pay $120,000 in gold, not to forget taxes, interest, and expenses; moreover, they could not afford to divert any of their laborers or finances to work on the Western Pacific anytime soon, especially since they were pressing to build over to the Truckee by the fall, and current expenses were up to $300,000 per month. Furthermore, he said, "our loans on collaterals having no established market value [were] already alarmingly large and dangerous in case of a financial panic." This was "an extra hazardous year," he said. It was plainly inexpedient. "It is not sound policy," he pronounced. "If we do it," he would warn, "it is only a question of time when we will be cornered and washed out." But because it was "a financial question," Hopkins was willing to submit to a vote with the majority ruling. In the meantime, he would sign the papers. Charley signed them. Hopkins invoked his power of attorney and signed for his partner in New York. He cautioned them to do nothing with the contract. And then he telegraphed Huntington: what was his vote?[15] Both Hopkins and Judge Crocker dispatched several long letters explaining their positions. Impatience mounted, especially in Stanford and Crocker. Day followed day.

All they obtained was silence. Ten days later, the puzzled Hopkins sent off another message: why no response? He finally got an answer after two more days, on April 22, but the question was hardly answered by Huntington's terse, baffling, four-word reply: "Do as you like."

Both Hopkins and Judge Crocker were mystified. What could he mean? They wired for a confirmation, each then dashing to his desk to compose extensive queries, explanations, and complaints. Hopkins had no confidence in Stanford's financial judgment, he wrote, but he had much less respect for the Judge's. "They were presumptuous enough to commit themselves unconditionally without first ascertaining our opinions," he griped. If Huntington wanted to do it then he would go along—but unwillingly, because it was a dangerous move. "For until I see where the money is coming from . . . I think nothing should tempt us to withdraw thought, labor or dollars from the Central Pacific."[16]

His hand was even and controlled as he wrote, for that is the way he was—and, tellingly, Judge Crocker's was shaky, his inkpot in his spleen. "Imagine *my* surprise when your *noncommittal* reply came, *'As you like'*—thus throwing the whole thing apparently back upon Hopkins," he scrawled. "Now I *know* you are in favor" of the purchase, he said—all of Huntington's letters indicated that. And while he had his friend's attention there was the matter of Huntington's eagerness to pick up the northern roads: "I have a very poor opinion of it," the Judge said. He now thought they should simply sell out whatever interests they had already acquired ("cheap—*very cheap*"). He warmed. "I feel myself in a very mortifying position, to go on and spend 2 weeks in fixing the details of the most important trade we ever made, & then to have it all unceremoniously kicked over, in a way that places me in the position of a man claiming *power,* & yet without any in reality. I shall never want to look McLaughlin in the face again." But here came the threat: "Nor shall I ever presume again to negotiate a trade with anybody," the railroad's attorney was saying. "I prefer to lay back & exercise the imperial veto power over trades made by the others. It will be far pleasanter."[17]

More days of tension and silence as Huntington continued to hang back. Stanford began to threaten to carry out the purchase on his own—which would have been an embarrassment to the united front they presented to the world.[18] And probably it was brawny Charley Crocker who broke the impasse, after his brother had forced him to read all of his letters from Huntington. He may have cast his memory back to boyhood on the Indiana farm and considered whether Ed would ever budge from his position. And then Charley somberly told Mark Hopkins that he could find no "*financial* objection" in Huntington's months-old letters. Hopkins acquiesced. "It would seem unreasonably fractious," he wrote the still-quiet Huntington, "or self-opinionated in me longer to refuse when all are for it."[19] The trade was officially accomplished on June 8. The Central Pacific absorbed the Western Pacific, its rolling stock, and twenty miles of rails previously ordered but undelivered; McLaughlin retained rights to sell off the land grants. In the next few years Leland Stanford would take a keen interest in the enterprise, which would accordingly be dubbed "the Governor's Road."

"Do as you like." What was Huntington up to that spring? His California partners repeatedly wondered about this in their letters, seasoning their progress reports and requests for materials with complaints and arguments. On the issue of whether the Central Pacific ought to take on a government mail contract, they all thought it was a capital idea; Huntington's dispatches

seemed to frown on it. Hopkins was particularly puzzled. Did not his partner realize that in gaining the mails they could with one swipe deal a serious blow to their competitor, the Wells, Fargo stage company, which crossed the Sierra by the Placerville route, and also to the hope of rivals who had been trying to confuse and divert investors from the Central Pacific in favor of a Placerville-to-Nevada rail route? Getting the mail contract was an economical tactic, Hopkins lectured—far better than engulfing rivals with money or credit at this tentative stage. "Our future power & influence as 'Rail Road men' depends more upon our real strength," he maintained, "than our rapid spread."

In his letters Judge Crocker repeatedly remonstrated, too. Why had Huntington purchased fifteen hundred tons of iron when they could not lay it for at least a year and a half? On May 8, in one letter to Huntington, he reported on good strides in the mountains—in the Summit Tunnel only 681 feet of granite remained between the headings, and the snow continued to retreat before the grading crews—but Crocker also complained about their communication. The last steamer from the East had delivered a package from Huntington containing the latest U.S. map of San Francisco and the lower end of the Sacramento Valley. "I notice," Crocker wrote, "the Central Pacific R.R. is entered on it, extending from Sac. to Goat Island," and it had been inked in Huntington's hand. But he could not imagine why the line deviated "materially" from the one adopted by the company, and his friend had not enclosed an explanatory letter. Nine days later he took pains to object to Huntington's exasperating vagueness in wanting "a report" to hand to Interior Secretary Browning and wanting "a report on an estimate"—without telling him what he needed them to say. "It puts me out of all patience when I think of the way you write for these things," he griped. For want of "half a dozen lines" of specifics, he had to study Huntington's words and read between the lines. There had to be a better way.[20]

Part of the problem had to do with the way Collis Huntington worked. He had no clerks, no fetchers and carriers. Much of his time was occupied with correspondence by mail and telegraph, his days spent in a small, shabby Manhattan office at 54 William Street. "Have had only one small room & their is allmost allways some one in talking to me," he wrote to Hopkins in his crabby hand and with his characteristic spelling, "and you know how difficult it is for me to write when I am disturbed by persons talking to me, so I hope when I get another room I shall be able to get my ideas on paper in better shape." He rented an adjoining room and hired a bookkeeper in May, and late in the year added a clerk. The comparison of his headquarters with Dr. Durant's—both were conducting business in the many millions—gave him no end of amusement.[21]

He did get out, selling Central Pacific bonds whenever and wherever he could. In March there had been an important trip to Washington; the congressional session was opening and legislation for the Goat Island easement had to be written and introduced. He also called at Interior Secretary Browning's office, on April 25, to hand over the Central Pacific's latest projected location map for Browning's approval. Based on Butler Ives's surveying notes, the route passed the California border and the Big Bend of the Truckee River, crossed the Forty-Mile Desert and skirted the Humboldt Sink; it followed the burnt-red Humboldt Valley eastward to the green meadows of the Humboldt Wells, the historic campground on the old California Trail where emigrants had paused to rest and fatten their animals, taking advantage of lush grasses and the dozens of clear springs called "wells" by the emigrants, who thought them bottomless. The new Central Pacific route was 370 miles long, climbing gradually from the Truckee Meadows' altitude of 4,500 feet to Humboldt Wells at 5,600 feet. Beyond were the Pequop and the Pilot Mountains—and the Utah border, only sixty miles away.

Browning was not in his office when Huntington called. He had been taken seriously ill with an abdominal obstruction two days before, and would not return to his duties until mid-May. Enfeebled for weeks thereafter, bogged down with the Interior Department's Indian policy and disputes with Secretary of War Stanton and with the army, Secretary Browning would delay his approval for nearly three months, costing the Central Pacific those three cooler months of time for location surveys. Toward the end of April, Judge Crocker impatiently urged Huntington to prod the secretary, not knowing how near death he was. The Judge fretted that they should submit a new map all the way to the mouth of Weber Canyon in Utah—and "if he says he does not know whether the line from the East will be run through Weber Canon, *then let him approve the location from Humboldt Wells to the upper end of Great Salt Lake.*" Browning, however, followed his own calendar—he would have even if he had been well—to the ever-present exasperation of Crocker and Huntington. There was, after all, much to contend with in Washington in the spring of 1867.[22]

Had there ever been so many beggars in the federal city? The streets were crowded with the unemployed, the poorhouses were impossibly full, large sections in all districts remained in ruins, shantytowns sprouted everywhere, many public buildings were still wrecked by wartime misuse, and mud ruled over all. Collis Huntington would have pushed through the sidewalk tumult outside Willard's, lucky to have found a boardinghouse room as accomodations dwindled with the opening of the Fortieth Congress. And, fighting the same throngs, Grenville Dodge was there to represent Iowa's Fifth Congressional District, at least for the moment. During the extra session held over the winter until March 3, the two chambers had overriden Democratic President Johnson's veto of the two bills establishing martial law in the South and forbidding the president from removing any Senate-approved official he had appointed without that body's permission. Another act vetoed and overridden transformed the Territory of Nebraska into a state. Also, the president was virtually deprived of command of the army, instructed to issue all military orders through the general of the army, Grant. Now, with the merge into the Fortieth Congress, two-thirds of the Senate and the House were now controlled by the radical Republicans, and one could not walk very far, indoors or out, without hearing the bitter word "impeachment."

Naturally because of his powerful political connections and his experience fighting in the Army of the Tennessee, Dodge was almost immediately invited to a dinner at the White House. Wide-eyed at the sumptuous repast—there were twenty courses "and as many different kinds of wine"—he could not get comfortable. Since New Year's Day, the president's dinners and receptions had become increasingly tense, though well attended; several weeks earlier so many guests had jammed into the White House that women fainted and police had to be called to restore order. Now, of the president's twenty guests, eighteen were radicals. Dodge must have stiffened when Johnson turned to him. "General Dodge, you knew me in Tennessee," the president said. "It seems to me we should not be so far apart now." But Dodge was already committed to the other side. He replied, "[I have] no ill feeling toward you, but I cannot sustain in any way your past course."[23]

"Johnson is looking to the future," Dodge had written home a year before, expressing his hardened opinions; "he believes that the Confederates and their northern sympathizers have more power than the true Union man and they have the most votes and ever since he was made President he has trimmed his sails to that breeze, has sought every occasion to alienate Union men from him. . . . I tell you that no part of the Republican Party will follow Johnson in his reconstruction policy." And now Dodge was at the center of things. He had drawn a good seat in the chamber for a novice radical; grinning old Thaddeus Stevens, unquestioned leader of the House and mastermind of the new Republican Party, sat on his right, tottering to his feet and hobbling around the desks on his cane, keeping his troops in line. On Dodge's left was the old barnburner Benjamin F. Butler of Massachusetts, leader of the growing impeachment movement and the man who would soon spearhead a farcical congressional probe into connections between Johnson and the assassination of President Lincoln. Butler, who was cross-eyed, frequently leaned in Dodge's or Stevens's direction to comment on proceedings. "Do you know," Stevens whispered to Dodge as he probably had to his predecessor, "whether he was talking to me or you?"[24]

Dodge would have been present as Congress drafted and quickly passed a supplemental Reconstruction act directing the military to enroll southern voters it deemed qualified. Conversations in the hallways of the Capitol orbited mostly around the issue of the South, prominently that month about whether Johnson might try to remove Dodge's former comrade, General Philip Henry

Sheridan, from his post overseeing affairs in New Orleans. But there were many other things on their minds—military matters, transportation issues, statehood questions, Indian affairs—and Dodge found that as chief engineer of the Union Pacific he was better known than nearly all the other newcomers, and he boasted to his wife that many prominent men had turned to him for advice. However, his chair near those two seats of power would in a short while be vacant—as committee assignments were taken up, as the House continued to enlarge and clarify Reconstruction of the South, as Johnson moved conscientiously to execute the laws passed over his veto, as Congress watched the president's every move and calculated its usefulness in removing him. As representative of the Iowa Fifth District and new point man of the Iowa Regency, Dodge found his days and evenings consumed by patronage demands, interviewing job seekers from home who clogged his office and spilled out into the corridor, answering the torrent of mail he received. For days and days he trudged from agency to agency—being forced to introduce himself and wheedle out spoils for constituents or make sure those in place remained.

Two weeks of this was about all he could bear. He wrote to his wife back in Council Bluffs, "[I have] had the blues ever since I left home." But so many crucial patronage matters remained that the engineer could not return to his railroad. When he could, Dodge lost no time getting out of Washington, leaving on April 6. "I have no disposition now to come here again," he told Anne Dodge.[25] Perhaps he would return in December to serve his constituents for the second session, or at least part of it.

When Dodge arrived home in Council Bluffs on April 19, the Missouri River was still out of its banks and Reed was frantically supervising the rescue of their tracks from the swollen Platte and its tributaries. Within a week, though, trains were leaving Omaha on a regular schedule and Reed thought his men would be ready to turn to new work in a few days. Everything westward still looked like a disaster, though—hardly the controlled, well-ordered machine they would have wanted when the brass arrived, which they did, by the steamboat *Elkhorn* from Council Bluffs, on April 27. Dr. Durant was there with Oliver Ames (though they barely spoke), and also their fellow Union Pacific executive board members John Duff Sr. and Springer Harbaugh, and muttonchopped company director Sidney Dillon. The stated reason for their visit was to officially accept the 305 miles of completed work from the Crédit Mobilier, a formality which might have been accomplished with less misery if they had stayed in New York and simply shook their own hands. The officials climbed aboard an inspection train with Grenville Dodge and rolled out toward North Platte.[26]

All wondered what would happen. The weather had become abnormally hot and uncomfortable. The inspection party went out to the end of track at O'Fallon's, alighted, and stayed there for exactly five minutes before setting back east. On the way Oliver Ames looked out at the hot Nebraska prairie west of Fort Kearney and despaired of its ever amounting to anything—what settler or farmer could want it? If the board were smart, it would refuse government land grants, even gifts. So much for the hope of profiting from land sales, he said. There would be little money in this until the project was completed and freight and passenger revenues picked up. Durant begged to differ—the only profit in railroading in their day and age would be in its construction, not in its operation. This set off Ames, which set off Durant. When on May 4 the tense officials returned to Omaha they looked over the accounts, mumbled cryptically among the tiptoeing employees, found time to tell Reed they were "well pleased," and indicated that there would be some changes. Webster Snyder was to be appointed superintendent of operations; Reed, retaining the position of construction superintendent, would relocate to North Platte, which disappointed him. Later that day, even that small certainty crumbled.[27]

"They broke up in a row," reported Samuel Reed, "and no one knows what will be the end." During the furious argument Oliver Ames warned Durant "about his lawless way of doing work taking the whole thing into his hands—& forbid his doing it without consultation." Two days later Ames and his supporters climbed aboard the *Elkhorn* and left for the East. Only the Doctor stayed behind—in a horrible mood. No work had been let west of the fourth hundred-mile section "and will not be until they come to some agreement among themselves in NY," said Reed, with Ames muttering that the only solution was Durant's ouster. "The Dr. is jealous of everyone," Reed told his wife, adding that he didn't dare do any work until Durant left for the East. "This fight places me in a very unpleasant situation," he wrote, "and I have a mind to resign my position"—but he decided to get back to work out at North Platte and wait a few days until he heard from New York. A day later it seemed less like a personal crisis. He threw himself into his job.[28]

Jack and Dan Casement had been hard at work driving their grading and tracklaying crews since the rains and the repairs had been finished. Getting a slow start on grading the fourth hundred miles, they were by early May making decent progress. They had a new contract for the season at $850 per mile, and laborers by the hundreds poured into North Platte to sign on. Dodge's former medical officer, twenty-eight-year-old Major Henry C. Parry, was headed out past Julesburg to Fort Sedgwick. "I found as I passed through North Platte," he wrote on May 16, "that the Indians had driven all the traders and miners in from the mountains, and at North Platte they were having a good time, gambling, drinking, and shooting each other. There are fifteen houses in North Platte: one hotel, nine eating or drinking saloons, one billiard room, three groceries, and one engine house, belonging to the Pacific Railroad Company. The last named building is the finest structure in the station. I observed that in every establishment the persons behind the counters attended to their customers with loaded and half-cocked revolvers in their hands. Law is unknown here, and the people are about to get up a vigilance committee."[29]

Just over a week later another commentator passed through North Platte. It was the twenty-six-year-old Welshman, Henry Morton Stanley, a rags-to-riches-to-rags adventurer who had found his writing vocation by accident, and who now was on assignment to travel west of the Missouri for the *St. Louis Missouri-Democrat;* over the course of seven months his dispatches from the Great Plains would attract the attention of James Gordon Bennett of the *New York Herald,* who would send him to Africa and commission him to find Dr. David Livingstone. When Stanley descended from the rail coach after a fifteen-hour journey from Omaha, he noted large quantities of freight piled up and covered with sailcloth, beyond which was a vast encampment of more than twelve hundred covered wagons—Utah-bound Mormon emigrants, settlers heading for Idaho or Montana. "The prairie around seemed turned into a canvas city," he marveled. But as he walked past the emigrants into the cacophony of North Platte, his breath was taken away. "Every gambler in the Union seems to have steered his course for North Platte," he recorded,

> *and every known game under the sun is played here. The days of Pike's Peak and California are revived. Every house is a saloon, and every saloon is a gambling den. Revolvers are in great requisition. Beardless youths imitate to the life the peculiar swagger of the devil-may-care bull-whacker and blackleg, and here, for the first time, they try their hands at the "Mexican monte," "high-low-jack," "strap,"* "rouge-et-noir," *"three-card monte," and that satanic game, "chuck-a-luck," and lose their all. "Try again, my buck; nothing like 'sperience; you are cuttin' your eye-teeth now; by-and-by you will be a pioneer." Such are the encouraging words shouted to an unfortunate young man by the sympathising bystanders. On account of the immense freighting done to Idaho, Montana, Utah, Dacotah, and Colorada, hundreds of bull-whackers walk about, and turn the one street into a perfect Babel. Old gamblers who revelled in the glorious days of "flush times" in the gold districts, declare that this town outstrips all yet.*[30]

By this time Jenny Reed had joined her husband for a visit at North Platte, and though the town was appallingly wild, the couple had one morning of reflection when an Episcopal bishop, D. S. Tuttle, came through on a journey west and lodged at their hotel, finding a room for the two women in his party but being forced to sleep with his male companions in blankets on the floor of a common room. After Reed made their acquaintance he arranged space for them on a flatcar in one of the westering construction trains, saving them fifty or sixty miles of staging and exposure to Indians; before they left, Tuttle conducted a Whitsunday service in the Reeds' room. "I read the morning service entire," Tuttle remembered, "except the lesson and the commandments, and we commended ourselves especially to the protection of our Heavenly Father. At North Platte no religious services of any kind are held on Sunday. Men work and trade and buy and sell just as usual, and gamble and quarrel more than usual."[31]

It was to be a busy season. The immense fields of covered wagons glimpsed at North Platte by Henry Stanley, with their cookfires and noise of livestock and shouting men and playing children and calling women, would be backlogged for weeks, their vanguard having already moved on in escorted wagon trains of twenty-five to thirty, raising dust and deepening ruts up the trails, driving off game in Western Nebraska, the Laramie Hills, and Powder River country. Three days' ride south of the line scored into Sioux hunting grounds by the trails and the Union Pacific, there was a similar scene of an emigrants' encampment on the Smoky Hill Trail from Kansas to Denver, bisecting Cheyenne and Arapaho territory with the approaching Kansas Pacific, Eastern Division, railhead punctuating the matter at Spring Creek; and farther south, of course, there was the Santa Fe Trail, which concerned the Kiowa and Commanche. Now, normally with springtime, with the buffalo and antelope herds swollen and distracted by their young, and forage blossoming, the tribes were stirring, too. Could there be anything but a collision, especially with a belligerent, foolhardy career soldier taking command at such a volatile moment?

All winter, back in Washington, closeted with his cabinet and pressed by so many matters, Andrew Johnson had watched the feud between Stanton and Browning—the war secretary was for total war after the distressing Fetterman debacle, and Browning was swayed by his Indian Affairs commissioner and a conciliatory plan involving commissions, treaties, and ultimately Indian settlement on reservations. Johnson, predictably, resolved in both directions: he authorized a commission to investigate the Fetterman engagement and made no utterance to hold back the army. Accordingly, General Sherman ordered a campaign into Kansas, a department that had been almost entirely peaceful in the previous year, under the command of General Winfield Scott Hancock. The forty-three-year-old Pennsylvanian, who had stood like a rock against Robert E. Lee at Gettysburg and had suffered grievous wounds, now intended to show the Cheyenne that "No insolence will be tolerated from any band of Indians whom we may encounter." Prominent among his expedition of fourteen hundred infantry and cavalry was George Armstrong Custer, reckless hero of Gettysburg and Yellow Tavern, "Boy General" of the Volunteers, now, at twenty-seven, holding the Regular Army rank of colonel and eager for more glories against the Indians.

At Fort Larned in mid-April, the local Indian agent tried to dissuade Hancock from approaching a large camp of Cheyenne and Sioux some forty miles to the west, on the Pawnee Fork River. He predicted that they were still touchy because of the Sand Creek massacre, and would run away—and strike out against the first civilians they encountered. Hancock went anyway. When a Cheyenne delegation met the soldiers, who were deployed in a line of battle, and pleaded for them not to approach the camp because the women and children were frightened, Hancock went anyway. At midnight, he ordered Custer to take his 7th Cavalry and surround the village. These actions could have had only one intent: it was hoped that the encircled Indians would start shooting, provoking another battle—even, perhaps, a massacre. But with the first light of dawn Custer's men discovered that the Cheyenne had all fled. Hancock thereupon retaliated by looting and burning the deserted Cheyenne village. And true to the prediction of the Interior Department's field man, the Indians rode northwest to the Smoky Hill Trail, convinced that they had narrowly escaped a replay of the Sand Creek massacre, and attacked and burned three stagecoach stations, killing three people.

In due time Washington declared that south of the Platte the Cheyenne, Arapaho, and the Sioux had made war against the United States. A peace commission would be sent out. Hancock warned the Cheyenne to clear out of the vast hunting range between the Platte and the Arkansas Rivers, knowing that they would not, and he soon withdrew to Fort Leavenworth with most of his force, leaving Custer's regiment to clean up the mess. At the end of that four-month Kansas campaign, the soldiers had killed two Sioux and two Cheyenne; Custer had been court-martialed and suspended from the army for a year for a grab bag of charges, which included ordering deserters shot, failing to bury soldiers killed in battle, and leaving his post and traveling 250 miles to spend a night with his wife.[32]

But before the ashes were completely cooled in the destroyed village on Pawnee Fork, Cheyenne warriors were striking all along the Smoky Hill Trail, running off graders and surveyors for the Kansas Pacific, and Sioux were riding to spread the news and to take revenge in the western Platte Valley and the panhandle and the Laramies, three days north. Fleets of prairie schooners were on the move. Surveyors' parties had edged out across the high plains and mountains. And twelve thousand graders, timberers, and tracklayers were at work on the Union Pacific.

Meanwhile, on the island of Manhattan in New York, there transpired a different sort of fight, a bloodless one, a war in the boardroom. By the time Durant had returned to the city from his uncomfortable tour, his fellow stockholders in the Crédit Mobilier were determined to throw him, the largest individual stockholder, with 5,558 votes, off the board. The Ames brothers, holding between them 6,015 shares, had important supporters, none more influential at that time than their Rhode Island friend, the retired woolen mill owner Rowland Hazard, who had not only invested heavily in the Crédit Mobilier but lent hundreds of thousands of dollars to the Union Pacific. Hazard had grown to hate Durant, whom he considered a crook. Moreover, in the weeks preceeding, Hazard had been trying to meet the Union Pacific's crushing needs by selling the railroad's first-mortgage bonds, and nearly every banker he had seen had said that he would deal with Hazard or the Ameses—

but not if the Doctor were to continue in authority. They were fed up. Hazard told Oliver Ames that "some radical change must be made." The annual meeting of the Crédit Mobilier had been set for May 18 in Philadelphia, and Durant had made it known that he was not bothering to attend. The setting was perfect for an ambush.[33]

Oliver Ames saw the unsuspecting Durant and his supporter Cornelius Bushnell on May 17 in New York and obtained their absentee proxies for the vote the next day. On May 21 Ames was back in New York with his allies to break the dramatic news: the Doctor had been deposed as president, director, and executive committee member, and Sidney Dillon had assumed those places. Hazard was now on the executive committee along with Oakes Ames's friend and fellow investor John B. Alley, who had just retired from Congress to tend to his Lynn shoe factory. Hazard also told Durant that he was going to have his financial "defalcations" investigated. The Doctor's reply was to slam his hat on his head and stalk out of the office—determined to get even.

The following day, at a Union Pacific board meeting, Durant sat inobtrusively until things were quietly under way and then nodded to his lawyer, standing by with some papers, who then handed the sheaf to Ames. It was an injunction—forbidding the execution of the Williams construction contract. The gathering dissolved in shouting and threats. When they resumed the meeting the next day, May 23, tempers were still high, and the Doctor pretended to listen politely to their complaints and accusations with the smugness of someone with the law on his side. But the Bostonians knew a few tricks, also. As Durant sat there, they resolved to hire the eminent attorney Samuel J. Tilden to get the injunction lifted. They also formed a Committee of Five to direct construction to the base of the Rockies, and another committee to settle with the Crédit Mobilier. Two days later, after recovering from the altercations and from stoically putting his signature on nearly a thousand railroad certificates and first-mortgage bonds, Oliver Ames sat down to inform Dodge (who had already heard snippets from Silas Seymour) what was transpiring in the wild East. The coup, he wrote with unbridled sarcasm,

> *has raised the very devil in that amiable Gent, & he has come down upon us with injunctions and proposes to visit us with every form of Legal Document to keep us honest. Such a lover of honesty and fair open dealing can't bear to see the money of the U.P.R.R. wasted on such scoundrels as make up the balance of the Board of Directors. I cannot understand such a change as has come over the Dr. The man of all others who has from the beginning stole wherever he had a chance & who is to-day we think holding stock and a large portion of his stock on fictitious claims and trumped up* [accounts]. *He is now in open hostility to the Road and any orders he may give you or any parties under you should be entirely disregarded.*[34]

Ames continued to redraw the lines of communication and control. He telegraphed Omaha on May 30 to tell Snyder and Reed that the Doctor no longer had authority to act for the company. And on June 6 he again wrote to Grenville Dodge. "I and all in connection with the road here have never been so sanguine of the success and great merit of this Road," he exulted, "as we are since Durant has been put out of its management. We are now selling from 15 to 20,000 $ of our Bonds Daily and are getting our money at much better rates than formerly. The Moneyed Interests here have now much more confidence in us and will I have no doubt soon be applicants for our paper."[35]

The financial situation had finally begun to improve—but no one in the East, not the Ameses or Hazard or Dillon or Alley, could entirely count the Doctor out. He still held his post as vice president of the Union Pacific, his stock portfolio still connoted strength, and his genius for manipulation was intact. The fight had simply entered another phase—as had, quickly and tragically, the trouble out west.

"I think this year is our crisis on the Plains," Tecumseh Sherman had written to Dodge on May 7. He soon had details. East of Fort Sedgwick, in the northeastern corner of Colorado where graders were preparing the line to Julesburg for track, Indians swept in over a ridge and drove off all the livestock of two subcontractors. They "scared the workmen out of their boots," Dodge would report to Sherman, "so they abandoned the work and we cannot get them back." It was only the beginning. An engineering party surveying in Lodgepole Valley was startled by raiders, who spared the whites' lives but took their two packmules and warned them to clear out; almost in afterthought, the raiders pulled up all their survey stakes. Another group working between Lodgepole and Sidney under the direction of an Arkansas engineer named C. H. Sharman watched as a Cheyenne party "swooped down on us and captured about seventy head of stock, mules and horses," Sharman recalled, "in spite of the soldiers who were supposed to protect us." A team of woodcutters in the Laramie Hills was attacked and driven off, as was a similar party on the Laramie Plains. Three stagecoach stations were burned, their stock taken. Farther west, at Rock Creek on May 14, a war party bushwhacked an engineering team led by Percy Browne, killing one soldier, and leaving a surveyor named Stephen Clarke for dead after riddling him with arrows, scalping, and mutilating him; Clarke was found alive by his friends before the wolves arrived, but died from shock in camp the next day.

To General Sherman, Dodge questioned whether the army's departmental commander, General Augur, had either the manpower or the judgment to protect the thousands of civilians along the line. Augur had announced he was leading an expedition north into the Powder River basin as a sign of determination lest the Sioux grow too bold there. But what about the railroad, the mail route, and the telegraph? "The mail will stop unless Augur will protect the stations," Dodge warned. "You know men will not run those routes with scalping Indians along them, unless troops are there to protect them. And we can not hold our men to our work unless we have troops, and Augur can not furnish them even after the road is built. Our station men will not stay at the tanks and stations, twenty miles apart, unprotected." The great difficulty, he wrote, was that Augur had only seven or eight companies of infantry to protect some three hundred miles of opened work, and only two companies of cavalry to patrol the whole line. "My engineering parties are driven into [Fort] Sanders, and Augur says it is now impossible to increase their escort, and they are working in the worst Indian country you have got." Augur should forget Powder River—his command ought to focus on the line, urged Dodge, and let General Hancock take care of the Cheyenne and the Sioux on the Platte and Smoky Hills. Dodge had "smothered all of the recent attacks and kept them out of the press," he said. "Augur and myself only know it, but should our men get at the real truth they will stampede. Stage agents, telegraph men, emigrants, tie contractors, and railroad men of all description out there are pressing for protection."[36]

Dodge, now convinced that the raids were part of a concerted action, had been suppressing the truth about the reopened war from his men while confined to his sickbed in Council Bluffs for two weeks, but he said he was determined to get out on the line; his letter was on its way to Sherman as the roving correspondent Henry Morton Stanley toured the wide radius of trouble from North Platte to Julesburg, interviewing Augur and his commanders, examining the army's stockades and counting heads, coming to a similar conclusion to Dodge's: it was a losing battle. "When the opportune moment arrives," he informed his readers back east,

> *from every sandhill and ravine the hawks of the desert swoop down with unrivalled impetuosity, and in a few seconds the post or camp is carried, the tent or ranche is burnt, and the emigrants are murdered. It is generally believed here that, if the present suicidal policy of the Government is carried on much longer, the plains' settlers must succumb to the unequal conflict, or unite in bands to carry on the war after the manner of the Indians, which means to kill, burn, destroy Indian villages, innocent papooses and squaws, scalp the warriors, and mutilate the dead; in fact, follow in the same course as the red men, that their name may be rendered a terror to all Indians.*[37]

Meanwhile, Dodge pressed Percy Browne and his engineers to return to the Laramie Hills, convincing him only by increasing his military escort to seventy men. "I think if they are vigilant active and will fight," he told Durant, "they can whip what ever comes against them." However, he added, "we must however expect loss of stock, perhaps some men this summer." Within a few days he had more grim news to back up this prediction. On May 25, a Sioux war party crept up on a surveyor's camp, killing one man and stealing the camp's thirty head of stock; another band rushed a train halted at the end of track, near present-day Brule, Nebraska, killing three men and wounding one, and rounded up thirty-one animals. East of there on the same day near the Overton station, a crew improving the track was attacked, with five out of six men killed and scalped. By then Durant had been alerted to the trouble and immediately wired General Grant in Washington. "Unless some relief can be afforded by your department immediately," he warned, "I beg leave to assure you that the entire work will be suspended."[38]

Dodge was able to go west when three government commissioners arrived in Omaha, expecting to look over the completed track before new federal bonds could be released. The inspectors included a Connecticut congressman, W. M. White, and two army generals, J. H. Simpson and Francis P. Blair; the latter, a former Missouri congressman, had commanded with distinction under both Grant and Sherman at Vicksburg and Chattanooga, and during the March to the Sea, and General Blair could still be considered their eyes and ears. He got an eyeful on May 27 out at Ogallala on the South Platte, where Jack Casement's men had spiked past mile 342. At noon, while all, workers and dignitaries alike, were on lunch break, their rifles stacked nearby, some one hundred Indians leaped out of a ravine. "In plain view of us," said Dodge, "they cut out several mules and horses and got away with them before the graders could get to their muskets." The commissioners, all three men of action, "showed their grit by running to my car for arms to aid in the fight." After the raid, which had seemed to take only seconds, Dodge lost his temper. "We've got to clean the damn Indians out or give up building the Union Pacific Railroad. The government may take its choice!"[39]

On that very day when General Blair and his fellow commissioners were persuaded in a lightning flash that the government had a crisis on its hands, Sherman, in St. Louis, responded to Dodge's pleas for help. He was, he explained, being "pestered from all sources" for troops—Kansas, Colorado, Dakota, and Montana, of course, but also New Mexico, Minnesota, and the whole South. General Augur had already been given "a very large fraction" of Sherman's command because of the importance of the railroad to national affairs, but Sherman promised to increase the troops "if I can get the men." However, Augur would still be dispatched north to go against the Sioux in the Powder River country. Finishing that unsatisfying letter, Sherman wrote to Grant to tell him of Dodge's worries. They were "a fair sample of what I get from every quarter, each party as a matter of course, exaggerating their difficulties and necessities and underrating that of others." But all could see that the region badly needed more soldiers, and Sherman was loath to call for volunteers: "it would stampede the whole country." He would immediately journey out to the troubled work camps and stockades to see things for himself. "I will go up the Platte," he wrote Grant, "& find out how the Indians can be so *universal*." As he headed west, though, newspapers in Omaha and other western towns began to pick up word of the slaughter of civilians out on the line, and their dispatches were reprinted in the East. Ultimately, as a political result, the Powder River campaign would be canceled.[40]

Scares and panics—real and imagined—became common across the region, even in bustling North Platte, where Samuel Reed recounted a false alarm on May 27: "Mrs. Casement & all the ladies in the North Platte house were badly frightened Sunday evening by some drunken rowdies raising an indian scare. Some men were as badly scared as the ladies."[41]

Other alarms were genuine. West of the tracklayers in Lodgepole Valley, beginning at dawn on June 2, surveying parties along a twenty-mile stretch were attacked. A young Iowa surveyor named Arthur N. Ferguson was roused by the cry "Here they come, here they come, boys!" and everyone in the tents scrambled out of their sleeping bags, grabbing rifles and bits of clothing and emerging to see a mounted raiding party cantering down upon them from the northern bluffs. "Some of our men were almost naked," Ferguson recorded in his diary. "I had nothing on but shirt, drawers and stockings." Rifle fire erupted all through the camp, wreathing it in blue smoke. "Here I saw one of the most daring Indians that I ever saw or ever heard of," said Ferguson. "He had dismounted from his own horse, and was endeavoring to pull up a 'picket pin' to which was tied one of our horses, and while trying to do this he was subjected to the fire from the muzzles of at least ten guns, and that within only two or three hundred feet of him." The raider succeeded "and in the meanwhile escaped as by a miracle from our shots. Now he springs to the back of his horse and endeavors to ride up the steep bluff immediately in front of us, when all at once a bullet strikes him, he sways to and fro in his saddle, as if the force of the bullet came very near to dismounting him; another shot strikes him, he reels and falls to the ground." One of the whites ran up "to dispatch him," but he veered away when warned that another raider was drawing a bead on him. "By this short delay we lost the chance of getting a fine scalp," Ferguson said, "and the other Indians got this wounded man away from us." All of the surveyor's livestock but four head were driven off. One of the engineers in the Lodgepole Valley "captured a white woman's scalp," Ferguson recorded, indicating that he had seen enough scalps from both sides to be a good judge of them. "It was quite green, having been killed but a few days, to judge from the appearance of it." From then on, whenever he stood at his instrument he had a rifle lying at his feet.[42]

Heading out toward the trouble, General Sherman got to Omaha only to be confronted by the sight of a party of junketing U.S. senators and their wives, who were about to climb on the same train to go out to the work. The Lincoln Car had been hitched up, presided over by none other than Dr. Durant, smooth as usual as he glad-handed his illustrious guests, and Citizen George Francis Train. Sidney Dillon was there, too, nervous about what Durant might do so soon after his defeats in the boardroom. Hub Hoxie sidled up to make conversation with the war hero and, anxious to make flattering conversation, he told Sherman that the railroad was planning to name one of its stations after him. "Where is it?" the general wondered. "Down here in Nebraska," replied Hoxie. "Oh, I don't want a water station named for me," Sherman grumbled. "Why, nobody will live there. Where is the highest point on the road?" Hoxie pulled out a map and pointed to a place in Wyoming, labeled "Altimont." Sherman grinned satisfactorily. "Just scratch out that name, and put down mine."[43]

They reached North Platte in a brisk twelve hours, which included one stop along the way while the senators were encouraged to take potshots at a herd of antelope grazing near the tracks. Samuel Reed happened to be on a passing Omaha-bound train and caught a fleeting glimpse here of Dr. Durant, who seemed "cheerful." During the stop for fuel and water at North Platte, concerned citizens surrounded Sherman to tell him about recent depredations. As if he needed "ocular proof," several even ran off to the cemetery to exhume some victims to display their wounds, but the bodies were too badly decomposed to be of use.

They moved on. About twenty miles east of Julesburg, the excursion train slowed to a stop. It was the end of track, for the moment, and the Casement brothers' men gave a demonstration of what they could do. Then there were the requisite senatorial speeches, and even the tight-lipped Sherman said a few words. Then it was Citizen Train's turn. He delivered a windy, jocular oration which set Sherman's teeth on edge. "Most of the frontier towns like war," Train told the audience. "It makes good trade; hence traders and military men become active. . . . Help me cheat the Indians and I will give you one half! The officer on small salary says 'extermination' and the war bugle is sounded." The general took his leave from the dignitaries soon thereafter, before he totally lost his temper at Train's ill-considered, half-baked ideas.[44]

Sherman arrived at Julesburg and Fort Sedgwick on June 6 and took stock of the situation. "The Indians are everywhere," he wired to the governor of Colorado Territory, who had been peppering him with entreaties for aid. "Ranchers should gather at stage stations. Stages should bunch up and travel together at irregular times." He was conferring with his commanders (including Custer, some weeks before his famous desertion to visit his wife) at Fort McPherson, one hundred miles east, when even luckless Julesburg, at that point merely an assortment of tents and a few frame buildings, was attacked, on June 10. About fifty Cheyenne galloped into the town in the early evening, meeting instant resistance. The shooting—and even the Indians' war whoops—could be heard inside the stockade at Fort Sedgwick, two miles away, and a cavalry company set off immediately. Major Henry C. Parry grabbed his surgeon's kit and followed, reaching Julesburg after the raiders had been repulsed. Parry treated five men who had been wounded by arrows.

"I never saw an arrow wound before," he noted, "and regard them as worse than a bullet wound." Two whites had been killed, scalped, and mutilated, while one Cheyenne had been killed and two wounded before being assisted off by their comrades. Major Parry recorded a gruesome sight: "One of the men killed was lying on the ground, pinned to the earth by an arrow through his neck," he wrote. "He must have been shot after he had been scalped."[45]

Back in Washington, the policy struggle within President Johnson's cabinet continued. Interior Secretary Browning had been absent from his office and from cabinet meetings during his month of illness, during which time peace advocacy had taken a beating. After one angry session on May 28, the still-feeble Browning emerged, complaining about Stanton and his generals: "The War Department seems bent on a general war and will probably force all the Indians into it." The commission sent out to investigate the Fetterman battle had yet to be heard from, but in weeks it issued the report, signed by three generals, including its chairman, General Alfred Sully, a West Pointer who had been a cruel campaigner against the Sioux in the high plains before the Civil War, and who had helped widen the trouble into the Dakotas in 1863 during the Minnesota punitive expedition. But Stanton and Sherman and the others would not like what Sully would say now: Hancock's aggressions against the Sioux and Cheyenne in the southern plains had inflamed the entire plains area to the point that the commission could not make peace with the northern tribes. The best course, Sully argued, was for the government to give up the Bozeman Trail as unnecessary, concentrate the army along the Platte route, and earmark eighty thousand square miles of terrain from the upper Missouri to the Yellowstone basin.

Sherman's characteristic retort was to support the general he had sponsored: Hancock's forceful actions in Kansas in the spring had actually improved a dangerous situation, he claimed. And if they rode away from the Bozeman forts it "would invite the whole Sioux nation down to the main Platte road." The reservation idea, however, was something he himself had urged.[46]

Meanwhile, the railroad workers needed support. During his inspection trip Sherman had made particular note of one of General Augur's units, organized a few months earlier, which would escort him at various times, and which had the express mission of protecting the Union Pacific and warring against the Sioux and Cheyenne. With the exception of its commander and a handful of officers, it was made up entirely of Pawnee from the camps along the Platte—and the recruitment took advantage of the long-standing emnity between the tribes. Pawnee hunting territory, occupied before the tribe's recorded history but according to traditions by conquest, spread across the Platte Valley, particularly around the Loup Fork. Pawnee were already there when the Sioux arrived, their lands bounded on the west by the habitations of the Cheyenne and Arapaho, on the east by the Omaha and the Kansa and

Otoe tribes; and the friction between Pawnee and Sioux was instantaneous, even fiercer than the frequent clashes with Cheyenne. This, however, was generally not the case with Pawnee as European explorers and traders, and eventually emigrants, ventured onto the plains, although from time to time Pawnee would bother wagon trains or stagecoaches, begging or shaking them down for "presents," or even attempting to collect tolls at river crossings. But as white incursions across tribal grounds increased, the Pawnee had decided that cooperation was preferable to resistance, in contrast to their immediate neighbors.

Treaties in the late 1850s had concentrated the Pawnee on reserved land on the Loup Fork, where they lived in earth lodges and hunted along the Platte; there were numerous attacks on the reservation by Sioux, especially into the 1860s, the Pawnee complaining that they were, like the proliferating white settlers downriver, getting insufficient protection from the U.S. Army. When, in early 1867, General Augur ordered the creation of an army battalion of Pawnee scouts under the command of Major Frank North, he was continuing a tactic of using Pawnee to war against their racial brethren, and he was choosing an officer who had lived among them and who spoke their language.[47]

Frank North's life would become inextricably linked with the progress of the Union Pacific westward into hostile territory, and consequently with figures of great historical note. His life was also the stuff of dime novels. Colonel William F. "Buffalo Bill" Cody in 1867 was reducing buffalo herds for the meat tables of the Kansas line's construction crews, but after their association began he would say admiringly that Frank North was the best revolver shot, whether standing still or on horseback, whether shooting at running men or animals, that he had ever seen.[48]

"Probably no man in the Government service," wrote Grenville Dodge, "was ever more worshipped than Frank North by the Pawnees." Dodge and North had probably met as early as 1856 in Council Bluffs or Omaha, when Dodge was surveying for the Mississippi and Missouri, and when the sixteen-year-old North was traveling west. He had been born in 1840 in Plymouth, Ohio, and, after his father was frozen to death in a storm while working as a land surveyor on the prairie near Omaha, Frank North moved with his mother and brother, Luther, to Columbus, Nebraska. They were near enough to the Pawnee villages to pick up the spoken Pawnee language as well as the sign language of the Plains Indians. Both Frank and Luther (or L.H., as he was known) worked periodically as mail carriers while still in their teens, and farmed and drove freight teams, and one winter Frank earned about $250 by poisoning wolves for their hides, an old Indian and fur-trapper trick. During the war, Luther joined a Nebraska volunteer regiment and later was on the Sully punitive expedition against the Minnesota and Dakota Sioux; Frank became a clerk and interpreter at the Pawnee reservation. In the opinion of

George Bird Grinnell, chronicler of the Plains tribes, "Major Frank North was undoubtedly more conversant with the spoken Pawnee tongue than any other white man has ever been." In late 1864, having been commissioned a captain, he organized and led a company of Pawnee scouts, which joined General Patrick Edward Connor's Powder River expedition. The association worked. Grinnell would find the key to North's success as leader of the Pawnee scouts in "the unwavering firmness, justice, patience and kindess with which he treated them," such qualities a distinct rarity even in the treatment of allies at that time. "He never demanded anything unreasonable," said Grinnell, "but when he gave an order, even though obedience involved great peril or appeared to mean certain death, it must be carried out. Their commander was at the same time their brother and friend. Above all, he was their leader. Going into battle, he never said GO, but always COME ON!"[49]

Paternalism, the attitude of the times, had its different aspects. In his book about the North brothers, Grinnell recounted Frank North's bravery and toughness, which in life-or-death situations edged into a semblance of a cruel, angry parent. "He expected in his men the courage and steadfastness that he himself possessed," wrote Grinnell. "In one fight when he and his scouts were in the open, exposed to the fire of concealed enemies and some of his men showed a disposition to retreat, Frank North said to them, 'I shall kill the first man that runs.' No one ran."[50] Though only twenty-five during the Powder River expedition, North was called "Father" by the scouts. When his brother joined the company just before it was mustered out in early 1866, Luther found that his brother's "children" could be terrifically brave and loyal themselves. During a winter cavalry ride from the Platte down to the Republican River in southern Nebraska, Luther North's scout detail was attacked by a much larger force of about 150 Cheyenne; the bluecoats sped toward nearby Driftwood Creek, intending to make a stand. "But the best laid plans sometimes miscarry," recalled North, "and when we were about a half mile from the creek my horse slipped on some ice and fell. I landed on my head and was knocked unconscious. Those 10 boys stopped, dismounted and surrounded me with their horses and while 9 of them fought the Cheyennes, the other one rubbed snow in my face and finally brought me out of it, but not until 4 of their horses were killed and 3 more wounded." Finally they drove off the war party, but it was nearly daylight before the scouts got their injured officer back to camp. "My boys were so well mounted that they could easily have ridden away if they had been willing to leave me to my fate," he recalled, "but with odds of some 15 to 1 against them, they jumped off their horses, formed a circle about me, fought it out and saved my life—which took cast iron nerve. Is it any wonder that I have always stood up for the Pawnee scouts?"[51]

After they were all mustered out, and Luther had gone to Michigan to study accounting, in late February or early March 1867 Frank North received his commission from General Augur and wired his brother to join him and the scouts. "On the way back I was caught in a blizzard," recalled Luther North. His train was stranded in Iowa for eight days, and when he finally got to Omaha he found the Union Pacific was not running. "I saw a work train being made up to shovel snow," he recalled, "and hired out as one of the crew for $1.50 a day. We rode in box cars, and it took nearly 24 hours to reach Columbus. I made a quick decision to leave that railroad job for the more exciting adventures of military service; so went home without the usual formalities, but in the meantime Frank North had enlisted his 200 Pawnees and gone on to Fort Kearny. I joined him there a few days later—in March '67, the month I became 21 years of age; and my duties as captain of scouts began. The U.P. still owes me for shoveling snow 24 hours; or, maybe, I owe them something, as the distance from Omaha to Columbus at that time was 91½ miles, and the fare 10c a mile—over $9. Anyhow I concluded to call it square with the company!"[52]

Major Frank North's Pawnee Battalion was comprised of four cavalry companies, each with a captain, lieutenant, and commissary sergeant, all whites, and the two hundred uniformed and mounted Pawnee, armed not only with old army-issue muzzle-loading guns but with their own bows and arrows and knives. Two companies were sent to Fort Sedgwick to exchange their muzzle-loaders for the new fast and accurate Spencer carbines, and then dispatched into the hills to escort surveyors' parties; two were ordered to guard tracklayers and graders. Captain Luther North's company, detailed to the end of track at Ogallala in late May, was trackside at lunch on the day when Grenville Dodge and the government commissioners witnessed the Sioux lightning raid on the graders' livestock. "We started in pursuit," North recalled, "and overtook them a few miles north of Ogallala. They abandoned the mules, and in the running fight which followed, one of the Sioux whose horse had been shot was running afoot." North spied a bow-and-arrow-wielding scout named Baptiste Behale, of mixed Spanish and Pawnee ancestry, who had seen the running warrior and quickly dismounted. "Behale ran up to within a few hundred feet of the Indian," remembered North, "and shot the arrow, which struck the Sioux in the right side, ranged forward and the point came out of his left side, a little toward the front. The Indian stopped, took hold of the arrow, pulled it through his body, fitted it to his own bow and shot it back at Baptiste. After taking two or three steps he fell dead, without having touched Behale, though it was a very close call for him." Later North discovered that the slain Sioux of extraordinary courage was the brother of the Brulé leader Spotted Tail.[53]

As effective as the Pawnee Battalion may have been in helping the work continue, four companies could not be everywhere, especially far in advance of the large concentrations of graders and tracklayers. The surveyors were at highest risk, isolated and strung out in small parties across the most dangerous stretch of the route. One crew led by L. L. Hills was working from the

Lodgepole Valley in panhandle Nebraska toward Crow Creek, some 140 miles, with the Laramie/Black Hills rising ahead of them from the plains. James Evans's team was working along a fifty-mile stretch across the high grasslands west of Crow Creek; to the south of them, the snowy peaks of Colorado could be seen; northward, there was a succession of pine-clad ridges. The Laramie River and its overlooking stockade fort was their summer's goal. Percy Browne and his crew were given the heaviest responsibility: a demanding 275-mile run northward across grasslands—with the Medicine Bow Range rising to the west and the bluish Laramie Hills standing to the east—climbing westward then to higher sagebrush plains and drier hills, running next along the rim of the Red Desert and imperceptibly across the Continental Divide, past buttes and sandstone hills to Green River. Beyond, in the Wasatch Mountains, T. H. Bates and his men were exploring alternative routes. Day and night all of the teams were in a state of high anxiety. Game trails crisscrossed their paths. Scrapes from travois poles emerged in the dust. Every mile could be seen from a thousand hiding places. Even with lookouts, even with armed escorts in this remote terrain, it was all too easy for members of the engineering parties to be separated or surprised, with tragic results.

On June 12, L. L. Hills and his team of ten surveyors, guarded by six cavalrymen, were on horseback working east from Crow Creek and seeing to final details, the deep blue mountains at their backs. Hills and a helper inattentively allowed themselves to fall behind the rest by about a mile. All of a sudden a mounted Arapaho war party burst out of a ravine, cutting off the stragglers. Most of the Indians occupied the larger group of whites, who leaped from their horses and, returning fire, took cover; other raiders rushed in pursuit of Hills or his assistant, who had spurred off in different directions. After the Arapaho gave up their attack and disappeared, the surveyors searched for five hours for their comrades. The groaning assistant, when found, had been badly wounded by a thrown lance, and then trampled by his own horse; nevertheless, he was able to crawl away to cover. But Hills's body was discovered with five bullet wounds and nineteen bristling arrows. He was buried there. Late in the year, the young surveyor Arthur Ferguson would be ordered out to exhume the body for reburial at home, a practice taken seriously by the Union Pacific as their advance men were picked off.[54]

"There seems to be a fatality attending many of the best men on the road," a depressed Samuel Reed wrote to his wife back in Joliet after he learned of Hills's death. "Last year many a good man connected with the road in various capacities died in Omaha—you know what a narrow escape I had" with his bad bout of typhoid. And this season there were all of these Indian killings. On top of this, it seemed almost too much when his fellow engineer James Evans lost his wife to typhoid fever soon after she had left her husband and other company wives in pestiferous Julesburg to return to Omaha. Reed had seen the bereaved Evans as he went through Julesburg to the railhead, where a special engine waited to speed him to Omaha in time for the funeral. As he wrote this, Reed's thoughts naturally turned to his own small children, whom he saw three or four times a year, and he asked after them. He was soon to head off to the mountains himself—Dodge was going, too. And although he was quick to reassure his wife that he was to be escorted by Major North and his Pawnee scouts and about twenty-five other armed railroaders, his anxiety was, understandably, about as heavy as the early summer's heat wave.[55]

25

"They All Died in Their Boots"

The trouble was this: early June, mid-June, late June, the Sierra snow would just not vacate the summit. And they had more work than they had men. The crews could only get so far above Cisco before they were confronted by the compacted, glacierlike snows filling the cuts and frowning above them from higher elevations. Even the goodly amount of sunlight served only to make the snow more rotten, threatening avalanches. In early June, Judge E. B. Crocker wrote to Huntington to report they had doubts they could reach the Truckee River with rails that year. "Our force is not now increasing," he worried. There was other work to attract the Chinese, principally in mining, which was not as hard. "We have proved their value as laborers," he complained, "& everybody is trying Chinese & now we can't get them." In the spring, Charley Crocker had raised monthly wages from $31 to $35; they hoped that as the word got out, and as new ships arrived from Canton, the pace would pick up. If the snow would ever cooperate, they would think about transferring some of the main tunnel crews outside to work on the craggy approach to the summit, where appallingly difficult stretches of hard rock awaited. Though it certainly was June, the Associates were very aware that the season for outside work at seven thousand feet was short. If any work was left undone by the time of the next snowfall, "it will be in the tunnels, where the snow will not stop it."[1]

Then, on June 25, as if a switch had been thrown, laborers removing rock on the western heights between Cisco and Strong's Canyon, two of the hardest miles of their entire route, all threw their picks and shovels on the ground and dissolved into a sullen mob, shuffling back to their encampments. Construction Chief Strobridge's attempts to bully them back to work went nowhere. He immediately telegraphed the bad news to Charley Crocker in Sacramento: the Chinese were not satisfied with their raise to $35; they now wanted $40 per month with their days shortened from eleven hours to ten. Both Hopkins and the Judge immediately informed Huntington in New York.

"This strike of the chinamen is the hardest blow we have had here," reported Judge Crocker. "If we get over this without yielding, it will be all right hereafter." His brother was leaving immediately to attend to it. "Of course," said Hopkins, "if they are successful in this demand, then they *control* & their demands will be increased." When Charley Crocker confronted some of the "leading chinamen" and told them he wouldn't pay a cent more than $35—he would stop the work first—the leaders smiled and bowed and said they would advise the men to go back. But the workers stayed in their camps, deaf to all threats. "The truth is," E. B. Crocker lamented, "they are getting smart."[2]

As the talks continued, the Judge and Hopkins both urged Huntington to do something about supplying laborers from the East—the specter of once-placid Chinese agitating for their rights was almost too much for them. They even thought they had come up with a winning formula: import black freedmen to finish their railroad. Earlier in the year, the Judge explained, someone named Yates had approached Stanford with an idea: "a colored man, formerly a steward on one of the S.F. boats, & a very smart intelligent man, left here some time ago for the East. His plan was to get a large number of freedmen to come to Cal. under the freedmen's bureau, & under the aid of the Government, that is a sort of military organization crossing the plains. He stated his plans to Stanford & wanted to know if they could get work on the R.R. & he replied Yes." As if in a timely reply to their urgent labor trouble, Hopkins and Crocker had found an item in the newspaper reporting that Mr. Yates was now in Washington promoting his idea to round up five thousand men—he was even identified there as an agent of the Central Pacific—and if he had not yet approached Huntington, they said, find him and help him. "The only safe way for us is to inundate this state & Nevada with laborers," wrote Crocker. "Freedmen, Chinese, Japanese, all kinds of laborers, so that men come to us for work instead of our hunting them up. They will all find something to do—& a surplus will keep wages low. It is our only remedy for strikers."[3]

Two days after the initial walkout, workers struck all along the line from Cisco up, over, and down to the Truckee Meadows. In reply to Crocker's threats, their demands actually increased—to $45 per month and ten hours a day. Hopkins fired off an urgent wire to Huntington: Help that man Yates! As he explained in a follow-up letter, "a Negro labor force would tend to keep the Chinese steady, as the Chinese have kept the Irishmen quiet." In the fullness of time Yates's remedy would fail as emphatically as the Judge's theory of inundation—though in another week their most urgent problem would be solved. Charley Crocker had begun to suspect that his heretofore faithful Chinese had been "instigated by Chinese gamblers & opium traders, who are prohibited from plying their vocation on the line of the work," and later he would decide that outside agitators working for the Union Pacific were behind it, as if there was no room for a legitimate grievance—but whatever the cause he was bound not to give in to them.[4] His solution was cruel but effective. "Their agents stopped supplying them with goods & provisions," reported the Judge,

> *& they really began to suffer. None of us went near them for a week—did not want to exhibit anxiety. Then Charles went up, & they gathered around him—& he told them that he would not be dictated to—that he made the rules for them & not they for him. That if they went to work immediately they would remit the fines usually retained out of their wages when they did not work, but if they refused, he would pay them nothing for June. They tried hard to get some change, as to the hours for a day's work, or an advance of even 25¢ per month. Not a cent more would he give. The great majority gave right up when he was firm, a few threatened to whip those who went to work & burn their camps—but Charley told them that he would protect them & his men would shoot down any man that attempted to do the laborers any injury. He had the sheriff & posse come up to see that there was no fighting.*

Thus confronted by the sight of the huge, clench-fisted Charley Crocker; the evil-looking, eye-patched Jim Strobridge; a strange man with a silver badge, and, behind him, an armed, deputized mob of whites, the hungry Cantonese strikers were, as the Judge said, "glad to go to work again."[5]

Indeed, everyone was relieved to get to work. The Associates had planned to have a dramatically increased workforce across the mountains in June, but here it was into July and they were just getting under way, really, and with half as many men as they wanted. But matters quickly turned around—perhaps something like a strike had been needed to rally not only labor but management and concentrate everyone's focus. "Since the strike the men are working hard & steady," the Judge wrote Huntington on July 10. "There is a rush of chinamen on the work—most of the fresh arrivals from China go straight up to the work. It is all life & animation on the line." The men were "thick as bees." Rock and earth work in the Truckee Canyon was rapidly finishing, and crews were being transferred back up to the summit. Both Jim Strobridge and Charley Crocker were encouraged to resume predicting that they would lay rail to the Truckee by the season's end. Moreover, in the second week of July the company was finally able to act on a notion prevented by the snows for the six or seven months since first contemplated: it began to ship heavy materials across the summit by ox-teams, including iron rails, components for platform cars, and, most dramatically, a locomotive. It had been owned by Charles McLaughlin's Western Pacific and would soon come in very handy. "It will be set up there immediately," noted the Judge in a flush of confidence, "& by that time the Iron will begin to move over. The grading and bridging on that 24 miles is well under way."[6]

Two weeks later, E. B. Crocker went up to see the operation for himself, the first time since the spring. He was glad to report that the laborers were working well and seemed to be satisfied with their wages, and that men were arriving from China in large numbers—they were initially ushered to the lighter work to acclimatize them to high-altitude life. With the retreat of snow, Cisco had become "the liveliest place in this State": long freight trains pulled into the yards at the rate of fifty to eighty cars a day, the periphery crowded with wagon teams loading up for Nevada, Idaho, Salt Lake, and Montana. The freight business had become so demanding—and so lucrative for the railroad and the Associates' wagon road through Donner Pass—that they had been hard-pressed to squeeze in any iron shipments for the eastern slope. But a wagon master with one hundred head of stock appeared, and contracted to deliver two thousand tons over immediately, which was two-thirds of their requirement. It now appeared that they would lay fifty miles of continuous track by the year's end, not only to the Truckee but out of the mountains onto the open plains beyond. The plan was then to take ten thousand tons of rails over to supply crews all the next winter. "By the opening of next Spring," the Judge wrote, "I have no doubt we shall be laying track East of the Big Bend of the Truckee, as fast as any road was ever laid."

The Judge watched from a safe distance as the men whittled down the deep, heavy rock cuts between Cisco and the summit, with the mountain appearing more cooperative once the blasters figured out a system. "The rock is full of seams," Crocker noted. "The men work the earth out of the seams with long hooked Iron rods, & then a keg or so of powder is fixed in them, which cleans out & opens the seam. Then 10 to 20 kegs are put in & the explosion sends the rock flying clear & out of the way." The company was using up three hundred to four hundred kegs of black powder per day. As soon as an open cut was finished and the graders had moved on, carpenters swarmed in to erect the framing for snowsheds. Other squads were at work below at wood shops at Emigrant Gap, framing two large bridges for placement seven miles above Cisco. A. P. Partridge, a member of this detail, recalled that they split their timber with handheld whipsaws, producing lumber "16 by 18 and 20 inches, and from 40 to 60 feet long." Then the carpenters were sent up to meet their handiwork as it was hauled in by ox-team, whereupon they framed it up in anticipation of raising the bridges in October. And then they were dispatched to the Truckee, where more bridge sites awaited them.[7]

All of this momentum naturally cast the Associates' thoughts eastward—to the passage across Nevada, and beyond those mountains, to furnacelike Utah and the natural exit through Weber Canyon. In early June, Butler Ives left to lead a party of sixteen surveyors to run a preliminary line from the head of the Humboldt River to the mouth of Weber Canyon, by way of the north end of Great Salt Lake. He had noted the line in the summer of 1866. "It is a new route, never explored before," he proudly wrote his brother William Ives,

"& the best through that desert country." By July first the team had worked its way up the Humboldt and ridden through the brown Pilot Range—and paused reflexively at the edge of the white-hot Utah salt flats, mirages shimmering ahead. They got their bearings and commenced.

Ives hoped to make from ten to thirty miles per day, and it would not be pleasant. For 150 miles they would have to haul water with mule teams to their "dry camps," and often the water was barely drinkable. Ives, who by then had become a bona fide desert rat, would find it "a good place to take the *romance* out of some enthusiastic young engineers I had in my party." He fully expected that they would not see a white settlement until the first of September, at Salt Lake, and beyond there was the wild Wasatch with its small scattered ranches. "We cross through Indian country all the way," he wrote. It would be, successively from the Humboldt to the Weber and beyond to Green River, the lands of the Northern Paiute, the Gosiute, the Shoshoni, the Crow, and the Cheyenne. He said he anticipated no trouble with them, "for we are well armed with Spencer and heavy rifles & are all pretty good travellers, when there are any *hostile indians behind.*"[8]

Judge Crocker, too, had been thinking about Utah in a big way all month, particularly as Butler Ives's reconnaissance reports from the previous summer were being copied to send to Huntington in New York, and he was anxious for Secretary Browning to support the northern route around Salt Lake. It was the only feasible one—south and southeast of the lake was a morass of sinkholes and mudflats, providing an effective barrier against an approach to Salt Lake City. "It may be that Young would prefer to have the road keep north," Judge Crocker mused, "as less dangerous to his harem, & system generally."[9] But it was important to try to gauge sentiment in the Mormon capital. Utah's Governor Durkee, in Sacramento on business in early July, promised Stanford and Crocker all the assistance he could manage. "He is one of our sort," the Judge observed. And Crocker dispatched a Sacramento businessman named Richardson to Salt Lake City on a sort of quiet political reconnaissance; he was "thick with all the stage men, & could find out a good deal through them." He was also "a shrewd businessman, & is pretty well posted about our plans of construction."

By midmonth Richardson had begun sending reports to both Crocker and Huntington, noting that Grenville Dodge had been expected to be paying court to Brigham Young by then, but had probably been "detained by Indians" in Wyoming. Having beaten the Union Pacific in that little race, Richardson had acquainted himself with one of Young's top advisers, a bishop, and expected "to find out through him what is going on." Within a short time, Crocker found out that the Union Pacific had not yet made any arrangements with Young; he would, moreover, form the opinion that Young was blissfully unaware that the railroads intended to bypass his city, or if he had heard rumors to that effect, he could not believe them. As a diplomatic gesture, as if to soften an approaching blow, the Judge sent a full set of stereographic views of the Central Pacific's Sierra work directly to Young himself. "He was highly pleased with them," Crocker told Huntington. "He had some from the Union Pacific, but they did not compare with ours."[10]

From the newspapers, and from Huntington's spy in the Union Pacific camp, they all knew of the boardroom turmoil which had been simmering for more than half a year now in the Union Pacific. He wished that a clear policy would emerge from the new men there. "If we could agree upon a point on Weber river near the mouth of Weber Canon," he mused to Huntington. "They ought then to agree to join with us in asking the Sec. of Interior to approve our location around the north end of the Lake."[11] This expressed, his conciliatory mood lasted exactly six days, until July 11, when he received a dispatch from Huntington. It said that the Union Pacific was leaking rumors that when they got within two hundred miles of Salt Lake they would send on a force some three hundred miles in advance of their completed line, "so as to preoccupy the ground & then take their time in building the road between." Crocker was livid. "Now I don't expect either they or we can gain anything by a trick," he shot back to Huntington. "Congress I think would promptly put a stop to anything of that kind. But even if Congress did not intervene it would avail them nothing." It was clear to Crocker what their response would be.

> *We will, of course, go right on in good faith, grading a reasonable distance in advance to keep the tracklayers steady at work. If in so doing we should meet a company of Union Pacific graders, or a portion of road graded by them, not connected with a continuous line, we would keep right on, grade our road right alongside their grading until we should "meet & connect" with their continuous line of track, get our Commissioners to report on it, & draw our Govt. bonds in the usual way. They would throw away some money on grading (which however might perhaps be used when a double track should be necessary) & would feel rather cheap besides.*[12]

That is how Crocker planned for them to deal with it. He must have had a crystal ball.[13] Prescient or not, he was ready for combat. Huntington had procured copies of the Union Pacific's engineering reports, and the Judge had studied them carefully. He saw that they declared that for the purposes of their government subsidies, the Rocky Mountains would be considered to commence at Crow Creek in the Laramie/Black Hills. He had traced their proposed route across maps and compiled a detailed table of distance and altitude, and he could see that Dodge's easy approach into the hills was only the beginning—the high but mostly level plains waited beyond. If the 1862 act approved a subsidy of $48,000 per mile for the "most mountainous and difficult" terrain (he quoted the pertient legislative section in his letter to Huntington), then it was "apparent to the most stupid" that the "most mountainous and difficult" region would not be reached until twenty or thirty miles past Green River—at the Wasatch (which could as much be considered a spur of the Rockies as the Laramies), threading over to the mouth of Weber Canyon in Utah. To compare the arduousness of what the Central Pacific was doing in the Sierra Nevada with what the Union Pacific proposed to do in the Laramie/Black Hills was a joke. But to make trouble for the Union Pacific, they could consider contesting their higher level of subsidies. And in any case, wrote Crocker, since the Central Pacific would claim "a direct interest" in the Wasatch mileage, it "would justify our appearing before the Sec. of Interior & the President, & contending for it."

With aggressive arguing, they might have the law on their side. And there was the additional reassurance of climatology. Crocker told Huntington to look at the table of distances and elevations he had prepared. For the Union Pacific, all the way from the foot of the Laramie/Black Hills to the mouth of Echo Canyon, a distance of 452 miles, "they are way up in the clouds, ranging from 5,535 to 8,242 feet high." Between the first of October and June, if not July, they would not be able to grade on account of snow. This left only three to four months in which to work. Construction crews would have to be discharged and then reassembled, costing more time. "I tell you," he assured Huntington, "we have got the Union Co. as tight as bricks—for all the way from Truckee to Weber Canon we can work at grading all the year round—never more than from 6 inches to 1 foot of snow, & then it don't stand long. We have got them whether we reach the Truckee this year or not. But we are going to make a desperate push for it, & I don't despair of doing it, for the men are coming on at the rate of 200 per day & the work is fast being covered up. So hold fast to Bridger's Pass as the place of meeting."[14]

Both Crocker and Huntington continued to be highly attentive to the company's relations with the press; the quiet arrangement brokered by Stanford with the Associated Press correspondent in San Francisco was still in effect, and the Judge had written at least twenty short releases about the road and its business, although he had no assurance that any of them had been wired to the East. He urged Huntington to send him any clippings he found.[15] However, at home the Central Pacific was confronted with a serious breakdown in relations with the *Sacramento Union,* which had faithfully supported every effort since the days when Ted Judah had been scouring the Sierra for a feasible railroad route and had finally convened the historic meeting at the St. George Hotel. Editor Lauren Upson, Judah's old friend, was still willing to give the railroad the benefit of the doubt. But he had been overruled by James Anthony, publisher of the *Union.* "Anthony is at the bottom of it all," the Judge wrote to Huntington. "It is all narrow-minded jealousy. He can't bear to see us prosper, especially as we don't drink & play cards with his crowd." After the big workers' strike, the *Union* had published an article riddled with what Crocker defined as "errors," but when they had complained the publisher refused to retract or print the Central Pacific's correction. "It don't hesitate to tell the dirtiest, meanest lies about us," the Judge complained. Years later, Charley Crocker had several other theories about why the *Union* turned against the Central Pacific—all personal, all petty, all regarding Anthony. It seemed Anthony felt that the Central Pacific should give him a greater share of its publishing of reports, broadsides, posters, even stationery and tickets; also, the railroad's news agent had angrily jousted with Anthony over credit for returned, unsold newspapers. But the last reason was best. "Jim Anthony . . . went out hunting," Charley Crocker recalled, "boarding a train with his dog. The rule of the company was that dogs and guns should be taken to the baggage car, but he did not want to have his dog taken there. The conductor, however, took him in nevertheless, treating him just as he would any other passenger. Mr. Anthony took great offense—and immediately the *Union* changed its tone, and assaulted the company in every way possible."[16]

The acrimony between the railroad and the *Union*—and, for that matter, the *San Francisco Bulletin* and the *Fresno Bee*—may have had many personal undertones but at base it was political, a decrying of corporate influences in the statehouse as well as the Capitol in Washington. Anthony had joined an alliance with the Central Pacific's other former supporter, Aaron Sargent, and with old-guard Republicans (that is, the Know-Nothing and Whig factions) in the state and "made war" on Senator John Conness and others. "We did this," Anthony explained later, "because we thought that these nominees had secured their nominations by improper influences, and were moreover pledged to railway measures which would burden the state with debt and greatly increase taxation of the people, without offering anything like corresponding benefits. We thought, in fact, that they would be more the guardians to great railway corporation's interests than of the people's." He threw his support, temporarily, to the Democrats, as a disagreeable alternative to "the railway radical candidates." Hopkins was worried about this splintering of the California Republicans—it put the state in danger of being delivered "into the hands of the disloyal Democracy."[17]

Whatever the reason or reasons, it hurt to be attacked on the Central Pacific's turf. And Jim Anthony was immovable. "He is a man of strong bitter prejudices, & being rich, he can afford to indulge his feelings & spit it out through his paper," the Judge would pronounce. "My plan is to buy him out as soon as the money can be spared."[18] But though it had become a constant irritant, rubbed in by friends at every social occasion, actual damage seemed to be minimal, especially in terms of business receipts and prospects. June earnings of $122,000 promised to be overshadowed by July and August; as Strobridge's crews now seemed capable of finishing the short, arduous stretch between Cisco and the summit, some $320,000 in federal bonds would become available; if they could finish the Summit Tunnel, lay track through it, and link with the eastern slope rails, another $1,152,000 in bonds would be released to them. Fisk & Hatch was advertising Central Pacific's California bonds at a very encouraging 95 percent of par in *Poor's Manual.* Six locomotives had been shipped from the East in April, and eight in May, all of which would be instantly put to work. The Associates had moved as soon as the Western Pacific acquisition went through and formed a new contracting partnership for it, comprised of just the five of them, calling it "The Railway Construction Company"; some months later the idea would be put into a formal corporate arrangement. And there was the prospect of real estate sales: 12,500 acres of federal land patents for the Western Pacific would arrive soon, although McLaughlin would have to be satisfied. The Central Pacific lands sold by August would bring in nearly $68,000—at an average bargain price of $2 per acre.

A raw and lively little town was springing up on the craggy, pine-shadowed banks of the Truckee, from which it would draw its name. And Stanford had gone over the mountains in late July to the Truckee Meadows, where he would look over a natural division site between the steep Sierra slope and some low brown hills, across the river from the trading post of M. C. Lake. For now the place would continue to be known as Lake's Crossing, but next year Charley Crocker would name the new town after the Union general Jesse Lee Reno, a West Virginian killed during the battle of South Mountain, whose many comrades from his Mexican War service had settled in the region and would urge the name on Crocker. Now, though, the Judge doubted there would be any future for a thriving town there beyond a few Comstock quartz-grinding mills built along the Truckee. "There are no big figures in town sites," he added, "because there is no country back of them to make a town." Reno, Nevada, with no big figures or future, it would be.[19]

It was a torrid summer everywhere. The Judge complained that "sweat like rain" ran down his face as he wrote in the Sacramento office, and Huntington complained that New York City was so bad that he longed to go into the country to find a cool place. But Crocker could lord it over his exiled friend, as Californians would always do—"we can take the Steamer & at San Francisco it is cool enough, or take the cars & in a few hours you can be walking over snow about the Summit." Indeed, as hot as it may have been in Sacramento, Crocker at least had the fresh memory of his five-day Sierra trip, even if it included a numb posterior from much horseback riding. A party had gone up to the head of the American River's North Fork to a nearby spring which had immense potential, he thought. "It is very fine mineral water," he reported to Huntington, "highly charged with carbonic acid, sprightly & pungent to the taste—clear as crystal, cold, & located in a beautiful valley surrounded by Summit Peaks, only 6 or 7 miles from the R.R."

The Sierra Nevada was beginning its next transformation, moving beyond the piled-up miseries of the Stevens-Murphy and the Swazey-Todd pioneer caravans, the frozen Frémont expedition, even the devastated Donners, building on the now-played-out auriferous era of Sutter and Marshall and the thousands of Forty-niners; the Sierra could be seen as a place of opportunity again—with their railroad, not a cruel barrier extracting its grim toll of lives, but a place where people might choose to go, and even linger. "It will, as soon as it becomes known, be a great place of fashionable summer resort," Crocker enthused. "The Summit is really the centre for people who desire to escape the fervid summer heats of the valleys. All around that point in every direction are points of great interest—& these mineral springs the most interesting of all. There is Lake Tahoe, & Donner Lake, close by. A few miles to the North are Lakes Weber & Independence, & all around are Summit Peaks from 9,000 to 12,000 feet high, easily accessible by trails, & even by wagon roads."[20] The tourism brochures of the future waited. In a few years Stanford and Hopkins would preside over the first of such area resorts: at Hopkins's Springs (later, Soda Springs), altitude 5,975 feet, genteel Victorian tourists could drink the healthful waters and take a short parasol-and-walking-stick stroll to view the prehistoric Indian pictographs of Painted Rock, with the once-frowning mountains standing all around them.

On August 3, at 11:00 A.M., deep in the west end of the summit excavation, after the smoke from a nitroglycerine explosion had cleared, Chinese laborers pulling out rubble from the site felt a strong current of air at their faces, which swept past them and up the central shaft. Lewis Clement lost no time in wiring Sacramento that they had broken through on the western end, and just as quickly the Judge's telegram was buzzing across the continent to Huntington. With the heading through in the west side, the bottoms could swiftly be dug and blasted out. Only sixty-five feet remained in the eastern end heading. E. B. Crocker left his desk and papers and sweaty office and rode right up to the summit. On the way to the tunnel, he told Huntington,

> *I was perfectly astonished to see the amt. done since I was there less than 3 weeks ago. The thing is done up scientifically. They work the rock up to a face, then go back 3 to 4 feet from the face, put in a hole 12 to 20 feet deep, fire it with powder, which is only powerful enough to "spring a seam," cracking the rock enough so that powder can be poured in. They then put in powder by the keg, from 1 to 50 kegs according to its size. The effect is to blow the greater part of the rock clear over the cliff & out of the way. It is a sight to see these heavy seam blasts go off. It makes the earth shake like an earthquake.*

It was beginning to look like a railroad there. The Chinese in the eastern tunnel face, though, were not progressing as quickly; the rock was harder, and the air got so foul in the tunnel that Clement soon withdrew the men working at the bottom so as to leave what little air there was for the crew drilling and picking out the rock in the heading.[21]

With the great encouragement of that first Summit Tunnel breakthrough, the tension was really on. Huntington pelted the Judge with requests for news, anxious to dislodge more bonds from Washington. On August 19, Charley Crocker came down to Sacramento, telling his brother that he could not predict when the tunnelers would get through—and, moreover, the pleasing pace in the open cuts below the western end of the tunnel had totally fallen off. "They have met with some infernal hard rock," the Judge told Huntington, "like Iron. Of course they are banging away at it as hard as they can."

Then, even with the slower work on the western gradient, at least on August 28 there was more good news: the light-headed workers in the eastern face of the Summit Tunnel had blasted through the last rock and were rewarded with a fresh surge of cool mountain air. "Toot your horn," Judge Crocker wired Huntington. "Locomotive on Truckee is in running order. Tracklaying commences Monday on the Truckee & will be finished through as fast as possible."[22]

Grenville Dodge, accompanied by several dignitaries, surveyors, two companies of Pawnee scouts, and two of infantry, moved westward past the place where, only three weeks before, L. L. Hills's party had been surprised by Arapaho warriors. About six miles west, following Hills's preliminary line on a large grassy plain to the place where Clear Creek trickled into Crow Creek, Dodge halted the column and they camped. They were soon joined by a Mormon wagon train of graders heading from the west, for railroad work; undoubtedly they were glad to see all the soldiers. The place Dodge selected was to be a Union Pacific supply depot and division point—someday to be quartered by the Pacific Railroad and its branch line ninety miles south to Denver—and Dodge predicted to those around them that the new city rising out of the plain would rival, even dwarf Denver. He called it Cheyenne.

First, Dodge and his crew began to lay out the streets to drive home that the Union Pacific had taken possession of this depot town. Just two miles north of town, the army would establish a post, Fort D. A. Russell, named after the Massachusetts General of the Volunteers, gallant hero of the Peninsula, Chancellorsville, and Rappahannock campaigns, killed by Jubal Early's artillery while defending Washington in 1864. The work of measuring and naming went only so far before it was time for a celebration—it was July 4, 1867. The combination of beautiful scenery and patriotism inspired one of their number. "Here on the rushing, clear waters of Crow Creek, flowing through a prairie adorned with beautiful flowers and rich, tall grass," he wrote, "with the towering heights of the rocky Mountains and the long range of Black Hills before us in the west and north, our national anniversary was not forgotten."[23] The railroad men and soldiers broke out libations for proper Fourth of July toasts. Of course there were speeches, cheers, congratulations.

The rest of that hot day and cool evening was spent depleting the spiritous supplies.

The next day, a Sioux war party attacked the Mormon wagon train. They had killed two men, driven off part of the livestock, and begun to ride circles around the encampment before the cavalrymen had so much as saddled their horses. The two Mormon emigrants were buried, their headstones to read "Killed by Indians," and they became, said Dodge, "the first inhabitants of that city."[24]

Dodge's expedition of nearly one hundred men, including the escort, had been necessarily planned for some time. He had been appointed land agent for the Union Pacific in late May, and despite the line being run through contested Indian territory there was the need to designate stations, junctions, railyards, and town sites; he was also to visit Salt Lake City in diplomatic fashion. The mission had taken on new urgency with Hills's death—there were numerous command-level engineering decisions to be made. As Dodge had made preparations in June to head west, it became known around Council Bluffs that he was going into the mountains instead of to Washington to represent his constituents of the Iowa's Fifth Congressional District, and some objected. Dodge, whose letters in late spring and early summer had referred to an unspecified illness he had suffered in May, now told his neighbors that he was suffering from an old war wound—presumably either the small pistol wound in his thigh inflicted when his own gun had discharged in its holster or the bad contusion he had received at Pea Ridge when knocked off his horse by a falling tree limb. The old wound, inflating even now in his telling, would serve many purposes, as his biographer has shown. Now it was keeping him from his electoral duties. Doctors had instructed him to recuperate in the Rockies, he claimed. "I regret very much that I have to go," he assured a newspaperman. "It is the first time in my life that I was ever away from my post, but there are plenty to do the work, and no close voting, and I consider it due to myself and my family that I should make an effort to get square on my feet again."[25]

Dodge would seem healthy enough during this long expedition, but not so his companion, Brigadier General John A. Rawlins, the adjutant to General Grant, who was suffering from tuberculosis and whom Dodge had agreed to include in his party at Grant's behest. Rawlins, a self-educated attorney from Grant's hometown of Galena, Illinois, and a possessive guardian of his commander's attention, was said to be the only officer who could influence Grant. Rawlins had a reputation for being irascible, but consumption had wasted him; he looked small in his uniform, already with the glow of the terminally ill, and he would be dead in two years. Rawlins brought along two Galena friends, one of whom was a newspaperman. And Dodge's party grew as it moved west: at North Platte he was joined by Samuel Reed, Jack Casement, James Evans, and, most disagreeably, Colonel Silas Seymour, whom Dodge was worried would meddle with his men's solid surveying lines as he once had with Peter Dey's. Seymour had already obtained maps and profiles through Dr. Durant's intercession, and he had made it clear that to him all routing decisions were still open, especially in terms of cost. Dodge winced at the idea of having to fight to resist Seymour's inevitable profit-seeking cost inflations. Colonel Seymour was escorting an Ohio engineer, Jacob Blickensderfer, who had been appointed by Interior Secretary Browning to determine the base of the Rocky Mountains; Dodge therefore had to keep a watchful eye on Seymour and also treat the government's appointee with kid gloves. Back in the East, even Oliver Ames had been uncomfortable about Blickensderfer's power and influence—over where the mountains began and the much-higher government bond rate kicked in. Ames had written to Dodge on June 13 to comment on "the very favorable aspect" of the Laramie/Black Hills line. "I am only fearful if the thing looks so very fine that it may influence the Judgement of our Mr. Blickenderfer in fixing the Base of the Rocky Mts."—the ease of ascent up Dodge's magic mountainside, and the easy work north across the Laramie Plains, might make the government contest the higher subsidy at Crow Creek.[26]

As preparations continued in Julesburg for Dodge's expedition, Casement's Irish legions spiked their rails through the buffalo grass and between dry, low, broken hills into town, barely pausing to acknowledge the settlement. "You can readily think how rejoiced we all were when we heard the shrill whistle of the engine, and saw in the dim distance its dark form come puffing toward us," the regimental surgeon Henry C. Parry wrote home. "Every cloud of its white smoke seemed to bring with it peace and civilization over the plains of the far West. Every ranch on the south side of the Platte has moved over to the railway side. Old Julesburg is no more, and a new Julesburg has been established." This marked, of course, the third "new Julesburg" to appear in that dusty corner of Colorado—there would even be a fourth, some years later—and the impermanence of it and the lives lived in it struck more than one commentator that season.

The unimpressed Henry Morton Stanley was there also, looking skeptically at the measly collection of tents and the half-finished restaurant of New Julesburg and reflecting on the life of the average shovel-wielding grader. For a Welshman of especially plain beginnings, Stanley had adopted the British-ruling-class view of the Irish with no vestige of irony. "Forty miles beyond . . . the careless fellow may be seen shovelling his contract, little thinking of the consequences resulting from his work," he wrote on June 25. "Step by step towards the Occident he throws the earth into a straight line over which will be placed the iron rails for the million travellers who will roll by without a thought of the navvies who toiled for them. The Irishman heaves his spadeful on the common heap, smokes his dudeen, eats his cooked rations at the 'shebang,' takes 'a smile,' has his petty quarrels, is reckless of the future, sanguine of the present, trolls out his carol day by day, and feels himself superlatively happy." Stanley turned away from his musings and from Julesburg, heading eastward for slightly more cosmopolitan St. Louis, passing the vanguard of a steady stream of freight trains bearing Julesburg's lively future: North Platte was packing itself up and relocating—Hell on Wheels was on the move.[27]

Dodge, Reed, and Casement would hardly recognize the town at the base of the bluffs when they returned. The expedition left Julesburg on June 28, swelled by the addition of the rich, smooth-faced young son of John Duff, who was being sent west to be toughened up, and by a geologist and a number of officers, including Dr. Parry. He had been detailed from Fort Sedgwick in such haste that he crossed the Platte on a raft at night, bedded down on the riverbank for a few hours and reached the main body as it was preparing to move out to the northwest. "Every man rode, either in wagons or on horseback," Parry recorded. "We marched rapidly. 'Reveille' sounded at 2 A.M. and 'unsaddle and go into camp' at 3 P.M. and sometimes at 11 A.M. Our course lay along the line of the railroad that is to be. Every ten miles we met grading parties, with their sentinels on the distant bluffs looking out for Indians, who frequently attack the graders."[28]

Samuel Reed had, like Dodge, developed a hearty dislike for Silas Seymour, and took pleasure at describing the colonel's pomp and posturings. "Col. Seymour was outfitted," he wrote Jenny Reed,

> *after the following style: First, the horse which he selected and paid a good round price for was, or ought to have been, the twin brother of old "Knockumstiff." On the horse he would have placed the saddle, attached to which was his carbine in its case securely strapped and buckled to be convenient in case of a sudden Indian attack; also his poncho, bed, etc., in bulk about a barrel, leaving very little room for the colonel. When mounted he would hoist his umbrella and leisurely follow in the wake of the escort or perhaps leading them a few paces. The Pawnees made fun of him from beginning to end.*[29]

Seymour may have presented a comic figure, but he still was capable of doing harm, especially after the expedition had arrived at Crow Creek and Seymour was left to prowl around on his own. Over the next several weeks Dodge was too busy looking after the line and creating the division town of Cheyenne to watch him. He had arrived only to find that L. L. Hills's men had done no more work after their chief was killed, leaving thirty miles of line to be located; this had the effect of putting him behind by three weeks. Afterward, with Dodge back in Cheyenne, down lines of wooden stakes in the grass would run the future avenues of Dodge, Reed, Ames, Dillon, Evans, Snyder, Seymour, and even Dey. Not surprisingly, there was no commemoration of Durant. Dodge kept 320 acres for railroad use—and the demand for land in the town began immediately, wagons trickling and then streaming in from points east on the intended railroad line but even more numerously from Denver, mightily encouraged by Cheyenne's prospects and by the close presence of the army stockade going up north of town. Putting up shacks everywhere, people had to be warned off railroad property by Augur's troopers. At an altitude of 6,100 feet, Cheyenne took some getting used to, even in midsummer. "In the middle of the day," Dodge recalled, "the sun burns, but at night I wore an overcoat and slept under two blankets."[30]

When the government engineer Jacob Blickensderfer went out on an inspection trip southward toward Denver, Dodge asked the seemingly idle Seymour to go with him. Seymour refused: it was, he sniffed, "none of his business to show him that country." Dodge had to spare a valuable assistant for the task. Then Seymour announced that he wanted to examine James Evans's route west of Cheyenne instead. Ah, now the trouble was manifesting. "Seymour is here for mischief and trouble—only finds fault," Dodge wrote to his wife. "I hope to get rid of him at Bridger Pass." It was irritating enough to get it on record back in New York, despite his shaky faith that the consulting engineer's alliance with Dr. Durant would discredit his interferences. "Seymour does nothing but complain of work done, lines &c. takes his ease and wants to be waited on," Dodge grumbled in a letter to Sidney Dillon. "I have no faith in his judgment. He does not know enough about the country to give a good opinion. He is first rate to criticize the work of others, but so far as any help to me, he is none." Dodge had lost his temper several times and while Seymour was gone wrote, "I expect he will pitch into me on his return, but I don't care; everybody along notices it and makes him a standing joke." Then, a few days later, Seymour rode back down to Cheyenne and boasted that he had found a better route than the Lone Tree Pass—it had a lesser gradient of eighty feet to the mile as opposed to Evans's ninety, and it eliminated, he claimed, the need for a high bridge across the Dale Creek gorge. Dodge had no respect for the man's engineering abilities, but when Seymour demanded a surveying party to confirm his figures, saying that he could save the railroad a quarter million dollars and was writing New York to alert the directors, Dodge had no choice but to give him assistance. Even the Ameses would listen to a quarter-million-dollar savings. He had his own schedule to follow, and although it worried him to leave Seymour behind to make mischief, Dodge put together his expedition to push westward fifty miles over the Laramie/Black Hills to Fort Sanders on the Laramie Plains; Reed and Casement and some others headed back toward Julesburg. It was July 21.[31]

Some 150 miles west beyond Fort Sanders, Percy Browne and his surveying party and escort were on the high, barren plain where the almost indiscernible Continental Divide ran somewhere through a basin of sage and cactus, beneath choking akali dust and sand. Leaving most of his team at the last known creek before the Red Desert, the engineer struck west with an assistant and five soldiers, looking for water and a good route. A colleague, Andrew Rosewater, would recall that "Mr. Browne was warned not to take these risks but he was full of ambition and zeal." Crossing near the southern rim of the Red Desert, an unpleasant expanse of sagebrush and greasewood where the changing angle of sunlight caused the red soil to shift colors as they rode, Browne's party advanced sixty miles in three days, in sight of remarkable sandstone buttes, always keeping watch for signs of Indians. On the morning of June 23, rising from his bedroll in a dry creek bed, Browne led the party down toward the aptly named Bitter Creek until midmorning, when he spotted a large party of Sioux in the distance. Raising an alarm, he led the men at a furious gallop toward high ground—but the warriors, three hundred in number, closed in. Two soldiers took charge of the horses when Browne and his assistant and the three remaining soldiers dismounted and formed a line, firing rapidly as they retreated toward the ridge. Mounted Indians galloped back and forth, yelling taunts and blowing whistles while others crept through the weeds, firing their rifles steadily. A furious half hour passed.

Abruptly Browne crumpled, screaming. He had been shot through the abdomen. He recovered enough to continue retreating a few hundred feet, but shock and blood loss got the better of him. He begged his companions to leave him behind after shooting him first, but they picked him up and struggled to awkward cover on the ridge, turning the horses loose in hopes that it would satisfy the Sioux. With Browne lying in agony, the men huddled around him, watching as the Indians returned after retrieving their mounts. The sun at its zenith beat down upon all, but the Sioux would not attack and finally rode away. Browne groaned at one point and whispered that he was sure to die—he had known only one man to survive a gut shot. He lay there while the others waited all afternoon and evening, but the warriors did not return. As night fell, they fashioned a litter with blankets and carbines and hauled Browne down the ridge to a road. Fortunately, after some distance, they encountered a wagon train, which carried them to the stage station at Laclede, fifteen miles away, but Percy Browne died soon after they reached it.[32]

Meanwhile, Dodge soon reached Fort Sanders, and it was there that he learned that the popular and capable Percy Browne had been killed. He "was one of the brightest of our young engineers," Dodge would write later, and his loss—which deeply affected the entire organization on all levels—was a severe blow to the engineering corps and to Dodge particularly. Browne, responsible for the division between the North Platte and Green River, was irreplaceable, and at this crucial time, Dodge reflected, "this put another party out of commission." From Fort Sanders he sent an urgent wire to Sidney Dillon. "I want answer to my dispatch to company about change of line and grade over Black Hills," he wrote, upset that his dispatches were not being acknowledged and that a proposal to improve the eastern grade from ninety to eighty feet per mile, at an additional cost of $200,000, was being ignored. Moreover, he had a sinking feeling that this loss of Browne might immolate his limited energies.

"I must push West. The Indians hold the country from here to Green River and unless I get out there, we will fail in all our plans for 1868. Brown, chief of party west of here killed yesterday after fighting and losing his stock." When the answer came, it was from Oliver Ames, and in effect, leaving sentiments about Browne to others, he simply told Dodge to try harder. "Durant and his friends are endeavoring to injure you and our standing before the country by calling us slow," Ames wired. In the name of speed, Dodge was to accept steepest grades of 116 feet to the mile, the maximum allowed by law, avoiding expensive and time-consuming rock cutting wherever possible. Later, after the line was completed, Dodge could have his milder gradients.[33]

It confounded the engineer in him to run on such a shoddy economy, but the politician in Dodge recognized that he would have to choose other battles at present—he still had to examine Evans's route to the North Platte, move farther west across Browne's division, pick up his team and egg them onward, and see to the canyon lines into Utah. It was a lot—perhaps too much—for one set of shoulders, wearied by the last few days. His party had crossed the hills to the Laramie Plains with no untoward incidents save several dreadful moments when, below Evans Pass with its altitude of 8,260 feet, the party startled a large herd of antelope "who seemed to be thoroughly frightened at the large body of troops following me," he recalled,

> *and instead of running away from us, they started towards us. I saw our danger and turned to the troops and told everyone there not to shoot at them and the men were so used to obeying orders that no doubt they saved us from being wounded. The antelope ran right among us so that the men could nearly put their hands on them. It was the most thoroughly frightened body of wild game I ever saw and it was the most astonishing sight to the party. I had seen antelope so frightened that they did not know what to do, but I never saw them run into troops, wagons, etc. There must have been one hundred of them. As soon as they had gotten away from us, I let the two hunters . . . go out and get two or three of them for our meals.*[34]

He had allowed himself one reconnaissance for pleasure, riding before dawn to the summit of the highest peak overlooking the pass, accompanied by Dr. Parry, a Pawnee squad, and several of the civilians. He could see Laramie Peak 100 miles north and Pine Bluffs 60 miles east; to the south he saw Long's Peak, 150 miles away; and 100 miles west, he saw the Medicine Bow Range, "with one half of the southern circle fringed with the ragged, snow-capped, Rocky Range, while the country east looked like a great sea." Dr. Parry looked eastward: "there was a faint tinge of orange color that gradually became yellow, then radiant with the rich golden hues of the rising sun," Parry recalled. "As the landscape and the cloudless sky lightened, the grass around us, bathed by the moisture of the night, sparkled like a sea of crystals, and quivered in the breath of a gentle wind." Both he and Dodge picked some flowers. "No matter how sterile or ragged this country," Dodge would write to his wife, "it is dotted all over with roses, leaves and grasses with singular stones and valuable mines."[35]

But on the far side of the hills, Dodge was confronted with perhaps the thorniest engineering problem on the whole line to date—the Dale Creek chasm. With Seymour's smug words in his mind, the saddlesore Dodge rode up and down the western slope from summit to the stream's edge numerous times over several days, taking figures on every possible approach, while others relaxed and fished for trout. On the hillside the wind had roared for three days, "a perfect hurricane," he noted, "and the sand and dust has filled every corner." Squinting down the line led him to one conclusion. "There was," he finally decided, "no possible way of overcoming this divide except by taking the heavy work at Dale Creek or else by long tunnels," the delays of which they could not afford. According to the notes, they would have to build a high trestle bridge some 1,400 feet long and 125 feet high over the creek. At least Evans's line and its ninety-foot grade would be free from curvature—making it easy to operate, which one could not say about Seymour's.

At Fort Sanders, then, reeling from the news of Browne's sacrifice, trying to make sense of his men's stacks of field notes and reports, learning in a letter received from Oliver Ames that Dr. Durant was poisoning the minds of the company's government directors, taking only momentary satisfaction from Ames's assurance of confidence in Dodge's judgment, the exhausted engineer wrote his wife on July 22 that he pined for fresh vegetables and fruit and that in all he did not bounce back the way he once used to. "I have," he said sorrowfully, "too much on my mind."[36]

Back in the East, Oliver Ames was in a similar state, with worries outweighing good news. For the first time, Union Pacific first-mortgage bonds had found a market, the success almost entirely due to the aggressive marketing campaign of Cornelius Bushnell. The thirty-eight-year-old financier's every prior business venture had turned base iron into pure gold, whether it was shipping between New York and New England, railroading in Connecticut, or even, most dramatically, military contracts—his shipyard in Fair Haven had backed, built, and relentlessly promoted the ironclad *Monitor*, winning after the *Monitor*'s victory over the Confederate's *Merrimac* off the Virginia coastline a generous contract with the navy. Bushnell was left, after the war, a fabulously wealthy and extremely energetic man in his early middle years, with a fortune at his disposal and connections everywhere. As an original director of the Union Pacific and as one of the three founders of the Crédit Mobilier, he had the most extreme motivation to see the railroad persevere—and make a little on the side. Given authority by the Union Pacific board in the spring to sell $10 million in bonds, a daunting task, Bushnell did exactly that—and in less than six months. "I went to work," he boasted later, "employed an advertising agent, and started advertisements in every leading paper in the Northwest and New England, and I sent travelling agents to every leading city." Driving up the price of the bonds, he was, by July, selling them at 95 percent of par.[37]

Despite the infusion of desperately needed hard cash, Oliver Ames could not celebrate—Bushnell was, from the beginning of course, Durant's man. And it seemed to him that tentacles were spreading everywhere he looked. The million-dollar loan which the doctor had negotiated at high terms with the banking firm of Clark Dodge and Company in late February had pended until May, after Durant had been turned out of the Crédit Mobilier board, when it became generally known that Clark Dodge was serving as a front for Cornelius Bushnell and others, chiefly the Delaware manufacturer and war profiteer Henry S. McComb, a Durant associate and original stockholder in the Crédit Mobilier. Also in May, the new members of the Crédit Mobilier executive committee, Rowland Hazard and John B. Alley, both of whom in any case despised Durant, had insisted that they could get better terms than Clark Dodge; thinking they had the support of Ames and the other directors, Hazard and Alley began to negotiate terms with other lenders. By late June, Oliver Ames found himself pushed into a corner. Bushnell and McComb demanded that the original agreement with Clark Dodge must be honored, and Hazard and Alley insisted that they had the right to dispense the loan. On June 21, Durant stepped out of the wings. His lawsuit to prevent the Union Pacific from writing a new construction contract was still tied up in the courts. Ames was desperate to get it moving. And the Doctor told Oliver Ames that he had to be readmitted into management. On June 21, Ames approved the million-dollar Clark Dodge loan. Its terms would balance out to a choking 16.5 percent.[38]

He could only have been thinking of placating Durant and Bushnell. But his move, of course, infuriated Rowland Hazard—who denounced the bestowal upon "an inside ring." A thick fight was promised for the July board meetings.[39]

July could be a dangerous month in a corporate boardroom, as Oliver Ames discovered. Many of the directors he counted on for their votes were away at resorts or touring Europe. For moral and political support, his brother, Oakes, came up from Washington, where Congress had reconvened—but the Ameses needed many more votes, and those votes were far away savoring the sunlight and the relaxing sound of water. The Union Pacific's five government directors were there in force. After two days of go-nowhere negotiations and tabled motions, though, Dr. Durant swept some of the government directors off to Newport for a sumptuous weekend. Ames knew it was for "corrupt purposes."[40] He could do nothing about it.

With the resumption of the meeting on July 16, Ames was confounded when a new contractors' proposal was placed before the board and the construction committee. It was for 667 miles—and it was submitted by Henry C.

Crane, who was Durant's secretary, and by John B. Alley, whom everyone had assumed was the Doctor's enemy. Then, as the resolution for approving the Clark Dodge loan was picked up, someone called a half-hour recess. When the directors filed back into the room, Henry McComb interrupted proceedings to move to compensate the government directors for their services—$5,000 to Springer Harbaugh, $3,000 each to the other four, including Charles Sherman, the brother of Tecumseh. The motion passed and the board adjourned. Oliver Ames saw what was up: "Govt directors evidently bought to support Durant's schemes," he noted in his diary.[41]

The next day, the Alley-Crane construction contract and the Clark Dodge loan resolution were both approved over Ames's meek dissent. But before Durant could get the meeting tidily adjourned, Charles Sherman seemed to have a change of heart, or at least he sensed a new angle: after moving to reconsider the Alley-Crane matter, he got the contractors' proposal tabled. And the board took the compensation of the government directors out of committee and then simply passed it.

In the days after that whirlwind meeting, the beleaguered Oliver Ames scurried to write a new proposal for a construction contract, intending to counter Durant's obvious move through Crane and Alley. But then the Doctor, who through Cornelius Bushnell stepped up his campaign for reinstatement into control of the road, edged in closer with an entirely new contractor's offer—in his own name. It even bid below the Alley-Crane terms. Ames could literally feel breath on his neck. For a few days there was some arm wrestling over calling a special meeting in mid-August, during which Durant with his superior votes would certainly prevail. To Ames's mounting distress, the Doctor's minions got the meeting on the calendar for August 15. Poised to bring a club to that conference, Durant arranged with a compliant judge for the hearing on his injunction against the company to be postponed until afterward.

Oliver Ames was no combatant. Despite saber rattling from the furiously upright Rowland Hazard, Ames called his brother up from Washington to fashion a compromise that they hoped would satisfy a majority—including, went the doleful reasoning, Thomas Clark Durant. During the afternoon and evening of August 15, and into the next day, in smoky conferences at the Union Pacific office and at the Fifth Avenue Hotel, the Ames brothers concocted a pie from which everyone seemed assured a piece. The Williams plan was set aside. The Alley-Crane plan was set aside. The Durant plan was set aside. Even the Crédit Mobilier was set aside. In their place, Congressman Oakes Ames became the contractor for 667 miles of road for a price of $47.9 million, with mileage compensation rising from $42,000 to $96,000. Details would be arranged later between Ames and a board committee. There were a number of problems the Ames contract did not immediately solve—Durant's injunction had yet to be lifted, and for the present he still had a commanding block of board votes; Rowland Hazard was still poised to sue Durant, and

Oakes Ames's months-long investigation into Durant's suspense account still inched forward. But money (or the prospect of it) was waved in front of, assuredly, everyone's nose, and the prospect of eventual subcontracts bought peace—for another month at least. Charles Sherman, his vote purchased for $3,000 and his loyalty for the assurance of as much as $20,000 of the new Ames pie, wrote to his brother the general to promise that everyone, including, many have assumed, General Sherman himself, would share in "a big thing." More would be clear in a month. But it was all so wearisome and tawdry, associating with those types. "Such a selfish, grasping set I never met before," he told his brother, "and therefore I do not rely on them."[42]

Out west, where even pretended scruples were not an issue in Julesburg, money and whiskey flowed freely. Samuel Reed had been in the mountains with Dodge less than a month before going back, and in the interim all of the raucousness of North Platte had rolled into town, and then some. "Julesburg continues to grow with magic rapidity," he wrote his wife on July 30, and

> *vice and crime stalk unblushingly in the midday sun. General Augur and staff returned here last Friday evening and nothing would do but they must see the town by gas light. I sent for Dan Casement to pilot us (I knew he could show us the sights). The first place that we visited was a dance house, where a fresh importation of strumpets had been received. The hall was crowded with bad men and lewd women. Such profanity, vulgarity and indecency as was heard and seen there would disgust a more hardened person than I. The next place visited was a gambling hell where all games of chance were being played. Men excited with drink and dally were recklessly staking their last dollar on the turn of a card or the throw of a dice. Women were cajoling and coaxing the tipsy men to stake their money on various games; the pockets were shrewdly picked by the fallen women or the more sober of the crowd. We soon tired of this place and started forth for new dens of vice and crime and visited several similar to those already described. At last, about ten P.M. we visited the theater and were asked behind the curtains to see the girls. From here I left the party and retired to my tent fully satisfied with my first visit to such places.*[43]

Henry Morton Stanley, still wandering about the plains, sent off a dispatch when he arrived at Julesburg. After a surprisingly tolerable dinner, he took a stroll through the streets. "[I was] really astonished at the extraordinary growth of the town, and the energy of the people," he wrote.

> *It was unmistakable go-ahead-it-ative-ness, illustrated by substantial warehouses, stores, saloons, piled with goods of all sorts, and of the newest fashion. As might be expected, gambling was carried on extensively, and the saloons were full. I walked on till I came to a dance-house, bearing the eu-*

phonious title of "King of the Hills," gorgeously decorated and brilliantly lighted. Coming suddenly from the dimly lighted street to the kerosene-lighted restaurant, I was almost blinded by the glare and stunned by the clatter. The ground floor was as crowded as it could well be, and all were talking loud and fast, and mostly every one seemed bent on debauchery, and dissipation. The women appeared to be the most reckless, and the men seemed nothing loth to enter a whirlpool of sin. . . . These women are expensive articles, and come in for a large share of the money wasted. In broad daylight they may be seen gliding through the sandy streets in Black Crook dresses, carrying fancy derringers slung to their waists, with which tools they are dangerously expert. Should they get into a fuss, western chivalry will not allow them to be abused by any man whom they may have robbed.

At night new aspects are presented in this city of premature growth. Watch-fires gleam over the sea-like expanse of ground outside of the city, while inside soldiers, herdsmen, teamsters, women, railroad men, are dancing, singing, or gambling. I verily believe that there are men here who would murder a fellow-creature for five dollars. Nay, there are men who have already done it, and who stalk abroad in daylight unwhipped of justice. Not a day passes but a dead body is found somewhere in the vicinity with pockets rifled of their contents. But the people generally are strangely indifferent to what is going on.[44]

Stanley had a ready prescription. "The only sure preventative of these murderous scenes," he wrote, "is martial law, or the ready strong arm of the *vigilantes.*"

Julesburg was indeed such an open town that the gamblers held a meeting and decided to contest the Union Pacific's right to the lots which Dodge had laid out and which were now being sold by a railroad land agent. They simply began building on them. The agent immediately wired Dodge, who was at Fort Sanders. Such vices as witnessed by Reed and Stanley were dwarfed by this outrage—and Dodge took the law into his own hands. "I immediately wired back to General J. S. Casement," he recalled, "telling him to take such of his force as was necessary and go back to Julesburg and clean the town out and hold it until the citizens there were willing to obey the orders of the officers I had placed in charge and pay for their lots. This was fun for Casement. I did not hear anything more in relation to what he had done there. When I saw him later, he said, 'I will show you what I did.' . . . He took me to a hill where there was quite a burial ground and he said, 'General, they all died in their boots and Julesburg has been quiet since.'"[45] Thievery and murder would continue in Julesburg, as Stanley would observe, but claim-jumping railroad land was no more.

While this vigilantism was transpiring in Julesburg, its author at Fort Sanders was preparing to head west; Dodge had been told, at least, that

Colonel Seymour would be returning to the East. "He is," Dodge complained in a letter to his wife, "the worst sneak I think I ever met; means mischief to our company—professionally dishonest and corrupt—and the quicker the company gets rid of him the better." Even Rawlins with his illness and his small entourage did not approach the bother offered by Seymour. "Gen. Rawlins has been of great aid to me," Dodge continued. "I believe he takes more interest in the road than Seymour, who seems not to care a damn whether the Indians get the road or not so long as he can play gentleman and have a big company to foot his bills. He will be my everlasting bitter enemy for I have taken occasion to give him some pretty strong licks at his actions; he cares about as much for our interests as we do for Japan and will sell any or all of us if he can better Seymour. He has some good ideas about engineering but the devil of it is no one knows when to trust him."[46]

Trying to put these cares behind him, Dodge and his expedition headed north from Fort Sanders up the wide, grassy Laramie Plains, with the 10,000-foot-high Laramie Peak rising deep blue and majestic to the northeast, and the Medicine Bow Range standing to dramatically define the western horizon. He had never seen such scenic country; "mountain piled on mountain," he wrote to Anne Dodge, "immense canyons, upland hills and perpendicular walls of stone." Some fifty miles from Fort Sanders, they crossed the Medicine Bow River—named because the tribes found mountain birch suitable for making bows along the stream (and frequently clashed with other, hostile tribes on encounters). They then followed Percy Browne's line veering westward through the Rattlesnake Hills. The pass, with an altitude of 7,124 feet, was named Browne's Pass in the engineer's memory. Dodge also indulged family pride by naming three mountains after his daughters, Ella, Lettie, and Little Annie. "Joined together like three little girls should be," he assured his wife, the peaks "will always be prominent points on the route." Every day, he said, they came across something new "to attract the attention, something to develop, and as I travel over it often think what scenery there will be from the cars."[47]

Beyond Percy Browne's pass and down the rock-ledged canyon of Mary's Creek, they reached the cottonwood-fringed North Platte, not far from the old campsite of John C. Frémont, who had briefly skirmished and then smoked the peace pipe with Cheyenne and Arapaho some twenty-four years earlier. (In a year the army would establish a post there named after General Frederick Steele of Delhi, New York, who had commanded alongside Dodge in Missouri and Arkansas and who had died of apoplexy after the war.) Dodge's camp, for two or three days, saw no evidence of Indians while the engineers combed the creeks of the Rattlesnake Hills and a downriver stretch of the North Platte for alternative lines; the loss of Browne and the disappearance of his field notes required Dodge's men to cover much of the same ground, though they found no route superior to the one they had initially traveled. While still at the North

Platte camp, the party was joined by Frank Appleton, Browne's assistant, who had taken charge of his men; looking for help, he had left them forty miles to the west—"unable to go forward for want of water," Dodge wrote, "with their horses gone, their escorts used up, and apparently with no alternative but to back out of the country that Browne was killed in while endeavoring to get a line through." They were, however, in good spirits, Appleton assured him, and Dodge felt assured that they "could soon be put on their feet again." He pushed westward, toward the Continental Divide, to aid them.[48]

A day later, Dodge grew concerned about the lack of water and ordered everyone to spread out across the sagebrush plain to search. Dodge and Rawlins, riding south, found a fresh, cold spring gushing from a rock. Rawlins tasted the water and called it "the most gracious and acceptable thing" he had ever found. He told Dodge that if the engineer were to name anything after him on this journey it should be a spring—not a pass or a mountain but a place like this. Dodge replied that this miraculous place in the desert would be called Rawlins Springs, as it was.[49] It would prove to be a godsend to the railroaders. Some miles farther west, Dodge's reassembled party found the rest of Browne's men at a stage station. They were, as Appleton had said, ready to get back to work. With supplies and fresh horses, they went off to finish the line eastward to Rawlins Spring and the North Platte.

The Continental Divide ran impressively over the Rockies past the famous Bridger's Pass and, northward at about thirteen thousand feet until it dropped some six thousand feet onto the Great Divide Basin where Dodge and his men stood—a dry depression nearly one hundred miles north-to-south, and two hundred miles east-to-west—whereupon the divide dropped another five hundred feet into a deeper basin and rolled on northward toward the Wind River Mountains. As Browne had discovered, the destitute plateau promised a relatively easy transit to the railroaders but offered its own set of problems—chiefly, water. Three days out across the Red Desert, the Dodge expedition ran out, staggering on two more parched days, finding only dry creekbeds—"shallow graves of deceased rivers," Dodge called them, though those courses could, during the region's rare torrential rains, wash out in the blink of an eye roadbeds which had taken months to build. At the edge of an undrinkable alkali lake, the crestfallen men rested while Dodge and Rawlins climbed a promontory to look around. Through his binoculars Dodge thought he saw Indians—but the distant travelers were leading teams of horses. It was the survey party led by Thomas Bates, pushing eastward from Green River.

They were, if anything, in worse shape than Dodge's group, sun- and windburned, blistered, with tongues badly swollen from nearly a week of no water. They exchanged notes and supplies, and Dodge sent Bates and his men to try another line north of Dodge's, hoping to find fresh water for a route. The main column continued westward. Past the dramatic sandstone mesa, Table Rock, across more sagebrush waste, they reached the strangely eroded gray sandstone ledges, Point of Rocks, looming a thousand feet above the alkaline Bitter Creek. Here Appleton was sent with an escort to examine the terrain southward; Dodge continued west between greasewood cliffs and sage hills toward the old Pony Express station of Rock Springs, until finally, 125 miles west of Rawlins Springs, they came to the place where they could look upward a thousand feet to Castle Rock and downward to the Green River in its valley. Sixty-five miles beyond, across more hills and sandflats and past the remarkably eroded sandstone cliffs called the Church Buttes, lay the worthwhile goal of Fort Bridger in its rich valley, the old corridor of the fur traders. Beyond there was the Bear River valley—and the climb to the "rugged, bold, and narrow" Wasatch Mountains. As strenuous as it had been, the Dodge expedition was far from over. And General Rawlins was still a boon companion, finding everything they saw worthy of exclamation, forcing the literal-minded Dodge to see the world with new eyes, reading poetry at night to the men by their campfires. "He is one of the purest, highest minded men I ever saw," Dodge would write to Annie, "and that he must die with that dread consumption seems too bad. This country when it loses him loses a great man."[50]

All during his ride from the Union Pacific railhead across the Laramie/Black Hills, the Laramie Plains and the Rattlesnakes, and the Great Basin all the way to Green River, Dodge had talked to two small parties that had been attacked by Indians—Hills's and Browne's—and he had witnessed only the one raid at Crow Creek. Dodge did not misinterpret this relative quiet, having fought off the earlier raid at the Casements' work camp on the South Platte; his large escort of infantry and cavalry undoubtedly had warned off any warriors from harassment. Elsewhere, though, the war continued to burn into late summer. Civilian traffic on the Bozeman Trail had slowed to a trickle, and rare settlers' wagon parties on this route were attacked, as were railroaders' teams on both the Kansas and Nebraska lines. If anything, these actions of smaller war parties were about to get larger, and more deliberative. In the earlier half of the season, tribal leaders had been divided as to a target for a concerted strike along the Bozeman Trail. At a large gathering around midsummer, during the Sun Dance ceremonies of the Lakota Sioux, there had been a furious debate over whether to strike at Fort Phil Kearny, to add another victory on top of the one seven months earlier against Fetterman, or to hit the lighter-defended Fort C. F. Smith, some ninety miles northwest of Kearny in southern Montana Territory, on the Bighorn River. When the Sun Dance was over, they had decided to do both: the Sioux would attack Fort Kearny and the Cheyenne would go against Fort Smith.[51]

It was the last post on the hated trail until the fast little mining center of Bozeman; Fort C. F. Smith had been named for a West Pointer from Philadelphia who had fought Mexicans and Mormons and who had won the affection of both Frémont and Grant; Charles Smith had coined Grant's famous phrase about "unconditional surrender" in the Tennessee campaign, but the warrior had quickly died of gangrene from a scraped shin in a minor, non-combat-related accident when jumping into a small boat. The ironic qualities inherent in this could not help but be attached to the little post on the Bighorn; during the winter, after the Fetterman fight, Fort Smith had been isolated and cut off from communication and there were widespread rumors that it had been overrun and all the inhabitants slaughtered; when a relief party had made it through the snows from Kearny in March, they found all at Smith to be blissfully alive, healthy, and unaware that their deaths had been exaggerated. Then, later in the season, the creator of the trail, John Bozeman, was killed on his way from Virginia City to Fort Smith, on a supposedly peaceable stretch of the route through Crow territory—and not by the Crows or the Sioux or the Cheyenne, all of whom he had inflamed with his creation, but by an unexpected wandering Blackfoot party.[52]

Fort Smith would be harassed all summer. Then, on August 1, some five hundred Cheyenne and Arapaho, with a few Lakota Sioux, approached Fort Smith across the tall-grass meadows along the Bighorn River. Sent, and possibly joined, by the leaders Dull Knife and Two Moons, the intimidating force encountered a soldiers' haying party three miles from the fort and attacked it. The soldiers and their livestock took shelter in a sturdy corral built out of logs, which afforded them effective shelter against the vastly superior attackers. For several hours the Indians' arrows and bullets found few targets, and a grass fire set outside the corral failed to dislodge them. A desperate Indians' assault on foot was turned back. Only a week before the soldiers had received their new Springfield breech-loading carbines; though they were single-shot rifles, their rapid fire and accuracy, which utterly surprised the raiders, finally forced the tribesmen to withdraw, having inflicted only three deaths and as many injuries among the whites. This engagement became known as the Hayfield Fight.[53]

Meanwhile, a force of eight hundred to a thousand Oglala Sioux, led by Red Cloud himself (and also by the important leaders High Backbone and Crazy Horse), were riding toward Fort Kearny. Scouts had informed them that a woodcutters' party was camped at their worksite some six miles from the fort. At 7:00 A.M. on August 2 the workers were already at their jobs, their previous day's cordwood on the way to Kearny, as an empty wagon train approached across a grassy plain. When the large force of Sioux struck, firing volleys of arrows, the two strings of wagons hightailed it for the stockade, leaving the woodcutters to take cover inside their own corral. Their commander was Captain James W. Powell—the officer who, eight months earlier, had been detailed to rescue the fort's haycutters under attack but who had been outranked and replaced by the eager Fetterman, who wanted to ride through the entire Sioux nation.

Powell's men were now sheltered in a corral constructed of fourteen sturdy wagon boxes, most of which were loaded with logs, ox-yokes, and sacks of grain and beans; the wheeled undercarriages had been removed to transport felled trees to the cutting camp. This afforded the twenty-eight soldiers and four civilians an excellent shelter, even when the Sioux began to shoot in fire arrows. The smoke from hay bales and manure piles made things uncomfortable, but the soldiers had also just been issued the Springfield breechloaders; they literally poured fire back at the galloping Sioux. The attackers had grown used to the whites' method of warfare and waited for its predictable outcome—they expected the usual volley of fire, and then a pause while the soldiers desperately reloaded their single-shot muzzle-loaders; then the Indians would seize the opportunity and rush their foes. This time, however, there was no lull. The first assault, by warriors on galloping ponies, failed, with dozens of casualties. The second, a frontal swarm on foot, got within five feet of the enclosure before breaking, with even heavier casualties. The Wagon Box Fight, as it would be called, continued until the Sioux had rescued their wounded and retrieved most of their dead—taking, as many commentators have noted, even greater risks than during battle itself. They withdrew at the sight of a relief party towing a howitzer out of the stockade. Three whites had been shot through the head, and two were wounded. Estimates of Sioux casualties vary widely, but Powell's number—60 killed and 120 badly wounded—was probably accurate. Both sides would claim victory. Red Cloud would admit many years later that he had lost the best of his warriors at Wagon Box. But the Sioux drove off a large number of horses and mules—and there was much prowess to boast about in the coming months before winter.[54]

The Hayfield and Wagon Box fights on August 1 and 2, as Lakota Sioux and Oglala Sioux engagements protesting the fortified Bozeman Trail through Powder River hunting grounds, had a profound effect on Washington's Indian policy. So did another act of warfare in Nebraska on August 4, which involved the tribesmen who had been routed only four months earlier from their friendly encampment near the Pawnee Fork in Kansas by General Hancock, who looted and burned the "insolent" Cheyenne village and sent Colonel Custer to chase them northward. The large number of Indians had split into smaller units and fanned out to elude pursuit.

As has been seen, the Platte Valley, with its high number of civilian emigrants and railroad workers, waited for the obvious and tragic consequences. Soon after the Pawnee Fork rout, one particular war party, led by Spotted Wolf, rode up from the south and saw the first railroad train they had ever seen. One warrior, Porcupine, who lived to be very old, told the Native American historian George Bird Grinnell, what transpired. "We looked at it from a high ridge," Porcupine recalled. "Far off it was very small, but it kept coming and growing larger all the time, puffing out smoke and steam; and as it came on, we said to each other that it looked like a white man's pipe when he was smoking." After the train had passed, the Cheyenne rode down to examine its tracks, being seen by emigrants and railroad workers but mistaken for the

Pawnee scouts of Major North. They rode north of the Platte to the encampment of Turkey Leg, whereupon they decided to return to the railroad tracks.

"Now the white people have taken all we had and have made us poor," they reasoned, "and we ought to do something. In these big wagons that go on this metal road, there must be things that are valuable—perhaps clothing. If we could throw these wagons off the iron they run on and break them open, we should find out what was in them and could take whatever might be useful to us." It was the same tactic used by numerous outlaws currently in knee pants or at that point unborn, from Big-Nose George Parrot and Dutch Charlie Burris to Butch Cassidy and Flat-Nose George Currie—derailment of a fast-moving train for the loot inside. Spotted Wolf's band rode down to the Union Pacific tracks near Plum Creek to see what they could do.

Porcupine recalled that with a friend named Red Wolf, he tied a big stick across the rails. They built a bonfire close by and sat down to wait. "Quite a long time after it got dark we heard a rumbling sound, at first very faint, but constantly growing louder," Porcupine said. "We said to each other, 'It is coming.' Presently the sound grew loud, and through the darkness we could see a small thing coming with something on it that moved up and down." It was a handcar—with five men pumping it. The railroaders saw the fire and the shadowy shapes nearby and began pumping frantically to get past them, but when the car struck the stick it flew high in the air. The men picked themselves up from where they had fallen and began to run away. One man, an English telegraph repairman named William Thompson, was overtaken by an Indian on his horse and shot in the arm; running farther, he was clubbed down. "He then took out his knife," Thompson recalled, "stabbed me in the neck, and making a twirl round his fingers with my hair, he commenced sawing and hacking away at my scalp. Though the pain was awful, and I felt dizzy and sick, I knew enough to keep quiet." The process seemed to take a half hour. "He gave the last finishing cut to the scalp on my left temple, and as it still hung a little, he gave it a jerk," Thompson said. "I just thought then that I could have screamed my life out. . . . It just felt as if the whole head was taken right off. The Indian then mounted and galloped away, but as he went he dropped my scalp within a few feet of me, which I managed to hide." He could hear the warriors moving about in the dark nearby and didn't twitch a muscle.

There were rifles on the handcar, and tools. The Indians picked up the guns and handled them, but the guns broke in two at the middle and they threw them away, saying it was too bad that they were broken. They were the first breechloaders they had seen. With the tools they found, the Cheyenne decided to see what would happen if they pulled up some rails; they pried out spikes and bent one of the rails about two feet upward. "Looking east over the long level plain," Porcupine continued, "we saw a small light close to the horizon, and someone said: 'The morning star is rising.' 'No,' said another, 'that is one of those things that we have seen.' 'No,' said a third man, 'the first one has gone out and another one is rising.'" They were looking at two approaching work trains, one directly behind the other. Spotted Wolf sent two men eastward on their fastest ponies to investigate the two lights, instructing them to yell and shoot and frighten them off, and then come back to report. As soon as the scouts realized they were trains, they wheeled about and galloped back, but the first train overtook them. They shot at it and tried to throw a rope over the engine, but their horses spooked and veered away. The engine sped up at the gunfire. From where he lay, Thompson heard "the low rumbling of the train as it came tearing along," but he did not dare flag it down. When the engine reached the gap it jumped high in the air, and the cars tumbled together with a splintering crash.

They immediately shot the engineer. The unsuspecting brakeman, carrying a lantern and swearing loudly, ran up from the rear. He, too, was shot, and both bodies were scalped. Four other men who had been back in the caboose slipped out into the darkness—three running down the track to warn the second train, the fourth hiding for a while beneath the car before following them, being pursued for the better part of a mile by two of the Cheyenne on foot. Meanwhile, the engineer of the second train had been alerted and stopped a good distance away; as he blasted his whistle, four or five men from his train walked up toward the scene, but as soon as they realized what had happened they ran back—and the train backed up toward the east. The Cheyenne made no move to attack the train or follow, being curious about the wrecked cars. Thompson could see by the light of their bonfire that the Indians had climbed into some boxcars. Still in shock, he watched, terrified but fascinated. While they were distracted he finally crawled off, clutching his scalp, and after he had made his way to the Willow Island station, and was taken by rescuers to Omaha, he told Henry Morton Stanley what next transpired. "They plundered the box-cars of everything that might prove of the least value," Stanley would write, "or what attracted their fickle fancy—bales of calicoes, cottons, boxes of tobacco, sacks of flour, sugar, coffee, boots, shoes, bonnets, hats, saddles, ribbons, and velvets. They decorated their persons by the light of the bonfire which they made of the rifled boxes; their ponies were caparisoned with gaudy pieces of muslin, and the ponies' tails were adorned with ribbons of variegated colours. The scalp locks of the Indians were also set off with ribbons, while hanging over their shoulders were rich pieces of velvet." Stanley's account was detailed and highly imaginative but basically borne out by the memory of Porcupine himself: they found whiskey and danced a dance of triumph.

The next morning, some of the railroaders returned to Plum Creek. With a spyglass they could see that the railroad cars were now all on fire, with some of the Indians galloping in a circle; others were carrying away the plunder. As evening began to fall, a company of soldiers appeared and went to examine the wreck and retrieve the charred bodies. These were transported back to

Omaha on a train which happened to also be carrying William Thompson. At the station there was quite a push to gape at the bodies—and there was the living spectacle of the Englishman. "People flocked from all parts to view the gory baldness which had come upon him so suddenly," Stanley reported. Thompson exhibited his other wounds. "In a pail of water by his side," said Stanley, "was his scalp, about nine inches in length and four in width, somewhat resembling a drowned rat, as it floated, curled up, on the water." Although Thompson hoped that a surgeon could reattach his scalp to his head, this was of course beyond medicine. William Thompson later sent the scalp, after it was tanned, to his doctor in Omaha. Ultimately it went on display at the Omaha Public Library, in the children's section, and now reposes at the Union Pacific Museum.[55]

Major Frank North was out at the end of track when he was summoned by telegraph to Plum Creek. Taking a company of Pawnee scouts by special train, he sped east the 220 miles and found the Cheyenne with much of the loot not far from the wreck; there were perhaps 150 by now, including women and children from the Turkey Leg encampment, their ponies heavily laden with packs of clothing and food. The Pawnee crept up the creek bank and fired a volley without warning, killing seven. There was a running fight for the rest of the afternoon and into the evening, during which a number of Cheyenne were killed and wounded. When the Pawnee scouts returned to their command, they had recovered at least part of the booty, and two prisoners besides—a woman and a young boy.[56]

The two prisoners were later exchanged by Turkey Leg for five white prisoners—three girls in their teens, a pair of twin boys aged six, and a baby — captured in raids in the spring. The exchange—which appropriately took place in the railway eating house at North Platte—was a crowded one, and it had all of the solemnity of an official, high-level war council. There were many important Sioux and Cheyenne leaders present, including Spotted Tail, Pawnee Killer, Spotted Bear, and Standing Elk. The opposite side included Generals Sherman, Augur, Harney, Terry, and Sanborn; Missouri senator John B. Henderson, who chaired the Senate Committee on Indian Affairs; and the Reverend Nathaniel G. Taylor, an old Tennessee friend of the president's and the recently named commissioner of Indian affairs. These, and several others, had been authorized by Congress into a Peace Commission a few weeks before the events on the Bozeman Trail and the Union Pacific; they had been sent up the Missouri and then to North Platte, and would next go to Fort Laramie. At North Platte, after the prisoner exchange, participants got a taste of how the peace mission would fare over the next month of its existence leading to a large council to be held in October near Fort Larned in Kansas. The Indians complained that the railroads in Nebraska and Kansas, on top of the wagon roads in Kansas and through Powder River country, were driving off all of their game; they were willing to live with the Union Pacific line, but the others—the Smoky Hill and the Powder River, and the whites who used them, would have to go.

At North Platte that day they were emphatic about such a trade—the Union Pacific in exchange for the Powder River and the Smoky Hill emigrant routes, along with all of their stockades. And that back-step was something Sherman would not accept. The centerpiece of Commissioner Taylor's proposal was ventured—it espoused setting apart two large reservations, inside of which the Plains Indians were expected to take up farming. Henry Morton Stanley, who seemed to be everywhere that summer, was present to record the response of one Indian leader. "Ever since I've been born," he said, "I have eaten wild meat. My father and grandfather ate wild meat before me. We cannot give up quickly the customs of our fathers."[57]

26

"There Are Only Five of Us"

He now spent one-third of his life in Pullman cars, excavating his way through mounds of paperwork or dozing, obtaining little meaningful sleep, the dark and dreary landscape speeding past his window unheeded, his little family neglected back in a small hotel suite, his friends some three thousand miles away. Business, however, was everywhere, and as he frequently said, "a man that goes to do a thing can do almost anything." With hardly any exception, each week he would spend four days in Washington when Congress was in session, always including a Sunday. Then two days in New York and one day in Boston. "That called," Collis Huntington would remember, "for four nights riding a week. I rode through the hot weather; I used to go over to Washington, stay in New York through the day and take the night train for Boston; worked there through the day, take the next day in New York and the next in Washington." During his fleeting visits to New York's Metropolitan Hotel, at Broadway and Prince Street, where Elizabeth Huntington tended to six-year-old Clara, he could enjoy what little domesticity his occupations would permit. Perhaps recognizing that he—or at least his wife and adopted daughter—needed more, in that early autumn of 1867 he would arrange to buy a townhouse at 65 Park Avenue in Manhattan, although he would insist on filing it in his nuts-and-bolts mind as a shrewd business decision affecting, of all things, the salability of their securities. "The Union Co. were using as one of their arguments that they were a New York Co.," he would explain to Hopkins, "that their Books &c. could be seen at any time, that the Directors lived here &c. and that the C.P. Co. were a Cal. Co. and that all the Directors live there." Notwithstanding his nearly five years residing in New York hotels, Huntington was about to become a property owner and eliminate that irritating criticism. "I now say," he would tell his California partner, "that I always expect to live here."[1]

Life at the Metropolitan Hotel may have had its drawbacks, but there were, to be sure, the obvious amenities—service, dining, a shoeshine boy in the lobby, a good cigar stand, the ready availability of all the newspapers. On September 7 there would be the satisfying advertisement of Fisk & Hatch, stretching the considerable length of one newspaper column and trumpeting in festive display type a not altogether truthful boast:

> *Across the Sierra Nevadas—The Central Pacific Railroad—the Western Half of the Great National Trunk Line Across the Continent, Being Constructed with the Aid and Supervision of the United States Government, is destined to be one of the most important lines of communication in the world, as it is the sole link between the Pacific Coast and the Great Interior Basin, over which the immense overland travel must pass, and the Principal Portion of the Main Stem Line between the two Oceans. Its line extends from Sacramento, on the tidal waters of the Pacific, eastward across the riches and most populous parts of California, Nevada and Utah, contiguous to all the great mining regions of the Far West, and will meet and connect with the roads now building east of the Rocky Mountains. About 100 miles are now built, equipped, and in running operation to the summit of the Sierra Nevada. Within a few days 35 miles, now graded, will be added, and the track carried entirely across the mountains to a point in the Great Salt Lake Valley, whence further progress will be easy and rapid. Iron, materials and equipment are ready at hand for 300 miles of road, and 10,000 men are employed in the construction.*

As of August 31, the brokers assured the public, the quarter's gross earnings were $487,579.64, the operating expenses were $84,548.47, and the net earnings $401,031.17—or, "at the rate of about two millions per annum, of which more than three-fourths are net profit, upon less than 100 miles worked." With such figures, with such future success, Fisk & Hatch was offering thirty-year first-mortgage bonds at $1,000 each, discounted at 95 percent—which would yield "nearly nine per cent upon the investment."[2] And Fisk & Hatch was now selling bonds from the line projected into Nevada, a change in policy which only underscored its faith. The ad was like a gold ribbon on the page—easily dwarfing the paltry notices of Union Pacific securities for sale by Wall Street bankers Henry Clews, John Cisco, and Clark Dodge. A careless reading of the Central Pacific notice would make an investor think the company was closing in on Salt Lake if not the Rockies themselves.

On September 9 the front page of the *New York Times* would report that President Johnson had decreed "a general act of oblivion," granting amnesty and citizenship to most southern rebels with the exception of Jefferson Davis and some of his cabinet. But there would be, if Collis Huntington paused to see them, two more inspirational articles. One announced the opening of the railroad line across the Alps between Germany and Italy across the Tyrol Pass, which was "the most direct and easy route across the main Alpine chain," which recovered for that "natural highway" all the "importance it possessed from the remotest Roman and German ages." Such a stirring and yet similar feat! And in an article a page or two away there would be a detailed description of a German wine factory and the particular qualities and merits of vineyards on the Rhine. In a few weeks Huntington would be writing to Hopkins about that very subject. "I have talked with many people (mostly German) about settling on the line of our road in Cal. to cultivate the Grape," he said, "and I am told by the Germans that have been in Cal. that much of the land on the Western slope of the Sierra Nevada Mountains is very much like the best vine lands of Europe, and I had thought it would be well to get up a pamphlet showing the advantage of these lands for grape culture, and circulate it in the wine countries of Europe." If Hopkins would get the manuscript written for such a pamphlet, Huntington said, he would get it printed in New York in German and perhaps in some other languages "and have them circulated."[3]

Such propositions occurred regularly in his mind, but overspreading them there was always the palpable presence of the competition, doors away from his office on William Street, and the question of how much of the line the Central Pacific could throw out until stopped. Judge Crocker had written that he saw no reason why the Union Pacific should oppose their locating north of the Great Salt Lake. That, in Huntington's mind, was missing the point. The Union Pacific was resisting them "not because we locate north of the lake," he patiently wrote to Crocker on September 6, "but they oppose any location at this time for this reason . . . that if they can prevent any location east of Humboldt Wells until they get, say within 200 miles of Salt Lake, they get already to ocupy the 300 miles that each Co. are entitled to under the law, then come in and agree upon our location, and go to work, and in that way get 100 miles west of Salt Lake. That would be one of Durant's smart tricks. It will not be done, but that is what they are playing for."

Huntington still thought, nonetheless, that they would prevail in building at least to the Lake, and as a way to do this he was of like mind with the Judge; recently he had talked with Oliver Ames, and "I have told him that all my Associates had fixed on Fort Bridger or some point near there, as the place of meeting." He was not going to leave that up to chance; a year earlier Huntington had begun to cultivate a friendship with T. T. Davis, a partner with Ames in the Cedar Rapids Railroad and the line's attorney. "I told him," Huntington informed Crocker, "if he would bring about an arrangement that would suit me, that I would give him $20,000 of our stock."[4] It now seemed that Davis was nosing in toward the bait. And Huntington would continue his discreet campaign to plant spies all around the Union Pacific directorate.

It was quite a network; a few weeks later he wrote Crocker that he had "a way of finding out what is done in the Union Co's office," and later in the year he would boast, "I have sent men that they did not suspect, to inquire." He pumped the executives of iron mills for intelligence about the company's orders and construction plans, and he even located a Mata Hari of sorts: "I called on a lady friend and she told me that C. S. Bushnell called on her the evening before and he told her that they had made up their minds to build 400 miles of road that year." And he paid an assistant doorkeeper in the Capitol building, one John Boyd, to watch for any meetings between congressmen and the Union Pacific brass. Soon Boyd would see Dr. Durant, Oakes Ames, and Grenville Dodge in the corridor as they entered the Committeeroom on Ways and Means. Boyd trailed along, innocuously straightening up the room, and overheard them all agree "to push the work without much regard to cost," and that Durant had muttered "that he would be damned if he would not prevent the Central Pacific from coming more than 200 miles east of California."[5]

The Judge, of course, had already enlisted operatives in Salt Lake City to keep him informed of events within the Mormon circle of power and of anything to do with the Union Pacific. In mid-September he picked up news from the city's Wells, Fargo agent about Grenville Dodge, who had pressed past Fort Bridger through the Wasatch and finally reached Salt Lake City. "He did not stop at the Townsend House," his informant wrote him, "but pitched his tent at Camp Douglass." While Dodge was roughing it, disdaining a comfortable city hotel presumably in an egalitarian gesture to his surveying party, he let it be known that as an engineer he favored the Weber Canyon passage and also the northern route around the lake. He called at the residence of an army officer and while there his host had shown him his collection of recent stereoscopic views of the Central Pacific Railroad's progress. Squinting at one after another, Dodge was eager to get some for himself and asked if he could procure fifty. Judge Crocker, when told of this by his informant, immediately sent a package with his compliments to Dodge care of the Omaha office, asking for a matching set from the Union Pacific. "It is well," he explained to Huntington, "to do the genteel with such a man as Dodge."[6]

Dodge himself was trying to learn all he could about the competition, as Judge Crocker learned from Butler Ives, at that point locating their fanciful line between Weber Canyon and Green River. "It seems that Gen. Dodge sent for him," Crocker wrote Huntington, "but as he was travelling over the Mts. the agent did not find him. It is just as well. Dodge could not have pumped him, but it is as well that he was not subjected to the trial." Having learned how to "shell the woods" at the feet of a real expert—Abraham Lincoln—perhaps Dodge would have managed a better interrogation than Crocker thought.[7]

There was steady progress to report since the Summit Tunnel breakthrough. Crews were laying track down the Truckee Canyon "as fast as they can," the Judge said, and they expected to get that twenty-four-mile section complete by the first of November, by which time his brother thought the Chinese would have spiked rails up and through the Summit Tunnel. "But still," he cautioned, "all these matters are uncertain. A single ledge of very hard rock would disappoint all our calculations—& for one I am determined not to again disappoint you by promises if I can help it." But he found it hard to rein in his enthusiasm. They were hauling iron across the summit which was destined the following spring to connect the Truckee with Lake's Crossing—the little trading post on the Truckee Meadows which would later become known as Reno—and it made sense to build their next freight depot there, he thought. In Nevada there were plans to build a branch line from there down to Carson City—"that road is in the hands of our friends," the Judge said, and if the Central Pacific could "pile up a big lot of Iron" at the summit before winter, they could be building toward the Big Bend of the Truckee all winter long. "Our men keep together well," he said, "& the force is increasing instead of diminishing. Every ship from China gives us more—but still not enough."[8] A great many Chinese returned home on the steamer *Great Republic* at summer's end, their pockets full of sufficient gold to see them into luxurious retirement in their back-country villages—but luckily, Crocker noted, "several leading men who have been furnishing us laborers were among them. They go to get a large immigration to come over & work on the road next year. They know all about the work & can explain it to their countrymen. I have no doubt they will be able to induce thousands to come over." They would be sending other recruiters over armed with a new printing of the Chinese handbills. "We want 100,000 Chinamen here," the Judge told Huntington, "so as to bring the price of labor down."[9]

All of the Associates were feeling the frustration as they peered at the increasingly smaller breaks in their line on the map and at the increasingly smaller number of pages on the season's calendar. As Stanford noted, even five hundred more men working that summer would have guaranteed reaching the state line by the season's end, saving them this tension, just as another thousand in 1866 would have gotten them to the summit a year earlier. "It is very trying to see such important results missed for the want of a few more men." A few weeks later, when it seemed that the season might fall short of a continuous line to the Truckee, Huntington was inclined to be philosophical. "I know all will be done that can be," he wrote the Judge, "and I do think that Charlie can do a little more than any other man in America, and if it is not done, I shall know it could not be." And he was in accord with his partners' desire to go all-out to encourage Chinese immigration. "It would be all the better for us and the state if there should a half million come over in 1868."[10]

Another matter consuming Huntington's thoughts was the prospect of a little more empire building—at a discount. All spring and summer, Colonel Charles Wilson had done everything he could to revive the California Central Railroad. It still ran from the Central Pacific's Roseville station north through the Sacramento Valley to its tantalizing standstill, only seven miles shy of lively Marysville on the Yuba River, with its hill-town trade. One scheme after another had failed, and finally the Judge reported, "Wilson has got to the end of his rope." Owing money to just about everyone, he had borrowed more from the Boston financier Charles Lambard to buy his seven miles of iron, giving Lambard as security three hundred California Central bonds, each valued at $100. When he was unable to meet one and then two interest payments, Lambard refused to deliver the iron. Technically this put Lambard in violation of his contract, but Wilson was in no position to bring suit. "He can't get anyone to help him out," Crocker commented. Wilson had appealed to Leland Stanford for help—but to goad him along, Crocker had a debt execution issued for the money owed to the Associates, and sale of the company was set for September 23. Stanford was all for proceeding, knowing that Huntington was eager to engulf the northern roads and eventually to enter Oregon, but would not take a step because of Wilson's outstanding debt to Lambard "and his friends." The matured interest coupons on the bonds had become due in January and July—but, as Stanford wrote and telegraphed Huntington, "Lambard would consent to cancel those and surrender those of January 1st, 1868, if by so doing responsible parties would take the road and build it." Moreover, the bonds were nothing but "stagnant property" in Lambard's hands; Stanford or Huntington could begin to buy them and others in New York and Boston at a discount. Even cautious Mark Hopkins saw this as an opportunity. Wilson's California Central, "like everything he has ever controlled . . . is so encumbered and entangled that nobody having the means to buy is willing to touch it—unless we do it," Hopkins wrote to Huntington. "Men of means enough to carry him on with his programs in this part of the world are few, & as no man has ever escaped from financial disappointment who has taken an interest in his affairs, nobody can be found to back him on an enlarged scale—hence his willingness to close out for the least price which will clear up & pay his confidential debts." Wilson, now a broken man, was eager to sell all of his interests in the California Central and the Yuba line, and in the franchise of the unbuilt California and Oregon; he asked Stanford for $200,000 but in a few days desperately lowered it to $85,000. Hopkins assured his partner that they would offer $75,000 in one- and two-year notes—and would meet Wilson's last terms only if all bondholders including Lambard forgave their overdue interest and if the Marysville town fathers agreed to give them $10,000 toward building the Yuba bridge and a depot.[11]

It seemed to Huntington that everything about the growing empire was falling into place, even in the right order. When Judge Crocker's letter of September 26 arrived with its suggestion that Huntington could begin buying California Central bonds in the East for as little as $50—half of their printed worth—he had already begun doing so. Later, when he discovered that the well-meaning Stanford was buying bonds in San Francisco at full price, he was livid. But the sale and the transfer of all interests began to proceed, although predictably it was a long and bumpy road until the end.[12]

For his part, however, the Judge was beginning to worry. "I dread the purchase of these northern roads," he confided to Huntington. "We are expanding too much—getting too much on our hands—too much of the grasping monopoly." He admitted that if they didn't go ahead they would probably lose about $200,000 lent to Wilson or already invested, and so he would have to go along with the others even if necessarily all of the legal tangle would drop on him alone. "I have hardly any leisure to think out & study up the important matters relating to our affairs," he complained a few weeks later, seeing that at this of all times the campaign for rights to a future deepwater terminus on Goat Island was going to intensify as disposition moved before the Senate. "My time is taken up with too much detail, & it is increasing all the time. Now to add on the work, the vast work of properly improving Goat Island, is too much. We will find we have too many irons in the fire."

Besides, he said, "we can't expect to monopolize all the good things in the State. We will make enemies by it." Crocker proposed to let others in—at least beginning with Goat Island—"to take this burden off our shoulders, expecting that they will furnish money, time & brains to carry it on. There are only five of us, & we can't expect to swing every thing here." It would take large sums to improve the Oakland waterfront and the hem of Goat Island, not to forget the island warehousing complex he himself had espoused in the spring, and "I have no objection to letting others assist in the investment take a share of the risk as well as the profits. When we get to be the principal owners of a R.R. from S. F. to Salt Lake, any pecuniary ambition will be fully gratified. It will be as much as all of us can handle or use properly the rest of our lives. I would like to see others getting rich, too." He had a number of San Franciscans in mind—Lloyd Tevis, Thomas Parrot, W. C. Ralston, W. W. Stow, and others, all wealthy, energetic men with political influence. Even the name of Lester L. Robinson, an avowed enemy, came to mind—money and the prospect of obtaining more of it had forged stranger alliances. He had known for more than a month that investors had paid more than $30,000 for adverse claims to Goat Island. Capitalists unfriendly to the Central Pacific had, in the spring, incorporated a tiny railroad between Vallejo (where the California Pacific, which had recently been reinvigorated with new money, began its run to Sacramento) and Oakland; the backers had even obtained from the state legislature title to submerged land between Oakland and Goat Island, and they had the temerity to call their corporation the Terminal Central Pacific Railroad. In his fatigued state the Judge might even have been open to consorting with them.[13]

In any event, Goat Island was in the hands of Congress. And as far as the Associates were concerned, there would be no new hands coming on to take a share of the burden. Had the Judge known that, he might have been driven to despair. One foot in front of the other, he moved onward.

Butler Ives, surveying for the Central Pacific's route in Utah's eastern mountains in September, had worked all the way up Weber Canyon and had moved into Echo Canyon, planning to run across the summit of the divide into Wyoming and Bear Run, and Fort Bridger beyond. He had been sent a copy of the Union Pacific's engineering reports from Sacramento, although his supervisor, Samuel S. Montague, had cautioned him not to rely on Samuel Reed's data but to develop his own. And Chief Engineer Montague was glad that his opposite, Chief Engineer Dodge, had not found Ives. "Am glad you did not go out of your way to meet Genl. Dodge," he wrote. "If he is extremely anxious to meet you he will probably send his escort to capture you, but if he only wants information he had better get it as we do ours." Montague warned Ives not to give away any of his own work—but that it would be good, if the Union Pacific people found him, to let them know that the Central Pacific company firmly planned to build to Fort Bridger.[14]

As it turned out, Ives was in no danger of encountering Dodge that season. He and Rawlins had camped outside Salt Lake City for a week, liking the place very little. The Mormons were, he thought, "mostly ignorant and superstitious and bigoted, without education, and doing so much better than they were in Europe, they are easily handled by so cunning an old rascal as Brigham Young and his lieutenants." Hating the idea of polygamy, he called it "abhorrent to the finer feelings of any educated person." And he reserved a particular pity for Mormon women—they lacked "that cheerful, genial, home-like air of others," he wrote to Anne Dodge. "They all look as if some great trouble was weighing them down." He could not deny that the people had done wonders with agriculture and industry. But with his large home and grounds, their leader seemed to be reaping most of the benefits. Brigham Young had granted Dodge an interview, and then talked him into a stupor. He did indicate to the engineer, however, that when the railroad approached Utah there would be no lack of laborers eager for work. When Dodge and Rawlins folded up their tents and rode out of the valley it was with a mixture of relief at leaving the City of the Saints and anticipation of new terrain ahead. Instead of doubling back through Weber and Echo Canyons, they went farther north, for Dodge was interested in the terrain after Brigham Young had reminded him of the potential for a railroad to Puget Sound; up the Bear and Snake Rivers the Dodge expedition went, crossing the Rockies, and then it went through South Pass, the Sweetwater Valley, and Seminoe Gap. At a camp west of South Pass, Dodge saved Rawlins's life when he was charged by a grizzly bear, shooting it between the eyes; Rawlins got the bearskin as a trophy. On October 4 they were at Fort Sanders. Two days later they rode into Cheyenne. Were it not for the familiar topography, they might have thought they were somewhere else—Cheyenne had been transformed.[15]

The population had shot to more than fifteen hundred, and in another month it would nearly triple. Saloons, gambling dens, shops, and hotels stood all up and down the dusty streets. "Three months ago it was nothing but bare prairie land," Dr. Henry Parry marveled. Finished with his detail to the Dodge expedition, he awaited orders for a dreary winter at Fort Fetterman. A newspaper, the *Cheyenne Leader,* had just come into existence; its first issue announced the arrival of the Julesburg Theatrical Troupe, by stagecoach, for a protracted engagement. Promoters cobbled together the largest building in town, some eighty by twenty-six feet, in a week, dubbing it the King Theatre, with "parquet, dress circle, private boxes and all modern improvement." Soon more establishments would join it—the Variety Theatre, Melodeon Hall, Beevaise Hall, the Theatre Comique, and so on—often combining under one roof a theater, gambling house, and saloon, sometimes all in one room. Every evening, as in Julesburg, bedlam descended. The *Leader* ran a daily column titled "Last Night's Shootings." Soldiers at nearby Fort Russell were on strict orders to stay away. While Dodge was there, however, he had some wide-eyed troopers come over to evict some claim jumpers—particularly desperate characters, he thought—from railroad property.[16] As soon as the bluecoats turned their backs on the railroad lots, though, newly arrived entrepreneurs swarmed in; their last port of call was swiftly becoming a ghost town.

"Julesburg is now an overdone town," pronounced Henry Morton Stanley, riding west on the Union Pacific from Omaha, "a played-out place. . . . It is now about to be abandoned by the transient sojourners, and many of them are shifting their portable shanties to some prospective city west—Cheyenne, or some 'prairie-dog town,' where cash can be made without work, and by any means that will not subject the operator to an indictment before a Grand Jury for obtaining money under false pretenses." His train had rolled past the site of the train wreck at Plum Creek—the charred debris was still piled at trackside—and paused at desolate Julesburg for lunch; trash heaps seemed to outnumber occupied buildings. By early evening the train had passed Sidney, Nebraska, a Hell on Wheels town for a space of mere weeks, and reached the end of track.

Stanley dined well with Generals Sherman and Augur and staffs, and the next morning he plunged into a brawny multitude in the tracklayers' boarding car of General Jack Casement, where he enjoyed "a most excellent breakfast." Afterward, Stanley watched the Irish lay rails "as rapidly as a man can walk" until Casement's superintendent decided to show off a little for the correspondent's benefit. "He gave his men the hint," Stanley wrote, "and in the space of exactly five minutes, as timed by the watch, they laid down the rails and spiked them, for the distance of seven hundred feet." At that punishing, showoff rate, Casement's crews could lay sixteen and a half miles of track in a day. Continuing his tour, Stanley was soon introduced to Major Frank North of the Pawnee Battalion, "who invited us to accompany him to his camp, where he would show us a lot of Cheyenne scalps obtained by his men." Stanley was curious and willing but he had to forgo the display; without seeming to see any irony, he was summoned to witness instead the preliminary peace talks at North Platte, and hurried off.[17]

General Augur planned to have North's Pawnee Battalion mustered out when winter arrived and the plains quieted down, but because of the scouts' success at defending the Union Pacific crews, he planned to reactivate the unit the following spring. "I have never seen more obedient or better behaved troops," he wrote the War Department. "They have done most excellent service." He asked Washington to consider organizing as many as three battalions of four hundred each "from the friendly tribes in this Department. It opens to those people a useful career, renders them tractable and obedient, educating and civilizing them more effectually than can be done in any other way. They are peculiarly qualified for service on the Plains; unequalled as riders, know the country thoroughly, are hardly ever sick, never desert and are careful of their horses." Indeed, the Pawnee seemed to be the best soldier material he had encountered—but Augur would not get his four battalions in 1868 or any other year. Major North would return in the spring to command just one. Augur's last words in his Pawnee Battalion report that September were particularly resonant, given the number of Regular Army infractions in the orbits of Cheyenne, Sidney, Julesburg, and North Platte. "I have never seen one under the influence of liquor," the general reported, "though they have had every opportunity of getting it."[18]

Samuel Reed was, in late September, back in his office in Julesburg, having gone to inspect the grading work west of Cheyenne. He wrote to his wife that they hoped to have those sections done before winter. While in the mountains, Reed told her, he had allowed himself some leisure, going out fishing in a stream above Cheyenne. He caught seventy "fine trout" in two hours, "the best sport I ever had of the kind," he said, giving no indication that he had thrown any back. He also took some souvenirs to Julesburg for eventual presentation at home—"two splendid mountain cacti" for his wife, and "a fine Pony for Anny and Mary. It is from the Blackfoot Indians, not much taller than Anny and as gentle as a lamb. It will follow the children like a dog and I know they will be pleased with it if I can only get it home." The little Indian pony would be on its way to his daughters via the Union Pacific shortly thereafter. But the news he had retrieved from the mountains was not so happy; part of the reason he had gone was to resolve engineering controversies that were slowing the graders, and Durant's "consulting engineer" was to blame. "Col. Seymour seems determined to delay the work as much as possible," he told Jenny. "The object is apparently to injure somebody's reputation. Genl. Dodge appears to be the scapegoat. Col. S. has been working to get a line with grades ten feet less per mile than the previous location and he can do it but the line will not be as good as Genl. Dodge's or as easily operated when built. I have written to New York and given a full statement of the matter, and trust the company will be fully posted about the facts in the case. I know they will as far as the delays are caused by changing the line."[19]

Dodge had been kept sketchily aware of Seymour's machinations during most of his trip, receiving forwarded mail and telegrams at successive army posts, but his firm letters and wires of opposition sent to Oliver Ames seemed to be getting him nowhere. Granted, Ames seemed to have accurately appraised Seymour's character as "an indolent man with a strong desire to criticize others work and do nothing himself." Moreover, Ames had assured Dodge in a letter received in Utah, that "[h]e has been kept more for the purpose of writing the Dr. Reports and doing his correspondence than meets the public eye and whitewashing his rascalities than for any real engr service he has done or will do the Co." But Dodge did not like what he next read. "I have told Seymour to make a survey of his proposed changes," Ames had written, "and report the comparative gain. If he has found a better line it is our duty to adopt it. If a 2 or 3° curve will throw us out of heavy work and hasten completion of the Road I should do it. You understand our views and for the present act up to them."[20]

Reed had wired Dodge in August that Seymour's interferences had thoroughly confused their engineers and paralyzed the grading parties on either side of the Laramie/Black Hills divide, and Dodge, feeling increasingly helpless at such a remove, had telegraphed Reed to choose the best line. But Colonel Seymour had not finished with his flurry of proposals. One week he telegraphed Ames to complain petulantly that any of his improvements to the eastern approach to the summit were now rendered untenable because of grading already completed; the next week he reversed himself with the spectacular news that he could, after all, reduce the eastern grade to eighty feet, save the company $120,000, and move the government engineer's fixed base of the Rockies some three miles to the east—to present the company with an additional $48,000 in federal subsidies. Ames was, predictably, dazed and confused.[21] He told Seymour to go ahead with the changes.

Reed had immediately alerted Dodge by telegraph. "It was clear what a fearful mix-up there was about changes in the Black Hills line," Dodge recalled many years later, "but I saw it was impossible for me to remedy where I was." James Evans, who had returned to the hills after an extended leave over his wife's death, wrote to Dodge on September 22 that "The folks are still sweating in the Black Hills. As I came along three separate lines of grading could be distinctly traced in places showing that some of the changes had been changed. . . . A considerable ingenuity is being used to avoid Dale Creek crossing, and as a consequence Evans' Pass. Alas, for my immortality." But of course it was more than that to Evans. Learning of this latest evidence of needless delays when reaching Fort Sanders on October 2, Dodge on his own initiative wired Seymour to cease all of his surveying in the Black Hills. Rather than replying, Seymour sent a protest to Ames, adding the alluring vision of a new improvement that would make the Dale Creek crossing easier and $200,000 cheaper.[22] Going carefully over the summit and down to Cheyenne, Dodge had been furious at how the good start of the season had been reduced to a shambles during his absence. Angry at the wavering support shown him by the Union Pacific's president, he headed east, girding himself for battle if necessary. After that, perhaps he might even make the next roll call in Congress. As he went toward New York he prepared to joust not just for himself and Evans's original line but also for Samuel Reed. Colonel Seymour had been busy poisoning several reputations, which Reed had perhaps not fully realized when he wrote his wife of his concern for Dodge. Oliver Ames had somehow gotten the idea that the blameless Reed was responsible for the delays in construction—he was guilty of a "want of system and Application to duty . . . he prefers doing any thing but his duty."[23] That had the unmistakable whiff of Colonel Silas Seymour to it.

Why was Oliver Ames acting so irresolutely? Anyone with an outside perspective would have seen Seymour's interferences as a delaying tactic to make the railroad, its employees, its contractors, and management look bad—indeed, to spoil the summer so completely that Seymour's puppet-master would have to step in, whether at the stockholders' or the government's behest, to put things right both for the railroad and for the contractor. Granted, Ames was not a good manager; the shovel works had represented the pinnacle of his executive skills, which, confined to that level, had worked well for the family and its employees; the brothers had inherited their father's business, after all, and benefited from building booms, western expansion, and of course the gold rush. But of course in the family factory he did not have Thomas Clark Durant lurking about. Ames disliked controversy, loathed open conflict. Durant thrived on it. Ames hoped for consensus. Durant with his extraordinary will and ruthlessness had become rather fond of the technique of the battering ram. Oliver Ames was a gentleman—a Bostonian—and Durant, from out in the hilly western Massachusetts provinces, had acquired a smooth veneer with money and power, but the Doctor still knew where the jugular, the kidneys, the groin, and the instep were located. He remembered how to use them to his advantage—and when. Durant had an exquisite sense of timing. And as Seymour had been sending back his collect telegrams to Oliver Ames in August and September, Durant had been watching the calendar. The hastily contrived construction contract of Oakes Ames, which meant so much to the Bostonians, had yet to be adopted. And the annual meeting and elections of the Union Pacific were set for the first week in October.

The two combatants readied themselves to unseat one another from the Union Pacific at the October 2 meeting, following that up with the seizure of the next construction contract. Ames hoped to outmaneuver the Doctor by winning his two longtime supporters, Cornelius Bushnell and Henry Mc-

Comb, over to the Bostonians' side, judging them as practical men who would listen to reason. "I don't feel that we should do right to put Durant in as director," he wrote to McComb from his home in North Easton, Massachusetts, on September 17, "unless he withdraws his injunction suits and submits to the will of the majority. He cannot hurt us half as badly out of the direction as he can in, and there is no pleasure peace safety or comfort with him unless he agrees to abide the decisions of the majority as the rest of us do." Ames's plan was to lure McComb and Bushnell into compliance by dangling an ingenious and ironclad trusteeship for the construction contract before them; it ensured profits, control, and momentum, and eliminated the roadblock of T. C. Durant and his vow to stall the Crédit Mobilier or any other contractor arrangement until he was included. In this plan, Crédit Mobilier would fund construction with 7 percent loans, with a small commission of 2½ percent, but the control of that construction was placed in the hands of seven trustees who would subcontract from Oakes Ames or else directly contract with the Union Pacific. The trustees would have to abide by majority rule, and they were empowered to unseat any member who made trouble; they would distribute profits on the construction contract only to stockholders owning shares in both the Crédit Mobilier and the Union Pacific, and only if the stockholders agreed to place control of 60 percent of their Union Pacific voting shares in the hands of the trustees. The shrewd plan was circulated, and over Rowland Hazard's angry objection Oliver Ames let it be known that Durant would be let into the trusteeship if he promised to recognize majority rule.[24]

Durant was eager to avoid that—unless, that is, he controlled the majority. On September 21, Ames was tipped off that the Doctor had made his move, just two days after his attorney, W. F. Allen, warned him that Durant was up to something "by which he expects to succeed or else to foil & defeat the election. I don't know what it is. He is hard at work & is unscrupulous & reckless." In a bold maneuver, Durant tendered offers for some sixty-nine thousand shares of Union Pacific stock, reaching out to commit the resources of all of his supporters and an alarming new face: James "Jubilee Jim" Fisk, the aggressive, thirty-four-year-old stock manipulator who had only the year before pushed onto the board of the Erie Railroad with partners Jay Gould and Daniel Drew and seemed eager to drain it dry. This offer of nearly $3.8 million—let alone the invitation to a maverick hired gun like Jim Fisk—as the sure path to controlling the election shocked the Bostonians, but it was a road they could travel, also. In little more than three days, calling on every resource imaginable, the Ames group met Durant's offer of 55—for an extraordinary $50 million in stock. The Doctor was dumbfounded but then realized the plan. John P. Cisco, the Union Pacific treasurer, had objected to Durant and Fisk's bid, saying that the law forbid him from accepting anything less than par, and he refused to take any draft. But if Cisco approved the discount, the Bostonians had Durant beat; if not, the Ameses would still win on sheer numbers. Nonetheless, both sides pulled out more ammunition, Ames objecting that Durant's offer, to be considered, must be fully tendered in cash, and Durant objecting that the Bostonians' offer was too late to secure voting privileges under the company bylaws. The Doctor swore that if he were not allowed to vote the $3.8 million in stock on top of what he already held, he would have an injunction issued—and stop the election, and the business of the company, dead.[25]

With the election just a few days away, all of the contenders swept into New York, prepared to do mortal battle. Oliver Ames was so disturbed that he called on Collis Huntington in his office and stayed two hours, seeking advice on how he might oust Durant. He unloaded all of the events to date—Huntington was delighted to hear it all firsthand, for once—and then Ames made a surprising proposal: would Huntington consider coming onto the board of directors of the Union Pacific? When Huntington got over his surprise, he later told Crocker, "I answered that I had as much as I could attend to." Then Ames turned to the subject of the Union Pacific election and asked another startling question: if Durant succeeded in beating the Ames brothers, would Huntington and his partners allow Oakes and Oliver into the Central Pacific, so that they could at least band together to fight Durant out on the plains as well as on Wall Street and Washington? The response to this breathtakingly desperate question was polite, cautious, and noncommittal. "I said to him that I would like very much to have him & his Brother work with us," Huntington recounted to Crocker, "but that my associates did not know them as I did & that as we were through our hard work & out of debt . . . I was quite sure that you would not commit to any changes in our board."[26]

The downcast Ames left Huntington's office empty-handed, intent to line up his lieutenants—and several prospective conscripts—behind the trustees proposal. He focused most of his appeals on Henry McComb. Desperate to push his campaign over the top, Ames did something which, under the circumstances, was as extraordinary as bidding to combine with his California competitors: he offered to resign the presidency of the Union Pacific in favor of McComb. It was as if, on the eve of Gettysburg, Lincoln had volunteered to leave the White House if Robert E. Lee would deign to move in. Even if Ames were certainly no Lincoln and McComb no Lee, it was a rash, even foolhardy move. McComb was after all in the Durant camp; Ames's stipulation that McComb would have to pledge himself to an honest administration would have little moral or practical force. Even if Ames had realized that he had no appetite for the job it would not diminish the shock value of his offer.

The first skirmish took place on October 1, when the present board of the Union Pacific met to appoint election inspectors and there was a tussle over who should have the power to accept or reject votes; the board adjourned without concluding, and everyone withdrew to battle rooms at the Fifth Avenue Hotel—one group to pursue action against the Doctor, and the directors to haggle over Ames's trustee arrangement. The discussions went late into the night. The next morning, the day of the election, the tense and bleary-eyed combatants sat down to face one another in the boardroom. Everything was a battle: identity of the election chairman, composition of the electoral inspectors, and within minutes all were on their feet, fresh injunctions brandished in each other's faces, leading to recesses and further conspiracies. It was "a Devil of a quarrel," said the gleeful Huntington when his spy reported to him. Finally the two sides began to fight while not acknowledging that the enemy existed; the Bostonians appointed their own inspectors and then elected a new board of directors in which Durant and his supporters were turned out and replaced by John Alley, Rowland Hazard, and John Blair of the Cedar Rapids road, who they hoped would step higher into management of the Union Pacific. Unfazed, Durant and his men appointed their own election inspectors and then their own board of directors, of course unseating Ames and company. Two sets of directors! Back in his own office, Huntington would savor the news. "So it is a question for the courts," he informed Crocker. "It would take ten pages to write all I have heard."[27]

The news would get better. On the evening of October 2, the Ames board met at the Fifth Avenue Hotel. Learning the distressing news that Henry McComb was defecting to the Durant camp after the Doctor offered him the presidency, the Bostonian faction chose officers: Oliver Ames as president, Rowland Hazard as vice president, and John Cisco as treasurer. The next morning, back at the Union Pacific offices, the combatants filed stiffly into the conference room. Still looking through Durant, Oliver Ames announced the election results and selection of officers, and he declared that last year's board was dissolved and the old members dismissed. Durant rose, barely hiding his anger, and strode out with the others. When the door had closed, John Williams rose to call the new board together, but he had only begun to speak when the boardroom door slammed open and Durant appeared, his face purple with rage, and seized Williams by the collar. He began hauling him toward the door, intent on keeping the Bostonian from convening the new board. For a few seconds everyone else was paralyzed with shock, but then the brawny Oakes Ames shouldered himself between them and forced the Doctor to relinquish his grip. Without a word Durant disappeared outside.

Shortly thereafter, the shaken board members were startled when Durant reappeared—but he was unarmed and seemed calmer. He apologized to Williams and Ames and the others—but then he was joined by Clark Bell, his attorney, who had hurried out to obtain another injunction against any further proceeding.[28]

Truly, as Huntington soon heard, in the face of these three days of quarreling the Union Pacific was "having men running to keep the quarrel out of the papers." Now, with the Ames board temporarily thwarted by the courts, the really nasty negotiating got under way, stretching exhaustingly for the rest of the day, long into the night, and into the next morning. Piecing together the story from his several sources, Huntington dispatched the intelligence westward. Either before or after the election, which had come close to dissolving into a big fistfight, said Huntington, the accounting books of the company had been in the possession of Durant, "but the Ames crowd got them, and in the night the [Durant] crowd stole them and would not give them up, nor would they tell where they were, and Durant swore that if he could not have his rights that the Books should never be seen in the office again." But then upon this petulance the Doctor revealed his next weapon, which just about froze his enemies solid. "Durant told Ames that the Co. had not done as the law required in any one thing," Huntington reported to Hopkins, "and that they had forfeited their charter ten times over, and that he and his friends had only $800,000 in the concern, while he (Ames) and his friends had about $5,000,000—and that he had made up his mind that he was willing to go to the block, provided he could see their heads chopped off with his." This cool threat of Durant's to sacrifice his fortune along with the Bostonians' much larger one chilled Ames and most of the others, although Rowland Hazard of all of them seemed unafraid. "Durant went on with much such talk," reported Huntington, "until he scared the Ames crowd, and they finally let Durant and a part of his crew in the Board."[29]

It was a costly defeat for Oliver Ames, who, in caving in to Durant, lost the trust of men he had worked with for many years, who refused to associate with the doctor—Hazard, Blair, and Williams. On October 4 the two board elections were voided. Three absentees—General Dix, Charles Lambard, and Cyrus McCormick, who were all in Europe—were unseated, and a fourth director, Charles Tuttle, resigned. Their places were filled by John Alley, F. G. Dexter, Benjamin Bates (who was president of the Boston Bank of Commerce), and William Glidden (clipper-ship partner of John M. S. Williams). Oliver Ames and Thomas Durant remained as president and vice president. "Durant says he has the votes," commented Huntington, "but I am satisfied that he has not, although I have not seen the Ameses."[30]

Now they could turn to the construction contract. As previously negotiated, the seven trustees consisted of Oliver Ames, Alley, Dillon, and Bates, representing the Boston faction, and Durant, Bushnell, and McComb. Enemies or allies, they commonly savored the prospect of a nearly instantaneous profit distribution; the Ames contract began way back where the Hoxie contract left off, meaning many miles of line completed the previous year. Together with what work finished by October 1, with the Casements' gangs having spiked past mile 435, the trustees would share profits on nearly two hundred miles. That would assuage at least the trustees' feelings, if not those now-alienated friends of Oliver Ames, who not only felt betrayed but were left out in the cold financially. "We had a stormy time of it, but are now all harmony and the Doctor agrees to go along in harmony and for the best interests of the Road,"

Ames confided to Dodge in an excess of self-delusion. "The Board of Direction is made up of nice Gentlemen, who will Look only to the True interest of the Road in their action."[31]

"Ha! Ha! what a time the Union Pacific folks have," laughed Judge Crocker in a letter to Huntington on October 14. "That is a trouble we do not have. We are all united. . . . We have just got out of the woods, in spite of their opposition. We are just bracing ourselves for a race over the plains, & have the means to go it alone. . . . Let them fight their own battles."[32]

Autumn weather on the Sierra summit had been kind to the workers of the Central Pacific, and most encouraging to the Associates. In early September, the Chinese in the Summit Tunnel had chipped and blasted all but about 500 feet of the bottom work, which diminished to 198 feet by September 25; on that date three remaining Sierra tunnels, numbers 6, 8, and 11, were drilled through with the bottom work to be finished by November 1. The Summit Tunnel would be done by then also, they hoped, and track would swiftly be laid through. In late September Judge Crocker had sent his completion report on the section between miles 74 and 94, and they expected that in early November they would release a report on completed miles 104 to 139 in the Truckee Canyon. It seemed unbelievable that there had been no snowfall on the summit. "I think fifty days good weather with undiminished force," Stanford told Huntington on October 11, "will let us make the connection with the track on Cold Stream," creating a continuous line. Within a day the mountaintops saw their first snow of the season, but it was only a foot of accumulation and it melted away in a couple of days. "The weather is splendid now," reported the Judge, "& the men are working better than ever." He assured Huntington that they still hoped to connect the separated lines, though no one planned to operate regular trains above Cisco in the winter season. "But the whole matter depends on a kind Providence," he said, "so address your prayers to him, being well assured that we shall do all in our power."[33]

A few days before, Stanford had taken Charley Crocker and Samuel Montague to show them the projected line of the Western Pacific, so that the latter two would be able to plan the engineering and construction details. They had ridden from Sacramento up the San Joaquin Valley toward Stockton, crossing the San Joaquin and the Calaveras Rivers and passing the old Rancho Campo de los Franceses, the hundred-square-mile ranch sold twenty-two years earlier to Captain Charles Weber for forgiveness of a $60 grocery bill. Then they had turned west toward the wooded San Jose Range, the Alameda Pass, the old Mission San José, and clusters of small farms at the base of the foothills. Crocker and Montague had never seen the country before. "Both were highly pleased," the Judge told Huntington. "It is a splendid agricultural country, well settled up, & will furnish a fine local trade." The Western Pacific road would not be costly. "We shall go right at work & have the surveys made," he said, "& our $800,000 of property there will not be idle much longer."[34]

That month, with construction plans, propositions, and pipe dreams cantilevering across several California landscapes, over alkaline Nevada and Utah, the Associates concluded to formalize their financial arrangements; until then, the Charles Crocker company had contracted to build the line in return for some cash and many discounted securities, which in turn were informally held for eventual distribution to all five Associates. The Judge researched the corporate laws of California, found them liberal as far as limits of liability and other important matters, and drew up a plan. "We have talked this matter over a good deal," he wrote to Huntington on October 25, "& such is the result." Since early in the year, the Judge had pumped Huntington to find out everything he could about the Crédit Mobilier—who was behind it, how were they tied to the Union Pacific, what were their arrangements—and the Associates decided to call their corporation the California Crédit Mobilier. Their names would not appear in the articles of association filed with the state, Crocker said, "but we propose to get our friends' names to them, & then after letting the contracts to the new corporation have been taken, they will transfer their stock to us & resign, & we will take their places & carry on the business & all business we may have under that name."

The next day the newspapers reported that "the French 'Crédit Mobilier' bubble had burst," Crocker would write, "so we concluded to drop the exploded name & have settled on 'Contract & Finance Company.'" The Associates planned that as the work continued the Contract & Finance Company would accept Central Pacific stock in lieu of cash. Over time it would become a large stockholder in the Central Pacific. "In that way," the Judge explained, "less stock will stand in the name of C. Crocker," which, after all, was a potential source of trouble since it was stock they all proposed to share. The capital stock of the corporation was an appreciable $5 million—"pretty large," Crocker admitted, "but the business will be large." Remuneration to the contractor was set at $86,000 per mile, which would be paid half in cash and Central Pacific first-mortgage bonds, and half in Central Pacific stock discounted by 70 percent. Probably the Contract & Finance Company would be building the Central Pacific, the Western Pacific, and the California & Oregon—"so it will have its hands full from the start." Huntington would later wistfully complain that he preferred the old informal arrangement—but it was obvious to all, in this season of dramatic growth, that they had to move to a new level.[35]

In late October, Mark Hopkins happened to be riding on the train as it pulled above the foothills into the heart of the Sierra, and he spied another passenger, a particularly prosperous-looking man who was gaping out the window at the yawning canyons and the daring railroad work. He went over to introduce himself. The man was William T. Glidden of Boston, partner with John M. S. Williams in the clipper-ship business, stockholder in the Cedar Rapids and Union Pacific and the Crédit Mobilier, and newly named trustee for the Oakes Ames contract. Hopkins conducted him like a celebrity up to the cab of the engine so he could see the view better. "He expressed his astonishment at the work," reported the Judge to Huntington. "If he reports & tells the truth, we need not be ashamed of it. I wish you could go over it, & get a full realizing sense of its magnitude & grandeur."[36]

It was grand, and as soon as they connected over the summit they would receive the magnificent sum of $320,000 in government bonds—which would, in Hopkins's words, "show the Railroad & financial world East & West that we are a party of progress & worthy of confidence." It made the four western Associates almost giddy. "For the first time this season Charles talks of making the connection this season with a reasonable amt. of fair weather," the Judge told Huntington. "He & Strobridge say that with 5 weeks—30 working days—they will connect through, which will bring it to the 5th of Dec. with all fair weather." And Charley himself wrote one of his rare letters. "The weather is splendid now," he said on October 30, "& the *Wise People* say we are going to have an open winter." This made him pause, for "they also said [that] last winter—from such a winter as last the good Lord preserve us." If they did not connect, he vowed, it "will be mortifying in the extreme."[37]

By the last week in October, the boardroom combatants of the Union Pacific were in Omaha for an inspection trip out to the mountains and perhaps even a truce of sorts. Durant and Oliver Ames were there, along with Sidney Dillon, Charles Bushnell, and George Francis Train, among others. Samuel Reed was in Omaha to arrange their junket and, having been told of the trustee arrangement for the new construction contract, got the mistaken impression that Vice President Durant's influence was about to be diluted. "T.C.D. will be *one* of seven to manage affairs the coming year," he wrote to Jenny Reed on October 29. "Everything looks bright ahead."[38]

The executives went west, stopping at Fort Kearny for some fun. "None of them knew anything about the west," historian George Bird Grinnell has commented, "but each one wished to kill a buffalo." Major Frank North and Captain L. H. North were detailed to take them south toward the Republican River, with a company of Pawnee scouts. The capitalists rode in ambulance wagons, with saddle horses at the ready once buffalo were sighted. Fifteen miles away from Fort McPherson, the scouts spotted a herd of about one hundred buffaloes. While the others watched from the wagons, Citizen Train, Dr. Durant, and Sidney Dillon mounted horses and were provided with revolvers, while Luther North explained the procedure. They rode as close to the herd as they could, and when Captain North gave the word, the chase was on. The ground was so dry that the galloping horses and buffaloes filled the air with dust—so thickly that it was impossible to see anything until one was upon it. North, way ahead of the others, cut a young heifer out of the herd and began to drive it out of the dust cloud so that the men in the ambulance would have a view, but his horse stepped into a hole, fell, pitched him off, and then ran away toward the excitement. As the captain was dusting himself off Sidney Dillon, his long white muttonchops flapping in the wind, galloped past. "He was not a very good horseman," said Grinnell, who heard the story from his friend North, "and had perhaps lost his stirrups at the first jump and when he went by Luther he was holding on to the pommel with one hand and in the other was the revolver sticking straight out to the right and the bridle reins were flapping on the pony's neck. It was doing its best to overtake the buffalo, but like Luther North's horse, it found a hole and went down and Mr. Dillon was badly bruised, but no bones were broken." North and Dillon limped off trying to capture their horses, which took some time. By this point the herd had scattered, but the scouts managed to maneuver a couple of old bulls back near the ambulance. A Pawnee by the name of Traveling Bear, a large and powerful man, shot one with two arrows; the first was driven up to the feathers, and the second went clear through the body—"a feat that he probably never duplicated," said Luther North years later. Meanwhile, Durant had managed to kill his buffalo, and the scouts a few more. When the hunt was over, some of the Easterners "expressed a wish to see some hostile Sioux," Grinnell reported drily, "and if possible, to witness a fight with them."[39]

Seeking to avoid that, several weeks earlier the peace commissioners had gathered on Medicine Lodge Creek in southern Kansas, about seventy miles south of Fort Larned, to meet leaders of the southern tribes—Southern Cheyenne and Arapaho, Comanche, Kiowa, and Apache—and most of them signed a treaty which allowed them buffalo-hunting rights south of the Arkansas River but concentrated them on reservations, dependent on annuities for thirty years. Then, however, the commissioners headed north on a similar mission but also nearly a complete failure. At North Platte, Spotted Tail and all the other leaders were away with the exception of Swift Bear of the Brulé Sioux. During the brief talks there, the commissioners listened with sympathy to pleas from the Sioux for rifles and ammunition with which to hunt buffalo, but apparently most of those proffered weapons were used within weeks to drive the friendly Pawnee away from their own hunt on the Republican River, sending them back home "without meat and greatly discouraged." The delegation moved farther north to Fort Laramie, on the eastern edge of the Laramie/Black Hills, for a council with the northern tribes. When they arrived at the fort on November 9, they were chagrined to see that the Lakota Sioux and the Northern Cheyenne and Arapaho had not bothered to appear; Red Cloud and the other leaders had sent their regrets, being too busy with the buffalo hunt and too insistent on the abandonment of the Powder River occupations. Only a few friendly Crow chiefs were there, eager for presents. The downcast commissioners headed back to Washington to write a report and plan for another try at peace in the spring.[40]

Meanwhile, the Union Pacific executives had made their way past the end of track—it had passed mile 490 on October 17, and was now closing in on Cheyenne—to see the sights in that raw new station town, and to examine the Laramie Summit route of Grenville Dodge. After all of the summer's delays with Seymour's "improvements," Oliver Ames took almost no time before deciding that Dodge had been right all along. He also stood at the edge of the Dale Creek gorge and approved the high bridge which Dodge and Evans had determined was the only crossing.[41] This accomplished, the capitalists returned to the East, leaving Reed and the other engineers to shake their heads in frustration.

Reed moved his base of operations to Cheyenne to focus on the hills west of town. He was as uncomfortable with the hellish nature of the raw settlement as with its predecessors on the Union Pacific line. "Vigilance Committee were busy last night," he wrote to his wife around this time.

> *When I went from office to breakfast there had been brot in three dead men taken from the place where the vig. comt. had hung them during the night. Three men are reported hung outside of the town not yet brot in. One of the parties hung was tried yesterday for murder and acquitted by the jury. Gamblers thieves & robbers do not make good jurymen, so think the vig. committee judging from their acts. Last night's work may have a good influence on the morals of Cheyenne for a few days at least. Love to all. Kiss the children for me.*[42]

The population had, over the past month, shot to four thousand; town lots that had sold for $150 in August were by early November earning speculators $2,500. Newcomers put up all manner of shelters—shacks, shanties, lean-tos, tents, propped-up wagon boxes, packing crates, even dugouts; there may have been as many as three thousand of these dwellings in Cheyenne. One commentator called them "standing insults to every wind that blows." The *Cheyenne Leader* noted the curious juxtaposition that on the same day, November 9, that Cheyenne's first public school was planned, there now being some two hundred children between the ages of five and twenty-one in town, John Shaughnessy and John Hardy fought a 126-round prizefight in an improvised ring downtown. The first school building would be dedicated on January 5, 1868, as the thermometer registered twenty-three degrees below zero. Prizefighting, too, would become a regular event. On November 13, the tracklayers spiked their way over to the the waiting station house, and soon after Cheyenne's first train whistled into town. Behind the locomotive stretched a long series of flatcars, upon which were stacked the broken-down shacks and dismantled frame buildings, tents, whiskey crates, brass beds, mattresses, looking glasses, barstools, and gaming tables of what was now a dusty, desolate ghost town in the northeastern corner of Colorado. "Gentlemen," called a guard as he swung down off the train into an expectant platform crowd, "here's Julesburg!"

That evening, on Eddy Street in front of the unfinished, unpainted City Hall, throngs of celebrants stood in the flickering light of many torches to mark the happy occasion. General Jack Casement climbed up before the crowd to deliver a short speech, followed by Sidney Dillon and several Cheyenne dignitaries. Nearby banners read, "Old Casement, We Welcome You," and "Honor to Whom Honor Is Due." Above the heads of the speakers hung a large transparency. It bore the motto, "The Magic City Greets the Continental Railway."[43]

On November 30, Stanford, Hopkins, and the Crockers rose before dawn in Sacramento and with their wives all met at the station to take a 6:00 A.M. train. A knot of bleary-eyed legislators and editors joined them—the *Union* and the *Call* were represented, and, for the *San Francisco Bulletin* there was Benjamin Parke Avery, who served as the Central Pacific's Associated Press correspondent-on-retainer. A single passenger car and locomotive waited to take them on a ceremonial ride to the summit—to watch a squad of Chinese tracklayers do their work through the Summit Tunnel. They were treated to breakfast at Junction, and to a lunch of cold turkey, sandwiches, head cheese, pickles, and champagne on the train, arriving at the summit around 2:00 P.M. All onlookers—executives and laborers, politicians and newspapermen—cheered joyously as the beaming Chinese banged the last spikes into place on the eastern slope and a locomotive, its whistle screaming, eased its way into the darkness of the tunnel and out the other side. Correspondent Avery was stationed at a jury-rigged telegraph hookup nearby and wired the news to New York as it happened—"the first ever from the summit," he later told his wife.[44]

"I have long felt that it would be a pleasant sight to reach a point where a train would gravitate toward the East," exulted Mark Hopkins in a letter to Huntington in New York. "For these years past gravitation has been so continually against us that at times it seemed to me it would have been well if we had practiced a while in smaller & shorter hills before attacking so huge a mountain. Often I have thought our Union Pacific Railroad friends were too highly favored compared with us, in having so long a stretch of easy work over meadow bottoms preparatory to their mountain work. Still we have worked on up the mountain—the labored and rapid puff of the Engines told how heavy & hard the work." Hopkins tried to convey the emotion of the moment to his faraway partner: "At last we have reached the summit," he said, "are on the downgrade & we rejoice. The operators & laborers all rejoice—all work freer & with more spirit—even the Chinamen partake of our joy. I believe they do five or ten per cent more work per day now that we are through the granite rock work & can trot along toward Salt Lake instead of remaining in each camp so long that they become sick & tired of it." If the weather continued to be clear, Huntington would have the completion report to the 114th mile. "But," Hopkins added cautiously, "weather is uncertain here in Dec."[45]

Even James Anthony of the *Sacramento Union* was moved by the occasion he witnessed on the windswept peak. "Thus has been completed the most difficult portion of the Company's labors," the *Union* had said that morning. "As compared to what have been encountered, the rest of the route to Salt Lake has no obstacles worth mentioning. The day the first locomotive passes through the tunnel will be a bright era not only in the history of California, but in that of the United States." That said, Anthony's short truce expired once again. The much friendlier Avery rode a pony through the tunnel, making notes, perhaps, for a later dispatch. "The scenery was of the grandest," he wrote, "yet of the most savage and desolate character. The very highest peaks were close to us, bare rock, except for a splotchy covering of thin snow and a few stunted trees. In a deep gorge to the left Lake Donner, to the right lofty cliffs of granite. The air was cold and the wind strong, but I had borrowed an overcoat at the hotel, and with my gloves on was very warm." Sore and a little lame from his unaccustomed horseback ride, the self-described "sedentary scribbler" rejoined the celebrants. On the way down to Sacramento, the engineer kept a slow pace above Cisco, where the track was only a few days laid, but then brought the speed up to forty miles an hour. Then, "the last 13½ miles to the American River bridge were made in 13½ minutes," marveled Avery. "Hold your breath and fancy us going through the night at that rate."[46]

The tracklayers had been busy all November on both sides of the mountains, pushing toward Strong's Ravine from the west and the head of Cold Stream from the east, and graders were busy in between. By midmonth there had been two snowstorms at the summit; then, a heavy, three-day rain late in the month had not only resisted turning to snow but washed away the earlier deposits. It had slowed the work, but not alarmingly so. It was with great relief that in the evening of November 29 they had laid rails up to the portal of the Summit Tunnel—and even with the symbolic victory celebrated with champagne the next afternoon, the Associates prayed for continued easy weather to connect the line all the way down the eastern slope to the Truckee.[47]

Judge Crocker's long, always-optimistic letters poured out to "Friend Huntington" at the rate of two or even three per week, keeping him apprised of the mountain work, of consistently high Central Pacific revenues, of financial conditions in the West, of their new railway acquisitions to the north and southwest of Sacramento, of matters pertaining to Goat Island, of the beginning of a new state legislative session with the Democrats in control, and of their developing equipment needs. Now that their emphasis was shifting higher into the mountains, the locomotive engineers had begun to tell the Judge that their equipment was just not powerful enough, especially after getting above Colfax. "The trouble is that the engines are not heavy enough to make the necessary traction on the rails," Crocker wrote, and "that the wheels slip on the rail before their full power is used." Their heaviest locomotives could only take up to nine or ten loaded cars on the high grade above Dutch Flat, and "this is too small a load for profit." Recent letters from Huntington informing him what locomotives were being sent by ship around Cape Horn were upsetting—he was not sending enough heavy freight engines, and even the biggest on the way were not equal to the Sierra demands. Crocker had read about a hard-duty British "double-jointed" engine and sent the specifics to Huntington. Despite their being forbidden by federal legislation from buying foreign iron, Crocker urged Huntington to procure one of the British locomotives in the name of the California & Oregon Railroad—not covered by the law—so that they could test it on the Sierra.[48]

Such an engine would have come in handy on December 4, when the first load of freight was shipped through the Summit Tunnel, off-loaded, and teamed across the seven-mile gap between the rails, and then reloaded on rail cars for the distance between Coburn's Station to Camp No. 24. The correspondent of the *Virginia City Enterprise*, paying even closer attention to the Central Pacific as it closed in on Nevada, was told by one of Charley Crocker's foremen that they would close the seven-mile gap between the Summit and Coburn's if the weather would just stay clear for another two weeks.[49]

Two days later, losing no opportunity to throw a rolling party for the public relations value of the opened tunnel, the Central Pacific arranged a Saturday junket to dwarf all previous excursions—nearly eight hundred passengers invited from the new Democratic-controlled state legislature, the government, the Sacramento Pioneers' organization, and the press, squeezed into ten passenger cars and two baggage cars arranged, respectively, as eating and drinking saloons, where "something warm" was plentifully offered—the first passenger train to cross the Sierra summit. To protect them from cigar smoke, splashed chewing tobacco, and other offenses, women and their escorts were given sanctuary in the last two passenger cars. The festive train stopped only for wood and water until it reached Colfax, where a second locomotive was attached and they chugged up the steepening mountain grade.

From there onward the car platforms were crowded with sightseers. The *Union* correspondent noted "cabins along the route, some moss-covered, roofless and falling into decay from age, and others more recently built, but now deserted and half torn down," which "furnished the subject of many little stories by Pioneers and others who had passed years of their California life searching for the precious metals." Cape Horn and the yawning gulf above the

American River, sixteen hundred feet below, caused many to gasp and hold on to each other, and passengers exclaimed at the thousand-foot cut at Hornet Hill, the six-hundred-foot embankment constructed with only dump carts at Blue Canyon, and the sights at rainy, misty Emigrant Gap, where the train paused for a moment as if to gather energy for what lay ahead and above. Here the train resumed with such a jerk that the bumper and passenger railing between two cars were torn off, and a man passing from one car to the other suddenly found himself being dragged up the tracks as he held on for dear life. He sustained only a twisted ankle. The car was switched off and another quickly substituted, and the excursion continued.

The weather grew colder. Rain turned to snow, and passengers saw larger patches of it on the way; trees grew stunted and sparse with increased elevation. The cuts through solid rock became too numerous to attract attention, "though the empty powder kegs strewn along the road spoke eloquently of difficulties met and overcome," noted the *Union.* Finally, at half past two they reached the Summit Tunnel and entered it to loud cheers and frivolity. Beyond, on the eastern slope, they went through two more tunnels and halted at the mouth of Tunnel 9. By then the snow was falling thickly and the view of Donner Lake was obscured, but those hardy enough to descend from the cars could stand in the swirling flakes and wind and gape at the majestic mass of rock known as Donner Peak on one side, and the dim, unfolding Donner Lake valley on the other. "It was interesting to note the expressions dropped by the visitors to this mountain spot," noted the *Union,* "many of whom had never seen a snow storm before, owing to a residence in the South previous to migrating to this State, and others who had not seen such a thing for years. Many of the excursionists walked on through the ninth tunnel; others climbed to the tops of rocks to gain a higher point from which to view the scene; while a more numerous body engaged in snowballing with great zest, much to the demoralization of high-crowned hats and immaculate shirt-collars."

Then it was time to return. The train re-entered the Summit Tunnel and got well inside—but then abruptly stopped. For some reason there was no draft through the tunnel as was usually the case, and smoke from the locomotive quickly engulfed the cars. Panic-stricken passengers shut the doors and windows, screaming for aid, and the air inside was quickly fouled. Conductors and brakemen groping through the dark tunnel discovered that someone had pulled the connecting pin from between the third and fourth cars, leaving the rear of the train deep in the smoky tunnel. After a quarter-hour of clamor and confusion the train was reconnected and the cars drawn out of close quarters. Then one of the passengers discovered that in the bedlam his watch and chain had been stolen, and it was generally agreed that the train had been stopped as a subterfuge for the pickpockets. Two men, who had got on the train at

Cisco without invitations, were seized and searched, and although no stolen goods were recovered, the train was halted out in the middle of nowhere and the two men put off.

The adventures were not over. As the train neared Yuba Pass, the fireman walked out onto the engine to oil the valves, slipped on the icy footboard, and fell off. The train was stopped immediately and the fireman, who had fallen on a soft spot, ran up and jumped back on. Farther down, at Colfax, another passenger car had to be switched off, turned around, and recoupled at the rear after it pulled free from the leader. Stanford and the Crockers, by now thoroughly embarrassed, assured the patient passengers that the two damaged cars had been brought up from the Lincoln road—thus, one supposes, spreading the blame to the hapless Colonel Wilson: what can you expect from such a man? "At five minutes before ten the excursionists got back to Sacramento, tired and hungry," reported the *Union*. "The ladies, who had kept in their cars for fifteen hours, on account of the stormy weather, were much fatigued. Notwithstanding all the drawbacks, however, every one was pleased with the trip."[50]

Collis Huntington found himself in Washington a great deal in late October, November, and December, seeing even for him an extraordinary number of legislators as they trickled into town after the elections. Waving "the bloody shirt," for the most part, had failed for the radical Republicans: after the fall election campaign was over, after the Democrats had swept New York, New Jersey, Pennsylvania, and California, coming close in Ohio, Kansas, and Minnesota, among others, after suffrage for African Americans had been defeated in various referenda, Washington in November was a settlement on edge—unsure whether it had just endured a war to end all wars or merely another battle in an unfolding, complicated, often baffling campaign. Thaddeus Stevens, rumored to have been dying only weeks before, showed the rumors to be true when he appeared among his congressional troops, pale, emaciated, and easily exhausted when the special session was called on November 21; even his home town had voted Democratic. It was a place of mixed moods; only Andrew Johnson, who interpreted his party's devastating defeat as a rebuke of the radicals, seemed to be eerily cheerful, waiting for his enemies in Congress to make a move over his suspension of War Secretary Stanton, in August, for treachery; by doing so against the will of Congress Johnson had violated the radicals' Tenure in Office Act. Trying to mitigate the removal by replacing Stanton with the reluctant and unreadable General Grant had surprisingly backfired, however, as the war hero was being villified for "going over" to the president; as far as Johnson was concerned, every day Grant seemed less distinct than his enveloping cloud of cigar smoke.

Huntington, at least, had formed a clear picture of every one of his enemies, and also his friends. In their campaign for Goat Island, Huntington would be the front man, just as Washington would be their first and most crucial test; the California Associates would quietly bide their time, staying away from the state legislature until Huntington obtained his federal grant for the island. As the weeks unfolded through the special session toward the opening of the Fortieth Congress's regular second session, on December 2, he had been in constant contact with his lobbyist, former congressman Richard Franchot, who had already sized up the new California delegation as "not worth much" and who was keeping his eye on two pieces of legislation involving rights to Goat Island. One, of course, was the Central Pacific's. The other was on the behalf of the upstart San Francisco and Vallejo ring calling themselves the Terminal Central Pacific Company—there were some old enemies of Huntington's behind that, such as J. Mora Moss (formerly of the Sacramento Valley Railroad), Alpheus Bull (previously of the California and Oregon, and determined to seize it back), and two Boston financiers, James P. Flint of Flint, Peabody, and his son Edward P. Flint. The elder Flint had earned Huntington's emnity several years before by promising to market Central Pacific bonds and then sitting on them; Huntington later discovered that Flint's greater interest was in rival California firms—and he retrieved his bonds with some heat. Now the son was in Washington making the rounds for Goat Island. Huntington had been there, too—by November 2, he told Crocker, he had already spoken with some seventy members of Congress.

"That young Flint is rather smart and is working like a Beaver," wrote Huntington to Crocker on November 29, "but he has undertaken too large a Job, and he will find it out before his little Ring gets possession of Goat Island." Flint and his allies had begun to boast that their group would control all of the northern California business as well as Oregon's, and thus deserved the Pacific terminus. But Huntington was confident that he had sewed up the Senate Committee on the Pacific Railroad; recognizing this, Flint had arranged for his sponsors to refer his bill to the House Committee on Public Lands and Railroads—but Huntington thought that while the latter committee was not as dependable as the former, it was reliable enough "so to hold *their* little bill long enough for it to have a long sleep." Huntington was feeling fit and sassy—between himself, Franchot, and Fisk & Hatch, he reported bond sales for the month of $590,000. Franchot's estimation of the Washington scene and the Vallejo ring excited him for combat—he was "quite sure that their swindle will not pass the Senate." But there were more politicians to see, and so Huntington went off, as he had every week, toward the Capitol.[51]

"There is a very strong combination against us," he wrote to Hopkins after sizing them up, "but I expect to beat them—but it will cost something." One weapon would be the promise of shares in the underwater property off Goat Island, after they had obtained a grant from the California legislature.[52] Another weapon was the distribution of favors; he had a sixth sense for what a senator or congressman required for his promise of support and any number of different favors lengthened his list of "friends." For Senator Henry Winslow Corbett of Oregon, for instance, it was the carrot of an eventual partnership of the Central Pacific with Corbett and his friends in building a rail line from the northern bend of the Humboldt River up to Eugene—too much to take on immediately, Huntington said to the senator (who had eagerly brought the whole matter up), but when they built to Green River and met the Union Pacific, the venture was a certainty. ("I hardly think it any thing that we want to take up," he privately reassured Judge Crocker.) For Congressman Donnelly on the Land and Railroads Committee, it was just a little something to tide him over: "Mr. Donnelly, MC, and a first-class man on the Land and Railroad Committee, and a good friend of the Central Pacific Road, was a little short of cash," he told Hopkins, "and I loaned him $1000. He says that Flint's Goat Island bill shall sleep in the Land Committee." For Senator James Warren Nye of Nevada, it was a nest egg: "He said that we ought to have [Goat Island] and that he would do all in his power to get it for us," Huntington reported to the Judge. "I told him that he ought to have an interest in the new city [Lake's Crossing, soon to be Reno] that must grow up on the Truckee." For Congressman Samuel B. Axtell of California, it was a quiet job on the side: "I have employed him as our attorney (confidential)," he wrote Crocker. For Senator William Morris Stewart of Nevada, it was no less than his entire re-election chest—"he said he hoped it would not cost over $20,000," Huntington wrote to Hopkins, "but it possibly would cost $50,000." It was a lot of money, but they would be needing his influence in Nevada as well as in Washington. Huntington told the senator that "I would not object" and told him to send his agent, Nevada Surveyor General Gage, to pick up the money from Hopkins.[53] His list continued to get longer and longer.

For Cornelius Cole, newly elected California senator, his requirement was the most intriguing. In late October he had appeared before Huntington to ask his help in a personal matter; the senator's brother, George Cole, had been arrested in a murder case and needed good representation; Huntington agreed to help and called on a prominent New York criminal attorney, James Brady—"he said he would take the case and do the best that he ever done in his life," Huntington wrote to Hopkins, "and that he would want $5,000 for his services, and although I did not agree to pay him, I expect that is what it will amount to." In return, Huntington told the senator that he expected his support on Goat Island, and Cole promised to do all he could. He repeated his pledge every time he saw Huntington over the next five weeks—but then suddenly began to balk; it became clear that Cole was beholden not only to Huntington, who was having his brother defended, and who had also promised the senator a monetary reward for his support, but he was also working for the other side. In early December, after "the Vallejo fellows" circulated the rumor that they had the support of all senators west of the Rockies, Huntington spent a couple of days calling on all of those legislators, and, he wrote the Judge, "they all told me that they would vote for us, and Corbet, Conness, Nye and Stewart will work for us." But Cole was another matter entirely: he had been "backing and filling." Huntington asked him squarely if he was still in their camp, "and after some twisting, he said he would do what he could for me," but not until Huntington promised to take care of his friends—"the squatters on the Island." Impossible, said Huntington—"I would have nothing to do with them," and "I had enough to do without having squatters running after me." Cole would get his money and out of that he must settle with his friends. Nevertheless, with Cole suddenly reluctant to introduce the Goat Island bill on behalf of the Western Pacific, Huntington asked Senator Nye to introduce it. But the reversal exasperated him, turning a mild dislike into something stronger. Later in the month, after Senator Cole had promised Huntington and Hopkins to help with a patronage job for a friend only to name someone else, the situation did not improve. "The fact is, there is no good feeling in Cole," wrote the angry Huntington to his partner; "his heart is cold, and his blood is white, and I do not nor *cannot* like him." Judge Crocker replied that "I cannot believe that Flint has bought him up." He liked to think that Cole was merely mistaken in his constituents' wants, but Huntington was more cynical. Politics had indeed come to a sorry place when a politician would not stay bought![54]

In both Washington and New York that season, Huntington could not help but encounter emissaries of the Union Pacific, and in the wake of all of the internal struggles over leadership, as the survivors continued to jostle for position, Huntington was a cheerful witness. In November, former congressman John Alley had called upon him in New York to run down Durant and to tell him details about the new contractors' agreement—that Oakes Ames was assigning to the seven trustees. "Alley said that I could come in if I would like to," Huntington wrote to Judge Crocker with some amusement. "I told him *No. I thank you.*" He suspected a calculating cleverness—"I understand they are going to Congress next Winter for more aid to help them through the Rocky Mts., which no doubt accounts in part for their cleverness." Some weeks later, a few days after Christmas, he encountered Vice President Durant on a New York street. "He told me that they had a tunnel 2600 feet in length and that it would take 2 years to get through it," he reported cynically to Crocker. "Of course it was a lie, and for a blind." And the next day President Oliver Ames paid him another visit. "He was very good natured," Huntington told Hopkins; "he said the fight was going on between him & Durant, and that he hoped that we would build as fast as we could and get the Road completed, as there was no pleasure to him in going on as they were going, &c. If the cohesive power of plunder was not so great, I should think that there would be a great blow up in the Union Co., but as it is I guess there won't be any."[55]

Oakes Ames was also well aware of the cohesive power of plunder, and of the character of his Washington colleagues. Earlier when he had gone among his friends and associates in and out of politics, trying to sell his discounted Crédit Mobilier bonds, he had found few interested. Now, though, with a sort of peace prevailing in the directorship, and a new contract let which would release past profits, the entire situation was changed. In a widely anticipated move, on December 12 the seven trustees of the new contract awarded Crédit Mobilier stockholders 60 percent in bonds and 60 percent in stock of the Union Pacific, a payment soon increased to 80 percent for the bond allotment. Crédit Mobilier also announced a dividend of 12 percent to cover the present year and also 1866, to be paid in Union Pacific stock valued at 30. All told, then, a Crédit Mobilier stockholder was to look forward to a profit division of 80 percent in bonds and 72 percent in stock. What this involved, of course, was not gold or currency but the promise of such, whether the securities were sold immediately or held until the Pacific road was completed and profits soared. Thus someone holding one hundred shares of Crédit Mobilier stock (worth, at par of $100, some $10,000) would be looking to collect $8,000 in bonds and seventy-two shares of stock in the Union Pacific. If he immediately sold the bonds, selling at $97, and the stock, at $30, it would bring the holder $9,920. To be sure, there were grumbles because the new trustee arrangement stipulated that vote proxies had to be signed over to the trustees—Hazard among several dissenters did not like sending any votes in Durant's direction. Grumbles, however, were easily drowned out; this was an exceedingly handsome return on investment: 99.2 percent.[56]

Suddenly Oakes Ames was the most popular man in the Capitol Building. Even before the dividend was officially announced, word must have somehow leaked out, if not by Ames then by someone close to him such as the former congressman Alley; legislators or their agents literally crowded around Ames, reminding him of earlier solicitations and announcing that they were now ready to lend a hand. "There were so many who talked to me," he said later, "about getting an interest in it, when they began to think it was a good thing. I cannot recollect all the names; I do not know whether they had all spoken to me before. I know that several of them did." As the spectrum of bright, happy, expectant faces passed before him, amid the hearty backslaps and hale handshakes, the cigars proffered and dinner invitations dropped, he was from first to last, he would say later when sitting in a witness chair, "influenced by the same motive—to aid the credit of the road." There seemed to be little thought about whether it might be wrong, or look bad, or that the arrangement might unravel, to be selling securities to the same legislators who would be voting on laws affecting the railroad.

What Collis Huntington had been simultaneously doing around the corner or at the next table was less easy to trace, more in keeping with the way things were always done in Congress (as if that excused it, of course) than this, with Congressman Ames carefully noting every transaction—or at least many of them—in a little black notebook. But Ames had been anointed with the hand of Lincoln on his shoulder and the words, "Ames, I want you to take hold of this," and finally, after all of those years of doffing his hat, he had his colleagues' eager attention. "We wanted capital and influence," Ames would later say in his defense. "Influence not on legislation alone, but on credit, good, wide, and a general favorable feeling. If the community had confidence in our ultimate success, that success was insured." To his fellow Crédit Mobilier trustee Henry McComb, he explained that he was disposing of shares "where they will do most good to us. . . . We want more friends in this Congress, & if a man will look into the law, (& it is difficult to get them to do it unless they have an interest to do so), he can not help being convinced that we should not be interfered with." Keeping his brother informed, Oakes wrote that he was selling the stock "to those M.C. [members of Congress] who will pay for it and not over much to any one of them." Some were inclined to piggishness, demanding fifty or one hundred shares, but Oakes's purpose was to spread the effect as far as he could. By buying in they would be "interested to prevent any legislation that will injure us."[57]

This veritable cornucopia of stock certificates, this cascade of wealth and influence, had to come from somewhere—and that is an interesting story. In June 1867, after Sidney Dillon had been elected president of the Crédit Mobilier, he had examined the books and discovered a block of 650 shares entered in the name of Thomas C. Durant—but never paid for. When queried, the Doctor claimed that he was holding the stock for unnamed persons who had agreed to buy it, but who had not paid him yet. While this may have been partially true—Durant had played loose like this before and would do so again, as would Oakes Ames—it was also the Doctor's poker style to slide some of the chips into his lap when no one was looking. Most of his fellow players suspected the latter.[58] But as soon as he heard of Dillon's mysterious discovery, Oakes Ames spoke up—he, too, had a string of unnamed politicians and capitalists to whom he had promised shares.[59] Dillon did nothing further about the stock through the summer and early fall, waiting to see how the succession of boardroom battles would transpire. In October, preparing to dispose of it, the Crédit Mobilier transferred the block from Durant's name to Dillon's as trustee. And suddenly it was even more alluring: as original stock the eventual bearer was entitled to take advantage of the capital increase announced early in the year, by buying an additional 50 percent of stock—if he owned a hundred shares, for instance, he had the privilege of purchasing fifty more shares, collecting the bonus Union Pacific $1,000 first-mortgage bond also declared in February for every $1,000 invested in Crédit Mobilier. Then the happy investor would gloat over the anticipated Crédit Mobilier dividend on all 150 shares.

Of course Dr. Durant and Oakes Ames renewed their claims for the 650 mystery shares—on behalf of their deserving friends and colleagues. Congressman Ames had the added urgency of all of the colleagues' names in his little black book, and presumably others. But then Henry McComb, original Crédit Mobilier stockholder, Union Pacific board member, friend to Durant, and perhaps after Durant the brashest of the lot, interjected a claim of his own: he, too, had obligated himself to others for some stock. He demanded his due, 250 shares. Dillon put the matter to a vote of the major stockholders: himself, the Ames brothers, Durant, Bushnell, Duff, Bardwell, and McComb. In the ensuing compromise, they voted to divide the bounty with 370 shares going to Durant and 280 to Oakes Ames; McComb went along with the majority—for the time being.[60]

The compromise did not last long. With his welcoming reception in Washington, Congressman Ames still needed more shares to distribute. Moreover, the disappointed McComb insisted that Ames must sell some of his stock to McComb's friends, such as Delaware senator Richard Bayard, Tennessee senator J. S. Fowler, and two Iowa representatives, William B. Allison and James F. Wilson. Ames had no choice but to ask his partners to give him ninety-three new shares of Crédit Mobilier, which were quickly issued. This, though, prompted McComb to renew his demand for 250 shares, including privileges (which would give him a total of 375 shares), dividends, and bonuses. To keep him happy, Oakes Ames lent him 500 shares; thus McComb would be able to add them to the 750 shares he had originally bought and paid for, to collect the dividends on all 1,250 shares when the profits would be paid, on January 3, 1868.[61]

Only this is known about the congressional customers of Oakes Ames: there were at least two U.S. senators and nine U.S. representatives, from seven states. One, Schuyler Colfax of Indiana, was Speaker of the House, and would soon be elected vice president; another was Representative James Garfield of Ohio, who in thirteen years would be elected president of the United States; another was Senator Henry Wilson of Massachusetts, who would become vice president in 1873. Iowa congressmen William B. Allison and James F. Wilson signed on; two Pennsylvania congressmen, William D. Kelley and G. W. Scofield, came in; two midwestern congressmen, John A. Bingham of Ohio and John A. Logan of Illinois, joined up; and two New England men, Senator James W. Patterson of New Hampshire and Representative Henry Laurens Dawes of Massachusetts, jumped aboard. Amounts were in even blocks of ten or twenty shares, though Senator Patterson took thirty. Congressman Ames was very accommodating about arrangements—flexible, even elastic, in fact. As "a convenience" but also probably for discretion, for the most part he held on to the stock, keeping it in his name. Henry Wilson, Scofield, Patterson, Bingham, and Dawes all paid cash up front, receiving full benefits; Colfax and Allison accepted their blocks without paying, waiting until the bonus was credited to them before paying for all at what amounted to a discount of nearly 70 percent; Garfield, Logan, Kelley, and James Wilson claimed to be short on cash, so Ames obligingly paid for their stock himself, holding on to the securities but sending them dividends. As events unfolded, three of the lawmakers had changes of heart in 1868 or 1869 because of possible suits or even ethical considerations, so Ames bought Allison's and the Wilsons' shares back for cost and interest.[62]

This was not the extent of it (nor, in fact, the beginning of it, as Ames had brought in Massachusetts congressmen Samuel Hooper and John B. Alley, since retired, as stockholders with 750 and 290 shares in 1865 and 1866, respectively, and Indiana representative James Wilson Grimes in a 380-share joint account with Ames in 1866). Durant had previously sold Pennsylvania representative Benjamin M. Boyer and his wife some one hundred shares in 1866, and now he found a new customer in the New York congressman James Brooks, the only Democrat to be so involved, who had been offered two hundred Crédit Mobilier shares a year before and turned Durant down. Brooks chased the Doctor down only after the December 12 dividend had been declared, but nonetheless he said he was ready now to buy the two hundred shares. Durant let him have one hundred. Then, despite having missed the dividend deadline, the congressman demanded an additional fifty bonus shares; Ames hesitated long enough for Brooks to go over his head to Sidney Dillon; Representative Brooks got his reward. But since the congressman had been named as a government director of the Union Pacific in October, which precluded him from owning company stock, he had all the shares entered with his son-in-law as the straw owner. Elsewhere, Springer Harbaugh had tried to sell up to three hundred shares to Representative J. K. Morehead six months earlier, to no avail. And some politicians had demanded shares but were actually turned down as insufficiently friendly or deserving, such as Tennessee senator Joseph S. Fowler. Others had been approached—by Ames, Durant, or McComb—at the Capitol but had declined: Delaware's James Bayard, Maine's James G. Blaine, and two more sons of Massachusetts, George S. Boutwell and Thomas D. Eliot. Later Blaine, who would become embroiled in his own damaging railroad scandal involving a land grant in Arkansas, would piously defend himself and Oakes Ames: "It never once occurred to me that Mr. Ames was attempting to bribe me."[63]

That winter Oakes Ames received a total of 373 Crédit Mobilier shares—with the option to acquire half again as many. Eventually it would come to light that he had disposed of 160 shares to eleven members of Congress, and 30 shares to a private citizen who has yet to be tied to any elected official. For twelve decades investigators and historians failed to do the elementary math; then, in 1987, Maury Klein in his official history of the Union Pacific, asked: What happened to the other 183 shares? "Two possibilities suggest themselves," wrote Klein, "each one fascinating in a different way. Oakes may have kept the remaining stock for himself, or he may have sold it to congressmen whose involvement escaped the notice" of the eventual investigating committee.[64]

That December, one particular congressman missed the train—though he arrived at the Capitol station in plenty of time and was well known to Oakes Ames, dispenser of tickets: the representative of Iowa's Fifth Congressional District, Grenville Mellen Dodge. Sharing a house with another Iowa legislator, James F. Wilson (who was complaining to Dodge at not getting fifty shares of Crédit Mobilier from Ames but only ten, on credit), Dodge was keenly aware of the stock distribution taking place. He had asked for some earlier, to no response, and in the feeding frenzy of December he asked again but was put off—perhaps because Oakes Ames was focusing on active congressmen whose work on legislation would be helpful, and Dodge was largely an absentee. It would take a personal appeal to Sidney Dillon, in March 1868, to dislodge one hundred shares, which Dodge purchased in his wife's name.[65]

For Dodge as well as the other railroad legislators looking after the concerns of the Union Pacific or the Central Pacific among the constant oversupply of special interests, to be sure, there was other business to transact in Washington at the close of 1867—attention to the administration of the conquered South, to Indian affairs, the military, the national budget, the championing of Stanton and the curtailment of the Johnson presidency. There was also the matter of a bill introduced on December 5 by a Wisconsin representative, Cadwalader Washburn (a retired general who had served near Dodge in Arkansas, Tennessee, and Mississippi, and who now considered himself a guardian of the public purse), to regulate—and reduce—the fares and freight rates of the Pacific railroads. It was assigned to Oakes Ames's Pacific Railroads committee—where most considered it would slumber. "They will not do anything with it this session," Collis Huntington reassured Judge Crocker.[66]

One man most emphatically being kept awake about freight charges that month was General Jack Casement, after being informed by the Union Pacific operations superintendent Webster Snyder in Omaha that henceforth the contractor would be charged the regular tariff for all the supplies he shipped on the railroad out to the work camps. Previously he had not been charged for these supplies—some of which were employed in the contracting, some of which fed and outfitted his gangs, and some of which went to the privately owned general stores Casement operated for his own profit at the track towns. The question of freight charges for the track contractor had never been emphatically settled, which had allowed the little general much maneuvering room—and higher profits. Contemplating another winter layover, he had ordered hundreds of tons of supplies brought out to Cheyenne. "As I made him pay us some $33,000 for freight that he expected to get free," Snyder wrote to

Dodge on December 13, "the genl is not one of my warmest friends." Although at that moment Dodge was seeking favoritism and a free ride regarding the Crédit Mobilier stock bonanza, he was inclined to agree with Snyder about Casement and wrote to New York to back him up.

However, there was another matter of disagreement: the efficiency of, and indeed the long-term prospects of, Construction Superintendent Samuel Reed. Back when Snyder and Reed were sharing an office in Omaha there had been a tenseness between them, a tinge of competition more keenly felt by Snyder, and as the supply line lengthened dramatically in 1867 and the line bore more regular and special traffic, there was a very evident friction between them. Adverse weather, constant raids by tribal war parties, and ambushes by Colonel Seymour with his "improvements," did not speed progress or make either superintendent's job easier that year, but the burden fell especially on Reed, who was often far from a convenient telegraph operator, who was not an inveterate complainer or a political schemer, and who disliked pinning the blame on others. Snyder could contain all of these qualities—but he, like Reed, was an honest man, though Snyder could be irritatingly smug about it. Being thoroughly convinced that no one could do as good a job as he could did not help matters either.

On December 5, Vice President Durant had enough of the complaints and also enough of the friction between the operating and construction departments, which was slowing progress during the final frantic weeks of tracklaying before winter stopped movement in the Laramie/Black Hills. "There must be no more delay in sending Reed iron and materials," he telegraphed Snyder, "even if you suspend all business except passenger and mail. Any further delay will be followed by an order giving Contractors control of trains for the present." Snyder, stung by the rebuke and pained by the operational headache this would cause him, could only comply. But he could also—obliquely—complain. "Reed dont keep up with his work—tanks not enclosed, cuts and embankments not in shape to run trains safely, not a bridge between Sidney and Cheyenne that might pass muster," Snyder wrote to Dodge a few days later, and Dodge, knowing that these complaints were buzzing over the wires to New York, had hastened to his fellow engineer's defense several times in November and December. He also wrote Reed to reassure him a few days after Christmas that he still had a job and that he would soon be writing Reed some suggestions about the coming year's plans; Reed should put more pressure on his assistants, keep "as pertinacious as possible" with Snyder, send all of his New York letters and telegrams directly to Henry Crane so that none were overlooked, and by all means, "steer clear of New York complexities, which we have no direct interest in."[67]

For his part, Dodge planned to stay in Washington and see to the backlog of patronage demands as the second session got under way. In the early spring he would be going back west past the head of the route across the

Laramie/Black Hills—the place he had named Sherman Summit after his commander and friend. Another old comrade was being transferred in by Sherman to assume command of the Department of the Missouri: tempestuous Major General Philip Sheridan, who had been removed by the president as military governor of Texas and Louisiana for his ties to the radicals and his overly harsh, even cruel dealings in those conquered provinces. In the West, Sheridan would oversee the states of Kansas and Missouri and also Colorado Territory, not to forget the Indian Territory (present-day Oklahoma) and New Mexico—given that tribal hunting patterns did not respect the whites' military divisions, the command would affect not only General Augur's Department of the Platte directly to the north but also, in a great way, the future of the railroads.

Immediately after taking command at Fort Leavenworth, Sheridan had begun an extended leave, spending much of the autumn and into the New Year in Washington, enthusiastically making the rounds of private parties and official receptions, seeing Dodge, Sherman, Grant, and other old comrades often. Out west, in 1868, Sheridan would rediscover himself, in the process coining a phrase that would outlive him: "The only good Indian is a dead Indian." To his former comrades-in-arms in Washington as the year 1867 waned and winked out, Sheridan likely voiced the strategies he meant to follow—the sentiments would find their way into his first official report, for that matter: the only way to guarantee peace was that the Indians "be soundly whipped, and the ringleaders in the present trouble hung, their ponies killed, and such destruction of their property as will make them very poor."[68] A war of attrition on the plains seemed a certainty.

As a note of interest, Sheridan's successor in Texas and Louisiana would be General Winfield Scott Hancock, burner of peaceful Cheyenne villages and igniter of the prairie war of 1867, rewarded for his efforts with a frustrating new job made more so by the fact that he was a Democrat with no sympathy for Reconstruction, and in any case could not have these subjugated people shot or driven off. In the face of the tragedy he had helped create on the plains, with deaths of noncombatants far outweighing those of bluecoats or warriors, the tracks of the Union Pacific had—miraculously, given the perils—extended some 245 miles from O'Fallon's Bluffs on the Platte to a point 20 miles beyond Cheyenne and just 13 miles shy of the summit, and, though few were complaining, 30 miles short of reaching Dodge's year's goal of the Laramie Plains.

Out in the rising winter winds the Casements put up a station there on the steep eastern slope—named Granite Canyon for the furrowed outcrop of hard stone they were encountering, a resistant, red-tinged rock which, when rendered as fine gravel, would henceforth delineate the roadbed. The tracks were emphatically into the terrain for which the federal government would pay the Union Pacific $96,000 per mile, and by year's end the railroad would have earned $3,840,000 in subsidy bonds for the work of 1867; the government business in freight, mails, telegraph, and troop transportation was $813,000. Its returns on operations combined with energetic land sales in Nebraska, Colorado, and Wyoming had brought in a profit of about $2 million. As Dodge would also report to the directors, the Union Pacific was operating regular train service over a 517-mile line, employing 53 locomotives, 17 passenger cars, 6 baggage cars, 838 freight cars, and 99 handcars—and this did not count what was being used to keep the contractors going.[69] Boldly the directors predicted that they would build 350 miles in 1868. The corporate future looked at least as rosy as the granite Casement's graders were removing, but as Jack Casement noted a few days before Christmas, they were now up against the hard stuff.[70] The notion of "blasting" had now crept into correspondence with New York, meaning higher costs for him and a pass-on to the company. More deep rock cuts, embankments, the towering Dale Creek Bridge, and even tunnels lay in the future. Stockpiling of materials continued even after Casement laid off most of his men, leaving them to seek warmth and comfort in Cheyenne, whose garish lights lit up the winter sky as never seen before. The Union Pacific was officially in high country, but far off to the west, the Central Pacific had all but completed its herculean trial over the Sierra.

Around dawn on December 13, a thick fog crept out of the deep river canyons onto the western slope of the Sierra and tenaciously held on. Only a few hours later there was a horrendous din—a Central Pacific construction train had collided head-on with a quarried-granite train, badly damaging the locomotives and splintering and heaving cars off the track. Several men were badly injured. The damage estimate was $8,000. "Thick fog was the excuse," Mark Hopkins wrote to Huntington, but an "incompetent train master was the cause." With all their dreams of a railroad empire, the Associates could not find someone to manage their train traffic. The company was "much in need of a *thoroughly competent & honest man* for 'train master,'" Hopkins continued. "We have none & can't find one. I wish you would send us one."

Judge Crocker explained the problem: the company's first train master understood the business well, "but was dishonest in many matters." He had been fired. His replacement did very well also, "but we discovered he was cheating us," and he was fired. One of the two had even set up a system to extort a percentage of his underlings' salaries, on top of his other perfidies. In desperation Hopkins had promoted one of their oldest conductors—but he had proved so incompetent that the operations schedule became a disaster, and through his carelessness those two heavy trains had been sent down the same stretch of track at the same time. "We are in continued fear that other accidents may occur," Crocker told Huntington. "It needs a man who has been brought up to the business of railroading, & knows all about it. We beg of you to look up a fit man there, & get him to come out. There are no doubt plenty there. We have not railroad men enough here to select from. By & by they will grow up on our road, but we cannot wait for that." Such were the perils of their growth and success—and both Hopkins and Crocker urged Huntington to find them "a first class railroad business man" to assume the post of assistant superintendent, an important job given the Associates' own lack of experience in running the road.[71]

On Monday, December 16, rain began to pour down on Sacramento. On the summit it took the form of the snowfall that would stop them in the Sierra for the winter. Charley Crocker, up there to distribute the payroll, telegraphed that six feet had fallen and was "too deep to handle." He ordered the whole workforce moved to the Truckee, leaving the seven-mile gap uncompleted. "It is too bad, to come so near it and fail," E. B. Crocker wrote to Huntington. "But we have done our best." He had craftily worded the commissioners' report on the completed line to skip over the inconvenient gap, prompting praise from Huntington, and now, even in the face of being driven out of the mountains by the storms, he saw no need to send in a correction. Fortunately, for the sake of construction progress in the winter they had managed to transport twenty miles of iron, two partially disassembled locomotives, and ten freight cars across the gap with ox teams before the storms had struck. Snowstorms continued to rage in the mountains for five days, but, as Judge Crocker boasted to Huntington on December 23, "the Central Pacific has come out of them with flying colors. Not a day has passed that the trains have not made the trip over the road. The snowplows, with the assistance of the snow sheds, have had no difficulty in keeping the track clear of snow up to Cisco," where the snow had thus far piled up to three and a half feet. He thought there would be no difficulty by the next winter, when they had closed the gap and the track had fully settled, in operating the road clear through. Huntington happily replied: "After some of these terrible storms it would be well to send or have sent over to the Associated Press here, how the storm came and beat upon the Central Pacific Road, but it harmed it not as it was founded on a *Rock*."[72]

Many years later the Central Pacific carpenter A. P. Partridge recalled the winter of 1867–68 (though he or his transcriber seriously erred, by more than three months, in noting when the snows had closed down the summit work, leading several careless historians to claim the onset of the Sierra winter was in August). Though fogged by time, Partridge's memory was detailed when it came to the rush and confusion down at the Truckee River when some four thousand laborers were hurriedly moved in from the stormy upper slopes. "Most of the Chinese came to Truckee and they filled up all the old buildings and sheds that were in Truckee," he recalled, indicating that even in the Truckee work camp they were surprised by the sudden amount of precipitation. "With the heavy fall of snow the old barn collapsed and killed four Chinese,"

he added. "A good many were frozen to death. There was a dance at Donner Lake at a hotel, and a sleigh load of us went up from Truckee and on our return, about 9 A.M. next morning, we saw something under a tree by the side of the road, its shape resembling that of a man. We stopped and found a frozen Chinese. As a consequence, we threw him in the sleigh with the rest of us, and took him into town and laid him out by the side of a shed and covered him with a rice mat."[73]

As Charley Crocker and Jim Strobridge worked to reorganize the labor force into grading parties to be sent beyond the Nevada state line, where some twenty miles of grade had already been completed, Judge Crocker and Leland Stanford tiptoed around the Democratic state legislature, which had opened its new session with a hearing into legislation to lower freight rates of the monopolistic Central Pacific, and quietly worked in both Sacramento and San Francisco to promote their interests in Goat Island. They put together a petition and memorial to the U.S. Congress on behalf of the Western Pacific which, they hoped, would persuade Congress to name their company as beneficiary of rights to terminal property on Goat Island; the petition involved some fancy footwork also, as Stanford persuaded a number of investors in the various railroad corporations in northern California to sign in their support—carefully avoiding sworn enemies in the top echelon of the Terminal and the Vallejo lines, but hoping to undercut their position by signing up some of their stockholders. The requisite local approval of underwater rights off the Oakland shore would not be sought until they got the go-ahead from Congress, which they hoped would persuade even the Democrats in Sacramento.[74]

Other political matters had reared up in that last month of 1867. On the complaint of a landowner in the mountains, a woodchopper working for the Central Pacific was arrested at his work in the forests and charged with trespass and theft; he had been cutting fuel for locomotives in an even-numbered section of mountain land, and the Central Pacific had been granted only odd-numbered sections in the congressional railroad laws for the purposes of land sales. It had grown out of a mere "piece of neighborhood spite," Judge Crocker told Huntington, but the complaint had drawn the U.S. District Attorney into the matter. Although the official was reluctant to proceed with prosecution, he wanted a ruling from the federal land office.

It was complicated by the fact that "little of the land had been surveyed, and that which has been the lines are so imperfectly marked that it is nearly impossible to trace them," Crocker explained in a letter which Huntington was supposed to forward to the Washington authorities. The company had suspended woodcutting until the matter was resolved, despite needing "great quanities of this timber for ties, bridges & snow covering, & for fuel" not only for the California stretch but probably for all the way to the Great Salt Lake. "You will at once see the importance of the question," he continued, "for as our wood choppers are mostly very ignorant men, and cannot tell one section of land from another, they don't know where the odd sections are, and when found they are often very inconvenient to get at." The letter of the 1862 Railroad Act clearly gave them rights to all timber within twenty miles of the line, he said, and the company was the victim of squatters and grabbers who had gone onto public lands near the railroad, "marking out immense tracks of choice timber land by merely marking and girdling trees or felling trees in line to make a rude fence, & claim thereby to be in actual possession of & are stirring up prosecutions against our employees." Crocker hoped that his letter would elicit a quick and favorable ruling.[75] Their hapless woodchopper and others of his calling were reluctant to go back into the forests if they were going to be dragged into the calaboose for their trouble.

Other natural resources were on the Judge's mind. Early in December he had talked with Clarence King, the geologist who was then in the midst of a geological survey of all the cordilleran ranges from California to eastern Colorado, a mammoth government project following the 40th parallel; King would move from that seven-volume undertaking to lead the U.S. Geological Survey, later writing a series of charming scientific sketches about the Sierra. When Judge Crocker consulted with him, King was just twenty-five, a few years out of Yale, obviously brilliant, and thoroughly versed in the commercial prospects of the western terrain. There were immense, high-quality peat beds along the Humboldt, he told Crocker, and the coal veins on Snake River were the best he had seen in the West; he had no doubt that the Goose Creek country would be rich in coal, and saw no reason to dispute the Central Pacific's surveyors, who had sent in a report of vast coal deposits in the Wasatch Mountains.[76]

Crocker passed the news on to Huntington, and a few weeks later he had a reply. While Huntington had been dining with California senator John Conness, the senator told him he had just heard from Clarence King that there were few indications of coal west of Great Salt Lake—but that he felt the coal beds in the Wasatch were "of immense value" to them. Huntington had recently pumped a Utah congressman about it, also, who agreed—but he said that "Durant and his party were *bound to control them.*"

To Judge Crocker, Huntington posed the question, Would the Union Pacific do it? If "one and only one" of the Associates in California were running the Central Pacific, Huntington wrote, "I should say no, they will not." But he was worried about the challenge the Central Pacific faced with its executives' divided responsibilities—previously the five Associates had each found a niche and performed extremely well, running by committee and propelling the corporation from an upstairs storefront office and a few shovelfuls of muddy earth to its place now, on the pinnacle of the Sierra, against all naysayers' predictions, looking eastward across the continent. But the view was toward a race with even greater stakes than they had ever contemplated back when they were mere storekeepers with ambitious dreams. Would the strength in numbers they had relied on for their success now prove their undoing? Did one of the men in Sacramento need to rise and take over the next campaign? "Few great battles were fought after a consultation of war," Huntington cautioned—but who in his eyes would be that great man? He left it unsaid. Not Stanford with his horror of detail. Not Hopkins with his disinclination to leadership. Not Charley Crocker with his lack of polish and subtlety. Did he mean Edwin Bryant Crocker? As was common in Huntington's letters, the line of thought played out like footsteps vanishing into the wastes of Nevada and Utah.

But it was, and is, only speculation. One had to shake off one's uncertainties. The future beckoned. Huntington had sold half a million dollars in bonds in December and fairly preened in satisfaction. Their entire completed total for 1867 had been only thirty-nine miles, but it had been across the roughest, most challenging terrain of the entire transcontinental route, and they had managed it without any federal mortgage bonds that whole year. Of course there was the galling seven-mile gap east of the summit, but they would overcome that—and the bonds would flow in. They had succeeded in engulfing the little northern roads and the Western Pacific. If Huntington's hunch was right, they would have their fingers on Goat Island soon. They had considerably increased their rolling stock: from 19 to 51 locomotives, from 199 to 443 freight cars, and from 6 to 10 passenger cars, the last of which would be sufficient to meet the limited traffic to Cisco but soon in need of correction. Their freight and passenger receipts hit higher plateaus each month. So Huntington shook off his misgivings about a divided command.

But he as much as issued a challenge to Judge Crocker. "I think," Huntington wrote on December 27, "if 1868 should be financially as 1867, I can supply you with what money you will need to lay 350 miles of Road, and if so, can you do it?"[77]

Part VIII

1868
Going for Broke

27

"More Hungry Men in Congress"

Not since the time of Theodore Judah and his Pacific Railroad Museum in the Capitol building had a railroader been assigned such a Washington lair: Congressman Grenville M. Dodge found his office more than convenient. It was in the warrens of the Interior Department building, with a telegraph and clerk within close reach should he need to communicate with his men in the West, or with the office in New York, or with Oliver Ames in Massachusetts. He took advantage of it frequently. "It was a fortunate thing that I happened to be in Washington," Dodge would record, "as a great many questions were brought up in Congress and the departments"—the Washburn railroad tariffs bill, for instance, and Indian affairs, and departmental approval of railroad routes, and acceptance of track sections, and granting of federal bonds, and certification of railroad land sections for sale. Interior Secretary Browning was not far away, though even with that proximity Dodge would have to keep his eyes open—"there was a great deal of friction," he continued, "on account of the attitude of the Central Pacific and the friendliness of the Government to them, and as I was right on the ground, I could generally meet and solve [all pertinent matters]." As far as Dodge would be concerned, that friction would all but send sparks up in his face.[1]

But until then he felt himself in the center of things, and not just in railroad matters. Sharing a house on F Street with Iowa representative James Falconer Wilson, he found his housemate shared many sympathies. Wilson's hometown of Fairfield was in southeastern Iowa and within the gravity field of the Mississippi and Missouri Railroad, Dodge's old line; Wilson, with his newly won Crédit Mobilier stock from Oakes Ames, was among "the believers" in Congress and in January pocketed $329 in dividends despite not having paid any money for the securities; as the Republican chairman of the House Judiciary Committee, Wilson had turned away the feeble impeachment charges against President Johnson the previous summer (most ridiculous, the evaporative links between Johnson and the Lincoln assassination) as having no hope of prevailing in a Senate deliberation, and had done so again in December, wanting a stronger case. But Wilson would serve as one of the managers (or prosecuting attorneys) in the third impeachment attempt. To the glee of Johnson's most rabid enemies (and to the relief of the defiant president, who was girded for the legal battle), the process began before too many weeks of the New Year had elapsed, and although Dodge was a minor player, he saw it all from his central vantage point.

Dodge's consumptive expeditionary friend, John A. Rawlins, became an emissary between the radicals and his commander, Grant—that unhappy interim tenant of the War Secretary's office since Stanton had been suspended by the president. Visiting Dodge's and Wilson's house almost nightly for political advice and instructions, presumably sharing an anecdotal chuckle over being saved from a charging grizzly bear by Dodge only a few months before, Rawlins ran Grant's messages by the congressmen before they were conveyed to Johnson. By now it seemed certain that Grant would get the nod from the Republican Party for the 1868 presidential campaign. He urgently needed to put some daylight between himself and Johnson. On January 10, the Senate Committee on Military Affairs vindicated Stanton's position as secretary of war, affirming the solidity of the Tenure in Office Act which Johnson had violated by removing Stanton despite congressional disapproval. Grant overlooked his promise to confer with the president before taking any action and passed the keys to his office back to Stanton. Having thus regained his command, Stanton defiantly refused all eviction notices from the president, actually moving into his office in the War Department, taking his meals and sleeping there, surrounding himself with a phalanx of radical congressmen.

Meanwhile, Johnson continued to try to persuade Grant to come to his aid, to no avail. On February 10, Grant tried to draw the politics-hating Sherman into the fray by asking his friend to prepare a statement endorsing Grant's position. It greatly upset Sherman. "I never felt so troubled in my life," he wrote Grant from St. Louis on February 14. "Were it an order to go to Sitka, to the Devil,—to battle with rebels or Indians, I think you would not hear a whimper from me, but it comes in such a questionable form that, like Hamlet's Ghost, it curdles my blood and mars my judgment. My first thoughts were of resignation, and I had almost made up my mind to ask Dodge for some place on the Pacific Road, or on one of the Iowa Roads, and then again various Colleges ran through my memory; but hard times and an expensive family have brought me back to staring the proposition square in the face, and I have just written a letter to the President."[2]

Within days Johnson—whom one critic would call "the nightmare that crouches upon the heaving breast of this nation"—would find a substitute to send in Grant's place in combat against Stanton and his congressional supporters.[3] The best he could secure under the circumstances was the elderly army bureaucrat General Lorenzo Thomas, a bibulous windbag who celebrated to excess his ascension to high office only to be awakened the next hungover morning by a federal marshal with an arrest warrant, for conspiracy with the president of the United States to violate the Tenure in Office Act, signed by Secretary Stanton. The confrontation between General Thomas and Secretary Stanton was one of high hauteur and low comedy (T: "I shall stay here and act as Secretary of War." S: "You shall not and I order you as your superior back to your own office." T: "I will not obey you but will stand here and remain here." S: "You will not." Etc.). It had a befuddled ending in a morning's bottle of whiskey, with Stanton doing the pouring, and with Thomas's grateful, relieved capitulation.

The next day, Sunday, February 23, a delegation of fifty legislators, including Dodge, crowded into Stanton's office to urge him to stay and fight, especially as the president had just handed them the legal means to impeach by sending Lorenzo Thomas to remove him. At their head was Schuyler Colfax, who handed the secretary the signatures of sixty other supportive congressmen and declared that Stanton was "the Thermopylae, the pass of greatest value to reconstruction by Congress," that he "represented more than any man of the day the policy of Lincoln and the spirit of the people who crushed the great Rebellion." Nine or ten others delivered similar sentiments, but Representative Dodge's was the fieryest. Stanton was "the greatest war minister ever upon the earth," Dodge thundered, "the victim of the hate and venom which characterized the official acts of the renegade at the White House toward loyal officers and people throughout the land." The noble Stanton had won a "world-wide reputation which would survive the reputation of the President."[4]

As Dodge spoke in that room thick with cigar smoke and rhetoric, the president and his lieutenants were at the White House scrabbling for legal precedents and a joust before the Supreme Court. And soon the Committee on Impeachment triumphantly forwarded its go-ahead to the larger body, scouring Johnson across eleven articles of impeachment; ten had been written by Stanton himself. It would take two full days of speeches—during which the city police had to be called several times to quell disturbances in the packed galleries—before a majority of the House voted for Johnson's impeachment and sent the articles to the higher body, on March 5.[5]

The grand inquest within the Senate, with Chief Justice Salmon Chase presiding, would unfold over the ensuing three months, and Dodge kept himself posted as a House member and frequently commented on the proceedings to family and associates back home. During this time the machinery of government would crank on, at quarter speed. For Congressman Dodge, however, there were also the demands of the Union Pacific and the coming year of construction. As supervisor of the land department of the Union Pacific, Dodge pushed to open up government lands along the right-of-way to settlement (planning on selling plots as cheaply as $2.50 per acre) so that the price of the railroad's reserved parcels would rise; he also investigated how other railroads handled their real estate records, hoping to improve upon their experience, and set in motion an inquiry into all homesteads settled within the one-hundred-square-mile limits adjoining the Union Pacific tracks, with the intent to contest squatters' claims, evict, and take possession.[6]

Dodge also made plans for even more roads. There was the branch line to Denver, the construction of which would begin by midyear. And there was Oregon. His report of his early autumn reconnaissance with Rawlins in northern Utah and southern Idaho, looking to a line along the Bear River and Snake River toward Oregon, had won approval and elicited a letter from Oliver Ames in January: the plan had "great merit," Ames said. "The Union Pacific Railroad Company will use their best efforts to secure the construction of this Road whenever the people on its line shall awake to its importance. It cannot be built without good aid, but our Co should not appear as applicants for this Charter, but it should come through the exertions of Representatives of Oregon. . . . I trust you will be able to enlist active workers . . . without cloging it with too many other interests." His colleagues from the Northwest would of necessity be recruited, Ames seemed to be saying, without recourse to Crédit Mobilier or Union Pacific inducements.

Aware that mechanisms were already moving in Congress, Dodge wrote to Samuel Reed to tell him that the main line was now not their only concern, that beyond the Denver branch would be the Portland branch. "If we get the legislation that the N.W. Coast ask," he said,

> *we will have that to build via Ham's Fork, Bear River, Port Neuf, Gap and Snake River to Columbia River. All the N.W. Coast, Oregon, Washington, Idaho, and Montana are working vigorously for it and even if we do not get it this winter, we will next. This will give us the best and shortest route to the Pacific, avoid the Sierra Nevadas, and make us a star of the through lines, and at the same time, kill the Northern Pacific. Our company will build the Portland Branch even if we get no Government aid. By it we can get to navigable waters of Snake and Columbia in 300 to 400 miles.*[7]

Dodge was still concerned about protecting Reed from snipers within the organization. Blessedly, Dr. Durant had taken his family and sailed for Europe and was not expected back until early March, but Seymour was still skulking about, and there was Operations Manager Webster Snyder in Omaha, who meant well but who had gone out on an inspection trip along the completed line and was expected to find fault which he would lay on Construction Manager Reed and forward to Oliver Ames. So Dodge sent a long list of recommendations to Reed in Cheyenne, some of which he had also presented in New York and some of which Reed had enacted on his own initiative. Over the winter some two hundred miles of iron was to be stockpiled west of Cheyenne at the rock-cutters' camp of contractor Lewis Carmichael; as Dodge noted, after tracklaying resumed they would thus not be dependent on the five hundred miles of road in the rear, and the opened line would be freed to concentrate on profitable passenger and freight traffic. He also urged Reed to "test the country in advance of track, dig your wells and get at the living springs of the country if there are any, and when you find water in abundance, accommodate the stations to these points instead of forcing the wells to stations arbitrally fixed as heretofore." Water would be the worst problem, he said, in the fifty or sixty miles between Medicine Bow and North Platte, and in the seventy or eighty miles between North Platte and Bitter Creek. Also crucial to speed and smoother management in the coming construction season would be the setting out of telegraph wires far in advance of even the grading teams—it should be done all the way out to Bitter Creek and beyond to Green River, as soon as weather permitted, and though the project would be unbearably expensive to the company, the cost was a bagatelle, Dodge said, "compared to its advantage."

But most important to Reed's future, and that of the coming season, was a team of assistants—"get good honest active" men, Dodge urged—three or four supervisors, each of whom would concentrate on one construction phase: bridge building, well drilling, road grading, tie cutting and transportation, and work-train dispatching. Back when the line was stretching to reach the 100th meridian it was manageable by one chief, but the previous season had proved Dodge's point: Reed could no longer do it all by himself, nor could any one man. "The company want to build a large amount of road," he told Reed, and "to do it they must give you more authority and allow you head men."[8]

Reed was all too aware that he had enemies—with Durant and Seymour he assumed that they saw him as an impediment to under-the-table profits, with Snyder he put it off to different personalities and individual styles, and with Oliver Ames he ascribed it to isolation and perhaps even weakness. Both Dodge and Reed saw that they would have the same trouble with Seymour and Durant now as in 1867, since questions of route never seemed to go away. "I have no time to fight for private interest or to keep track of mistakes made by enemies," Reed wrote, going on to thank Dodge "for pitching in on my behalf." Already he had begun the practice of leaving a wide paper trail of all he did, even though it slowed him down and irritated him. "My whole time is devoted to the work," he explained, "and I had flattered myself that the progress made was satisfactory"—they had, after all, laid track to mile 549. He thought that the grading contractors would be done by the end of January.[9]

On January 13, a severe snowstorm struck the entire Union Pacific line, dumping two to three feet of snow in a few hours, with subsequent high winds burying embankments and filling roadcuts. The eighteen hundred graders slunk back down the Laramie/Black Hills into Cheyenne, their work finally ended for the season. "I was delayed two days on a trip to Fort Sanders," Reed wrote his wife. "The wind blew a gale & a large amount of snow fell and was blown furiously over the country, being deposited in high piles behind rocks, in ravines canyons & railroad cuts. After the storm it was intensely cold, thermometer sank to 30° below zero."[10] But despite the halt, Reed was able to project an optimistic note a week later to Henry C. Crane, the corporation's secretary in New York, who would be on the receiving end of Reed's stream of reports and accounts. Reed had built a temporary warehouse for the winter's tools and supplies at Carmichael's Cut near the Sherman Summit, and he proposed building a permanent warehouse in the spring at Fort Sanders and hiring a supervisor there. He had appointed an iron master already, and although the two feet of snow had killed any prospect of adding to the present stockpile of bridge timbers and some 120,000 ties, the contractors, Davis and Associates, had employed nearly four hundred men for cutting and transport; their considerable work was now buried under two or three feet of snow up on the Big Laramie River, but it would be there when the snow melted. Quarrymen and masons were nearly finished with the footings and approaches for the Dale Creek Bridge. Precut bridge timbers had begun arriving in Cheyenne from mills in Chicago, wanting only a delayed shipment of bolts from New York to be ready for assembly. And even an outbreak of labor problems had been solved with a get-tough remedy: "My stone cutters have done but little work, only 270 yards of stones were cut when all the cutters struck for wages and I would not let a single man continue work even at reduced prices," Reed reported. "I was glad of an excuse to get rid of them for the winter. They paid their own transportation home from Cheyenne."[11]

Such labor relations would have raised the pulse of capitalists in the East, though the harsh terms of this midwinter layoff increased the shiftless population of Cheyenne and probably affected the already-high crime rate. While Reed was making his report, Webster Snyder was heading west on a series of freight trains, stopping at every railyard, bridge, water well, and station to inspect. Trains were moving promptly and the road itself, he said grudgingly after he reached Cheyenne, was "in first rate shape" or else rapidly improving under the continuing efforts to upgrade facilities to bear heavy traffic. Most of the bridges in Nebraska were now acceptable, he thought, though to correct shoddy conditions west of Sydney and into Wyoming he had ordered over ninety thousand feet of timber. Assisting Reed in moving the next season's iron into the Wyoming hills, Snyder was also stockpiling coal all along the line between Grand Island and Cheyenne, which they were forced to import from the East as the coal on the eastern slope of Wyoming was not thought to be suitable for locomotives. Luckily, the temporary trestle bridge built over the icy Missouri River from Council Bluffs to Omaha had made that importation not only possible but advantageously priced. The Missouri River bridge, as a matter of fact, had turned into a great boon—not only in getting coal and iron, he reported to Dodge, but there was now "no change of cars at Omaha. Tariff on Chicago & Northwest cars will pay for the bridge, and crossing our own material will save us twenty thousand dollars and over."

Snyder's heart always quickened at such economy—and moved him to an already familiar topic: insisting that contractors like the Casements pay for everything they shipped on the Union Pacific. During his inspection tour he learned how prevalent the practice had been. "I find that Carmichael and others working by the day for contractors, have been keeping stores for their individual profit and getting their freight free. It is not just to continue that style of operations if they wish to show fair earnings for the road." Moreover, the Casements, Carmichael, and the others were unfairly competing with the railroad's paying customers, "not a safe operation for the road." But this insiders' finagling was dwarfed, he told Dodge, by a corrupt supervisor in Snyder's own office in Omaha who was eliciting kickbacks and also diverting department funds to himself. "I propose to remove Mr. Bean, Master of Transportation, this week," Snyder wrote.

> *I am very poor and have a great deal of humility, but I am neither poor nor humble enough to occupy the position of Supt. and have as rotten an administration as Bean's under me if I am able to prevent it. One of us must quit the road at once. If the Directors would like to make Mr. Bean Supt., I will remain with him one month and impart to him all the information I possess relating to the affairs of the road. If they wish me to remain, I must have the power to remove any person who disobeys orders. Had you been Supt. Mr. A. A. Bean would have becn dismissed three months ago. I have borne the affair much longer than I ought, and as long as I can. If my head goes off, it is with a clean record and I am ready to start on some other road, as a warehouseman, where I commenced, as I have the muscle, if not the brains, to earn an honest living.*[12]

With the situation being analagous to that of President Johnson's disputed right to fire his own secretary of war, it is interesting to see that Dodge threw his support to Snyder, especially after the latter appointed Dodge's Iowa crony Hub Hoxie to take Bean's place. But Snyder's action drew a complaint from Oliver Ames, who endorsed Bean's fitness for the position and suggested that he merely be transferred to superintend the road west of Cheyenne. Snyder could not put up with this. He dashed off a protest to Dodge—including his and Hoxie's letters of resignation for Dodge to use as ammunition if need be. Dodge went up to New York to try to reason with the board of directors. With his own experiences with Seymour and Durant fresh in his mind, he told Ames and the others that it was a terrible time to be allowing honest and competent men to slip away from them, especially in favor of people who seemed to be larcenous. To underscore his argument he brandished a letter from Snyder and began reading it aloud. "In the first place," Snyder had written, "the Directors know but very little of the affairs of the road here." There had been so much bickering and intrigue in New York that the supervisors out West had been left to hew their own ways. The company ought to be sending investigators to Omaha to see "whether the road has been managed properly, honestly, and economically," to "probe every department, see where the leaks have been, if any, and discharge at once incompetent or dishonest employees." Snyder had urged such a program upon Ames during his last visit to the field. Nothing had happened. "The directors ought to know," he continued,

> *that old R. R. thieves from all parts of the States are flocking here. They look upon this road as a Government institution, and come here to steal all they can. It is not necessary to stock this road with men known to be thieves, as others can be procured. My policy has been to clean out those characters as soon as their sharp practice was manifest. If a man comes here with a dime, borrows money to bring his family, has no income but his salary from the road, and in six months on wages as conductor of $3 per day saves money enough to live at the rate of $12 per day, and loan money to his friends, it is sufficient evidence to me that there is some stealing. His discharge from another road on same ground only confirms my belief. And I don't believe in splitting hairs or paying $1,000 to detectives to work up the case. It is safe and cheaper to get rid of the man at once.*

Complaints from such men, which of course had reached New York, ought to be discounted as much as those from Omaha businessmen who charged that Snyder was a despot. "If the directors think that all the business men of this country are Christians," he snorted, "then it is perfectly safe to run the road without a head. If to make myself popular with all I give every man who asks it a pass and ship his freight at half tariff, there would not be much money for our pay rolls. If these men growl at my exactions for the road, may it not be presumed that I am looking out for the interests of the Company? If I have any 'pets' to whom special favors are shown, or grant any favors on account of locality, religion, politics, relationship or for any other cause, it can easily be ascertained by having an examination made." As to President Ames's complaint about the firing of the Master of Transportation, Snyder said,

> *since Bean was removed we have had a different administration on trains. The loafers have been removed and everything on the road has improved. Whiskey is no longer the principal motive power. Train men understand their duty now and know they have got to perform it. They have more respect for themselves, and treat patrons of the road respectfully. There is*

now some character to the transportation department, and the travelling public appreciate it.

Let the question be decided at once, Snyder had written, and Dodge echoed that urgent challenge to the board of directors. Snyder's last words underscored the rising crisis he saw between the complacent, ignorant directors and the savvy operators crowding to get aboard the Union Pacific. "They can offer me no salary or inducement that will keep me here in charge," Snyder wrote, "and responsible for the management, if my hands are to be tied and the road plundered under my own eyes."[13]

To Dodge's surprise and relief, the government directors were the first to reply—and they insisted that Snyder and Hoxie be retained, urging a new attentiveness on the other board members. "This was," Dodge recalled, "about the first time they had taken a stand in these matters; they generally followed whatever the Company desired. This was rather a surprise to the Company, and they evidently saw there was a new factor in all these questions." He doubted, however, that much would change about either standards of competence or honesty. When he wrote to Snyder and Hoxie he urged them to try "to get along with everyone, not to criticize, but to adapt themselves to the positions they found; I thought this would avoid a good deal of friction in New York." Somehow everything they said was repeated in New York in a different light, which only reflected badly on the complainers. He begged them to bide their time.[14]

In a year they would almost be in sight of the end, and it did not make sense to quit without seeing it through; there would be, too, other roads to build—and, as Dodge hoped, other fortunes to be made on the side, whether or not they met the Snyder probity test. His former comrade-in-arms from Powder River, Patrick E. Connor, was still out in Utah, now running prospecting enterprises for silver, gold, coal, or iron mines, partly in a partnership with Dodge, who had agreed back in 1866 to help underwrite expenses. Connor was also scouting for suitable sites for foundries to be passed on through Dodge to interested capitalists, for fees or interests, and he was expecting to form a foundry of their own at the best site he discovered.[15] Even more recently, Samuel Reed had suggested that he go in with Dodge "and others to supply the Company with coal as soon as the track is laid to the coal on the Laramie Plains."[16] Most of the others in management were similarly casting out nets. Dr. Durant had granted a contract for ties and timbers to Davis, Sprague and Company, in which he was a secret partner; Harbaugh had a contract for axles and McComb one for train wheels, and McComb was surely not the only one to arrange for contracts to be let—in true congressional fashion—to firms in his hometown.[17] Undoubtedly there were other sidelines (such as Hub Hoxie's interest in an Iowa coal company), especially when one considered Durant's particular genius. Most of these businessmen had been war profiteers; was it any surprise that they had only redoubled their efforts with the outbreak of peace, such as it was? With such an example at the top, was it any surprise that out on the line a conductor would pocket a percentage of his fares, that freight would arrive at its destination lighter than its initial weigh-in, that clerks would take a payoff to look the other way, that supervisors would shake down their subordinates for kickbacks?

Oliver and Oakes Ames had a complex web of business interests from Boston Harbor to the Missouri River and westward into California but seemingly the Crédit Mobilier was their predominant sideline springing out of the Union Pacific. Such, though, kept them occupied on top of the railroad's administration and managing a heavy traffic of demands ranging from the merely irritating to the outrageous. The worst came from John Adams Dix, still living it up in Paris as U.S. minister but mourning the loss of his no-show job at the helm of the Union Pacific (which had paid him $8,000 a year) and his ouster from the board of directors in October. The old general who had once ordered a subordinate to shoot any civilian secessionist who tried to haul down the American flag was not about to give up with a whimper. He sent word to the Ameses that, if they didn't want him, he wished to sell out his five hundred shares of Union Pacific stock. Much or all of these had been paid for by Durant, but what was more galling was that Dix demanded them bought at par of $100; at the time, they were selling at around $20. It was "a kind of black-mailing," Oakes knew, but as Cornelius Bushnell later testified, Dix's attorney threatened that "the Union Pacific bonds would meet with unfavorable reception in Europe" unless Dix got his price. He might even have to obtain an injunction against the next stockholders' meeting. Oliver Ames offered to buy the shares for $32,000, but Dix held firm. Then Bushnell agreed to pay the $50,000, convincing the rest of the board to go along. It was "a complete swindle of the company," Oliver Ames grumbled. "They paid him $50,000 to get rid of him," Oakes Ames would say in summary, "and to get rid of the claim, or the threat, or whatever you call it." Coming at such a time, it "was a hard thing on the company."[18]

In Washington, Representative Ames suspected that behind the demands of John Adams Dix lurked the shadow of Dr. Durant, who had met with Dix as soon as he arrived in Paris in January. Even after Durant returned to New York in March and claimed that he did not like the settlement, Oakes Ames remained doubtful as ever about the Doctor's virtue. But that was not the greatest betrayal he would suffer in the early winter months, as he finished the distribution of Crédit Mobilier shares among his colleagues and as the first and second dividends, 80 percent of the value of the shares held, were paid to the worthy recipients. He had placed them "where they will do most good to us. I am here on the spot, and can better judge where they should go," he wrote on January 25 to Henry McComb, who had been advising him about recipients. In Delaware, some impulse led McComb to pause before carefully putting Oakes's letter in a safe place; he sent off more advice about distribution and obliquely warned Ames to be careful. Almost as expected, Ames puffed up a little with pride over the neatness of his scheme. "I don't fear any investigation here," he wrote McComb on January 30, stepping a little farther into the noose by going into greater detail about his rationale for choosing beneficiaries, and linking such developments as Representative Washburn's tariff-reduction bill to their decision to go all the way with "our bond dividend in full" despite some discreet reservations in the board: "We can do it with perfect safety." McComb would put this letter away with the first—but not until he conferred with Oakes Ames in Washington. During that meeting, McComb would later testify, Ames read him a list of names from his little memorandum book—twelve members of Congress thus far, and one reporter for the *Philadelphia Inquirer.* McComb copied the names at the bottom of Ames's letter of January 30 and filed it safely away.

Three weeks later Ames responded to McComb's demand that he sell him some of his Union Pacific stock by writing a third time. He grudgingly agreed to do so but said that none of a new distribution of Crédit Mobilier stock would go to McComb. His appetite was proving insatiable. "I want that $14,000 increase of the Crédit Mobilier to sell here," Ames wrote. "We want more friends in this Congress, & if a man will look into the law, (& it is difficult to get them to do it unless they have an interest to do so,) he can not help being convinced that we should not be interfered with." Ames's third letter joined the first two and would wait—like all tools of self-destruction, even those unwittingly delivered—until the time was right, in this case some four years down the line, when the stakes had grown even higher than they appeared in the winter of 1868.[19]

Even at its great remove from the furious doings in Washington, the state of California kept abreast of President Johnson's arm-wrestles with Congress and even with two incidents in faraway Britain. One involved the continuing dispute over how much the United States should be compensated for war losses inflicted by Confederate cruisers built and outfitted in England, most dramatically by the infamous warship the *Alabama;* Secretary of State William Henry Seward had been thwarted at seeking redress. The other incident concerned George Francis Train, the noonday mystery, whose lecture excursion to the British Isles had been cut short at customs in January when "seditious" literature linked to Irish independence had been seized out of his luggage during his entry. Train was thrown in jail, into a cell bare of bed or chair, a burly constable detailed to watch his every move. His appeals to the American legation heated up editorial inkpots on the homeside of the Atlantic. It seemed a good time, given Andrew Johnson's general situation, for the president to mutter darkly about serious trouble between the two nations.

This was, in the *Sacramento Union*'s eyes, "a mere demagogic expedient of Johnson, of which he knows he may escape the responsibility while getting whatever credit (very little) may attach to it. Whether the proposed demand upon Great Britain is in regard to the *Alabama* claims or George Francis Train we are not advised. In either case there is no occasion for an appeal to war, which will take the public by surprise. The American people are not of the sort that will follow blindly or agree submissiviely to the freaks or follies of their executive officers. On a question of war at least they will insist upon being consulted." Only Congress had the power to declare war, the *Union* grumbled.[20]

In California, talk of war was not confined to the newspaper columns. Collis Huntington had written to both the Crocker brothers and to Hopkins in the first week of January, reporting that his spies in the Union Pacific camp told him that the company was determined to build no less than four hundred miles of road in 1868. Durant, Seymour, Dodge, Reed, and the board had met; it had been vowed that four hundred miles of iron would be in Chicago by June 1, and they intended to meet the enemy west of Great Salt Lake. This challenge had to be answered. "I would make a mightier effort to do it," Huntington told the Judge, "and I would commence now to organize my forces and I *would do it,* and I would commence with laying 400 miles of Iron in 1868, and I would not say anything to the newspapers what I was doing, and I would have it understood by the superintendent foreman that there should be no unnecessary talk, but that the work was to be hushed." He went on: "I would build the cheapest road that I could and have it accepted by the [Commissioners], so it moves on *fast.* I would not put one man on the Western Pacific or Northern Road that could be worked on the Central." If there was to be a final battle it had to be east of the lake: those abundant coal seams in the Wasatch pulled him like iron filings to a magnet.[21]

His letters poured out to his four associates, full of exhortations. In a couple of days Huntington learned that the Union Pacific had ordered forty locomotives and he dashed off another missive to Judge Crocker about meeting the four-hundred-mile challenge. "You write that you think the Union Co. can not build 350 miles this year because they only built 200 miles last year," he said, "but I think they will, as they will be better organized, and Durant had next to nothing to do with the construction last year, but he will this year as the Co. have appointed a committee of three to have all to say about construction & Durant is one of the committee, and he will push on the work with reckless energy, and I sure think they will build more than less than 350 miles in 1868, so buckle on your Armor for the fight—for the ground once lost is lost for ever." The breathless sentence went on. And Huntington was ordering twenty-four locomotives, three hundred flatcars, one hundred boxcars, and would get them aboard ships by the first of May, and, as he wrote Hopkins, iron was "going on board ship very freely."[22]

The Crocker brothers fired back their replies. "We are going to do all that it is possible to do, you can depend upon that," Charley Crocker wrote in one of his rare letters. "Every man that can be had we will employ & if you send the Iron & machinery to move it with, 'By the Eternal' as Old Jackson said, it shall be laid. If the Union Co. lay more track in the year '68 than we do I will pay the damage. . . . [W]e will overcome all obstacles & beat the Union Pacific Co. to Salt Lake. Stick a pin there." Brother Ed was tickled to see his doubting brother and partners "all fired up" by Huntington's challenges, which had now increased their year's goal from 350 to 400 miles but had been picked up for a change by eager hands. "*[T]hey* regard me as over sanguine & have laughed at my ideas of progress," the Judge wrote on January 16, "so I have been careful what I wrote you." But after the recent letters "we have been looking around to see what can be done—& *they* now see that we can do a good deal more than they ever calculated on. The laborers are all at work on the Truckee & they cover the work from the 138th mile to the 160th as thick as they can work to advantage, & no cut will take more than 60 days to get out, & they are now satisfied that by spring the whole line will be graded to Big Bend of Truckee."

The Summit had been under thirty days' continuous assault from snow, Charley told Huntington—between five and eight feet of snow fell in the first few days—and although they had ceased work on the gap in trackage at Cold Stream on December 17, they still had not moved all of the men to lower camps. Moreover, he said, "it is an immense job to get the carts down from those mountains, scattered as they were for 6 miles through such deep snow. But our good feature in the thing is that the men do not scatter as they did last year. The Chinamen all stick [though] the White men have scattered some." Charley was proud of what was done—"the same number of men & horses never accomplished more work than was accomplished on the C.P.R.R. in the year 1867," he said, and he was ready to do better in 1868. "But will the Rails come on as fast as we want them," his brother wondered. "That is our greatest fear. Don't fail to get that Iron off as fast as possible."[23]

"Make it cheap," Huntington urged them from New York—satisfy the commissioners, but do all to push forward. "Run up and down on the maximum grade instead of making deep cut & fills, and when you can make any time in the construction by using wood instead of stone for culverts &c., use wood, and if we should have now and then a piece of Road washed out for the want of a culvert, we could put one in hereafter." Hopkins was of a similar mind. "We don't expect to build a road of the character we have been building through the mountain & deep snow line," he assured Huntington. "We expect to build the cheapest road we can make answer to purpose. Undulating in grades, wooden culverts where rock would delay, trestle whereever it will tend to more rapid progress. In short, to build road as *fast* as possible of a character acceptable to the commissioners. And we *know* the commissioners will readily accept as poor a road as we can wish to offer for acceptance." As Huntington approvingly replied, "let the paint and putty man come afterwards—that is our policy."[24]

If Judge Crocker had, as he claimed, truly reined in his enthusiasm when writing to Huntington over the past year—and it is hard to detect evidence of restraint—then what could Huntington have made of an audacious, wholly impractical scheme which for a year had been keeping the Judge up nights and making his California partners roll their eyes in exasperation? The proposal concerned the parched Nevada desert east of the Truckee and probably ever eastward toward the Salt Lake, a certain logistical nightmare requiring endless mule-driven caravans laden with water, food, timber, and iron. "My idea," he finally disclosed to Huntington, "is to have the graders *behind* the tracklayers, so that they & all can receive their supplies of wood, water & provisions direct from the cars—& save the immense amount of teaming them ahead. Just as well to lay the track right on the natural surface & fill in behind to raise it up 2 or 3 feet for drainage, & when a gully or water course is reached have the timbers ready to put up trestle or timber over, & at cuts run around them, until the graders behind come up, take it out, & move the track through it." In other words, lay ties on the alkaline flatness, spike on the rails, gingerly roll work trains forward, and then either dismantle the trackage so the earthmovers could get their gradient in or else somehow lift up the twin irons and ties like ribbons and shove the embankment underneath like children with sandcastles. The idea was as untroubled by engineering principles or physics as it was mad—perhaps as mad as Crazy Judah's hundred-mile-an-hour juggernaut locomotive, double-width cars, four-track system, and forty-hour transcontinental trip. Jules Verne, at that point working on *Twenty Thousand Leagues Under the Sea,* could have dreamed up no better a scheme than Judge Crocker or Ted Judah. "Those are my ideas, & the rest are working into them," Crocker told Huntington, who strangely enough did not comment or encourage his friend, and little more was ever written about it. "Engineers will fight it," the Judge admitted, "but they will have to do as they are told, for we don't intend to have any foolishness from this on."[25]

Nonetheless, sobering enough to dampen almost anyone's enthusiasm, that seven-mile gap at Cold Stream stood, despite all their denials to one another, as an almost mocking reminder of all they could be doing were it open and they were free to send the iron over toward the state line. Huntington told them about meeting Jack Casement in Washington, and he used the little general as a way of a dare—"they have a man that can lay more track in a day," he said, "than any other man in the United States." Casement boasted that he had laid four and one-fifth miles in one day by daylight, "and he says he can lay it day after day." Oliver Ames had asked Casement to begin working also at night by torchlight, but he refused, saying he could still "lay the track to Salt Lake this year and not work nights either." To Hopkins, Huntington urged that the Cold Stream gap be closed as early as April and hang the weather: "I

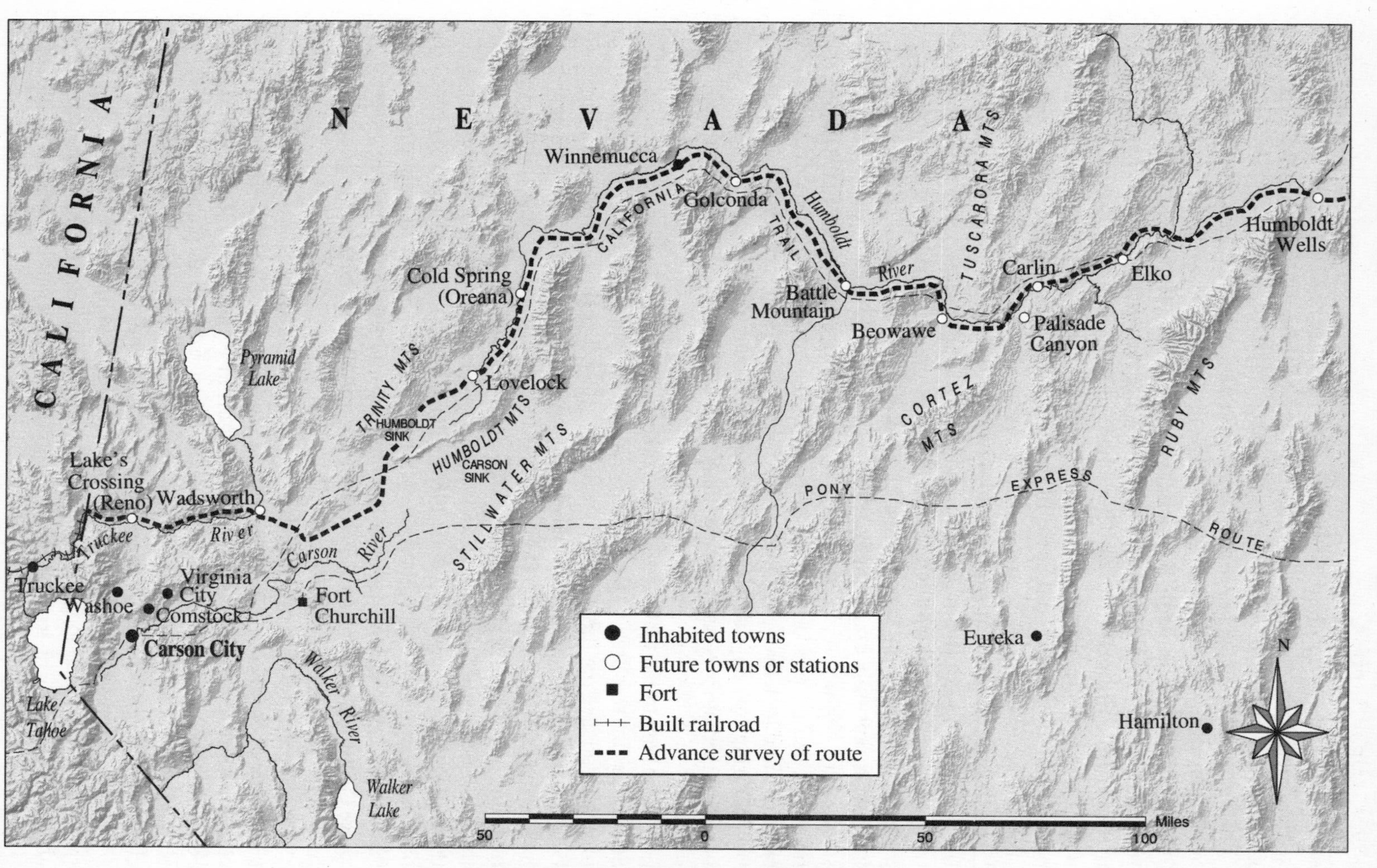

Central Pacific Route Across Northern Nevada, 1868

would make that connection in April if I was compelled to melt the snow in tin kettles," he wrote, "for as I understand it, you have about six days work there if the ground was bare."[26]

Hopkins tried to convey just how many kettles might be needed after going out on the work with Charley Crocker in the second week of February. Their train from Sacramento reached Cisco at noon, and Hopkins saw that the snow had reached eight feet, deeply burying the now unused tracks leading up toward the Summit Tunnel. They changed to a stagecoach outfitted as usual for the season with metal runners, and over the next twelve hours they passed over the summit and down to the Lower Truckee Canyon—"the best sleighing I ever saw on the coast," Hopkins assured Huntington, "& as good as I ever saw anywhere." Snow was only six inches deep at Hunter's Station on the Truckee, though the temperature had grown colder the farther east they traveled, down below zero by the time they put up for the night in the hotel there. Hopkins slept on "a wild duck's feather bed with ten blankets & a sheet over me & was only comfortably warm—the thermometer in the morning stood at 10° below zero." The construction chief Jim Strobridge and his engineer took them the next morning to examine the anticipated Lower Truckee crossings and look over the route on the forty-five-mile desert beyond the Big Bend. During two days they agreed on some expedient changes and the site bosses immediately began to stake out the new line and locate culverts and masonry.

"How long must it take to reach the desert with the track," Hopkins asked Strobridge. By the first of June, the reply came, "with my present force of men—if you give me the iron and ties to do it." By that time, he thought, the Cold Stream Gap would be closed. Twenty miles of iron had already been hauled over to the state line, Hopkins reported to Huntington, and the order had gone out to haul over 40 more. "We know we can do that," he said. "As to the ties, it will be difficult to get 135,000 ties so early in the spring as we want them. But Charley says he will do it & I believe he will." When Hopkins got back to Sacramento and conferred with his partners, the word went out to send a hundred shovelers, several locomotives, and the huge snowplow out above Cisco to get the line reopened to the summit. When the weather got a little warmer they could tackle the gap.[27]

Steamer mail was taking from two to three weeks to make it from the West to the East Coast and vice versa, so Judge Crocker's series of letters with their calls for more iron were reaching Huntington in New York in January and February just as the Hudson and the East Rivers froze shut, paralyzing pier traffic for more than three weeks and creating such a shipping backlog that Huntington's goals for January and February were badly lost. On top of nearly six thousand tons sent off before the freeze, he wrote to Judge Crocker on February 21, "I shall be loading 7 ships next week that take in the aggregate about 10,000 tons of iron." He added that they would sail by the middle of

March and that another forty thousand tons—rails, spikes, fish joints, and other hardware—would ship by the first of June. "Tell Charley to let out another link," Huntington wrote. They would get all the rails and spikes they could use, he vowed, "if Old Ocean don't get her back up, so as to break up or drive our Iron-laden ships off their courses." He assured his partners that rolling stock would be going soon, too—although it took some three months to build a locomotive, and they would not ship as quickly as they wanted. He left unsaid the probability that with this slow start—the Associates had hoped for forty thousand tons of rails to leave the East in March, not June, and they had immediate need for the heavy-duty engines—they might at worst run out of iron in August or September no matter what the urgency of battle.[28]

Compelling them forward was the specter of new competition from a quarter they had failed to anticipate: Grenville Dodge's Snake River route proposal, obvious to anyone as a way to edge control of a western terminus away from California and the Central Pacific, had been written into a House bill. As Huntington told Judge Crocker near the end of January, "the Durant Crowd" had tried to put the bill before the Pacific Railroad Committee. Normally Huntington felt this committee was friendly to the Central Pacific, but with Thaddeus Stevens absent with his terminal illness, Huntington felt his influence weakened and feared that Durant's representatives would exploit that weakness. "After a little fight with them," Huntington boasted, "we sent their bill to the Land Committee, and then got that Com. to refer it to Donnelly [Ignatius Donnelly of Minnesota], and in his hands it will have time for a long sleep." Huntington preferred slaughter to slumber for this bill, knowing more about the Union Pacific plans after learning that Samuel Reed was visiting Washington; he sent the young banker Edward P. Flint, with whom he was developing a cautious strategic alliance over the Goat Island matter, to question Reed. Soon Huntington knew almost as much about Snake River as Dodge or Durant.[29]

Intelligence gathering was expensive—but hardly as much as buying congressional loyalty. Demands during the feeding frenzy as the second session of the Fortieth Congress opened had not abated by the early months of 1868, for it was an election year and campaign funds had to be raised. Two senators went directly to him for help—Aaron H. Cragin and James W. Patterson, both of New Hampshire. Huntington gave them $1,000—"and it is a good investment," he wrote Judge Crocker, "as it will give us the N. H. delegation as we want them." Another delegation for sale, apparently, was Iowa's, even with its strong ties to the Union Pacific. Senator James Harlan wrote Huntington several times asking that the Central Pacific hire a certain lobbyist to take care of its land matters; told that the railroad already had a good lobbyist, Harlan pressed the matter by saying that "it would please all the Iowa Delegation &c," as Huntington related to Hopkins, the senator adding that "it would do us much good." Though he had only sporadically received Harlan's support in the past, Huntington agreed to pay the lobbyist $3,000 a year, "which I think somehow goes to Harlan," he said—and the Central Pacific would keep its original lobbyist.

"There are more hungry men in Congress this session than I have ever known before," Huntington marveled. "Several Senators are very short." Senator William Stewart of Nevada needed campaign help—which was given. "Stewart will do anything for us that I ask him to," Huntington assured Stanford, extending the senator a further courtesy by paying for a set of law books for a partner back in Nevada. Another notable example was Senator Roscoe Conkling of Utica, New York, whom Huntington came to realize would require either an interest in their construction companies "or in some city property somewhere. He is no common man. He is about the greatest man in the Senate." Conkling was barrel-chested and pugnacious—after Charles Sumner had been caned on the Senate floor, Conkling had stationed himself next to sharp-tongued Thaddeus Stevens to discourage southern "criticism"—and his leather-lunged vitality had once moved James G. Blaine to jeer at his "haughty disdain, his grandiloquent swell, his majestic, super-eminent, overpowering, turkey-gobbler strut." Collis Huntington now liked Conkling for his severities with the Kansas line's legislation, and over ensuing years he would be rewarded with both public and behind-the-scenes chores for the Central Pacific; at that point the New York senator was consolidating his control of the state political machine and the New York City customhouse; publicity over the entrenched corruption there would boil over in a decade in a battle between Conkling and President Rutherford B. Hayes over the patronage job of Port Collector Chester A. Arthur, who, of course, was later to succeed the onetime Crédit Mobilier congressman, Ohio's James A. Garfield, when that president was assassinated in 1881. The web of influence was indeed tangled during the Gilded Age![30]

Conkling's bare-fisted political style had won Huntington's admiration, but his liking for Representative Samuel Axtell of California was doubling by the week. Already an off-the-books "attorney" for the Central Pacific, costing $5,000 annually, Axtell was proving his value in Washington. "I have had several longer talks with him," Huntington told Judge Crocker, "and I am satisfied he will stay by us." In March, as Axtell prepared to make a reelection campaign visit to California, he promised Huntington to see Hopkins while in San Francisco, and Huntington urged his partner to give the congressman support. "He is the best man to manage a bill in Congress that we have ever had from Cal," he said, "& while we would have a republican, all other things being equal, I must say that for myself I would rather have Axtell."[31] Huntington, that founding father of California Republicanism, that stalwart Union man, was beginning to learn the dawning capitalist tenet that there are several different kinds of loyalty but the most useful was of the green variety.

There were many more demands on Huntington—so many, in fact, that he thought it was necessary to bring former New York Congressman Richard Franchot in on permanent retainer of $25,000 a year to help him keep track of their political interests.[32] But even with all of his backroom dealings he afforded himself a little outrage against the "Union Pacific folks": they "divided 3,000,000 of their first mortgage Bonds a few days ago amongst the stock holders of the Crédit Mobilier," he had written Charley Crocker on January 24, "and they feel strong and are disposed to push the Snake River Bill through Congress. . . . [T]hey have several members in Congress that are directly interested in their Road." Behind it all, of course, was "Old Ames"—"he is an old skunk," Huntington exclaimed. "I hope to hole him." And his righteous indignation grew. "As the Union Co. are so very corrupt," he complained to Judge Crocker on March 28, "it has to a considerable extent demoralized all the Road being built under the Pacific R. R. Act—and I should not be surprised if Congress should order a committee to overhaul all the Co's., and while I know everything is All Right with the C.P., I would be very careful that the Co's. Books should make it so plain that any one could so see it."[33]

Indeed, how was one expected to get one's legislation passed (and the enemy's stalled) with such skunks fouling the burrow? Beyond the Union Pacific's Snake River bill before the Fortieth Congress, Huntington was still feeding and watering his Goat Island bill. So avid was he that he overlooked the written opinions of Judge Crocker and Mark Hopkins, sent to him late in December, that the California associates had decided the terminus project on that unappetizing island was not worth Huntington's lobbying time, and certainly not anything like the $50,000 he had thought he would have to pay to shove the bill through the Senate. "We have lost interest in it," the Judge had reiterated on February 6, a week later acknowledging that the army's expert on fortifications, General Henry Halleck, would report on Goat Island and urge that it be retained for strictly military purposes and that was another reason to look elsewhere in San Francisco Bay or Mission Bay. But in any event, "We have enough to do now on the C. P. without taking our time over matters at the Bay."[34]

Nonetheless, Huntington continued to nudge the bill along for some time—though he allowed himself time to stay abreast of many other Washington matters. Keeping an eye on Washburn's slumbering tariff reduction bill, he considered Mark Hopkins's suggestion to take the pressure off by getting the original Pacific Railroad Act amended; the law's language reserved the right to reduce rates whenever dividends exceeded 10 percent on its capital stock, but Hopkins advocated a clever substitution—"the rates *shall not be reduced* until the Company shall pay a dividend exceeding 10 per cent"—which he thought would guarantee all the railroads "peace & security" from agitations.[35]

Beyond the tariff issue, Huntington watched their competitors for congressional attention. He counted fifty railroad companies in town asking for land grants and federal bonds, most interestingly the Union Pacific, Eastern Division (UPED). Their graders and tracklayers may have only been concerned with the terrain between Kansas City and Denver, but the Kansas executives had their eyes focused on the faraway Pacific—and a land grant and federal support to enable the new line southwest of Denver down across New Mexico and Arizona into southern California. Concerning this, Huntington worked both ends of the track—working throughout the winter under cover to prevent the UPED from getting any aid and manipulating Ames and Dodge to oppose them out in the open, turning the Kansas people against the Union Pacific and giving the appearance that the Central Pacific was friendly.

In early March, the president of the UPED, John D. Perry, sent word to Huntington that they should meet. Perry offered the Central Pacific the grant to build the California portion of his proposed road. Huntington surprised him and refused outright—saying he wanted no part of "a small feeder" line—and countered with an offer to build the entire western half of the route between the Pacific and Denver. It was the only way for Perry to get Huntington's support, he said—and his opposition to the scheme would kill any bill they introduced, despite any personal feelings of friendliness he felt toward the Kansas people. He left the meeting confident that he had set good bait. On March 21 he saw that he had, and that he seemed to have hooked not only the UPED but the Pennsylvania Railroad: the next meeting included Perry and a number of other "first class" businessmen, including Thomas A. Scott, a director of the UPED and also vice president of the Pennsylvania Railroad. Scott, it turned out, had been quietly working with Perry for nearly a year to create a true transcontinental line. In their minds the Pennsylvania would link Philadelphia to St. Louis, a line would be flung across Missouri, it would join the UPED, and the line from Denver to southern California would then turn up toward San Francisco: board the cars on one Market Street, near where the Delaware commingled with Atlantic brine, and emerge on another Market Street, a continent away, meeting the salty Pacific breezes, and all under the aegis of one railroad company! Huntington kept his head.

"Their proposition was that we come in with them and build the road under one organization," Huntington wrote the Judge. "I of course refused." When they offered the California section, he said that he had evidently come over laboring under a mistake—that he had already ruled out such a meager portion, that he "was not aware before that they had anything to give west of Denver," and that furthermore the Central Pacific had enormous resources in and out of their state, had locked up California, and controlled the approaches to San Francisco. Perry and his associates hemmed and hawed, Huntington told Crocker, and said that they had to consult with their engineer when he came to Washington in a day or two. "But Tom Scott, who is very sharp, came to see me a few minutes after we had this talk," he continued, "and said if I would give a certain party a small interest in our part of the line he thought he could carry it with his people." Now Huntington knew he was really on firm ground. He replied that he did not think his California partners would go along with any further division of control—but that Huntington might convince them to part with a portion of the construction proceeds. "He then said he wished I would not mention it to anyone," Huntington chortled. By the end of the month he was crowing to Hopkins that they would wind up holding the southern line between San Francisco and Denver.[36]

This fit so neatly into the Associates' long-range aspirations as they had developed that it was a marvel—but its sudden appearance at that time also greatly alarmed Huntington's California partners, especially Judge Crocker and Mark Hopkins, and to see how two such contrasting paragons of optimism and pessimism, respectively, had arrived at the same stage of nervousness, we must backtrack (as the expression goes) ever so slightly. In that first quarter of 1868 E. B. Crocker and Mark Hopkins had been very busy. Beyond the frantic planning for the coming construction year on the Central Pacific—going out on the line, estimating matériel needs and placing orders, managing regiments of workers, negotiating with the labor importers for more legions of Cantonese, worrying about the responsibilities of running regular passenger and freight service (and the threat of more collisions due to mismanagement), peering at terrain maps, printing and handling securities, keeping the private and public sets of company books, seeing to all legal exigencies, and of course handwriting those twice-weekly four- to eight-page letters to Huntington—there was the Western Pacific from Sacramento down to the bay, there was the braid of short northern lines up the Sacramento Valley pointing the long, long way toward Oregon, there was the question of locating some inexpensive wetlands for a Pacific terminus, and there was the opportunity to look southward toward capturing all of the possible trade of the lower state. Stanford, particularly, had become enthusiastic about a branch from the Western Pacific line at Stockton up the San Joaquin Valley. With Hopkins and some engineers, he had ridden up into the valley to examine it. "We had to carry our supplies for both horses and ourselves," he would recall, "and we prepared to camp out. You could ride miles and miles through that valley and see nothing better than a sheep herder's little cabin. It was almost entirely unoccupied."[37] And so he filed the incorporation papers. These other concerns were like a classroom of children all shouting "Me! Me! Me!"; they were like a corridor filled with messengers all seeking admission at once. They were like a half-dozen hands stretched toward them, each of the hands clutching so many greenbacks that some were sailing lazily to the floor, and there was the sound of clinking gold coins in the air, and each hand beckoned seductively to be grasped. And more hands appeared.

Before the Judge was used to writing the new year as "1868"—he changed the date in afterthought in a letter to Huntington on January 8—he was reporting on the entity called "the Southern Pacific." This paper railroad had been chartered in 1865 and had won federal land grants in July 1866, its incorporators hoping to build down the California coast from San Francisco to San Diego, then swinging eastward toward the border and a presumed meeting with another paper company's projection building from the East. Some time later, though, the founders decided to change the route away from the coast to run up into the San Joaquin Valley. This opened up the problem of the original land grants—but in any case, nothing more could be done, for few bonds were sold, and now the founders were about to forfeit their federal land grant for not having completed fifty miles within two years. The Southern Pacific had become, at least for the Associates of the Central Pacific, merely an object for polite scrutiny—like an ant stumbling and limping nearby with broken legs, its aimless mission perhaps to be ended with a firm bootstep, or perhaps helped; but there were more immediate concerns. In January, however, Crocker learned that the Southern Pacific had been taken over by two powerful and possibly useful businessmen: the shrewd San Francisco speculator Lloyd Tevis and Oakland's baronial Horace W. Carpentier. Tevis had been suggested by the Judge last year as a possible partner in the Goat Island makeover; Carpentier controlled most of the Oakland waterfront and his support had been entertained in 1867 by Stanford and Judge Crocker both in the contemplated northward extension of the Western Pacific from Niles and in the plan to aquire Goat Island's potential for deepwater facilities. In a formidable partnership, Tevis and Carpentier proposed that the Southern Pacific be built jointly with them and the Associates. Crocker and Stanford declined, the Judge told Huntington, "because we have enough on hand."

But then Tevis countered by proposing that they all buy the fledgling little operation, the San Francisco and San Jose Railroad, and this was a temptation. Doing so would gain Tevis and the Associates three and a half miles of waterfront on Mission Bay. Then again, Crocker wrote to Huntington, "I would prefer, & I think Tevis will, to buy the San Jose for ourselves and own the *butt end* of everything, giving the Southern Pacific a favorable running connection. But we are already so deep in railroads," he continued, "I don't like to extend any more. Still this seems a kind of necessity. What is done must be done quickly, as several schemes for getting this water front will soon be before the legislature." They could, after all, get "a magnificent grant" for waterfront and submerged lands for "a grand terminus." Sooner or later, he thought, they would want to build the Western Pacific from Vallejo Mills across the upper neck of the bay to just such a place. Such a network, and such a facility, made Goat Island and its Oakland waterfront approach look as uninteresting as they were, in all probability, unattainable. To Mark Hopkins, purchase of the San Jose line seemed preferable to opposing it somewhere down the road, and the notion of a "tidewater termination" appeared both "practical and profitable." But as Hopkins confided to Huntington in a letter also dated January 8, "a monopoly of the entire RR services of the state is liable to prejudice the public against us, I fear, and by Legislative restrictions, curtail the profits."[38]

These new concerns, on top of everything else, rolled forward. In the statehouse, legislators scored easy points by rising to denounce the growing monopoly. A rumor sprang up that the Central Pacific had already engulfed the Southern Pacific and the San Jose; Stanford soberly went before the press to deny that any such connection existed.[39] The *Sacramento Union* was unsparing in its venom. By late March, Tevis and Carpentier had gone ahead and signed a three-year option to buy the San Jose, and the California Senate had passed a bill granting thirty underwater acres in Mission Bay and an access strip to Point San Bruno, with the House expected to go along. Lloyd Tevis took this opportunity to tell Judge Crocker that he and Carpentier were not getting along well, and that the Associates should join him in taking hold of the Southern Pacific. Then, by coincidence, the Judge was paid a visit by the chief engineer of the Kansas Line, General William Jackson Palmer, sent West by Perry and Scott; Palmer traced their proposed entry into California on the Judge's map. His finger crossed the blistered Mojave Desert, wound up and over the narrow pass in the Tehachapi Range and down into the San Joaquin Valley. From there it was just a peremptory slash of the finger up to San Francisco. Judge Crocker ventured the opinion that since the Associates intended to build all the way up the San Joaquin from Stockton, it made sense for the Central Pacific's San Joaquin branch to meet General Palmer in the saltweed hills at the California line. Palmer haughtily replied that he intended to build straight up to San Francisco himself. "I, of course," Crocker wrote Huntington, "had no reply."[40] In the same paragraph of his letter, however, he had linked the Central Pacific, the Southern Pacific, the San Joaquin Valley, and the Union Pacific Eastern Division. Obviously they were coming to be all connected in his mind, as—three thousand miles away, that same day—they were in Huntington's, as he prepared to meet that second time with John Perry and Thomas Scott. And when Huntington had gleefully emerged from his private huddle with the Pennsylvania Railroad's Tom Scott, predicting that one day they would control the line from southern California all the way to Denver, both Hopkins and Judge Crocker reflexively began to clamp down on the brakes.

Hopkins was the first to speak out against the precipitate purchase of the Southern Pacific. He did not want to do anything "that brings new partners in the Ring," he wrote Huntington firmly on March 21, and besides, they were so close to owning a California railroad monopoly that they were in grave danger of heavy sanctions. "We own enough now," he said.

> *We are committed to the building of enough now. No matter how desirous the public along any & all routes may be to have Rail Roads built, when once they get the road built all must go "deadhead" [free] for passage & freight, else they join in the cry of demagogues, fools, & knaves to reduce the rates of the "monopoly"—and denounce it a "soulless & bloated corporation."*

Let other folks have some, he pleaded.[41]

The Judge's alarm was expressed more personally, immediately upon hearing "San Francisco to Denver" from Huntington. Crocker admitted they could do such a thing

> *without much trouble, . . . but is this business of expanding, extending, & buying up Railroads never to end. Hopkins talks very decided against it, & I must say that I don't much like the idea of slaving myself to this railroad business for the rest of my days. Stanford, I believe, would buy up every road in the state, & you don't seem disposed to stop. This buying up of roads & starting new ones is working against us. You see how the* Union *abuses us for it. It frightens the people, & scares the legislature. They actually fear that we are becoming too powerful for the legislature & people.*[42]

The brakes had been applied—but was this a runaway?

The *Sacramento Union* suspected so, as Hopkins and Crocker had uncomfortably read for themselves. The paper was "so hostile and malignant," Hopkins complained to Huntington. It had said that the Central Pacific must be controlled "before it grows more powerful than the State, as has been the case with others of a like character in New York and New Jersey. In a few years this corporation is likely to be the owner of thirty millions of dollars in railway property, and to have controlling interests in most of the railways of the State. If they are allowed to take the reins in their own hands now in the infancy of their power, what may not be apprehended from them when all their machinery is rounded out and completed?" The first step was to control rates in the legislature—as another critic, the *Auburn Stars and Stripes*, had remonstrated: "At the present rates for freight, we are advised that the cost of sending a barrel of flour from Sacramento city to Cisco is greater than sending it on shipboard from the foot of J street to the docks in Liverpool. . . . [S]uch rates must be sensibly and promptly reduced." Ten dollars to ride up to Cisco, said the *Union*, was an outrage. And beyond the tariffs issue was another bill introduced by the Central Pacific and their shadowy partners, "the naked grant" of millions of dollars of waterfront and submerged real estate—"all the greatest real estate interests will be in the hands and under the control of this delectable partnership, of which the Central Pacific Railroad Company would, of course, soon become the chief."

Early in the legislative session James Anthony, the *Union* editor, seemed confident that the will of the people would not be oppressed against what he would call "a corporation already dangerously powerful and meddlesome in political affairs." By the second and third weeks of March, however, he commented on the sudden and confounding evaporation of support for controls—sworn enemies in the statehouse and at the editorial desks of numerous newspapers sat down and were silent. "We have heard," wrote Anthony, "that one of the chief stockholders in the Central Pacific, some time ago, remarked substantially that, though the company was then compelled to submit to the exactions and restrictions imposed by sundry interests, the time would soon come when it would be a great power in the land and be able to do as it pleased. There is but one way in which this hope can be realized, and that is by forestalling the will of the people and defeating the public interests through the aid of money and a venal press."[43]

Seeing such a reversal was unbearable. The editor stoked his fires. Judge Crocker, who wished that the others had listened to him in the fall "to buy out Anthony's interest in the *Union*, & get some first rate man in his place," wrote optimistically to Huntington that "the old fellow is so terribly abusive that it destroys the effect against us." They had sustained many bruises during the season's battles, but probably would survive to fight again, even if Hopkins had been thoroughly shaken. "All the gall & bile that Anthony has been gathering up for years," Judge Crocker commented on March 20, "he is pouring out without stint. But he like all the others will find it hard to buck against the locomotive."[44]

Three thousand miles away, scrambling to get iron aboard ships and to figure out the enemy's next moves in Washington, Collis Huntington could not offer much. "I have but little trouble to get favorable things said of us," he told the Judge, half in sympathy but half in boast, "by giving the local Editors a box of Cigars. But we have grown, as that old [James Gordon] Bennett of the *Herald*, and others like him, think we are fat enough for them to gather tribute of—and so they forbid anyone about these papers saying any good-natured thing of us."[45]

Even with the bad publicity in the West, by March 31 Hopkins was much relieved to write Huntington that all efforts in the legislature to reduce railroad tariffs had failed—"& we have peace for two years to come." By then he hoped they would have sold their bonds and have no need to make such an effort at controlling news and information, let alone lawmakers. The underwater land grant in Mission Bay had been approved—thirty acres each to the Southern Pacific and Western Pacific, and rights-of-way, "of much value," he reported; also, the Oakland negotiation was continuing, "which promises to be of still more value." Hopkins and the Judge savored the victory. "We feel," wrote Crocker, "that we have come out first rate because the legislature has done us *no harm*." But still they were both exhausted. "It has been a heavy load to carry through this legislature," Hopkins admitted. "It has cost us a few thousand dollars to do it."[46]

Certainly it was time to look ahead. For more than six weeks bond sales in the East had been extraordinary; Huntington had sold some $500,000 to German investors, $640,000 to Philadelphians, and $700,000 to New Yorkers, despite the fact that he was raising the selling rate. Even so, on the day after he raised the bond rate to 98, "supposing it would check sales," he sold another $100,000. "They no doubt would bear 105, possible 110," he told Hopkins, but he was going to be careful and not get too greedy. "We took the wind all out of the Union Co's sails, when we put our bonds at 98," he exhulted, "which was 8% above theirs, & when they raised the price to 95% I did not see any mention made of it in the papers, only the 90 changed to 95. They changed the 2nd day after we did, and I suppose they thought it looked too much like following us." The money would help pay for the tonnage now being loaded in New York—and for the renewed energy being focused on both slopes of the California mountains, where the six thousand laborers at work in the Lower Truckee in late February were now being augmented with fresh companies of Cantonese. Huntington had reached the point where he could legally sell no more bonds—though perhaps he could sell *the promise* of more—until the Associates closed the Cold Stream gap and made the line continuous.[47]

This had been suddenly pushed back in March; as the legislature had wound down, the snows had piled up. Five feet of snow already stood in Cisco, but then in three days another thirteen feet fell. For the next week, days of high, snow-piling wind alternated with more storms. Train service had halted below Cisco at Blue Canyon, but the giant snowplow, eight engines, and three hundred men cleared upward to Emigrant Gap, hoping to meet an equally large force working downward from Cisco. Service to Cisco was finally restored after five days, and they prepared to resume sending iron over the summit on sleighs, though it was difficult hauling iron "on such a depth of soft slumpy snow," Hopkins told Huntington. He hoped it would be the last storm of the season.[48]

If it was the last storm, then it left a lot to remember it by. Charley Crocker, who had been stranded in Cisco for several days but had finally escaped on snowshoes to Emigrant Gap, where he caught a train, returned to Sacramento with bad news. "There is no place between Cisco & Cold Stream Gap with less than fifteen feet of snow lying on the track & line of uncompleted work," he reported on March 29. Some places had between fifty and a hundred feet "drifted or slid in." An avalanche on Black Butte Mountain had taken out one bent of trestling, and "a dozen snow slides between Cisco and Emigrant Gap raised hell generally."

"Now some people may think that is a nice country to railroad in," Charley Crocker told Collis Huntington, "but while up there *I did not 'see it.'* "[49]

On the east bank of the Laramie River, at the southeastern edge of the cold, windswept, and barren Laramie Plains, Grenville Dodge's surveyors pitched their tents a mile or two north of the stockade of Fort Sanders in February. They began to lay out the streets and building lots of Laramie, Wyoming Territory, the town, plains, river, and overhanging mountains named for the French Canadian fur trapper, Jacques La Ramie, who nearly half a century before had set his trap lines along the tributaries of the North Platte until he was killed by Indians.[50]

The commander of Fort Sanders, General John Gibbon, noted that the surveyors were locating the railroad line, the depot and yards, and a portion of the town on land reserved for the army post; Fort Sanders had been established by Dodge when he was still in service and marked off with a ten-square-mile section of grassland, only a small part of which was currently in use. General Gibbon, an artillery expert and former commander of the stylish Black Hat Brigade who had been wounded at Fredericksburg and Gettysburg, and who was a favorite of Grant's, had nothing but sympathy for the Union Pacific and Dodge, but he had to notify Grant and Stanton of the technical trespass. Dodge, still in Washington, had "no idea" that the government would object, but the incursion made a complicated series of letters necessary involving Dodge, Generals Gibbon, Augur, and Sherman, and War Secretary Stanton. All wanted the railroad put through as speedily as possible, and saw the land acquisition as a relatively simple exchange of equal properties—but the military men were justifiably worried about the proximity of depot and town to the fort and its effect on military discipline and morale. "I want you to understand," wrote Sherman to Dodge at one point, "that both Augur and I are more than friendly to your great enterprise, but of course, don't want to plant a dirty little town right along side of our military posts."[51]

Two miles to separate the whiskey shops and gaming tables from the young bluecoats was finally deemed sufficient, although General Gibbon was still understandably eager to keep order for the good of the army as well as the railroaders. Recently though, a ruling had been handed down by the Interior Department which threatened this. "Frequent complaints have been received from your R. R. people that these whiskey sellers squat along the line of the road make their hands drunk and interfere terribly with their work. I have always on these complaints sent and destroyed the whiskey and in some cases arrested the sellers," he wrote Dodge, but with the entry of the railroad into the territory, the Interior Department had decided to cease designating it as "Indian Country." With a stroke of a pen, the area was now considered "public lands" of the United States. If it was not Indian Country, Gibbon said, the law did not permit him to roust the vendors of whiskey. Perhaps Congressman Dodge could arrange for the law to be rewritten, giving the army jurisdiction over all public lands outside organized towns, within the limits of the railroad's granted lands. "Without some such provision," he warned Dodge, "I am afraid your work for the next year will be sadly interfered with."[52]

Law or no law, the railroaders would continue to suffer from camp-followers. From Cheyenne, Samuel Reed reported to Henry Crane in New York on March 13 that while he was out inspecting work on the Dale Creek bridge, rustlers stole three mules and two horses from Carmichael's graders' camp nearby. "The night watchman discovered them as they were leaving the yard," he wrote, "and as soon as men could saddle up ten good men followed and overtook the thieves, demanded them to surrender which they paid no attention to [and] they were fired upon. Men and stolen stock returned to camp. Thieves did not follow the party back to camp." Several weeks after he reported the theft and executions, Reed again wrote from Cheyenne, this time to Dodge, saying that he was contracting out the grading for station grounds at Laramie. But he regretted "exceedingly," he added,

> *that nothing can be done to suppress the whiskey traffic along the line. A few nights since two men in Carmichael's Camp were shot (badly wounded). Welch, a contractor, was robbed of $1100, and nearly killed contiguous to a place on his work called Robbers Roost. One man was shot dead through a window at Creighton's Camp. Horses and mules are frequently stolen from some part of the work. If these depredations are to be continued, it will soon be worth a man's life to go over the work. Let us have martial law if necessary to keep off the whiskey. Every pay day the men lose several days; the work is materially retarded in consequence.*[53]

What Cheyenne and the camps were like to a greenhorn was related by Christen Nelson, a Swedish American blacksmith's apprentice who journeyed out from Omaha to arrive in Cheyenne that same month of March 1868; all the way he had heard stories of the rough frontier people. As the train pulled into the station there was some hesitation to leave the coach, but conductors said they should "fear no danger from the crowd that met us there." Outside, with the false-front town sprawling off from the track yard and the noise of saloons engulfing them all, he found people unlike any he had seen at home. "Nearly everybody there wore a long blue overcoat with brass buttons, the regular United States soldier uniform left over from the Civil War, with one or two revolvers strapped at his side," Nelson recalled. He had no sooner stepped down off the car when a dozen or more rough characters approached—they were hotel touts rustling up business. "Here," one said, "let me take your grip. I keep the best hotel in town, I set the best table and have the best liquor at my bar. I have two fine girls waiting on the tables and they are full of fun. If you have a pair of blankets you can sleep on the floor." The wide-eyed young Nelson went along, soon getting a job in the Union Pacific shops; he bought a blanket and a buffalo hide, for a while sleeping in the shop on the bellows and cooking his meals on the forge. In due time he and another employee took possession of a dugout in the bank of Crow Creek "that someone had occupied but apparently deserted." He saw a lot of life and death in Cheyenne, some of the latter a by-product of the Indian wars and of course some of it right in the streets of town. "Cheyenne was referred to as the wonder city of the plains," he wrote many years later, "with a climate where man could live forever. No one ever died a natural death. They generally had to shoot a man to get the cemetery started."[54]

A few days after Reed reported to Dodge about the shooting of two men at Carmichael's graders' camp, he wrote to Henry Crane in the New York office to tell him about the pernicious "whiskey ranches" which kept springing up at trackside. In retaliation for the shooting, Carmichael's men had swarmed out of their camp and demolished the offending whiskey ranch—but now the emboldened vendors were threatening to swear out a warrant against Carmichael. Reed was desperate for help—could not the War Department or Congress somehow prevent the debauchery? "As soon as a party of men come on work a lot of tents are put up in the vicinity and whiskey furnished to the men. Robberies and murders are of frequent occurence."[55]

In spite of such pervasive drunkeness and lawlessness in these late winter weeks, however, Reed could mark real progress on the line—even with two large snowstorms, the only serious snowfalls of the winter, which had stopped everyone for a few days in the first week and the last week of March while they cleared the tracks and the grades and the cuts. Seven contractors commanding large pick-and-shovel brigades were scattered along the line from just below Sherman Summit on the eastern slope across and down the steep mountainside to Dale Creek; some outfits were so close to finishing their portions that they were sending advance teams west of where Laramie would rise, to begin work on the northward turn up the plains. Bridge contractors had completed most of the trestles for the Dale Creek Bridge, and the upper truss work was being sent from Boomer's works in Chicago in two large shipments. The Davis firm had, after delays, put a large number of teams hauling ties and timber to trackside. Reed had caused a well to be dug in the granite of Sherman Summit, struck water at ten feet, and ordered a large tank and an even larger windmill and pump erected overhead. Supplies of iron and coal from the east, however, had slowed to a trickle in the second week of March when the Missouri River had broken up, sending the temporary railroad bridge floating hazardously downriver in pieces toward St. Joseph. Agitation in Omaha for a permanent railroad bridge swelled in response to the loss. "Everybody here appears to be crazy on the bridge question," wrote Webster Snyder to Dodge on March 26.[56]

Back in Washington, Congressman Dodge had begun to hear from his far-flung advance parties in Wyoming and Utah. Jacob House had been sent from Cheyenne into the headwaters of Lodgepole Creek and Crow Creek to prepare for aqueducts to supply the town and tracks with constant fresh water. James Evans, after locating at Laramie, had taken two location parties north from Fort Sanders, being too impatient to tarry for a military escort that was waiting for horses and wagons; scouts had reported Indians in the vicinity of the North Platte, so Evans kept his parties bunched until the soldiers caught up. Three weeks later, after struggling across "considerable snow" on the Laramie Plains, they were at the North Platte with the escort, and despite lingering fears about the Indians, Evans was ready to dispatch location men west toward Dodge Summit and Bitter Creek. "Push location," Dodge wired on March 20. "Indians will not trouble you for a month. Get all parties to work on the ground. The company are pushing me. They want line to Bitter Creek as soon as possible."[57]

Dodge had also heard from his engineers farther west, and he hastened to keep them busy. Fred Hodges, whose assignment that season would be leading surveys west of Great Salt Lake, had been in touch with his opposite number in the Central Pacific, Butler Ives, who against orders was communicating with the enemy. There was, after all, a brotherhood of engineering—which sometimes transcended company loyalty in the consuming interest of understanding and conquering the obstacles of terrain. Ives claimed to have found a route through the Wasatch Range which was thirty miles shorter than the Echo Canyon route. Dodge was thereby bound to investigate it, detailing his new engineer, Jacob Blickensderfer, to organize a party at Salt Lake City and work eastward across the mountains. Ives's letter, moreover, "was a clear indication" to Dodge that the Central Pacific planned to build east of Ogden. Blickensderfer had arrived in Salt Lake City on March 7 by sledding over a three-foot snow base in the Wasatch from Fort Bridger. He got a slow start at organizing parties; the quartermaster at Fort Bridger refused to supply him with horses, and then a sudden snowfall deposited a few inches on the Utah Valley but several feet in the mountains. He wrote Dodge that he would be sending parties up Weber Valley, the Cache Valley, and Box Elder Creek, but "how we shall succeed in getting over the snow in the Wahsatch, I do not know." He got an impatient telegram from Dodge, setting firmer and more urgent priorities:

> *Get location in at Green River and head of Echo as soon as possible. Company are starting forces out there. Then over rim of Basin and through narrows and Weber Canon. After that close gaps if you have plenty of help and men. Put a new party running line north of lake over Promontory Point, and across Bear River arm of lake around Promontory Point, so as to determine best line to locate curve.*[58]

Dodge was not happy to hear the reply, that another storm had deposited some four inches in Salt Lake City and that Blickensderfer feared "in the mountains the fall has been quite heavy." Either Dodge—or the fear of Dodge—spurred him to go out anyway. A week later he was reporting that one party was making but slow progress, and although the roads were breaking up all along the way to the mouth of Weber Canyon, wagons were getting stuck everywhere. "Have hired a team of 8 cattle to keep along," he explained, "and today with 4 mules and 8 bulls to one wagon, had to rest every 50 feet." Dodge promised to send more engineers as soon as they became available but after hastening to locate through Weber and Echo Canyons, Blickensderfer should move quickly to locate north of the lake; when there was time they could, for the sake of thoroughness, survey the route south of Great Salt Lake, even though "for grades, distances, water, work," and to avoid the extensive quickmud flats, "the north line is best."[59]

Dodge may have been convinced of this but as far as the public was concerned he was still saying the opposite, for clearly political reasons. Shortly before, queried by a journalist doing a story on the Union Pacific, he had sent a highly detailed account of the planned location from Sherman Summit all the way through to the Salt Lake basin—and this is how he ended it: "We open out on to the broad table lands of the Weber Valley and soon strike Great Salt Lake Valley at the mouth of Weber Canon; from here we skirt the Great Salt Lake 35 miles to Great Salt Lake City, 214 miles from Green River and 1055 miles from Missouri River, elevation of Great Salt Lake City 4245 feet above the sea." He knew they would do the contrary, but his story, until it collapsed, would keep Brigham Young happy—and quite possibly confuse and confound Mr. Collis Potter Huntington.[60]

On March 13, Andrew Johnson's impeachment trial opened in Washington, although the president did not appear on that first day; his attorney asked for forty days to respond to the articles and was given ten. Tickets for the Senate trial had become the capital's most hotly desired item—but nonetheless Senator John Sherman was surprised because the trial had "so little effect on prices and business." Life and legislation went on.[61] Orville Hickman Browning added the duties of attorney general, filling a temporary opening, to those he performed as interior secretary; one of his first actions was writing the draft of a presidential veto of a recently passed law taking away the jurisdiction of the Supreme Court in habeas corpus cases. Johnson sent in the veto on March 25, two days after his impeachment trial resumed. The 25th, a Wednesday, was also the day in which Grenville Dodge rose to make his solitary speech in Congress. It was against a new version of the Washburn bill to reduce railroad tariffs, in which the Wisconsin representative relied heavily on information supplied him by John Richly, a correspondent in Columbus, Nebraska. The data had seemed suspect to Dodge, who had written to Webster Snyder in Omaha. Snyder's reply had denounced Richly—"as notoriously dishonest as any man who ever crossed the Missouri"—and it gave Dodge enough ammunition about freight rates and logistics to demolish Representative Washburn's hearsay argument—and bury his bill. "We are very greatly delighted with your success in the controversy with Washburn," Oliver Ames hastened to write. "It shows the necessity of having some one there that thoroughly understands the whole subject."[62]

But the Union Pacific's most sterling Railroad Congressman was already gone from Washington in his mind. On April 1 he planned to be on a train heading west to inspect the line and huddle with his engineers; in May there would be the national Republican convention in Chicago, during which he would work all-out on Grant's benefit within the Iowa delegation, deliver a vice-presidential nomination speech for colleague Schuyler Colfax, and let it be generally known that Dodge would not seek a second congressional term. He would return to Washington only to apply for a leave of absence from finishing out his term, having delivered to his constituents no more than twenty weeks of work for his two-year term. There is no question, however, that the Union Pacific was satisfied with his labors in the capitol.[63]

Even though as he prepared to leave Washington, Dodge's thoughts could not wholly embrace the thought of terrain in western Wyoming and eastern Utah and the engaging list of problems the dry lands posed him, for a new and consuming irritation had been presented to him the week before his tariff bill speech, one which would be a burr beneath his saddle blanket all the way west, and certainly for the remainder of the year. On March 16—the day in his office at the Interior Department which also marked Secretary Browning's last abbreviated appearance in the building before moving over to the Justice Department for the hearings—Dodge received a letter from Oliver Ames. Dr. Durant had returned from his two months in Europe, and the board of directors of the Union Pacific had voted to appoint him general agent of the road, "with power to assent to a change of the grades and location of the road, as provided in the contract with Ames, and do all other things necessary to expeditite the construction of the road and telegraph in connection with the contractors."

Power to change grades and location? "I saw that this meant trouble," Dodge wrote. He realized that he would also "have to fight to maintain my lines," that there was the danger of "another demoralization of the forces." Lower down in Oliver Ames's letter the Union Pacific president had repeated the obvious, that Dodge's location lines had always been subject to the board's approval and since the past year had fallen under the purview of the board's committee on location and construction. The Doctor had been only one member of that board and committee, even with his cunning ways and extraordinary willpower. But here he was being given the whole package. "I hope you will not consider this assumption of authority by Durant as a final settlement of this question," Ames had gone on to say, as usual trying to have it both ways. "If he could by these annoying exhibitions of his peculiar character drive us all out of the road, he would do it, but I don't intend to be driven out or coaxed out, but will adhere to the strict interpretation of the Resolution, and if he abuses the power we must repeal the resolution conferring it."

Dodge could not believe what he was reading. He also saw the hand of Silas Seymour in it. Immediately he scribbled off a letter to Ames—a "great mistake" had been made; it "would not be a month before Mr. Durant would be changing things all along the lines and it would bring on a conflict with the Government." Dodge would not stand "for any changes which were made for the purpose of saving work and hurt the commercial value of the road"; the government "depended on me to see that the road was located and was accepted by them as built, and that Mr. Durant, nor the Company, had any right to change these lines after we had submitted our maps and profiles to the Government." His fury over this new turn of events must have helped speed his letter northward toward Massachusetts, but not before he heard from the newly appointed "general agent":

> *Please forward to this office as early as practicable detailed copies of maps and profiles of the different routes run between Ft. Sanders and Great Salt Lake, over routes that may be regarded as competing for the final location; also a summary of the different lines showing their relative length, cost, maximum grades, &c. with such other characteristics as should be taken with consideration by the company in deciding upon the final location. Also copies of the reports of the division engineers for the year 1868, and your instructions to them for that year. Also names of division engineers employed at the present time, their assistants, salaries paid and copies of your instructions to them.*[64]

Such a request would keep an officeful of clerks busy for weeks copying out the information. The notion sunk in that every twist and turn of the railroad from Cheyenne to Salt Lake was going to have to be defended as vigorously as was the best route across the Laramie/Black Hills in the autumn. Realizing that the Doctor was applying pressure to exasperate him into quitting, Dodge's answer to Durant came as clear and simple as the Sherman Summit line: "I have no duplicate copy of these reports," he told the Doctor. Durant could find them in the New York office. He added that the line "was definitely fixed upon Ft. Sanders to the North Platte" (even though he privately knew his men were at the moment actually doing the final location across the snowy Laramie Plains) and that "any changes in the line would delay the work." He sent a copy of his reply to Ames.

Oliver Ames did not reply to Dodge's entreaties until March 22, and his reassurances that they "do not intend to have your lines interfered with" did not ease the engineer's worries. Ames continued in a similar vein. "I did not think that the resolution authorizing the Dr. to change the grade and location with your assent would give him any power to disturb your lines," he wrote, "except in some cases where it might greatly expedite the construction of the road, and in this case Reed might change the line to get the road along. I know that the Dr. is for assuming all the power whenever he has a chance, but I trust that as this power was given solely for the purpose of advancing the road under this Ames contract and for this summer alone, you will not find it annoying."[65]

But the chief engineer was already powerfully annoyed. A week later, as Dodge was seeing to his last tasks in Washington before heading west, Ames wrote again, continuing his attempt to placate him. Durant and Sidney Dillon were going west too, he announced, and if Dodge felt anything but sickly at the prospect of contending with the Doctor when he was trying to get his engineers organized and motivated for the extraordinary push needed this spring and summer, the next words of Ames's could not have improved his disposition. "I think I have never seen the Doctor more pliable and anxious to please everybody than now," Ames assured him. "It may, however, be for the purpose of getting power in construction of road," he found room to admit. "I hope, however, he will be as he now pretends only anxious to push forward the road."[66]

Dodge contemplated Ames's words. Dr. Thomas Clark Durant "pliable"? "Anxious to please"?

In a pig's eye.

28

"Bring On Your Eight Thousand Men"

From the Donner Lake Wagon Road it stood above in baleful reproach, that mere seven-mile gap separating the ends of track between Strong's Canyon and Cold Stream. Buried under unknown amounts of snow in the clear bright days of early April, it was both painful to the eye and infuriating to the soul. A seven-mile downhill walk—it would have taken the impatient James Harvey Strobridge much less than two hours to tramp from his boyhood farmhouse in Albany, Vermont, down the tumbling Black River to, say, Irasburg; and for bull-like Charley Crocker, seven miles would have gotten him from the town center of Albany, New York, to his childhood house in Troy. Seven miles was nothing. Seven miles was humiliating. But they were far from home, grown into men with terrible responsibilities, with thousands of workers to command and millions of dollars hanging over their heads as clearly as the late-winter clouds in the sky. These seven miles in the High Sierra may as well have been in Mongolia for all the trouble they represented.

"The weather has turned off warm & the snow is settling fast," Crocker had written to Huntington in New York on April 4 before leaving Sacramento to see the mountains for himself. At that point Crocker was more worried about getting ties fast enough, once they had commenced tracklaying. "I have fretted & fumed more on that head than anything else," he confided, adding that the best solution was for himself to go into the sawmill business on the side and "show these fellows how to make ties." A few days later, on April 10, some four feet of snow fell on the summit and the east face of the Sierra, causing two large avalanches on the exposed line, recently graded and tracked, and on the stage road, delaying travelers for eighteen hours and paralyzing the little work trains supplying graders and trackmen lower down on the Truckee. Now, seeing the gap and ordering Strobridge to begin moving his graders back up the canyon to assault the covering snow, it was plainly too early for Charley Crocker to count it out. Strobridge would have five hundred men wielding shovels at Cold Stream by the end of the second week in April, and a thousand halfway through the third. But they were confronted by an immense amount of snow. And although the days were lengthening, the sunlight reflecting off the mountainsides took a cruel toll on the diggers whether they were working slowly up the gap or pressing higher on the western slope above Cisco toward the Summit Tunnel, where hundreds were still at work. "The snow has blinded a good many," Edwin Bryant Crocker reported to Huntington, "& we have had to buy up all the cheap goggles in the market. Only think of that. What an item in railroad building."[1]

One-eyed Strobridge was still feared among the Chinese legions when they saw him bearing down on them in the mountain glare, and even domestication apparently had little effect on him; earlier in the winter, his wife had joined him at the end of track in the Truckee Canyon, moving into a boxcar which had been converted into a comfortable little one-bedroom apartment with curtained windows and a small verandah recessed in the side where Mrs. Strobridge could rock and knit in the sunny midday. A newspaperman visited them and gallantly pronounced it "a home that would not discredit San Francisco." Other boxcars had been connected to the Strobridges' and similarly converted into offices, a kitchen and store, and sleeping quarters for the supervisor's assistants. Every several days as the track lengthened down the canyon, the cars were pulled down to a new siding and reconnected to the telegraph cables overhead.[2] Tracklaying down the canyon continued slowly toward Lake's Crossing and the infant town of Reno, at that point a small collection of squatters' shacks, tents, and many wooden stakes poking out of a light covering of snow. The town would be formally born in a few weeks—at auction. Bidders appeared well before auction day on May 9 and camped on the ground, buying up all the available blankets against the still-cool nights and eating all the food. The first lot sold brought the Central Pacific $600, and about two hundred more lots were taken that one day. In a month Reno would boast a hundred houses.[3]

Grading across the meadows and the lower canyon to the Big Bend of the Truckee was nearly completed—in the late-afternoon light the workers began to think about knocking off when the steep Virginia Mountains shone red above them—and at that place where the Truckee veered north, the company planted a division station called Wadsworth, altitude 4,077 feet, near an old trading post and the remains of a Paiute village. The local Indians had almost entirely withdrawn up into the foothills of the Truckee Range, especially after a short, furious series of battles eight years earlier which had been sparked by the abduction of two Paiute women by some whites, who kept them imprisoned in a trading post on the Carson River. A tribal party, learning where their women were held, attacked the post and killed five abductors. Some days later an angry force of more than a hundred miners swept over from Virginia City and down the Truckee looking for retribution. In a wooded draw the posse was ambushed and mostly cut to pieces, although survivors made it back to the mining camps. A larger force, bristling with Regular Army troopers, met the Paiutes in another battle, but the Indians soon vanished into the mountains above Winnemucca Lake. Very infrequently there would be a raid on a stage or wagon train, but for the most part of five or six years there had been peace. Crocker and Strobridge had instituted a policy by which every Paiute would receive a railroad pass, which all but ended any mischief along their line. But now, as the graders worked toward completion of the bed to Wadsworth, Stobridge had no idea how terrified his Chinese were of Indians, nor that a new kind of Indian trouble, this one wholly imaginary, was beginning to cloud his immediate future.[4]

Judge Crocker's concerns were as substantial as iron, however. George M. Gray, the consulting engineer hired in the autumn to bring a greater professionalism to the construction as well as the operations arms, had just completed an inventory of the rails and other iron equipment on hand. The Central Pacific had 122 miles of iron on hand, he said, and the Western Pacific had 20 miles which could be borrowed for the main line, and there were 10 miles stored at Freeport—152 miles of iron in total, less the 7 needed to close the gap at Cold Stream. And sidings had to be figured into the total. "Charles says he is quite confident of completing the connection by June 15," the Judge wrote Huntington in early April, "& that he can lay 3 miles per day after he gets his forces fairly settled at the work, say by July 1. At that rate the 145 miles would be laid for 49 days—or say by Sept. 1. By that time we trust your shipments will be on hand so that we can keep up the gait." But he still fretted, wishing that they would have a hundred more miles on hand on June 1 and another hundred by the first of July—only then would he feel they were fully insured. More consultations with his brother and with Strobridge led the Judge deeper into pessimism. While it was true that they could lay rails all the way to Great Salt Lake by the end of the year, he wrote Huntington on April 13, "it is evident now that we shall get all the Iron here laid before the new supply reaches us. It is not your fault, as the frozen streams East prevented you from getting it off early—but remember we have storms & hindrances here which sometimes delay progress & cannot be overcome readily. . . . Keep sending Iron therefore as fast as you can get it off—until we have enough to lay to the mouth of Echo Canon, on the Central Pacific, & the 100 miles on the Western Pacific." Totaling up the amounts sent was nothing short of depressing. "It chafes us terribly when we look over the shipments of Iron," he confessed to Huntington on April 21, "to see how little was got off in Jan. Feb. & March—Jan. 3,698 rails, or only 8½ miles. Feb. 5,856 or 13 miles. March about 21,000 or about 48 miles, not enough to supply the tracklayers one month."[5]

Shortly after this was written, the Judge's load of cares were increased by one which even a fleet of freighters would not drive away. "I have had a slight attack of paralysis," he wrote Huntington on April 27, the day after he had observed his fiftieth birthday. "The physician tells me it is caused by too much confinement, lack of exercise—so I am going to start next week with Charles & Montague up the Humbolt to 12-Mile Canon, & perhaps to Humbolt Wells." He would be gone two or three weeks from the office, he added—"Must do it."[6] The level of medical knowledge about diet, blood pressure, stress, and heredity being as primitive as it was, the small stroke E. B. Crocker had just suffered was only an anticipation of a larger, ticking time bomb waiting to go off, and a few weeks' vacation in Nevada would only temporarily lessen his stress. The work would pile up in his absence and wait for his return.

Overwork and health were looming in Huntington's mind, too—for days he had suffered from a blinding, "fearful headache" earlier in the month ("I can hardly think and much less write," he had told Judge Crocker to apologize for one brief and sketchy letter); the pain was as familiar a companion as the small stack of bills of lading for iron leaving New York for the slow sail around Cape Horn to California. On top of what the Californians knew about, Huntington had sent two thousand tons the first week of April and as he wrote Crocker, six strong, "first class" ships were loading in the Hudson and would depart in a matter of days carrying another ten thousand tons. Though with chagrin he realized this would keep the Chinese busy for only ten or twelve days, he jocularly promised "you will have plenty of Iron in the fall" and fumed at the chain of misfortunes beyond his control which kept him from meeting the company's demand. "The Devil seems to have been among the Rolling Mills," he complained to Hopkins on April 13. One factory had been put out of commission for two weeks when a shaft broke, and another lost ten days when one of its rollers disintegrated. A third mill had been rolling seventy-five tons per day and was nearly done with its contract when it burned to the ground. Such bad luck made Huntington suspicious. "I have sometimes thought the Union Co. had something to do with mills Breaking & burning," he told Hopkins, "but I know nothing certain." He did not blame Durant or Ames when a freighter loaded with twelve thousand tons of iron was disabled and returned to New York, requiring the load to be transferred to another ship. But the thought must have fleetingly occurred to him. He could, however, lay direct blame on the Union Pacific for spoiling his timetable for supplying locomotives. "They bought Engines before we did," he admitted to Charley Crocker, "and took the capacity of nearly all the best shops for some time, so it will make you short of motive power for the early fall business." But he hoped to have thirty-seven locomotives on the way "by the final or by the middle of June."[7]

When in Washington, he could see that their legislation—all legislation, really, and there were now eighty railroad bills under consideration—had slowed to a walk, such were the concerns about the president's impeachment trial, which continued across April. Material needs of public servants did not diminish with the reducing pace, however. Judge Crocker was having free railroad passes printed for all members of Congress. That, they recognized, was coming to be a normal part of doing business. Of course it only began there. Representative Ignatius Donnelly of Minnesota appeared in the office to ask for $10,000 "to control some paper in his state." He was on two important committees—the railroad and land committees—so Huntington gave him $5,000 in a check made out to intermediary Richard Franchot, "but I did not like the way he called for it, and I was sorry I gave it to him." That same week the postmaster of the House of Representatives kited a check for $1,000 against Huntington in some transaction with a New York congressman, then went to his office "and begged so hard that I finally paid," he wrote E. B. Crocker, "but I told him never to do it again, and he never will."[8]

If he was feeling taken advantage of by some sorts in the capital, he kept alert when it came to the Kansas and Pennsylvania crowd. Securing the western end of the UPED's transcontinental line through Denver, still as complicated and still as important as it had seemed earlier, depended on Tom Scott and John Perry in the East (for that franchise across Arizona and Colorado) as much as it did the Southern Pacific's Lloyd Tevis and Horace Carpentier (for an unimpeded run from San Francisco and Mission Bay up the San Joaquin Valley). It was quite a seesaw balancing act, with Stanford and the Judge working the western end and Huntington guiding the eastern. He had felt the UPED people stiffen palpably upon the return of their tough chief engineer, General William Jackson Palmer, who had been facing Kiowa and Apache in Kansas and snorted at the Central Pacific's power and stature in the West; when earlier there had been negotiations about the Californians building either to Denver or at least to the Colorado River, Palmer returned from the West insisting that the California land was the most valuable to the company. Sensing no promise with them, Huntington stopped off in Philadelphia to see board member Scott. "I could do nothing with him," Huntington reported to Hopkins,

> *and I got up to leave, at the same time remarking that I should follow the instructions of my associates and put in a bill asking aid of bonds and land for the San Joaquin Valley & Arizona and make the bigest fight on that that I was able to. He then started up as though he did not want it done & said if I would give a friend of his an interest he thought we could do something.*

Scott promised to see Huntington in New York the following week—and, true to his word, brought Perry with him. They huddled at the Fifth Avenue Hotel until midnight on April 16 and again the next day at the Central Pacific's office, and finally agreed to meet their lines at the Colorado River and unite in one congressional aid bill. This doubled the urgency, in Huntington's mind, of buying control of the Southern Pacific, though there were many interests to satisfy. Work on the bill's several drafts consumed the rest of the month, requiring consultations in Washington and Philadelphia, but finally, on April 28, Huntington wrote to Judge Crocker that he had negotiated a document they could live with, putting it in the name of the San Joaquin line. "I believe we can pass it between this and the 4th of next March"—presidential inauguration day—"and if we could get control of the Southern Pacific I think (with this Bill of ours passed into a law) that it would be worth much money to us." Realizing that they would take some heat from antagonists in San Francisco, he said that they should get control of the S. P. secretly "and then sell out the San Joaquin Valley to the Southern Pacific. I shall go to work today to get the New York papers to say good natured things about this government aiding a Southern Pacific Road. Shall not have them say anything about the S.J.V. at present."[9]

Huntington was in Washington a week later when he learned some news about the Southern Pacific which altered his thinking about Tevis and Carpentier and what the Associates' strategy should be. Senator Conness's friend George B. Gorham, who had recently failed in his campaign for the governor's chair in California, was in Washington in the Senate Secretary's office; Gorham was also a significant stockholder in the Southern Pacific. He told Huntington the surprising news that Tevis and Carpentier were purporting to control the railroad but actually had "no money or next to none in it." It would be better to combine with Gorham and another stockholder, the San Francisco lawyer William B. Carr, and "it would take but a trifle" to buy the others out. "I have but little doubt but what a bill can be passed in Congress, giving us large aid," he wrote Judge Crocker, "in fact a grant worth many millions, and I am inclined to think that it will be the Road that will do most of the through business between N.Y. & S.F."

And that was not all he was thinking, Huntington told the Judge. He was going to see the president of the Northern Pacific, chartered four years before and expected to one day construct a line from Minnesota to Oregon. Somehow Huntington would "get him to hold his matter over another year and then we go in with him and control the west end of it." As if in reply—which it wasn't, as it required two to three weeks for mail to make the steamer passage between California, Panama, and New York—E. B. Crocker's letter arrived a couple of days later: "Do you ever stop to think where you are leading us, with all your schemes?" "I do not expect to lead you at all," Huntington replied smoothly. One day there would be three great transcontinental roads, he wrote, not just their one, and by letting others build the northern and southern lines they could take the western ends and get "the cream of the business."[10]

Other business on Huntington's mind concerned coal—those rich deposits in the Wasatch Mountains just waiting to be exploited, coal to fire locomotive boilers, coal to send both east and west on the transcontinental line, coal to forge a rich anchor to their railroad at its eastern extremity. But his spy network was reporting great leaps on the part of the Union Pacific—the bold announcement to build 350 miles in 1868, the certainty that such could be done under a dynamo like General Jack Casement (whom Huntington met in the late winter in Washington, and who had shaken the Californian's confidence in his own team). Butler Ives had seen the Union Pacific engineers in the Wasatch Mountains. Grading contracts were being taken across Wyoming Territory, and iron was rolling westward. If Durant got the Union Pacific within a hundred miles of Ogden, then by law he could send graders three hundred miles in advance of their continuous track. This would plant the Union Pacific in Humboldt Wells, Nevada.

"I see no way to keep them from occupying the ground for say, 100 miles west of Weber Canon," Huntington wrote to Judge Crocker on April 17, "but for us to go on and occupy the whole distance into Weber Canon as soon as we get within say, 400 miles, of course claiming that the distance is inside of 300." His suggestion was to "get a lot of teams and men and material and send them on the Upper Humboldt—but before they stop let them reach Weber Canon, and so as to have everything look fair I would start the work on all the hard points between the Hard work on the Upper Humboldt and Weber Canon, and I would make up my mind to hold all the line to Weber Canon as arbitrarily as though it was held by the great I am."[11]

Huntington's bid for the Wasatch coalfields and the trade of Salt Lake City—couched in a statement sent to Secretary Browning by way of Huntington's secret lobbyist, Thomas Ewing, who was Browning's former law partner—conveniently overlooked the seven-mile gap in the Sierra, and alleged that the Central Pacific tracklayers had reached the Big Bend of the Truckee when in fact only the grading parties were there, and it tabulated the distance between the Big Bend and Humboldt Wells as 240 miles, shaving 60 miles from the actual figure. It asked the secretary to approve the Central Pacific route north of Great Salt Lake to Monument Point. He sent it off to the Interior Department and settled down to wait for a reaction. A few days later he discovered that Durant had gone west "to hurry up the work." With both Dodge and Durant out of town the chances of their discovering Huntington's scheme before Browning ratified it were lessened.[12]

Meanwhile, there was a strange thing to report. Charles Tuttle, a director and the secretary of the Union Pacific, had appeared at Huntington's office on May 7. He offered to buy $1 million in Central Pacific bonds at their regular price of par plus interest and a 1 percent commission. "He stated that he wanted for a party in Europe & that he would make a small commission on them," Huntington wrote to Judge Crocker. But they were no longer hard up for money and the offer was just too suspicious, so he had politely refused. "I have a man on the track of this," he wrote, and hoped to know if it had been legitimate or someone's subterfuge.[13] The Doctor was an obvious suspect.

Dr. Durant and Grenville Dodge stood at trackside in a small crowd of laborers, soldiers, and dignitaries in the thin, nippy air of Sherman Summit on April 16, 1868; a locomotive and a few cars waited nearby as Durant raised a sledgehammer and pounded a ceremonial spike to close the connection of Dodge's magic crossing over the Rockies. It is not recorded whether Dodge felt a stab of resentment at the moment the hammer rang out—at having the executive hound the credit for a route which had nearly cost Dodge his scalp to Crow Indians once upon a time, and which Dodge had had to defend two seasons before against Durant's and Seymour's attacks. But there were champagne corks soon popping, and congratulations from Sidney Dillon, Webster Snyder, Casement, and Reed, and the less-welcome handshakes and pleasantries from Silas Seymour and the Doctor, who then had to bustle over to where a patient telegrapher sat inside the little station house—ready to tap out a thumb-in-your-eye message to the Central Pacific.[14]

"We send you greeting," Durant's telegram to Leland Stanford read, "from the highest summit our line crosses between the Atlantic and Pacific Oceans. Eight thousand two hundred 8200 feet above the tide water. Have commenced laying the iron on the down grade westward."

The reply from Sacramento shot back almost immediately, although Durant did not get it until the next day in Cheyenne. "We return your greeting with pleasure," Stanford wired, "though you may approach the union of the two roads faster than ourselves, you cannot exceed us in earnestness of desire for that great event. We cheerfully yield you the palm of superior elevation. Seven thousand and forty-two feet (7042) has been quite sufficient to satisfy our highest ambition. May your descent be easy and rapid."[15]

Dodge, Reed, Snyder, and the Casements might have echoed Stanford's last sentence to Durant with a more pointed and personal meaning. But it would have been hard not to feel a sense of eager anticipation after the symbolic event at the summit. Casement's men had spiked past the Sherman station on April 5, some 549 miles from Omaha. They moved quickly on the downgrade past the beckoning tents of the whiskey ranches which had bedeviled Carmichael's graders; by Thursday, April 16, carpenters working under Hezekiah Bissell and L. B. Boomer completed the dizzying pine-timbered Dale Creek Bridge, 135 feet high and 650 feet long, anchoring it with guy wires to steady it against the assault of gale winds. Casement's tracklayers barely missed a beat, on April 21 swarming across what was then the highest railroad bridge in the world and moving into a deep, curving granite cut toward Laramie. "Great load off my mind," Samuel Reed wrote his wife on April 23.[16]

Before the tracks were run past the new station at Laramie, a tent town had sprung up on the riverbank, populated by speculators and entrepreneurs and other fast-buck artists. On auction day, the railroad men could barely record the sales quickly enough. Some four hundred plots were sold within a

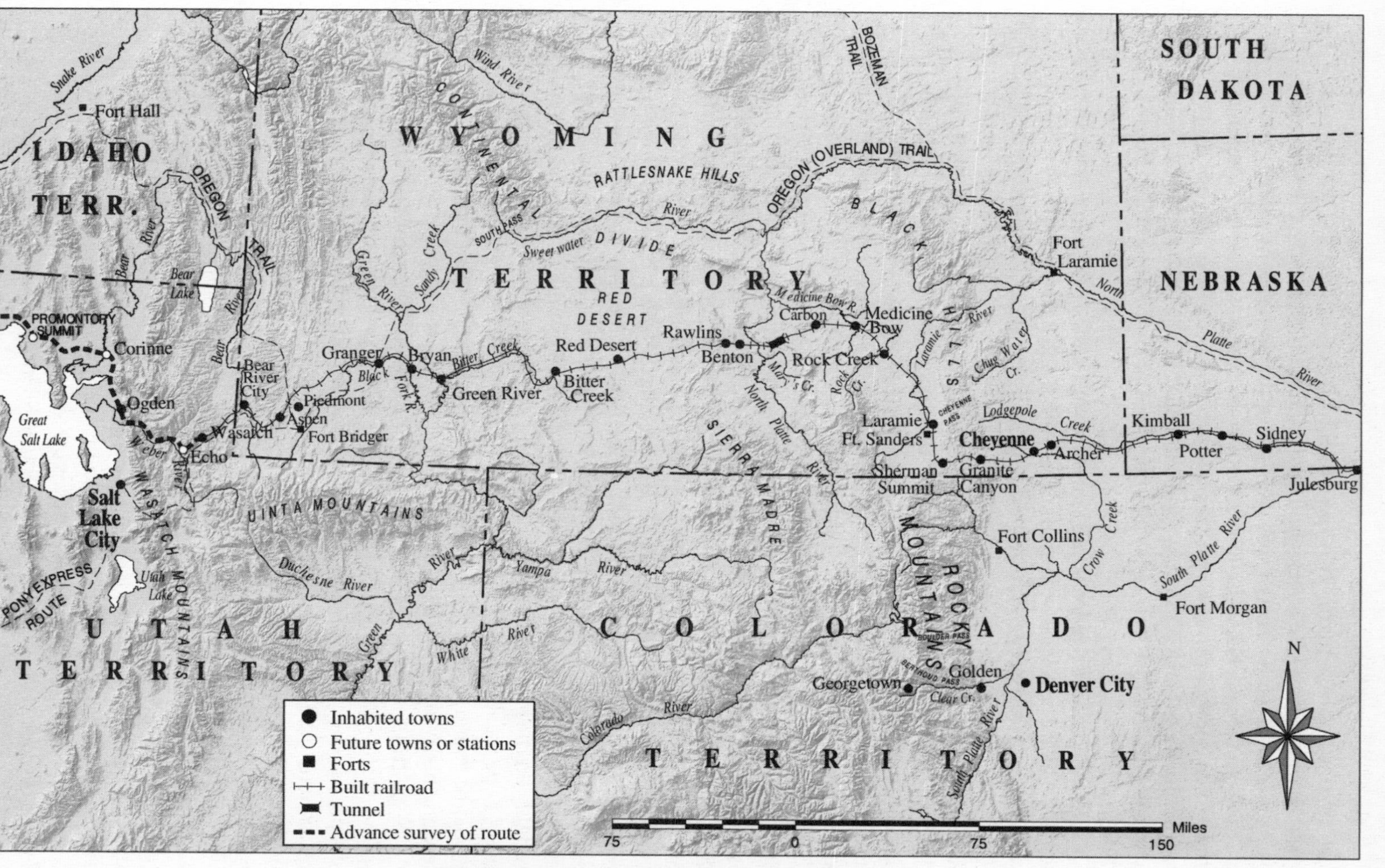

Union Pacific Construction Across Wyoming, 1868

few days at prices ranging from $25 to $260, and in another ten days no fewer than five hundred shacks had been slapped together. The first regular passenger train would ease its way slowly over the new, raw mountain grades on May 9, its coaches raucous with saloon keepers, gamblers, peddlers, tradesmen, brothel owners and their "prairie flowers," the flatcars spilling over with all of their various paraphernalia and with towering stacks of dismantled building sections. Hell on Wheels had advanced a little farther into the West.[17]

New merchants needed some kind of civic solidity, no matter what they sold for a living or tolerated from their patrons, so in early May they formed a provisional government and elected a mayor and other officials. Unfortunately, in organizing they were beaten to the punch by the lawless element, which had formed a union of its own. The legitimate government lasted only three weeks—before resigning in fear of their lives. Anarchy would prevail in Laramie throughout the summer and into early autumn, with no meaningful measures being taken by the commander at Fort Sanders other than to enforce a strict off-limits policy for his troopers. Robberies occurred in broad daylight, and nighttime murders were commonplace. On paydays and for several days thereafter, the mayhem would increase dramatically. In such an atmosphere even vigilantes were afraid to strike back.[18] Cheyenne, meanwhile, had been partially depopulated at the opening of Laramie, though things were still lively there. C. C. Cope, a brakeman on the regular passenger runs across Nebraska to Cheyenne, boarded at the Rollins House, where "the partitions between the rooms were all boards and papered over; was almost a city of tents at that time and all gambling houses. I can hear the gamblers yelling now, 'Roundo wins, Coolo loses.'" The vigilantes in Cheyenne continued to ride roughly on the most outrageous miscreants. "Nothing strange to see a man or two," Cope recalled many years later, "hanging on the telegraph poles as we would be coming east on the train in the morning. The telegraph poles were low and no trouble to swing them over."[19]

As with the previous years, violence erupted on other fronts within the sound of the Union Pacific's train whistle. In April, the brakeman Cope stepped off the train at Sidney, Nebraska, for dinner, where he encountered the conductors Tom Cahoon and Wilkes Edmundson. They told him that they were going fishing down at Lodgepole Creek. "Looking over towards the bluffs we could see quite a few Indians," he recalled, but they must have decided they were friendly Pawnee so close to the line. Cope climbed back on his train and Cahoon and Edmundson went off to get their fishing poles. "On leaving Sidney," Cope wrote, "I opened the side door of the baggage car nearest the bluffs, threw down a trunk at the door and sat down. All at once, *'Bang'*—there was an arrow sticking in the trunk between my legs just about six inches too low to get me. I didn't wait for the second arrow to come, but closed the door, went over to the locker where we kept about fifteen Spencer rifles to use in case of attack, took out a Spencer which was loaded, pushed the door open just far enough to stick out the end of the rifle and let go all the shot that was in it."

Cahoon and Edmundson were not so fortunate. Footing it a mile and a half out from the station, they were cut off at the creek by a Sioux party. Cahoon was shot and scalped. Edmundson, with several arrows in him, brandished a revolver and kept the Sioux away; his shots drove them off from Cahoon's body before his companion could be killed. Both survived their wounds. Cahoon worked for the Union Pacific for many years, wearing his conductor's hat well to the back of his head to cover his bald spot.[20]

At around the same time, some two hundred warriors attacked a construction party at Dale Creek not far from the bridge, killing two and wounding four, and rustling off twelve head of stock. To the east, at Elm Creek in Nebraska, five track repairmen were surprised and killed. The alarm was up—before the army was prepared to react, much less prevent. Major Frank North had just reorganized his two companies of Pawnee scouts, but the first attack at Sidney was a hundred miles west of his command, which had been placed to patrol the hundred miles of track along the big dip of the Platte between Willow Island and Wood River. The troops nearer to Sidney, at Fort Sedgwick, were unable to track down the raiders who had attacked the Union Pacific train and the conductors, but were now on heightened alert—as were the bluecoats to the west at Forts Russell, Sanders, and Halleck in Wyoming and the new post about to be erected at the North Platte crossing, Fort Fred Steele. Dodge was particularly worried about the tracklayers moving up the southern Laramie Plains, and the graders working northward above them, after learning of the second attack so close to Fort Sanders; he obtained assurances from the commandant, General Gibbon, to deploy escorts and scouts immediately. He also conveyed his trepidation to Oliver Ames. "I have feared this trouble with the Indians," Ames replied immediately, "and see no way to avoid it unless the Government will feed them or give them such severe punishment that they will not feel that they can rob with impunity. I see nothing but extermination to the Indians as the result of their thieving disposition, and we shall probably have to come to this before we can run the road safely."[21]

With a bravado meant to reassure his wife, Jack Casement wrote home from the end of track on the Laramie Plains north of Fort Sanders on April 28. "The Indians are on the Rampage," he said, "killing and stealing all along the line. We don't apprehend any danger from them [because] our gang is so large." Indeed, the ever-expanding force of Irishmen seemed to work with heightened vigor as the days warmed. "We are now *Sailing,*" Casement wrote a few days later, "& mean to lay over three miles every day." He had taken on a new enterprise beyond tracklaying, signing a grading contract for a section 150 miles ahead of track—now the commander of "about 500 teams and over a thousand men." Lewis Carmichael took the next stretch and left in a huge caravan of wagons, teams, and three hundred men bound for the desert and the continental divide. A late-season storm in the second week of May blocked the tracklayers for three days. "I never saw a worse storm," Casement wrote home, but in the same breath he boasted that the four hundred shovelers were nearly done and that his men could "lay 4 miles a day if we can be kept at work." Over forty miles of track were ready to be examined by the government commissioners. Just weeks later, Casement had freed up enough men on the Laramie Plains to wire Durant for another grading contract west of Green River.[22]

Before he had returned east, Dodge received reports from his engineers in the field. Dispatches announcing progress would follow him as he went. James Evans finished his final line from the North Platte all the way across the desert to Green River. "We save considerable in distance and altitude both over the preliminary lines," he wrote Dodge on May 7. Further west, J. O. Hudnutt had staked the route over the rim of the Wasatch into Echo Canyon, and Jacob Blickensderfer was locating through the mountain canyons toward Great Salt Lake. For Dodge, the urgency of avoiding trouble after seeing Dr. Durant on Sherman Summit was only intensified by having so many hundreds of laborers at work on one line with one will; with a clear scientific line beckoning them on into Utah, it was paramount that the company avoid self-defeat. Before he had gone, he had "a very plain talk" with Durant and Seymour at Cheyenne; Sidney Dillon was there to diplomatically underscore the need for amity. "Dr. Durant assured me that he had no desire to interfere with the work or delay it, but he only wanted to help," Dodge noted. "I told him we were well organized; that the lines had been well thought out; that all the engineers were very able men and that nobody could go over their work superficially and change it, and Mr. Dillon agreed with me."[23]

With Durant's promises to cooperate in his ears, Dodge finally headed east—there would be time for only the briefest home visit in Council Bluffs and a pass-through at the state Republican conference in Des Moines, where delegates were to be selected for the national party convention, before he had to hurry to Washington. The Republican back rooms of Des Moines, like their counterparts in state capitals all over the nation and indeed in Washington, were in a furor as the party faithful turned on a furious behind-the-scenes campaign to influence senators deliberating on President Johnson's fate in Washington. Then, delegates like Dodge were on their way to Chicago when the news came that their nominal leader in Washington had been acquitted of impeachment charges on May 16. On that day seven radical Republican senators joined to vote for Andrew Johnson's acquittal, finding that the prosecutors had failed to prove that the president's constitutional challenge of the Tenure of Office Act could in any shape or form be judged treason. The radicals thus were one vote short of conviction. In Chicago, the joy and expectations raised by General Grant's accepting the presidential nomination, with

Schuyler Colfax squeaking in on the fifth ballot as vice-presidential candidate, heaped more earth over the bitter contentions. As written, the party platform largely reiterated the articles of impeachment rather than set out a program of peace and prosperity, and the convention adjourned after only two days. But Grant now stood at the top. It was time to move on, win the election, and get down to business—business as usual.[24]

"I truly regret that we did not wake up sooner than we did," mused Collis Huntington in a letter to E. B. Crocker on May 9. He was thinking about the iron not shipped and the locomotives not ordered in the wintertime, and the host of other factors combining to slow their progress—and their promise—now as the summer construction season was truly in earnest. On one score, at least, he had caught the Union Pacific napping. Still waiting to hear from Secretary Browning about the fraudulent route map filed from the Truckee River to Weber Canyon, he had learned on May 5 that the Union Pacific had not filed a location with the Interior Department farther than fifty miles west of Cheyenne. The same day, Thomas Ewing wrote to say that he thought Secretary Browning would approve the Central Pacific location as far east as Bear River that week. "I have written him that we want it to Weber Canon," Huntington told the Judge.[25]

Three days later, Oakes Ames was in Browning's office when the secretary let slip that the Central Pacific was claiming to be within three hundred miles of Great Salt Lake and petitioning to be allowed to occupy the Utah line deep into the Wasatch Mountains. The congressman could not believe his ears. "Can this be?" he wrote his brother in New York. Oliver sent the news back down to the capital to Dodge that the Central Pacific was attempting to "lap over to the East side of Salt Lake. This should not be granted and I think my Brother feels that you will be able to check its adoption." Furthermore, they had to meet the enemy's challenge by immediately posting location engineers west of the lake. By the time the letter arrived, Dodge had already heard rumors about Huntington's scheme while in his temporary office in the Interior Department, and had huddled with Oakes Ames. They went directly to Secretary Browning to complain.[26]

"Dodge & Ames are fighting us very hard," Huntington informed Judge Crocker on May 13. However, "the Secy. has agreed to approve any location as far as the north end of Salt Lake, which I told Genl. Ewing to accept. Ewing said that Dodge told Secy. Browning that they would build to Humbolt Wells before we got there." Dodge cast aspersions on the C.P. surveys—they had been only preliminary barometrical readings done twice by Butler Ives last year, he knew—and he began talking with great familiarity about the terrain north of the lake, "and he finally got the Sec. to agree to give him three weeks to make further surveys," Huntington related, "and Genl. Ewing says he has got the Secretary to almost say that at the end of three weeks he will approve our location, if there is no further light on the subject."

Huntington said he wanted to guarantee that—suspecting that the Union Pacific might close in on its own three-hundred-mile proximity that soon. "I told Ewing that if he would get the location within 3 weeks," he confided to Crocker, "that I would give him $10,000 cash and $20,000 stock which I think will bring the location this week." Sure enough, three days after Huntington promised Secretary Browning's old friend and former law partner that $30,000, Secretary Browning approved the Central Pacific location—as far as the north end of the lake. A jubilant wire went westward that day, May 16, to Stanford, even as the cocky-feeling Huntington moved on to write Hopkins. He urged his partner to have a location map drawn for the mileage east of Monument Point down to Weber Canyon—it wasn't even necessary to base such a map on any further surveys, he said. He felt confident that Browning would approve the new map. And they would then gain possession of the extended line into Echo Canyon—if Charley thought his men could reach that point before the Union company. "They are pushing that point with a terrible energy," he warned, and they clearly still intended to hold the entire Utah line "and build at their leisure." This Huntington vowed to prevent. "With money enough," he assured Hopkins, "we can get the line located to any point that we want, and if so located we would take and hold all we wanted."[27]

Not until Friday, May 22, as the brass bands were blaring forth at the Republican Convention in Chicago, did Ames and Dodge learn that the secretary had ratified the Central Pacific location to Monument Point. "They were very mad about it," Huntington crowed to Hopkins. "They said Browning promised them not to ratify it for 4 months or until they had time to make further surveys &c." But ratification did not necessarily mean possession, he cautioned. "It looks to me as though the Union would reach a point within 300 miles of Salt Lake long before we do and then they will take the 300 miles and hold it, and I think we should take all we want calling it within the 300 miles & then get within that distance before they could prove it, more as it seems to me that we ought to be able to build three miles in the plain while they are building one in the Wasatch Mountains."[28]

Dodge was more than angry at having his engineering words disbelieved by Secretary Browning. He promised they would beat the Central Pacific across Utah. Huntington had taken the Union Pacific's engineer's measure at Browning's office and it had reminded him of his scorn for the man who had gotten him into railroading in the first place. "There never were two peas more alike than Genl. Dodge & T. D. Judah," he wrote to Judge Crocker. "If you should see Dodge you would swear that it was Judah, and if you had anything to do with him you would be more than satisfied. The same low, thieving, cunning that he had, then a large amount of that kind of cheap dignity that Judah had. So keep him as far from Sacramento as it is possible to do." The Doctor, though, was even more furious than Dodge, and much more dangerous an adversary. Huntington took care to warn his partners. "You have no ordinary man working against you on this end of the Road," he wrote, "& Durant is a man of wonderful energy, in fact reckless in his energy." It was clear to Huntington that the Union Pacific would get to Salt Lake before them—and that the only way they could prevail over them and their own problems with iron and the unclosed Sierra gap was to continue to lie about distances. Now, Huntington had just received a dispatch that Durant would have 640 miles of road done by the first of June. At the rate they had laid the last forty miles, Huntington said, they would be at Salt Lake by the first of December.

"I expect to go to Washington tomorrow night," he told Hopkins, "and if $100,000 will get the line located to the end of our line as per map in the Interior Department, I shall get it done."[29]

Trumped by a storekeeper! Congressman Dodge knew that the Central Pacific route had been ratified illegally. In the twilight of this administration, officials were beginning to look toward the day when they would not be on the public payroll, and Dodge knew that Huntington had hired General Ewing, Browning's former partner, and that "inducements" had been passed over. "It is putting a block in our way," he had written Henry Crane in the New York office, "that it may trouble us to get out."[30]

But there was even more distracting trouble brewing out west. It was not difficult to sniff out the source: Durant had lingered out in Laramie and Cheyenne after Dodge had returned east, and behind the Doctor there was most certainly the consulting engineer, Silas Seymour. Inexplicably, Durant had ridden into Cheyenne and announced that he was moving the Union Pacific division from Cheyenne to Laramie, meaning the much-anticipated branch line from Denver was to be relocated, and the big Union Pacific machine shops would go up in Laramie. He said he would do it even if it cost him $500,000 personally—although in all likelihood a scheme for personal enrichment lay at bottom. The announcement had paralyzed company land sales in Cheyenne and thrown new landowners and hopeful entrepreneurs into a fury: what kind of double-dealing was this? Durant's willful change of mind recalled the trouble when he threatened to move his eastern terminus from Omaha to Bellevue; it cost the company thousands in lost revenues and engendered ill will everywhere. The order to change the division would be rescinded—in this case the Union Pacific board would have to step in—but not before it sorely cost the company.[31] By then, probably, Durant would have cleaned up, buying depressed Cheyenne real estate.

This was a mere annoyance, and for a few days in early May, at least, Dodge's suspicions may have been lulled by his own engineer, James Evans, who had written Dodge to report that Durant had questioned him closely about the newly located line between the North Platte and Green River. Although the Doctor was "full of notions," Evans convinced him that "so far as the line to Green River was concerned, he could bring on his eight thousand men as soon as he pleased." Evans was "quite satisfied that the only course insuring comfort in dealing with him is to put on a reasonable amount of assurance. I gave him to understand that we could locate the line faster than he could march men and transportation over the road. After that everything was right. Seymour kept in the background and didn't trouble any. I don't know what use he is here unless it is to drive teams for the rest of them."[32] If it was amusing to think of a mustache-twirler like Seymour as a dusty, manure-booted muleskinner, the next letter Dodge had received from Evans changed everything.

It was dated May 11, and Dodge had received it as he was readying for the Chicago convention. Evans said that on May 9 he had been handed a notice published by Dr. Durant. No copy had been sent to Dodge. It was dated May 6 at Fort Sanders and titled "General Order No. 1." "For the purpose of facilitating and perfecting the early location of the line between this point and the Great Salt Lake, and enabling the company to place the large construction force and supplies which are now moving westward, upon the most difficult points without unnecessary delay," it read, the following points were ordered: Consulting Engineer Seymour and Chief Constructing Engineer Samuel Reed were to examine the entire proposed line to Great Salt Lake. The division engineers and chiefs of engineering parties would furnish Seymour and Reed "all maps, profiles and other information in their possession," and assist them in every way as they inspected. While the division and constructing engineers would make decisions in all cases of questionable location and choice of routes, all matters were now subject to approval of Colonel Seymour, who was also empowered to make further changes in the line and grades if he deemed them necessary. All locating engineers and their corpsmen were henceforth to report to Reed in the construction division, and they would be needed to plan "temporary tracks, either around or over the most difficult points," to hasten the work and avoid being stopped by tunnel or bridge projects.

One final point was the biggest slap in Dodge's face. "In order to prevent unnecessary delay in the work during the absence of the chief engineer from the line of the road," Durant ordered, "the consulting engineer is hereby invested with full power to perform all the duties pertaining to the office of acting chief engineer, and his orders will be obeyed accordingly by every one connected with the engineer department. Any orders heretofore given by the chief engineer, conflicting with orders that may be given by the consulting engineer during his absence, are hereby rescinded."[33]

Dodge digested all of this as he continued Evans's letter—how upon being handed the circular by Seymour he had instantly resigned, how he agreed to stay on a few days or weeks to finish his office work (but only if he were not subject to the General Order), and how Durant had telegraphed to offer him the temporary post of assistant construction engineer between Fort Sanders and Green River. Evans refused—and disclosed his reasons to Dodge: "Everything connected with it is chaos," he complained, "and they are building so fast and the work is so light that there is no time to organize it properly." By the time he could get it organized, Carmichael and the Casements and the others would have rumbled on through. "You can't hardly imagine how much I have desired to have you on the ground during the past two weeks," Evans wrote. "Reed is the weakest backed man I think I ever saw."[34]

What is going on with Samuel Reed? Dodge may have asked himself. Was the fifty-year-old playing it safe for his job, thinking of his wife and two young girls and the farm in Joliet? He was at base a good man, even if he quailed at the hint of a fight. But Evans had to be retained, and without a doubt others of Dodge's engineers would be walking off the work in protest. He immediately wrote Evans and said he must accept the supervisory job and try to get along with Reed, that Dodge needed him in the field and that it was important for Evans to obey Reed's orders.

The next letter arrived from Oliver Ames in New York, who said that Oakes had obtained a copy of the Durant order and wrote him about it. "The whole circular is one of those peculiar exhibitions of character which Durant everywhere exhibits and which shows the impolicy of giving him power which he is sure to abuse always," Oliver said in his smooth Bostonian way. "I think at our next meeting, we should definitely fix up the powers we intended to give him or repeal altogether the resolution making him agent for this work." Durant had no power over the location other than in the smallest instances, he wrote, and his moves "will not be sustained by the Board of Directors."[35]

Dodge went into a frenzy. How could Ames just blink and smile his patrician smile and talk about eventually doing something about this emergency? "You have allowed such matters to go on for harmony until one after another men leave until no one is left . . . who has one drop of manhood in them," he scrawled to Oliver Ames, his anger reducing his usual barely legible handwriting to near gibberish. Dodge wrote that he would resign before he would permit his men to be directed by "a man who has not an honest drop of Blood in his veins, who is connected with the Co. for the sole purpose of bleeding it and who the Co. say they cannot discharge for fear he will Black Mail it.

"It is your duty," he told Oliver, "to *promptly decidedly countermand that order.*"[36]

By May 14 the track on the western slope of the Sierra had been cleared of snow to the lower end of Summit Valley, and Hopkins hoped that in another week they would clear it to the Summit Tunnel. "At this time it averages 7 to

12 ft. deep & almost as solid as ice," he told Huntington. "Nearly all the men from the Truckee have been brought back & are at work in the snow, shoveling." Hopkins had just gone over the entire mountain line and could hardly contemplate the crushing task facing his laborers. On the eastern slope at the Cold Stream gap, the snow was so much deeper and so compacted that the workers had to cut in large, descending steps, or benches, down toward where the tracks would be. The deeper the gangs dug, the more snow had to be flung upward to the next higher bench and thenceforth up to the next—"in some places six times over from bench to bench," Hopkins said, "to get it up to the top of the snow cut & out of the way." Stanford would recall there were places where a snowfall of 63 feet "was pressed down, perhaps, into not more than 18 feet, but packed as hard as ice, and requiring the pick and powder to make a passage." By the end of May, the west-slope shovelers uncovered the entrance to the Summit Tunnel; aided by the big snowplow, they worked through the tunnel and began on the downgrade eastward. Ahead, other gangs had worked their way deeper to uncover the headings of rock tunnels they had blasted and chipped out the year before; inside those tunnels they found, to a depth of two or three feet, solid ice, which also needed to be removed.[37]

Charley Crocker predicted that the force would clear the gap and build the connecting track by June 15—"& then as soon as we get thoroughly organized and drilled for tracklaying," he assured Huntington on May 20, "we will go after Mr. Genl. Casement's best work, not for one day but steady every day work & you shall see what you shall see in that line. We have no men on our line with high sounding titles—Generals colonels &c. &c. as we hear of on U. P.—but I think we can build R. R. as well & as fast as they." He had been all through the Humboldt Canyon with Strobridge—"gone 15 days among the Indians & sagebrush"—and was much encouraged. The stretch of 250 miles would take his graders no more than thirty days to prepare for track, so his strategy was to send no men forward until they had laid rails past Reno and the Big Bend of the Truckee, and past the Forty-Mile Desert and the Humboldt Sink, to the stagecoach station at Oreana. There, in the portal between the Trinity and Humboldt Ranges, they would begin to raise some alkaline dust—sending diggers and blasters far ahead to Emigrant's Pass and westward, where short but very heavy work was required, while the tracklayers hustled up toward them along the Humboldt—"we will lay the Iron as fast as it arrives here up to the upper end of the Canon," Charley promised.

Hopkins echoed this a few days later, adding that "'where there's a will there's a way.' I believe we shall lay all the iron we can get here previous to 1st Oct. and from 50 to 90 miles a month after that. During the long days, Strobridge thinks he can work two shifts of men, eight hours each by day and during clear moonlight nights a third shift—& with horses to haul forward the ties & part of the iron, he thinks he sees how he will be able to handle 300 or 400 tons of iron & the required ties from the end of a track daily." As the Judge had suspected, for great distances they could actually lay track on the ground after clearing brush away, and graders would follow to raise the track to a proper businesslike level.[38]

The question was the iron. How they could build all the way to Weber and Echo Canyons by the end of the year if they were going to have a month in August with no rails to lay before those slow-sailing ships began completing their 120-day voyages around Cape Horn? First they debated paying the higher freight charges to send iron by steamship. But by mid-May it was clear to all that Huntington would have to send some iron down to the isthmus. It would be hauled across Panama and reloaded for the voyage up to California, to cover those slack weeks. It would be fearfully expensive—to use one of their favorite expressions—but, as Hopkins wrote, "We think it may not cost as much per ton as we have paid to haul 50 miles of Iron over the summit." Huntington found that it would cost more—$30 in gold per ton of iron for the isthmus, as opposed to $20 per ton for the Sierra summit—but the difference was immaterial. On May 21, after a rush of telegraphed messages, he put the order through for ten thousand tons to go over Panama. To reduce the cost he had them locate California merchants and shippers willing to send a freighter down loaded with grain, at which point the iron and the grain would change places.[39]

All five Associates were greatly relieved—although Huntington's headaches would continue to pound as further disasters struck his iron suppliers. The worst concerned the big Scranton Mills, where a main balance wheel pulled loose at the end of May and flew into pieces—"one piece went through two puddling furnaces and one piece, about half a ton, went out through the Roof," Huntington moaned to Crocker, "and I won't pretend to say where it went." The mill had been producing 150 tons per day but now would be stilled for at least three weeks. Huntington scrambled to fill that gap.[40]

Knowing iron would take the short cut across Panama removed one care from Judge Crocker, too, but of course there were a host of others. To shake off the scare of his short attack of paralysis he had actually listened to a doctor and taken two weeks away from his desk, and the long rides through the Humboldt Valley with his brother, Strobridge, and Montague had given him a welcome respite. Five minutes back at his office in Sacramento erased it all. The waiting stack of letters from Huntington with their dreams of an empire in the West signaled that they would be tied "down to slavish work the rest of our lives," he complained to his New York partner. "The detail of our already immense business falls heavily on me, & I am breaking down under it. . . . So you see it is not strange that I hesitate about taking on my shoulders any more work. I don't like to say anything but Stanford & Hopkins don't like to stay in the office any more than they can help—& when I am here they know I will not neglect it, & so it goes on. The mental strain, & the lack of physical exercise will soon use me up, & then what good is it all?" Two days later he felt himself sinking even deeper. "I give up," he scrawled to Huntington. "I am hopelessly committed to work enough for 3 or 4 lives. I hope to be able to live through it all, but if I do, I shall be a very old man." His next consultation with the doctor left him with a warning to spend as little time in the office as possible during the summer.[41]

Hopkins had confided his worries about the Judge to Huntington. "I am fearful we may suddenly lose him some day," he had written a month earlier after seeing the portly but suddenly frailer Crocker off on his working vacation up the Humboldt. "Though it is possible that others of us who feel more secure of our hold on life may yet go before he does. He is a strong feeder & with healthy strong digestive process that supply an amount of animal vitality requiring abundance of open air & physical exercise. Such natures can't stand as much confinement as feebler [organisms] who run their machines with a less press of steam." Whether Hopkins's concern went beyond lip service is as unrecorded as any offers he may have made to come in to work more often or to take on some of the Judge's load of responsibilities, just as there is no way to ascertain whether E. B. Crocker partially brought stress upon himself by being unable to delegate responsibility. The fact remains, however, that the Judge stayed at his post longer than he should have throughout the summer, hastening to apologize if he took a few days off. Aside from his administrative duties he remained Huntington's most vital link with the organization in the West. "I am very sorry to hear that your health is not good," Huntington wrote him on June 9. "We can't spare you. Not yet. I know you must be very busy and I can sympathize with you, for with attending to matters in Washington, buying and getting together R. R. material, attending to financial matters &c. &c., I get very tired and sometimes I think that a man can wear out."[42]

In the wearing-out business, Huntington gave as much as he received. This was especially straining on the Sacramento partners when it came to the great urgency about Utah in late May and early June, as Huntington sent his cables and letters westward with accounts of his schemes to use the interior secretary as a means to drive a railroad all the way across into the Wasatch coalfields. The Union Pacific's answer signal—its intent to begin grading across the same territory—drove Huntington from his usual bold and aggressive self into something rasher, even dangerous to the Associates' common cause, in the opinions of Judge Crocker and Mark Hopkins.

Huntington's urgings to call four hundred miles three hundred in order to get their own graders within sight of the coalfields—and then "build before anyone could prove to the contrary"—alarmed them. They had been bending truth and geography for years—since they had moved the base of the Sierra twenty-one miles to the west—but this was a lie of a more hazardous order. Let the Union Pacific people do such a thing, wrote the Judge on May 23.

> *It is in violation of the act of Congress, & is a piece of sharp practice on the part of Durant—which seems to me will not be sustained by Congress or the Sec. of Interior. It places them in the wrong. Will we not be equally in the wrong if we follow suit, & set men at work more than 300 miles ahead of completed road. Bear in mind that ours is a national work, that the public are watching us all. Our strength lies in building railroad* fast*—& keeping within the law, & acting fairly & honorably, free from all sharp practice.*

Better to send a protest to Secretary Browning, he argued. They should then simply grade alongside the Union's line, confident that their work would get the approval of Congress. "Your dispatch has troubled us a good deal," he went on. "Of course we can set men at work over there, though at a great disadvantage, & at heavy expense. We cannot send them from here, but would have to send agents, & hire men there." The morning's papers had carried a note that Brigham Young had taken a contract to grade from Echo Canyon to Salt Lake City. "It cannot be strictly true," he said, "because Young don't take contracts in his own name. The mere letting of contracts amounts to nothing, it is the *doing of the work* that violates the law."[43]

Hopkins sent his own cautionary letter eastward, but they all agreed on one thing: they needed a presence in Utah. "I have been fearful that we have made a mistake in not keeping a good man at Salt Lake for the last, say, 2 years," Huntington confided to Hopkins, "as I think we want a large quantity of men to work this season in the Wasatch Mts and necessarily a large amount of Provisions to feed them." Charley Crocker would be the one to send—but he could not be spared. Hopkins and the Judge decided to delegate Stanford—he was the most able-bodied, and as a former governor he would have diplomatic standing both with the territorial leaders and with Brigham Young.

Stanford delayed his trip for a fortnight, however—his wife delivered their first child in eighteen years of marriage, a son they named Leland Stanford Jr., on May 14. The anxious new father finally left, reluctantly, on June 1, after signing over a packet of Contract & Finance Company stock worth some $9 million to Jane Stanford in case something happened to him on the mission. George Gray and Utah governor Durkee had preceded him by a few days; the territorial governor would be an intermediary for tie contractors on the Bear River and for buying coal properties. "We hope," said Hopkins, "they will be able to get a Mormon or Gentile force at work there—perhaps let a contract to Brigham Young or his representative for 50 or 100 miles from Salt Lake westward towards Humboldt Wells."[44]

There would be no waiting impatiently by the telegraph key to learn of

Stanford's success or failure; there was too much else to do. Huntington kept up the pressure on his partners to get control of the Southern Pacific. He urged them to see who really controlled the line and how they operated, if financier Lloyd Tevis was not truly a front man; where, too, stood William Carr in the picture? Before Stanford had left for Utah he had gone overnight to San Francisco and returned with a roster of directors and a strategem. "None of them are anxious to sell," Judge Crocker reported. "If any of us were to approach them their ideas of its value would be up in the skies. We can only work through Carr and Tevis. So you see it is not an easy thing to get the control of the S.P.—but we are after it." The directors had expended very little money, he noted, but they would cite the value of their submerged thirty-acre grant in Mission Bay, the worth of their thirty thousand acres of land grants out to their railhead in Gilroy—between $2 and $5 million, theoretically. "They treat these as so much overplus assets on hand capable of being divided," the Judge complained, "making no account of the expenditures necessary to obtain these grants. With men so crazy as to figure that way, it is hard to deal, & time must be allowed for the blood to cool." But of course, other local railroad men were beginning to show interest, and "it is getting complicated."

Hopkins had concluded that the price would remain too high, but they all assured Huntington their blood was still up; Stanford and Crocker thought it might be best to get Tevis to their side by giving him an interest in the San Joaquin Valley Railroad, and to this Huntington agreed. But perhaps he could drive the price of the Southern Pacific downward by directing their congressmen to enter a bill for the San Joaquin Valley Railroad. "I have not thought it policy to have the S.J.V. Road come to the surface," Huntington explained, but if it shook the arrogance of the Southern Pacific directors into worrying that the Associates would run right up the valley and over to a deepwater terminal, bypassing them altogether, it would be worth the inevitable press assaults on the monopoly.[45]

For his part, Huntington had been seeing to other aspects of that monopoly. Complete control of Colonel Wilson's California Central line up the Sacramento Valley eluded them as long as there were outstanding bonds to buy up, and in New York the last directors balked at signing Huntington's indemnities to protect against any of their outstanding hidden debts; it was a tedious and frustrating chase, Huntington said, but he hoped to run down all of the leads soon. Vastly simpler was the California and Oregon Railroad business. Since getting a hold on the charter in the fall, they had all been aware that it called for construction of twenty miles by the summer of 1868 or they would forfeit the charter. With no rails to lay and too many irons (as it were) in the fire, their only course was to have the law concerning the California and Oregon changed, extending its time limit. Huntington had gotten the bill introduced in the late winter, and with so many friends in the Senate the extension had passed that body late in May. He pulled out all stops to get it through the House before members would evaporate in July.[46]

The Goat Island bill continued to navigate through hidden shoals, suffering from sudden adverse winds. In late May, while in the capital, Huntington heard that ordinary real estate concerns such as the grand isolation and breathtaking views of San Francisco Bay might be affecting his cause. "I have been told that some of the military officers have some fine cottages on Goat Island," he told Judge Crocker. "Very likely that is the reason why it is so necessary for military purposes." He also had news on the Pacific Railroad tariff regulation bill: it would rest in the Senate committee, "as Morgan, Conness, Stewart, Harlan & Drake are all right on it." His circle of helpful public servants was expanding by the week; Senator George H. Williams of Oregon stayed at his house in New York for three days, during which time Huntington plied him with hospitality while keeping up a running commentary on business. "He has been rather slow on Rail Roads," he told Hopkins, "but I think I added something to his R. R. education. He is a first class man and can do us much good." When Senator Williams returned to the capital he did so as a friend upon whom Huntington would count when their San Joaquin and Southern Pacific bills surfaced later. Congratulating himself on this among many other things, Huntington wrote expansively to Leland Stanford on the eve of the latter's mission to Salt Lake City to let him know just how well they were doing. "Our bonds are selling very freely," he said, "so much so that it gives us a large surplus of money, and if the sales should hold up for the next month as they have for the last 3 or 4 months, I shall advance the price to 105." In fact, he put the price up to 103 in only two weeks.[47]

With some of that surplus of money—the $100,000 he had told Hopkins and the others about—he intended to take the train down to Washington on the night of Sunday, May 31, to see Thomas Ewing and Secretary Browning about the Utah business, but more trouble with iron delayed him several days; another of their principal suppliers halted when its employees struck for higher wages, leaving Huntington with four ships in port with only partial cargoes. He got one launched fully laden, promising the captain a bonus of $250 if he made the run in 125 days. Huntington finally caught his train to the capital on June 3. "I rode all night in the cars," he told Hopkins, "was at work at Washington until 3 o'c in the morning thursday and returned friday night, and am about as much used up today as I ever was in my life." He found official Washington, however, draped in mourning and closed in honor of ex-President Buchanan, whose funeral was on June 4, although Huntington did gain a special audience before the Senate land committee and succeeded in forestalling a land grant about to be awarded their adversaries in the Goat Island case—"I *saw* two of the committee before they came together," he told Hopkins after his return to New York, presumably with his satchel a little lighter.[48]

If Huntington did not see Ewing or Browning that week, he at least cooled down considerably—even to the point of reversing himself. The short trip may have been exhausting, but it gave him time to reconsider his hasty orders sent out to his partners, especially in light of their telegrams and letters calling for caution. "I telegraphed you yesterday not to take possession of the line more than 300 miles in advance," he explained to Hopkins on June 10. He recognized that the Union Pacific had gained possession of the canyon lands for the present, so he had left letters for Ewing and Browning, making a formal objection about the Union Pacific being allowed to continue westward. "I thought we had better take *high ground* and confine ourselves to the law until we are where we can make more to break it than keep it," he said. Replies to his Washington messages arrived shortly. Browning ventured the unofficial opinion that the Union Pacific would beat them to Great Salt Lake, leaving Huntington to fill in the obvious blank as to the outcome of a claims fight. But Ewing's letter was more positive, encouraging them to stay on the higher moral ground but to build into the Weber orbit as soon as possible and then contest the claim. "If the Genl. is true to us I think it can be made to work, and I have no reason to suspect him," Huntington wrote Judge Crocker,

> *except that he is connected with the Shermans by marriage, and they are very close to the Union Co. Charles Sherman is one of the Government Directors in Union Co., but I have no fear of [Ewing] going for the Union Co. unless he gets as many dollars from them as he does from us. But we can only get to Weber Canyon before the Union Co. by pushing our work as work was never pushed before.*[49]

The direction of his "high ground" scheme took on more definition in a letter to Hopkins: they would contrive a location map with their line as far east as Echo Canyon, and they were to indicate the need for a tunnel just east of their line in Echo; then they would wait until Strobridge and Crocker had gotten the track within three hundred miles; Huntington would then try to convince Browning that the Union Pacific should confine itself to the eastern end of the tunnel location until the Central Pacific had finished blasting it through. But a week later, on June 19, Huntington found his confidence eroding—not only in this plan but in whether he was actually accomplishing anything for all of his frenetic activity and the whirlwind of ideas forever swirling in his head. "I have been so much perplexed about our failure to get Iron, as per contracts," he fretted to Hopkins, "and in getting ready so much more material than we expected to want early in the year, and in getting it shipped, that I have realy had to neglect our business in Washington." How hard it was to do a competent job on all fronts![50]

Feeling the same, though a trifle relieved to have one or two of their New York partner's schemes deflated a little, Hopkins and Crocker saw another way to burnish their "higher ground" position—by claiming better engineering and construction than the Union Pacific, and certainly greater comfort and convenience on the Central Pacific cars. The Judge was researching how they might break an eastern monopoly on the right to build sleeping cars, which when their service moved into Nevada they would require; any tinkerer in California could solve the problem of converting upright daytime passenger seats into horizontal bunks, but to manufacture any kind of convertible car in California would be to collide with existing patents. Crocker had presciently gotten bills passed in the legislatures of California and Nevada which would invoke the principle of eminent domain in patent law, allowing them to have a patent condemned in state court. "It is rather of a novelty in railroad matters," he confessed to Huntington, "but I think I can make it stick—for a patent right is no more sacred as property than a man's house." In those freebooting times such a notion might actually prevail long enough for them to move forward, or at least use the threat of a lawsuit to blackmail the eastern manufacturer into granting a cheap franchise.[51]

Mark Hopkins's contribution was the news that the Union Pacific in its greed and haste was building terribly. A friend returned to Sacramento after an overland journey by rail to Laramie and thereafter by stagecoach to say that "the Union Co. are building road rapidly," and that he thought "they will reach Salt Lake before we do," Hopkins told Huntington. "But he says their road is an inferior one compared with ours—that it is an undulating grade, the alignment crooked, their ties are one third less in number than ours & not exceeding one half the size of ours, and none of the road ballasted. He says it was impossible to read a newspaper while riding on their road, but it was easy to read while riding over the newest & roughest portion of ours." Gone, apparently, was the Associates' heady resolve to build as cheaply and expediently as the Union Pacific. So Huntington would adopt the claim of quality and safety when he renewed his campaign against the Interior Department.[52]

Never, however, could the Associates expect their own ride to be smooth. Late in May the Chinese began quietly filtering out of the work camps and heading back toward civilization. It was not a walkout—Strobridge and his pick-handle had been keeping that problem at bay. The foremen thought at first that it was simply out of loneliness or an unwillingness to advance out across the Nevada desert far from established Chinese communities and supply lines, though some of the laborers said they were not making enough money with their constant moving of campsites. Then Strobridge got to the bottom of it. "Worthless white men have been stuffing them with stories," Judge Crocker reported, "that east of Truckee the whole country was filled with Indians 10 ft. high who eat chinamen, & with big snakes 100 feet long who swallowed men whole." To counteract this Strobridge organized a junket for fifteen or twenty men—"& we sent good men along with them to show them 100 or 200 miles of the country where the road was to be built. . . . The fact is there are no Indians on the line until we reach Winnemucca, & then they are harmless like the Piutes on the Truckee, & the Chinese despise them. We have a ticklish people to deal with—but manage them right & they are the best laborers in the world. The white men we get on our work here are the most worthless men I ever saw."[53]

Dodge, stuck back in Washington after the Republican Convention, wanting to be nowhere but out with his engineers laying claim to Utah lands north and west of the lake, tried to maintain his vigiliance within the Interior Department building against another coup by Huntington. At the same time he attempted to control the plunging morale in his own corps, long distance by letter, and maintain a firm position with President Ames and the Union Pacific board. Everyone but Durant seemed to be operating in a fog—Dodge included. The chief engineer had seen no maps or profiles of the terrain west of Green River, Wyoming, on the eve of this battle with Durant and Seymour over the Wasatch rim line and the locations through Echo and Weber Canyons. His teams were out in the field, unsuspecting should Colonel Seymour appear to make demands with his spurious authority. And none of the executives in the east had seen a copy of Durant's "General Order No. 1" with a name signed to it, which led some of them to wonder if it was all just a hoax. Meanwhile, the Doctor and his consulting engineer could operate freely.

Granted, Oliver Ames had replied to Dodge's bruising letter demanding action, explaining again in his diplomatic way that "This order of Durant as far as it confines itself to construction and a change of location to facilitate construction may be within the scope of that Resolution. But when he interferes with your authority as Chief Engineer and the control of your parties he is entirely beyond his limits, and should not be recognized. I will write him at once." Dodge immediately had written to warn Ames that he would not pay attention to Durant's or Seymour's orders and had so instructed all of his employees. He had followed this with a warning to Jacob Blickensderfer out at the head of Echo Canyon, reiterating that the chain of command through Dodge remained; he should do all in his power to aid Samuel Reed and the construction bosses, and that they were permitted to change lines or grades—but only temporarily. Under the contract they must "build the line we turn over to them," Dodge had emphasized. "You will therefore turn over to them the permanent location as fast as possible and endeavor to turn it over so fast that they shall make no complaint for want of line."[54]

Samuel Reed and James Evans, too, had needed attention. In a benign-appearing move, Durant had ordered Reed and Seymour to continue through the Utah canyons to Salt Lake City, with Reed to arrange with Brigham Young for grading parties. This made sense to all, since Reed had spent time in Salt Lake City and had established good relations with Young and his people. Evans had been appointed to temporarily take over Reed's duties as chief construction engineer; he had his doubts, as he had written Dodge from Fort Sanders: "I hope [it] will be short, as it is no sinecure. Track, grading and bridge all in a pile together. I think it will be difficult if not impossible to prevent delays. . . . Everything, of course, is being done in an extravagant manner."[55]

Reed had found respite and success in Salt Lake City. "I have not felt so free from care and anxiety since leaving home," he wrote his wife. When Samuel Reed obtained an audience with Brigham Young, the Mormon leader was eager to discuss obtaining good-paying work for his faithful. In the valley there had been, memorably, plagues of crickets and grasshoppers, but now, with the Saints' empire firmly established and blooming, there were locusts; for three years running the farmers' crops had been affected. What surplus there was of hay, oats, and potatoes, Young knew, they would sell to the railroaders. Moreover, as an original shareholder in the Union Pacific, he savored the trains' approach, still blissfully convinced that the Pacific Railroad could never avoid running through the City of the Saints. Reed had been instructed to be noncommital on which way the railroad would turn upon reaching Ogden. The strategy was correct, Dodge thought: "I knew the importance of upholding that friendship," he would write, "especially as the Central Pacific were catering to it and trying to influence it." On May 21 (ten days before Leland Stanford would depart Sacramento on the same mission as Reed) at the Continental Hotel in Salt Lake City, Young signed a contract for $2,125,000, encompassing all grading, tunneling, and bridge masonry from the head of Echo Canyon to Ogden, the onetime fur-trappers' camp at the junction of the Weber and Ogden rivers, now a burgeoning town in itself.

Young subcontracted the work mostly in small parcels to bishops of various wards; larger and more specialized work was subcontracted to his son, Joseph A. Young, and his elder partner, Bishop John Sharp, a Scot, who was assistant to the superintendent of public works in the city; another local man, Joseph F. Nounan, a Gentile, was another principal subcontractor. All of the subcontracts, large and small, granted a tithe to the church. Later it would be said that Brigham Young profited handsomely from his $2 million railroad construction contract; after his death in 1877, administrators would find that the profit was far lower than popularly thought: $88,000. Though Young's personal finances were often indistinguishable from church finances, there was in this case little confusion regarding his estate; the railroad project was considered "official," and those profits went to the church.[56]

When he signed, Young was grateful for one particular concession granted him by Durant, who wired Snyder in Omaha to "transport passengers on [Young's] orders or those of his agents, at the same rate charged contractors." (In New York, Huntington's spy in Durant's office apprised him of this. "I am told," he would write to Stanford, "that Durant gives all the Mormons passes over the Road and I suppose that will give him some hold on Young, as there is . . . quite a good many Mormons coming from Europe this year.") After the deal was sealed, Reed had gone up to work with Blickensderfer; the line would roughly follow Reed's preliminary survey. There, up in Echo Canyon, Reed was to share a tent with Silas Seymour as the colonel began to look over Blickensderfer's work, and there, Dodge would soon learn, the trouble for his men and himself would begin to increase.[57]

Before Seymour's complaints in Echo Canyon began to resound, Dodge saw that his outift's morale problem was getting serious when he next heard from James Evans:

> *There will be a great pressure brought to bear to keep me on construction. What the result will be, I don't know, neither do I care. Rest would suit me better than anything else. The last year has been particularly hard. I feel it most sensibly the work could have been easily gotten along with, but there are other things constantly with me, and I am quite sure that it will lead to a breakdown very soon. As long as I can, will try and do whatever there is to be done. This I suppose is the duty of all of us, if not to ourselves, to those who come after us.*[58]

Dodge could respond as a commanding officer and try to buck up spirits. And three days later, on June 1, Blickensderfer's letter arrived, requiring more support and guidance from the chief engineer.

The head of Echo Canyon, where he was camped, with its silver-green and aromatic sagebrush everywhere and with the unsettling way in which noises reverberated from the high rocky walls, was far from the fifty-two-year-old Blickensderfer's native Ohio. His consultations with Colonel Seymour, too, were unlike those he had had previously on the Ohio Public Works Board or the Steuvenville and Indiana Railroad. They were unworldly. As for the setting of their debate, deep, narrow, boulder-strewn Echo Canyon wound for nearly thirty miles, its dusty wagon road bumping alongside the creek below squadrons of diving and swooping swallows and the high, red rock walls; the Harlan-Young wagon train had been the first along it in 1846, followed shortly by the doomed Donner-Reed party. The Mormons' advance party under Orson Pratt had gone through in 1847, their leader noticing how the overhanging rocks "were worked into many curious shapes, probably by the rains," the most fantastic being Castle Rock, looking like something right out of a storybook, and others soon earning names like Steamboat Rock, Sphinx, and the Giant Teapot from subsequent emigrants. Samuel Clemens had ridden through Echo Canyon in a stagecoach on his way to Nevada. "It was like a long, smooth, narrow street," he recalled later, "with a gradual descending grade, and shut in by enormous perpendicular walls of coarse conglomerate, four hundred feet high in many places, and turreted like medieval castles. This was the most faultless piece of road in the mountains, and the driver said he would 'let his team out.' He did. . . . We fairly seemed to pick up our wheels and fly."[59]

That Mormon-maintained wagon road may have provided a smooth descent, but Jacob Blickensderfer was being thwarted in setting out anything like it for the railroad from the crest of the Wasatch Divide down toward the junction with Weber River. From an engineering and construction standpoint, he had to worry about physical hazards of steep canyon walls, possible landslides from above, the likelihood of snow obstructions in winter, bad rock cuts, expensive embankments, the serpentine nature of the course, and the possible requirement of tunnels. He had to hope to avoid crossing the creek. He had to tabulate every physical feature of a projected line in terms of how the laborers might run temporary track around it if it posed a problem or delay. And he had to run his lines at an acceptable grade.

Blickensderfer reported to Dodge that he had isolated three possible routes down Echo Canyon, each at a gradient of ninety feet per mile, each with serious drawbacks, and that he had sent his assistants down to develop the three lines more fully so that a clear choice could be made. Unfortunately Seymour and Reed had appeared. Blickensderfer could tell that Reed agreed with all he had done thus far, especially the premise of a ninety-foot grade—"but he does not press his views so strongly as the Colonel does." Seymour was "a strong advocate of the use of higher grades, and the cheapening of the cost of construction by that means," Blickensderfer commented. "This strikes me as singular, being the very reverse of his views strenuously urged last season in reference to the Black Hills location." Seymour told Blickensderfer to run the three lines down the canyon again at a hundred-foot grade, and then to compare the two sets and calculate the savings—and then submit his recommendations. This Blickensderfer agreed to do, but it would cost at least a week of time.

He had gone to give instructions, he related to Dodge, returning to the base camp a week later only to find that Seymour and the acquiescent Reed had already elevated Blickensderfer's grade through Weber Canyon and Devil's Gate by ten feet. Seymour sent a messenger to recall surveyors James Maxwell and Thomas Bates away from the important tasks they had been doing and put them onto this relocation. Both engineers had thus lost a week in changing position—bridges had washed out and travel was difficult—and the changes in location would cost at least another week.[60]

Immediately after reading the letter, Dodge fired off a wire to Blickensderfer and told him that Seymour had no control over his men and no right to interfere. He also wired that he wanted no grade to exceed ninety feet. In New York, Durant got wind of Dodge's telegram; stepping back once and planning to advance twice, the Doctor wired Seymour himself and countermanded his orders, for the moment supporting Dodge. This game only heightened Dodge's frustration. "All this conflict of orders was demoralizing to parties, made delays in the work, and was being felt along the entire line," he said later, "and it was with the greatest difficulty that I kept up the discipline and everything moving."[61]

Twice in that one week Dodge went up to New York to contain the damage being done out west. Durant was there—turning the argument away from his general order and the interferences to complain that Dodge had failed to send all of his instructions and preliminary profiles and maps to the New York office; the Doctor seemed unimpressed by the explanation that several of the field reports had been submitted late and that the short-handed department in Omaha had been working overtime to get the material completed, and were often called away for other company business; at present his field engineers in Wyoming were under orders to give their new location maps directly to the grading contractors to avoid delays. Dodge could tell that Durant's complaints were being lodged as ammunition to get him fired.

Back in Washington on June 4, he wrote Blickensderfer to warn him that Seymour had been sending telegrams to New York filled with objections and dire predictions, but that Dodge had persuaded the board that a ninety-foot maximum grade was best, even if more expensive by half a million dollars. "Col. S. says that the adoption of your line is suicidal policy," he reported. "Its great cost, deep cuts, high banks, walls, &c. with liability from snow and snow-slides, make a location in the present and prospective, that the Co. will not sustain." Dodge emphasized that he was leaving the matter to Blickensderfer's judgment. "If the work is very heavy," he added, "we will have to put in a temporary track and go around it, say with 140 or 200 ft. grade and thus not delay work on location or delay progress of track. I have thought that track could be laid cheaply, right on Reed's old line; a short 200 ft. grade would bring it in to the valley that Reed ran down and avoid the very heavy work which it is feared we cannot take out before track arrives there."[62]

As for Reed, Dodge did not suspect him of perfidy in remaining silent at Seymour's side, as opposed to the hot-tempered Snyder back in Omaha. Reed had shown himself to be resistant enough to Durant's previous machinations to earn the Doctor's emnity. "While Mr. Reed was a quiet man, and said very little," Dodge would write, "he was very much opposed to all their movements and all their interference with the work in the West. They not only upset my matters, but his also, and he was in the habit of writing me confidential letters." Dodge's understanding of the new duties was that James Evans, reporting to Reed, would attend to the construction campaign from Laramie to Green River, while Reed supervised the important work in Utah.[63]

The pressure was being exerted upon Blickensderfer from all sides—above, there was the brusque puffball Seymour; eastward, the track was moving up the upper Laramie Plains and about to veer west through the mountains and the Great Divide Desert toward Green River and the Wasatch; and westward lay thirty-mile-long Weber Canyon with its own problems for grading. At the confluence of Echo Creek and the Weber River, the latter canyon presented itself as broad-floored and passable, but within a few miles the rock walls narrowed dramatically, rising in places to four thousand feet, and the bottom of the canyon was filled by the river. It had been named after an early fur trapper, John G. Weber, who pronounced his name to rhyme with "fever." In 1846 the Harlan-Young party had fought their way through the lower Weber Canyon through endless tumbled boulders, winching their teams and wagons over nearly vertical protrusions of mountainside; later emigrants like the Donners and the Mormons had therefore turned off from the Weber and worked their ways down Emigration Canyon—itself with its own problems. Above the lower Weber there were the strange rock formations called the Witches for the facelike formations topped by witches' caps, and the Devil's Slide, two colossal, parallel limestone reefs affixed to the vertical canyon walls which in more innocent times had earned the formation a name for its resemblance to a children's slide; raunchier trappers, desperadoes, and their successors on the railroad would crane their necks at the sight and make reference to a woman's anatomy.

Some miles downstream, where it narrowed to the impassable V-shaped gorge called Devil's Gate, the river would soon flow out between sagebrushy foothills toward the plain and the slate blue great lake beyond. The Promontory Mountains could be seen to the northwest. This uncompromising terrain would soon—the company hoped—begin to yield under the powder, picks, and shovels of Mormon graders. When Dodge finished his letter of June 4 to Blickensderfer he had said that the Union Pacific executives "seem to think that by Sept. 1st track will be at Green River. If so, every foot of line to the valley ought to be covered with men now." It wasn't, just yet. Both Dodge and his man in the field knew that they had already lost precious time to make the final location. "I regret that Col. Seymour was sent out," Dodge wrote to Blickensderfer, "for I was certain that he would cause trouble. There never has been a man on the road that he agreed with; what Mr. Reed thinks of line I have not heard."[64]

While Dodge was writing his engineer, Durant, still steaming over their arguments, was composing a wire for Dodge:

> *I have telegraphed to Weber that all division engineers not complying with general order number one will be discharged. I do not know as the Board will sustain me but I shall not stay in the company if they see fit to keep you in their service to run politics to the neglect of your duties as Chief Engineer.*[65]

The noose had been draped on the ground—and Durant was expecting Dodge to step into it. Notifying Jacob House in the Omaha office that he was leaving Washington on June 15 to resume work along the line, Dodge told him to prepare a camping outfit for a party of about fourteen. He would show Oliver Ames and other board members, and government director Jesse Williams, the terrain and the location lines and have them talk with his surveyors and the construction chiefs, "telling them," he intended "there was no other possible way for them to meet the continual interferences." He then wrote Oliver Ames to deflect the barrage of criticism leveled at him and his men by the Doctor. "On behalf of my parties and my chiefs," he wrote on June 8,

> *I challenge the world to show an amount of work done by any one before that has been done by them. Summer and winter they have faced all the dangers, steadfastly: two have been killed by Indians, others drowned in crossing streams; [others have] frozen to death and suffered everything that man is heir to in that wild country; and while I have had to neglect some of the details, I have endeavored to so shape their course and work that they would meet most effectually the wants of the company.*

He explained the reasonable delays in forwarding maps and profiles, as the original copies had been furnished to contractors, as ordered. Besides, he said, "last year I sent in my report with a box of profiles and maps and as they were never opened for six months and you then said it was not necessary for me to submit all the matters to the New York office . . . they could be kept at Omaha." He had "labored under a very great disadvantage," he added, "in not being appraised in time of what the Co. really desired done. I have had to act almost exclusively on my own knowledge of matters without orders and thereby endeavor to anticipate the demands of the company." Now, suddenly, rules were changing almost daily, edicts appeared and threw his engineers into turmoil, and Dr. Durant was complaining that Dodge was not only derelict in his duty but that he was disobeying orders. "As a soldier," he said, "I had to obey orders fully, faithfully and in the spirit given, and to the best of my ability, and I believe I had not varied from that course since I have been in the employ of the company—and if Mr. Durant reflects one moment, he will be convinced I have done all in the power of man."

President Ames's reply, as usual, glossed over the source of his chief engineer's anger—instead of reassurance he passed the treacle; he praised the Blickensderfer findings in the canyons sent earlier by Dodge and treated the whole controversy about the chain of command as if it were a dispute over place settings. "I am quite satisfied we shall be pleased with the course you have taken, in having the parties report to you," he assured Dodge. "I have never seen the Dr. so courteous and confiding as he has been since I have been here these two days and I should think from Blickensderfer's letter that Reed and Seymour had not actually interfered with his line but simply asked for aid from his parties in preparing line for contractors." Ames closed by saying he was looking forward to meeting with Dodge out in Omaha the next week.[66]

The trouble was of a vastly higher order, Dodge knew, and might not wait until he had gotten out in the field; from each letter and telegram he unwrapped, out popped a new problem. In the few days left to him in the capital, he readied a location map to send to Interior Secretary Browning. It showed the proposed eighty-mile Union Pacific line from the mouth of Weber Canyon to the north end of Great Salt Lake. His accompanying letter asked for acceptance of the new line. He did not resist, of course, pointing out that Browning had accepted the Central Pacific map from Humboldt Wells to the same point north of the lake. "We are," he said, "nearer that point with our completed road than they are." Then it was time to close up his congressional office. Once he was out in a tent again in the mountains, face to face with friends as well as adversaries, perhaps he would regain control—before it became total chaos.[67]

For three months now, several thousand shovel-wielding Cantonese and a small number of whites had labored under the harsh springtime sun to move snow off the deeply covered railroad tracks on the eastern Sierra slope—"all the men who could be found," Mark Hopkins wrote to Huntington on June 16, "willing to *work themselves blind & their faces peeled & seared as though they had been scalded in the face with scalding water.*" Down the grade, other teams worked under similarly hard conditions to lay track over the now-bare seven-mile gap. Hopkins had hoped that the closure would be made by June 17 at noon, but it had been snowing and raining on the summit for twenty-four hours and a day's delay was possible. As it was, he thought, the track would probably be finished within hours of when the snow brigade would chip out the last of the solidly compacted white stuff. Strobridge's men, however, pushed themselves to the limit, and that night of June 16, Judge Crocker sent a wire to Huntington in New York: "The track is connected across the mountains. We have one hundred and sixty seven continuous miles laid."[68]

The first locomotive through would end an excruciating necessity—hauling iron, ties, hay, grain, provisions, and other necessary supplies from the summit for two miles on ox-pulled sleighs and transferring the goods to freight wagons headed down to Coburn's Station on the Truckee, where the track resumed—even with seventy teams they could manage only half a mile per day. No wonder the section to the Big Bend of the Truckee was still uncompleted. The government commissioners would begin examining the stretch between the 94th and 114th mileposts; by the time they reached the Cold Stream gap it would certainly be closed.

Two days before, Charley Crocker had surprised Hopkins by appearing on his doorstep on a Sunday, greatly agitated. One of the three government commissioners, Frank Denver of Virginia City, Nevada, was apparently ready to block federal approval of the twenty-mile section, and their bonds, in a shakedown. A Sacramento steamship captain supposedly friendly to the Central Pacific had told Crocker, and offered to "fix it up" for them on the best terms possible—he thought $50,000 would do it. Crocker and Hopkins decided that the intermediary was a partner in the scheme. "[W]e would not consent to be blackmailed—whatever generous things we might do with our friends," Hopkins told Huntington; "we certainly never would 'stand & deliver' to a highwayman when we were armed & could defend ourselves." The next day in Sacramento Crocker saw Nevada governor Bigler, also a government commissioner, who expressed shock and dismay at the scheme and promised to quietly halt Denver from doing any damage or else have him removed. The governor was in a hurry to leave for the Democratic National Convention on July 4, so he compliantly signed most of the Central Pacific inspection papers in advance of his trip, collected the commissioner's fee of $250, and left a happy and virtuous man. "Denver is a *Bummer,*" exclaimed Hopkins to Huntington, "of little ability & no standing in Virginia City . . . and if Union Co's money or anybody else could be put into his pocket by acting adverse to our interest, he would readily do it."

Hopkins could give no news, however, on Leland Stanford out in Salt Lake City. All he knew came out of three telegrams: that Stanford had arrived, he had gone up to see Weber Canyon, and he had left for California on June 14. "Nothing good or bad to gratify our impatience," Hopkins groused. "Nothing satisfactory."[69] Huntington was more than a little irritated that Stanford had been sent on the mission in the first place. "We need a bold, sharp man," he wrote, meaning the opposite of Stanford.[70] What they should have had at Salt Lake City for the past two years was "a first-class man . . . that is up in the morning and that does not knock off until afternoon. A man that [can] handle the Mormons." Apparently the Governor liked to sleep in and go home early.[71]

By the time Stanford and George Gray had ridden back in view of the California Sierra, they found that the Cold Stream gap was plugged, temporary bridges were about finished over the Truckee at the Big Bend and over the Humboldt above Stark's Ferry, and a small force of graders was completing the stretch to the Humboldt Sink, forty miles beyond the Big Bend; flatcars loaded with rails and ties had begun to be pulled over the summit. But Stanford returned to Sacramento empty-handed, telling them all that Samuel Reed had beaten him to the Mormons and had left the city with a signed grading contract before Stanford had even arrived. Reed and Seymour did return to Salt Lake City as soon as they learned that Stanford was there—but by then the Governor had obtained an audience with Brigham Young.[72] The leader was initially "cold and close," Stanford told Hopkins in a much-delayed letter from Salt Lake City, which trailed Stanford back to Sacramento, *but I have, I think, got pretty near to him. He and everybody here was dead set for the southern route. How to meet this bothered me a good deal, but this afternoon being pressed I was able to find good reasons why they would be most benefited by the northern route. They do not seem, any of them, to be aware of the location from Humboldt Wells to the north end of the Lake. I have not thought it advisable to enlighten them.*[73]

Stanford and Gray "did much to counteract the influence of the Union company," they claimed, and "convinced Brigham that it was his interest for both Co's. to meet at Weber or Echo Canon—a point he had not seen." Stanford said he arranged with one of Young's sons (and thus with Young himself, although this would remain unacknowledged) to put together a Mormon grading force to work east from Humboldt Wells—on top of the proposed contract, "giving him 10 cents per day for all he will furnish up to a certain number," Judge Crocker reported, "& increasing as the number of men increase." Stanford said the agreement would commence "as soon as the supplies can be sent from here"—he had not yet signed a contract. It was enough, he thought, that Young now appreciated the Central Pacific's position. "I did not think best to offer more to Brigham as that was satisfactory," he explained to Huntington a month later after Huntington had complained. "But I think we can manage to get what men we may want from Salt Lake. We do not want any yet." Of course Huntington would disagree—but he was three thousand miles away.[74]

The first passenger train across the width of the Sierra rattled up to the summit on June 18 and halted at the great tunnel for several hours while "a swarm of Chinese" (in the *Alta California* correspondent's words) cleared snow and boulders from a small avalanche just east of the far portal. "The water pours down in torrents from the numberless crevices and seams in the granite walls and roof of the long, dark, cavernous tunnel," he wrote, "but we struggle through on foot and anxiously inquire after the prospect of getting through." Once under way, the train was forced to halt again and again to be accommodated by the shovel brigades, but finally it rolled beyond the snow line and descended along the Truckee. "As the first through passenger train sweeps down the eastern slopes of the Sierras," the newspaperman continued, "John [the Chinese worker] comprehending fully the importance of the event, loses his natural appearance of stolidity and indifference and welcomes with the swinging of his broad brimmed hat and loud, uncouth shouts the iron horse and those that he brings with him." The train rolled across the California line and on into bustling little Reno just after 8:00 P.M. "Reno is a lively place, and one month from the day the lots were sold could boast about 200 buildings, stores and dwellings," the Central Pacific chief engineer Samuel S. Montague wrote to his surveyor, Butler Ives, out in Utah. "The day I passed through there on my return, an opera company was to entertain the good folks. On Sunday last I saw a circus tent in process of erection—a circus company performed there last week to an audience of *1,000* people."[75] Montague was as understandably excited by the railroad's entry into Reno as the editor of the *Sacramento Union* was unimpressed. "There is a novelty in this event which must for the moment excite general curiosity," James Anthony wrote,

> *but further than that, the announcement is of no importance. Long ago the public have justly regarded this much-lauded and patronized transcontinental enterprise as a merely private affair to be used for the enrichment of a very few individuals, without anything like compensating returns to the liberal public, by whom it was started, and by whose money built. The speculators and their snobbish pipers of the truculent press will, no doubt, make a great ado over the event, and herald it far and wide. . . . But reflecting people, who know better, will be apt to regard it as merely another link completed in the chain whereby a couple of selfish corporations are endeavoring to prove to all the world that gratitude is weaker than avarice, and that it is dangerous to trust any private individuals with the management of such vast public interests.*[76]

In Washington, at least, similar sentiments were beginning to be heard, to Huntington's great chagrin. With Senators Stewart and Conness he had drawn up a bill for the San Joaquin Railroad from Stockton up the valley and over the Arizona line to the Colorado River, and for the UPED from there to Denver; the bill had been introduced on June 6—only to be denounced in the most angry terms in both chambers of Congress. That many of the critics were in the thrall of the Union Pacific—which had no interest in seeing a southern transcontinental route helped along—did not escape Huntington's notice. The bill would go down in defeat.

Another failure was his beloved Goat Island bill. Through his cheerleading and other influences the Senate had passed it, 28 to 8, but when it was deliberated by the House it ran aground; at one point Huntington even appealed to Oakes Ames for his vote, which the representative agreed to give. After the bill died, with Ames's "no" vote helping to bury it, Huntington went looking for him. He cornered Ames at the Willard Hotel. "I told him he was a treacherous old cuss and that I would follow him as long as I lived," Huntington told Judge Crocker. "I got him mad, as I wanted to, thinking if I did I would get some truth out of him; and he went on and said that he did not work for me; that he thought the Union Co. would want a part of that island. . . . I really unearthed the old skunk." As he suspected, Ames and Dodge were already thinking of how they might build down from the Idaho stretch of the Utah-Oregon route, enter California and head for Goat Island themselves.[77]

Stalls, false starts, fitful progress, and galling retreats continued. The bill granting an extension of deadline for the California and Oregon line, to begin building its first twenty miles up the Sacramento Valley, inexplicably began to lose momentum. "I have written Franchot almost every day for weeks on the subject," Huntington told Hopkins, "have also written Stewart and Conness, as also Higby & Axtell, and they have all answered that it should be done, and that they thought there would be no opposition &c, but it has not been done. . . . [I]n the mean time you had better prepare to do what is necessary to be done to save the Franchise without it." Then, as the Sacramento partners struggled to react to the defeat, in the eleventh hour the extension was passed after all.[78]

There may have been reverses—but somehow, even those became part of the greater forward momentum, which could especially be felt along the thirty-mile stretch of the Truckee River between Reno and Wadsworth, where sunburned and dazzled shovelers released from snow removal in the Sierra began to concentrate and regroup under Strobridge's evil eye into tracklaying crews for the push toward the Forty-Mile Desert. Meanwhile, Butler Ives was four hundred miles to the northeast, urging his small surveying crew away from Humboldt Wells into western Utah, feeling that same forward push. In his case it was coming from Chief Engineer Montague, who had ordered him on June 16 to begin the locating survey toward Salt Lake. Soon another team, under Lewis Clement, would begin working westward from the lake toward Ives. "The necessity for *pushing ahead* will compel us in many places to sacrifice good alignment & easy grades for the sake of getting light work," Montague had written. "Make temporary location, by using sharp curves & heavy grades (keeping within the max.) whenever you can. Make any material saving in the work. The line we construct now is the one we can build the soonest, even if we rebuild immediately."[79]

So Butler Ives pushed east to locate, just as Montague and Charley Crocker moved forward in his footsteps with a force of five thousand Cantonese tracklayers; economy and speed were most urgently on their minds, not the least of their motivation being the completely dreadful, unfriendly terrain ahead.

In Sacramento, Edwin Bryant Crocker could not shake his nervousness over what would transpire in thirty to forty-five days when the available iron would be used up. "We want that iron across the isthmus *as early as possible,*" he told Huntington. "If we run short, though, I tell our folks, we must beg, borrow, or steal all the Iron we can lay hands on, & that failing lay down *wooden rails*—& replace them as soon as possible. But the locomotive must move on."[80]

29

"We Are in a Terrible Sweat"

Four summers before, Samuel Benedict Reed had found the outdoor life in the deep, shadowy Weber and Echo gorges to be magnificent. He drank in the scenery, feasted on the abundance of mountain trout and wild berries, in the cool nights slept on buffalo and beaver robes and felt, upon awakening and emerging from his tent for the next day's survey work, remarkably collected and clear-headed. Now, in June and July 1868, nothing was clear; he was deep in a canyon—a canyon of confusion, only partially of his own making. Despising controversy, he was in the center of it. Distrusting passion, it rose around him like angry smoke from a campfire. Desiring above all, as an engineer would, to always know where he stood, he could not shake nagging suspicions about people he had thought he knew, about plots which seemed to rise at hand and at a distance, about his part in the scheme of things. His tentmate, Colonel Seymour, filled his ears with grandiose imaginings. Only had he been blind could Reed have escaped noticing the resentful expression on the face of Jacob Blickensderfer as Seymour spun out his orders like measuring tape off a spool. Only if he had had no feelings could Reed not have felt a flicker of guilt as he said nothing. And only if he had been elsewhere, out of this hole, would he have dismissed letters from former co-workers back in his superintendent's office in Laramie such as the one from S. S. Benedict, a relative from Joliet, which arrived in mid-June to make the pebbly ground beneath his feet turn to quicksand.

Benedict was writing only out of a sense of duty to Reed and his family, he said—but Reed was being undermined in all manners of ways. His main nemesis was Operations Superintendent Webster Snyder, who from his office in Omaha had never missed an opportunity to clash with Reed nor, apparently, to complain about Reed behind his back. Because of Snyder, James Evans had been appointed acting construction superintendent in Reed's place—and now Evans was using his authority to prosecute a "reign of personal abuse" against Reed's loyal staff. One draftsman identified only as Erastus, who had worked with Reed for years after Reed had brought him out from their hometown, took as much as he could and then decided to go back east. When a request for a courtesy railroad pass went up to Snyder, he at once issued special orders to conductors to watch out for Erastus and "to collect fare of him or suffer the penalty, an immediate discharge." A conductor took up the man's pass on the eastbound train though it had been signed by Reed, and extracted full fare on threat of throwing him off the coach platform. Another employee on the way over the hills from Laramie to Cheyenne to count bridge timbers there had his pass confiscated by a conductor because it was signed by Reed and also had to pay the fare. Still another employee detailed to keep accounts of iron being shipped toward the end of track was let go in favor of one of Snyder's men.

Everyone was certain his was only the first of many firings to come. "Such are the abuses your men have heaped upon them every day," Benedict said. "The Const. Dept. here is entirely in the hands of Mr. Snyder who has found, and I say it truly, a willing tool in the person of J. A. Evans. . . . Our stay here now is far from pleasant. I might say it is very humiliating, knowing that every thing was being done to injure you and your men." James Evans, he warned "is not your friend by any means, by his conduct here I shall always think he accepted the position merely as a stepping stone to occupying it permanently." Benedict said that at present Evans seemed to have other things than advancement on his mind—but, he finished, if Reed planned to remain in Utah much longer, "please excuse me from further duty."[1]

Apparently others in the department were sending warnings to Reed, along with copies of orders signed by Durant. One directive given to Evans particularly slighted Reed, even in his newly appointed temporary jurisdiction in Utah. "You will procure and furnish all supplies, teams and men," the Doctor had written Evans, "and have them forwarded, as the Construction Engineer has so much to do that he cannot attend to the detail." As to the departmental employees, Durant told Evans, "organize your forces to suit yourself."[2]

Dispirited, Reed wrote his wife. "There is so much jealousy and hard feeling . . . both in New York and on the line of the road," he told Jenny back in Illinois, "that there is not much pleasure in trying to advance the work." He told her he was submitting his resignation to Dodge, who by then was on his way west. Jenny Reed, who gave birth to their third child that season while her husband was in Utah, took matters into her own hands. She wrote to Dodge herself, begging him to intercede before her husband was forced out. The letters from husband and wife reached a surprised Dodge when he arrived at his home in Council Bluffs. He instantly sent a wire to Reed and told him he was mistaken—and that Dodge would fix what was wrong. "I knew the importance of keeping him on the work and in charge of it," Dodge recounted later. When he had sorted things out he partially blamed "mischievous interfering of men who pretended to be friends of Mr. Reed's but wished to make trouble," but of course what loomed higher in the trouble was the Doctor's shadow. Fortunately, Oliver Ames had come out west also and was staying at the Dodges' house. When Dodge read him the letter, Ames immediately telegraphed Reed to reassure him and asked to meet him at the end of track, by then closing in on Rock Creek, on the first of July.[3]

That there were so many ill feelings rampant within the company's staff was, however, beyond Dodge's ability to heal, or perhaps even to appreciate. Certainly it was beyond Ames's. "I was so busy," Dodge said long after, "that I did not pay much attention to it." Part of the friction between Snyder's and Reed's departments was natural—neither the construction nor the operations arms appreciated what the other was up against, especially when their work was adversely affected by inevitable slowdowns due to weather or inefficiency as the organization grew so quickly. And Dodge simply did not believe that Evans would conspire. But one had to add other ingredients to the recipe: competition (healthy or not) and ambition, and abrading tempermental differences, and probably finish it with the almost universal element of greed; it was so common that nearly everone suspected everyone else of it.

Despite a complete lack of evidence, Snyder was still convinced that Reed was crooked. "If the contracting firm west of Green River is not Young, Reed & Seymour," he had complained to Dodge, "then I dont know the men, or Brigham knows them too well." His distrust had been conveyed to Sidney Dillon in New York, who recirculated it to Dodge and to Ames. "I think that Seymour and Reed have an interest with all these parties that have taken work at Salt Lake," Dillon wrote Dodge. "When you see any of them if you work it right, you can find out all about it. Make them all believe that we are all glad. I would give most anything to know if it is so. . . . Be sure to speak of it." As Dodge and Ames headed west on the rumbling train across Nebraska, and then up over the Wyoming hills, they each had their suspicions. It was no secret by then that Durant and Seymour were up to their eyelids in conflicts and secret side deals injurious to the railroad. But Dodge, at least, knew that Reed "had no interest whatever" in the Utah work. "He was very careful in that way."[4]

With such issues of probity on his mind, Dodge got to the end of track at Rock Creek, forty miles up the grassy plain from false-front Laramie, and rode up into the hills to inspect the coal mines he had opened at a place he called (at a loss for something better) Carbon; he was satisfied to see that miners had uncovered an eight-foot vein there and no doubt savored the prospect of retailing it to the railroad, seeing, of course, no conflict. He circled back to Laramie then, to adjust the troubles between Ames, Reed, and Evans. Snyder would be there, too, but perhaps a reconciliation on that front was too much to hope for.[5]

Reed appeared, tight-lipped, unusually assertive, and ready as never before to unburden. "I have never known Mr. Reed to be so positive and decided," Dodge recalled—and as the location maps were spread out before Dodge and Oliver Ames, Reed announced that here was the final location to Salt Lake, with the exception of short stretches at the head of Echo Canyon and at the Narrows in Weber, where tools had been delayed in reaching the parties. He certified Blickensderfer's line as the best and said that if "they" would only let the engineers alone, there would be no trouble, but "they were continually changing and interfering and it was almost impossible to meet the views of the Company." Every change and every order, he said, "interfered with the operations on the ground, was costly to the company, and was detrimental to the surveys, and only kept the engineers and contractors undecided as to what they should do." In addition, he said, Dodge's engineer corps had "the active support of the military authorities and the Mormon church." As he listened, Dodge could only bask in his surprise and pride. Reed then told Oliver Ames that "it was very embarrassing to him to have men from New York right with him recommending things as coming from New York which he did not like to oppose but which were against his judgment. The only object of Col. Seymour," he continued, "was to get rid of the Chief Engineer, if he could."[6]

That said, Samuel Reed subsided—although as the engineers conferred, Evans defensively spoke up to claim that he meant no one any harm, plainly referring to Reed, and Reed snarled out his disbelief. During the presentation Dodge was impressed, but somehow Ames was not. As averse to controversy and face-to-face confrontation as poor Reed, Ames uttered platitudes, telling Reed how valuable his efforts were for the company, mollifying the engineer enough for him to agree to pick things up in the Utah canyons. Ames went with Reed and Snyder back on the cars to Cheyenne, during which they continued their plain talk and Ames, like a noncommital doctor, only listened. And as it turned out, the boil had been lanced—but the infection, unattended, was not leaving. "After the conversation," Snyder wrote Dodge, doing his part to spread the poison, "Mr. Ames told me and told others that he was more than ever convinced that Reed was not competent to run a very large machine." For the moment, though, Dodge could do nothing more: he was on his way west to inspect the engineers' lines between the Laramie Plains and the rim of the Wasatch—and to look for more coal mines.[7] The trouble would get even more personal, and rise to envelop him by the time he reached Green River; Colonel Seymour was still abroad, and still active.

"You are paying at least a million for bad engineering between head of Echo & mouth of Weber," Seymour had written to Durant from Salt Lake City the week before, on June 23, complaining that Dodge's interferences were being adopted as company policy though his expertise, "you know as well as I," the colonel continued, was "not worth the snap of your fingers." Thinking he had the silent Reed's support, endorsing Reed's cause in the disputes with Evans and Snyder, Seymour said that "their" plans might have saved Durant "a good deal of time & money"—but Dodge had stepped in, he complained, and meddled with Durant's "General Order No. 1." He would "give anything if you were here to see how matters work." Blickensderfer had stopped speaking to him or acknowledging his letters. And the engineering corps was mobilized with graders, obviously committed to the route around the north end of Great Salt Lake. "I believe there is more in this than you think," he said. It was clear to him—and Seymour said he had just written Oliver Ames about it—that a southern route through the city and up the west side of the lake only made sense. And while he had Durant's attention, the colonel asked if he could have a piece of the Davis, Sprague contract to provide ties: "I can make a little money if they will allow me to subcontract for them from Bear River west at 80 cts. [per tie]—won't you arrange this for me? I have a half interest in the best coal mine in the country for you & myself if you wish it." The wily colonel knew just how to evoke a response from Durant—find a conspiracy, an insubordination, and an insult in one place, suggest a reversal and a surprise move in another, propose a deal on top of a deal, promise a profit. He was a genius at this kind of choreography, and he had just proposed a breathtaking series of turns. "I understand that Dodge is on his way out," Seymour closed. "When are you coming? You cannot come a minute too soon if you intend to get the track here this Fall."[8]

A few days later, having ridden up to red-walled Weber Canyon to find that Reed had gone to consult with Ames and Dodge, Seymour helpfully sent a telegram, collect, after his letter to Durant. "If Mr. Reed's absence imposes any additional duties upon me," he said, "please specify what they are." The next day, he sent another. "Been waiting several days to know where to put [workers'] shanties," he protested. Reed had left without getting the profile of the head of Echo Canyon, he alleged, and now "several hundred of Brigham Young's men have left line dissatisfied & tired of waiting." He got no response to either wire, and on July 2 impatiently telegraphed Henry Crane in New York. "When is doctor coming here?" he complained. "Best part of season being wasted by delay in location and no one here with power to remedy it." He waited for some kind of answer.[9]

Meanwhile, Grenville Dodge was riding west with a small party (including for part of the journey his ally Jesse L. Williams, the government Union Pacific director). From the new log palisades of Fort Fred Steele on the North Platte, they threaded through rolling, short-grass hill country to Rawlins Spring and beyond to the broad, bleak Continental Divide plateau. Grading contractors along the way gladly interrupted their hot work and told him they would be done with the section to Bitter Creek by the end of the month. Indians had killed two men at Rawlins Spring two weeks earlier, and the night before they had attacked Reed's party not far away, but the engineer had lost no men and no stock. On July 8, Dodge crossed the barren divide, camped overnight near a promising eight-foot coal vein in a sagebrushy bluff, and then joined the Bitter

Creek Valley for a fifty-mile stretch, "remarkably direct and well built," he noted, "banks high above snow and very few cuts. Coal shows itself in the bluffs, and looks like good coal." The line down the creek crossed it often, he noted, because the "valley is crooked and torturous but we get good line by taking advantage of its bends; as we get towards Green River [we] avoid the Creek more than we do above. In many places we change the channel."

Interviewing contractors who had taken ten or twenty-mile sections and were fairly well along, noting sites for iron truss bridges and trestles, Dodge found more promising places for coal in the weathered gray-green bluffs between Point of Rocks and Rock Springs. He opened numerous veins of the black rock. He established stations called Separation Creek, Continental Divide, Red Desert, Clay Buttes, and Bitter Creek, all aptly named. ("It is not a very inviting spot," he admitted about the last station, "but the best to be had.") More would be marked down, as Black Buttes, Point of Rocks, and Salt Wells. He saw that before being transferred to Utah, Samuel Reed had provisioned for water at Benton and Rawlins, and Dodge ordered men to begin drilling wells at each of the dry new station sites. "I know we will get water," he would report, "but we ought to have been at it long ago; it may lay deep and through the rock. Several of the wells show water but it is bad and we must get below the surface water, and we cannot give too much or too careful attention to this." Dodge was glad to see that Williams, normally a severe critic, was "much pleased" with the adopted route, saying it was "a bold, judicious" line, "the ground evidently studied and good advantage taken of it."

With the thousand-foot Castle Rock in sight rising from sandstone cliffs above the banks of the Green River, Dodge climbed from the Bitter Creek Valley through heavy, unfinished rock cuts which were wholly innocent of work gangs. He did not like what he learned at Green River. Lewis Carmichael, the contractor, had remained obdurate when his laborers complained that they deserved more pay; all hands had struck on July 4, and no more work had been done in a week. It was the second strike since they had started work two months before. With Dodge there, all three hundred went back to work. It seemed to Dodge that they would be finished the heavy work before the tracks arrived—but not so the contractor working the eight miles of greasewood flats and sparse hills toward blue-black Church Buttes. Malloy by name, he had been on-site with a paltry crew for two weeks and worked four days, complaining to Dodge that the easy side-hill cut before him needed at least five hundred men and supplies. "The great trouble with him," noted Dodge, "is want of tools, cars, scrapers and corn." This was, he added, the same problem with Brigham Young's crews working—or waiting, rather—in the bare red hills westward across the Bear River Valley. Six wagon trains mostly loaded with provisions had made the thirty-day journey from Laramie to Green River, and the Mormons had come down from their hills and all but emptied them for their own grading work. Eleven more wagon trains were on the way. "The light work can be easily gotten along with," Dodge reported, "and as soon as tools arrive a large force can be put on the heavy work. I guess they are pushing matters all they can; but 300 miles teaming of everything is tedious."

At Green River Dodge found that squatters had eyeballed where the trains would go and had formed a private organization to sell town lots. "They really have no right," he noted, "but I see no way to help it." Besides, the engineer continued, they had established the town site on the low valley bottom. "Green River sometimes covers it four feet," he said with no little satisfaction. "It is now within a foot of going over it." Using a little more common sense, he would lay out a small town, but would save the large site and shops for the better-situated plateau above Black's Fork of the Green River, fourteen miles farther west, as Reed had recommended. On July 14 he camped near the Church Buttes at the deeply cut Black's Fork, saw that they would need four bridges in the vicinity, and then climbed to the top of the cathedral-like outcropping to catch the view. Standing there, with sharp-shinned hawks sailing overhead in the blue sky and swifts swooping through the air at eye level, he could see many miles in every direction—the Pilot Buttes, the Uintah Range, the rim of the Great Basin, the Medicine Buttes, Aspen Hill, and Bridger Buttes—in a clear, undeceiving landscape.[10]

At Fort Bridger a large packet of mail was awaiting him. One letter was from Schuyler Colfax, announcing that the Speaker of the House (and the presumed next vice president) was bringing a large excursion party west to the end of track, wherever that would be, and asked for cars and accommodations and stagecoaches for a side trip from Cheyenne to Denver; the *New York Tribune* correspondent Albert D. Richardson would be with them. Dodge made arrangements for the luminaries. Another letter, from Webster Snyder, said that the end of track was now at Benton, 660 miles from Omaha. Dodge had located the train shops there, with the nearby North Platte River offering fresh water and a highway for upcountry timbers and ties, and he had seen many building lots sold before pushing off. Within two weeks of the arrival of tracks, Benton would have a population of three thousand. "The streets were eight inches deep in white dust," the former country lawyer and itinerant journalist John H. Beadle would soon observe about the raw new town, "as I entered the city of canvas tents and polehouses; the suburbs appeared as banks of dirty white lime, and a new arrival with black clothes looked like nothing so much as a cockroach struggling through a flour barrel; the great institution of Benton was the 'Big Tent' . . . a nice frame building 100 feet long and 40 feet wide, covered with canvas and conveniently floored for dancing, to which and gambling it was entirely devoted."[11]

Another dispatch for Dodge was from the home office in New York. It was signed by Charles Tuttle, secretary of the company. Dated July 17, it was addressed to Grenville M. Dodge, but bore the greeting "To the Officers and Employees of U. P. R. R."

Mr. T. C. Durant, Vice Prest., has been appointed to take supervision of the affairs of the Company along the line of the Road, including surveys now being made. Officers and Employees are subject to his directions and will act accordingly. He leaves tonight for the west.[12]

Incensed but determined to avoid losing a minute of work to useless worrying, Dodge pushed westward, knowing that trouble was now following him. It is possible that before he left he sent a reply to Tuttle threatening he would resign under such circumstances. In one of the four accounts he left of this time, Dodge claims he resigned—but if he did, he did not take a step away from his work, as if he were confident of being sustained.[13] And there was much work to do. On July 18 he reached the sunbeaten and treeless rim of the basin with surveyor J. O. Hudnutt, soon to be joined by Jacob Blickensderfer, who had closed up all his work to Salt Lake. Blickensderfer brought word from surveyors Fred Hodges and James Maxwell—Hodges had reached the Humboldt, and Maxwell was near Promontory Point at the north end of the lake. Together for the better part of two days Dodge and his men hashed over two troublesome lines from Muddy Creek to the summit, finally deciding on the easier sixty-foot-grade line. Dodge was next heading over the divide into Echo and Weber Canyons to see the location advocated by Blickensderfer and Reed, but, trusting them, he had the satisfaction of knowing that they now had stretched a true line from the Medicine Bows through the Wasatch, close enough to kick a pebble into Great Salt Lake.

A messenger brought him news from home on that day, July 19, which could not have pleased him. Writing from Council Bluffs, his wife told him that Hub Hoxie had paid a visit that morning. "He told me the company had telegraphed you to return with profiles, &c.," she said.

Something is to pay with the Central Pacific. I think it will be too bad for you to return over the hot plains and then go back, and I hope you will not do it. I hope to hear from you what is the matter; as to being ordered about everywhere by Durant, I would not do it, and I have thought lately you could do better off of the road than on. How can you like being gone so much and having so many times to fight men like Durant? Still I dislike saying anything for fear you will think it is from selfish motives. It seems to me you have harder work than any man who ever built railroads before. They are trying the same game with you that they have with Reed, and I think there is more in Seymour's being out there now than you think. I don't believe any small business would keep him out there so long. He is cunning and crafty, and you had best look out for him. I don't think much of the men that keep him, and it would be just if you left the road to him and them, and take care of your own business interests or build other railroads.[14]

Dodge had heard nothing of a summons. He worked west.

When still commanding from his office at 20 Nassau Street in New York, Durant had been facing problems on all levels, not just the difficulty of getting Dodge and Reed, for starters, to resign and give him more maneuvering room. Relations with James Evans were prickly—the superintendent had clashed with Seymour over the latter's demand to be given the franchise to build all stationhouses and tanks between the North Platte and Green River; in less eventful times Evans's truculent wires to Durant might have gotten him fired, but he was crucial to the enterprise now, so the Doctor had been close to conciliatory in telling Evans to let Seymour have the plum. He was less than polite a few days later when dissatisfied with their progress. "Our instructions were to complete grading by July tenth," he wired Evans, "and we did not leave it discretionary with you to extend the term. By delay the whole program of the company is interfered with and it may make a difference of millions to us." On June 18, Evans wired him with a new shock: Casement's tracklayers had nothing to do—"We are out of ties this A.M." He followed up. "So far as ties are concerned we are working from hand to mouth. I think delays in the future will be common, having no stock ahead something must be done to give us more ties—or it is useless to keep up so large a track laying force." This involved the tie contractors Davis & Sprague—with whom Durant was heavily involved—so the Doctor told Evans to "give them a fair chance" to improve, but that if they could not keep up, Evans should take them over himself: "Show them this and act promptly."[15]

Then, upon learning that the grading contractors in Echo Canyon were idle for lack of everything from picks and shovels to the horse-drawn scrapers, it brought forth a series of angry telegrams between the Doctor and Evans: "Telegraph Reed fully and at once," Durant said, "and do all you can instantly to save us as much as possible from damage in consequence of the required amount of tools not having been sent to Echo." To Evans's retort that "We are sending and have been sending tools & supplies west" and that soon "everything will be right," Durant exploded. "Send Reed at once detailed statement of what you have sent to Echo and Green River," he wired on June 25. "Neither he or myself have any data to go on and don't know if you have sent one tenth the amount required, and fear very much the work will be delayed on this account. I think you had instructions to do so before."[16]

Durant could not blame Evans for the next brush fire, which broke out two days later when graders working for contractors Reynolds & Dowling struck and Evans telegraphed that it was likely the strike would spread. "The cause assigned is that they want pay while en route to Green River. If this is conceded it will be something else. The best way is to commence paying them off as fast as they want to leave & throw them on their own resources." Within three days a large crew working for Lewis Carmichael, also heading toward Green River, halted until all their back pay was handed out—and then nearly all the graders between the end of track and North Platte went off, demanding an increase of wages from $2 to $8 a day and board. "We understand," the Laramie *Frontier Index* disclosed on July 1, "the contractors have been ordered to discharge and pay the men off."[17]

It just showed that agitators had to be thrown out before they caused trouble. But what about Reed and Dodge? That score had yet to be settled, even if Reed had still resisted leaving and had survived the confrontation with Snyder, Evans, and Oliver Ames; as for Dodge, he was heading for Utah, conferring with engineers all the way, which could only mean more problems. But Durant could not do anything about that yet; there were legal difficulties in New York. The Pennsylvania businessman Duff Green, from whose hands George Francis Train had purchased the entity renamed the Crédit Mobilier, was now suing them—probably to get into the action. And the pirate "Jubilee Jim" Fisk, who the year before had helped Durant in the Union Pacific election fight, had yet to recoup the mere $3,200 he had advanced in buying up stock. In June, Fisk had discreetly purchased six shares of Union Pacific stock, paying $240, by this joining the ranks of legitimate stockholders. Then he, too, filed suit, on July 2, to secure the twenty thousand shares he had attemped to buy on margin the previous fall—failing that, he would "get what he could" from the company. Fisk obtained an injunction preventing the Union Pacific from doing any business with the Crédit Mobilier until he was satisfied. He wanted $75,000 to settle, though Cornelius Bushnell thought he would cease for $50,000.

To avoid having the work out west shut down altogether, Durant called an executive committee meeting for July 3. Present were Durant, Bushnell, Dillon, and John Duff; Oliver Ames, still out west, was absent, as were Charles Lambard and government director James Brooks. The Doctor had no trouble having the contract with Crédit Mobilier dissolved—which temporarily met the Fisk threat and served Durant's own purpose. He also had the committee pass a resolution giving him authority over constructions and surveys, putting Dodge again under his putative control. He allowed himself the luxury of two weeks' preparation for his trip out to Wyoming before telling his secretary to notify Dodge toiling out there near the Utah border that he, his staff, and his surveys were emphatically subject to the direction of Vice President Durant. He climbed aboard a train on July 18, savoring the showdown to come, confident of his imminent victory.[18]

Dodge was camped at Yellow Creek, a few miles shy of the Utah border, on Tuesday, July 21, when he received a dispatch from Durant ordering him to meet the Doctor at the end of track by Wednesday or Thursday—an impossibility. Dodge cabled his regrets and, instead of beginning the trip east, diffidently went in the opposite direction; in retrospect it seems as if he were stalling for time. Later that day, though, he met Samuel Reed riding up over the divide. Reed was dutifully responding to Durant's summons to the end of track; Dodge did not accompany his associate back eastward, as one might expect, but sent him on ahead after learning that all of the grading work between Green River and the mouth of Weber Canyon was now contracted and progressing. A day later, Dodge was thirty-two miles west on the Weber, pronouncing Blickensderfer's route over the Wasatch rim to be as wise as he had supposed.

Then, at Weber on July 23 a wire sent from Durant and Sidney Dillon was handed to him—and this was an invitation he would not ignore, to a party he did not want to miss. It told him to attend a meeting at Fort Sanders on Sunday, July 26, to settle the question of authority. Durant and Dillon would be there—and the meeting would also include some old friends: Generals Grant, Sherman, Sheridan, Augur, and others, who were out west to take a working holiday and to do a little campaigning for Grant along the settled routes. With such old allies to be present at this fight, Dodge lost no time in taking the next stage east. He would never admit to having invited his sponsors himself to the company conference, but one wonders; in any event, he arranged to meet the Grant party ahead of time at dusty Benton on July 25.[19]

Grant's excursion was primarily intended for recreation, although some politicking was inevitable. The presidential candidate of course wanted Sherman's support—something a military officer such as Sherman would studiously avoid—so Grant enlisted his friend on the trip, knowing it was enough for people to see them standing side by side. And Sheridan, too—he had escorted Sherman and presidential candidate Grant across his new command in Kansas, taking the train westward from the Missouri and stopping frequently for short, rear-platform speeches to fascinated crowds. "I return my sincere thanks," Grant would call to the cheering people, and then turn away. At St. Joseph, in an unusually voluble mood, he had added that he wished to be excused from further campaign comment. "I am fatigued, weary, dusty, and unable to address you." Most of the crowds had seemed happy enough to see the heroes in the flesh, although there were occasional hecklers supporting the Democratic candidate, Horatio Seymour of New York. At Fort Leavenworth, on July 17, Sherman convinced Grant to extend his trip longer than he had intended—he had taken the Kansas line; why not return by way of the Union Pacific? "This will probably be the last chance I will ever have to visit the plains," Grant wrote his wife, explaining the holiday extension, "and the rapid settlement is changing the character of them so rapidly, that I thought I would avail myself of the opportunity to see them."[20]

Privately, correspondents accompanying the party found the usually taciturn officers to be in good humor, full of stories, with Grant seemingly not drinking. At the end of the Kansas line they had boarded a stagecoach for Denver. Sherman had equipped the officers with Spencer carbines, and they fired frequently from the stage at grazing antelope. "We killed two," Grant wrote his wife proudly. After a brief pause at Denver, they found their luggage to have been delayed, so the officers visited some of the mountain settlements such as Golden City, Central City, and Black Hawk, before retreating to the mining camp of Georgetown, up between the Park and the Front Ranges and not far from Berthoud Pass, where they loafed in civilian clothes for a week, relatively unnoticed by the locals. Another coach took them up the Wyoming Trail to Cheyenne. Waiting for a train to take them over Sherman Summit and up the plains to the end of track and their meeting with Dodge, they wandered around the streets "with so much of an unassuming and quiet air," the *Cheyenne Leader* commented, "that, had not everyone been on the *qui vive* to see the renowned visitors, they would not have been distinguished from any ordinary gentlemen strolling around the town."[21]

As if it were not arranged, as if they did not have an appointment, as if this reunion party unprecedented since the late Civil War was drifting northward like an unconscious tumbleweed to the unlikeliest by-chance meeting place in the West—cosmopolitan canvas-walled Benton, depth of dust, eight inches; population, three thousand roustabouts, gamblers, saloon denizens, and painted ladies—Generals Grant, Sherman, Sheridan, Augur, Harney, Gibbon, Dent, Slemmer, Potter, Hunt, and Kautz encountered General Grenville Dodge at Benton on the day before battle, and the warriors listened as Dodge told what he knew.

If Dr. Durant felt that he held any cards whatsoever on the next day back at Fort Sanders, it is not recorded. The extraordinary assembly crowded into the one-story log structure used as the officer's club, found seats where they could, and everyone turned to regard Durant. He avoided looking directly at Dodge—Seymour was there, too, also evading Dodge's eyes—as he let loose a litany of complaints. Because of Dodge's interferences the company owed the idle graders in Utah, and even some of Casement's teams, some $6,500 with no work to show for it; the lines down Echo Canyon from the rim and around Weber's various bottlenecks were "bad ones"; Dodge's impudent and disloyal engineers had not even finished the final location. Seymour was called on as an engineering authority, but he, too, was far out of his element in this blue-coat-and-brass crowd—that was *Colonel* Seymour, you said?—and his words sounded as empty as his honorary title. As he spoke, Dodge could tell by Sidney Dillon's expression that he thought Seymour was talking through his hat, and even more satisfying, that Durant was realizing "that statements made by Seymour had nothing to base them upon."

Then it was Dodge's turn. He met every complaint, refuted every statement, demolished every assumption. He announced that he would not submit to such interferences as were being enacted not for the benefit of the company but simply "for the purpose of driving me off of the road." Durant and Seymour "knew they could not have their way while I was on the road and watched my every movement." It was clear he was hinting at the corruption which constantly threatened to boil out of the enterprise and into public sight. He ended by promising to resign if he did not have free rein.

Then it was time for the next president of the United States, and the next general of the army, to be heard. Grant and Sherman "took very strong ground" with Durant and Dillon. They said that the government would not stand for any change in Dodge's lines—and that they insisted that Dodge stay on the road. And the meeting did not break up until Durant and Dillon had promised that all orders changing locations would be withdrawn, and Dodge had agreed to stay at his post until the railroad was finished.[22]

They filed outside like military jurors to pose for the New York photographer Andrew J. Russell, who had been hired early in the year to document the Union Pacific work and distribute stereograph prints commercially. Russell had them line up along a picket fence in front of the log building, with straw-hatted Grant in the center. Dodge stood behind and apart from the group in the doorway, grim and gaunt from his weeks of hard desert riding, and he glared defiantly right into the camera lens. Durant, off to the right and trying to look casual, popped himself up onto the fence, his legs dangling like a schoolboy's. When Russell snapped and then told them to hold for another while he moved his camera back for a wider shot, Dodge moved around the edge of the fence to stand next to Dillon and Sheridan, certainly a symbolic sight, as Durant eased himself off his awkward perch onto firm ground, and the moments of confrontation, triumph, and defeat, still resonating among the group, were captured on Russell's film, in stereo. Then, the day following that momentous event, Dodge swept the entire Grant party back down the line to Omaha and Council Bluffs, accommodating Grant, Sherman, and Sheridan at his house, where they were free to kick their boots off, put their feet up, fill the air with cigar smoke and whiskey fumes, and reminisce about all of their battles and triumphs—that is, whenever Dodge would take a breath from filling them in on the rest of the railroad's story, and what he would do now. Not until the eminent guests had returned home would Dodge begin to realize that whatever victory he had won, it would be, at best, fleeting—as far as his day-to-day official duties were concerned.

On July 7 Charley Crocker burst into the Sacramento office of his brother with startling news: he believed the Central Pacific "would soon be out of spikes." He had set men to counting the supplies. "We have not looked into it before," E. B. Crocker wrote to Huntington, "because we supposed that you were sending & had sent the spikes with the iron. Is this so? If we have not spikes for the supply of iron now on hand we are in a horrible fix. It made my blood run cold when he spoke of it. If true, it is a terrible oversight, equal to that of the Vallejo

Co. who forgot to order chairs until after their iron arrived. I hope Charles is mistaken." Even before the men were finished counting it was obvious that he was not mistaken—the Judge instantly wired Huntington to send a thousand kegs of spikes across the isthmus.

Only a few days before, Crocker had been complaining that they were "lamentably short of Locomotives as well as cars," fretting that they could not keep materials moving quickly out to the end of track. Now spikes—and then they found a shortage of fish joints, too. "We are in distress," he said. Another day of counting prompted an urgent wire: they were short twenty-two thousand splice joints and as soon as the next shipment of iron rails arrived they would be down thirty-two thousand. "Track laying will stop unless we receive via Isthmus fastest freight possible straps & bolts," Charley Crocker dictated frantically, "the bolts are equally short with straps. Spikes are also largely short." He explained to the Judge that Huntington had probably been misled into thinking twice as many of the connecting parts had been shipped with the rails. "It is appalling to think of this state of affairs," the Judge then wrote to Huntington. "We had supposed all the time that you were sending the spikes, bolts & fish bars needed along with the rails, for the rails are useless without them." They would always need these in excess of the ordered rails "because they are liable to get lost, injured, & are also often defective. We are now working in a sandy country, & in the tiring rush of laying track these small articles are liable to get lost in the sand. If a fish bar proves defective or don't fit well, it is thrown aside in the hurry & is liable to be lost." They were moving quickly—the rails were ten miles beyond the Big Bend of the Truckee and Charley was preparing to double the tracklaying force to lay four miles per day. The graders were working seventy-five miles ahead. As far as the Judge was concerned, "we can go right on & lay track until we meet *their* tracklayers. If you can delay them any by proceedings at Washington all well & good—& we will do all we can to help you—but the main thing is to *lay the track*."[23]

Some quick experimentation out there in Nevada led Charley to decide to continue to lay track even with the deficiency. He had the men spike seven out of every thirteen ties, which saved a third of the spikes, and when they ran out of the joints he would have a tie placed beneath the joint and a fistful of spikes hammered in around it to keep it approximately in place. Of course they could not run passenger trains over such improvisations, but they could creep the work trains forward at a slow walk, hoping not to derail. And he had not spoken of it before, but because Huntington had been buying iron from so many different mills, it arrived to confront the laborers with seven different lengths of rails and a baffling assortment of fittings, which would work on one type but not another. "I tell you," he groused to Huntington, "when we are straining every nerve for distance these things are great drawbacks." Two days later, on July 17, the Judge wrote again. "We are still in a terrible sweat about spikes, bolts & fish plates," he said. "Yesterday 25 miles of Iron arrived in S. F. with only 12 miles of spikes & bolts." It truly felt as if they were all working twice as hard for half as much effect.[24]

Huntington's telegram on July 16 that he was shipping five thousand plates by fast steamer in no way mollified the Sacramento associates; "it will be 2 weeks yet before she arrives," the Judge responded, "& then they will only supply us 4 or 5 days." He felt bound to make Huntington squirm. "It was a *terrible mistake* in not sending fish plates & bolts as fast as rails," he said. "We could be laying now 4 miles per day, had we a full supply of these *indispensable* articles—but as it is we cannot lay more than 2 or 2½ until we get a supply. We have raked the county over, & are getting along very well with spikes, though short. The fish plates must be put on & at least half bolted." Charley and Strobridge had split the men into two forces and organized them into two shifts, each working 8 hours, from 4:00 A.M. to 8:00 P.M. "Think of it," he wrote, "we have over 100 miles of Iron on hand here, without a strap or bolt to lay with." There was some consolation that the Union company was not progressing as fast as they expected, according to what the Judge had heard. A little more delay, he said, and "they cannot reach Echo Summit before the snow will block them—& then they will have a little of our mountain experience." But Crocker knew nothing for sure about the Union men. "I tell you," he told Huntington, "we have got a big elephant on hand—& being so far from the Eastern workshops is a great drawback. We are battling away & doing our best—& that is all any man can do."[25]

On Wednesday, July 15, a through train drawn by the spanking-new locomotive *Ogdensburg* left Sacramento at 6:00 A.M. and took the first passengers some thirty-five miles beyond Reno to the raw new village at Wadsworth, in the elbow-bend of the Truckee, arriving precisely twelve hours after it had set out. "That town at present consists of about twenty houses," the *Sacramento Union* noted, "of not very beautiful architecture, but more are being built as fast as lumber can be obtained." Wadsworth would become the supply base for the five-hundred-mile stretch of track ahead. And it was bleak out there—not even one inhabitant for every ten miles, all the way to Bear River in Utah. "With the exception of a few cords of stunted pine and juniper trees," Engineer Lewis Clement would recall, "all the fuel was hauled from the Sierra Nevada Mountains. Not a coal bed on the line of the Central was then known, and the only one yet discovered is a poor quality of brown lignite. . . . There was not a tree that would make a board on over 500 miles of the route, no satisfactory quality of building stone. The country offered nothing." Therefore, at Wadsworth piles of sap-smelling railroad ties would rise next to stacks of iron, and wholesale food would be warehoused, almost requiring armed guards because it was so expensive—"barley and oats ranged from $200 to $280 per ton, [and] hay $120," Clement wrote. Strobridge agreed. "Supplies cost enormously," he recalled. "I sent a wagon load of tools from Wadsworth to Promontory, and the expenses for the team and trip were fifty-four hundred dollars. I found a stack of hay on the river near Mill City [in the Humboldt Valley], for which the owner asked sixty dollars a ton. He said I must buy it as there was no other hay to be had. The stack was still standing in his field when we moved camp and it may be there now for all I know. Another settler had a stack of rough stuff, willows, wiregrass, tules and weeds, cut in a slough. I asked him what he expected to do with it. Not knowing that he had a prospective buyer, the man answered, 'Oh! I am going to take it up to the railroad camp. If hay is high I will sell it for hay. If wood is high I'll sell it for wood.'"[26]

Wadsworth would mark the terminus for passenger service until spikes and splice joints began appearing to make the line eastward safe for the riding public. Travelers would be able to board the overland stage at Wadsworth for points east in a few days after trains began running, when service began. As stagecoaches rattled out over the desert, kicking up great clouds of corrosive dust, they would soon begin to pass the host of dirty canvas tents spread out alongside the new tracks beneath the unforgiving sun, a little farther on marking the sweating laborers laying out the heavy wood and iron and hammering it all into place. With the Ragged Top Mountains rising to the north in the tricky, shimmering light, the workers moved beyond the desert hot spring—the old Spring of False Hope where emigrants' wagon-teaming oxen would go mad at the scent of water and lunge into the scalding pool—and slowly the railroaders pushed northeast. The swampy Humboldt Sink would appear in the distance, where Peter Skene Ogden had trapped beavers forty years before and commented on swarms of pelicans overhead; the Carson Sink was just beyond. Samuel Clemens had passed by during one of the prospecting frenzies in Humboldt County, his party camping for two days near the Humboldt Sink. "We tried to use the strong alkaline water of the Sink," he wrote, "but it would not answer. It was like drinking lye, and not weak lye, either. It left a taste in the mouth, bitter and every way execrable, and a burning in the stomach that was very uncomfortable. We put molasses in it, but that helped it very little; we added a pickle, yet the alkali was the prominent taste, and so it was unfit for drinking. The coffee we made of this water was the meanest compound man has yet invented."[27]

To supply the daily needs of more than five thousand workers and four hundred horses, and the requirements of every steam locomotive, the Central Pacific had a number of deep wells bored in the desert east of Wadsworth, the engineers being encouraged at striking abundant water between twenty and forty feet below the sandy surface. It was perfectly clear, apparently good, and so they had filled a locomotive tender with it. It "foamed up so the Engine could not even move itself," the Judge wrote Huntington; "they had to haul it off with another engine & take out all the water & even wash it out with pure water before they could use it." For the present, the company would use a fleet of flatcars with water tanks built above, and take on water from the Truckee;

with one of these tank cars an engine could run across the Forty-Mile Desert and back again to Wadsworth. They would try to burrow deeper for better water out there in the desert, but with eons of leached minerals souring it, it would not serve. Lewis Clement would later recall that "thousands of dollars without result were expended in well boring; tunnels were bored into the mountains east of Wadsworth to develop small springs, and when water was found it was carefully protected and conveyed in some cases over 8 miles in pipes to the line of the road."[28]

It was indeed a nasty place to be at work. Years before, during the surge of overland travel at the first gold rush, one worried sojourner recorded his observations as he pushed west from the Humboldt Sink across the Forty-Mile Desert. "Even the very wagons seem to know that we are off today for the great adventure—in sand, volcanic ash, alkali, furnace heat, and the stench of putrid flesh," he wrote. "We crossed along the edge of an immense baked plain with the fetid stinking slough for a guide, although the wreckage along the way almost paved our route. . . . It must have been here that one emigrant said he counted a dead animal every 106 feet."[29] Of course much of the wreckage, both animal and mechanical, was gone from the desert by the time the Central Pacific worked through, but it contributed its own debris. And in the heat tempers frayed and patience shriveled, especially when it came to Charley Crocker. On Saturday, August 1, his men laid four miles and 180 feet, "without crowding or extra exertion," but the record unleashed his frustration almost like never before; the feat, he wrote Huntington, only occurred because they had been so short of materials the day before that they had only laid a mile and a half before being idled for 9 hours, which would have driven Crocker crazy. By Friday night the missing materials had caught up with the workforce—so on Saturday, with two days' supplies, they had easily passed four miles.

Crocker reminded Huntington that when he had issued his call for four hundred miles in 1868, "I telegraphed to you to do it we must have plenty of rolling stock. If we had 10 more locomotives & three hundred more flat cars now on hand in addition to what is coming we would show those U. Pacific people how to do it & Genl. Casement would be called on to yield the title of the fastest track layer in the U.S. The fact is, Huntington, we can lay all the track there is to lay if we can only get the Iron & ties forward & we can't do that without the machinery." He could hardly believe his eyes now to read in Huntington's letter that "you knew we were short of fish bars & bolts & spikes & did not sooner ship via isthmus to make up the deficiency—when at the same time you have been urging progress in tracklaying &c. &c." They had put down forty-five miles half-spiked, and he feared they would not be able to resume normal procedures for another hundred miles. He was tired and disgusted. "I notice your remark that you would not work so hard another year for the whole road," he told Huntington, and

> that's me exactly. *I have felt like giving up several times. It is not when things go smooth but sometimes everything seems to get at sixes & sevens & as every particle of work connected with the running & construction of the road devolves on me—everything in fact except drawing a check on you monthly, & buying Rail Roads & talking over big things in the office. We have 85 miles of Rail road built & 65 more graded that Stanford or Hopkins have never seen or been over. Well it is all right. I will hold my grip until the C. P. is finished—but I give fair warning that there is to be a fairer division of labor when the other roads are to be built—which have been planned & talked over so lavishly. My time & paper have both given out.* Burn this—*I am ashamed of the complaining spirit manifested.*[30]

Self-pity was common to all of the Associates at times, of course, but Charley Crocker had reason to feel an extra burden of worry and fatigue when contemplating some of those other roads—especially, now, with the Southern Pacific. Negotiations between Stanford and Judge Crocker and the principals in San Francisco had continued through the month of June, and obstacles seemed to begin falling out of their path by July 2, when the Judge wrote Huntington that Lloyd Tevis had detected what he thought was their waning interest—and suddenly he was sure they could get control of the Southern Pacific for $300,000. On Saturday, July 11, the trade was consummated—for $350,000—giving them a majority of the stock and the seats on its board. It was done completely in secret; they put the financier W. C. Ralston down as purchaser, with only Tevis and Carpentier in the know. "All are to keep still," Stanford wrote Huntington, "and we will change the board gradually so as to avoid attracting attention." They would salt the board with "confidential friends." Tevis was bought out with the others, "but I think we will keep him in the board." Stanford's letter went out on its tedious trip to New York by steamer. A more expedient telegram sent to Huntington on the purchase day was, unfortunately, indecipherable, as the Governor had left his code book back in Sacramento.[31]

In Washington the weather had grown insufferably torrid, with many in the capital going down with fever and all looking forward to the congressional adjournment on July 27. Apparently on the eleventh Stanford's strange telegram arrived—and Huntington, impatient as ever with Stanford's inefficiency and pomposity, waited for a clarification. While he waited Huntington made an unhappy discovery. He learned that Secretary Browning had told subordinates he was canceling the federal land grants to the Southern Pacific; the company had, after all, shifted its location from the original southbound coastal line to the interior, up the San Joaquin Valley, and done so without consultation or permission. Minutes after hearing this news, Huntington's frantic wire to Judge Crocker was heading west. The reply came back just as quickly, and nearly stopped his heart: they had already closed the deal, on July

11, purchasing control of the company and board for $350,000. The Southern Pacific was theirs—and they now had no land to build it on.[32]

Blame ricocheted across the continent like artillery. "If we had got this before the trade was consummated," the Judge pointed out in the rebuke most likely to wound, "we could have bought it for $100,000 less without a doubt, but it came too late." He spread out the disagreeing Southern Pacific route maps from the Articles of Association (threading down the coast) and the now-defunct federal land chart with its thousands of acres of withdrawn valley territory. He was spending less time in the office with his infirmity, and a lassitude sometimes seemed to overtake him. "There is a land fever here," he glumly told Huntington, "& speculators are entering land largely." The federal agent had "entered 200,000 acres principally in the San Joaquin Valley." The Judge mused about perhaps swinging the Southern Pacific up into the Salinas Valley—but back in Washington, Huntington did not intend to give up easily. He laid siege to Secretary Browning's office. And Stanford was thinking in concert. "We don't know exactly what to make of the withdrawal of reservation of S. P. lands," he wrote Huntington, "but I think you will straighten it if it needs it. I have a hope that it means aid to San Joaquin V. Road. With Govt. aid for both those roads, we would have a great thing." They could run the roads up the valley in parallel, each hugging a flank of the enfolding mountains.

"You can," Stanford continued, underlining each word almost deep enough to tear the paper, *"well afford to pay for such aid."*

And the interior secretary paid attention to Huntington's inducements. The Southern Pacific land up the San Joaquin Valley—its gentle, sparsely settled terrain for the present entertaining only scattered herds of peaceful sheep and a few lonely homesteaders, its thousands of acres of lush orchards and crowded vegetable farms, its packing plants and distribution centers, its chains of prosperous communities and burgeoning populations hardly a gleam in the eye of the railroaders at the moment—was, with a stroke of Orville Hickman Browning's pen, restored to the Southern Pacific.[33]

Major Frank North's Pawnee scouts battalion was continuing to patrol the Union Pacific line in central Nebraska when, in July, the major took a detachment of fifty scouts to ride south to the Republican River. There, North's troops encountered their fellow tribesmen out on the annual Pawnee summer buffalo hunt in the Republican Valley, and North decided to join the hunt to bring some fresh meat back to the scouts still patrolling the railroad. Early one morning at Mud Creek, North's party of nine had cut off and surrounded a number of buffalo; the hunters were individually moving in to finish off the prey when they saw about one hundred riders heading their way—who soon began firing at them. They were Brulé Sioux, from Spotted Tail's camp up near North Platte, and, as it happened nearly every summer, when enemy tribes ran into each other during their buffalo hunts, they turned to hunting each other. North's men retreated to a sheltered ravine and, though surrounded, kept their attackers at bay for hours; one of the scouts slipped through the Sioux cordon on his pony and galloped off for help—but he was seemingly cut down.

Meanwhile, a much larger party of Sioux had attacked the main encampment of Pawnee—thinking they would surprise the unprotected women and children and elderly—but most of the Pawnee warriors were nearby and rallied to the defense, and began to drive the Sioux off. North heard the sounds of a pitched battle moving closer over the next several hours. By then the soldiers were nearly out of cartridges, three were wounded, and six of their nine horses were dead. They were suffering terribly from lack of water. Suddenly, the Sioux cleared off—North's scout had gotten through, and the Pawnee had worked their bloody way toward the ravine and rescue. They recovered the body of the one scout killed and scalped at the onset of the battle—he would be the only Pawnee scout killed while serving the North brothers—and they limped back northward toward the railroad.[34]

In comparison to the summer before, though, scrapes between whites and Indians were dramatically declining. In March, Grant had ordered the army posts in Powder River and Bighorn country closed to appease Red Cloud and the other hostile Northern Sioux. In April, when the long-awaited Laramie Treaty was presented for signing, the northern tribes had remained aloof, waiting for the bluecoats to leave, but the Brulé and Oglala Sioux of the plains lying between the Union Pacific and the Kansas lines had come in and made their marks. Bearing the piles of presents always proffered during such occasions, the chiefs took their people south to hunt buffalo—not realizing that they had overlooked an obscure provision in the treaty, which ceded their hunting rights on the southern plains and ordered them to relocate to a reservation up in Dakota, on Whetstone Creek in the Missouri River Valley.

It was not until July when intermediaries told the southern Sioux of Spotted Tail and Swift Bear, who were hunting in the Republican forks of Kansas, the astounding news that they were bound to move north. To reinforce this, the Indian Agency, with its source of supplies, had closed. Annuities promised under the Laramie and Medicine Lodge treaties, and others, had still not appeared by midsummer; Congress, so hypnotized by the bright, simplistic light of the impeachment trial, did not take up important business such as the approval of those treaties until the end of July, in the rush to leave Washington on time. "The poor Indians are starving," General Sherman wrote his wife on July 15. "We kill them if they attempt to hunt and if they keep within the Reservations they starve. Of course [the peace commission] recommended they should receive certain food for a time . . . but Congress makes no provision and of course nothing is done. I wish Congress could be impeached." The tribal leaders had a terrible time explaining the treaty and its overlooked codicils to their young firebrand warriors, but with Augur's troops menacing them from the Union Pacific country, and Sheridan's men attending them in Kansas, there seemed little choice but to acquiesce. They slowly, grudgingly moved northward across the Platte Valley. The younger warriors did not understand—they had not, after all, been defeated in battle—and a number slipped away from the main body of Sioux to prove to unfortunate settlers and wayfarers that they still had fight in them.[35]

Up nearer to the main flurry of railroad activity in Wyoming, which on July 25 would be deemed an official territory of the United States when carved from parts of Dakota, Utah, and Idaho, the friendly tribes which for years had camped around Fort Laramie up on the Oregon Trail were suddenly ordered north to their new reservation. They, too, moved away from the whites, perhaps taking with them the smallpox germs which wiped out many of their number during the journey. Farther up the North Platte River and closer to the hunting grounds of the recalcitrant northern tribes, where the Union Pacific tracks crossed from the Laramie Plains before turning westward, the young surveyor Arthur Ferguson was sitting in his tent one day in late July when "a bullet entered the canvas, making quite a hole in it." He assumed it was a potshot from an Indian and not a stray from an altercation between whites because a few days before, on Sunday, July 19, Indians had made an attack on some railroaders, and "killed and scalped 4 men besides severely wounding another," he recorded in his diary. "On Saturday a band of 500 Indians made their appearance before Fort Sanders and killed 2 men." Still, the vast prairie between the two great railroads was steadily being swept clean as he wrote, and the Iron Horse was opening up more country to settlement and exploitation. "The time is coming, and fast too," he observed, "when in the sense it is now understood, THERE WILL BE NO WEST."[36]

Whatever it was and would be, it was still wild. Not ten miles west from Ferguson's tent, Benton still rocked. The dividing point between the Union Pacific's freight and passenger divisions and the construction division, the station saw two immense freight trains arrive each day to disgorge all the materials earlier hauled over the plains, which were now reloaded onto wagons bound for Utah, Montana, and Idaho. "For ten hours daily," the traveling correspondent John H. Beadle would note, "the streets were filled with motley crowds of railroad men, Mexicans and Indians, gamblers, 'cappers,' and saloon-keepers, merchants, miners, and mulewhackers." At sundown, from the tents the "lively notes of the violin and guitar" called the citizens to "evening diversions," he said. For the *Cincinnati Commercial* he described the interior of the "Big Tent":

> *[T]he right side is lined with a splendid bar, supplied with every variety of liquors and cigars, with cut glass goblets, ice-pitchers, splendid mirrors, and pictures rivalling those of our Eastern cities. At the back end a space*

> *large enough for one cotillon is left open for dancing; on a raised platform, a full band is in attendance day and night, while all the rest of the room is filled with tables devoted to monte, faro, rondo coolo, fortune-wheels, and every other species of gambling known. . . . Fair women, clothed with richness and taste, in white and airy garments, mingle with the throng, watch the games with deep interest, or laugh and chat with the players.*

Somewhere in town, always, a fistfight was erupting. Drunkenness was so common that sobriety was remarked upon, and the lawless element found a roost. The mild Massachusetts editor, Samuel Bowles of the *Springfield Republican,* passed through and made haste to get out: it was "a village of a few variety stores and shops, by day disgusting, by night dangerous, almost everybody dirty, many filthy and with the marks of lowest vice, averaging a murder a day, gambling and drinking, hurdy dancing and the vilest of sexual commerce." E. C. Lockwood, the young brother-in-law of Dan Casement, was working by then as a payroll and commissary clerk. "It was frequently my duty to go to the end of track to carry the mail for the men and currency for the superintendent's use," he recalled many years later. "I often carried as high as $3,000 and was never once held up, although when the office was at Benton station, I started out one morning at three o'clock with a mail bag full of mail and $2,000 embedded within. A man followed me so closely that I turned and faced him. I had in my hand a silver-plated six-shooter. The starlight was brilliant and shone on the barrel. The man evidently saw it, for he quickly disappeared in the sage brush. The train I was to take stood on the side track a quarter of a mile away, so I boarded it in short order and arrived safely at my destination."

Benton, observed a Wyoming newspaperman, "like the camps of the Bedouin Arabs, is of tents, and almost a transitory nature as the elements of a soap bubble." It would last about as long. As the Union Pacific crews pushed away from the dirty little town it began to fade—and the businesses unscrewed their portable buildings and folded their tents and moved after them. When Beadle passed back through he found only rubble and wreckage, a few skeletal chimneys, and—of course—a weed-choked cemetery, all of it coated in the bitter white dust.[37]

The long stretch from the North Platte to Green River was no place for loitering, especially during the hottest part of the summer. "This is an awfull place," Jack Casement wrote home, "alkali dust knee deep and certainly the meanest place I have ever been in." Many men came down with sun poisoning and heat exhaustion; others wheezed from the choking dust, coughing and spitting blood; bandanas soaked in water and masked over the mouth did little good. "We haul all our water," Casement said, "about fifty miles on the cars." Mules keeled over in the heat—"Six nice fat ones died in less than an hour today," Casement complained—and they were dragged off from the grade for the buzzards and coyotes. Correspondent Beadle passed by in August and noted the terrible contrasts in temperature: "one of my mules twice fell exhausted with the heat; that night ice formed in our buckets as thick as a pane of glass." But Dodge's engineers had given the graders and tracklayers a relatively straight and level path, and much of the iron was arriving in thirty-foot lengths, which at more than five hundred pounds each rail made them more difficult and dangerous to maneuver but resulted in satisfying mileage at the end of a shift. Casement's men averaged 2.3 miles per day, six days per week; some days they managed between 3 and 7 miles, which usually earned rewards of alcohol or cash from the contractors.

In that empty and unfriendly terrain, their only problem seemed to be gawkers—there was an endless parade of excursion trains filled with curious newspapermen, politicians, army brass, ministers, educators, their engines chuffing up slowly behind the laborers' backs as if to chide them on, well-dressed Easterners stepping out of the cars to watch the sweat-glistened Irishmen work at a pace a little quicker than comfortable in order to impress the dignitaries. Charles Dana of the *New York Sun* and Albert Richardson of the *New York Tribune* came through with a host of other scribes; Schuyler Colfax took a party through, complaining later that he did not receive treatment royal enough for a vice presidential candidate; Roscoe Conkling went through also, any complaints being unrecorded. "Today," Jack Casement complained to his wife from the end of track on August 17, "all the Professors of Yale Colege and a lot of Rail Road men with their Ladies have been here it is a great nuisance to the work Yet we have laid over four miles of track today." A few weeks later, as the track met and ran parallel with a stage road, the stream of visitors even increased—"Since I have had a tent," Casement wrote, "I have been alone but one night"—and because there was no place to stay for excursionists as well as unconnected travelers, the railroaders could hardly refuse accommodations. Women were given their own quarters and men crowded into the contractor's tent. Among the visitors was Jenny Reed. "Mrs. Reed, her Pastor, four other Ladies and several Gentlemen bound for Salt Lake staid with us Saturday night and yesterday," he told Frances Casement. "Mrs. Reed left her baby at home with a wet nurse and milks herself while away from the Baby."[38]

Casement was taking on another grading contract for a sixty-mile section west of Salt Lake—"I am very busy fitting out parties to put on it"—so visitors, no matter who they were, disturbed his preoccupations. One traveler who had somehow missed him but who had written Casement from home in Pittsford, New York, upon his return at the end of July, would have been an interesting visitor there on the Wyoming desert. It was Dr. Hartwell Carver—who still could give away old pamphlets from his campaign against Asa Whitney more than twenty years before, and who still claimed to be the originator of the idea and the project General Casement was now supervising. "I have spent my fortune and many years in the prime of life in getting the Pacific railroad where it is," he told Casement. "I am the first man who ever thought of the thing and have got the printed documents to prove the facts beyond any possible doubt." He was now eighty and "in hopes to live to cross the road when done." But that was, he admitted, "uncertain."[39]

Between Benton and Rawlins, Casement's men had been briefly delayed by trouble in the first tunnel on the Union Pacific Railroad, which was some 680 miles west of Omaha near St. Mary's Creek; it had been commenced on the last day of April and worked from both ends until early June. Dodge had seen it and not liked it, and before the track crews arrived, his engineers detected a soft spot in the west end of the tunnel; there was no way to quickly reinforce it from falls, so the rock above the headings there was carefully blasted away, creating an open cut and reducing the tunnel length to 215 feet. Not content to wait until the demolition was done, Casement had his men put in a temporary track around the bottleneck. They would not have to contend with another tunnel for nearly three hundred miles, all the way to Echo Canyon, where, in July, work began on the longest bore on the Union Pacific line—it would be 772 feet long when finished.

For a month, workers contended with heavy cuts only through clay. In August, though, when they finally struck rock, it was an exasperating type—hard and requiring drills and blasting powder, but when exposure to air had dried the material out, it cracked and crumbled, noted a civil engineer who examined the work and gave estimates to Reed, "like lime in slacking. These qualities made the work very expensive; rock prices had to be paid, and earth slopes taken out." The entire length would have to be lined with timber. Tunnel 2 would require every bit of ten months to complete, causing no end of worry as the track crews drew ever closer. Down in Weber Canyon, Tunnels 3 and 4 would curve and descend through harder black limestone, and quartzite nearly hard as granite, and would be commenced in September. As predicted by Reed, Blickensderfer, and Dodge, all three Utah tunnels would hold up the work, making short, steep, temporary tracks necessary.[40]

Dodge got back out to Utah around August 4. On the way he had conferred with Durant and Dillon in Laramie, and it was probably there, on August 2, that the Doctor handed him a copy of his "General Order No. 7," which was dated two days before. It was soon clear that the showdown at Fort Sanders, even with its eminent audience, had not settled anything regarding authority out on the line; the only concrete result of the confrontation was that Dodge had personally promised the company and the next president of the United States that he would not resign until the project was done. Letters received from Oliver Ames back in Massachusetts would show that even Grant's and Sherman's firm reinforcement would not move the company president to stand up and really support Dodge. "I think the Doctor coincides with you in regard to what should be done to locate and construct road," Ames had written on July 26 in a stunning avoidance of reality. A day later, he added that he expected Dodge to behave like a good boy. "As our great object is to complete the road," he wrote, "we must as far as possible set aside all these annoyances and let no ordinary thing turn us from this object. I am glad that our line is so well located and is ready before the graders are really ready with their tools to take hold of it."[41]

To this maddening incomprehension was added "General Order No. 7": Dodge was to turn over to Durant "all maps, profiles, surveys and notes" for the line east of Weber Canyon; he was to give Durant "one or more" full locating parties to be used "in perfecting the location"; he was to give "personal attention" to the line between Weber and Humboldt Wells—which would put him far away from the action in the Wasatch canyons, where Colonel Seymour was still lingering; Seymour would "continue to make himself familiar with all engineering operations," and he would "from time to time make such suggestions, and give such advice to the engineers"; Durant and the compliant Reed would command "immediate surveys and estimates" based on Seymour's suggestions from the mouth of Weber Canyon north to the Bear River confluence with Great Salt Lake, at maximum allowable grades. Dodge had already heard from Jacob Blickensderfer that the Doctor had been asking him for "estimated quantities" of earth and rock removal at the head of Echo Canyon, obviously to have figures with which to object to their line. How little had changed! Dodge left Durant and Dillon at Laramie after pointedly insisting on written orders—saying, in other words, that the superiors could not be trusted and that Dodge would be stockpiling his own ammunition for future battles.

His letter to Oliver Ames, dated August 8, spelled out Dodge's new position. "You can put such construction on the action of Durant and the Executive Committee as you deem best," he wrote.

> *I know what it means and where it will end. He has accomplished his end. Seymour has supreme control here and he in New York. If the country knew it today more than one injunction would be served on you. Nothing is being done on repairs and the orders of the Vice President is to skin and skip everything for the purpose of getting track down. Your temporary bridges will now hardly stand to get trains over them and winter will close in on you with nothing done. Your immense subsidy will be spent in dividends and what few men you have among you who have a name or reputation will be, in the eyes of the country, disgraced.*
>
> *I am sent out, here under orders to look after 200 miles of line. Seymour is sent here with orders to break up the finest location that was ever made by putting in 116 ft. grades and 8° curvature and I am not even under the order allowed to say a word or give an opinion; however, the company may go on and do all this, but I doubt whether as a mile of road will be accepted, with such a location, as the first principles of a first class rail-*

road is a first class location and you may rest assured the country will know it—such things cannot be hid. I say to you that this location that they are endeavoring to change, to save a few thousand dollars, is the best location for the money to be had, and the work can all be done before the track reaches here. I made Reed admit that yesterday.

Dodge reminded Ames that in the spring, everyone had told him to "hire the ablest engineers in the country" and promised him that there would be no interference—"that you wanted no more Black Hills operations." His men had produced the most superior and unassailable line on the continent.

What is the result? Col. Seymour under order No. 7 has instructions to change, alter &c. it whenever he deems best. He knows It is a fine location and will resound to the credit of all of us who were engaged on it. No grade to exceed 65 feet going west. None, going east, to excel 90 ft. and no curvature to exceed six degrees, and very few, if any, undulations and this in a country where it was thought 116 ft. grades would have to be often used. While this line was being located, we were abused, lied about and unmercifully attacked because we were slow, all of which were unmitigated lies, and men were waiting here for tools not for lines and in six weeks we located 240 miles of difficult line, including such points as Green River, Rim of Basin, Head Of Echo, Narrows and Weber Canon, all of which were as difficult to locate as any line in the United States; but how changed things are now. There is plenty of time to make changes, plenty of time for Col. S. to make all the surveys he wants and they move as though they had all summer to do it in. I came here to see the lines they had run. I have seen them and if any of the high grade lines are accepted, somebody will have to answer for the swindle.

All of Seymour's "pet lines" were steeper, longer, more expensive, and required more time than the ones adopted by Dodge's team, no matter what the colonel claimed. It was hard to say what was the worst alternative proposed by Seymour, but the change already instituted at Devil's Gate stood out—"the 116 ft. grade, with the additional bridge, costs more than the 90 ft. grade."

My parties are now all at work locating west of Salt Lake and I will be at Humboldt Wells with entire location by Sept. 1st. I desired to make my Oregon reconnaissance but my orders I construe as positively prohibiting it. . . . I expect to remain here until my lines are located, and I trust for that short time, I will not be interfered with or further hampered. After that I shall be at liberty and the men I brought here and whom I consider I was bound in a certain degree to protect, will be through and I shall feel then at liberty to take such course as circumstances may dictate.

No one—not even Oliver Ames—could mistake the threatening tone of this letter. It "caused a great many comments in Boston," Dodge was to admit many years later; copies somehow made their way to the government directors of the Union Pacific, who were enemies of Durant, and they in turn complained bitterly against the Executive Committee. Dodge would be on to other business during the two weeks it would take for the government directors to make their own threats and cause President Ames to call a meeting of the board of directors, set for September 2, and then send a letter back to Dodge. Ames promised that the Committee on Location would certify Dodge's lines over Seymour's. "Though the line of the road will be fixed all along the route where your engineers have laid it," he told Dodge, "it is exceedingly annoying to have to fight for it all the way and for the purpose of satisfying Dr. Durant that he is a power and to be consulted in all matters." Seymour's arguments had seemed compelling, he admitted in partial apology, and they had no way of judging them; "your telegrams of cost," he wrote, "have fixed this so that there can be but one opinion as to line to be built." And when the committee had fixed the route, they could move some iron. "We are ordering some 20 more additional locomotives to get our construction material ahead," he said, "and are doing everything possible to get our road on to Salt Lake this year, and 200 miles west of there next year if possible." The Union Pacific was now within three hundred miles of Salt Lake, and they had the right to work three hundred miles ahead of the track—"we can now put our men on the heavy work west of Salt Lake at once."[42]

Dodge was already moving to put this new phase into effect by the time Oliver Ames's letter reached him. On August 14 at Salt Lake City he met and compared notes with Lewis M. Clement of the Central Pacific—the first such official conference between engineers of the two railroads, although there had been informal trades of data earlier. Dodge had been committed to the northern route around Great Salt Lake for some time, and Clement told him that the Central Pacific had come to the same conclusion: the southern way was longer and had more obstructions. But now, with this finally acknowledged, it was time to disabuse Brigham Young of any lingering illusions that the companies would run through his capital.

When Dodge obtained an audience and told the patriarch, Young was, in Dodge's words, "greatly disappointed and much dissatisfied." It is supposed that he may have threatened to throw all of his support to the Central Pacific, giving them primacy in Utah and all of the business of Salt Lake City; one Dodge biographer has asserted that the engineer dissuaded Young by giving him the news that Clement had confirmed the Central Pacific intent to go north. But Young was too astute a businessman to see any benefit in hostility or obstruction; moreover, he had already discussed construction contracts with Leland Stanford by that time. He may have shot appeals eastward to the Union Pacific board of directors, as Dodge later claimed—but if so, Young was told by Ames that the government would not support any decision to add such an expensive detour—probably the Union company would build a branch line to Salt Lake City once the lines were joined.[43] But all evidence supports a reaction of disappointed acquiescence. Ten days before he met with Dodge, on August 4, Brigham Young was writing to an associate:

> *Work on my railroad contract is progressing rapidly; several jobs are already completed, and nearly all the light work would have been done ere this, had the work been staked out in time. The Western Company [that is, Central Pacific] are wishing me to contract to grade 200 miles for them, which I expect to begin as soon as the stakes are driven. These contracts give us many advantages, besides furnishing money for labor to those whom the grasshoppers have left but little to do, and who could not well otherwise supply themselves with food until another harvest.*[44]

Young made it known that he would address the faithful in the New Tabernacle on Sunday, August 16, on the subject of the railroad. Earlier in the season he had challenged his Mormon workers to work so hard and well "that in the halls of Congress it might be announced that no part of the national railway was completed more satisfactorily." Now he rose in that grand vaulted space and began to tell them the new situation. In his autobiography, written many years later, Grenville Dodge said that he was seated with his wife and children near Samuel Reed and Colonel Seymour, and he listened as the president of the Latter-day Saints severely denounced him in a markedly aggressive manner, implying that the railroad "would not be safe" if it ran around the northern end of the lake; Dodge claimed that Young seemed to be embarrassed—after all of his assurances to his people that the rails would join at his capital—and that this moved him to tell the audience that he "had a revelation on the matter," sanctifying the railroad engineers' disappointing reports. Dodge went on to say that Young's tirade "alarmed my people very much and they were very anxious for me to get out of Salt Lake, but it did not alarm me any."[45]

Such a stance would have made Dodge seem heroic—but the threats and the tirade seem to have been wholly invented by Grenville Dodge himself. They have been repeated without question in most subsequent accounts, growing with each embellishment. Actually, on August 16, according to the *Deseret News* (which published Young's full text), the president of the Saints reserved his spleen for "careless and unconcerned" church elders who frittered away their time and neglected to attend meetings and services. "In my experience," he thundered, "I never did let an opportunity pass of getting with the Prophet Joseph and of hearing him speak in public or in private. . . . I notice that even my own natural brothers when they come into my office, which is very seldom, if there are important matters on hand—when I am teaching the brethren the principles of government, and how to apply them to families, neighborhoods and nations, will leave the office as though it was a thing of no account. And this is the case with too many of the Elders in the Church. This is mortifying to me." He gestured over to the empty seats of high councillors and elders—and then, his temper apparently cooling, Young spoke about sundry matters before addressing the qualms of some in Zion who feared the railroads for the outside influences they would bring, who wished their society to remain in isolation. "Are we isolated?" he asked.

> *No I do not think we are. We are right in the great highway from sea to sea. And instead of the railroad being any detriment to us, all I have to regret is that they tried to get it on the north side of the Lake; we want it in this city where it belongs. And that is not all, the attempt to carry it in that direction is an insult to the people of this city, for in doing they have tried to shun us. They would not have had a telegraph or railroad across the continent, and coaches would not have run as they do now for one generation yet if it had not been for the Latter-day Saints; and for them to try to take it away from us I look upon as an insult. We do not care about it, we are in the habit of being insulted and imposed upon. Far from wishing not to have a railroad it ought to have been built years ago. When we came to this valley we never traveled a day without marking the path for the road to this place. We anticipated it, and if they had done as they should have done, instead of going to war and killing each other, we should have had a railroad long ago.*

"These are my feelings with regard to the railroad," Brigham Young finished, "and whether it comes through this city or not it is all right, because God rules and He will have things as He pleases."[46] After this mild criticism, and after Young's additional hour of sermonizing about missionaries, about polygamy, about the splendor of life in this sanctified city, about the fond hope that "the Latter-day Saints will be Saints indeed," Grenville Dodge would have risen stiffly from his pew, with his wife and restless children, feeling considerably better about himself and his railroad than the poor, inattentive, hooky-playing elders who had suffered the effects of true wrath on that warm Sunday afternoon. No wonder that in a few days, when Dodge called on Brigham Young to say goodbye before going back out on the line, Young greeted him "as cordially as ever," and welcomed Mrs. Dodge and their eldest daughter, Lettie, and took them home with him so that their children could play together.[47] As Young had written to an associate,

> *this contract is viewed by the brethren of understanding as a God-send. There is much indebtedness among the people, and the territory is drained of money, but labor here and coming we have in large amount, and this contract affords opportunity for turning that labor into money, with which*

> *those here can pay each other, and import needed machinery, and such useful articles as we cannot yet produce, and those coming can pay their indebtedness, and have ready means with which to gather around them the comforts of life in their new homes.*[48]

If only the railroads would pay their bills, all would be Edenic. If only they would pay their bills.

When Dodge had written to Oliver Ames with his unveiled threat to be "at liberty" to disclose malfeasance and misappropriation along the nation's railroad, when he muttered about the "swindle" and that "the country will know it—such things cannot be hid," that at such time "more than one injunction would be served on you," the chief engineer was just stepping into a crowd as truculent, self-serving, and dangerous as any Hell on Wheels barroom throng, minus only the ready pistols and lynching rope. By late summer the climate of threats and counterthreats in New York City had grown nearly as bitter as the alkali atmosphere in Benton, Wyoming, and pressure was mounting against both Oliver Ames and Thomas Durant. If they had been seated around a card table with a collection of cutthroats and cardsharps, one could have looked at the two of them—the Doctor, straw hat pulled down to shade his heavy-lidded eyes, languid and tense at the same time like a coiled snake, and the King of Spades, sweaty and uneasy in his itchy three-piece suit, nervous at the coming confrontation—and found few bet takers that the Doctor would fold first.

The others at that table would have increased the pressure and tension as well as the stakes in the fanciful Hell on Wheels boardroom poker game. There would be Jubilee Jim Fisk, whose suit and injunction "to get what he could" out of Durant and his friends sought to recoup his $3,200 stock investment by 2,300 percent; the threat of contempt citations in New York would cause Durant and Dillon and other principals to skip town for the duration of the summer. Oakes Ames would try from outside the circle to halt the Fisk threat by writing a bill which passed into law before Congress adjourned. In cases of suits against corporations with federal charters, it would remove them from malleable local jurists and courtrooms and put them in federal courts. It would have as profound an effect on American business as the Crédit Mobilier–inspiring notion of limited corporate liability. But as far as the immediate threat by Jubilee Jim, it gave the lawyers more to do.

Another player upping the stakes would be Government Director Jesse Williams, who in communications with Dodge and Snyder, among others, had been alerted to the unsound bridges, shaky trestles, settling and eroding fills and embankments, lightweight rails, and substandard, badly fashioned, and mismatched wooden ties, which the miracle machine of General Jack Casement had left behind in the hurry toward Utah. Williams also sided with Dodge over the Wasatch line. After he inspected the railroad himself, Williams told Secretary Browning that the Union Pacific should create a $3 million reserve fund of government bonds to cover repairs and finish work. The board members, shocked and dismayed that such a sum would be put into unreachable limbo, only grudgingly agreed to a reserve fund if it was drawn from the less-desirable first-mortgage bonds. Such would be adopted in September.[49]

Still another player would be the tenacious leather merchant and Crédit Mobilier stockholder Henry McComb, who continued to threaten to file suit over his failure to get more out of the previous winter's Crédit Mobilier distribution. McComb, however, was in for the long haul—keeping his presence (and his muscle) felt as the others dealt and raised and raised again. It would be the grandfatherly Rhode Island Yankee, Rowland Hazard, righteously indignant over the gross veniality of Dr. Durant, who called—over the still-nagging question of Durant's mysterious, unexplained "suspense account"—and caused a dramatic little flurry in that high-stakes game.

All summer Hazard had tried to corner Durant to disclose details about the princely sums charged to the account. All Durant would tell Hazard's auditor was that the money should be entered as a Union Pacific expenditure; he had obtained it from the railroad for purposes too delicate to be disclosed in the company's books, and for that reason it had been charged to the Crédit Mobilier under the Hoxie account. Durant refused to say what he had done with the money, though he assured Hazard's people that it was used for the company's benefit. By late August, Hazard had lost all patience. Durant had fled New York City to escape Fisk's citations, lodging for a while at Saratoga to attend to his private pet project, a railroad into the heart of the Adirondacks. Then the Doctor had gone to his other favorite watering hole of the wealthy, Newport, and it was there, in Hazard's home state, that a lawsuit was sprung. Charging Durant with "intent to defraud" the Union Pacific by seizing funds "for his own wrongful use," Hazard was only able to get Oakes Ames to supply an affidavit—the others were too scared of Durant to step into the light. But the word of a Massachusetts congressman and a Rhode Island captain of industry was enough for the judge to order detectives to appear at the side of Vice President Thomas Durant at that patrician seaside, to put firm hands upon him, and escort him off to the calaboose.[50]

The Doctor was too smooth to stay behind bars for very long—he delighted his guardians by bribing them generously for his release—and he was too slippery to be pinned down by Rowland Hazard and his lawyers. Back in New York in the first week of September, he joined the executive committee in authorizing a $6.5 million payment of securities to the trustees of the construction contract for work done out on the line, and after dangling that juicy plum before their noses for a few days, Durant had the executive committee withdraw it. He also had the committee vote to retain counsel on the Doctor's behalf. And he arranged for a new pending construction contract in the name of Oakes Ames to be postponed. Like a Hell on Wheels gambler, he looked around the table at the empty faces as the significance of his cards sunk in.

Magically, wonderfully, all fell into place for him. Oliver Ames, as a member of the board committee authorized to inquire into the suspense account, joined John Cisco in certifying their acceptance of Durant's expense "vouchers"—which no one, including them in all probability, ever saw. The executive committee thereupon charged the amount to construction and closed the matter. The new Oakes Ames contract went forward again. The $6.5 million went out to the trustees. And Dr. Durant, exonerated and triumphant, went west with newly bestowed power to "remove and appoint" all officers while out on the railroad line. "It is one of the miseries of our Road," feckless Oliver Ames had written Dodge, "that we have a man in it who is so desirous of Power and so jealous of every act that does not coincide with his notions."[51]

On August 12 Leland Stanford left his wife and infant son and rode the cars 250 miles to the end of track and then took a stage eastward to Salt Lake City. As it happened he arrived on Sunday, August 16—the day of Brigham Young's temple sermon about the railroad—but did not see the leader for a few days as Young went north to visit the settlements. Stanford avidly followed him, and when they met, Young told Stanford after some discussion that "it would not answer" for him to contract with the Central Pacific directly. He said that he had a good deal of unfinished work in Weber and Echo Canyons, and also that he was afraid to offend the Union Pacific; free passage for all able-bodied men on the railroad, and reduced rates for their families, was too great a boon to the Perpetual Emigrating Fund to be trifled with. Young assured Stanford that he was "very anxious" to have the two roads meet at the mouth of Weber Canyon. He arranged for three influential associates—Ezra Taft Benson, leader of the faithful in the Cache Valley; Lorin Farr, a successful miller and the mayor of Ogden village; and Chauncey W. West, who lived in Weber County—to take the contract. "All three of them . . . have large property interests to be affected by our reaching Weber," Stanford wrote Huntington. For approximately $4 million, they agreed to grade one hundred miles westward from Monument Point in the Promontory Mountains above the lake to Humboldt Wells. If the contractors performed well, they would have the right to grade the one hundred miles eastward from the Promontories to Weber.

By the last week in August the contractors had put together eighty teams and mule-drawn scrapers and sent them north. Stanford was optimistic after talking with Chauncey West and others in the canyon settlements that the Union Pacific would not advance over Echo Summit before the rails were halted by snows. At the same time he thought the Central Pacific could build to the mouth of Weber in time to lay track to the junction with Echo Canyon.

"This would if possible be a great thing for us," he told Huntington. "It would take us nearly to the coal mines which are about 5 miles up Weber beyond where Echo comes into Weber. Echo Cannon is high, cold and subject to drifting snows. The U. P. is welcome to operate it."[52]

Benson, Farr, and West advertised to pay their Mormon hands $2.50 per day and board and $9 for mule teams, which would "board themselves." The pay, higher than that offered by the Union people, might even attract laborers from the competing work. Indeed, Stanford chased a promising lead to hire 180 teams working for the Union Pacific east of Echo Summit, dropping it when he concluded the parties were acting in bad faith. But he told Judge Crocker that "there is great dissatisfaction among the laborers on the U. P.—that when work is started on our line, hundreds will leave to come over to us."[53]

While the Mormon teamsters began to scrape their way down through the bare and unpleasant Promontories, Strobridge's tracklayers had finally begun to measure satisfying progress up along the Humboldt in Nevada. Spikes, bolts, and fastening plates sped over the isthmus had reached them, and while crews went back over the badly cobbled section of line to hammer down the missing hardware, freights rolled in with new batches of rails which Charley Crocker pronounced "splendid." Even the splicing bars "fit beautifully," he complimented Huntington, who had not been hearing many compliments lately. "It takes no longer to lay two 30 foot rails than two 24 foot rails," Crocker said, "& the difference on a day's work is big." On August 19 he wired New York and Sacramento that they had laid six miles and eight hundred feet in one day. It was done, he wrote more expansively from Sacramento on August 22, "by no increase of men—but merely by having a full supply of material all day & the Boys putting in for glory even the Chinamen went in for a big day's work & to beat the U. P. big day's work. We commenced at 4 A.M. and quit at 8 P.M., working 16 hours. I gave the Boys a few kegs of lager after they got through. It was a terrible hot day & the gnats or sand flies were present & on duty by the millions."

By then the rails had progressed past the tule swamp of the Humboldt Sink and, swinging northward up the river between the Trininity and Humboldt Ranges toward the great bend, they were closing in on the 288 milepost at the optimistically named Humboldt City, a dusty little mining supply town. Charley Crocker had wired Huntington to send twenty ready-made locomotives "the earliest day practicable." "This question of locomotive power is *the* great question with us now—the ability to get the material to the front," he wrote. "We have proven to our own satisfaction that we can lay it down after it is there."[54] His brother soon echoed him. "If we fail to lay all the track we calculated on," the judge wrote to Huntington, "it will be mainly for want of locomotives." They were working at full capacity—but newly arriving iron had actually begun to pile up on both sides of the Sierra because they could not move it out to the Humboldt Valley fast enough. With a full supply of motive power, Judge Crocker wrote, "we could lay 4 to 5 miles per day as easy as 2 or 3," which had become their average. Huntington was invited to think what they could attain at such a pace. Even so, by the end of August the line would be stretched to three hundred miles. "We have track laying reduced now to one of the exact sciences," Charley boasted, "& can beat the world at it."[55]

Not only were locomotives in short supply—the company still drastically needed men to drive them, along with, in Charley Crocker's words, "first class men as Division Superintendents, Train Masters, Engineers, Conductors &c. &c. The fact is, *we have not the men.*" The year before, when Huntington had recruited George Gray as consulting engineer and Erastus Corning as assistant general superintendent, the extensive experience of both had improved matters greatly. But Corning, retired from the New York Central, was elderly and in poor health. "Every time he exerts himself by going over the road," Charley said, "the jar of the cars & excitement of business affects his spine & he is threatened with paralysis. He has been laid up several times since he came here & is now away at San Francisco resting in order to recover." It was a crucial void to fill—but the organization was shot through with holes. "We had a collision this morning on the other side of the mountains," Crocker noted on August 31, "& smashed up two Engines just for the want of the right man in the right place at the right time. There is no sense or economy in trying to run a Rail Road with incompetent men—first class Rail Road men don't come here of their own volition seeking work." Huntington had to get out and beat some bushes for "men of experience & capacity—energy and force of character who can control & organize others."[56]

The Judge was also ill, slowly mending after his paralytic stroke and under his physician's orders to spend as little time as possible in the office. Typically, he solved this by doing "a good deal of our business" in San Francisco and on the line, as if this was what the doctor meant. True, if he were not in the office, work would not devolve upon him; but he remained busier than he should have been—taking less time to sit down and write his long letters to Huntington in New York, which at least were relaxing, instead seeing to more "crucial" matters. In one letter, though, the Judge found a little time to commiserate with his faraway partner. "You complain of the effect of confinement & overexertion on you," he wrote. "You know something how I feel. I have been trying for 2 months to find time to go to the end of the track & get a little relaxation, but have not found it yet."[57]

During one of his trips to the bay city, Judge Crocker had closed a new deal in their expanding empire by purchasing the majority of stock in the San Francisco and Oakland Railroad Company, paying $261,375. This worked out to 75 cents on the dollar—and it was an even better bargain because the railroad property in Oakland included ferry slips and four ferry steamers.

"Why," Crocker exclaimed to Huntington in a letter on September 10, "one of the steamers is worth a good deal more than $75,000." He moved on into negotiations for the Alameda Railroad with the same seller—and as with the Southern Pacific acquisition, it was all kept a secret; the Associates handed the seller their notes, keeping him as the straw owner, and arranged to rake in the profits themselves. More from other regions were anticipated, and with the Central Pacific's capital stock raised to $100 million by unanimous vote at the annual meeting in midsummer, the Associates began to discreetly seek safe places to park their money; Judge Crocker, for instance, had Huntington arrange a trade of $53,000 in gold coin for $75,000 in 6 percent, thirty-year U. S. savings bonds.[58]

Keeping track of such trades and distributions was of course the job of E. H. Miller, who calculated in the first week of September that Huntington had now sold more than $15 million in first-mortgage bonds—almost too much of a good thing. He warned Judge Crocker, who immediately wrote Huntington. "It seems to me it is time to be a little cautious about the sale of bonds," he said. The Central Pacific was entitled by Congress to sell those bonds pegged to their actual mileage plus 100 miles in advance of the end of track. Under that legislation the company was within $300,000 of reaching the limit—and "this is sailing too close to the wind," the Judge warned. If there were so many eager buyers out there, Huntington had better switch to selling their government bonds; it would be tragic if they connected with the Union Pacific and actually had to refund the government for first-mortgage bonds sold over their limit.

Ed Miller flagged another delicate accounting matter that same week and was instructed to write Huntington directly. "In your statements of account, there have been some amounts entered as loaned to [Senator] Stewart and [Congressman] Axtell," the treasurer wrote dryly. "These I have entered on our books to account against them. Judge Crocker and Mr. Hopkins think their names should not appear on our books, since they seem to have an idea that the loans will never be paid. Is it so? Should the names be taken off our books here?" Knowing the answer, Miller suggested that thereafter Huntington's clerk not enter charges such as "June 26. S. B. Axtell account: $5,000," but rather mark them as "June 26. Expense (or some other account): paid S. B. Axtell $5,000," so that the entry might be given a proper burial in the Sacramento books.[59]

Meanwhile, the tracks reached the great Humboldt Bend and Winnemucca, with the Santa Rosa and Eugene mountain ranges rising north and west and the Sonoma Range, including fabled Auld Lang Syne Peak, standing south, still plentifully seamed with gold and silver and supporting numerous prospecting claims. The place had been settled in 1850 by a French trader and named by emigrants, ever logically, as French Ford, and when mineral strikes transformed it from a wayside stop to a destination in itself, with entrepreneurs dreaming even so far as trying to dig a canal up into the hills, it was renamed Winnemucca after the Paiute warrior chief, son of Chief Truckee, who had beaten off the vengeance-burning force of whites up near Pyramid Lake in 1860. The Central Pacific was keeping its government inspectors busy, with Judge Crocker dispatching twenty-mile certificates every week or two. At Winnemucca the Chinese hammered in their 336th milepost, and soon the Judge was sending off the latest timetable, Sacramento to Winnemucca, for Huntington to see for himself.[60]

By this time matters in Utah had grown to the point that the company needed a full-time agent. Stanford appointed George Gray, who left for Salt Lake City on September 14, promising to send reports frequently. "If he meets with difficulties that require my presence," Stanford assured Huntington, anticipating objections to delegating such an important responsibility, "he will telegraph and I will go immediately." Within days of getting out there, Gray had met with Chauncey West, the Mormon contractor who would actually supervise the work up north; his people had organized into shifts working late into the night by sending men out into the hills to gather sagebrush for bonfires.

It was well that they appreciated the urgency. Gray learned that Brigham Young's subcontractors, Benson, Farr, and West, were nearly done grading in Echo Canyon and were beginning to mass in Weber Canyon. Moreover, he heard that the Union Pacific had sent laborers to work east from Humboldt Wells. "It seems to me foolishness," Stanford wrote Huntington on September 25. "There is no more sense in it as I can see than there would be in our grading east where we know they would be first with their track." Of course Judge Crocker had the year before exuberantly recommended that if need be they should grade in parallel across Utah, expecting in his lawyerly way that they would ultimately prevail and be awarded bonds for their own mileage. If Stanford was truly missing the message and the intent here, it did not matter to Huntington or to the Judge—but it was important for them to know that the companies were now working the same hundred-mile section of desert in Utah from opposite ends; the graders would have to meet, and pass, before too long. Strobridge was instructed to have a man investigate the rumors about Humboldt Wells, and he soon telegraphed that there was an encampment of Union Pacific engineers up there on Nevada soil; the invaders seemed to be waiting for supervision, and Gray had heard that Samuel Reed, Colonel Seymour, and even Dr. Durant were heading west. But the practical Union presence, at least for the moment, was limited at the Wells to only six teams and their drivers, scraping gamely eastward.[61]

However, competition began to flare as hotly as those sagebrush fires up in the Promontory Mountains. Gray dashed off a bulletin to Stanford and Huntington: the Union Pacific had sent operatives "to demoralize our contractors; it requires extraordinary exertions to keep up their courage." Whether actual threats were exchanged at this point he did not record—but Gray did note that "they are offering and contracting 42 & 45 cents [per cubic yard] for earth work right alongside our contractors, while our contractors only get 42 cents. If this is not met by a corresponding advance on your behalf the subcontractors cannot be sustained on our work." Even posting touts to induce the cash-poor Mormon farmers to jump their work for the competitors' was a serious threat at that point; who knew where it would end? Because Chauncey West had assigned nearly all of his line to small, independent teams working at most sections of a mile and a half long, the isolation worked against him.

Within days the bidding worked both ways, often backfiring, causing innumerable headaches for George Gray. After West moved to subcontract a larger portion of his hundred-mile line to a firm called Carter & Hall, which had just completed a grading job in Weber Canyon for the Union Pacific, work was suddenly delayed when creditors attached liens to their tools and teams. Carter & Hall could not pay them because the Union Pacific withheld its pay for the Weber work—undoubtedly to blackmail the firm from going over to the other side. West was forced to buy all the subcontractors' tools and teams himself to keep from losing ground. He would not discover for some months just how deep a hole he was digging for himself.[62]

Gray sent another alarm when he found compelling evidence that the Union Pacific would build south and west of Great Salt Lake instead of going up over the Promontories. "It may possibly be that they intend to build road South of the Lake," the Judge commented to Huntington, "intending to play a sharp game on us—but they will slip up on it, for I don't believe the Sec. of the Interior will go back on that approved line." The sharp game, however, was only in the inventive rumor-mongering going on in Salt Lake City—a Union Pacific campaign to lead the citizens of Salt Lake City into supporting it over the California line, the by-product being to confuse and confound the Central Pacific. In that aspect it was not dissimilar to the one practiced by the Central Pacific only weeks before. A government director of the Union Pacific by the name of T. J. Carter arrived in California and went out on his competitors' line, going so far as to study tracklaying east of Wadsworth. "He seems a very fair & candid man," the judge opined, "but is undoubtedly here as a spy." Carter turned up in Sacramento and San Francisco next, inquiring about Goat Island. Judge Crocker entertained him before turning him over to his brother and Strobridge, who took him back out over the road and filled him with misinformation, that "our material came so slow that we could hardly build to Humboldt Wells by next June," the Judge chortled, "& the roads would probably meet there." Strobridge told Carter that he was so frustrated he wanted to quit; Charley Crocker said he almost agreed. Out in the Forty-Mile Desert, Charley stopped their stage to show Carter the water tankers being hauled out toward the work. "I took him to the stage station," Crocker laughed later, "and gave him a glass of water," which was represented as the water they were forced to use, and "he was unable to drink on account of the alkali, of which it was full. Cattle would not drink it at all. Horses would drink a little of it, but not much." Carter left the desert grade convinced that the Central Pacific would not build more than 150 miles under those conditions. Still later, the Union Pacific director was found to be inquiring if the San Francisco and Oakland Railroad was for sale. The owner kept his silence about his recent under-the-table sale to the Central Pacific, stalling Carter by saying he would write him at the Union Pacific headquarters soon. "Now if they are so anxious to reach San Francisco, & are so rich as Carter says," joked Judge Crocker, "why cannot we sell them the Central Pacific?"[63]

The iron continued to advance up the Humboldt, passing the tiny supply town of Golconda and dodging between the Battle Mountain Range to the south and the Hot Springs Mountains to the north, the high flanks visibly scarred by silver-mine openings. Trying to keep the Chinese legions supplied, numerous freights hauled iron, supplies, and water to where the sweating men labored in smothering white dust near Valmy. In addition, the work required thirty carloads of wood daily, all of it hauled from the Sierra. Regular train service would not be extended past Winnemucca until the rails reached another fifty miles to Reese River. But on September 29, the line passed a legal landmark, the fact immediately wired east to Huntington. At a place some 160 miles shy of Humboldt Wells, they were also now within 300 miles of Monument Point in Utah. The Central Pacific was entitled to ask Secretary Browning for permission to grade east of the point down to Weber Canyon and beyond—up to the mouth of Echo Canyon. As Judge Crocker exulted, "It is a big point gained." With the Union Pacific facing them down at Humboldt Wells and in the Promontories, they could now be really bold and push toward the promise of those coal seams in the Wasatch. At such a time it was all the more urgent that the Central Pacific increase its motive power. Off sped another wire from Charley Crocker to Huntington: they needed more locomotives—and regardless of cost, he should ship them across Panama. "They ought to have been shipped that way 2 months ago," Judge Crocker commented to Huntington, "& then we would have had them at work by this time. Hundreds of tons of freight have accumulated in this city. . . . We would have been within 50 miles of Humboldt Wells before this if we had been able to move our material faster."[64]

There was certainly no doubt about that. "If we only had 20 more big locomotives," the Judge wrote, "we would knock the socks off the U. P."[65]

After she said goodbye to her husband at Salt Lake City late in August, Anne Dodge returned to Council Bluffs by stagecoach to the Union Pacific station at Benton. "The trip home was quite tiresome and terribly dusty as you know," she wrote Grenville Dodge.

As we came past Rawlins Springs Station, the construction cars passed going to Benton—with the Seymours on board, and when we went in to the U. S. Restaurant for breakfast they were there and came in with us. At Benton when I took my sleeping section I drew the curtains down and laid down; pretty soon I heard Seymour and J. D. Casement in conversation, Seymour scolding and C. swearing. The principal subject was the water tanks and the management of water west of Cheyenne giving Evans fits. Seymour asked Casement if he would take charge—which was accepted and S. said he would see the Doctor, &c. You can well imagine it was rich. Seymour hinted that his object in going East was to get the Doctor out here in September and I thought from what he said that the Doctor needed some representations to come out. I fancy there will be lively times this fall.[66]

Dodge was finding it lively enough for his tastes north of Great Salt Lake, where he roamed the Promontory Mountains east and west of Monument Point with his engineers. One of his chief tasks was to locate up the eastern slope of the mountains with lighter work and grades, but he could find no way to improve over a ninety-foot grade and heavy work through limestone outcroppings. All during the last week in August, he followed his crews' stakes, pausing at every difficult section and calculating how to run around it. Regarding the eastern approach, he wrote Durant on August 27, "I have no question but what we have adopted the cheapest lines in a commercial point of view, to be had in that country." Before he could ride westward to begin examining the territory to Humboldt Wells, he received a letter from Jesse Williams, who had been over it carefully. "You are no doubt aware," Williams told Dodge,

of the contracting of 100 miles west of Promontory Point by the C. P. R. R. Co. . . . The entire 100 miles is very light work and if our company do any work west of Monument I should not go west of summit of Toans Pass. Up to that point which is 200 miles west of Weber, work is all very light, beyond that there is some rough work in crossing drainage of table lands. Heavy cuts and fills in gravel with chopped up grades and crooked lines. The C. P. R. R. can without doubt meet us there or even at Monument Point, if they show proper energy—Monument Point is just halfway between the ends of the two tracks. I know nothing of the company's intention and under this new phase of the matter, but if you had the teams, you could cover the 100 miles with graders, even before they could. They have now only about 25 miles located, and that right over the easiest portion. They commenced staking out at Monument Point yesterday. . . . The work on east slope of Promontory will be good winter work, and I suppose there is no hurry about it as the C. P. R. R. abandon everything east of Monu-

ment Point and when they commence to grade will cover our stakes though sometimes we are a mile apart, at other, right on the same ground.[67]

Dodge soon saw the Central Pacific stakes, with an identifying "L" painted on them, threading across the mountains, and he used his old spymaster skills from the Civil War days to find out all he could about Governor Stanford's arrangements in Utah. Stanford's shifting the Central Pacific engineering parties west, to get their lines down for grading teams, was signaling something important, Dodge thought, "virtually acknowledging that [Stanford] would not contend for that ground, and making the debateable ground still west." This showed, he wrote Oliver Ames from his engineering camp at Wilton Springs on September 4, "considerable sagacity" in selecting Monument Point as a place to begin grading, and he guessed "on what ground he will contest his right to build to it, whether he gets there first with his track or not": the point was equidistant between the two ends of track, about 345 miles east and west; Secretary Browning and President Johnson had approved the Central Pacific line to that point; by the time the contractors began grading, the Central Pacific would be within three hundred miles. Led by these deductions to believe that Stanford and his associates had decided to abandon all hope of entering Weber Canyon—which was anything but the case in September 1868—Dodge applied common sense to this new phase of competition across the bare and rocky Utah soil. "Now, it seems to me," he told Ames,

that it is very poor policy for two companies to go on and build or grade two lines over this 100 miles. I could have made an arrangement with Gov. Stanford for our company to grade to a certain point and his to do the same, and the one who got there first with road was to go right on and use the other's grading and that is the sensible way to do it; the lines over this flat, broad country are a mile apart, and an equally good ground. Again, I can see no good ground why the C. P. R. R. Co. cannot reach Monument Point by the time we do; they have less work, less snow to contend with and should we fail to get over Wahsatch, with work before winter, they certainly can reach there as soon as we do.

If Ames wanted the Union Pacific to grade west of Monument Point, Dodge continued, building upon Williams's notes and his own, they should concentrate only on the first, and inexpensive, flat stretch of 120 miles, "mostly like Platte Valley work," which would have the effect of stretching the Union line some 200 miles beyond the mouth of Weber Canyon with only one difficult and costly section on the east slope of the Promontories. The only hazard would be losing any Union grading ruled extraneous should the Central Pacific be awarded that country. Under this plan, Dodge recommended that they agree with the Central Pacific to meet at a place like Red Dome Pass, 112 miles west of Ogden,

> *and thus avoid the possibility of any loss by either party. You do not want to be grading 100 miles of road on an uncertainty and my plan would make a certainty of it. Another thing, you must remember is that the C. P. R. R. are laying as much track per day as we are, and in their letting of work were not grasping, commencing only at a point halfway between the two roads, with the light end of the work with them.*

Thus lulled into thinking that Stanford and Huntington were no longer striving to head them off before they could poke their rails out of Echo Canyon, Dodge moved his camp to the precipitous and broken eastern slope near Promontory Point—"six miles of it," he complained to Ames, "the most stubborn refractory country on our entire line." Give him two weeks to polish it a little, he said, and "I know I can save $50,000, perhaps more if I can only have some time and not be driven to death."[68]

Back home in Council Bluffs, a controversial matter was finally moving a little toward resolution. The issue of the Missouri River railroad bridge had been a sore one among residents of both Omaha and Council Bluffs, stirring no end of debate over its location as well as design. Seymour and Durant had waded into the debate late in the spring, as anyone might have predicted, and they argued for a cheap, low bridge; spring floods, winter ice jams, and a shifting, soft riverbed all argued against Seymour's design, and not until Government Director Jesse Williams interceded was the sensible high bridge design and location adopted. Finally the Union Pacific agreed to construct it—if Council Bluffs would issue $200,000 in bonds, and Omaha $250,000. The bond issue went before the voters that summer, promising an economic boon to the region. Insiders were assured of great benefits; by the time it was approved, Dodge and several business associates in Council Bluffs had bought up most of the land along the Iowa approach to the bridge.[69] Although most of these transactions were accomplished by the end of the summer, Anne Dodge was certain that more could be done, and wrote her husband as excitement about the bridge rose all around her. "There are many strangers coming in this fall," she said.

> *The papers and telegrams this morning say that the bridge contract is let to Boomer of Chicago, one of your friends, I believe, and so if you want to get in it you better hurry home. Can't you take part in that bridge next year and make enough at home to pay for leaving the road? And then we could build our house, too, or get ready for it. I wish you would. The bridge is going to make this place, and next spring property will be high and sell well, and I think right here in the course of next year you can make a fortune if you are*

only free from the R. R. and can give your attention to your own interest instead of the Company's.

In the end, Dodge's multitude of interests connected with the road would serve him far better than if he had left the Union Pacific in the final phase of this, the largest endeavor of his career, to attend to mere real estate transactions in Council Bluffs. Through his associates, the crucial purchases had already been made. And as it would happen, those bridge investments would take a sharp punishment when Durant's excesses prevented the Union Pacific from building the bridge; seeking to get the project moving after a two-year stall, Dodge would go to unseemly lengths, exposing himself to controversy and public embarrassment at home. Thus it is fortunate that in early September 1868 he avoided digging himself a deeper hole and remained with the Union Pacific—though his wife would continue to hope that he would quit. "We have beautiful weather now, cool and pleasant," she wrote, wanting to lure him away from the moonscape of northern Utah, "and I would like to take some rides, so hurry home. I hope you will not be careless in coming through Indian country. The papers now are full of stories about them and I guess there is some trouble. Have your pistols with you and be careful."[70]

Far away from where Dodge read this letter near Red Dome Pass in Utah—back at the old Indian ford on the Platte River, at Plum Creek—some of the young Sioux warriors had slipped away from their tribes riding into exile up in Dakota. They were determined in their retribution for the humiliating loss of the rich hunting lands won in years of battle with the Pawnee and Crow, now ceded—or so they were told by the elders—by baffling marks on paper, and through no decisive fight. At Plum Creek, where the year before a repairer's handcar had been thrown off and most of the whites killed and scalped, the warriors rode down toward the tracks of the Iron Horse. About midnight an eastbound passenger train passed Plum Creek. "The sky was clear," the Union Pacific baggageman C. C. Cope recalled many years later, "with the moon shining bright. Looking out of the side door of the baggage car towards the Platte River we could see quite a band of Sioux Indians headed towards the track, but not close enough to do us any damage." They met a freight train at Wood River, stopping at the switch to notify the engine crew to "look out for the Indians at Plum Creek."

By this time the Sioux had used metal tools they found lying around to pry up the twin rails, bending them backward. They made a pile of wooden ties. To be certain, they shinnied up poles and tore down the telegraph wire, using it to bind the ties together, laboriously knitting the mess around the jutting-up rails. If the train crew was watching out for trouble, they did not see this obstruction as they rushed upon it, throwing the engine and tender high into the air. "The first two cars back of the engine were flat cars loaded with bricks," Cope said later, "and most of them went off the cars as though shot out of a gun. For years after one could see pieces of brick lying along the right of way." When the locomotive came down it crumpled, killing the fireman outright and pinning the engineer, Brooks Bowers, against the red-hot boiler. The head brakeman, Fred Barker, climbed into the wreckage to try to extricate the screaming man, but it was no go and Bowers died in agony. Indians were now closing in on the train, so Barker ran into the dark grassland north of the tracks and did not stop until he had run a mile. Then he lay down in the brush to wait for daylight.

Back in the caboose of the wrecked train, conductor Billy Kinnie and rear brakeman Fred Lewis climbed down onto the gravel roadbed as soon as they saw the Sioux. They carried lighted lanterns beneath their coats to avoid being seen, and high-footed it back down the track to flag a train that was following them. The men boarded the stopped train as it began backing up to Wood River, where word of the wreck was wired to Omaha. When a special train was sent out to the wreck, baggageman C. C. Cope went along, intending to get the remains of his friend Fred Barker back to his family. "I was glad to find him very much alive," Cope recalled.[71]

Back in Utah, Dodge had moved westward from the Promontories, camping at bisecting creeks for a few days and going over the lines through the sagebrush and alkaline hills with his assistants Blickensderfer and Maxwell; on September 9 he noted in his diary that he rode fifty miles and back in a day, a great feat for his horse, certainly. On September 21 the mail pouch carried a letter from Samuel Reed at Echo City. Back in Wyoming, the track had reached the looming gray thousand-foot cliffs, Point of Rocks, eighty-five miles beyond Rawlins and some forty miles before Green River, and the cuts and grades on either side of Green River would be done within three weeks. Farther west, the work to the head of Echo Canyon was progressing well, with a deep cut at the rim of the basin to be finished by October 10, and with the Echo tunnel headings driven sixty feet already. Three large grading parties were on their way to cover the hundred-mile stretch between Humboldt Wells and Monument Point. "My orders are to do the grading on that portion of the road as quick as possible," Reed said. "Shall have five hundred or more teams on the work very soon." There was a greater urgency in the air: Dr. Durant, detained in Chicago for some days because of illness, was now on the move west with his party—coming to take personal charge of business. "Work night and day," Durant had wired. "Put on all the men that can work. The tracklaying will be increased to four miles per day. The great point is to get one hundred miles more track soon as possible. If you are caught on grading you better give up railroading."[72]

There was also a letter from Jesse Williams, now back in New York and pressing his campaign to earmark $3 million for bringing the road "up to a proper standard including more rolling stock, shops, water supply, etc.," as well as better bridges, trestles, grades, and culverts. In submitting his report to the Interior Department, Williams had assumed that Dodge expected the two lines to meet ninety miles west of Weber Canyon at Monument Point. This would, he calculated, give the Union Pacific some 1,110 miles of line. Whether Dr. Durant and Collis Huntington would agree on that was still very much an open question.

Then Williams presciently brought up one of the most farsighted matters posed during this epic project, showing that at least someone was thinking of the long-term consequences of the Pacific Railroad beyond the certain changes to be wreaked by increased traffic and commerce, and settlement, and the curtain coming down on the traditional life of the Plains Indians, all of which had become subjects of comment in the press and in private correspondence by witnesses and participants in this continental enterprise. All along the way, as towns were planted and game driven off, as hillsides and river bottoms were stripped of their trees, as streams were dammed or diverted, radical, unalterable changes had begun to take place. Williams had ridden through the Wasatch Canyons and over the foothill passes, across the grassy herder's country in the eastern Salt Lake Valley and up around the hilly, sagebrush-covered northern shore of the lake. Something in what he saw—and foresaw—nagged at him. "Before you leave that country," he wrote Dodge, "I think it would pay to get the exact level of Salt Lake. If you lay your line near its level you had better keep up pretty well, for I have a theory that when the country gets settled and ditched out and tramped by stock up the numerous valleys, a far greater proportion of the rain and melted snow will run into the Lake than heretofore, and its surface may gradually rise for fifty years." Jesse Williams would be right, and the effects of settlement and exploitation would continue to be felt—all along the line of the Empire Express, and for much longer than fifty years.[73]

30

"A Man for Breakfast Every Morning"

Back in the insufferable month of August in Washington, Orville Hickman Browning had struggled to push as many developing matters off his desk, out of his office, before fleeing the capital for a cooler place. He supervised the removal of the Plains Indian tribes to the upper Missouri, wanly hoping that they would have time to plant corn and construct shelter before the frost; he had responded to telegrams from General Sherman about Indian outrages along the railroad line; through his agents and by keeping abreast of cable traffic, he had kept his eye on General Sheridan out in Kansas, hoping that the peace would keep. He had dealt with sundry matters in the customs office. He attended the president's cabinet meetings and noted the eroding attendance as officials made their excuses and departed. (There was no mention in his diary of the final passing of the disagreeable Thaddeus Stevens, on August 11, a day in which Browning recorded "Nothing of special moment.") He had even presided over the emptying of his house on Georgetown Heights and the transferral of all his furniture to a rented house on I Street, fondly anticipating a return to private life after the next president took office.

Before he could leave for vacation with his family, however, there was the matter brought up by his emissary, Union Pacific government director Jesse Williams, about the makeshift and corner-cutting work being perpetrated across Wyoming Territory, with no objections being raised by the lackluster government inspectors. On Friday, August 14, in the cabinet meeting, when Secretary Browning conveyed Williams's misgivings, it was agreed to ask Attorney General Evarts if they might withhold bonds in reserve.

No further action was taken, although cabinet members agreed that a reserve fund was only prudent. On August 26 Browning, his wife, daughter, and a servant departed for White Sulphur Springs, Virginia, for a five-week rest. At a welcoming reception in the resort's ballroom, Browning met a number of former Confederate leaders, notably General Robert E. Lee. "He spoke of the crippled, impoverished and distressed condition of the South," Browning would record, "and expressed his earnest wish for the restoration of peace and fraternity, and the revival of industry." Several nights later, when outlying cottages in the resort mysteriously went up in flames, Browning and all the other guests—northern and southern men alike—joined in the bucket brigades to save the main structure. It seemed then, and in the days following as tinges of autumn began to be felt, as if all was in accord. He felt positively agreeable.

By Tuesday, October 6, Browning was back in the capital, in a cabinet meeting with all present but the attorney general. A new twenty-mile section of the Union Pacific line had been approved and submitted by the commissioners, but the cabinet was now alert and suspicious. News of the power struggles within the company, and of Dodge's threatened resignation, were as impossible to ignore as the bad reports of Jesse Williams. The cabinet decided to hold up the government bonds until a special board of three engineers inspected the whole line; Browning was also to order the government commissioners to thoroughly examine the company books—especially "whether the Company had declared and divided dividends, and if so to what amount, and from what sources the dividends were derived." Before the cabinet adjourned for the day, Browning noted that he had approved reports on the two last twenty-mile sections of the Central Pacific given to him by Mr. Huntington, "that road being in no default," he would say, "so far as I am advised."[1]

On that day, October 6, President Leland Stanford put pen to paper in Sacramento to write Secretary Browning directly, notifying him that the Central Pacific had laid iron and strung telegraph wire to the place in Nevada within three hundred miles of the easterly end of the line approved by Browning at Monument Point. Stanford thereby requested, in language so diplomatic as to obscure the boldness of the plan of Collis Huntington, that the accepted line be extended—past the mouth of Weber Canyon (where Browning might expect the Central Pacific to petition, at a natural meeting point in Ogden), past the mouth of Echo Creek (where Judge Crocker thought they had a chance), indeed all the way up to the head of Echo Canyon. This, Stanford explained smoothly, would allow them to prepare for and then commence tracklaying over the winter months—to eliminate delays and keep the laborers from disbanding, and (of course) to shorten the time of completion of the Pacific railroad.

He sent the letter off to Huntington, who was in New York, preparing to storm the castle of Orville Hickman Browning, and who most impatiently wanted Stanford back in Utah as soon as possible. "Some one of *us*," he had written Hopkins on September 14, "must be there soon after the 1st of October to take possession of the line . . . up to the west end of the tunnel" of Mr. Durant's railroad, before they broke through or hastily ran their track around it.[2] In Huntington's mind, as soon as he had persuaded Secretary Browning to approve their new line, he would send a code telegram to Stanford in Salt Lake

City: "Go and see him." This would spur the governor to unleash Mormon grading crews all along the length of the two Wasatch canyons, ignoring the all-but-completed Union grades. Possession, and influence in Washington, would do the trick—at least that is how Huntington envisioned it.[3]

But despite urgings Stanford was still in Sacramento on October 6. He would postpone his trip, in fact, for another two weeks, explaining to Huntington that he had concluded that the Union people had apparently given up hope of reaching the mouth of Weber Canyon "before next spring" and by spreading rumors and sending their men out to Humboldt Wells were only trying to intimidate and embarrass them. Everything was well in hand out in Utah. "Brigham Young assured me confidentially that he would stand by our contractors," he told Huntington. "I think I got pretty close to him and that Durant will find it difficult if not impossible to change him."[4]

Out in the City of the Saints, however, there was some disagreement about this. "I am trying to do all [that] can be done here," a frustrated George Gray wrote Huntington on October 10, "but fear we are too late in the field to meet our own desires or your intentions." Had his advice to Stanford been followed back in June when the governor had made his brief appearance in Utah—Gray had urged him to press Young for a vigorous response, and to ride the contractors to work their utmost—then "the grading would now have been completed from Humboldt Wells to Monument Point. Brigham Young would [have] put on 3,000 men as he agreed to do, but the plan was broken upon return to Sacramento by whom I won't say but I can guess." Of course he meant Stanford who, by inaction, had shattered the schedule. "When the Gov. returned here the second time" in August, he continued, "Brigham had other views—Union Pacific had presented him with $10,000 stock free, & made him a director & promised him all the work he could do." Now they would be extremely fortunate, Gray angrily finished, if they reached Monument Point.[5]

The Mormon contractors were at least making progress on the hundred-mile contract west of Monument Point, he said, despite pressures from the Union Pacific's competitors, including a higher pay scale, and in spite of a continuing shortage of tools. The road would follow a gradual downgrade for some fifty miles to the northwest corner of Great Salt Lake, from which it would move over barren alkali flats and salt beds. Gray had heard that the contractors' supply wagons commissioned by James Strobridge in Nevada had been so heavily laden that they had cracked and broken every bridge they had crossed, and they had repeatedly sunk up to their axles in the alkaline bottoms, but they were expected that day.

Meanwhile, the locating engineer Lewis Clement had just wired that he would soon have another seventy miles ready for the contractors, once they were finished their present work. Gray was feeling the pressure. "Durant, Col. Seymour, Reed, Gen. Dodge &c. &c. arrived here two days ago," and they were scrambling around for new contractors' parties, he wired both Stanford and Huntington in cipher. And Durant wanted to get graders on the stretch from Weber Canyon up to Monument Point; Brigham Young had agreed to take the Union Pacific job but had expressed reservations about fulfilling it in time. So Durant was going to fill out the work with a Gentile force. Gray also learned that Reed and Dodge were trying to convince the doctor, with little success, not to grade parallel to the Central Pacific lines north and northwest of the lake.[6]

Apparently Dodge and Reed budged him a little. The next day, October 11, when the Doctor approached the Central Pacific officers, seeming almost conciliatory, his manner instantly put all on guard. Durant sent a message to George Gray for forwarding to Stanford and Huntington by telegraph, in which he proposed that the two companies avoid parallel grading over the one hundred miles between Monument Point and Humboldt Wells. Durant would grade—"and let either party pay at the average cost for as much as they lay the Iron on," he said, "or each party if preferred to grade an equal portion of the distance." He gave the Central people three days to respond.[7]

Telegraph traffic between Judge Crocker and Huntington was thick and expensive in the wake of the proposal, and subsequent letters filled out the debate. "We studied it over carefully," Crocker said, "& came to the conclusion that D. found himself in a bad fix, especially in pushing his graders out to Humboldt Wells. It is expensive to him, & he must feel sure that he can never lay Iron on it until after the connection is made." Durant's motive was to just halt them in Nevada, they decided, or somewhere up in the unappetizing northwest corner of Utah, far away from Salt Lake City business. "We are not to be caught that way," Crocker wrote. "If D. sees fit to grade parallel to us for 100 or 150 miles, why he is throwing away the money of his Company." There was no way that he could lay track on their side of Monument Point before the Central Pacific reached there. Their success lay, the Judge wrote, "in laying track as fast & as far as we can, & devoting our whole energies to that." The Wasatch canyons were unoccupied for the present, as far as they were concerned. They would let Durant's proposal expire—unanswered.[8]

The pressure in Utah, of course, was rising palpably, and George Gray felt it keenly. His own conviction was to add even more force to what Judge Crocker advocated. "Our only hope," Gray wired Huntington, "is now to push on the Iron rapidly and occupy any road bed we find east of Humboldt Wells, whether graded by Central or any other party."[9]

Such rashness and purposefulness had its mirror in New York, where Huntington was trying to meet an alarming, temporary slump in bond sales by ordering Fisk & Hatch to pay par for all Central Pacific bonds that appeared on the market—an open-ended order which could greatly increase their indebtedness. On October 13, in a mood of high urgency, he took the train down to Washington with a route map and profiles, based on Butler Ives's preliminary surveys in Utah and elaborately doctored to seem grounded on much more extensive planning and measurement. Huntington had also rewritten Stanford's letter to Browning, puffing a little more air into the company's claims. The Central Pacific, he said, was logging three new miles each day in Nevada; tracks had extended to Reese River, and when they had advanced to Humboldt Wells and moved onto the Mormons' gradient, they would go ahead with even greater momentum. "I think it quite safe to say," he wrote, "that the CP will be in operation to Salt Lake within 80 days from this date."

In reality, no one in the company—not even Huntington in his wildest fantasy—thought they could reach Monument Point by the middle of January in 1869. And the tracklayers were lucky in those awkward, frustrating autumn weeks if they managed two miles per day. And even if they had not spiked up to where the Reese flowed into the Humboldt, service to the station Charley Crocker would call Argenta would open in a few weeks' time. Great and small fibs spun out pridefully; as if patting and stroking the fraudulent maps, a sanguine Huntington finished his letter. "If the Honorable Secretary will approve the location as per the maps here presented, so that the company can get men and materials on the heavy work before winter sets in, the road will be completed to Echo Summit by the 1st of June."[10]

When they met, Huntington lost no opportunity to impress the secretary with the high character and quality of the Central Pacific constructions—and the shoddy, temporary nature of the Union Pacific's. Browning told Huntington that he intended to carry the portfolio into a cabinet meeting on Friday, October 16. He knew that the question of acceptance of the railroad location was his own decision. But it was an unsettled month, with many minds on the looming presidential election, and now there was this problem over the Union Pacific's fiscal and physical fitness—it seemed proper and prudent to include the cabinet. It might be in Huntington's interest, he said, to call on a few of the other secretaries, such as Hugh McCulloch of the Treasury and Alexander Randall of the Post Office. And his former law partner, Thomas Ewing, one of the most influential men in Washington and still collecting a retainer as Huntington's adviser, would agree to go see President Johnson. "What form their arguments took," says Huntington's biographer, David Lavender, "is anybody's guess."[11]

No clear opinion emerged on October 16 when the matter was first brought up in the cabinet meeting. Huntington had not made all of his rounds yet. His days in Washington were truncated because he did not want it discovered by Durant's spies that he was paying court to the cabinet; for that week he went home every night to New York City and made himself seen the next morning at his office before slipping out to make his Washington train. Browning noted at the meeting that Attorney General Evarts had still not returned to the city to give advice over the Union Pacific reserve fund, let alone this new matter. The cabinet officers present briefly considered appointing an engineer to end any possible controversy about a meeting point for the two railroads—but then they discarded the notion as creating an inordinate delay. "If the Cabinet shall not reach a conclusion at its next meeting," Browning wrote in his diary later that day, "I will decide the question myself."[12] Over the next several days, Huntington and Ewing did what needed to be done to ease the decision-making process.

Meanwhile, clerks in the Interior Department received and processed a packet from President Oliver Ames of the Union Pacific Railroad and sent it upstairs to Browning. "The Road is completed for 847 miles (near Green River) and grading and masonry will be completed (except for tunneling at head of Echo Canon) to Salt Lake in November," Ames had written to Browning on October 15. "The location of the line has been definitely made to Humboldt Wells."[13]

Waiting to hear some shred of news from Huntington, Judge Crocker wired him on Monday, October 19, to report that Stanford had finally left Sacramento for Salt Lake City, taking Chief Engineer Samuel Montague with him. Hopkins, who had gone out on the line, stopped at Reno and dispatched new maps and profiles, a commissioners' report on the sections between miles 350 and 370, and affidavits for government bonds.[14]

The next day in Washington, the cabinet devoted much of its meeting to railroad business. Secretary Browning blandly announced that he was accepting a twenty-mile section of the Union Pacific, which completed the line's eight-hundredth mile, and was releasing those bonds; he also ordered the government commissioners to examine and report on the next twenty-mile section. But the main discussion, with all participating, regarded the Central Pacific's location to Echo Summit, and the conclusion was unanimous: "the location should be accepted so that the work might progress this winter," Browning noted, adding that the company had asserted it would reach Monument Point "with its completed road, in 80 days from this time."[15] For fairness' sake he would appoint a three-man commission for the Central Pacific to look over construction in California and Nevada.[16]

The progress report from Oliver Ames, with its petition for acceptance of its line from the Wasatch canyons to Humboldt Wells, was quietly filed and seemingly forgotten. And Oliver Hickman Browning agreed—at Huntington's insistent urgings—to delay telling Ames or Durant of his decision for as long as possible.[17] Gleefully, Huntington went back to New York. *"I did it,"* he wrote Crocker. "*The line from Humboldt Wells to Echo Summit* is approved and is the legal line for the road to be built on and a road built outside of it will get no Government bonds." He trusted that "it will not take long" to "have men on the whole line between the end of the completed line and the tunnel." In afterthought he added the admonition that the whole scheme might fall apart if they didn't actually get inside the three-hundred-mile distance from the Central Pacific end of track to the tunnel heading at the head of Echo Canyon. "By

God, Charley, you must work as man has never worked before," he scrawled. "Our salvation is in you. Let out another link. Yes, a couple, and let her run. Don't mind Durant's blowing. *Say nothing, lay the iron on the approved line.*"[18]

But the most dramatic messages he would send were the shortest, tapped out on the spot in Washington on October 19 when Browning had told him the good news. "Cover line to Echo and hold it!" he wired to Judge Crocker. "Go and see him!" he wired Stanford within moments, the telegram speeding out over the wires toward Salt Lake City.

The governor's carriage would not roll into that place until October 31.[19] While George Gray waited tensely for him in Salt Lake City, he could not help but take some pleasure from hearing that Dr. Durant, who was buzzing angrily between the lake and Echo Canyon, was suffering more—over the sudden bright light of scrutiny being trained on the Union Pacific's business and construction by the federal government. "Durant is in a towering passion over the new commissioners," he wrote Huntington on October 23, "and thinks it unfair the other roads are not included, particularly ours. Let him dare, but in this madness they are building railroad very fast—beating us very badly. I think if this good weather continues to hold the Union will get the rails laid to 'mouth of Weber Canon' by the 1st of January. The grading will all be done by that time, *sure,* and they are laying and bedding cross ties on every foot of bank, fast as it is ready for them." He could not resist casting a little blame. "This is what I feared and predicted when I returned from this place last June," he continued, "yet Charlie Crocker laughed at me & said it was nonsense to put the Mormons at Humboldt Wells grading as he could grade it himself before a force could get on the ground. So much for over confidence."

Gray had heard that Stanford was finally on the way. "I hope he will come soon," he said, "as there are some very important things which need the attention of a person with power to act. I know something can be done to stop the Mormon grading at the Wells, but it can't be done effectually unless done promptly. I have not the power to act. I could buy the contractors over to us for I think $50,000 bonus. . . . The Union Co. could not put another force there before we should have the Iron laid."[20] All Gray could do was keep his ear to the ground to listen for what Durant would do next while he hoped, day by day, to hear the sound of Governor Stanford's footsteps. Then, surely, Gray's bold plans would be adopted, and they could respond to the Union Pacific's challenges in quick and inventive ways. And so Gray waited, and listened.

Grenville Dodge spent the first two weeks of October examining location lines west of the Promontories and then inspecting the work in the eastern canyons. His meetings with Durant in Salt Lake City and Ogden had showed that the Doctor was still obdurate about grading to Humboldt Wells regardless of the waste, and so Dodge grumpily rode across the Wasatch past a number of noisy graders' camps, scowled at the slow, disorganized progress of bridges at Black's Fork and Ham's Fork, and finally came down to the raw new depot town of Bryan and the end of track, fourteen miles of sagebrush, rabbit brush, and sand flats west of Green River. Work on machine shops and a roundhouse was progressing, at least, and the dusty place bustled with stagecoach and wagon traffic. Colonel Seymour was there, ostensibly seeing to company buildings and watering facilities while attending to his town lots. He airily told Dodge that he had no doubt that Congress would rip up the special government commissioners' report, "whatever it might be." Dodge's response went unrecorded. He climbed aboard a Union Pacific coach and rattled eastward across Wyoming and Nebraska toward Council Bluffs and a few days' visit with his family. He would check off his absentee ballot for General Grant and Speaker Colfax, of course, and exult when his sponsor was elected president on November 3 despite a slim popular majority, but with a healthy plurality in the electoral college. The Fourteenth Amendment, adopted in July and granting citizenship to African Americans, was clearly seen as the thing that had pushed Grant over the top, with over seven hundred thousand new voters going for him.[21]

Dodge celebrated the election back out west on a special inspectors' train—the examination was ready to begin, and the chief engineer had to be with the special commissioners every inch of the way. He and the other company officials had, by the time he had reached Council Bluffs, learned the identity of the special commissioners. Jacob Blickensderfer, asked by Jesse Williams and Browning to take a leave of absence from the Union Pacific to become a government watchdog once again, would provide the kind of engineering expertise that reassured Grenville Dodge and infuriated Dr. Durant; the two others were also trained engineers who had coincidentally both been commanders at Little Round Top at Gettysburg: General James Barnes, who had been the oldest and perhaps least effective senior officer at Gettysburg, and General Gouverneur Kemble Warren, another military foot-dragger, who had surveyed in the West before the war in the Army Topographical Corps. "I understand their instructions," wrote Oliver Ames to Dodge on October 19, "embrace not only the examination of the road, but they also have authorized them to decide the location west of Salt Lake, and fix the line on which the two roads shall run, so that no grading may be lost."

If this was true, no wonder Durant was fuming! "Blickensderfer knows these lines and is just the man to decide this question properly," Ames went on, adding that he hoped Dodge would make a favorable impression. He did not have to say, in afterthought, what was obvious to Dodge: they wanted the line located far enough west of the lake "to take in the Oregon branch. . . . You will have them with you a number of days and can show them your superior line and they must come to the conclusion that our line is the best and adopt it." It was very important that they make an early report, he said. "Our bonds are being withheld for the report of this commission and we need them for the work on the road, and are now raising all the money being used from private sources—which can't hold out long."[22]

He received other words of caution and urgency from Jesse Williams, who asked him "to note carefully any changes in location or grade from your final location" while out with the special commissioners. "I am apprehensive," Williams explained, "that without any formal and reported changes the constructing engineers in the hard cuts will quietly change the grade so as to increase it above the maximum on that particular division of road. In returning from Salt Lake, I found that the grade on the summit west of Twin Lakes had been increased to 62 feet with some changes at other points. So at Rattlesnake Summit the grade had been made 66 instead of 65 feet." This sort of change would be more than simply irritating or embarrassing; it would put the senior engineers and the responsible company officers into hot water after the ones responsible were long gone, their pockets stuffed with unearned cash, and it would likely result in their having to do the work all over again. Williams was not happy about his current assignment, stuck in New York examining the books of the Union Pacific for malfeasance, feeling bored, unappreciated, and demeaned, and earning nothing but resentful looks from the office people: "I have no other calling," he complained to Dodge, "that amounts to a row of pins."[23]

The Doctor was still nipping at the heels of his contractors between Echo Canyon and Salt Lake City, urging them to greater speed, when he briefly entertained a novel and ironic proposition from California. It came from O. D. Lambard, a Sacramento businessman who always found himself competing with the Central Pacific, and Daniel W. Strong of Dutch Flat, whose connections to the Central Pacific had remained severed since the death of his partner and friend Theodore Judah, and his own falling-out with Stanford and the board. Lambard went to Salt Lake City to see Durant, and after some negotiation he extracted a contract, dated October 10, from the Doctor:

> *I authorize D. W. Strong, of Dutch Flat, to contract with Ab Him or Ah Coon, or any responsible party, on the part of the Union Pacific Railroad Company, for any number of Chinamen, not exceeding two thousand, to be delivered to our agent at the terminus of the Central Pacific Railroad Company's track on the following terms of agreement, viz:*
>
> *The Union Pacific Railroad Company to furnish, free of expense, necessary transportation for tents, cooking utensils, and camp fixings; to furnish them with work on this road east of Humboldt Wells, and to pay them in United States gold coin $40 per month for twenty-six working days, and allow their pay to commence from the time they report themselves to our*

agent. The railroad company to protect them from Indians with soldiers if necessary.

To be used on receipt of telegram from me.

THOS. C. DURANT, *Vice-President*

Strong could have used the money, and Lambard, who ran up considerable expenses waiting to see Durant, waited for some weeks more for the deal to be closed. "This is hell for me," Lambard complained at one point to Strong, "for I shall nearly die in this stupid town." But the confirming telegram from "the big Indian of the U. P.," as he called Durant, never came—local talent from around the Salt Lake basin, augmented with imports from the Wyoming camps, would satisfy the Union Pacific's needs.[24]

As the Casements' tracklayers worked their way toward the rise of the Wasatch and the Utah border, the Irishmen struck a new record for a day's work: on October 26, they put down seven miles and 1,940 feet of iron, being rewarded by triple pay by the Doctor and congratulated by Oliver Ames, who wrote that it was "the achievement of the year"; telegrams of a like nature were read to an uproarious throng of celebrating workers.[25]

A similar cheerful mood (though not watered by spirits inbibed by the tracklayers) affected the Mormon graders working over the rim of the Wasatch and down the head of Echo Canyon. W. C. A. Smoot, a farmboy whose father was away on Mormon missionary work in the southern states, had taken a job with the contractors for the ready cash that would help his family. "Our camp was made up of a far different class of workmen" than the railroaders, he recalled many years later. "The Utah work, east of Ogden, was done by Mormons, who were law-abiding, conscientious and peace-loving men. They rested from all labors on the Sabbath Day and heeded religious worship. And they also asked a blessing upon their food and closed each day with singing and prayer." One of the campfire songs, "which was often sung by three hundred lusty voices after a day of hard work," he preserved:

At the head of great Echo, the railways begun
The Mormons are cutting and grading like fun
They say they'll stick to it 'till it is complete
When friends and relations they're longing to meet.

CHORUS:
Hurrah, hurrah, the railroad begun,
Three cheers for contractor, his name's Brigham Young.
Hurrah, hurrah, we're honest and true.
And if we stick to it, it's bound to go through.

Not far from Smoot's camp was a large camp of some five hundred Gentile graders, mostly Irish. "They were good workers and a jolly bunch of men," Smoot recalled, "but often got the worse from drinking bad whiskey and when they laid off work for a good old Irish wake, they sure let everybody know it for many miles around." He recounted the story of when a well-liked Irish gang boss died and the camp proceeded to give him a first-class send-off. The deceased was placed in a carelessly made coffin, which they loaded appropriately onto one of their two-wheeled dump carts, drawn by "a big Missouri mule." The coffin was much too long for the little cart. "After some ceremony and much drinking of bad whiskey a long procession was formed," he wrote,

> *the mule cart and body taking the lead, and a start was made for the cemetery. It was a lonely spot about two miles away from camp, over undulating hills, covered with stunted cedars and sage brush, where the man was to be buried. When the grave was reached, further ceremonies and much drinking were indulged in. While all this noise and confusion was going on, the mule, growing impatient, started for camp across country. Meantime dissension arose among the boys as to who should speak, offer toasts, etc., etc., which precipitated a free-for-all fight. The sun went down, the mule made camp with the empty cart. The corpse was missing and no one knew whither he had gone.*

Several days later, very late at night, young Smoot was out riding to bring in stray horses, when he unexpectedly came upon the coffin at the bottom of a three-hundred-foot cliff, where it had come to rest. "My frightened saddle horse threw me to the ground and bolted for camp five miles away," he recalled. "This left me alone with the corpse, but I was not long in parting company with him because the coyotes which infested that lonely region, commenced to howl. It was several days before the funeral was completed."[26]

These doings, at least for the present, were rather mild compared to things across the line in Wyoming. Back in Laramie, lawlessness had reigned for some five months after the town's provisional government had been chased out of town. H. W. Guy, a railroad carpenter, was building the mechanic shops and quarters then, and recalled it as a wild and alarming time. "We built rooms in the upper part of the shop" for the family of the master mechanic, he remembered; when they had finished, the master's wife came out on an evening train from the East. The next morning, "she looked out of the window and saw a man hanging by the neck at the old log house across the tracks." (Instead of leaving on the next train she stayed, and before the summer was over organized the first Sunday school in Laramie.) In August a committee of perhaps twenty shopkeepers and railroad workers took a first violent step toward order and abducted and lynched an outlaw who called himself "The Kid." In retaliation, his gang members took over the town. By October the number of vigilantes had swelled to five hundred—and they were ready to declare war. The battle plan called for simultaneous raids on several outlaw dens. On October 29 it began badly, due to miscalculation and mischance—but the vigilantes rallied outside the notorious saloon Belle of the West and, during a pitched gun battle involving nearly every resident of the town, five men were killed and fifteen wounded. Four desperadoes surrendered and were strung up from telegraph poles. When the Vigilance Committee had run out the gamblers and robbers from Laramie, H. W. Guy recalled that it was a thrilling night, "but it cooked them."[27]

By then all that was left of the railroad town of Benton was dust, debris, and the graveyard, but the miscreants fleeing Laramie had other goals down the track. Struggles between outlaws and the authorities were common. H. Clark Brown, a Union Pacific bridge carpenter from Iowa, was working west of Green River in the fall of 1868 and stepped off a train intending to get something to eat at the roadhouse. A tent city had sprung up not far away, "perhaps the roughest place on the railroad at that time," Brown said many years later. "There was an officer with a man by his side that had been arrested," he recalled. "They were about ten feet in front of us when the man with the officer said he wanted to go in to wash and he started for the door. The officer told him to come with him. He did not do so and the officer drew his revolver and shot him."[28] "Every town had its share of crime," recalled Robert Lardin Fulton, then a young train dispatcher,

> *but Bryan had a surplus. Hundreds of Wyoming's worst citizens congregated there and it was one of the bad places of the system. More murders were committed there on less provocation than at any place of the same size. In one week there was found on the street, what the slang described as "A man for breakfast" every morning for five days in succession. One ruffian stabbed a companion within a few feet of the spot where a friend and I were passing. We rushed up to stop the assault, but too late. As he was seized the murderer whined, "I was forced to do it, gentlemen. He called me a vile name."*

"Human life was held as cheaply as on a battlefield," Fulton said, "with tragedy so common that death seemed almost a matter of course. Desperadoes fought at a word, almost at a look, and were accountable to no one. Whisky or opium, one or both, caused most men's downfall, and were the immediate cause of most of the crimes." More than fifty years after the fact, Fulton wrote:

> *I heard a death yell once and only once, but I can never forget it. Half a block away a noisy quarrel was in progress one quiet evening, when a blaze a foot long flew out of the mouth of a revolver, apparently a big one. It was a fatal*

shot and the stricken man let out a yell, half squeal, half roar, so filled with terror that it seemed hardly human. It was a negro who had had words with a white man off the grade, drew a razor and cut the latter across the wrist. Of course the man was "heeled," everybody was, and drawing a big Colt's navy he fired at short range, dropping the negro in his tracks.[29]

Ammunition was expended on the greater level of the Laramie battle at another quickly appearing Hell on Wheels town, Bear River City, or "Beartown," which sprang up in forty-eight hours in the red hills bordering the riverbank, along a tributary aptly called Sulphur Creek. A grader named Theodore Haswell watched the place fill up, just as he had seen back at Benton, "with saloons, gambling places and questionable characters of both sexes." Entrepreneurs quickly hammered together shanties and structures of only slightly more permanance than the dirty white canvas tents, which as usual, predominated. An itinerant newspaperman named Leigh Freeman, a rabidly Democractic Southerner, established a sheet called the *Frontier Index*, which railed at Republicans, foreigners, Indians, and outlaws in equally vitriolic terms, encouraging vigilantes to take matters into their own hands. Most people believed Freeman was a vigilante leader—and he seemed like a likely target of the lawless element. Haswell was in the streets on the day a man who had just committed a murder was hanged by the vigilantes. "This greatly incensed his friends at a nearby grading camp," Haswell wrote.

They formed a mob of about 200 armed men and attempted to annihilate the town. They succeeded in burning the [Frontier Index] *printing office and set fire to other buildings. Twenty-four of us gathered in a storeroom for defense and were quickly surrounded by the mob. One of our group, who was acquainted with the most of the men in the mob, went out and attempted to reason with them, but returned shortly, convinced that it would be impossible to accomplish anything in this way. Just as he returned, a shot was fired from the mob and he fell dead. Then the mob opened fire on our group. Smith, the leader of the mob, with a revolver in each hand, ran up to the front of the storeroom we were in, and emptied both guns at us. Then we opened fire. Smith was severely wounded and five of the mob fell dead, and others were wounded. Almost immediately, the mob withdrew. The man who was killed was the only casualty in our group. With nearly every window of the building shot out and the woodwork torn with bullets, it seemed an act of Providence that we were not all killed.*

A company of troops from Fort Bridger dispatched to Bear River City arrived about twelve hours after the gun battle and imposed martial law. Editor Freeman, whose newspaper had so dramatically helped to raise the level of passion in town, had escaped on horseback into the Wasatch Mountains, going on to similar pursuits in Utah. "Enough dead bodies were gathered up in the street," recalled train dispatcher Robert L. Fulton, "to fill quite a graveyard on the hillside, where most of them still remain." Before the soldiers left, the Union Pacific tracks arrived, passing by the Hell on Wheels town without acknowledging its existence with station, siding, or slow-down sign, and within a week or two Bear River City was as gone and forgotten as Benton.[30]

The noise of the tracklayers' and gamblers' camps at Granger may or may not have been loud enough to disturb the slumber of General James Barnes's special government commissioners, aboard their sleeping car on a siding on the night of October 30, but the next day, accompanied by the watchful Grenville Dodge, they began their inspection trip eastward. "I am glad that you are on hand to escort Barnes out," Oliver Ames had written him. "We did what we could here to impress him favorably in regard to the road, and I feel sure that you can satisfy him that everything on the road has been done fully up to the generality of new roads. Barnes says he does not expect or desire that we should take out the trestle work we have until it becomes unsafe, but it may be proper to lay aside a sum sufficient to make the culverts and piers of bridges of stone, and generally to make it a first class road as rapidly as the decay of our wood work requires."

Judging from such preexamination lenience, Ames had prepared Barnes well. And as the train worked its slow way across the desert and the Continental Divide and out toward the Medicine Bow Mountains in the east, Dodge found nearly as many faults as Barnes and Warren, and even Blickensderfer, who had reason to be suspicious before this trip, was surprised—locations and track curvature had been changed and grades made heavier than ordered, sometimes, apparently, by lazy or cost-cutting subcontractors, but more than once by Seymour or Durant.

At one place, between the Medicine Bow and Lookout stations, Seymour had ordered some twenty miles of extra curves put in, which at $48,000 in government mortgage bonds would give Durant and the other main contractors nearly a million dollars. Barnes and Warren wanted to blame Dodge, as chief engineer, for the boondoggle, but Blickensderfer had a copy of the originally located line, approved and signed by Dodge. It took both of them to argue that the graders had followed Seymour's lead while Dodge was in Utah, and when he discovered the change and protested, Durant and the contractors had persuaded the board of directors, wrongfully, that any change at that point would cost the company three months of progress. They moved on, the examiners grumbling that for subsidies of $48,000 per mile, the company could have hewed a straighter course. There were more problems on the Laramie Plains and in the hills, and indeed all the way to the Platte Valley. They "did not seem to be satisfied," Dodge recorded. He did all he could to minimize the seriousness of most of the defects and convince them that the major flaws could easily be corrected. "What report that would make on it," he wrote, "I have no idea."[31]

No matter how he tried to see things from Huntington's perspective, Mark Hopkins found himself in firm disagreement with his longtime partner over the Wasatch canyons and parallel grading. By nightfall on October 23 he had been wired that the end of track was now 388 miles from Sacramento—the Chinese had spiked 14 miles in five consecutive days, and in three more days would make that record 23¼ miles—but at their best they were not managing more than 60 miles per month, and winter was fast approaching. Each successive gain would give them more to worry about in transporting material, wood, and supplies. Parallel grading in Weber and Echo Canyons made sense only if they could lay their track to the mouth of Weber—"otherwise," he wrote Huntington, "it appears to me a doubtful expenditure."

Learning from Gray that the Union graders were almost down to the bottom of Weber, Hopkins said the best he thought they could hope for was grading and laying track toward Salt Lake with a punishing winter falling on the Union Pacific's heads in the canyons. "I do not feel so confident of being able to do it as Charley & the Judge express themselves." But even Judge Crocker was feeling doubtful. "You tell us to '*cover* the line approved by the Sec. to tunnel in Echo Canon & *hold* it,'" he wrote Huntington on October 27. "A thing easy said but not easy done." He was clear that it was best to put a thick force of men from the Nevada border to the north end of the Lake, and in fact even farther, to the place on the northeast bank of the lake where Bear River emptied in, and Crocker could even see the merit of grading to Ogden at the mouth of Weber Canyon. But, he continued, "hold it" to the head of Echo? "Do you mean that we must drive off Brigham Young's men that are at work along there? I think not." Should they grade parallel in the canyons? "I would say yes," he wrote, but only if they could lay down iron first. Whoever put down tracks had the odds of getting the government bonds, he thought—regardless of Huntington's coup in the Interior Department.[32]

Charley Crocker, on whose thick neck lay the weight of the construction gangs and every locomotive and platform car he wished they had, was feeling very doubtful. "You do not seem to realize the great need of Engines & Cars we are laboring under," he complained to Huntington after his partner wrote to remonstrate that he could not understand why they laid track so slowly. "Our Engines are kept running constantly with double sets of men day & night & don't get cold until they have to go into the shop for repairs—& we cannot spare them from the road long enough for thorough repairs." The day before, he had gone over to the city wharf where three freighters had just tied up with thirteen locomotives on board. He bribed the stevedores to take them out ahead of all the other receivers. But they were all the ready-made small-cylinder engines Huntington had snapped up at a "bargain" and was continuing to ship—they were much less efficient at burning wood, and besides, one larger-cylinder ten-wheeler was worth any two of what Huntington was sending.

As to matters in Utah, Charley Crocker said he agreed with Huntington about the urgency and the lofty goals,

> *but there is no use talking. Stanford* will not *stay long or go away willingly from his Wife. If it was Washington, N.Y., London or Paris, all would be right; he would go immediately & stay indefinitely. As to work he absolutely succeeds in doing nothing as near as a man can. He spends an hour or two per day at the office if we send for him—& acts only in an advisory capacity, never will take charge of any work or power, any responsibility or act independently. His plea is "don't understand."*[33]

With Stanford gone something was unleashed in his three California partners. Their complaints—none of which carried a hope of a solution—poured out to Huntington. Judge Crocker said that he had been talking to Lloyd Tevis and Horace Carpentier, their partners in (among other things) the Oakland Waterfront Company, and he was told that Stanford had carelessly promised them an interest in Goat Island if and when Huntington procured it from Congress. "I am afraid he had some loose talk of the kind," the Judge said, "which they have worked up in their minds as an agreement or understanding." Some time later, when Crocker upbraided Stanford for his carelessness, the Governor would admit to giving Carpentier an opening—but it was the other's fault for being "a troublesome man. He is like a duck called the diver," Stanford said. "You can't tell when he goes down where he will come up. He is smart and the most tricky man I ever knew." It was just possible that people were recognizing that with five chiefs only occasionally in the same office at the same time, sometimes conferring sporadically but often not at all, they could take advantage of the Central Pacific company, quoting statements never uttered or handshake agreements never occurring. Still, the Judge would say, if they could not figure a way to correct the impression, they would lose 70 percent proportionate to whatever Goat Island interest was demanded for the Oakland partnership—just because of careless talk.[34]

Hopkins found, to his horror, that Stanford had mismanaged the Central Pacific's relationship with its chief banker, the Bank of California. The alert had sounded after the region's money market had tightened, and cashier W. H. Ralston had "flattered or squeezed" a million-dollar draft out of Stanford "to suit their convenience," Hopkins complained. The Governor had signed the draft, dashed airy notes off to Hopkins and Huntington, and left for Utah, thinking the bank was dealing generously with them while the company officers were forced to scramble to cover the unexpected draft. After that "favor" had been done for the bank, Ralston began tightening the screws; first, he held up a Central Pacific telegraphic money transfer with no notice, forcing Hopkins to suspend paying bills and payroll to meet a tax installment and avoid a truly punishing penalty. Next, instead of the earlier practice in which the Central Pacific could deposit gold in New York City and "float" the amount in San Francisco upon notification, the bank abruptly signaled that henceforth they could write no more drafts; the bank now required five days' notice for telegraphic transfer from Huntington—during most of which it had free use of the money.

"It is a worse state of things than ever before since we began Rail Roading," Hopkins despaired. "We have never before seen the time when plenty of securities failed to bring what coin we must have." Ralston ascribed the new policy and the tight money market to everything from an unforeseen good agricultural harvest, to dry weather in the mountains which had reduced the gold yield from hydraulic miners, to an earthquake causing some $3 million damage in San Francisco, but when he began blaming weather and geological mishaps in Panama, Hopkins blew a gasket. They were being taken advantage of on every front, he complained to Huntington, and the Bank of California was the worst offender, making the Central Pacific jump through hoops like a trained monkey, not only in its finances but in the considerable freight business it did between Nevada and California for the bank. It really came down to a question of power. For Hopkins, November was a humiliation he would not ever forget. The whole relationship offended him—"and if it didn't suit Governor Stanford's lazy way as a good fellow doing a large business with clever fellows in a clever manner," he said, "it would please me very much better—and I think it would give real dignity to the finances of our concern at this end of the route—which is something it has been sadly in want of."[35]

With Stanford off in Utah on such important business, all three left in Sacramento soon had something else to complain about: as always, the Governor neglected to write them. "We have received but one letter from him since he got there," Hopkins complained to Huntington on November 16. "I hope he has written you more frequently."[36]

On October 31, Stanford had ridden into Salt Lake City after a stately inspection tour of the graders' work east of Humboldt Wells and a conference with Chauncey West, their principal Mormon contractor; West agreed to put more men on new adjoining sections to the east and west of his existing contract. As Stanford had traced the line he noticed that the Union Pacific surveyors crossed and even followed the Central Pacific stakes—at least down to the eastern side of Great Salt Lake, where, to spite their rival, the Central Pacific engineers had pulled up their stakes all the way down to the mouth of Weber Canyon. "I am told they have copies of our maps filed in Washington," Stanford wrote Huntington on November 4. "I think you will find they will file nearly exact copies with our line as their own laid down as we have it." It was a worse situation away from the level of the valley, up in the narrow canyons where in many places there was hardly room for two engineers to stand side by side, let alone their trackbeds.

The Union Pacific graders had rolled down these parts, blasting their path from one side of the streams to the other. "They are all over our old line," Stanford cautioned, "often on and across. There I shall do nothing as they can with good weather have their rails down before we could make a showing." He was going up there again, he said casually—to meet Dr. Durant. "He telegraphed from some where on his line . . . that he would come out to Echo if I would meet him," Stanford continued. "I don't expect much from meeting him, but have thought it best to give him an opportunity to have his say." Luckily for Mark Hopkins and the Crocker brothers, they would not hear of this summit until after the fact—or they would have burst from anxiety at the prospect.[37]

It would surprise them when Stanford did not give away the store. He met Durant at Echo City on November 5, and the Doctor immediately offered to sell them the grading already done east of Humboldt Wells. "I think he would like to get out of that work . . . but I do not think it policy to help him out of the scrape," Stanford told Huntington. In the meeting both executives were too cautious, too distrustful, to come to an agreement about anything. Sizing up Durant, the Governor took to examining the seals on every letter he received, and he told Huntington not to trust cable traffic: he was sure the Doctor was tapping into every transmission they sent. Stanford's several meetings with Brigham Young were more fruitful; the Mormon leader repeated his willingness for new grading contracts with the Central Pacific east and north of Salt Lake—even though he would be doing parallel work for the Union Pacific. He solemnly promised to treat both companies equally. Stanford came away from the meetings persuaded that Young planned for the competing lines to merge at Ogden—after he had collected for grading two lines from Weber Canyon to Monument Point.[38]

Stanford may have been thoughtlessly neglecting to keep his California partners abreast of his activities—which Huntington compounded by not passing on much when he did hear from the Governor—but Samuel Montague returned with news to Sacramento on November 12, "& from what he says," Charley Crocker told Huntington, "I judge Stanford will do nothing about occupying the line to Echo Summit which you have urged with so much persistency." While this showed some appreciation of legalities—the Central Pacific end of track, after all, was still some four hundred miles short of Echo Summit—it seemed to them hardly the time for probity. A few days later the local papers reprinted a Salt Lake story alleging that Stanford had agreed with Durant on a meeting place north of the lake; later it was denied and disproved, but it caused some agitation in the Sacramento office. The Central Pacific track had extended on November 11 to Gravelly Ford on the Humboldt; that sandy crossing, overshadowed by reddish brown mountains, the ever-present sagebrush relieved by willows on the riverbank and occasional cottonwood groves turning yellow in the nippy fall air, was the site of the deadly quarrel within the Donner Party which ended in a murder and the exile of the killer, James A. Reed—who not only survived but pointed the way for the first rescue party over the wintry Sierra. Grading along the Humboldt—through the deep and curving, formerly impassable gorge ahead of the track, named for the almost perpendicular Palisades which towered hundreds of feet overhead—was nearly complete, Crocker reported. But progress would be temporarily slowed as his tracklayers caught up with bridge builders and then the graders.

Crocker was feeling unusually hopeful—he had sent Strobridge to chase down a rumor of a place beyond Humboldt Wells, and within twenty miles of the road, which had timber enough for 120,000 ties. "If it proves true," he wrote happily, "I will send all the men we have in the Sierras now covering track, as soon as they are driven out by the Storms, over there & get out ties. It will help us out amazingly in transportation if we can get that number of ties so far out." Moreover, he had sent some prospectors about 12 miles from the head of Humboldt Canyon, and they reported striking a vein of coal 12 feet thick. "It is almost too good to be true," Charley exulted.[39]

Hopkins was soon feeling a little better too, with company earnings for October posted at $300,000, with government reports on miles 390 to 410 sent east for collection of new bonds, and with an emergency transmission of $500,000 from New York clearing so he could resume normal business. To feel more secure, in a few days he dispatched the report on miles 410 to 430 prematurely, dating it ahead and telling Huntington to "deface the figures of the postmark on the envelope as not to show the awkward discrepancy in dates, in the event that you should show the envelopes at the [Interior] Department." Despite these minor forgeries, Hopkins was alert to larger issues, especially after hearing Huntington's proposition that they settle all the accounts of their dummy construction company, the Contract & Finance Company, with Central Pacific bonds at the market price. This went against the contract, which stipulated cash payments, and was dangerous, Hopkins countered, especially in light of the Crédit Mobilier suits in New York: there was nothing inherently illegal about the Contract & Finance Company—but it could still blow up in their faces. Because they were absorbing so many companies in California, they had to think ahead to every contingency, including lawsuits and bad publicity. Bonds, he explained, "might be traced & identified." They might put the contractors (themselves) to "inconvenience, delay & expense of a legal explanation," he said, "& cash cannot be traced & identified."[40]

Meanwhile, Leland Stanford was camping north of Salt Lake in the Promontory Mountains, returning after six days to his comfortable hotel on November 20. Stanford had examined the Central Pacific lines with George Gray, he told Huntington, and the locating surveys which had taken their engineers two months in the summer now seemed to require too much heavy rock work. "I was much disappointed," Stanford said. He ordered new surveys to give them quicker, lighter grading work. The Governor was probably basing his opinion on that of the more experienced Gray, but the air of executive assertiveness in his letter may have caused Huntington to blink in surprise. As Stanford had returned from the north, he said, he found that Chauncey West had put teams between the Promontories and Monument Point for the Central Pacific, and also above Ogden—working north and west over fertile, canal-irrigated lands toward the orchard-ringed and bee-keeping village of Willard, the great blue lake and the western slopes of the Wasatch always in view. Other Mormon work gangs contracted by the Union Pacific were grading only a quarter-mile west of the new Central Pacific line above Ogden. Farther north the lines crossed frequently, and in many places the stakes had been hammered in side by side.

The man Stanford had sent to look over the Union Pacific presence in the canyons had much useful news to report. One fact would slow them down: the Union Pacific line called for no less than ten bridges, with spans ranging from thirty to one hundred feet; eight of them were in Weber Canyon alone. All were to be of prefabricated timbers shipped from Chicago and hauled by teams over the Wasatch rim. In addition, one of Brigham Young's sons had contracted to supply ties in Echo Canyon, for the royal price of $4 per tie. "There is considerable talk here lately that the U. P. is not prompt in their payments," Stanford told Huntington with some satisfaction. "I am told they are two months behind with many." The work in the canyons, he had heard, was as cursory as the line in Wyoming: they had bridges resting on ties, and not enough culverts—"it was certain to be washed away in many places. . . . [T]he spring freshets will await much of it in Echo, I can say from my own observation."[41]

Huntington was still intent on getting their laborers up there; they had just a month to play with until Secretary Browning's pledge of secrecy would expire and he would tell Ames of the approved Central Pacific line up through Echo Canyon. Huntington dashed off a letter to Stanford on November 13. "If it is in the power of *God*, man or the devil to get our rail laid to within three hundred miles of Echo by, say, the tenth of December," he raged, the words fairly shouting from the page, "then I would take possession of the line [immediately]." As soon as Stanford had his graders up next to the Union Pacific tunnel, "let me know," he urged, "and I will have it in all the papers here. I think if I could tell it today, I could break the Union Pacific." At this exhortation, Stanford dug his heels in a little deeper. What did Huntington expect him to do? "The whole force of the U. P. are on the 45 miles East of Echo Summit and in Weber Canyon," he said. Was he supposed to shoulder past all those work gangs and engineers in Weber, plant himself below the Union Pacific tunnel heading, and invite those unwashed minions to swarm over the mountains and down on his head? And besides, the canyons were full of obstacles and difficult engineering problems; as far as operating regular train service in them, he could see little value. "If it were not for the coal mines near the mouth of Echo," he said, "I would not consider the line up Weber of any value to us."[42]

Helpless to intercede from his office in Sacramento, Mark Hopkins urged that Stanford and Huntington break their impasse by meeting halfway, in Omaha, face to face—an idea to which the Governor gave lip service but claimed there were too many pressing matters in Utah to allow him to leave. Again he went north to camp with the graders in the Promontory Mountains, aiming to push them to greater speed. His associates wished for something perhaps more presidential. "It is," Huntington admitted to Hopkins on November 18, "a fearful position to be in when one has to trust vital, material things to Providence only."[43]

Jack Casement sat in his tent in Granger, Wyoming, on October 31, 1868, enjoying a few minutes of quiet. Dodge had just left, escorting Blickensderfer and the other special commissioners out on the line—and Casement could finally write his wife Frances, missing her and his two small boys, and apologize for not writing oftener. There was so little time, and his tent was usually full, and when not, he was on the move. With winter about to blow in, his men had constructed one warehouse stocked with "provisions enough in it to weather one small storm" (they already had one short storm) but he worried that he could not get enough food for the tracklayers—"it is hard to get anything up the road except ties and iron." Durant was still lurking about—"trying to hurry things up but he only creates delays & we have held him level by keeping the track onto the Bridges and our men idle more than half of the time—but as he claims to love to pay for those things I propose to give him his fill of it."

They were, he wrote a few days later, "straining every nerve to get into Salt Lake valey before the heavy Snows fall—thirty more days of good weather will let us do it." Two weeks later, worried about his ailing mother back in Painesville and more homesick than ever after hearing his brother Dan talk of the Ohio autumn during his furlough, General Jack had all he could do, needing to move his supply base westward to Bear River and manage his large force of tracklayers. "The Indians are on the Rampage East of us," he reported, "and are detaining trains." Probably they were Cheyenne and Arapaho—driven northward by General Sheridan and Custer, the latter finally loosed from his court-martial; the bluecoats were sweeping across the southern plains of Kansas, aiming to end the pattern of raids, rustling, and abductions. Dodge had seen one of the Union Pacific's railroad bridges in the Platte Valley in smoking ruins while out with the commissioners. For the tracklayers in western Wyoming, though, it was peaceful and the weather had continued to be favorable. As Jack Casement wrote his wife, "we have been delayed by Bridges and want of material or we would have been in to Salt Lake valley before this time." With good providence, things would be under control in early December and he could look forward to some time off with his family.[44]

Ahead at Bear River, James Evans had decided that he could put up with the pressure as supervising construction engineer no longer. He wrote to Durant on November 19 that even with some anticipated delays the track would be at the end of the division assigned him—and he wanted to take that opportunity to turn in his resignation. "The work is taken up so hurriedly," he explained, "that we find it in bad shape," and "if I go any further it will be a very constant and persistant fight and I would rather be out of it." He said that a bridge contract would suit him better, and Durant, satisfied that track across the Wyoming division was indeed close enough to Reed's territory in Utah, gave Evans a contract.[45]

Webster Snyder was also sick of the pressure and the rottenness, and he wanted out of operations. He complained to Dodge in late October that he would leave as soon as he opened the road to Bryan, that he "would not stand the fight and be crippled any longer." Hub Hoxie told Dodge that if Snyder resigned, he would go too. Again, Dodge had to plead with them to stay the course, but Snyder was clearly very troubled and on the edge. His department owed over $1 million but New York would not give him the money, pleading poverty with the government bonds still being withheld; he was spending large sums for new machinery, shops, and fuel; all the contractors were still hauling material and not paying him the freight bills. He said he did not believe the Union Pacific would be a paying proposition once the lines were connected. Everyone seemed to be stealing the road blind, Snyder groused, and he could not even slow it down. Even men from the government were joining the frenzy: a new federal commissioner, Cornelius Wendle, arrived on the job in late October to examine 25 miles west of Green River. He told Snyder and the other officials there that he "did not intend to examine it in the cars," the usual practice being to ride the special examiner's car and inspect while rolling at a slow speed. Wendle proposed "to walk over it and find such defects" as he could—he was there "for business." The word "business" was given special emphasis, and all present knew what the commissioner required. Before Snyder could get the section accepted, he was forced to pay Wendle $25,000.[46]

When Snyder wrote Dodge to tell him about the shakedown, Dodge wrote back "that under no circumstances should he pay a cent as it would ruin us and it would give the politicians such a hold on us that we would never recover from it." A similar case had happened to Dodge himself with a government commissioner named Snow who returned from his examination of the line—"his report was brought to me for sale," Dodge recalled many years later, but

"I refused to pay a cent upon it. It was finally offered to me for $1,000. He had two reports—one was a very severe criticism of the road and the other was a favorable opinion of it. Finding that he could not obtain anything from us for his report he filed it in the Interior Department. I immediately filed . . . a protest, demanding that it should not be used." He told the department that "it had been for sale and was a blackmail report, and if the Interior Department used it, we would take proper action in that matter."[47]

Back east, Oliver Ames was feeling as if he was being held up also. The money market was even tighter in New York than it was in San Francisco, and both Central Pacific and Union Pacific securities were taking a beating. One of the key economic debates during that season's presidential campaign had been whether to pay the burdensome national debt in gold (as said the Republicans) or greenbacks (the Democrats). For Oliver Ames's company debts, with no government bonds even on hand, the question was how to keep the Union Pacific from going under. "We have now 80 miles of road done," Ames despaired in a letter to Dodge, "on which we have rec'd no bonds nor can we issue our 1st Mtg. Bonds until we receive the Govts." To keep the road moving along they were forced to seek as much as $5 million in the market, a hard proposition. Then Cornelius Bushnell stepped in nearly at the darkest hour with a temporary salvation: the Union Pacific would borrow $2.5 million from its stockholders for 120 days, at 7 percent and a commission of 2.5 percent. Don't worry about our running out of money, he assured Durant out in Utah—in fact it was too late "to fear Huntington. You are bound to beat him to death. . . . We will borrow or Steal all the money you need."[48]

Indeed these were terms to warm the Doctor's heart. Almost at that moment he was hatching yet another scheme. When Casement's crews spiked past milepost 914, alongside the Big Muddy creekbed one mile west of Bridger, the Oakes Ames construction contract expired. Only Durant was paying attention. Congressman Ames had moved to extend the contract a few months before, but the board was not meeting while the Fisk suit was pending, and Ames had apparently forgotten about it. Using James W. Davis, his front-man partner in several railroad-tie and coal-mining enterprises (who happened to be the young brother-in-law of George Francis Train and the son-in-law of a longtime Durant associate, George T. M. Davis), Durant wrote a new construction contract on November 1, covering the rest of the mileage to the joining of the lines. As Hoxie and Ames had done before him, Davis signed the contract over to Durant—he was to be paid $5,000—and the Doctor's fellow trustees in the Ames contract; terms were the same as the earlier contract with the only difference being that the trustees now represented stockholders of the Union Pacific—the Crédit Mobilier was represented no longer. Not only the terms of the contract but also its very existence were kept secret by Durant for four months.[49]

Blessedly ignorant, the Ames brothers found their worries multiplied. Henry McComb began to fulfill his threat to drag down Oakes Ames and the directors of the Crédit Mobilier by filing suit on November 11 for the disputed 375 shares he had been demanding for a year; that case would be lurking on Ames's doorstep for years to come. For that matter, Duff Green's injunction on Crédit Mobilier was continuing to skulk in the courts, Jim Fisk's injurious nuisance suit was continuing, and the upright Rowland Hazard still kept his suit against the Doctor alive, causing some difficulties and bad feelings in the Union Pacific executive suite. Oliver Ames called for harmony in management if they were to finish the great project and wrote a stern rebuke. "Your Suit disturbs this harmony and tends to retard its progress," he scolded Hazard. By the time Hazard replied, he too had climbed on a high horse: he learned that President Ames had approved Durant's notorious suspense account. "I happened to be present," Hazard wrote, "when you completed an examination of the accts & stated as your full belief that he had got all his stock and 200,000 dolls. besides out of the companies and your brother still says Doct D declined to account for the monies received by him."[50]

While Oliver Ames was occupied with writing letters and giving depositions regarding all these lawsuits, another part of his mind occupied with the worry over the stock market and their lack of negotiable government bonds, the nervously anticipated government report was submitted to the Interior Department by General Barnes and almost immediately released to the press. Surprisingly, it was generally complimentary. Dodge, his engineer's nose rubbed in those shocking high grades and sharp curves sneaked by him by Seymour and Durant, had forgotten how an outsider from the east might be impressed by his magic route across the Rockies and by the sweat expended across the unforgiving Wyoming deserts. (For his part, Oliver Ames thought that having an old friend detailed to the commission as secretary may have helped.) Looking at the entire route, Barnes reported, the Union Pacific was "well constructed" and the line "exceedingly well selected." He estimated that the weaknesses would cost some $6.5 million—certainly nothing to sneeze at—but the problems could mostly be forgiven except for the high number of needless and even dangerous sharp curves: it was inevitable, he said, that many flaws were endemic in new construction.[51]

The Barnes report had enough in it, however, to fuel criticism in Congress and the press about profiteering and corruption, but more objectionable to Dodge was the financial report on the Union Pacific submitted by Jesse Williams—which he felt underestimated the corporation's operation expenses by millions. Williams would complain that he could not even get adequate data from the Union Pacific, for he was unable to work his way upward to the highest and most secure business records kept secret by the board of trustees. (Neither, for that matter, was the company's own chief engineer ever able to penetrate that far.) Dodge, though, living in the actual world of railroad work, was including among more legitimate costs the extraordinary figures a stickler could only charge to fraud, corruption, and mismanagement.[52]

Naturally the Union Pacific people hoped that with a generally forgiving report in hand, Secretary Browning would release their bonds. It did not show good timing, therefore, when the Doctor decided to respond to bad publicity from the report and to the predicament the suspension of federal support had put them in by firing off salvos against the Interior Department and its appointed inspectors. Jacob Blickensderfer drew much of his ire. At Durant's behest, Colonel Seymour began giving statements to the newspapers rebutting the critical phrases in the report and alleging that the Union Pacific had fired Blickensderfer for incompetence and dishonesty—and that this was his way of revenge. Blickensderfer read the charges in the *Chicago Tribune*—he was in Omaha, having been directed by Browning to examine the Kansas and the Sioux City lines—and was greatly upset. He wrote Dodge, who had originally urged him to take the leave of absence and accept the temporary government appointment, asking him for advice and intercession before his name was entirely ruined by Seymour and Durant. "That Seymour and those whose pockets are likely to be affected by the report will do all they can to injure me is not singular," he said, "and I presume it is not singular either that they should resort to falsehood if it tends to further their ends." He begged Dodge to help correct the lies about his Union Pacific performance and release for this official duty.[53] Dodge, however, like Oliver Ames, was more concerned with any high-level reaction to the Doctor's outbursts; those bonds meant all.

Durant had also sent his complaints directly to the White House, complaining to President Johnson that Browning was biased and had been almost incompetent in appointing someone like Jacob Blickensderfer to oversee the Union Pacific, and petitioning the president to replace the engineer; such a letter (it had been mailed on October 29), whether read aloud in a cabinet meeting or simply slid across a table without comment, had hardly pleased Secretary Browning. It was the same kind of aggressive defense reaction that had irritated Jesse Williams—a few months before, when the government director had relayed some of his misgivings to the Union Pacific board, Durant had snarled at him and derided him as a mere "watchdog" for the government, as if that were something of shame. But at this point, in the last week of November 1868, Browning held all the cards. He spent a good portion of the week writing his annual report to the president, in which he would note Jesse Williams's estimate that the Union Pacific had profited by some $17.7 million on its first 710 miles, after expenses (an overstatement, though understandable). However, seeing as the three special commissioners had estimated it would require $6.4 million to make the line "first-class," he believed the railroad should correct the deficiencies—before any more bonds or land patents were released. On Tuesday, December 1, during the cabinet meeting, Secretary Browning read his report and recommendations. The president and cabinet unanimously endorsed the move.[54]

On that same day, Dr. Durant, who had come back east, picked up his Crédit Mobilier congressman from New York, James Brooks, and hopped on a Washington-bound train. Dodge planned to go to the capital anyway for the opening of the congressional third session, and he would join them, as would Oliver Ames—but first Dodge would be detained until after the Chicago reunion of the Grand Army of the Republic, where a multitude still able to squeeze into their old uniforms in those fattening postwar years would do so, and cheer the prospect of one of their own marching into the White House. Once together in the capital, Dodge, Durant, Ames, and their representatives would use whatever influence was still available to them against this lame-duck administration; they would get in to see Secretary Browning, and they would try to gauge the depths—and the source—of his unfriendliness. Most important of all, they would, all of them, try to pry those bonds from his grasping fingers.[55]

On November 21, Charley Crocker returned to his office in Sacramento from a week's trip to the end of track to learn that the three-man team of special government commissioners had been named—and would be ready to begin their inspection in three days' time. Miraculously, with the ready supply of enemies and competitors in California, the identities of the men seemed most favorable: the name of railroad entrepreneur Lloyd Tevis, unknown to anyone outside their circle as their secret partner in the Southern Pacific and other ventures, was a friendly, shining beacon; the second, Sherman Day, was a former California state engineer whom Huntington had known for more than a decade since the Sacramento storekeepers had begun to support a Sierra wagon road; the third was Lieutenant Colonel R. S. Williamson of the Army Topographical Corps, who had been surveying and mapping California since the great army surveys for the Pacific railroad back in 1853. Crocker had yet to learn how friendly Williamson and Day could be, and he confessed to Huntington that he was "afraid of trouble with them. Tevis says they talk of going to the front & then returning over the line on horseback at the rate of 15 miles per day & examining every inch of the road." Tevis had been appalled at such a rigorous and exhausting assignment; he was prevented from declining the post by the Crocker brothers: "we have persuaded him," Charley said, "to go & prevent any such damned folly."

Still he could not shake his anxiety. Some portions of the road beyond Wadsworth on the bitter Forty-Mile Desert would not presently bear close scrutiny at ballast, culverts, and trestle work—but between Sacramento and Wadsworth, he said proudly, "I will put it up against any road in the United States." Then his solution came like a beam from heaven above: "I will have them out examining culverts, ballast & bridges between here & Wadsworth—so that they won't want to hear of culverts or anything of the kind beyond there." On November 24 Crocker and Montague escorted them out on the line in a special car fitted with beds, lounges, tables and chairs, and a kitchen. "It is not a *palace* car," the Judge admitted, "but has all the comforts of one."[56]

Over the past several months the *Sacramento Union* had been pointing its editorial finger at precisely the bad stretch of Nevada track, east of Wadsworth, which had worried Charley Crocker. The newspaper advised passengers not to travel over it, warning that they would take their lives in their hands. The attacks were difficult to ignore—and the three commissioners certainly referred to them as a reason for their caution. After the special train glided comfortably over the Sierra and down toward Reno, the commissioners were properly impressed with the work—and Charley Crocker's ruse to tire them out worked. By the time the special train reached the shaky stretch of road, Crocker used a cardsharp's old sleight-of-hand trick to distract them. "I said to the commissioners that we were approaching the point where the *Union* had said it was unsafe to go over the road," he chortled many years later,

> *and I would tell them when we passed it. "And," said I, "Here is a tumbler of water which I will set on the floor of the car. Now, gentlemen, take your watches: here is the station, and from this station to the next is so many miles. Note the time we leave this station and when we arrive at the other, so that you will know the rate of speed at which you have gone." I instructed the conductor to tell the engineer that I wished to go over that piece of the road at 50 miles an hour, and they made a little better time than that. The tumbler was still standing and but little water had been spilt. "Now, gentlemen," said I, "there is your Sacramento* Union.*" And the commissioners all laughed and said that was the strongest proof that had been given. They did not ask for any more; did not want to get out and look at the culverts; they were satisfied that if we could rush over the road at 50 miles an hour and did not tip over a glass of water it was a pretty safe road and well built—and so it was.*[57]

Having been snookered into staring at pocket watches and a water glass instead of looking at loose ballast, soft fill, and cobbled culverts, while the hale, smooth-talking host rushed them past what he had privately admitted was shameful, hasty construction, the commissioners wired their report to Secretary Browning, who read it on December 4. "The road is being constructed in good faith," they said, "in a substantial manner, without stint of labor or equipment, and is worthy of its characteristic as a great national work." In New York, Collis Huntington sent Charley Crocker his gleeful congratulations. "I think," he said jocularly, "you must have slept with them."[58]

After the inspection trip Crocker was back at work in Sacramento without even breaking stride, impatient for his tracklayers to do better than the mile and change per day they had been completing for the past several weeks; work through the three deep canyons of the Upper Humboldt required an inordinate amount of curving, each rail needing to be hammered and bent by brute labor. The total length of the line passed 440 miles on December 1, and Charley vowed to his partners that the progress would pick up once they were out of the canyons. As before, he and his brother and Hopkins could, in the absence of news from the Governor out in Utah, assume the worst. "We have not heard from Stanford in ten days," Charley told Huntington. "Don't know what he is doing. *I* guess nothing—in fact I never knew him to do much himself—he is awful lazy & never attends to details—wants somebody to come along after & stop the leaks & do the work." Hopkins agreed, sending a coded telegram to Huntington: "We think Stanford intentionally avoids meeting you." They had sent Darius Mills out there to persuade Stanford to go east and meet Huntington halfway, to resolve the debate about the Wasatch canyons once and for all, but Stanford had demurred: he was, he said, too busy. "I fear he is ashamed of his past management of Salt Lake matters," Hopkins commented, "& therefore dreads to meet you there." Hopkins was also irritated that Stanford had taken Montague out with him to Utah, delaying the engineer's signature on mileage affidavits and thus the commissioners' reports and the government bonds. The Governor could seemingly do no right.[59]

Leland Stanford would of course disagree, and as the volume of complaints issuing from his associates entered the audible level, he stiffened, grew slightly haughty, and defended his earnest efforts in the letters he sent Huntington every week or ten days, when convenient. His latest expedition to the Promontory Mountains had good results, he bragged on December 4—"We have at last got a line thru on which I am willing to go to work," he said. The surveyors had laid out a new half-mile line at one point which obviated the need for a tunnel—a savings in time that would keep the Union contractors from seizing and occupying the gradient. He had felt them nearly breathing down his neck while up in those mountains; they had the superior force, with the Central Pacific graders scattered all along the line, and without stretching Stanford could see them at work as close as a hundred feet away from the Central Pacific stakes.

But it was still not as apocalyptic as the excitable Huntington was making it out to be in his barrage of telegrams, such as the one in which he repeated Durant's boast that the Union Pacific track would be at the mouth of Weber Canyon by December 10. Stanford replied that he doubted Durant's men would be at the mouth of even Echo Canyon by the first of the year—he was still confident that with an ordinary winter the unprotected Union Pacific track would stall materials in the Laramie Hills and paralyze them in the Wasatch.

Tantalizingly, Stanford toyed—merely toyed—with the notion of their sending graders up into Weber and Echo. Of course they could do it, he admitted to Huntington. "If the weather will permit I can put forces in there." He had talked it over with Brigham Young, and the Mormon leader had even offered to supply all the men they needed. He dangled that a bit in front of Huntington's nose. But of course that was "mere speculation," he went on, oblivious of how this would raise the blood pressure of his partners. "But my idea, Huntington," he finally pronounced, "is that we will not strike a lick East of the mouth of Weber Cannon and perhaps we may not care to go more than a couple of miles east of Ogden, though I think we ought to go up into the Cannon far enough to just give the U. P. just a chance to peek out and get a glimpse of Salt Lake Valley." All by himself he had decided how to get around the problem that they were still over three hundred miles from the head of Echo Canyon: knowing that the federal government was granting the Pacific railroad its right-of-way, the farmers in the valley and the canyons "may well object to having their land cut up by two lines of road and say that the right of way exists for only one." The Governor's idea, he wrote both Huntington and Hopkins, was that he would go to each and every farmer—and *purchase* the conveyances.[60]

The idea set off explosions inside every one of Stanford's impatient partners. What "a waste of money," Judge Crocker responded sharply—the 1862 Pacific Railroad Act gave the Union Pacific "a *clear legal* right of way over all lands which were public lands on the 1st day of July 1862, & our company has the same right." If any plan was going to work, it was Huntington's scheme to take advantage of Secretary Browning's still-secret ruling. His veneer of brittle politeness to Stanford crumbled when commenting about the Governor's proposal to Huntington. "It made me boil," he fumed, "when I read his letter." Thinking to buy off the settlers—"expecting thus to block the wheels of the U. P. locomotives. Of course he might as well throw a straw before an engine."[61]

Darius Mills came back to Sacramento, dejected. Stanford and Huntington were just not talking in the same language—the governor did not comprehend what Huntington wanted him to do, and Huntington had no understanding of topographical matters in Utah, let alone the reality of politics out there. He was baffled by how to get them together. Stanford had said only an order from the board of directors would move him to change his mind. Hopkins and the Crockers, none of whom had set foot in Utah during this crisis, could think of no solution, their impatience over the continuing slow tracklaying progress in Nevada making them feel even more impotent. "Nothing but the best of systematic organizing energy can overcome the difficulties," Hopkins wrote Huntington on December 10, "foreseen & unforeseen, which meet us from day to day."

They were at mile 452¾, he noted—this was some 200 miles away from

Monument Point—and a new 20-mile report was going out by express. The weather on the Humboldt was getting colder, the temperature ranging from nine degrees above zero to six below when measured at 6:00 P.M. every day. "Such weather brings difficulties which we are not altogether prepared to meet," Hopkins said. "Streams freeze up, & Water pipes. Fires have to be built of sage brush under the cylinders of Engines standing still—& the thoughtlessness or carelessness of one man often delays many. All these things tend to make progress slower than our sanguine people expect and promise."[62]

People standing around on the streets outside Leland Stanford's hotel in Salt Lake City on Tuesday, December 13, may have seen the bolt of bright light shine down from the celestial vaults and through his window, finally striking him. "Your views and mine seem to be directly opposite," he scrawled out to Huntington, the exasperated, plainspoken letters from his California and New York partners in hand. "One or the other must certainly fail utterly to comprehend the situation. I have at last got things here in such shape that I think I can leave them in safety. I go tomorrow to Ogden and if every thing appears right, I shall telegraph you to meet me at Omaha. . . . I do not see that we can understand each other without a personal interview."[63]

Huntington left New York with his wife on December 17, going by train to Omaha, where Governor Stanford was waiting on the twenty-second. Both of them a little grayer in the temples and the beards, their impatience with each other out in the open with this late controversy—how would they have begun talking, and what might they have said? Particulars went unrecorded. Then they climbed aboard a private Union Pacific car to go back over the competitors' line. Operations Manager Webster Snyder provided them with all courtesies for the trip across Nebraska and up into Wyoming. A bad snowstorm which had in the first week of December blockaded the road in many places was followed on the seventeenth by a worse blow; track crews had been idled for several weeks, but apparently by the time the Central Pacific executives had rolled past Laramie the worst was over.

There would have been little to see from the cars over the little spine of the Rockies and perhaps no view of the Dale Creek Bridge, and no Indian war parties to trouble them, for the tribes had been pushed into winter camp on the distant margin of their hunting lands, and most of the former Hell on Wheels camps were now just cold, rude gravestones hulking under the snow, and the desert's furnace heat would have been long gone. When the train had glided to a stop at Green River, the ride would be over for Stanford and the Huntingtons, for Snyder had been cautioned not to let them see the miserable track work farther west. Though service had just been extended to Evanston, 110 miles beyond Green River, Stanford and the Huntingtons had to board a special stagecoach for the journey to Salt Lake City. The Governor had arranged relays of horses for a quick journey, and as the coach had thundered down the famously smooth chute that was the wagon road in Echo Canyon, past Castle

Rock and the sudden opening of Sawmill Canyon, past the old Mormon breastworks—a light snow over the red walls—and down along the Weber to the high-canyoned Narrows and the squeeze through Devil's Gate, with Stanford pointing out the Union Pacific grades all the way, the camps, the swarms of men, the impossibility of seizing and holding this terrain.

And as they would have emerged from the mouth of Weber Canyon into the openness of the Great Salt Lake Valley, with the smoke plumes of Ogden village rising ahead and the cold gray-blue of the lake beyond, Huntington would have certainly begun to concede that Stanford had had a point about the Utah canyons after all, and behind this dramatic change of heart would have loomed the weighty dark red palisades of the Wasatch, the humbling immensity of Wyoming in wintertime with those puny iron ribbons running across it, and the huge piles of snow pushed and shoveled from the right-of-way. Ogden did not have the coal seams of the upper Weber, but it did have trade and freight and mails and agricultural business, and there were probably short-line possibilities to the north, south, and east.

Huntington would have begun to formulate a fallback position to present to their associates as he and Stanford conferred in Salt Lake City for two or three days. They observed Christmas there surrounded by Saints. Then, while the governor stayed behind to push the work, Collis Huntington and his wife, Elizabeth, boarded another special coach for Nevada—Stanford had wired ahead for relays of horses to be ready all along the route. The Huntingtons sped westward, not going by the hard roundabout way up to Humboldt Wells to trace the Central Pacific line, but following the old express wagon route across central Nevada, the "washboard way" across no fewer than seven mountain ranges. At Reno, with the wintry Sierra frowning close by, he would have escorted his wife onto an elegant private car on the Central Pacific Railroad, gilded words he would have seen only on stationery and securities over the past several years, and Huntington would finally get to look on what Charley and the Judge and Mark had been boasting about as the big engines hauled them up the Truckee, then steaming through a succession of long snowsheds and through extraordinary deep cuts and past towering retaining walls, frequent tunnels blacking out the views of snowy mountainsides and steep valleys falling away, and as winter would begin to slough off them, as the train rolled down into Sacramento Valley, any slight lingering deference at seeing how the Union Pacific had crossed the Rockies and the Continental Divide, and almost conquered the Wasatch, would have shrunk like a miner's cheap long johns on washday—and Huntington would have swelled with pride over what he and his associates had done. And what they would do.

Still, he felt a great loss, was dizzied from the breakneck trip but even more so from the frantic momentum about the Wasatch scheme having slammed against those narrow canyon walls, and Huntington reeled from making the necessary mental adjustments to start a new campaign under wholly different conditions. In the reunion with his old friends and business partners in the Sacramento office, there would be time for congratulations—as well as recriminations.[64]

Dr. Durant and Congressman Brooks spent Thursday, December 3 making the rounds of government offices in Washington, and the next day Secretary Browning must have felt their presence in town keenly as he read aloud his opinion of the condition of the Union Pacific to the cabinet; he must have anticipated some lessening of political pressure when the cabinet agreed to release subsidies on forty miles more of its road. But by Saturday, Durant had received a letter from Oliver Ames, dated December 3. "By Section 8 of the Act of 1864," Ames wrote, "we have a right to call upon the Govt to advance us ⅔ of the cost of work done in advance of the track."

It was a codicil that had also escaped Huntington's gimlet eye, to his everlasting dismay. "We have now expended three millions in advance of the track and we have nearly 4,000,000 of iron ties and superstructures on hand and most of it paid for," Ames said. The commissioners should forthwith go out and look at their grading and bridging work in the Wasatch canyons, and go all the way out to Humboldt Wells. Lighting up at this dramatic, promising new approach, the Doctor quickly pressed the request on to Browning. On Tuesday, December 8, Browning passed it to the cabinet with Durant's further proposition "to give the lands, to which the Road is entitled, in pledge for the completion of the road as a first class road in all respects, and thereby to release the subsidies to which to road is entitled." No doubt feeling more tense as Durant and the Ameses stepped up their pressure, counting the hours and days until he could finally inform them that he had already given away the Utah canyons to the Central Pacific, and dreading the moment when they would hear that news, Orville Hickman Browning came down with "a slight attack of Cholera Morbus" on the night of December 10, and therefore was mostly at home and unavailable until Tuesday, December 15—when he could finally let the Huntington cat out of the bag. But whom would it scratch?[65]

On that Tuesday, Browning wrote the awful news to Oliver Ames: that on October 20 he and given his consent and approval to the Central Pacific location from Monument Point to Echo Summit. "It is necessary, under the law," he informed Ames,

> *that your road and the Central Pacific shall be so located as to united and form one continuous road. You are therefore requested to cause the Union Pacific railroad to be located to Echo Summit, so as to unite and form a continuous line with the Central Pacific railroad as already located to that point, and forward the necessary map and profile for file in this department.*[66]

In a few days Browning could wince over the first response, when, on Saturday, December 19, the agitated Oliver Ames and Grenville Dodge were ushered into his office. Did Dodge pace? Did he raise his voice? Did he smack his palm with a fist? Did he thump an index finger down on his location map? We shall not know the actions, only the words he recorded soon thereafter. The Central Pacific "had never located the line," the engineer complained angrily, "it was merely on paper that curves had never been run in." They were "pretending to lay on the same or very nearly the same grade that ours is now building upon." That Browning had accepted it "is an outrage—and cannot be justified by any reasoning." Not only was the Central Pacific line merely on paper—using unnecessary curvature and steep grades, he could not help adding—"over 100 miles of the ground it pretended to lay upon has been graded and properly bridged by the Union Pacific." When Huntington had filed his fraudulent documents, the Union Pacific end of track "was nearer Monument Point than the C. P.," and "the 300 miles that we are allowed to work in advance of track has covered this ground." The Central Pacific's end of track was not even within three hundred miles of Echo Summit when it filed its report. Now the Union Pacific was nearly to the head of Echo Canyon—a week before they had crested the divide between the Bear and Weber drainages, five miles over the Utah border; their tunnelers were burrowing away, "and early in spring," he finished defiantly, *"we will cover the entire distance."*[67]

Did Browning wince, murmur excuses, lapse into legalese? Under Dodge's onslaught he retreated—filing a map, he said (obviously making it up as he went along), gave no right to one company over another in the building of the road. The secretary had only approved *a line.* He only wanted to avoid the problem of two companies building toward a meeting point at different elevations, grades, or curvatures, making the actual splice impossible, he said smoothly. Did he succeed in mollifying them? He had no idea. But that night, while the exhausted Browning attended a handsome reception at the house of Secretary of State William H. Seward (who was still embroiled in the *Alabama* war damages claim with Great Britain), across town Grenville Dodge went to war. He dashed off a letter to the new secretary of war, John M. Schofield, who like Dodge had spent time fighting rebel guerillas in Missouri and commanded under Sherman in the Atlanta campaign. He, too, had just attended the big army reunion in Chicago (which Secretary Browning had vaguely called "some sort of a soldier's convention" in his diary). Schofield was an able administrator with a philosopher's temperament, and Dodge appealed to reason in begging his former comrade to see things from his point of view before the next cabinet meeting on Tuesday, December 22. The chief engineer complained that the special government commission had included in their report some ninety miles of line before the Union Pacific was ready to show it to outsiders. "We had built on this 90 miles few permanent bridges, no stations, shops or tanks," he said, "and which made a vast difference in the estimate" of raising the railroad to first-class status. The regular commissioners had now seen that mileage, after it had been improved, and had sent in good reports. Moreover, he continued, "what I look upon as unfair, unjust and very detrimental to us and to the government is that one commission of able, critical, railroad engineers should examine road on this side of the continent and another commission should examine the C. P. R. R. assuming, as they have, on an entirely different basis. . . . [L]et us have all the roads examined upon the same basis under the same instructions and by the same minds, so that we may arrive at the true merits of the question." If he had but known about the demonstration with the water glass!

Be reasonable, Dodge urged his former comrade—though the Union Pacific was the "best built, equipped and appointed road in all its departments, we are the only road that has had any trouble in obtaining its bonds." The Kansas line and the Sioux City road retained "three times as much per mile as on ours," he complained. "The Central Pacific goes free and use the report of our commission to harm us in the money market, while the other [special] commission send in a telegram that [the Central Pacific] is or will be built within the law."[68] Wishing that he could remain in the capital and see Schofield, Dodge left that Sunday to confer with the Union Pacific board of directors in New York. "The great fear," he wrote, "was that [Huntington] might get bonds issued on the line" in the Wasatch canyons. Secretaries Browning and Evarts seemed in the pockets of the Central Pacific. General Schofield would, he hoped, favor him. McCullough of the Treasury did not like Huntington or his friends. There the count veered into darkness not unlike a railroad tunnel; what lay on the other side?

What waited for them all was a compromise which really satisfied no one. On Tuesday, December 22, the cabinet met to approve it. "It was agreed," Browning noted in his diary, "that all the lands granted to the road should remain unpatented, and that one half of the first mortgage bonds which the Road should hereafter issue should be deposited with the Secy of the Interior until the amount so deposited should amount to $3,000,000, to be held by him as security for the completion of the road." During the next two days he endorsed over two 20-mile sections to the Union Pacific, withholding half of the mortgage bonds. It would not be quick enough to pull the company very far back from the edge: on December 28 a million-dollar note would have to be signed at a regrettable rate just to get the Union Pacific through the night. "Money is awful tight," Oliver Ames wrote to his brother, Oakes, "and we have large amts to pay. We hope to get through but things look Blue."[69]

Headlong efforts out in Wyoming and Utah exacted the inevitable toll in the New York office when bills poured in from all directions—from Salt Lake City, Ogden, Echo City, Green River, Laramie, Cheyenne, Omaha, Chicago—all those graders', tunnelers', tracklayers', tie contractors', and bridge builders' invoices piling up toward the ceiling, sending Oliver Ames's heart plummeting. He dreaded to think how badly they were being cheated. "The demands for money are perfectly frightfull," he wrote to Durant, who had gone back west.

> *Some how the Road must be costing us very much more than we are getting for it, or every one out there is stealing. Awful stories come down here from outsiders of the competition of the Contractors Engineers and everyone in connection with the Road. If something is not done to stop it at once and discharge all supernumaries on the work; we shall be largely in debt when the tracks meet. I believed up to within two months [ago] that at the end of our work we should have a handsome surplus. It now looks as though we should have a large floating Debt. Do something to stop the thieves from stealing our last cent and making the Road suffer. We have Snyder asking for 700,000$, [Construction Cashier William] Kennedy for 1,500,000. This will take a large portion of our Govt and with the 900,000 wanted for interest on our Bonds will take 100 miles of our Road. We owe outside of this 1,500,000 more on which we have no securities, the 5,000,000 we owe on a pledge of our 1st Mtg Bonds leaves us with almost nothing to finish up the Road. It looks to me as if Kennedy was drawing or asking to draw for double the amt needed. . . . The Tunnels should be pressed. The Central say they are going to meet us at the Big Tunnel. . . . You know how to get through the Tunnel as well as any one and meet them at Head of Salt Lake. The work at Promontory Point is costing or will cost for 5 miles there 100,000$ per mile—if there is danger of their getting up to that point leave it for them to do.*

"Cut off all useless expense and economize everywhere, where it will not delay work," he begged Durant. "We dont want any surplus material on hand when the work is done." His alarm would have trebled if he knew that the Doctor had ordered Snyder to find a thousand more laborers in Omaha and pack them off for the Utah work, and wired Reed to work their men in night shifts down Weber Canyon.[70]

That night in Washington, Oliver Ames saw the president and Secretaries Browning and Evarts at a gala wedding reception, in which most of the upper strata of the capital participated. "The most sumptuous and elegant banquet I ever saw," Browning commented. "A marvelous amount of foolish extravagance." The affair calmed Ames a bit. Something in their meeting seemed to remind Ames of a private arrangement with Thomas Ewing—Browning's friend and partner who had, that autumn, acted as a conduit for Collis Huntington to influence the Interior Secretary. "[Were] you to give Ewing ten [shares]?" he wrote Durant the next day. "I suppose he wants it and if so write me."[71] Less than a week later Ewing told him plainly he wanted $10,000.[72]

On Christmas, Secretary Browning worked for an hour at the office, and then distributed presents of silver to personal servants at the department—a cake basket to his usher, a butter dish to his driver, a goblet to his barber. In succeeding days during that quiet holiday week in Washington there were several parties at the White House. During one business meeting in the Christmas season, President Johnson and Secretary Browning determined to discharge the corrupt railroad commissioner Cornelius Wendle for, presumably, excessive and unseemly greed.[73] It was on Christmas, too, that Dr. Durant arrived at the head of Echo Canyon to whip the forces into shape. The track had finally threaded over the rim—bypassing, as planned, the inflexible, unfinished "Big Tunnel" by way of a steep jackknifing temporary "switchback" track. "The ground is frozen about 2 ft. deep," Samuel Reed wrote home from his new campsite, forced to work through the holiday because "all the forces have to be sent west & placed on new work." He wrote Jenny Reed again a few days later. "You do not know how I want to be home," he said. "If anything goes wrong the Hell is to pay, the same if all goes swimmingly. I do not care whether I remain another day or not." Workers struggling with the frozen ground often had to blast it as if it were rock, and there were times, as Reed stood by helplessly, when Durant would order Casement's men to put down ties and track on ice and snow. When Dodge would finally extricate himself from Washington, where he was studying the fraudulent Central Pacific location map and continuing his siege of Secretary Browning's office, he would see out in Utah the results of Durant's hurry-up orders to put track on ice and snow; he would witness a train and its track actually begin to slide sideways—until the train rolled off the bank and landed upside down in a ditch.[74]

Financially the company may have been in a ditch, but on the opposite side of the balance sheet was the matter of sheer accomplishment. Putting together his final report, Dodge would note that the company had built 64 stations, 73 water tanks, and 15 coal houses; that it now operated 124 engines, 21 first-class coaches, 10 second-class coaches, 81 cabooses, 16 baggage and mail cars, 8 sleeping coaches, 520 boxcars, 1,734 flatcars, and 100 coal cars. The year's mileage (not counting sidings) was a respectable 446 miles; when the track stopped at the end of the last day of 1868, it was 982 miles from Omaha. But on the personal side in those last days of the year, with Durant hectoring everyone in earshot down there in Echo Canyon, with the wind whipping snow and sleet into their faces, winter—most assuredly—was doing its best to humble them all.[75]

Out in Nevada, where the burnt-red Ruby Range to the southeast was now obscured by snow and veiled by freezing fogs, the similarly cold and subdued Chinese gangs had stretched the line onto the wide north–south valley drained by branches of the Humboldt. Past the mining-supply and freighters' town of

Elko (altitude 5,063 feet), in J. H. Beadle's view a "pretentious and lively city," they were at milepost 468 and climbing toward the next depot, which would be called Halleck Station in recognition of nearby Fort Halleck—named for the brilliant, owlish West Pointer and former general-in-chief Henry W. Halleck, who between the Mexican and Civil Wars had been an attorney in San Francisco and briefly secretary of state in California. The valley the tracklayers had reached had the dubious distinction of being one of the coldest spots in the nation; frigid winter winds blasted down from Idaho, concentrating and intensifying before walloping the eleven-thousand-foot Ruby Range and coming to a frozen standstill over their heads.

"I imagine you are chafing with impatience at our Slow progress [in] track laying," Charley Crocker wrote to Huntington on December 15, knowing that he would be facing his associate before the letter was read. "I assure you I am about discouraged. It seems to me that I have exhausted every resource & failure seems to stare me in the face." Weather had been troublesome, of course—"thermometer going down 8 & 10 degrees below zero, froze up some of the mountain streams we obtained water from, which laid us out 3 days." And they had been plagued with mechanical difficulties with four locomotives, which "have never yet succeeded in getting through to the end of their route with their trains—something has broken or given out & they have been compelled to leave their trains & come back to the shop for some trivial repair that could not be done, however, out on the road." But the worst thing was green wood—fuel cut less than six weeks before, "& the result is an Engine can do but half work & cannot be depended upon for that even. Trains get behind time & delay others & the result is we are doing but little."[76]

Hopkins elaborated: "can't make steam," he wrote that same day, "our largest locomotives with 14 cars, all day getting 49 miles—out on the Humboldt, having to stop every half hour to get up steam enough to move slowly on, consuming great quantities of green pine wood to little purpose—& all this green wood has to be hauled in this way to supply the road from the Truckee 350 miles." They were transporting thirty-five cars of wet, green wood per day. "It is discouraging," Hopkins told Huntington. "I am not Rail Road man enough to see the way to overcome this difficulty, & it seems Charley was not Railroad man enough to avoid it, at the time & in the only way it could have been avoided—that is, to have had wood enough cut last summer for use this winter."

Erastus Corning had called Hopkins's attention to the problem early in the summer, and more than once Hopkins had mentioned it to Charley Crocker, but each time he had said "the choppers wouldn't chop it," though he offered a high price, and he "couldn't spare Chinamen then, but he would have them at it in time to meet our wants." The result, Hopkins said, "is we go into the winter with only green wood, & green wood won't do good work, but it does dishearten & demoralize all the men & render all systematic organization an expensive farce." He hated as much as Charley to now bear such bad news. "I have before now, more than once perhaps, told you," he wrote Huntington, "we had more talent for getting out of difficulties than we had for avoiding them. It often seems so to me. But I don't quite see how we can laugh or pettifog ourselves out of this one."[77]

But miraculously, only a week later Hopkins's black mood had lifted—partially because the money market was dramatically easing and partially because the newspapers were full of the Union Pacific's troubles with money and politicians and bad publicity. Rich mining districts in Nevada had just been discovered and publicized, which before too long would be nurturing their railroad. But even more good news had returned with Charley Crocker from the front, where he had gone to see how they could improve upon the Sierra's "green, sappy pine tree tops" they had been forced to burn. He had found a quantity of scrubby nut pines and junipers, along with a considerable quantity of dead timber, which could be hauled inexpensively down from the foothills—"as good fuel as seasoned pine wood from the Sierra mountain large timber," Hopkins wrote. "The consequence is that every Locomotive can do the work reasonably expected of her, & so many Locomotives & cars are released from hauling wood that much more material can be placed within reach of the track laying gang—who are all the time able to lay down all the track we can give them materials for, up to five miles a day."

The track had made 463 miles on December 21. It was now just thirty-six miles from Humboldt Wells. Ten miles beyond that point waited an abundant supply of good wood already cut from the slopes of the Pequop Mountains and stacked along the empty grade. They hoped to reach it in three weeks or less. "It is," Hopkins wrote to Huntington, "a real pleasure to relieve you of some of the apprehensions I felt & expressed to you when I wrote you last on this wood question."[78]

Indeed, perhaps things were not so bad after all, Hopkins and the Crocker brothers were thinking as the curtain on 1868 began to fall, with Huntington expected any day now in Sacramento and Stanford continuing to bumble on in Salt Lake City. There was a great deal of business to anticipate: land sales for the California and Oregon and the Southern Pacific, for instance, but also even those measly eastern Nevada townsites where the new mining interests would be drawn for freight and supplies; there were indications that Ben Holladay, the stagecoach king, was flexing muscle over an entity called the Oregon Central Railroad, which he might plant in the Willamette Valley and then sell, perhaps to them. Politics seemed to be brighter; every Republican was looking forward to a new start with the inauguration, three months away, and most Californians (including the Associates) were hoping that Senator Conness would be named to a post in the Grant administration—did they dare hope for the interior secretary? Other members of local delegations were falling into line, such as the Nevada representative-elect, Thomas Fitch, who had borrowed $500 in gold from Hopkins to pay steamship passage to the East for his wife and sister-in-law, and even more happily, California senator Cornelius Cole, who had finally promised to end his obstructions regarding the Southern Pacific, Goat Island, and the several waterfront enterprises. Cole may have been a "brazen faced fool" in Hopkins's view, who was demanding too much for his support, but at least he was now pledged to do everything they wanted. Huntington had lent Cole $5,000 before he left New York; presumably he entered the amount in the company books so as not to alarm the bookkeeper with a loan to stay "outstanding" until Hell froze over. To smooth their way in Washington even more, Huntington had hired the indefatigable and influential Senator Roscoe Conkling of New York as their attorney in the East.[79]

From fair Sacramento to the icebound Humboldt, they could boast of 7 engine houses, 6 repair shops, 140 locomotives, 13 passenger cars, 6 baggage cars, 2 mail and express cars, 1,333 freight cars, 95 dump cars, 79 handcars, 63 section cars, 10 yard cars, 34 iron cars, and 6 snowplows. They had capital stock of $100 million, with nearly a quarter of that subscribed; they owed nearly $38 million. They had receipts of $2.3 million for passenger, freight, and mail service. They had paid taxes of more than $122,000. The figures went on—and if they were staggering, they were also impressive.[80] The editor of the *Reno Crescent* had commented on the personal side of the extending business, only three or four weeks before—on the Central Pacific's effect beyond the Sierra's eastern slope. He marveled at the bright new timetable of service from Sacramento to Argenta. "Think of that ye Sage Brushers," he exhorted,

> *who used to occasionally span the Humboldt Deserts on the hurricane deck of a mustang, famished for water, eaten up with alkali dust, chilled by Washoe zephyrs or petrified with blasts like simoons, think of getting into a close, comfortable carriage and speeding away, at a pace which would kill a Norfolk in an hour, on and on, in a single night passing all that dreary stretch of country this side of the Humboldt, that country paved with the bones of animals, that died from exhaustion in the great exodus to California, that country that still lingers in the memory of former emigrants as the dark and fearful ground.*

"Verily the Railroad is a benefactor," he finished, "a benificent engine of glory and strength."[81]

Perhaps it was not such a bad year after all. And praise like that would be handy comfort in times awaiting.

On Tuesday, December 29, at the Union Pacific offices in New York, Oliver Ames convened a meeting of six of the seven trustees of the Crédit Mobilier construction contract—Ames, John Alley, Dillon, Bushnell, McComb, and Benjamin Bates, with Durant proxied from the Wasatch—to discuss the question of a new dividend on their investments, the sixth such allotment. Not all were in favor of such a munificent award for each of them to give themselves and the lesser shareholders of Crédit Mobilier; the Union Pacific, was, after all, in grave trouble, with the obligations still pouring in, no clear way to pay them all in sight, and much expensive work to go during the final push—through the Utah canyons and up around the north of the lake as far as they could push or intimidate the Californians. John Alley in particular was livid about such a grabbing scheme, and would soon resign from the trustees in protest. Bushnell and Ames may have had lingering doubts. But they stifled them, put on resolute faces, and voted, along with the absentee doctor and the others, to issue a 200 percent dividend anyway. Showing, at least, a slight prudence, it was payable in Union Pacific stock—rather than the earlier dividends which had been paid in cash. "We supposed," Cornelius Bushnell would explain lamely to investigators later, "we were going to make some money on the rest of the road." Out in Utah, though, the Doctor was in a breakneck race with the Governor—and, with the contest now getting so personal, there would assuredly be more to pay.[82]

In Washington there was a quiet, merry gathering at Secretary Browning's house on New Year's Eve. "The Almighty has been kind and good to us, far beyond our deserving through the past year," he wrote in his diary on Friday, January 1, 1869, "and I implore his mercy and blessing for the year upon which we are just entering. I beg that we may be forgiven for all the errors and sins of the past," he continued, "and be guided in the paths of wisdom, and delivered from the temptations of the future."[83]

Part IX

1869
Battleground and Meeting Ground

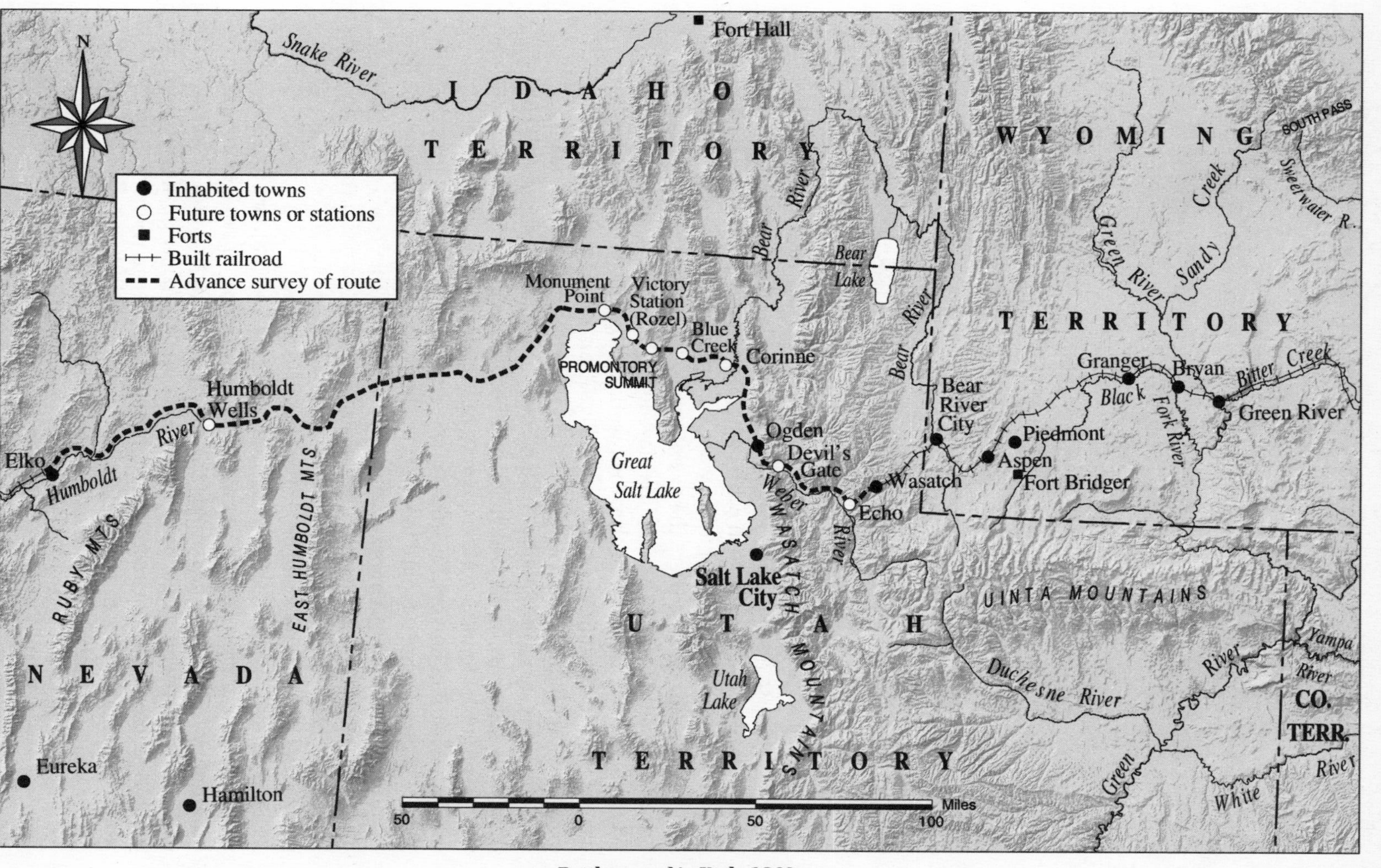

Battleground in Utah, 1869

31

"A Resistless Power"

California detained him only ten days—and Collis Huntington would boast many years later of that pellmell race across the continent by rail and special stagecoach relays—"only gone 31 days from New York," he told a biographer, and "it cost me some money. They ran a race night and day." On the way west with Stanford, Huntington had spied a party of woodcutters delivering ties down to the grade in Echo Canyon, and ordered the stage halted to see what they were up to. "I asked what the price was," he recalled. "They said $1.75 each. I asked where they were hauled from, and they said from a certain canyon. They said it took three days to get a load up to the top of the Wasatch Mountains and get back to their work. I asked them what they had a day for their teams, and they said ten dollars. This would make the cost of each tie more than six dollars." Huntington knew his railroad was taking a similar bite in getting ties down to Nevada from the California mountains, but during his return trip, which included another conference with Stanford in Salt Lake City, he found a sardonic end to his anecdote. "I passed back that way in the night in January," he said, "and I saw a large fire burning near the Wasatch summit, and I stopped to look at it. They had, I think, from twenty to twenty-five ties burning. They said it was so fearfully cold they could not stand it without having a fire to warm themselves." Huntington resumed his way back to the East—having had the pleasure of having seen some of Durant's and Ames's dollars swiftly and brightly going up in smoke.[1]

He hoped to reduce their hopes in Utah to embers, too. Stanford had been right about the impracticality and waste of building in the Wasatch Canyons, but Huntington's fallback plan would buy time by insisting that the Union Pacific be held to a continuous, permanent, first-class line going through—not around—those three time-consuming tunnels in the canyons; Stanford would document where the Union Pacific crossed their preliminary lines up Weber and Echo and wherever their completed grades were sliced over by the graders, and their engineers would sign supportive affidavits, preparing for

Huntington to obtain an injunction. One way or the other—and on this Huntington was resolute as he boarded an enemy train in Wyoming—the Central Pacific was going to get at least to the village of Ogden.[2] That is, he thought, if his associates did not let him down.

Stanford, meanwhile, had gone north to examine the work—and he returned with glum reports. "There was not as much work done as I expected or as had been represented," he wrote Huntington. The ground was so frozen that it defeated the small farming plows employed by their Mormon graders, so Stanford wired for fifty large plows to get the ground broken. Even with the slow work he thought Utah matters were such that as soon as possible he could go back to see his wife and baby boy in Sacramento for a few days.[3]

Out in Nevada, the tracklaying gangs had yet to pick up speed—the supply locomotives, still hampered by green wood and low steam, were supplementing fuel with expensive Mount Diablo coal, inching toward those waiting stacks of good wood upriver. Huntington had sent Charley Crocker a telegram from Salt Lake City, upbraiding him for delays and inefficiencies. It arrived at the Humboldt camp at the worst time. A few laborers turned up with cases of smallpox and it sparked a panic. "The small pox completely demoralized our track laying force," Charley told Huntington, "& they could not have laid much more Iron if they had it, as very nearly all the White man left the work & most of our best foremen also. We are breaking in Chinamen & learning them as fast as possible. They have much to learn but are apt." Then, he said, someone handed him Huntington's telegram. "Strobridge sick with a very bad cold & afraid it was the small pox as the symptoms were very similar. Men running off scared out of their senses. Two cases of small pox among the wood choppers at Elko. Thermometer 10° below zero—& everybody in a demoralized condition—on top of all your dispatch that 'if we failed it was my failure *as I had the means*' &c &c." Huntington's telegram was "*very encouraging,*" Crocker wrote with undisguised sarcasm, "was it not?"[4]

As Huntington raced back toward the East, his mind full of schemes, the situation in the capital was changing rapidly. When Grenville Dodge had stalked out of the interior secretary's office on December 19, with Browning's platitudes in his ears, he immediately launched his campaign to make sure his company prevailed over the Californians. His letter to his former comrade, Secretary of War Schofield, had been only the opening salvo. Every government dignitary he met in the closing weeks of the year and the opening of the next heard his diatribe against Browning and Huntington. By December 30 he finished a twenty-five-hundred-word response to Browning's rulings, in crushing detail, to go out over Oliver Ames's signature. The Union Pacific's location through Echo and Weber Canyons was superior in every way and had been certified as such by the special three-man commission, he said. The grading was all but completed, tunnels were well under way, the Union Pacific track had advanced to the temporary bypass of the big Echo Canyon tunnel. They could therefore not comply with Browning's written demand to join the Central Pacific line up at the rim of Echo Canyon. "The Central Pacific railroad location must have been accepted by the honorable Secretary through a misrepresentation of the facts," he wrote, "and as their map in no sense complies with the rules and regulations of the department, under which rule the maps of both companies should be filed, we do not admit that it is, in any sense, the final location of the road, or should in any way control our location or affect our rights." Moreover, he continued, the Union Pacific "received no notice from the Interior Department of the filing of such a line." Dodge repeated what he had pressured Browning to concede on December 19—"that filing a map gave no right to one company over another in the building of the road." He urged reconsideration, "and our location adopted to a point as far west as we have a certainty of building, or, say, to a point equidistant between the ends of the two roads. . . . This would be fair, impartial, and just to both companies."[5]

Oliver Ames signed the letter, but for some reason he held it back for five weeks, until February 10. However, Dodge had made multiple copies; in the interim he evidently shared either the letter or the information and arguments it contained with key congressmen and other members of President Johnson's cabinet. Knowing that a number of press correspondents were working on stories—it was, in fact, as if they were camped all over the capital, huddled over their own mysterious campfires—he poured his own propellant of news, accusation, and rumor, not knowing (perhaps not caring) how large a flame might result, how explosive it might be, and whether or not he himself might be burned. Snyder, Hoxie, Reed, Evans, and now Blickensderfer had all told him they didn't care what happened to themselves anymore. Everyone was exhausted, demoralized, fatalistic, and not a little crazed. So Dodge poured and stepped back to see what would result.[6]

Amazingly, at first Secretary Browning seemed to retreat toward his earlier agreement with Collis Huntington. On January 6, he insisted once again that they join the Central Pacific line at the Echo summit. Ames entreated Dodge to do something. "Is there no way for us to avoid this, and are we to lose our subsidy?" he anxiously wrote his engineer on January 8 from New York.

> *If this is to be so, we better give up our road where it is and stop our work. I have no idea of doing this as Browning desires. He evidently wants to force us to give up our grading and take that of the Central and build our road on their line or lose our subsidy. The old hypocrite! I thought when he was saying to us that the location of this line in advance of ours gave them no rights he meant what he said, and would simply ask that the roads should be joined when they met.*

You must get some immediate action of Congress to have this matter put right and not let our line be sacrificed in this way. The idea that men like Browning are to sit in their office and fix the line on which these roads shall run when they have not seen or examined the line of our road nor know anything of it or either of them. I see no way for us to act if Browning's action is sustained but to withdraw our forces and wait for Central to build the road. We can't go on without the subsidy and if the subsidy is to be applied to their line we must pull up our track and put it on their line or quit.

Later that day, after Ames had taken a train up to Massachusetts and had time to think about the matter, he wrote Dodge again. "It would be infamous for such action as Browning is taking to be sustained," he said, "and your documents can't be used too quick—to show him up in Congress. Our hope is in Congress. The cabinet will be too deferential, to the head of the Interior Department just now, that its corruption is being exposed, and this action of Browning shows that the head is corrupt."[7]

Dodge's campaign against Browning was already well under way. Many senators and congressmen tied to the army, or to the radical Republicans, or to the net of influence woven by the Union Pacific, were already beginning to speak up. "I was utilizing it very effectively," he boasted, and as he continued to circulate his long protest letter to the secretary he encouraged the whisper campaign about Browning's probable crookedness, and he let it get around that the secretary's order for them to abandon their nearly completed work for the Californians' imaginary line "was simply nonsense"—and, he vowed, "I proposed to bring it up in Congress unless an adjustment was soon made."[8]

Other changes were being made out in Utah, and as Dodge learned about them during those few heady days, his temper rose toward the boiling point. Durant and Seymour were huddled over location maps in Ogden and, to save time and much money, had begun ordering alterations in the approved line up into the Promontory Mountains, which Dodge's engineers had run with eighty-foot grades. Durant ordered the considerably steeper one-hundred-foot grades. Divisional engineer Jacob Blickensderfer was in Salt Lake City at the time, trying to put his office into order, anticipating a move to a less frustrating post, and he refused to go along. "This makes a cheap line but a higher summit," Blickensderfer would write Dodge, "and about 3 miles longer and more curvature. Commercially the line is so inferior to the revised 80 foot grade that . . . I told him it would not answer." The Doctor's reply was to fire the dissenting engineer. "You will please consider your services for this company at an end," he wired on January 2 from Ogden, "from and after the time when you left your work to accept an appointment under the Government." It was insult upon injury, so typical of Durant: on top of the dismissal, to deny Blickensderfer, a family man, nearly two months' pay due him during his sanctioned leave of absence to serve on the commission. When word reached

Washington—Jesse Williams probably learned first and alerted Browning's staff—the Interior Department was infuriated.

Dodge heard and immediately wired to countermand the order. On January 6 he followed with another telegram telling Blickensderfer to remain in Salt Lake City and finish the paperwork he was trying to do. Blickensderfer had intended afterward to ride north to inspect the Union Pacific's surveying and grading work, and even to see if he could document that the Central Pacific line to Echo summit was "an actual falsehood." Now he wondered if it were worth it. "I claim to be a friend of your road," he wrote Dodge, and

> *my sympathies are all with you. Your location I know to be good and that of the C. P. Company, as far as I have seen it, is inferior to yours. I think they have simply imposed an untruth on Mr. Browning, and if so I would be glad to possess evidence by personal knowledge, which I believe exists to prove this to the satisfaction of your friends or the cabinet at Washington if necessary. These being my feelings, I am not disposed to let the bad treatment received from Durant and Seymour influence me, but I am inclined nevertheless to do all for you that I can, and go East fully armed with the best information I can collect. On the other hand I am not inclined to go out on the line, ask questions and drum up facts, with the liability of being snubbed off by any subordinate hireling with the question, 'What business have you here?' For I know pains have been taken to let it be known that I am stripped of authority.*[9]

Operations Manager Webster Snyder was out on the line, and his dispatch reached Dodge on January 4. "There is so much to say about the work out here that I can't do the outfit justice in writing about it," Snyder told him. "In construction the waste of money is awful. It is the last part of Reed and his outfit and they are making the most of it. The track west of Aspen is not fit to run over and we are ditching trains daily. Grading is done at an enormous expense by day work under supervision of Company's men and the government subsidy in this section of country will not begin to pay cost of road." Snyder was ready to jump ship. "Personally," he said, "I am about worn out and if I had money enough to support my family six months, I would quit now. The company can't stand such drafts as I know the Construction Department must be making."[10]

Such news did not have to go through Dodge to reach Washington in those early weeks of the year and increase the pressure on Secretary Browning, or to surface in the press. Dodge was actively leaking information, to be sure, and quite possibly so were Blickensderfer, Snyder, Jesse Williams, even some of the rebellious investors such as McComb or Hazard. Whatever their sources, critics were popping up in print all over the nation. "What guaranty has the Government that the companies will ever complete their roads?" a

Washington correspondent of Horace Greeley's *New York Tribune* had written on January 5. Congress had granted subsidies without independently determining costs—"hence the evident fraud upon the government."[11]

If skulduggery was beginning to be unveiled by Congress and the press, the heaviest, most far-back curtain was skillfully pulled aside early that month by Charles Francis Adams Jr., the thirty-three-year-old Harvard man and grandson of President John Quincy Adams. Adams was trained at law by the eminent attorney Richard Henry Dana, who himself had earned a literary reputation nearly twenty years before with a book based on his year at sea as a common sailor, *Two Years Before the Mast.* Not long after young Adams had been admitted to the bar he hearkened toward the journalistic side, beginning his career in 1860 with an article in the *Atlantic Monthly,* "The Reign of King Cotton." The war had interrupted his rise; Adams enlisted in the state militia as a captain and by the end of the war had been promoted to colonel of the 5th Massachusetts Cavalry, a negro regiment. After the war, his health broken, Adams had spent nearly a year in Europe recuperating, but he returned to the states determined to make his mark as a writer.

The most likely area of specialty for an enterprising correspondent, he decided, was the railroad industry. His first article for the influential *North American Review*—a monthly founded by the father of his mentor in law and journalism, Richard Henry Dana Sr. On the "wretched history" of the Erie Railroad, it appeared in April 1868, driving thorns into the sides of Cornelius Vanderbilt, Jay Gould, and Jim Fisk, among others. He would return again and again to the Erie, but soon after that opening salvo, Charles Francis Adams Jr. turned his sights on the Pacific Railroad, for, as he would write, "The Pacific Railroad is already a power in the land, and is destined to be a power vastly greater than it now is." Already it "numbers its retainers in both houses of Congress, and is building up great communities in the heart of the continent. It will one day be the richest and most powerful corporation in the world; it will probably also be the most corrupt."

Adams confessed it was not pleasant to level criticism at the "able and daring men who are with such splendid energy forcing it through to completion." Undeniably it was important and daunting work, and the men at its head "incurred great risk, and at one time trembled on the verge of ruin." But the enterprises were riddled with mystery. Of the Central Pacific, he said, "absolutely nothing is publicly known. Managed by a small clique in California, its internal arrangements are involved in about the same obscurity as are the rites of Freemasonry." But the Union Pacific was closer to home, and rumors were as rife on Wall Street as they were in Washington, and Adams had been listening. Rumors talked about, he disclosed, "a new piece of machinery" called the Crédit Mobilier. No longer would the obscure words be whispered. "The Crédit Mobilier is understood to be building the road, he continued, but what this Crédit Mobilier is seems to be as much shrouded in mystery as is the fate of the missing $180,000,000 of capital stock of these roads." Now there was litigation over its proprietorship, he noted.

> *Whoever originated this anomalous corporation, it is currently reported to be the real constructor of the Union Pacific, and now to have got into its hands all the unissued stock, the proceeds of the bonds sold, the government bonds, and the earnings of the road,—in fact, all its available assets. Its profits are reported to have been enormous,—reported only, for throughout all this there is nothing but hearsay and street rumor to rely on. Sometimes it has been stated that the dividends of this association have amounted to forty per cent a month, and they have certainly exceeded one hundred per cent per annum; at any rate, it has made the fortunes of many, and perhaps of most of those connected with it. Nor are these profits temporary; every dollar of excessive dividend of the Crédit Mobilier is represented by a dollar of indebtedness of the Pacific Railroad, with both principal and interest charged to income, and made payable by a tax on trade.*

And who, then, he continued, constitutes the Crédit Mobilier?

> *It is but another name for the Pacific Railroad ring. The members of it are in Congress; they are trustees for the bond-holders, they are directors, they are stockholders, they are contractors; in Washington they vote the subsidies, in New York they receive them, upon the Plains they expend them, and in the Crédit Mobilier they divide them. Ever-shifting characters, they are ever ubiquitous—now engineering a bill, and now a bridge,—they receive money into one hand as a corporation, and pay it into the other as a contractor.*

Here, he continued, "is every vicious element of railroad construction and management; here is costly construction, entailing future taxation on trade; here are tens of millions of fictitious capital; here is a road built on the sale of its bonds, and with the aid of subsidies; here is every element of cost recklessly exaggerated, and the whole at some future day is to make itself felt as a burden on the trade which it is to create, and will surely hereafter constitute a source of corruption in the politics of the land, and a resistless power in its legislature."[12]

It was a brilliant piece of investigative reportage—a true classic—and an eye-opener for the general public. Adams backed his arguments with as many facts and figures as he could muster, given the mysterious circumstances of both sides of the Pacific Railroad, and he called for a new bureau of transportation to be created within the Interior Department, to educate Congress and the country. "We might then hope to know how large a tax is annually levied on business under the head of transportation, and how large a portion of it is applied to the payment of dividends and interest on paper capital," he argued. "We might then hope to know how much our railroad system has cost, and by what securities that cost is represented; it might then some day become difficult to deluge the market with forged certificates of stock, and call the so doing 'a financial irregularity;' it might even become questionable whether a railroad potentate had the right to double the nominal cost of a public thoroughfare without adding one dollar to its value."[13]

Other angry voices were rising, notably that of General Henry Van Ness Boynton, Washington correspondent of the *Cincinnati Gazette;* Boynton was a hero at Chickamauga and now a respected Republican voice who more than once over the years would team with Dodge in party maneuvers. His January articles on the pattern of frauds scrutinized several prominent and obscure companies in Kansas and Iowa, and even the Central Pacific's Sacramento and San Jose Railroad, which was all but unknown in the East. Horace Greeley reprinted Boynton's articles in the *Tribune,* praised them, and took up the call to find about more about the Union Pacific, the Central Pacific, and this mysterious Crédit Mobilier.[14]

This clamor in the columns was impossible for Orville Hickman Browning to ignore. Dodge himself may have shown him a path, sending the secretary an entreaty on January 11: if, Dodge said, the Union Pacific examiners had been too harsh and the Central Pacific examiners too lenient (indeed, he complained, it was "a whitewash"), the department should "have only one commission examine *both* roads again, in simple justice to both." It was a howl of protest from a punctilious engineer—and it would dismay the Ameses, who knew that every delay or new level of oversight was going to bury the Union Pacific deeper in debt—but a few days later Browning decided he could delay a public response no further, and edged himself back into safer terrain while pushing the Dodge idea into more stringent territory. He created a new commission to "examine both roads and determine on a point of meeting, *even if it takes a completely new location.*" So these commissioners would scrutinize it from west to east, and they would have the final say about a meeting point and the approach routes. An enormous amount of work by both companies was, with one stroke, imperiled. Browning appointed Jacob Blickensderfer and General Warren from the Union Pacific special commission, and Lieutenant Colonel Williamson, who had, likewise, inspected the Central Pacific.[15]

Dodge was beside himself. On top of the worry about Durant changing his lines north of Great Salt Lake, he had this commission—which might make "radical" alterations. He sent an anxious bulletin to Oliver Ames on January 15. Oakes Ames had gone to see President Johnson, the only one who could overrule the cabinet secretary, but the congressman was hardly in a position to sway the president. Oliver Ames was afraid that his chief engineer might run to the newspapers about Browning or Durant or otherwise rock the boat, and quickly reassured him not to fret on either front. "I have telegraphed [Durant] as you suggested that no alteration be made in the line of road as finally located," Ames said,

> *and have also written Durant a letter enclosing him a copy of the instructions of Browning. I think he will do what he can to remedy any changes he has made in line. Durant has been as furious in his demonstrations as any one could well be generally but has till now made no important changes in your line. . . . I hope you will feel that though the Dr. may want power and exercise it without judgment frequently, yet the board of directors are strongly your friends and I hope you will not let your feelings against Durant lead you into any demonstrations against the road.*[16]

That day, the *New York Tribune* printed a scorching editorial against the crimes of the newly unveiled Crédit Mobilier, the corruption of congressmen who had evidently sold their votes, and the greediness of the whole class of railroaders—"all willing to build railroads from the Sun to the Moon, *provided they can have subsidies.*"[17]

Collis Huntington arrived in New York City early in the morning of January 19. Waiting for him in the stack of office mail was the outrageous letter from Secretary Browning—telling him that the new commission would examine the whole line and decide who got the approval and the bonds in Utah. There was no time to consider the makeup of the board beyond Williamson, whom he knew—the inspectors were already on their way west, and he had to react quickly. "The Union Pacific has outbid me," he concluded.[18] Fuming, he hopped on a train to Washington, scrawling off a protest to President Johnson. The Central Pacific, he wrote, was entitled to the subsidies by the previous approval and by the unscrupulous, "temporary and superficial" construction of the Union Pacific company; Ames and Durant had placed their forces "from 50 to 350 miles in advance of their permanent work," seeking to take advantage of the Central Pacific's delays in the Sierra Nevada and its more responsible building. By doing so it would swindle the subsidies, steal the trade of Salt Lake City and the region, and snatch the Wasatch coal deposits.[19] He posted the protest in Washington and got in to see Secretary Browning—but the Interior Department's climate was now considerably chillier than before. Browning informed him loftily that it was out of his hands.

"For God's sake," Huntington implored his associates, "push the work on. . . . If I was there I would not take off my clothes or change my shirt until the rails were laid to Ogden City." He told Hopkins that he was going ahead with a lawsuit against the Union Pacific for trespass upon their approved line in the canyons. "I think by commencing this suit we can cripple them," he said. The Union Pacific was, after all, skating on the edge of bankruptcy, and

Huntington was now in the mood for pushing them all in. "I have felt for the last 100 days," he explained, "as though I would like a hell of a fight with someone."[20]

Stanford was miraculously still in Salt Lake City when the new commissioners got to Utah; Blickensderfer, of course, had been anonymously in town closing down his company duties. "Now can it be possible that we are to lose our work simply because we do not cover it immediately with the rails?" Stanford asked Huntington on January 24. He stayed, looked up the commissioners, and pumped them for all the information they were worth. It seemed to him, he warned Huntington, that they would approve the permanent Union Pacific line in the canyons up to the two-thirds subsidy for advance work, and save the issue of the undone tunnels for later. The Governor ventured to blame himself. "I fear [Huntington] is having a hard time in trying to save what a want of foresight has jeopardised if not lost," he wrote to Hopkins the next day. "I tell you, Hopkins, the thought makes me feel like a dog; I have no pleasure in the thought of Railroad. It is mortification."[21]

On January 30, Huntington received Oliver Ames and Dodge in his hotel room in Washington. Ames got to the point: he proposed the two companies take the distance remaining between their ends of track and meet halfway. Huntington's response was characteristically unvarnished: "I'll see you damned first." He named the mouth of Weber Canyon. Now it was Ames's turn to lose his composure. He refused and stormed out. There was much pushing him to a solution, though, and he went back to see Huntington. Then the Californian gave a small concession: Ogden, he said, but no farther. Here the parley, if one could call it that, broke down.[22] But Huntington was pulling ammunition out of all his pockets: he had wired Judge Crocker to send all the regular twenty-mile commissioners' reports before the first of March, knowing that the world for them would change for the worse when Grant was inaugurated on March 4. All of the lame ducks in the Johnson administration already had their minds elsewhere—how could he gain advantage?[23]

In Salt Lake City, General Warren, acting as chairman of the commission, asked Stanford for all the Central Pacific maps, profiles, and reports for the contested land lying between the ends of track. The Governor stalled him; they were all in Sacramento, he said, and would be copied. The Central Pacific continued to deem Secretary Browning's approval as final, he added. It was rapidly laying track and hastening the grade work. He said, too, that he would accompany the inspectors when they left on February 1.They would take the central Nevada stage route westward and, as if to remain unsullied or undistracted, would even go by coach over the Sierra. The plan was to commence at Sacramento, going all the way to Omaha—and see what they could see.[24]

The Governor would be leaving Utah with work on the Promontory now well covered. By the time the commission worked its way back there was even the assurance of new accomplishments. Happily, life in eastern Nevada for supervisors and tracklayers alike had changed enormously for the better in only ten days. The smallpox epidemic was almost over; they had established hospitals on wheels, so-called—there is no evidence that any doctors were present. "Only one new case there last week," Hopkins rather coolly reported to Huntington on January 31. "Nearly all died who went into the pest cars, & those who did not die increased the panic among the men more than those who died and 'told no tales.'" But at least the men and their locomotive boilers were sufficiently warm. "We have now [an] abundance of good wood on the Humboldt Division," Hopkins wrote, "plenty of motive power & no complaint of not having cars enough." Some twelve hundred flatcars were rushing material forward, and Hopkins hoped they could now improve upon previous records—they had laid 38 miles in January, he said, making their total 514 miles from Sacramento. Perhaps they might now reach twenty miles per week—Charley Crocker was going back out to raise the pressure—but Hopkins doubted they could do better, given their experience and organization. Now, though, they were but nine miles from Humboldt Wells, and there was enough iron on the eastern side of the Sierra to stretch all the way to Ogden.[25]

The commissioners arrived in Sacramento on Saturday, February 8, a day on which Hopkins would gladly note that he had sent Huntington two more twenty-mile reports, and that they had breezed across the snow-whitened meadows at Humboldt Wells and reached milepost 532. When he met Stanford and the newcomers, though, his mood darkened. The Governor was impatient with alarming news to send to Huntington: on the way from Salt Lake he had discovered how disinterested one of the inspectors—the severe one with the barely pronounceable Dutch name—would be. "Jacob Blickensderfer," he wired, "was the first assistant engineer & assistant in locating the line of the U. P. R. R. from Ogden to Humboldt Wells. Justice demands that L. M. Clement, our first assistant & locating engineer, be added to the commission." That night there would be a dinner, and the commissioners would observe Sunday there in the capital and then go see San Francisco for a week (during which Williamson had pressing personal business to transact) while the Central Pacific draftsmen finished copying their maps. It would be a useful hiatus while Huntington raised his protest in Washington. Even so, Hopkins was worried. "Warren & Williamson appear like pure minded, good men," he explained to Huntington, "& competent for the service required." However, "Blickensderfer is a different appearing man—apparently of puritanic conscience, scrupulous, cold & distant. At the same time a sinister something about his look & manner *warning you to beware.*"[26] The words might have issued from Thomas Clark Durant.

"The Doctor himself, I think," wrote Samuel Reed to his wife on January 12, "is getting frightened at the bills. He costs hundreds of thousands of dollars extra every month he remains here and does not advance, but retards the work."[27] A plentiful supply of railroad ties lay out in the woods, frozen and buried under several feet of snow after Durant had rashly redeployed his tie contractors' woodcutters into distribution duties at trackside; graders had been similarly shifted before the frost. Now, in January and February, Durant was paying huge sums to secure yet more ties and to blast frozen ground. It was a shocking, needless waste. Oliver Ames begged him over and over to save them. "It would be an eternal disgrace to us, and to you in particular as the manager of the construction," he had written earlier that month, "to be forced to suspend for want of funds to continue the work."

The bills kept pouring in. Joseph A. West, the son of the Mormon grading contractor Chauncey W. West, recalled how tension increased during the parallel grading in the Salt Lake Valley. "Competition for men and teams . . . became so great that the companies began to bid off each other's men by increasing wages," he wrote, "and thus the construction cost became enormously heavy, especially towards its close. Transient labor, too, became the masters of the situation instead of the employer, and as usually happens where the wages are high, the service rendered became very inefficient and undependable."[28] Oliver Ames's alarm rose with each passing day. "Every thing depends upon the Economy and vigor with which you press the work on Construction," Ames wrote Durant on January 16. "We hear here awful Stories of the cost of the work and the thieving of our Employees."[29] Durant would not—indeed, now he could not—pay such admonitions any attention.

And the Doctor's explosive temper was not helping matters at all—not even with the tunnels. For that, finally, newfangled nitroglycerin with all of its dangers would answer. Up at the head of Echo Canyon, the 772-foot bore was moving so slowly at the end of December that Durant had Samuel Reed import a small army of men fresh from the Wyoming rock work to relieve the Mormon contractors and address not only Tunnel 2 but the two smaller ones down in Weber Canyon, which were, respectively, 508 and 297 feet long. Reed put three shifts to work on all. Brigham Young admitted he was glad to be relieved of the pressure. "I could not have asked Dr. Durant to confer a greater favor," he had written an associate on January 5. But the doctor's experiment fizzled within a month. "The big tunnel which the company's men took off from our hands to complete in a hurry, has been proffered back again," Young noted with satisfaction, a good point having been proved.

> *They have not less than four men to our one constantly employed, and, withal, have not been doing over two-thirds as much work. Superintendent Reed has solicited us to resume it again. We were well pleased to have the job off from our hands when it was, as it enabled us to complete our other work on the line; but now that it is so nearly complete, probably we shall finish the tunnel. Bishop Sharp and Joseph A. Young are using the nitro-glycerine*

for blasting, and its superiority over powder, as well as the sobriety, steadiness and industry of our men, gives us a marked advantage.[30]

The headings of Tunnel 2 were blasted through on January 30, though more than sixty days would pass before the bottoms were cleared out. Not until later in the month of February would the Gentile contractors still working on the two Weber Canyon tunnels be moved to try nitroglycerine on that tough black limestone and blue quartzite. One in five of the laborers immediately went on strike. They were simply fired—nitro would hasten progress enough to let the contractors get by with only two shifts, and the company saved tens of thousands of dollars in the bargain.[31]

Brigham Young and many of the faithful had been worried about the moral pollution of their settlements when the railroad began importing men from the tough Wyoming camps. The town of Wasatch, altitude seven thousand feet, had become the winter headquarters of the Casement brothers' tracklayers and, with a population of some fifteen hundred cold and restless souls, was getting fairly wild. J. H. Beadle, who had bought the newly established and struggling Gentile newspaper *The Salt Lake Reporter* from former General Patrick E. Connor, visited in January and stayed a week, during which time the thermometer ranged from three to twenty degrees below, never rising to zero. "During my stay," he recorded,

the sound of hammer and saw was heard day and night, regardless of the cold, and restaurants were built and fitted up in such haste that guests were eating at the tables, while the carpenters were finishing the weather-boarding—that is, putting on the second lot to "cover joinings." I ate breakfast at the "California" when the cracks were half an inch wide between the "first siding," and the thermometer in the room stood at five below zero! A drop of the hottest coffee spilled upon the cloth froze in a minute, while the gravy was hard on the plate, and the butter frozen in spite of the fastest eater.

This was another "wicked city," he noted. "During its lively existence of three months it established a graveyard with 43 occupants, of whom not one died of disease. Two were killed by an accident in the rock-cut; three got drunk, and froze to death; three were hanged, and many killed in rows, or murdered; one 'girl' stifled herself with the fumes of charcoal, and another inhaled a sweet death in subtle chloroform."[32]

Echo City had been swelling in population for several months now; the track reached it on January 15, greeted by the usual brothels, gambling houses, and saloons plying their trades. Bitter cold weather had discouraged a number of laborers from outside railroad work, and though they resolved to go home had wound up penniless in Echo City with nothing but trouble in their hands. Often the sound of gunfire bounced off canyon walls in the clear, cold night air. Holdups, murders, hangings, and unexplained disappearances became common. Beneath a trapdoor in one saloon putative lawmen found a large hole in which seven unidentified bodies lay amid the tin cans, empty liquor bottles, and other refuse.[33]

Another Hell on Wheels town would soon take root in the Salt Lake Valley north of the Weber mouth, on the west bank of the slow and brackish outlet of Bear River. There Gentile speculators snapped up lots as soon as Union Pacific surveyors had laid out the town of Corinne some twenty-five miles northwest of Ogden, the first non-Mormon town in the territory, and within two weeks there rose more than three hundred frame structures, shacks, lean-tos, tents, and combinations thereof; at least nineteen saloons and two dance halls opened for business, catering to the rough graders imported late in the previous year and to the traders, teamsters, drifters, and ne'er-do-wells who naturally flocked in. "At one time," J. H. Beadle noted, "the town contained 80 *nymphs du pavé,* popularly known in Mountain-English as 'soiled doves.'" The population soared past 1,500, giving the town founders visions of grandeur—Corinne as the beginning of a Gentile empire in Utah, as possible meeting point for the competing Pacific Railroads instead of the Mormons' Ogden, as eventual successor to Salt Lake City as the territorial center. Former General Patrick E. Connor—onetime scourge of peaceful and unsuspecting Indians, still as anti-Mormon as ever, who was busy now with a tie contract for the Union Pacific and whose hopes for a mining industry in the Utah mountains were beginning to succeed—would become a major force in the flowering of the weed which was Corinne. His aspirations—and the town's—were destined, however, to wither.[34]

Corinne was the last "formal" Hell on Wheels town, if such a qualifier may be used, but it was not the last place of iniquity on the line; some thirty-five miles away, up on the Promontory at a place called Blue Creek, a tent town would briefly become, in the words of Mormon contractor Chauncey West's son, "the toughest place on the continent." To J. H. Beadle it was, "for its size, morally nearest to the infernal regions of any town on the road." Many called it Robbers' Roost. It was too close for comfort for the pious Mormon laborers. "On more than one occasion," West recalled, "this rough element assembled in broad daylight, with the avowed intention of raiding the camp of Benson, Farr, and West, where they knew large sums of money had to be kept with which to pay off the men, who invariably demanded coin for their services. This constant menace necessitated the employment of a large force to keep watch over the camp." A company of cavalry from the Territorial Militia had to be stationed there until the work was over.[35]

Accident reports mounted in the divisional office. The stretch of track across the Utah border had been laid on frost and, in Hub Hoxie's dolorous report, "goes down all the time—4 miles per hour is the maximum speed allowed," he wrote Dodge, "and then we are off the track about half the time.

The iron will be worthless by spring and there will be no road left."[36] Due variously to bad weather, poor organization, unsound engineering, hasty construction, insobriety, and simple bad luck, the railroaders were indeed plagued that month by a series of accidents, bad breaks, and near misses in the Wasatch canyons. The smaller ones—crushed limbs or lost fingers due to frozen or slippery switches or couplers—were simply recorded; it was becoming either a badge of honor or a rite of passage for track crewmen to be maimed in such ways, but these small dramas were occasionally greatly overshadowed. One night between Wasatch and Echo City, a drover was hauling a sled of freight and hastened his team to hurry across a grade crossing before an oncoming westbound supply train arrived. "It was a beautiful moonlight night in midwinter," recalled paymaster Erastus Lockwood, the Casements' brother-in-law,

> *and the snow was very deep. One of the sleds was about to cross the track when the runners settled down in the snow and the sled box containing the freight rested squarely on the rails. Our engine struck this sled box, tearing it to splinters, and scattering over the snow great quantities of baby shoes, destined for the Mormons at Salt Lake City. These tiny shoes caused us great trouble as they threw our cars off the track, right and left. Fortunately the engine remained on the track and was dispatched to the end of the track in Weber canyon for a wrecking crew, which was brought back in short order and the track cleared, for under no circumstances must the track laying be delayed. All this occurred about two o'clock in the morning; by seven o'clock the track was entirely cleared.*

Another accident involved a supply train of some sixteen flatcars, which was on a relatively level segment at the top of Echo Canyon when the last four cars became unhitched; the main part had pulled about a half mile ahead when a trainman happened to look back to see the four trailing and fully loaded flatcars, which were now finding the advantage of the grade. He yelled at the engineer to "go like Hell"—and, with throttle open and whistle screaming, the work train fled before the maverick cars. Two brakemen were aboard the pursuing cars but they were fast asleep. Courting derailment at such high speed, the engineer blew the signal to open switches and clear the track as his train rushed through the night. The trainman had by then worked his way to the back car and ordered some workers to hurl stacked ties down onto the track behind them to stop the runaways. Sure enough, when they struck they catapulted high in the air. Somehow the two brakemen landed in snowbanks and were unhurt.[37]

General Jack Casement was gone for most of the month of January, on business in Washington with a brief pass through home in Painesville, Ohio. His brother, Dan, had his hands full keeping the men going—until he had his own serious mishap. He was on a train between Echo City and Wasatch with a party of men when the train was overtaken by a fast, furious snowstorm. The engine stalled, and the men thought they could walk to the Wasatch station. "It was too much for Dan," Lockwood recalled; "the snow too deep, so he begged the others to go on and leave him. But a Captain Alford, who was one of the party, picked him up and carried him the rest of the way."[38]

Jack Casement by this time was en route back to Utah to relieve Dan; he expected that Dan would be gone for a month and return, but apparently the near-death experience was so harrowing that once he got home his brother would answer none of Jack's letters or telegrams until mid-April. General Jack wrote to his wife while passing through Omaha on February 8. "I am afraid the Union Pacific is in a bad way," he told Frances Casement. "They owe an awfull amount and as we are running a big machine that would run us out of money and in debt besides. I was in a great hurry to get out to the ground—don't be alarmed for I don't think we will go to the Poor house." Three days later he was in Echo City. "The company owe Millions of Dollars," he told his wife, "and as Congress and the might of Government is working against them and in the interest of the other end of the Road it makes matters look blue. Dan had collected more money than I thought he could, and if the company don't quarrell too hard amongst them selves we will all come out right." Not everyone was allowing even that cautious measure of hope. An exhausted Samuel Reed wrote his wife that week to confess that "I wish the last rail was laid; too much business is unfitting me for future usefulness. I know it is wearing me out." When they were finally done, he said, "I shall want to leave the day after for home, and hope to have one year's rest at least."[39]

As if in reply, a blizzard swept in. First it covered most of Wyoming, filling the great trough that was the Laramie Plains and blocking some ninety miles of track between Laramie and Rawlins for three weeks, stalling trains at isolated stations, marooning passengers, holding up many others whose trains were simply canceled outside of the stricken area—even Dr. Durant, on his way back to New York, who was stuck east of Evanston at the Aspen depot. Dan Casement, heading east on a plow train, intending to keep going all the way home, was awed by the storm's power and telegraphed Snyder to think twice before sending trains west. "You can't get trains over this division by sending a snow outfit ahead with provisions," he warned—the only way to move was to bunch a regular train right on the heels of the big plow and engines, "and as soon as you get through a cut have train follow. Have seen a cut fill up in two hours that took one hundred men ten hours to shovel out. Train west is well organized, but can't more than keep engines alive when it blows." James Evans reported that Laramie was snowed in, "having had but two trains from the East and none from the West in six days." He had tried to work his way west but gave up. "I am afraid," he wrote Dodge, "that we are going to have some trouble from now on until Spring."[40]

Laramie quickly filled with hundreds of impatient westbound travelers, but much graver trouble was up at the Rawlins station, where two hundred passengers on one train were stuck; a number had come from the West in plenty of time, they had thought, to attend the presidential inauguration. Food and water soon ran short. The station restaurant and other establishments in the raw, windswept hamlet took advantage of the situation, charging exorbitant prices, such as $1.50 for a piece of bread and molasses. The train crew consoled themselves with whiskey and refused to stir. Not until ten days had passed, after the infuriated passengers banded together in an "indignation meeting" and sent protest telegrams to the Congressional Pacific Railroad Committee and other dignitaries in Washington (even Collis Huntington received a wire from stranded friends from California), did the crew agree to push forward—but only if the passengers shoveled. In desperation they agreed, at one point clearing a drift a thousand feet long. But when that was open the engineer had only enough steam to carry them into the deepest part of the drift, where the locomotive stalled again. A telegram from divisional headquarters to Rawlins forbade any further sorties. At this point the crew went on a two-day drunk. About fifty passengers then left on foot for Laramie. They arrived there in four days after terrible suffering. "They denounce the road and its management," a newspaper reported, "in unmeasured terms."[41]

Snow had piled up seriously on the eastern Utah border and blown over the rim and down Echo Canyon, but the tunnel crews were of course under cover and Casement's tracklayers were safely down in Weber Canyon, having passed the thousand-mile post and finally worked their way, literally at the heels of the graders, to the Devil's Gate; fortunately enough food and construction supplies had been stockpiled by Reed and Hoxie to last the laborers while the track was cut by storms. Reed estimated on February 27 that they had food for perhaps twenty days, but after that, he wrote his wife, "we would be starved out." Colonel Seymour had taken lodging in an Ogden boardinghouse, and since there was no snow down near the lake, he could not understand why Casement was not already in sight at the mouth of Weber Canyon. He wired Durant to complain that the daily work rate was too slow. "Can't you induce Casement," he said, "to strike a three mile gait to Bear River?"[42]

The Central Pacific was blocked for five or six days in mid-February when a heavy fall of snow and an avalanche took out part of a trestle in Butte Canyon near Cisco. A correspondent of the *Virginia City Enterprise* was stuck on a snowbound train above Colfax for several days, watching as five locomotives behind a huge snowplow gained a few miles before proving unequal to the growing drifts. An army of shovelers from Sacramento appeared, and after they had been at work most of a day, the reporter decided with a few others to walk ahead nine miles to the Cisco station. "Pits had been dug the first four miles eight feet apart and four feet deep," he wrote, describing a quick method of work which would eventually allow the plow to barge through the remaining snow, "so it was into one, a scrambling out, into another, and so on, the hardest work possible. Four hours tedious walking and the bridge, a mile west of Cisco, was reached. Here a tremendous slide of snow from the mountain had come down and carried away four 'bents' of the bridge, at which a hundred men were at work rebuilding. . . . The train we left behind us did not get through to the bridge until Tuesday at ten o'clock, making the time from Alta just one week, and every man on the road working his level best."

Ten feet of snow had fallen on Cisco in just two days; small wonder the railroad had been overwhelmed. The storm demonstrated one thing: "From Truckee to Alta," the correspondent said, "the Central Pacific Railroad must be shedded—nearly every rod—to be rendered practicable in the Winter. Wherever the sheds are, two engines with a plow can clear the way; in other places, ten are inadequate to the work. We predict the road will be entirely shedded before another Winter."[43] Judge Crocker tended to agree, although when he wrote to Huntington on February 23 he stressed how the covered track had endured. "Our snow sheds were a perfect protection & worked splendidly," he said proudly. "We must put up a few miles more at each end this summer & we shall be all right."

The Judge had received a telegram from the Union Pacifics's Webster Snyder saying that service would open to Ogden on March 20. But Crocker doubted it—it "is a blind," he scoffed, adding that from what he had heard, the Wyoming snow blockades showed no promise of opening for supplies. Stanford and the commissioners had left on the examination, with Lewis Clement along, empowered by a telegram sent by Secretary Browning to Stanford on February 14. The Judge felt optimistic about what they would find, especially with two Californians now on the board: "Williamson feels a deep interest in sustaining his old report," he believed. "Warren & he are military chums, & will probably go together, & with Clement to draw the report, it will probably be all right. Blickensderfer is very particular, & I guess is an off ox generally."[44]

In those high-pressure weeks the Central Pacific also had its share of accidents. A new locomotive called *Blue Jay* was on the return trip of its inaugural run over the Sierra—the *Reno Crescent* had called it "prettier than a spotted mule, or a New York schoolma'am"—when it suddenly and violently caught up with a stalled lumber train. "Bruised, broken, and crippled," the *Crescent* mourned, "it was then taken limping to Sacramento for repairs. . . . [F]ortunately nobody was killed or even wounded." A decoupling similar to the Union Pacific episode in Echo Canyon occurred on the long, curving downgrade west of Reno, involving an eastbound construction train; the maverick cars overtook the front part of the train, crushing two brakemen and wrecking eleven cars. Another crash a few days later on the same stretch of track occurred after a heavily loaded eastbound freight overshot the depot at Verdi by a mile because it lacked enough brakemen; while the superheated brakes cooled down and the engine was lubricated, a construction train rounded the curve and crashed into the freight, reducing a dozen cars to matchwood.[45]

Meanwhile, back in Washington, Collis Huntington was avidly pursuing a new goal. In January he had happened upon the codicil of the Pacific Railroad Act allowing the companies to draw two-thirds of their subsidies for grading and other advance costs; probably he had pried the fact that Oliver Ames had been chasing the notion for six weeks, from Browning or someone else in the Interior Department. Wiring Stanford for a progress report on the eighty miles of work between Ogden and Monument Point, the governor had replied that the contractors were not far enough along to support such a claim, especially "considering the fuss that is being made just now about Pacific Railroad matters." Given that even Stanford had heard in Utah that the Union Pacific was pressing its own claim for advance reimbursement for grading in the canyons, Huntington would not let it go. Assuming that in a few weeks their contractors would have more done, he went ahead and passed the claim into the Interior Department. Apparently someone there tipped off Dodge; on February 10, Oliver Ames filed a protest directly to Browning; he appended Dodge's exhaustive letter, which Ames had himself signed but not sent. Browning contemplated the insistent pile of documents, considered the railroad controversy airing in nearly every morning's newspaper, and nervously stopped the Central Pacific claim dead. He did not have to wait long for the angry Huntington to be announced at his office door. Under such pressure Browning agreed to present the application to the entire cabinet at the next meeting, February 26.[46]

By then Huntington had learned that the cabinet had granted the Union Pacific bonds on completed work to the thousandth mile, in the Wasatch canyons, regardless of the unfinished tunnels, after the company agreed to put up nearly $1.7 million in first-mortgage bonds against its promise to become a "first class road"; it also promised to post half of all subsequent subsidies into the same fund. On the afternoon of Friday, February 26, Huntington was told that while Browning and three other secretaries favored releasing the Central Pacific bonds, the cabinet had decided to table the matter over the weekend while Attorney General William M. Evarts studied it. Huntington spent an anxious few days. But late on Monday afternoon, March 1, the cabinet unanimously agreed to release the bonds for advance work to Ogden. Treasury Secretary Hugh McCulloch was instructed to issue them immediately.[47]

Huntington had but two days before the secretary returned to private life on inauguration day, March 4. But for some reason McCulloch declined to release them. Huntington (the sole source for this assertion) said later he suspected that the treasury secretary had been reached by Ames and agreed to delay until Grant took office. So Huntington went straight to McCulloch's office and vowed he would stay until he got his bonds. The secretary fretted while Huntington simmered in the antechamber—but finally McCulloch sighed and agreed to let them go. "I went out," Huntington said with great satisfaction, and after eating dinner, "by eight o'clock I found the bonds in my room." They were in two "untidy" packets; Huntington had agreed to deposit $500,000 from each lot into a Treasury Department "security fund" like the Union Pacific's, so he was left with a total of $1,399,000.[48] Nonetheless Huntington was joyful—he would keep these bonds secret from the Union Pacific people as long as he could, and if this was not an endorsement of the Central Pacific's right of way to Ogden, he was hanged if he knew a better one.

Had he been delayed twenty-four hours or even twelve, Huntington would not have succeeded in extricating the bonds. The next day, Thursday, March 4, Ulysses Simpson Grant rode to his inauguration alone; President Johnson refused to ride with his betrayer, or even to attend the ceremony. Grant took his oath. One of his first actions was to announce his largely undistinguished cabinet, which caused some perplexity and dismay. "They will bring no strength to Genl Grant, and inspire no confidence in the Country," Browning would comment sourly. "It is a singularly weak cabinet, and in my opinion will go to pieces ere long." Before the day was out President Grant stepped into railroad business and issued an order to the Treasury and Interior Departments: no more bonds were to be issued to the railroad companies. This came as no surprise; on December 22 it had been reported that Grant had told Oakes Ames he was opposed to "granting any more money subsidies to the Pacific roads until the finances of the country were in a better condition than at present."

But more than money was at stake. A few days before his inauguration, Grant summoned Dodge to a private meeting in which the president-elect voiced the hardly surprising news to Dodge that there was evidence of "a great swindle" in the recent work done for the Union Pacific. They had talked about such matters earlier in the year, although with the exception of a privy few, that meeting had been a secret. Then, as now, the engineer saw clearly that a federally mandated reorganization of the Union Pacific was in the works, and that at least one head was going to roll within his company—a doctorly head.[49]

In the waning months of the Johnson administration Dodge had seen that, as he put it somewhat diffidently many years later, "the administration and departments had lost all confidence in Mr. Durant, and many of the decisions against us came on account of his interference and statements."[50] The whispers of yesteryear about the Doctor's side deals, hedge bets, stock jobs, hidden interests, secret funds, had risen to outraged shouts, and it was obvious in nearly every wire from Snyder, Hoxie, Blickensderfer, and the latter's successor as divisional engineer, Theodore M. Morris, that Durant was taking every advantage in this final contest in Utah to emerge richer than a king. Seymour was close behind; as Hoxie—the now respectable, even virtuous former construction contract shill—had warned, "The presence of Silas Seymour at Durant's heels and putting all kinds of foolish notions in his head will yet ruin the company. Millions of dollars spent for naught to simply gratify Seymour."[51] If Webster Snyder were to believed, Samuel Reed was trying to catch up; Dodge would forever doubt that, and Jesse Williams agreed with him: "Contractors' engineers the world over are apt to become a little corrupt," he commented to Dodge, but he believed Reed "merely passed it through," knowing "it was wrong," but concluding "to say nothing about it." But otherwise there was no shortage of crooks out on the railroad—and at their head was Thomas Clark Durant. In the overall feeding frenzy, in fact, the Doctor had been acting close to unbalanced.

"Durant was crazy on his last trip," Snyder had written to Dodge in mid-February, "& discharging me daily. Seymour and J W Davis dared not let him get out of reach of their voices fearing somebody might expose their operations." With the paymaster some two months late, the common laborers had been "growling, striking, and generally demoralized," Hoxie had warned. But thanks to the Doctor, all the field supervisors, and, top to bottom, all the engineers, had been walking on eggshells, uncertain where orders were coming from and which ones to obey. Firing Blickensderfer over Durant's cost-cutting and dangerous changes in the Promontory lines on January 2 had only been the beginning. The Doctor threatened the replacement, Theodore Morris; sacked two construction engineers, Thomas H. Bates and Major R. J. Lawrence; subjected Hoxie and Snyder to inhuman pressure hoping they would resign; and even blamed Colonel Seymour for delays caused by a bad landslide in Weber Canyon: "You were left with instructions and power that it was supposed would prevent delay," he raged, and because of such preventable delays, "you are now at liberty to return to New York." Seymour, though, knew better than to leave at such a time.

But Durant shone his fiercest light upon Dodge, back in Washington and increasingly unresponsive to the Doctor's provocations; Oliver and Oakes Ames had worried in mid-January that Durant's threats and insults would goad their chief engineer to resign and go to the press, but the following month the Doctor became positively unleashed. On February 3 Dodge received a summons—"I propose," Durant wired, "to have a line on the east slope of Promontory located for the best interest of the company without regard to former surveys. Can you come here next Tuesday?" The engineer paid no attention to the telegram, carrying on his pressing business with the Interior Department and his field correspondence, and when he did not appear in Utah on the appointed day, the enraged Durant fired off another summons and a rebuke.

> *I telegraphed you . . . not to send instructions west without first submitting the same to me. You are away from the work attending to other business and are not supporting parties. If you can't find time to report here I shall of necessity be obliged to suspend you.*[52]

What else but some form of derangement would move the Doctor to threaten to suspend the chief engineer for the Washington lobbying that was keeping the company from ruin in those crucial opening weeks of the year? What else, when, after he had returned to New York and turned up the pressure on Dodge in Washington, Durant would then attempt to pin blame for rampant corruption (including, of course, his own) on his chief engineer—within days of the inauguration of Dodge's mentor and friend as the president of the United States?

> *You have so largely overestimated the amount to contractors in January that it becomes my duty to suspend your acting as Chief Engineer until you give a satisfactory explanation of the same; a mistake of a trifling amount might occur, but when it gives contractors hundreds of thousands of dollars, it creates suspicion that all is not right. Your immediate attention is requested in order that if you have an explanation to give it may be done before the report becomes public.*[53]

To Oliver Ames the whole situation may have begun to resemble a tightrope walker's act as seen from the remove of the bleachers—Dodge up there in the dimness at the crown of the tent, and perhaps the whole Union enterprise and all of their financial hopes on that quivering wire, with the malign Durant up on the high platform shaking and rocking it back and forth, and all an observer could do was hope that the walker could steadfastly put one foot in front of the other and gain the far platform—for there was no net.

Nonetheless Dodge was almost cocky with certitude that it was the Doctor who would topple. But what would dislodge him? The righteous press? The suddenly upright Congress? The victorious new administration? Or the wobbly-kneed Ames brothers and the cliquish company directors with their constantly shuffling alliances, intrigues, and plots?

Perhaps it would take that entire army, and all its ammunition. Early in the year, during Dodge's confidential talk with Grant, after they had touched on matters similar to the ones discussed just six months before in Benton, Laramie, and Council Bluffs, Dodge got the president-elect's permission to write Oliver Ames and Sidney Dillon to say that "all we had to do was to hold things steady until after the 4th of March," after which time the whole rotten outfit "would have to get out." The letter was shown to Rowland Hazard—who, as a steadier, proven, more dependable enemy of Dr. Durant, had already begun to foment a directors' coup; Hazard wrote to Dodge in the first week of February from New York to say that he had "a plan pretty well matured" which he would be discussing with Ames and Dillon, their attorneys, and a few others, for a maneuver to take place at the annual meeting of the Union Pacific in New York beginning on March 9. Therefore Dodge knew that at least some opposition was rising—even if it were less direct than his own—but for the present "they kept secret all the inside information they had."[54]

On the day, February 4, that Dodge was reading Hazard's letter about the new and mysterious plot, Webster Snyder was sending Dodge some astounding ammunition. "It is news to me but may be old news to you," the operations manager wrote from Omaha, "that J. W. Davis . . . is the contractor for building the road west of the Oakes Ames contract." It was certainly news to Chief Engineer Dodge that young James W. Davis—whom he knew to be their principal tie contractor (though probably in secret partnership with Durant), and who was the brother-in-law of George Francis Train—was now the overseer for the millions in construction money expended from the end of the Ames contract in western Wyoming across all their Utah division. Snyder had heard it dropped casually by Durant in an unguarded moment when the Doctor was not attended by the protective Colonel Seymour, and apparently he wrote, "we assigned the contract for the 'Trustees for the Contractors,' whoever they may be—all arranged by Thomas Clark Durant." It might be, he thought, a repetition of the Doctor's dummy contract with L. B. Boomer back when they were getting started in Nebraska.[55]

Such a secret seemed to have a life of its own, and it wanted to get out into the light of day as quickly as possible; before the incredulous Dodge could alert New York, Durant himself disclosed the contract to John B. Alley, who told the Ameses. The next Union Pacific executive board meeting was in a few days, on February 25; when Oakes Ames and the others confronted the Doctor, contending that they had all just "assumed" the Ames contract had been extended, Durant merely assured them an official explanation was forthcoming, though of course, he said, he had been acting with full power to make such an arrangement. And at the meeting it was indeed like old times; Durant began talking immediately and seized the initiative, unleashing a dizzying flurry of financial matters which put the other executives on the defensive.

The meeting carried over to February 26, when Durant handed them the surprise contract and what seemed to be a letter back-dated to November 27. "I found it absolutely necessary," Durant smoothly explained therein, "in order to carry out the wishes of the board, to commence work on this portion of the road at once. The present organization, with its large outfit of teams, tools, and men, presented the most available means of doing the same. To have created an entirely new organization would cause much delay." No one was fooled; no one was mollified. Durant did not even bother to explain why he had kept it a secret for three months. Befitting the rules, the contract was referred to a committee for submission to the stockholders. Anyone who troubled to begin reading the contract while the Doctor talked on would see that figures for what the line would cost the Union Pacific were not there; not even Davis's remuneration was specified. It was simply a blueprint for wholesale robbery—and as Durant continued to raise unrelated financial issues, meaning to keep them befuddled and distracted, there may have been a few seconds of doubt or indecision in the boardroom as to whether he was going to prevail yet again, but one after another in those executive minds rose a date—March 9, the annual meeting—and then a variation on the image of the Doctor's head, on a silver platter.[56]

Strobridge's tracklayers followed the Central Pacific grade across the chilled and dry reaches of easternmost Nevada in the early days of March, with little to look at but sagebrush and bleak hills and mountains lightly furred with pinion and juniper: the heads of the Pequop Mountains and the Toana Range rose to the south; immediately north and west, between ranges of hills, was the opening of the Valley of the Thousand Springs on the old Fort Hall emigrant trail. On March 9 the rails passed milepost 556. Ahead the railroad grade passed through a wide natural corridor between the Goose Creek Mountains standing to the north and the Pilot Range to the south. The salty, sandy Utah border was about twenty-five miles away. "Cold weather has been a serious hindrance," Mark Hopkins wrote to Huntington. There had been little snow, but operations had been hampered by frozen water sources, and for more than forty miles to the east all creeks were so alkaline as to be useless. Eventually engineers would lay wooden pipes in deep, frostproof covered trenches up into the hills, but for the time being as the track was going down, the big, rolling water tankers would have to serve. They had been slowed to a crawl when they caught up to a grading party, but Charley Crocker now thought their rate would soon pick up to between three and four miles per day—and that would continue until they reached the Promontories, and probably Ogden.

The new government commission had recently passed by the end of track, having been augumented by Lewis Clement at Wadsworth. The inspectors reached Ogden on March 8, and were to spend several days in the eastern valley before commencing up Weber Canyon. Chief Engineer Samuel Montague would stay with the party all the way to Omaha—Stanford had become deeply suspicious of Warren and Blickensderfer the longer he spent with them, Charley Crocker thought Clement too "untested" to be left unsupervised with such types, and both the Governor and George Gray thought that the hostile Warren either wanted "a bid" for his favoritism "or that he has been *seen* by the other Company," as Hopkins termed a Union Pacific bribe. Hopkins reminded Huntington about Warren's lackluster military record ("superceded on the Battlefield by Genl. Sheridan & sustained by Grant") and his dependence on the ex-president's patronage ("has since been hanging around Andy at Washington waiting for something to turn up"), hoping these black marks might help the Central Pacific's case before the Grant administration. In a few days Stanford wired Huntington in cipher from Salt Lake City, urging him to see President Grant and have the commission removed before it issued a report, which it would write in Washington—the influence of the Union Pacific was simply too great. Failing that removal, it behooved them to build as much track as possible—Stanford desperately hoped they could reach Ogden—before the commission submitted its report.[57]

There was a new development in the situation north of Ogden, Stanford told Huntington in another coded telegram: the Union Pacific engineers had run their final line between Bear River and Promontory, managing to cross the Central Pacific no less than five times and at different grades—varying as much as fifty to eighty feet. It was, Stanford complained, "for the purpose of embarassing us"—at some point he expected his graders would be ordered off this defiant new right-of-way. But he had decided not to make any legal trouble about it; after all, Huntington had already pocketed the bonds for this part of the route, though Stanford did not understand why his partner was still keeping it a secret. When their graders advanced up to this disputed ground north of Bear River, though, "we shall probably have to show our entire hand."[58]

On Sunday, March 7, the peaceful and righteous village of Ogden with its fifteen hundred faithful souls entered the modern age—when the tracks of the Union Pacific moved out of the portal of Weber Canyon and over the plain, and were spiked past the log and adobe houses with their struggling little shade and fruit trees, to the place where a depot would stand. Three months before there had still been some debate about whether the railroad would locate to Ogden, as opposed to another town—possibly a Gentile settlement such as the still-nonexistent Corinne. But Brigham Young had convened a meeting of property owners on the western end of Ogden and asked them to sell him their five-acre lots in support of the railroad locating there. All consented, and he bought some 133 acres for yards, repair shops, and the depot, at the price of $50 per acre, and, on behalf of the church, presented the package to the railroad.

The exchange—of land and quiet privacy for business and the disturbing influences of modern society—was not made lightly. In "the old days"—even as recent as some five or eight years before in the time scale of the West—strangers were feared and distrusted in Ogden. "The emigrants parked along the street in their covered wagons," recalled one of the original citizens, "and all the people were afraid of them and what they might do. The grocers would hide what little money they had in coffee cans and anyone who had any valuables would get them out of sight. They would let the people stay so long, then the sheriff . . . would gather a posse and drive down the street and fire several shots with their rifles, and that was a signal for the emigrants to pack up and start moving again." Mayor Lorin Farr had for some years presided sternly over his village citizens, with his small police force hauling in all gamblers and whiskey purveyors who mistakenly set up in town; the law frowned on drunkenness, profanity, pilfering, fist-fighting, and other lamentable practices. Only fifteen months before the arrival of the railroad, Bishop Chauncey West predicted that "the time was not far distant when the police would have plenty to do," and citizens soon saw this would be true. The past August a policeman reported grimly "that there was some lose women in our town that would bear watching," and in December the officers arrested six men for "fast driving and loud hollowing in the streets."[59]

With these advance alarms, Ogden began to awaken from its peaceful slumber. Less than two years before the councillors had found it necessary to pass laws forbidding "cattle, horses, mules, sheep, calves, swine, or goats" to run at large within the village limits. Ogden still paid a bounty on wolves, and only recently, taxes had been payable in stock or produce.[60] Now, to the tune of a steam whistle and a clanging bell, the world began to chug in. Celebration for the arrival of the Union Pacific tracks on Sunday, March 7, was sensibly delayed until Monday, when the faithful could participate. On the day Young would inaugurate a new railroad company to build between Salt Lake City and Ogden—beneath welcoming banners and given a merry prelude by a brass band—Ogden speakers marked the arrival of the "national highway" to Utah. There were residents still living in 1940 to tell Works Progress Administration historians of that day "when the entire populace gathered around in Sunday finery to see the iron monster. Suddenly the engineer blew the whistle, yanked a steam valve, and announced he was going 'to turn the train around.' A wild scramble for safety ensued and many ran pell-mell through a nearby slough in their fright, ruining their Sunday clothes. Some terrified children were not found until evening."[61] An artillery salute punctuated the affair and, with it, any lingering hopes for solitude.

Trains from the East would not begin arriving immediately, however; though it felt springlike in the Salt Lake Valley, snow still blocked the road in Wyoming. "We hope to get trains through this week," Jack Casement had written his wife on March 3. "The weather here is very fine Birds are singing in the valey like Spring and farmers in the valey are plowing," he told Frances. He confessed to being homesick for her and their little boys as he progressed across this "wild looking country" so alien to conventional beliefs. "I have not been to Salt Lake City," he wrote, "and don't know as I will I have no desire to toady Brigham." He warmed to his anti-Mormon fever some days later with the opportunity to show a real discourtesy. "Brigham Young sent me word that he would like to see us lay track," he chortled. "I sent him word all he would have to do would be to come where we were at work and open his eye. He came here but I think so little of him and his pretensions that I did not stay to receive him."

Nonetheless, thanks to the alacrity of Young's grading and tunneling crews, Casement was in Ogden in the "delightful" weather, dealing with springtime mud instead of being somewhere high in the Wasatch Mountains kicking through still-deep snow—and he had ample change in his pockets. "We are getting along nicely," he wrote on the day they reached Ogden. "I have had much better luck collecting than I anticipated and if the Company pay in New York the Drafts they have given me here, I can send home fifty thousand dollars as soon as the Road is clear again. . . . I flatter myself I have some tallent for collecting."[62] A few days later, after but one train had struggled through the laboriously shoveled cuts and drifts across Wyoming, the biggest storm of the season commenced—in three days covering all the work again, blocking and stopping all movement between the Wasatch Mountains and the Platte Valley. Casement's crews had perhaps eight miles of iron left.[63]

In New York City on Tuesday, March 9, one day before the Union Pacific stockholders' annual meeting, the board of directors met. Most were in a mood not unlike that of warriors before a pivotal battle. Grim Roland Hazard and John Alley, the dour Ames brothers, the saturnine government directors Jesse Williams and Congressman James Brooks—all assembled with the others, including the stooping, standoffish Dr. Durant, ready for a fight but disagreeing, even at this late hour, about tactics and protocols. They were also all on heightened alert because of a new development revealed the previous Thursday—about which several had spent a long weekend worrying what could done.

On that Thursday, March 4, a New York State Supreme Court justice had refused to approve the Union Pacific petition to move Jim Fisk's nuisance suit out of state jurisdiction into federal court, following the letter of the legislation sponsored by Oakes Ames and passed eight months earlier. And Fisk's suit had grown far beyond his initial effort to parlay his $3,200 investment on Durant's behalf in the fall of 1867 into a $50,000 or $75,000 windfall. Now he clearly intended to grasp those disputed twenty thousand shares, not just compromise on "damages." Now he may have even been aiming to jostle his way onto the Union Pacific board; now, even worse could be contemplated. Barely three months earlier, in October 1868, using surprise proxies and other subterfuges, Fisk, Jay Gould and Frederick Lane had finally seized control of the Erie Railroad from the rest of its board; they had exiled the outcasts from all governance of the company, then quickly and secretly issued and sold more than $20 million in stock, pocketing the proceeds between the three of them; then they had successfully resisted all litigation by the outmaneuvered directors and bilked stockholders.

Only in their dreams could Durant, Dillon, Bushnell, and the Ameses—

and, for that matter, Huntington, Hopkins, and the Crockers—so quickly and effortlessly assemble an empire. Fisk and Gould had then gone on a well-publicized spending spree, buying theaters and opera houses and palatial mansions on Fifth Avenue. Some of the Erie profits had presumably been used to smooth the way toward procuring the large block of Union Pacific stock. What was now especially worrisome for the Union Pacific directors on the eve of their annual meeting was that the superior courtroom in which the Erie litigation was found for Fisk and partners was the same one in which the company's petition had just been denied. That placement—and the exquisite timing—was no mere coincidence.[64]

The judge in both cases was George Barnard, who served under the pleasure of the powerful Tammany clubhouse and the Tweed Ring; significantly two Tammany leaders, William Marcy Tweed and Peter B. Sweeny, were now on the board of the Erie Railroad and working closely with Fisk and Gould, so Judge Barnard's sympathies could be easily guessed. Hazard and the Ames brothers had strong suspicions that behind "Jubilee Jim" stood the wily Doctor; Durant had, after all, conspired with Fisk a year and a half earlier in the boardroom fight, and he had used Judge Barnard two years ago to snare them with an injunction. Their suspicions sharpened when Charles Lambard confided that Durant had recently offered to go in with Lambard and buy out Fisk's claim for $75,000. Associations were growing before their eyes like poisonous vines, and indeed, once one began counting and measuring them, the greater the danger seemed to loom for the Union Pacific in its vulnerable state—unmanaged, adrift, plundered close to bankruptcy, yet edging so close to completion.

It is questionable whether by this time the Doctor still felt he could control Jim Fisk, like him a shark but whose teeth and appetite dwarfed his own. Canny Cornelius Vanderbilt and Daniel Drew, for all their wealth and power, had been badly chewed up in the Erie struggles, and could offer a cautionary. Even more alarming, in January Charles Francis Adams's article had touched on the political horror of the vast wealth of the railroad corporations—Adams more feared the Erie than the Union Pacific—serving "the power and patronage" of an entity like Tammany. "Imagine," Adams said, "the Erie and Tammany rings rolled into one and turned loose on the field of politics and the result of State ownership of railroads will be realized." Later in March, the *New York Times* would quote those words while delivering the terrible news that Tammany and Erie were already as good as one; Tammany had made the New York governor, the New York City mayor, two state superior judges, a senator, a city chamberlain, and had even sponsored presidential candidate Horatio Seymour—and Erie was now, warned the *Times*, its treasury. Suddenly New York did not appear to be a safe or healthy location for the headquarters of the Union Pacific.[65]

Awareness of much of this, or at least of confusing, disconnected parts of it, swirled around all the directors at 20 Nassau Street during their board meeting on March 9, and in this atmosphere of suspicion and intrigue the directors accomplished very little during their meeting but the appointment of election inspectors for Wednesday's board election. Conspiracies continued into the night.[66]

The stockholders' meeting convened at ten the next morning, and the voting began almost leisurely, until James Fisk himself strode in. He was flanked by several beefy men who were soon revealed to be sheriffs sent by Judge Barnard. Fisk was being called "the Prince of Erie" in the newspapers for his roguish seizure of the Erie Railroad, when he was not being called worse—but he had a genuine danger about him which could not be missed. Three days before Christmas he had arranged for a judge to have the *Springfield Republican* editor, Samuel Bowles, arrested and jailed in New York for libel, bribing officials to make themselves scarce so that Bowles could not be bailed out. Supporters vied to get the editor released from Fisk's clutches—the list included Schuyler Colfax, Horace Greeley, Henry J. Raymond, and John Jay, with the aggregate offered sum going well over $3 million, including one offer from the *Cincinnati Commercial* editor, Murat Halstead, "to buy Ludlow Street Jail." Bowles was released on $50,000 bond the next day, vowing to explore Fisk's antecedents in a forthcoming issue of his newspaper. Soon Fisk would be hauling the *New York Times* financial editor, Caleb Norvell, into court for libel, demanding $100,000 damages; Norvell, too, would be bailed out for $50,000. The evil glow of Jubilee Jim flared brighter.[67]

And now here was the Prince of Erie striding into Union Pacific headquarters. When Fisk attempted to cast the ballots representing not only his token six legitimate shares but the twenty thousand he claimed from the company, the tellers refused to let him vote. He left with a markedly smug air; the sheriffs stayed behind, and after all the votes had been deposited, they stepped forward brandishing an injunction against the election pending a hearing of Fisk's grievance. Unsure how to react, the directors debated and brought several motions to move the meeting out of the judge's jurisdiction, to Washington or Boston. The sheriffs countered by producing arrest warrants—the directors were now in violation of Judge Barnard's order. Bedlam and outrage: Oakes Ames immediately shouldered out of the room, claiming congressional immunity, while the towering Sidney Dillon, his face purple against his luxurious white muttonchops, forgot the veneer his millions had bestowed upon him and tried to give the bum's rush to one of the sheriffs. He was deterred and taken in hand, as was President Oliver Ames—who managed to adjourn the meeting. The sheriffs announced that all the directors were under arrest and that they would be conveyed to the Eldridge Street Jail. There was a tumult of outrage—but after some negotiation the sheriffs agreed to a compromise and led away only Oliver Ames and Dillon, not to the hoosegow but to confinement in one of the parlors at the Fifth Avenue Hotel.

They were released the next morning after an appearance before Judge Barnard and posting $20,000 bail.[68]

Durant still professed innocence and ignorance when the directors met soon thereafter on that day, March 11, and the stockholders' meeting was then reconvened long enough to adjourn for a week, after which time it would resume in Washington. Then a resolute Oliver Ames opened a directors' meeting to deal with the Doctor. With his sixth sense Durant had foreseen the showdown; he immediately tried to deflect their wrath by resigning the powers given him by the executive committee on July 3, 1868, but his adversaries moved past him to revoke authority of the executive committee itself—it had become, of course, a tool of the Doctor's—and the supremacy of the board of directors was again unquestioned. In Durant's place as construction superintendent, the board appointed a committee of Dillon, John Duff, and Hiram Price, who, they hoped, would see the Union Pacific to its meeting place—wherever and whenever that would be. The coming board election, when the Doctor could be removed for good if Hazard's plan worked, was still left hanging. So was the Fisk lawsuit: the Union Pacific attorneys would again try to move it away from Tammany to the federal courts. So was the question of the current construction contract: was it to be Durant's jack-in-the-box Davis contract, which left out all the Crédit Mobilier people, or a renewal of the Ames agreement? No one knew. Perhaps federal intervention was their only hope, and a wholesale transfer of their offices to Boston or Washington, where, presumably, Fisk's tentacles did not yet reach, and where the Doctor's influence was now greatly reduced, or so they hoped. "We have had a lively time of it," Sidney Dillon admitted in a letter to Dodge, describing the tussle with Durant, "but we have beaten the enemy so far & have Bearded the old Lion in his den & if we all stand firm he will have to remain there. He seems very tame at this time yet he may be preparing for another leap So we must be careful to watch him close."[69]

Newspaper reports of the maladroit meetings, the scuffles, and the genteel jailings were enough to make a starched shirt wilt, and though Jubilee Jim had lately been portrayed in the editorial columns as the devil incarnate, he found himself in the unusual position of being cast as a hero, albeit a roguish one, sword in his fighting hand and lantern in the other. "There is a promise of light to be thrown on the dark doings of some of the Pacific Railroad Companies," opined the *New York Herald* on March 13, calling Fisk "Young Erie" and chuckling over his maneuver to imprison the Union Pacific board. The face of his complaint seemed just, the editor said sardonically. "If there is any underhand work going on why should he not have a finger in it?" Meanwhile, the newspapers and their readers could sit back and watch the capitalists fight. The *Herald* offered advice to both sides: "if the Union Pacific is a bird of the same feather with the modern Erie, we advise them to keep Fisk, Jr. out, on the principle of preserving character. If the directors have made the millions report has assigned them they can afford to compromise with Fisk, and if they will not come down à la Vanderbilt and Drew, we advise Fisk, Jr., to 'keep pegging away.'"[70]

Grenville Dodge was still in Washington, sizing up the situation with the new Grant administration and establishing influences in the new Congress, when Oliver Ames, Dillon, Jesse Williams, and other directors arrived in the capital to assist him in lobbying, and there, the annual meeting of the stockholders was extended almost day by day to keep it active. Dodge had some useful contacts in the cabinet, though many editors had been unimpressed with Grant's appointments: Samuel Bowles had thought them "a little obscure," the *New York World* called them an "absolute oddity," and the *Boston Post* had complained the men were unexciting.[71] But from Dodge's point of view there was his pale, breathless, tubercular comrade John Rawlins as secretary of war, and, as the new secretary of interior, Jacob Cox, the Ohio school administrator and lawyer who had been a brigadier general of volunteers. Cox had served well in Western Virginia (where Jack Casement had fought), Tennesee (for a time, Dodge's assignment), and North Carolina (where Cox had rendezvoused with Sherman); for one brief term after the war he was governor of Ohio.

The new secretary of the treasury was also a familiar face to both Dodge and Ames: it was George Seward Boutwell of Massachusetts, like Dodge just retired from Congress. Grant's original appointment to preside over the treasury had been the merchant prince, A. T. Stewart, whose politics were obscure and whose sole selling point had been that he had been extremely generous to Grant and his election campaign; Stewart had so many conflicts and disqualifications in trade and commerce that he was forced to resign within a week of his installation. Many of Grant's appointments gave sinecures to retiring politicians who had earned merit in his backers' eyes, and Boutwell had been one of the congressional framers of President Johnson's impeachment. He had also deserved, in Oakes Ames's eyes, an offer for Crédit Mobilier stock earlier in the last term—but Boutwell had judged it a poor investment and declined. Now, at least, Ames's former fellow delegate led the treasury.

As far as influencing Congress was concerned, Ames hired the former attorney general William M. Evarts as a lobbyist, and retained two eminent Washington attorneys, Caleb Cushing and William E. Chandler, the latter through Dodge's persuasion. Now the task of keeping old congressional allies and creating new ones from the freshmen began. The first test of influence would be a joint resolution to endorse the Union Pacific's desire to move out of the Tammany courts into federal jurisdiction and to relocate its corporate headquarters from Tweed's New York to Boston. Immediately Senator William Stewart of Nevada (with Collis Huntington somewhere behind him) rose, critical newspapers and the Adams account of the Crédit Mobilier in hand, to denounce any such scandalous machinations in Congress. But as

Stewart thundered away, Dodge and Ames and their lobbyists encountered surprising resistance in the Senate corridors, even among supposed friends. Cornelius Bushnell had stayed in New York, but when he heard of these unexpected changes of heart he suspected that the Doctor was behind it all—and wrote a very candid and private letter on March 20, in which he untidily lumped several of their antagonists together. "I discovered by this . . . your hand," he said to Durant, "as well as Fisk-Huntington & Co." Whatever Durant's private aims by deterring their removal legislation, Bushnell warned, he was going to cost them the bonds west of Ogden and play them all right into the hands of Huntington—if not Young Erie as well. "I hope," Bushnell wrote cautiously, "you have arranged to get Fisk out of the way."[72]

Whatever the sources of this new resistence in the Capitol, in a few days Dodge encountered a new problem and a rude shock. Whether he learned from Rawlins, Cox, Boutwell, or one of their subordinates, suddenly on March 26 it was revealed that Collis Huntington had slipped out the closing barn door of the Johnson administration, the prized bonds to Ogden clutched in his fist. Dodge dashed off a complaint to President Grant: the Central Pacific had fraudulently grabbed bonds which rightly should go to his company. "We are now determined to expose the entire fraud," he wrote to his wife, Anne, that day, "and have the world know what they are." He would work "day and night" to get new legislation through Congress charging the Central Pacific with fraud—it would be a real "stinger." That night he got in to see Grant at the White House—and subjected his old commander to three hours of invective against "old Huntington" and his servants in the discredited administration.[73]

Less than seventy-two hours later, Congressman John Bingham of Ohio—who in 1867 had bought twenty shares in the Crédit Mobilier from Oakes Ames and shortly exchanged them for a like value of Union Pacific bonds—rose in the House chamber to denounce Huntington and Browning and call for an official investigation by the Committee on Pacific Railroads. Dodge's newest affidavit sped straight to that committee. The Central Pacific map filed with the Interior Department was "false and fraudulent," he charged. "It has no topography, no stations, no courses, no angles, no scale, nothing by which any line could be identified by it on the ground." Their mere trial line, which he himself was unable to find in the Utah canyons, directly violated the Interior Department instructions. Therefore, he said emphatically, Huntington had no claim on those advance bonds west of Ogden because by right—and by dint of their completed track through the Wasatch—they belonged to the Union Pacific.[74]

While Dodge and the lobbyists and lawyers were thus occupied in the capital, another contingent of attorneys were sweating up in Manhattan in Judge Barnard's courtroom. Jim Fisk and team kept them and their clients busy. On March 11, Fisk had submitted an affidavit calling for the Union Pacific to show cause why a receiver should not be appointed: the Crédit Mobilier ring was depleting the dregs of resources of the Union Pacific, putting the railroad at dire financial risk and cheating its stockholders. The judge had overridden all objections. He appointed a scion of the machine—William Marcy Tweed Jr.—as receiver, and young Tweed instantly and officiously departed for 20 Nassau Street. Durant and Bushnell, their secretaries, treasurers, and accountants were there, their faces imperturbable masks, and when Tweed demanded all assets, property, trusts, bonds, and proceeds of the Crédit Mobilier Company, he was sent from office to office like a puzzled messenger and told that the company had no assets or bonds in the state of New York other than some securities pledged for a loan held by a local bank; they got rid of him by sending him to the bank, where officers refused to discuss, divulge, or divest. Tweed slunk back to court, under the impression that the Crédit Mobilier had already slipped out of town.

In a new affidavit Judge Barnard learned from Fisk that the joint resolution to grant the Union Pacific relief had just been introduced in Congress—it was evident to Fisk that the defendants intented to evade the judge's jurisdiction and remove the property of the Crédit Mobilier and the Union Pacific from the state, and Barnard agreed. He grimly appointed a court referee to elicit subpoenaed testimony from Durant, but the Doctor had refused to be sworn on advice of counsel; the Union Pacific disputed the right of jurisdiction in the state court proceedings. An angry Judge Barnard hauled Durant into court on March 20 and, in the presence of all the opposing attorneys and Jim Fisk himself, threatened to jail him on contempt of court if he did not swear. Finally the Union Pacific lawyers told him to go ahead, and after he had taken his oath, Durant tersely described the relationship of the railroad to the construction company, and his investment in them, and testified that the Crédit Mobilier had no property in the state, and the Union Pacific very little. When he was directed to produce the books of the Union Pacific, Durant replied, "Well, that I can't do. They are not in my custody."

Judge Barnard turned to the affidavits in hand. Fisk claimed that he had been told by some of the defendants of a "secret service fund" of about $700,000, "which they, in substance, admitted was used for corrupt purposes, although they suspected some of their own members of having privately appropriated a large part of it to their own use. " He had also been told that over $500,000 in bribes to members of Congress had been expended in a single year, and that a number of members were given shares of Crédit Mobilier "in their own name or in the names of relatives, friends, or servants, taking the same in trust." Then a Fisk lieutenant, Adin H. Whitmore, had sworn that from an interested party now out of state he had learned that the directors of the Crédit Mobilier and the Union Pacific were one and the same, that he had been told of the construction company's long detours, sharp curves, and steep grades, its careless usage and damage of equipment, that it had charged the railroad up to three times the actual construction cost, that the directors had formed fraudulent wood and coal companies, signing fifteen-year contracts for fuel at double the open market price, that a commissioner had been bribed and others taken over dangerous, substandard sections while drunk and asleep.

Reading these affidavits, and the Union Pacific responses, took the entire day. Then the judge ruled in favor of the Fisk complaints: receiver Tweed's duties were enlarged to "protect" not only the Crédit Mobilier but the Union Pacific—all bonds, proceedings, assets, and properties were to be seized, along with all company records. Tweed was empowered at the sufferance of the court to pay out all debts or expenses of the company from such seizures.[75]

Two days later, on March 22, they were all back in the courtroom, Tweed with crates of seized papers, little of them helpful. Sidney Dillon testified that the Crédit Mobilier had been incorporated in Philadelphia and since the previous December had removed out of 20 Nassau Street, out of the city, and out of New York state; it had done no business in the city and had no records or assets there. Furthermore, all of Fisk's and Whitmore's affidavits were "absolutely false and untrue." John Cisco, Charles Lambard, and Thomas Clark Durant submitted similar statements. Judge Barnard and Fisk's attorney questioned the Doctor closely about a number of related matters, and just before lunch, the judge ordered Durant to produce the books. When he heard the refusal, Judge Barnard told him if he did not produce them by 1:00 P.M. he would be jailed. After the lunch recess, Durant testified that he had gone to his offices, that the safe was locked, that the safe key had been in possession of the deputy sheriff, but that the combination was in the possession of his assistant auditor, Benjamin Ham, who was absent. Where, then, was Mr. Ham? Durant replied blandly that Mr. Ham was in New Jersey.

Attempts both by mail and messenger to persuade Mr. Ham to leave his home were for some reason unavailing. Over the course of the next week, as the Court waited, it admitted a copy of the complaint of Henry McComb vs. The Crédit Mobilier of America, at that point pending in the Pennsylvania State Supreme Court, in which, among other things, McComb alleged that there was still a large, undivided surplus of profits held by the construction company, equal to 500 percent of its capital. On March 23 Judge Barnard issued a new injunction barring the Union Pacific from receiving U. S. bonds or issuing new mortgage bonds, and again enjoined it from removing records from the state. And then, on March 30, the judge told the court that he suspected Mr. Ham was being encouraged or prevented to remain out of state. He ordered receiver Tweed and deputies to break or blast open the safe and bring him the contents.[76]

This was all reported gleefully, and in exhaustive detail, by the newspapers. "The inquiry . . . into the affairs of the Union Pacific and Crédit Mobilier jobs is a very good thing," the *New York Herald* said in an editorial on March 31, "although it does not spring from a very good motive." The paper urged

Fisk to "keep pegging away at the Union Pacific Railroad Company until the cancers which afflict it are brought to light and cured, before they shall have consumed the whole work." The *New York Times* agreed. "A great deal of Congressional wisdom has been wasted in pretended efforts to investigate the affairs of these railroads," it said on April 1, "but two or three days' work before a Court of justice has sufficed to lay bare one of the most monstrous frauds that was ever perpetrated upon any Government." The *Times* was awed by the directors' systematic stripping of the railroad "of every penny of its funds, so that in a short time its property may be sold under foreclosure of the first mortgage bonds, leaving the United States to look where it may for its paltry loans." And the efforts of Congressman Ames and his Senate friends had to be resisted. "We think it is about time that Congress should leave this matter alone," it said.

> *Bad as may be our State Courts, there is no evidence or appearance in this case of any injustice having been done to these directors; and our Courts have done good service to the public in throwing light upon these transactions. The attempt to transfer the case to the Federal Courts is not inspired by a love of justice, but by a desire to protect gigantic frauds from exposure.*[77]

The next day was Friday, April 2, and the safe at 20 Nassau Street awaited young Mr. Tweed. It was a very large, heavy, regular bank safe, walled in on three sides by massive brickwork, with a burglar- and powder-proof combination bank lock. "There was the safe," a *Herald* reporter would write, "behind the iron doors of which the wealth of the company may be hid; in [Tweed's] hand was the key; and yet, without knowing the combination, it was as useless to attempt to unlock the doors as to try to pry them open with a chisel from a toy toolchest." When Tweed appeared at the start of the business day he was accompanied by a deputy sheriff and an even dozen special deputy sheriffs. Five muscular, aproned artisans of the iron works which had manufactured the safe swelled the crowd; they carried immense sledge hammers, chisels, and steel augers. The purpose of their call announced, Dr. Durant stepped forward and told Tweed that he and his men were all considered trespassers and would be personally held responsible in civil actions for damages, as well as criminally for the trespass. Tweed merely smiled and ordered his men to begin.

"The men at once took off their coats, rolled up their sleeves and at it they went with their sledge hammers," said the correspondent, "striking blow after blow at the place where the lock was, and at each blow the sparks flew about as if it were redhot iron." At one point Tweed wandered out of the noisy, thick-aired room into the corridor when Charles Tracy, the principal Union Pacific attorney, appeared and shouted at the top of his lungs: "I order all you burglars to cease this work and leave the building within five minutes or I shall proceed against you both civilly and criminally!" This threat succeeded in temporarily emptying the place of the accumulation of idlers and onlookers, but not the iron workers or Tweed, and the work continued for more than three hours. Despite all this effort the men were able to penetrate only two layers of steel plate and produce a shallow opening about the dimensions of a book. At 2:30 P.M. the sweating laborers put down their hammers at the appearance of the owner of the ironworks; he heard the Doctor's threat to prosecute for civil and criminal damages and pulled his men off the job.

"Therefore," the *Herald* correspondent said, "the youthful but indomitable Tweed, Jr., like the son of Ossian, or Ossian, Jr., true to the instincts of his race, vowed . . . that the coveted prize within that safe would not be safe there another hour." He huddled with the sheriff and deputies and soon sent for a squad from another ironworks. Not long after they had begun their assault, they were through. The inner door had been left unlocked. Tweed rushed in. The sheriff and deputies rushed in. The ironworkers rushed in. The Union Pacific defenders rushed in. Later it was said that there was a melée in which papers, books, records, and even bonds went flying, some disappearing forever, but according to the *Herald* witness, "a sort of amnesty was proposed," some company witnesses came forward "with flag of truce brought on the ground, and the youthful conqueror of the safe, with courteous address and chivalric bearing, invited them to enter and be cognizant of whatever might follow from his achievement."[78]

While the siege had been shaking the walls of the Taylor Building at 20 Nassau Street, an attorney for Fisk had been upstairs in the meeting room questioning Durant, Bushnell, and Tuttle before a court referee. Their unresponsiveness was like layers of iron plate; so many of the answers were "I don't know" that the plaintiff's counsel began to falter, but then he pressed on against Assistant Treasurer Tuttle as gamely as young Tweed several floors below:

Q. Has the company any assets left?
A. Yes.
Q. What?
A. The road.
Q. Has the whole capital of the company, and all that it has received from the government and from its own first mortgage bonds, been already expended?
A. They have.
Q. Has the whole gone into the pockets of the contractors?
A. I do not know whether it went into their pockets or not.
Q. Has it been received by them?
A. Yes.

The exciting announcement that they had cracked the company safe set the torpid room buzzing, and Tweed's appearance soon thereafter with the Union Pacific's subscription book, cash book, and contracts book interrupted proceedings. The lawyers, witnesses, and referee sat there for some minutes as he thumbed through the ledgers, trying to make sense of them and respond to questions.

"I don't understand these books," he finally said.[79]

One block over, at the offices of the Central Pacific at 54 William Street, Collis Huntington was enjoying the spectacle. It would have been hard to miss, with full transcripts of the court hearings appearing in the daily *Herald* and *Times*, and it would have been hard for him not to enjoy the notion of Durant or Bushnell blinking under the light of official scrutiny, or gulping at the prospect of jail food, with the Tammany judge overruling every single objection of their counsel and the Fisk legal team reading reams of accusations and characterizations into the record for the obvious enjoyment of the newspapers. "Mr. Dillon and certain other officers of the Company had absconded" would be read out before the judge, and the Union Pacific attorney would leap to his feet objecting to the malicious and slanderous "attempt to calumniate in the newspapers," that Dillon "is a venerable, high-minded and most respectable gentleman. He is a fair dealing man and 'abscond' is not a word to use in regard to him."[80] Indeed it would have been hard for Huntington not to enjoy the enemies' suffering—but in enjoying the bonfire one had to take note of the wind lest oneself be burned.

Down in the Capitol, Senators Stewart and Nye and their other supporters were doing a fine job against the rising temperature, and despite the heat raised by Representative Bingham to investigate the bonds' release, some newspapers were nonetheless responding well. During one Stewart blast, the capital correspondent of the *Sacramento Union* noticed that "one of the boss men of its Crédit Mobilier, Oakes Ames, of the House, sat hard by and seemed to enjoy it, as any one could afford to who had made at least a round million out of it." The *New York Times* report on testimony in the Senate Railroad Committee on March 31 could have been written by Huntington himself: there had been "no over-issue of Government Bonds" to the Central Pacific and "the Union Pacific Road is far from being completed to Ogden, as alleged." But while Hopkins continued to send their twenty-mile reports of completed track at a pleasing pace—to milepost 570 on March 17, to milepost 590 soon thereafter—that final report exhausted the last $1.5 million of their $50 million federal appropriation, and it looked as if they might not soon collect those bonds. His Washington lobbyist, L. E. Chittenden, had seen Interior Secretary Cox and was informed that no more bonds would be issued until the special commissioners completed their report and gave their recommendation for a final meeting point and the routes thereto. "You should not lose a moment in advising your California directors of this," he urged, "that they may pay proper attention to [the] committee whose action may be of grave importance."[81]

"Proper attention" might be out of the question at this late date, though anything was possible. But Stanford was in Salt Lake City and reported on March 21 that a train from Omaha had supposedly reached the Wasatch station the day before. "If the road is open and shall remain so long enough the U. P. will undoubtedly shove the material through to take them to the Promontory," he said. Samuel Montague had looked over the Union Pacific line to the head of Echo Canyon and estimated that the unfinished trackwork and tunnel would be finished in about forty days. Once Casement's men picked up speed west of Ogden they would first cross the Central Pacific grade in twenty-five miles—Stanford urged Huntington to get their lawyers ready for an injunction to stop them. Meanwhile, Charley Crocker was spurring their Cantonese to higher attainments; Charley thought they would reach Promontory in mid-April. On March 22 they crossed the Utah line at the edge of the barren, gray flats of the Great Salt Desert and established a station called Lucin at milepost 597, within days opening regular train service to that unappetizing place. Fifty even emptier miles ahead was the western shore of the great blue lake. But from then on they had to be more economical than ever. On March 30—the day their tracklayers passed milepost 614—Huntington wired Hopkins: "Call on me for as little money as possible until the road is completed to Ogden. We will get no bonds until that time." Hopkins of course agreed, but wondered how they were going to prevail. "Unless we are able to make loans here, I don't see how we can get on with less than $750,000 a month," he said. "We will see what can be done to make it less—for I comprehend that without Govt. Bonds or 1st Mortg. Bonds to sell, it is not easy to raise large amts. of gold & meet your bills falling due there. I comprehend how our own mistakes have cut us short of millions of Govt. Bonds to help us to the end of the route. I think we all see it now & feel it."[82]

Collis Huntington was never one to resist an opportunity to say, "I told you so," especially to close friends like Hopkins and Crocker, but there would be plenty of time for that later. He had a lengthy statement to give to the Senate's Pacific Railroad Committee, which would defend the Central Pacific on many disputed fronts, such as the bonds, and lambaste the Union Pacific for its multitude of sins.[83] And the germ of a new enthusiasm had taken him, one that would give him control of a new metropolis which would more than make up for the loss of the Wasatch coal mines. He was determined to keep his federal mortgage bonds to Ogden, of course—but why should that flyspeck Mormon village become the junction city for the two greatest railroads when he could cause a new one to be built somewhere north of it, a true national city through which the commerce of the world would run? The Mormons would, if Huntington played his cards right, gullibly sell enough land to finally give the Central Pacific the kind of real estate business it deserved; being less than fifty miles from the coalfields might still make that fuel accessible if they bought up the right stakes. One of Stanford's assignments from Huntington had been to scout the area north and west of Ogden. Then, as Huntington was readying himself to go to Washington, on April 5, Stanford's telegram came with its bad news. "It may be important for you to know," the Governor told him,

> *that there is not a sufficient supply of suitable water for a terminal station at any point between Rosebud Creek forty miles west of Monument Point and Bear River, where it must be pumped and which at a low state is strongly alkaline. At Ogden is an abundant supply of pure water, a good elevation. Prior to the commencement of our work there was not and is not now a resident along the line between river & the western boundary of Utah excepting those brought here by railroad work. Brigham Young has organized a company to build a railroad from Ogden to Salt Lake city this station.*[84]

All that empty, unpopulated land but no water; they were thus tethered in some fashion or another to Ogden. But Huntington felt he still had room to maneuver. In a fighting mood, he took a train down to Washington, where the new interior secretary had just urged that the Pacific Railroad Act be amended to strike out the provision about collecting advance bonds; Huntington had his own amendment slipped in, fixing the meeting point at Ogden. But Dodge was alert and had the motion blocked; not resting there, he sent out a flurry of telegrams to press agents and newspaper editors urging that the companies be allowed to build until they met.[85]

But it had become clear to both Dodge and Huntington that they could either stand toe to toe and slam each other until one was dead or compromise before blood lust had the onlookers climbing over the ropes to get them both. The Union Pacific and the Central Pacific had better fix their own meeting point before Congress did it for them; such amity might even appease the scandalmongers. On the evening of Sunday, April 8, Huntington met Dodge in Washington, appropriately at the house of Massachusetts congressman Samuel Hooper, who knew them both and who faithfully held stock in the Union Pacific, the Crédit Mobilier, and the Central Pacific. The negotiations wound through the night—Dodge had wired New York to alert him immediately if Judge Barnard uttered any indication of not releasing their case from the New York courts, which would strongly affect his position—and lasted until nearly dawn. After a few hours' rest they took up again and negotiated late into the evening.

At one point a messenger delivered a wire to Dodge from Oliver Ames in Boston, who had convened a directors' meeting; they ordered Dodge not to agree to any point "East of the Eleven hundredth mile Post." But the engineer well knew how close the whole organization was to collapse. And the Californians already had the bonds. Cox and Boutwell in Grant's cabinet had told him that the Union Pacific could be enjoined from crossing the Central Pacific's grades.

He put the telegram away. Then he made the best arrangement he could. Huntington agreed to let the Union Pacific build as far as Promontory Summit, about fifty miles away from each company's end of track. Such would certify (at least in the engineer's mind) that Dodge's grades were superior to the philistines' lines in his own company, as well as to his competitor's; this being an unimportant debate for Huntington, he went along. As they worked it out, the Central Pacific would buy the track from the Union Pacific, for its actual cost, from Promontory down almost to Ogden. At that place, five miles from the village, Huntington would, with the Union Pacific's aid, build his new Gentile city which would shut Brigham Young and his followers out of the real estate markets and commerce. Of course Huntington and Dodge would keep such a detail secret. The pact would merely stipulate purchase of 47.5 miles of track—to the edge of Huntington's fantasy metropolis which would, in the course of time and events, never be built—and lease of the five-mile stretch, for 999 years, from the dream to the quiet reality of Ogden. And finally, Huntington and Dodge would arrange for their respective champions in Congress to pass a law officially fixing the point at Promontory Summit, so the bond question would be resolved.[86]

Telegrams of the agreement sped west to Utah and California and north to New York and Boston. Stanford, in Sacramento, instantly telegraphed Ogden to suspend all grading work east of the Promontory. Crocker was at the end of track, about to go back home for three or four days' rest; he felt as if he were breaking under the strain of this final push across the desert, nearly treading on the heels of his rock-cutters northwest of the lake and impatiently egging everyone on. "I have not time to spin a long yarn," Crocker wrote Huntington wearily, "but will merely say that I am doing all in my power to hasten track laying—which does not yet reach your wishes or expectations, nor mine." Then the wire from the East arrived and helped lighten a crushing load. "While I was building the road, I had all I wanted to do," Charley would explain.

> *I didn't think of anything else. I used to lie in my car, travel over the road, backwards and forwards; I would lie until morning before I could get to sleep. When I came home to Sacramento, my wife used to say, "Why, Charley, what's the matter, why do you toss about in bed?" "I don't know," said I, "unless it is that I am getting old. I cannot sleep as I used to." And we both decided that it was old-age. When Huntington telegraphed me that he had fixed matters with the Union Pacific, and that we were to meet on the summit of Promontory, I was out at the front. I went to bed that night, and slept like a child.*[87]

In North Easton, Oliver Ames received his notice of the agreement from Dodge. His response might have surprised Dodge, who deserved thanks—Ames's letter was more strongly worded than anything he had ever sent to the Doctor, who had made his life miserable for several years. "That part of the agreement giving the Central Pacific bonds on the road we build beyond Ogden is an outrage upon us and ought never to have been consented to," Ames complained, sounding more like Durant or Seymour than his usual genteel self.

> *We have burdens enough to bear to have some little help in bearing them, but for us to give the Central these bonds and let them pay us for the road when they get ready will I fear break us down. I can't conceive how you ever should have consented to it. If you had known the condition of the company you would not have done it. Certificates for these bonds have already been sold and it calls upon us to raise money at once to redeem them. When if you had stood for bonds of Government on all the road we build we should have gotten them and had part of our pay, while as it now stands we shall have a quarrel with the C. P. to get any pay out of them.*[88]

If Dodge felt any charge of resentment at Ames's intemperance, it would have been mitigated by another paragraph of the letter: Ames had sent Dodge a letter discharging Colonel Seymour from duty on the line of road, "to be sent to him or handed to him as you should see fit." Dodge—of course—preferred a face-to-face meeting. He was going out west to see the final act anyway, and this task would be one of the most agreeable ones of his service. Oddly enough, after he had gotten out on the line, Dodge would decide to hold the letter in abeyance; work came before pleasure, and even Seymour had some small value in the last weeks.[89]

On the morning after Huntington and Dodge had made their pact, their men in Congress quickly produced a joint resolution on the Promontory Summit connection and an officially sanctioned terminus "at or near Ogden," with the stipulation that the Central Pacific would buy the tracks as agreed. The resolution also authorized President Grant to appoint a board of five "eminent citizens" to inspect the completed line and note all deficiencies; the president would withhold enough bonds to guarantee that the national road would be brought up to "first class condition." Finally, the Union Pacific was authorized to move its corporate office to Boston and hold a board election there on April 22. Dodge wired the good news to Sidney Dillon, whose excuse for "absconding" from New York to evade assault charges for pushing a deputy would be that he was urgently needed in Omaha and Utah as a newly appointed member of the board's final construction committee. Getting the Union Pacific out of the Tammany courthouse had been a particular urgency for them all—it "put us out of the hands," Dodge would say, "of the New York Blackmailers and pirates."[90]

In Judge Barnard's courtroom, young receiver Tweed had proudly reported on April 3 that he was busy with an inventory; among the large number of books and papers he had found in the Union Pacific safe were "bonds and coupons to the value of hundreds of thousands of dollars." His time for savoring the prospect of dividing that fortune between Fisk and other creditors deemed worthy by the judge was very brief: Charles Tracy, the principal Union Pacific attorney, told the court that "There were some bonds found there, but they were the bonds of the Company which had never been issued, and, of course, are worth no more than waste paper until they have been put upon the market. They are worth about seven cents a pound as old paper—nothing more."

Much later it would be revealed that the Crédit Mobilier records so avidly sought had been presciently moved by Sidney Dillon in January to safekeeeping at the headquarters of the New Jersey Central, where he was vice president. And in a few days the case of James Fisk versus the Union Pacific Railroad was pried out of Judge Barnard's courtroom by a federal judge responding to directions from Washington, who removed it to federal jurisdiction, where the case would drag on for years. The litigation had given the railroaders immense bad publicity at a crucial time—it also caused the value of company securities to plummet—and scandal helped to set the public mind against not only the scalawags but the whole industry as well. "The truth is," commented the *New York Herald,* "there is cheating on the grandest scale in all these railroads, and it is only when sharp managers quarrel over the spoils that the public get at the facts." Before the case was seized away from the angry and petulant Judge George Barnard, there would be time for one good telling finger in the eye: in the transcript published in the newspapers was a list of stockholders, which took up half a column of tiny type. One could, with a little research, turn up some revelations in such a list.[91]

A Sacramento-bound train was crossing the Sierra on the morning of April 9, when a mishap and an adventure occurred which seized the public imagination on the western slope. Accidents occurred every day on a railroad track somewhere in the nation, and they never failed to excite newspaper editors and fascinate readers. On the morning in question, a messenger for Wells, Fargo named Michael O'Connor was asleep in his compartment in the express car, which was filled to capacity with freight. He awoke to a sensation of heat and suffocation, and found the forward end of the car in flames from, perhaps, the lamp. In the thick smoke he attempted to smother the fire with empty mail sacks but it was too large. He looked for the bell rope to signal but it was already burned. O'Connor leaned out the door only to see that the train was passing through a snow shed and there was no one to signal. Onrushing air was rapidly fanning the flames, and he saw that not only the express car but the whole train was in danger. Somehow he climbed to the roof. "Any person who has seen one of the express cars," the *Union* reported, "can readily imagine that the feat of climbing from the door to the roof is a somewhat difficult gymnastic undertaking when the car is standing still; with a train moving rapidly and swaying heavily from side to side, passing close to the timbers of a snow-shed, it was perilous." Up in the locomotive, the fireman happened to look around to see the amazing sight of a crazy man crouching on a car roof, gesticulating wildly and managing to duck beneath the snowshed timbers only inches away from his head. The train halted as soon as it could on the downgrade, the blaze was extinguished, and O'Connor was, briefly, a hero.[92]

Meanwhile, the Central Pacific tracklayers were moving bravely under the stern pressures of Strobridge and Crocker, who wrote Huntington that he thought they would reach the rock cuts on the eastern slope of the Promontory Mountains on April 25, and the summit perhaps on the twenty-seventh; Stanford left Sacramento for Ogden on the twentieth, telegraphing Huntington of his speedy arrival only two days later. Charley had told the governor that they would reach the connecting point on the summit on the first of May—but would have to wait at least ten days for the Union Pacific to get up the eastern slope.

Stanford, though, discovered a new crop of troubles just when they all should have been feeling the pressure lessening. Going up to Promontory with the intention of persuading the Union Pacific people to adopt the nearly completed Central Pacific line over the summit, he met Sidney Dillon, who had taken charge of construction. Dillon refused to tell his men to lay track on the completed Central Pacific grade; he said he had no authority, but did not seek it, and kept his contractors busy cutting and leveling the Union Pacific line for track. But the company was not following its own line as dictated by the approved route map, Stanford learned. In their haste and carelessness the graders seemed to be making it up as they went along—Stanford found four unacceptably sharp curves and two lengths of trestling "of a very imperfect character," he wrote Huntington on April 22. Shaky and insubstantial, one trestle was some 85 feet high and 400 feet long, with the other 20 feet high and 100 feet in length. Any kind of train would have to go over them at five miles an hour, and heavy freighters were out of the question. Stanford's engineers glumly estimated that getting this work accepted would cost more than $50,000 in repairs and improvements—whereas the Central Pacific work required less than $10,000 to perfect.

By this time Dillon had gone to Echo City. Stanford wired him to object to the shoddy job which did not adhere to the approved line or to the agreement between Huntington and Dodge. "I told him," Stanford wrote Huntington, "we could not be expected to pay for temporary work." The next day in Ogden Stanford ran into Silas Seymour, who seemed sympathetic to his objections. The governor accordingly telegraphed Huntington: "Seymour advises U. P. to take our line" if the government approved. "Time & money both saved by taking our line."[93] But when Seymour wired Dodge en route out west with the proposal, the chief engineer rigidly rejected it out of hand—such was his distrust of the colonel and of the Central Pacific.

Trouble still burned in the Hell on Wheels towns, even in burgeoning Laramie. There, a party of railroad tie cutters on a bender began to break up a saloon. They were arrested and placed in the town jail. A posse of other tie cutters decided to release them, and it seemed that they were about to succeed when a force of vigilantes came to the aid of the beleaguered constables. Gunfire blazed out and wounded a good number on both sides; one policeman was mortally wounded. It would take days to sort out culpability and the charges. Law-abiding citizens worried that their tenuous hold on civility was slipping.[94]

In the East, Union Pacific credit and securities were also diminishing. Thanks to the Fisk suit and publicity, first-mortgage bond prices fell from a level of 102 to 70 or 65, and the roar from the West for money showed no indication of abatement. In Boston, Oliver Ames had convened a directors' meeting on April 9 (this was the same meeting in which Dodge had been instructed to settle the meeting point west of Ogden and Promontory); the directors had created a new executive committee of five—Glidden, Bushnell, Brooks, Bates, and Oakes Ames—whose task was to get the company out of its distress. On April 10 the committee published a circular offering stockholders first-mortgage bonds at 85 percent and land-grant bonds at 55 percent, the limit pegged to their current stock holdings, on strictly cash terms; unbought bonds would be offered to those stockholders who had purchased their limit but who would pledge to quickly transact for the new securities. Oakes Ames was nearly out to the edge of his available credit, but he manfully bought all entitled bonds and offered to take more if others did not step forward. Few did—Glidden and Ames were reduced to begging some of the directors and stockholders (who had gladly taken profit allotments in cash) to now help give the company a push. The committee hoped to raise at least $7.7 million in cash for the $11.3 million in bonds at par value, but the money only trickled in, scarcely assuaging the great thirst for cash out west.[95]

"I am still here on a gridiron," Jack Casement wrote to his wife from Echo City, worried for the future but willing to grit his teeth and hold on, doing his part.

> *Things look awfull bad and I cant hear a word from New York. I can close up the work now and have some money left if the Banks dont burst. We owe*

> *our men about 90,000$ and have 160,000$ in the Banks at Omaha and Cheyenne. The banks are loaded with UPRR paper and if the company dont send some money here soon they will bust up the whole country. The company owe us over 100,000$ but we can wait on them if there is any show of helping them out.*

Bad as things were for his contracting company, Casement thought that "we are in the best fix of anyone and can't help but come out something ahead." His men would be finishing their tracklaying very soon, "but the Lord only knows when we will get paid for it."[96] Other outfits were indeed doing poorly, their men complaining and threatening to strike—Dodge wrote Oliver Ames that "men will work no longer without pay & a stoppage now is fatal to us." Material was still scarce. By April 22, Dodge had reached Council Bluffs on his way out to the work, and Webster Snyder wrote him to despair of the new building committee under Sidney Dillon—which to Snyder promised business as usual. "I judge that Seymour and Reed have gotten hold of Dillon," he said,

> *and are making the most of it. He takes their word for various matters without examining for himself and is telegraphing me to do various things which I know are not right, and which he would not do if S. and R. were not writing dispatches for him. I wish you could go out and look through the outfit. It ought to be done quickly or there will be nothing left. If T.C.D. comes here with any authority, I propose to quit at once.*[97]

That day in Boston, Thursday, April 22, the postponed annual stockholders' meeting convened at the shipbuilding offices of Glidden & Williams, on State Street. From wherever they happened to be, Dodge, Snyder, Blickensderfer, Jesse Williams, Secretary Cox, President Grant, Charles Francis Adams, Horace Greeley, and all the rest, expectations rose about the long-expected election of officers, and the ouster of Thomas Clark Durant. But after a few resolutions confirming the move to Boston and the creation of the new committees, the meeting adjourned without holding the election. This did not sit well with everyone. Glidden and John M. S. Williams wired Dodge to "Please hurry Duff and Dillon back and all Government directors, yourself with them, so that we can have an election. We are not safe until we do." But a majority of directors had decided it was too dangerous to go up against Durant until the project was done. That might be, the rationale went, only a matter of a few weeks—Dodge's wire, received that day, reported "Central Pacific Railroad, eighteen miles from Promontory Summit. We are twelve miles from Summit."[98]

Charley Crocker was hell-bent on entering the record books, and no general or colonel or other lofty-titled gentleman was going to show him up. The Union people had crowed over putting down four miles in a day, and Crocker's men had gone better with six; then Casement had spiked eight. "Now we must take off our coats," Crocker had told Strobridge, "but we must not beat them until we get so close together that there is not enough room for them to turn around and outdo me." In the last week of April the time had come, and Crocker told Strobridge they would lay ten miles in a day. "How are you going to do it?" the supervisor wondered, "the men will all be in each others' way." Crocker laughed. "You don't suppose we are going to put two or three thousand men on that track and let them do just as they please? I have been thinking over this for two weeks and I have it all planned."

The feat was announced for April 27, but a locomotive derailed and postponed it twenty-four hours. At dawn on the twenty-eighth, though, Crocker's army was impatient to begin. Dodge, Reed, Jack Casement, Dillon, Duff, and Seymour were invited to watch the work. Crocker had indeed planned everything, and although Dodge and Casement would grumble later that the Central Pacific had cheated a little by embedding its railroad ties beforehand, what ensued over the daylight hours was a miracle for the era. One by one, platform cars dumped their iron, two miles of material in each trainload, and teams of Irishmen fairly ran the five-hundred-pound rails and hardware forward; straighteners led the Chinese gangs, shoving the rails into place and keeping them to gauge while spikers walked down the ties, each man driving one particular spike and not stopping for another, moving on to the next rail; levelers and fillers followed, raising ties where needed, shoveling dirt beneath, tamping, and moving on—"no man stops," Charley Crocker directed, "nor allows another man to pass him"—and no man, Irish or Chinese, did more than one task, each a cog in that large, dusty, sweating machine advancing up the incline toward the summit. At midday Crocker sent the last train forward, "two train-loads with two engines upon it," he commented proudly, "and it was the heaviest train that ever went over the road," and "it went over the very six miles we had laid in the morning, and went safely, and if the track had not been good, it could not have gotten over." He offered to relieve the tracklayers with a reserve team—but they were adamant about finishing out the day themselves. All would get four days' pay for the feat.

A detachment of soldiers was nearby, and came out to watch the work. "Mr. Crocker," the commander exclaimed later that day, "I never saw such organization as that; it was just like an army marching over the ground and leaving a track built behind them." He had walked his horse alongside the men and measured their progress as "just about as fast as a horse could walk . . . a good day's march for an army." The afternoon's pace even included the necessary bending of rails for curves, as always done by brute strength with sledges. Other Central Pacific workers kept the telegraph wire unreeling during the day, stringing it up on freshly planted poles, and when they were all finished the news was flashed to Sacramento. "We got our forces together and laid ten miles 185 feet in one day," Crocker would chortle later, "and that did not leave them room enough to beat us on; they could not have done it anyhow, I believe, because [when] they laid eight miles they had lanterns, and we did not use them, we laid our ten miles by daylight."[99]

Dodge might have quibbled over petty details like distributing the railroad ties ahead of time, but the commander in him appreciated what Crocker had accomplished while the manager in him appraised the Cantonese tracklayers and found them eminently desirable as employees—"very quiet, handy, good cooks and good at almost everything they are put at," he admitted to Anne Dodge. "Only trouble is, we cannot talk to them."[100]

Stanford was also out at the end of track, and on the ten-mile day he wired Huntington in a final plea to get the Union Pacific officers to stop their steep grading and sharp curving on the eastern slope, and their temporary constructions, and just adopt the completed Central Pacific grade. It would be a hopeless cause, despite Huntington's wired assurances to both Stanford and Judge Crocker that the agreement called for adoption of the Central Pacific line to Ogden; "they send no orders to the contractors," the Judge complained from the Sacramento office. In this matter as in so many others inertia dictated; with Dodge and Ames against it, apparently the government commissioners and the interior secretary had decided on a status quo policy which would "not require serious action by the Government," as a Union Pacific lobbyist wrote Dodge.

All of the Governor's work adjusting the grades on the eastern slope would be for naught, though there was still ample reason to complain about the quality of the newer work—"the track passes over a piece of trestle which if we had possession we would not attempt to run over," the Judge protested, "but would immediately replace with new trestle or fill." And Stanford's reports home about prickly conditions on Promontory Summit also prompted the Judge's commentary. "The U. P. people at end of track act ugly about everything," he told Huntington, "& talk that we will never get possession of the track to Ogden. Of course this settles nothing, but shows the animus of the ruling spirits." Crocker had been trying to negotiate with Webster Snyder in Omaha about setting through rates, with some difficulty: the Judge proposed setting the rate from Sacramento to Omaha at $125 in greenbacks, divided equally between the two companies. Even $100 for first-class passage was acceptable, and when the steamship companies responded by lowering their fares Crocker could see dropping "down to 90, 80, & even to 60, when we get cars enough to take the crowd, & if it should prove necessary." But he expected resistance to the idea of equal division of tariffs; the distance from Sacramento to Promontory was 690 miles, and to Ogden, 744 miles; between Ogden and Omaha it was 1,032 miles, or 1,776 in total. "The great cost of surmounting the Sierras, and operating without coal or cheap fuel, entitles us to an equal division between the two roads," he argued. Truly, this would be a controversy not easily settled, and if Snyder was not the one with the power to negotiate he hoped Huntington would get someone sent out before this last late work was done.[101]

As the company's major task was nearly completed, the Associates looked toward the growing empire—and to new horizons. Work had continued, albeit at a slower pace through the late winter and spring, on the Southern Pacific (with a twenty-five-mile report ready for the commissioners) and the Western Pacific (with twenty miles); the latter had been slower because of the timber shortage and the number of bridges required between Sacramento and Stockton, and the many swollen freshets which had made construction difficult at that time of year. Consolidation of the California & Oregon with the Yuba Railroad was still held up by evasive Yuba bonds and unresolved debts, which Hopkins and the Judge were not willing to absorb. Crocker was also looking for ways to improve their Oakland waterfront property; the nearest rock for landfill was, ironically, on craggy Goat Island, and although the local army commander said he had no military objection to their mining it, the Judge knew this was something Huntington must handle in Washington.

With his reduced physical strength, Judge Crocker's office time was mostly consumed with these corporate interests in the state, but he—like Stanford and Hopkins—was enthusiastic about what might be a significant sideline to their railroad business. Lloyd Tevis and D. O. Mills had come to the Associates proposing they all form a Pacific Express Company, with a capital of about $8 million, with the Associates having a majority of the stock but none of the management duties. "One thing is certain," Crocker wrote Huntington, "the express business between here & Salt Lake & on this coast will be of vast magnitude . . . while the business between Salt Lake & the East will be comparatively small." Of course cooperation with the Union Pacific would be necessary. "In fact," the Judge said, "Durant & Stanford agreed that the Cos. would co-operate in the matter & do nothing without mutual consent." Time was growing short, but it seemed important that they get something in place before the two companies joined track at Promontory.[102]

Quickly, of course, this became only a matter of days. On Saturday, May 1, Charley Crocker scribbled a telegram to New York. "Our track is completed," he informed Huntington jubilantly. "The U. P. will meet us on the Eighth or Tenth of May."[103]

The four special commissioners—Warren, Blickensderfer, Clement, and Williamson—were still in Washington at the end of April, expected to finish their report and recommendations on both roads within days. But Dodge had already found a distressing amount of faulty construction in western Wyoming and Utah which was certain to raise flags; some of it was so bad that he would never let a train cross. The principal bridge contractor, L. B. Boomer, had sent Dodge a written protest refusing to put any more prefabricated bridges on their masonry, which had been done by subcontractors working for Dodge's former assistant, James Evans. "Three of his bridges have gone down, the masonry failing to support their weight only, no cars having run over them," Dodge wrote Oliver Ames glumly. "Two others are giving way." He had seen bad work like this back on Lodgepole Creek, but said he was going out himself to inspect. "Though closely watched it seems we cannot trust masons who have had the reputation of being No. 1 and honest, unless we employ an engineer to every structure to stand right over them while they put in a drain that will hardly cost as much as the engineer's salary." A few days later, having reached western Wyoming to see things for himself, he found things in shambles. Seymour had on a whim moved the location of a bridge over Black's Fork, and Evans had done likewise on the span over Green River, both now requiring a train to lurch and veer onto the new tangent. Dodge ordered the temporary track and bridges left in place and wired he would not accept the permanent structures for the line. Worse than the bad engineering was the outright thievery. "The contractors who put in masonry on Sulphur Creek should not be paid," he wrote angrily to Sidney Dillon, who was now deciding—with a logic known only to himself—which bills would be honored.

> *The abutments of five bridges are breaking and tumbling down. The Masonry on Bear River is worthless, the backing is dirt and free stone set on edge, and I doubt if there is a bond in any one of the piers. They are now delivering red sandstone there to put a course on top to cover their miserable work. I do not believe the masonry will hold up a truss and the placing of heavy sandstone on the piers will only help crush them. I should put truss on trestles and rebuild the entire work.*[104]

Similar problems became evident along Bitter Creek. A. P. Wood, the construction engineer working for Reed, had been ordered earlier in the winter to put up permanent bridges when the government directors threatened to hold up their reports on the sections. "I want you to go there and select the best rock you can find," Reed had told Wood, "and get a force of quarry men and masons at work." As Wood later recalled,

> *Any man of judgment knew that Bitter rock was worthless for bridge masonry. But what were we to do? An emergency existed. We could not build the railroad without money. I did as I was told to do. I soon had two or three of the smallest bridges completed and the track changed to the main line. This satisfied the directors and they allowed the money to flow again, and all went well until warm weather came, when the masonry dissolved, the bridges became unsafe and the road was put back on the temporary crossing.*[105]

Expediency was one thing, but what Dodge had been astounded to find along the line was clearly of a lower order. Snyder had been warning Dodge of the corner-cutting and the plunder, but to his discredit the chief engineer had been too busy to pay full attention, and unwilling to suspect some of the people he knew personally as being corrupt. To the end he would vouch for James Evans and Samuel B. Reed, for instance, and most historians have gone along, probably with good reason. Meanwhile, the complaints poured into the Operations Department, and Snyder obligingly forwarded them in typical high dudgeon. Excess material and tools stood all along the line until employees or passersby darted in like hungry fish to get a nibble. In Utah, Silas Seymour had seized a stack of company lumber and had set carpenters to work on a two-story house on a Union Pacific–owned lot, and it seemed that even Jenny Reed—who had come out to Echo City with her children some months before—was readying the company-provided quarters for shipment back to Joliet. "Seymour the Colonel looks at these things as his private property," Snyder was told, "while Mrs. Reed marks all furniture, bedding etc. with S. B. R.—may they long live and prosper." Snyder was ready to burst with indignation. "I am heartily sick of this outfit," he ranted to Dodge on May 4, "that talks so much about cleaning out thieves & yet weakens when in presence of the thieves & will let thousands be stolen under their own eyes while looking after old flaws & scrapes."[106] He would not be surprised, and in some measure would be relieved when, in less than three months, Oliver Ames would sack him, blaming the messenger instead of the miscreants.

In New York on Friday, April 30, Jim Fisk served injunctions on all the banks where he thought money belonging to the Union Pacific or Crédit Mobilier might still be lurking, "but," as Oliver Ames told Dodge, "we were too sharp for them. Bushnell was advised by one of Fisk's men who had been watching us for weeks, that the injunctions were to be served, and so he was prepared for them." The Fisk operative said that his employer would not pay him as agreed so he offered to work for Bushnell. Meanwhile, things were still lively at 20 Nassau Street; despite the fact that all signs for the Union Pacific company were removed from the lobby and the upstairs corridors and the offices were vacated, Durant had craftily rented the rooms for himself. Incomprehensibly, not all the records had been removed when the doctor took possession; he gleefully seized the trustees' books. His private secretary, Henry C. Crane, refused to give them to the Union Pacific, although he permitted treasurer Benjamin Ham to at least look at them and make copies. On some level Ames had been warned by the proceedings in Judge Barnard's court that his company's accounting practices were in criminal disarray—contractors' and suppliers' drafts by the hundreds, representing millions upon millions of dollars, appeared in New York and were paid without question or oversight approval of a company officer, the casual revelation of which dumbfounded most of those hearing it in the courtroom. It was, though, worse than anyone might have expected. As he peered at Crane's handwriting, Ham found one mysterious entry after another: drafts paid out with cryptic, dubious, or nonexistent explanations, duplicate and triplicate payments, with Crane unable to explain any of them. One $10,000 draft had been paid on a note Crane failed to reclaim, but he said he could not remember whose note it was or when it was paid. Taken together, the holes in those accounting books amounted to a good-sized grave for the corporation, and Ham as well as the Ames brothers despaired of ever completely plumbing their depths. "When we can get our Books away from NY and cleaned out from that sink of corruption," Oakes Ames grumbled to Dodge, "we shall feel safe and not until then."[107]

Any vision of safety Congressman Ames might have experienced was purely illusory. And he could not fathom what had overtaken the usually steady Sidney Dillon out west; his wires were verging on the hysterical, and his demands for money went beyond comprehension. Oliver Ames wired him and John Duff to cease drawing on the now-nonexistent New York office and the bank accounts they were rushing to close. Dillon's telegraphed reply was like a howl from the wilderness: "We must have five hundred thousand dollars to pay contractors' men immediately or road cannot run." Oakes Ames wondered wanly if Dillon knew what he was doing. "He wants to draw on N. York and Boston both," he commented on May 4 to Glidden and Williams in Boston, "and for all kinds of sums and [I] think he must be confused in his operations and the large amounts he wants to draw rather surprises me."[108]

By this time Dr. Durant was on a westbound train, intent as vice president on being the ranking Union Pacific officer at the joining ceremony—and taking credit for the whole gigantic enterprise. President Ames had been advised to stay in Boston by company attorneys because of outstanding legal matters, and told by them that as long as he commanded from the new home office, the Doctor could usurp no more powers; the edict had already gone forth that only Dillon or Duff was authorized to approve any more expenditures. Durant, however, was never short of schemes.

On May 6, a Thursday, Grenville Dodge wired Ames that the two companies would connect their rails on Promontory Summit on Saturday, May 8. Everywhere in the nation, citizens readied themselves for a tumultuous outpouring as the Atlantic and Pacific coasts were finally, after decades of expectation, linked by twin bands of iron, at unimaginable cost and only barely foreseeable benefit.

32

"We Have Got Done Praying"

Dr. James D. B. Stillman arrived at the Central Pacific depot in Sacramento on Wednesday, May 5, 1869, as the regular eastbound train was getting up steam for its departure at 6:00 A.M. Behind the scheduled train was a special—locomotive, tender, supply car, and Charley Crocker's superintendent's car. Dr. Stillman was an old friend of Mark Hopkins, who in 1849 had been with him on the horrendous Argonauts' voyage from the East around Cape Horn; he had kept up the association after Hopkins had gone into trade and Dr. Stillman had resumed practice and founded Sacramento's first hospital; currently he served as the San Francisco medical examiner. Dr. Stillman now had been invited to go out to Utah to see the ceremony joining the rails of the Central Pacific to the Easterners' line.

He was a good witness. Born near Albany, New York, he had stood in his father's dooryard as a little boy in 1825 as the Erie Canal opened, "saw the rippling flood as it flowed for the first time over the sandy floor of that stream . . . that immortalized the name of Clinton, and opened the great lakes and prairies of the west to the commerce of the Atlantic." He never forgot the sight of a group of older, bare-legged boys standing down at the edge of the "big ditch" as the moving water splashed by them, one boy leaning down and snagging a flopping, disoriented fish and carrying it away triumphantly. Not long after, his elementary-school classroom turned out to see early experiments with a locomotive for the new fifteen-mile-long Mohawk and Hudson Railroad. Everyone, even the doubters, laughed delightedly as the engine pulled away from a knot of competing riders on wild-eyed galloping horses; "one dog alone contested the race," he remembered long after, "bounding and barking on till lost in the distance, and on the long vista, where the paralleled lines met, the black speck disappeared, leaving a film of smoke to float away among the pines." Stillman was a superb writer with a keen eye for telling detail, and a contributor to the avidly read *Overland Monthly*. Periodically he had gone up into the Sierra to watch the progress of Crocker's legions. He was thrilled to be included in the excursion to see it finished.

There was room to spare in the superintendent's car. Mark Hopkins and both Crocker brothers all declined to leave town, contenting themselves with the prospect of a decorous carriage ride among old friends in the Sacramento procession on Saturday. So Leland Stanford would preside over the excursion car—which, for such a momentous occasion, was remarkably free of dignitaries: the Central Pacific's three government commissioners, Sherman, Haines, and Tritle; California's Chief Justice Sanderson, Governor Safford of Arizona, and Tax Collector Gage of Nevada; and a few others whom, as Dr. Stillman modestly admitted, were "like myself . . . not particularly distinguished but born to good luck." Alfred Hart, the official Central Pacific photographer, wrestled his bulky camera, tripod, and processing equipment on board. Their car was not especially luxurious; it had served Charley Crocker well, and had been considered good enough for the special commissioners' inspection tour, and although it was arranged with a kitchen, dining room, parlor, and private bedroom, and with fold-down beds in the public area accommodating eight, it was not a palace car, though it was amply provided with victuals, spiritous drink, and cigars.

After the regular train had pulled out from the station, Governor Stanford bowed from the rear platform to an assembled crowd, and the excursion began—across the Sacramento plain, where Theodore Judah had planted his instruments and pulled out his little engineer's notebook; over Arcade Creek, where Leland Stanford and a few scientists had moved the Sierra Nevada range; through the deep Newcastle Cut, where Strobridge had sacrificed an eye; up the miraculous, continuous divide toward Dutch Flat, where the grumpy Doc Strong still dispensed his patent medicines and ill opinions about the Central Pacific ring. Passing the beetling cliffs of Cape Horn, where adventurous farm boys from Kwangtung Province had woven reed baskets and swung out in the numbing air, the locomotive pulled them up through somber forests toward the snowy peaks, every cut contested by the hard-hearted granite, and marked and unmarked graves still visible along the roadbed. "We plunge into the bowels of the mountain," Dr. Stillman wrote—menacing nitroglycerine oil tested and proved in the tunnels—"and out at once into the sunlight and past the cheerful dwellings of men." He went on:

> *We are cribbed in by timbers, snow-sheds they call them; but how strong! Every timber is a tree trunk, braced and bolted to withstand the snow-slide that starts in mid-winter from the great heights above, and gathering volume as it descends, sweeps desolation in its path; the air is cold around us; snow is on every hand; it looks down upon us from the cliffs, up to us from the ravines, drips from over head and is frozen into stalactites from the rocky wall along which our road is blasted, midway of the granite moun-*

tain. We are in pitchy darkness in the heart of the mountain—the summit of the grade; out again into the light; on, on through wooden galleries mile after mile; a sylvan lake flashes out from its emerald setting among the mountains."

It was Donner Lake, and in a few days during the tumultuous celebration in Sacramento the whole city, including thousands of graying Forty-niners, would stand and fall silent for an entire minute to remember "the Pioneers of the Pioneers—the Donner party," mindful indeed of *all* the emigrants' suffering to surmount the Sierra or endure the rough seas and reach the promised land. Sacramento in May 1869 was filled with the scent of roses, but in the mountains the air streaming into their car smelled cold and moist, like wintertime, like pine—less threatening, less like death, with every climbing train.

Past the spanking new settlement of Truckee—where every board in every house, Dr. Stillman noted, "looks as if just from the saw-mills, so fresh and bright," where the streets were crowded with "great, healthy-looking, bearded men"—the train plunged down the Truckee Canyon.

Where the track ran between the foaming green river and the overhanging slopes, Dr. Stillman wrote, "we came near driving our last spike." Some Chinese woodcutters were far above them on the mountainside, and had seen the regular train go by as expected, and, being unaware that a special train was following, had shoved off an immense log down the steep slope toward the Truckee. The log landed partly on the track—it was some fifty feet long and as much as three feet in diameter—just as the special locomotive rounded a sharp curve and bore down on it. One excursionist, W. H. Harkness, editor of the *Sacramento Press*, was riding on the engine pilot (or "cowcatcher"), enjoying the best available view of the scenery while wrapped in buffalo robes, and may have been the first to see it. He leaped from the train, in his tumble sustaining only bruises and cuts, as the pilot picked up the log, "or did its best to do it," as Dr. Stillman commented, "and went through bankruptcy; but the force of the blow was not lost, for the heavy frame of the engine tripped the log and landed it where there was just room for it, yet did not prevent it from clearing away the steps of the starboard side of the train from stem to stern." Leaving the wreckage and the log—broken through the middle where it was at least three feet in diameter—the train limped down past Reno to Wadsworth, where the locomotive was unhitched for repairs and the excursionists' cars were made fast to the regular train.[1]

The sun was going down behind the mountains as they set out again across the Forty-Mile Desert. "Several of our party were among the overland emigrants" of 1849, Dr. Stillman wrote, "and they pointed out where, one by one, their animals perished, where they abandoned their wagons, and where their guns—the last article they could afford to part with—were planted, muzzle downward, into the hillocks in the desperate struggle for water and life." Governor Stanford, enthroned amply and wreathed in cigar smoke, had no such story, of course, having journeyed to California three years after the gold rush by Pacific steamer to join his brothers in the grocery business, but as the speeding train moved off across the darkening Nevada desert, never out of sight of snow capped peaks, raising a cloud of ash-colored alkali dust to settle over old animal bones, wagon fragments, mileposts, and occasional gravestones out in the sagebrush, he could call for more iced champagne from a liveried attendant and put their minds toward what lay ahead. "It was," Dr. Stillman said, "a country that one could not travel over too fast."[2]

The sumptuous Pullman palace car of Vice President Durant rolled westward from the Missouri—down and around the notorious oxbow and out to the Platte, clattering past the now well-established Nebraska towns and radiating farms, past sad heaps of buffalo bones out on the prairie, past abandoned encampments and old battlegrounds, past vanished hell towns and their rude, tumbleweed-snagging graveyards; the train paused in burgeoning but still rough-hewn Cheyenne and Laramie, taking on westering passengers who inevitably tried to peer into the windows of the elegant private executive car, and it moved smoothly up the Laramie Plains corridor between snowcapped mountains, pistol potshots of the riders responding as with every train in daylight hours whenever the dwindling game of the plains showed itself near the tracks; after Rock Creek, with the Red Desert and the Continental Divide ahead, the train would have slowed cautiously to accommodate the rotting culverts and subsiding grades, all the hurry-work and get-it-while-you-can contracting, but though slowed, the westward movement would have continued, day through night through day, with seemingly nothing to impede it.[3]

Intrepid correspondent Albert Richardson would be tracing this route from the East in a fortnight, having a happy encounter with Charley Crocker in Omaha as the prideful Californian took his family in their private coach on their first transcontinental railroad trip; Crocker "delighted us," Richardson would write, "with blooming flowers, and feasted us upon strawberries, oranges, and luscious cherries from California, brought upon Alaska ice 1,800 miles through the green valleys of the Pacific slope, and through the lingering snow-drifts of the Rocky Mountains. It seemed like a story from the Arabian Nights." Richardson would have just seen Dr. Durant in a Chicago hotel lobby crowded with "trunks and valises piled up like cord-wood, and variously labled 'San Francisco,' 'Salt Lake,' 'New York,' and 'Bangor.'" And Richardson would have beamed with pleasure at the sight of the Doctor, whom the writer never failed to praise in print as a near-superman, "the motive power" behind the Union Pacific—"one of the most marked, original characters of all this throng," yet "one of the quietest and least noticeable. There he sits, chatting carelessly in low tones, a rather tall man, in middle life, his hair and whiskers beginning to show streaks of gray, and his worn, mild, thoughtful face shaded by the limp brim of a low-crowned brown hat." Richardson would regard him from a distance and reflect on the many controversies, the competitiveness and recklessness, the hard work "like a galley slave," the unflagging drive, and then he would approach with greetings. "At last," he would comment,

> *after his every nerve has been strained for four years, he is foot-loose once more. "The rails are laid," he says, with a quiet smile, "and now I don't care whether school keeps or not." As he gets up for a stroll, we see the chief mark that his terrible labor has left on him; his frame is bowed, and he looks like a modern Atlas, a little surprised to find that his heavy burden has rolled off. He has done the work; let him have the credit of it. He is said to own one-fourth of the entire road. Now he will devote himself to his private affairs, which have taken care of themselves during three busy years. Perhaps, for this summer's recreation, he will build the plaything of a railway to the Adirondacks, in which he has a controlling interest, and where he owns half a million acres of land more or less. Where will his indomitable energy next find vent?*[4]

Subtract the hero worship and financial exaggerations, mark the stooping carriage to bad posture, too much time behind a desk, and middle age as opposed to something mythic, but still it recharges an image to place inside the palace car with its polished walnut parquet, its lush velvet draperies, its cut-glass decanters, its brass. His brass. John Duff the elder was with him, on the ascendant within the company directorate, and the train, heading for what was being called "the marriage" of the two continental lines, moved past the towering buttes of western Wyoming, crawled over all the temporary bridges around and west of Green River, and began the rise toward the Wasatch rim.

At the desolate station of Piedmont, near the Aspen summit, the train abruptly stopped and the wholly unexpected sight of at least three hundred angry and armed workmen surrounded the train. Railroad ties had been piled up on the track, and the private car was uncoupled from the rest of the train and sidetracked, with both switches locked. Conductors, rebuffed at gunpoint, raised their hands and retreated. And spokesmen surrounded Dr. Durant to tell him that they were tie contractors who had not been paid in months, that they were desperate men who would keep him a prisoner there until they were paid. It was Thursday, May 6, the festivities at Promontory were still set for Saturday, and they wanted at least $200,000—or else. Another westbound train filled with dignitaries heading for the ceremony rolled up to Piedmont and it too was stopped. No more trains would pass until they were satisfied, said the workers.

Durant telegraphed east and west for money, reaching Oliver Ames in Boston and Grenville Dodge in Utah. Dodge's first response was to call in the troops—he wired to nearby Fort Bridger for a company of infantry to rescue the executives. The message flashed—as was normal—up the railroad line in the Wasatch canyons, over the rim, and stopped dead in the Piedmont station with the sympathetic telegraph operator there, who gave the message to the ringleaders. They sent a wire back to Dodge: send the army and the hostages would suffer; no more trains would be going through, and within twenty-four hours they would raise a general strike from Omaha to Ogden. "Men are getting desperate against Durant," Dodge wrote to his wife on May 6, "and I do not blame them much. Duff is in a bad box but we will get them out today."[5] Then, in Echo City now with Sidney Dillon, Dodge's telegram went off to Ames. "Tie outfit at Piedmont hold Duff & Durant under guard as hostage for payment of amount due them," he said. "You must furnish funds on Dillon's call."[6] Dillon wired for $50,000 as a token first payment. While the money was being gathered, Dodge and Dillon tried to keep the news of the abduction quiet—but Echo City was already in commotion and news was spreading rapidly among the laborers and contractors. Dodge sent a follow-up telegram to Ames:

> *Trouble at Piedmont will cause trouble in running department unless Snyder gets immediate help—if you wait until trains are stopped [all along the Union Pacific] it will be too late to release them until we are forced to pay in fact everything due on line—half million at once will relieve necessities and enable us to keep moving—actions at Piedmont known everywhere & all know company was obliged to pay before officers were released.*[7]

After $50,000 was handed over, Durant and Duff were allowed to go. But something nagged at Dodge and Oliver Ames; the rebellious laborers worked for Davis Associates, in which the Doctor held a strong interest. "There is no doubt," Dodge would write afterward, "this was an arrangement made by Durant for the purpose of forcing the company to pay." It turned out that Samuel Reed had refused to accept a great deal of the timber and ties the Davis contractors had delivered to trackside. After the fact, Dodge even grew suspicious of Sidney Dillon, the Doctor's sometime ally, who responded to the threats of the mob by ordering a trainload of soldiers "to go through without stopping," Dodge said, "which was a great mistake, but I did not hear of it until it was too late to stop it. There was nothing left to do now but to furnish money to pay off these contractors to release Duff. I knew Durant would be released any time he wanted to."[8]

"Their seems to be no relief," Oliver Ames wrote Dodge, in his anxiety misspelling perhaps even more words than usual,

> *and we feel that the vortex out there will swallow all that can be raised out of our securities—and then perhaps the mobs on line of Road will stop the*

trains and the next thing we shall hear is that the trains have been stoped and Passengers Robed to pay starving men. It would have been better to have called out the Military and stoped this 1st mob and then we should have had no more trouble. I am informed that Davis & Associate men were the parties stoping the train. Could it be one of Durants plans to have these men get their pay out of the Road and we suffer for his benefit[?] Durant is so strange a man that I am prepared to believe any sort of rascality that may be charged agnst him.[9]

It was raining hard in the Wasatch Mountains when the Doctor's train, and the ones behind it, got under way, and as the lead train inched and backed over the temporary tracks around Tunnel 2 at the head of Echo Canyon, and rolled down along the swollen and tumbling creek to the Echo City station, news began to circulate that there was a new obstruction ahead: snowmelt and spring rains had washed away a bent in the new Devil's Gate Bridge, rendering it impassable, and another bridge at Strawberry Ford was dangerously undermined. Crews under Leonard Eicholtz had no sooner finished the 450-foot trestle up near Promontory Summit when they had to rush back down to Weber Canyon to address the Devil's Gate emergency; a new 50-foot truss would be required, and some hasty foundation work, presumably better than the previous job. Down beneath the sagging tracks, icy water swirling at their feet, the crews worked without a break.

The train carrying Leland Stanford's party rolled eastward along the Humboldt through the night. Dr. Stillman did not rest well. "The novelty of a spring-bed in a railroad car was too great to allow of sound sleep," he wrote; "it was too much like being tossed in a blanket all night; and with the first light of morning I was up. The air was cold, and snowy mountains were in sight—one is never out of sight of them. A volume of steam in the distance indicated hot springs." After the sun had climbed well above the Ruby Range the train halted at Elko, where most of the regular passengers hurriedly got off to change for stagecoaches heading south to the White Pine mining fields, a hundred miles away, where silver surface deposits could be picked up in chunks and lumps, where some thirteen thousand claims would be registered around the high-altitude and high-pressure town of Hamilton, and from which at least $22 million in bullion not lost to highwaymen would be freighted up to the Central Pacific in the next several years.[10]

Governor Stanford's party resumed, straining through the day toward and past Humboldt Wells and the eastern ranges, rolling across the Utah border in darkness, the dreary desert crossing blessedly eclipsed for most of the excursionists by sleep. At dawn, however, the tracks were now breaking through hills and then skirting the vast white salt flats, bumping over trestles and catching near and far views of the great blue lake and some of its craggy, still-snowtipped islands. The train edged past, on sidings, "the novel sight of a town on wheels," Dr. Stillman recorded, "houses built on cars to be moved up as the work progressed." Then there were the acres of tents and the smoky cookfires of "the Chinamen who had built more railroad in a given time than was ever done before by any people," he marveled, meaning not only the "formidable difficulties of the Sierra Nevadas," not only the desert country where "they passed like a hurricane," but also the well-publicized feat, the week before, of ten miles in a day. Soon Albert Richardson would pass over this same stretch, seeing "hosts of Chinamen, shortening curves and ballasting the track. Nearly four thousand are still employed in perfecting the road. They are all young, and their faces look singularly quick and intelligent. A few wear basket hats; but all have substituted boots for their wooden shoes and adopted pantaloons and blouses. . . . They are tractable, patient, and thorough; they do not get drunk, nor stir up fights and riots."[11]

On the rainy and dreary morning of Friday, May 7, assuming their party had reached the "marriage point" with a day to spare, the people in the excursion train felt themselves eased to a stop in a high, level, almost circular valley ringed with mountains—the Promontory summit valley. On the sandy gradient, several lengths of iron rails were missing, but then the track resumed, heading northeast toward a break in the mountains. Casement's men had come there a day or so after the ten-mile-feat and proceeded to build eastward with material teamed up from below; Casement had been worried that Crocker would keep working eastward, and even now the Union Pacific men were rushing to close the yawning gap in barely finished rock cuts on the eastern slope. On alighting from their coach, Stanford, Dr. Stillman, and the others were suprised to see that there was no other train, no party of celebrants, no weary gangs of laborers, and few hangers-on. Two or three tents "were pitched in the vicinity," said Dr. Stillman, "for the rendezvous of those ruffians who hang about on the march of industry, and flourish on the vices of men." They found the desultory telegraph operators at the end of their respective lines, only a few yards from one another, who could explain nothing about the lack of population and activity at Promontory, but the Union Pacific telegrapher tapped out an inquiry to officials down the line. "We were informed," Dr. Stillman wrote, "after some delay, that it would be impossible for them to arrive before Monday."

The Californians were aghast. Tens of thousands of celebrants were pouring into Sacramento and San Francisco for the programs the next day. They quickly wired the bad news home. After a further, understandable delay, the reply came back: the celebrations would take place according to the original schedule—Sacramento and San Francisco could not afford to put off or turn back the multitudes coming into town. They would try, though, to keep the festivities going continuously into Monday. "We all felt the embarrassment of our position keenly," Dr. Stillman recalled, "but we tried to make the best of circumstances we could not control."

In their best hats and tailored suits, some attired in light linen dusters, watch chains gleaming from their waistcoats, burnished boots beginning to scuff among the pebbles and bitter mud on that drizzly day, the excursionists looked around anxiously. What they saw was not reassuring. "Promontory is neither city nor solitude, neither camp nor settlement," Albert Richardson would say in a few weeks. "It is bivouac without comfort, it is delay without rest. It is sun that scorches, and alkali dust that blinds. It is vile whiskey, vile cigars, petty gambling, and stale newspapers at twenty-five cents apiece. It would drive a morbid man to suicide. It is thirty tents upon the Great Sahara, sans trees, sans water, sans comfort, sans everything." How could the Californians make the best of this setback? Their spirits sank. "To spend three days in this desolate spot," Dr. Stillman said, "surrounded with sage-brush, with only such neighbors as would make it dangerous to venture away from the car, lest we have our throats cut on the suspicion that we might have a spare quarter in our pockets, was not charming."[12]

Half of the Californians approached one of the layabout tents and hired a wagon and muledriver to take them down to the Union Pacific camp. There, the sympathetic General Casement put them on a work train heading for Ogden, where the excursionists could find weekend rooms among the ambivalent Mormons. Governor Stanford, Dr. Stillman, photographer Hart and the rest left stranded on Promontory Summit spent the night locked safely in the superintendent's car, until a Union Pacific special came partway up the slope to collect them, with General Jack in command. He gallantly put the train at Stanford's disposal so they could look over the canyon lines, "and we ran down to Weber Creek station, and an opportunity was enjoyed of viewing some of the finest mountain scenery in the world," Dr. Stillman wrote. "The Wasatch Mountains rise from the plain on the west shore of the lake to the height of six thousand feet above its surface, or ten thousand feet above the level of the sea. They are the very ideal of inaccessible snow-covered mountains, set off by the green fields and blushing tints of the peach orchards just coming into flower."

At Ogden, from the railroad grade, Hart excitedly took a number of views of the majestic peaks. Standing there, Dr. Stillman pondered the slopes and the five distinctive horizontal lines marking ancient beaches (called benches in that vicinity) of Lake Bonneville, which had inundated much of the valley in the last ice age fifty thousand years before; the highest shelf carved and pounded by waves was a thousand feet above the present level of the lake. The benches, and the growing bustle around him in Ogden, made him mindful of another obvious confluence. "The tide-rip is well marked," he commented, "where the currents of traffic from East and West meet—where the barley from the West greets the corn from Illinois, where paper is currency, and coal takes the place of Juniper trees as fuel. We feel, while looking about, that we have met half way."[13]

As promised, California proceeded with jubilees on Saturday, May 8, beginning in Sacramento perhaps earlier than most people expected. At 5:00 A.M. a special train from Reno glided to a stop at the Central Pacific depot; it was full of the firemen of Virginia City and Gold Hill, many of them earlier residents of Sacramento, all looking magnificent in their uniforms. Their band members mirthfully took up their instruments on the railroad platform and heralded the dawn of this new day—"the music of their band woke up the slumbering city," the *Union* reported, "with the piping notes of the swallows." The firemen were not alone on the platform for long; trains from Colfax and Lincoln pulled up, their sixty-five coaches disgorging vast numbers of military and civilians from Yuba, Nevada, Placer, Sutter, Colusa, and Butte counties. Next was the Folsom train, thirty cars in all, conveying delegations from El Dorado, Placerville, Shingle Springs, Diamond Springs, Latrobe, and Folsom. Engines and cars from the direction of the San Joaquin Valley brought some six hundred from Stockton, Woodbridge, Hicksville, Liberty, and the upper valley. Citizens from Vallejo and the farming settlements of Solano, Napa, and Yolo, at least twenty-five hundred strong, arrived next. "By nine o'clock the city was crowded in all the principal streets," the *Union* said, "with the largest, most orderly and eager number of people ever collected here at one time—and still they came, from farms, roads, river, in boats, cars, and every conceivable style of conveyance, till the sidewalks of J and K, Second and Front and Fourth streets were too small to hold the throng."

At eleven o'clock the extraordinary procession had been formed in ranks of eight divisions, each with its band and complement of dignitaries' carriages, floats, wagons, marchers, marshals. There was the First Division, with band, the Sacramento Light Artillery with four-horse guns and caissons, the Zouaves of Sacramento and Grass Valley in their billowing costumes, the city guards of Sacramento, Placerville, Nevada City, and Grass Valley in smart uniforms. More than a hundred California Pioneers led the Second Division, and behind a brass band there were the nine carriages of the Central Pacific—the Crocker brothers grinning in triumph, Hopkins acknowledging the cheers, knots of happy clerks and engineers in their Sunday best, waving and hallooing; next, there was a gigantic banner depicting the United States, with railway trains crossing the plains toward one another over the motto "May the Bond Be Eternal." Two hundred machinists and iron workers marched up; behind them were marvelous floats: coppersmiths standing around a miniature engine; boilermakers posing next to a partially completed locomotive boiler with smokestack rampant and painted with the motto "United We

Stand—the Central Pacific Railroad a Success;" blacksmiths in their leather aprons banging almost in unison upon their anvils; and finally, a four-horse wagon containing eight of the ten Irishmen who laid ten miles of track in one day. Close behind was a wagon festooned with red, white, and blue bunting, containing a printing press of the *Sacramento Daily Record* which, as the procession unfolded, printed the day's edition for the celebrants.

Next there was the brawny, heroic Third Division, comprised entirely of firefighters' corps and military bands, brass bands, and cornet bands, accompanying their gleaming, horsedrawn engines, service wagons, and hose carriages, the uniformed men bristling from the conveyances or stoutly marching behind. The Fourth Division, escorted by the Knights Templar of Sacramento and Placerville, brought the twenty-five carriages of dignitaries, including Governor Haight, orator of the day; Albert Hart, reader of the poem; the Reverend J. A. Benton, chaplain; distinguished invited guests; and state, county, and municipal officers. There were the Fifth and Sixth Divisions, with their bands; hundreds of marchers comprised of the Improved Order of Red Men, in body paint and feathery regalia; the Independent Order of Odd Fellows and members of the Grand Lodge; and the Mexican and Chilean Clubs, in traditional costumes, with flags of those nations brandished alongside the Stars and Stripes. And there were the Seventh and Eighth Divisions—twenty-nine omnibuses, carriages, and wagons filled with schoolchildren, and wagons and floats of the Atlantic and Pacific Telegraph Company, and Wells, Fargo, and the Pacific Union Express company, decorated with banners, transparencies, flags, and bunting. One of the last float decorations was a drawing of a large hand, the fingers of which were labeled Charles Crocker, E. B. Crocker, Mark Hopkins, C. P. Huntington, and Leland Stanford, "the Big Five," or, as the caption put it, "The Right Hand Men" of Sacramento.

Past sidewalks crowded with thousands upon thousands of cheering citizens, the procession followed a circuitous route through nearly every street of the business district, everything decorated with flags, garlands, signs, evergreen branches, and triumphal arches. "United We Are, United We Will Be," read a sign in front of one store, "by the Hand of Industry and the Flag of the Free." "The Wild Grand March Is Done!" read another, "The Guarded Ways Are Won, from Sea to Sea!" Another banner celebrated "The Enterprise of California, the Energy of Sacramento—the Pacific Railroad." And there was the waggish legend, "Seventeen days to Europe by the 'Dutch Flat Swindle!'" And perhaps the single open dissenter was the grouchy Atlantic and Pacific States' Telegraph Company office, with its sour-grapes "Untiring Opposition to an Oppressive Monopoly." About midway in the long and winding parade, at 11:00 A.M., though the true uniting of the continental rails was not to happen for another forty-eight hours, the ceremonial cannon, "Union Boy," boomed out a simulated message of completion, answered nearly simultaneously by twenty-three locomotive whistle-blasts on the levee and the ringing of all the bells in town. "This deafening clamor," said the *Union,* "lasted fifteen minutes."[14]

In San Francisco, a wire from Sacramento unleased a similar barrage of sound—"the bells rang," the *Bulletin* said, "and the steam whistles at all the foundries and machine shops, on many vessels in the harbor and the Mint were sounded, and sent forth a deafening peal of screams, continuing nearly an hour." Tens of thousands stood on the sidewalks to cheer the procession of mounted police, mounted trumpeters, artillery bands, regiments of infantry and cavalry, Zouaves and Hussars and Dragoons, batteries, California Pioneers, phalanxes of veterans of the Civil War, Mexican War, War of 1812, firefighters, Garibaldi guards, Italian, German, Mexican, and Chinese benevolent societies, while down in the harbor, all classes and sizes of vessels displayed their flags, national emblems, and company insignia. "Thousands of these gay colored banners floated out in the winds of the morning," said the *Bulletin,* "wreathing the entire front of the city, from Black Point to Point Avisidero, with a crown of glory. It seemed a fitting tribute by the commerce of ocean to the commerce of the land, which today achieves its latest grant triumph over the mighty barriers of time and space."[15]

Meanwhile, back in Sacramento, finally the first marchers began to amass at the Central Pacific station. Under a hot sun the ranks of celebrants and onlookers waited until all had joined them in the square. There was a prayer, a poem, a speech, another speech, another and another and another and another under the sun, each one punctuated by a tumult—but, the *Union* said happily, the crowds behaved themselves—"everything was conducted in a pleasant and orderly way, and that nothing of a public character disgraced the event. There was little unseemly dissipation and no fighting." After Governor Haight and other distinguished speakers had finished, several of the Central Pacific directors rose to speak. First it was bluff, braying Charley Crocker, whose blacksmith's brawn and iron will had forged the Central Pacific labor force, who would be the first of the Associates to triumphantly cross the nation on a well-deserved family excursion to the eastern states, to "Old Home." And next it was genial, somewhat wan Edwin Bryant Crocker, whose acceptance of the brutal weight of running the company had come close to finishing him, but whose optimism had kept his associates from despair. In just one month from the time he stood before the city to accept its accolades, the Judge would be conducting important, intricate, and extremely taxing company business in San Francisco when, upon retiring to his hotel room at the Lick House, he would suffer a paralyzing stroke from which he would never recover. Unable to move, barely able to express himself, he would retire from business forever, spending several years with his family in Europe and with them using his railroad millions to build an extraordinary art collection which he housed in a gallery next door to the Crocker mansion in Sacramento. Because of his debilitating illness and his early death in 1878, and the willingness of Stanford and long-lived Huntington to hog credit due others, his paramount role in the enterprise would be obscured for well more than a century.[16]

After the Judge had smiled and bowed and sat down, after another succession of speeches, and after a swarm of resolutions, toasts, and dedications to the California pioneers, to the Donners, to Asa Whitney, "the originator of the idea of a railway across the continent," to Theodore Judah, "conquerer of the Sierra," the vast crowds dispersed to private parties in every hotel, restaurant, saloon, and streetcorner in the city, and the downtown was brightly illuminated long into the night, and people strolling down Fourth and J Streets were delighted to hear the melodies of the Stockton Cornet Band floating down from the balcony of the St. George Building until long past what would normally be considered a respectable bedtime.

The Stanford party returned to its train at Promontory Summit on Sunday and rolled back to Monument Point, where there was easy access to the lake, and where they spent several hours on the shore. Some men, including Alfred Hart with his camera, clambered to the top of a hill to look down on the train and the shore and gaze far out over Great Salt Lake. While they were there, a wagon train of emigrants appeared from down the eastern shore. Hart lost no time taking a photograph of the canvas-topped wagons passing the locomotive and cars; the symbolism of the moment would make the published stereograph one of his most popular. In his stroll along the beach Dr. Stillman noted the curious "grasshopper line" along the high water mark—"the waves cast up masses of the remains of insects which have perished on its surface," he wrote. "A few fish in the lake would allow no grasshopper line along the shore, but here the insects are pickled when they perish, and are finally blown ashore." Squinting out over the bright blue lake at the islands, he recalled reading in Frémont's and Stansbury's diaries that myriads of birds bred there. On this day, he said, "we saw no living thing within or above its waters." At that time in the spring, lake gulls may have been flocking to the fertile valley, following the Mormon plows to gorge on uncovered insects, and the great white pelicans of nearby Gunnison Island must have been occupied elsewhere. However, Dr. Stillman said, their day's big meal was considerably improved after a foray into the wetlands: "Our steward with his gun procured a mess of snipe from a marsh where a fresh water brook lost itself in the sedge at a distance from the lake—among them was a rufus-headed Avoset!" The party returned in late afternoon to Promontory to sup and sate themselves in preparation for the next day's excitement.[17]

Down near the mouth of Weber Canyon, Union Pacific bridge crews wearily jacked and bolted the new truss into place after some four days of feverish work; rails were quickly rejoined, and Dr. Durant's train, and two backed up behind, were pushed and pulled across with ropes, all at trackside and on board holding their breath that the Devil's Gate bridge would support a load. Finally, the trains were allowed down to Ogden, where impatient passengers who had gone ahead by wagon to boardinghouses could be found and collected, and the excursionists could ready themselves for Monday morning's roll up to Promontory.[18]

The night in the summit valley was cold; water froze in footprints, puddles, and ruts, and Stanford's steward kept the coach's stove going all night. Very early on Monday morning, May 10, 1869, the Californians looked outside their car and beheld, down the line on the eastern side of the remaining gap, the surprising sight of a large number of Union Pacific tracklayers finishing what was obviously an all-night job; a complete siding and Y-track had been thrown down in what would be noted as the last and shortest race of the transcontinental railroad. Simultaneously, obviously by competitive reflex, Grenville Dodge and James Strobridge had decided to vie for the rights to the Promontory Summit terminal by dint of putting down the rails just before the ceremony. General Casement sent his gangs up overnight, and the side tracks were quietly and expeditiously done by light of torches and sagebrush bonfires. At dawn, when the Central Pacific work train drew up behind the locomotive *Whirlwind* to make quick work of the job, Strobridge saw the irritating vista of a fait accompli.[19]

With the Union Pacific the victor of that small skirmish for the barren, uninteresting dust of Promontory, Strobridge's Irish and Chinese workers descended from their cars to mix with Casement's proud, understandably smug laborers. There were no problems—the competitiveness was good-natured. Years later, colorful myths would be written that as the graders and tracklayers from the West and the East drew into Utah, the stretches where parallel lines were built saw open warfare between Strobridge's Chinese and Casement's Irish—fistfights and rockfights escalating even to explosives. There is, however, no evidence in all the telegrams, letters, reports, journals, and contemporary newspapers to support this myth of corporate race warfare, so gullibly repeated in many accounts. Besides, much of the parallel grade work was done by Mormon contractors who were happy at the prospect of being paid by both competitors. Now, at Promontory, Chinese, Irish, and Mormons mingled in shared relief and anticipation.[20]

Despite a moderate breeze, the air warmed with the rising sun; it was to be a clear day. More construction trains crammed with workers swelled the crowd. Enterprising camp followers quickly rode up to the valley, and soon they had expanded the temporary settlement of Promontory to fourteen tents, with signs advertising "Red Cloud," "Red Jacket," "Blue Run," and other potent distillates; the sound of hammers on nails announced that some planned to stay awhile. At 8:20 A.M. a Union Pacific train from Ogden pulled up, disgorging many sightseers including Andrew J. Russell, the Union Pacific's official photographer; Russell had lived in his darkroom-wagon or in construction camps for nearly two years, documenting the great work along the railroad, and publishing two editions of stereoptican sets of the views, which were brisk sellers; an art album of fifty "imperial" views, entitled *The Great West Illustrated,* had just been issued in April. Russell, commissioned to provide views of the ceremony and commentary for *Frank Leslie's Illustrated Newspaper,* was joined by his assistant, S. J. Sedgwick, equipped with his own camera, and by a friend from Salt Lake City, the photographer Charles R. Savage. Soon the three were out and working with Alfred Hart, taking no notice when a view included a competing photographer hunched behind his own big box camera.[21] Within half an hour, they heard a train whistle and a Central Pacific train appeared, half passenger coaches and half platform cars piled with telegraph poles and the final set of railroad ties.

And within minutes they heard another shrill train whistle, from the east: Dr. Stillman and Governor Stanford, who had grown visibly impatient, saw that it was the long-awaited Durant party, in two deluxe coaches being pulled by Union Pacific Engine No. 119. "We went over at once to meet them," Dr. Stillman said. "In a superb piece of cabinet-work they call a 'Pullman car,' we met Vice President Durant, of whom we have heard so much, with a black velvet coat and gay neck-tie, that seemed to have been the 'last tie' to which he had been giving his mind, gorgeously gotten up. General Dodge was there, and he looked like business. The veterans Dillon and Duff were there to give away the bride. General Dodge on the part of the Union Pacific, and Edgar Mills on the part of the Central Pacific, were appointed to arrange the preliminaries."[22] The Doctor was suffering from a splitting headache, and delayed himself in the Pullman as long as possible (whiskey was a common remedy for headache, even for teetotalers), but Silas Seymour, Samuel B. Reed (with his wife and daughter), and Jack and Dan Casement, among others, gamely emerged in the bright sun. They all posed alongside the coach for Russell, doffing hats at his command so their eyes would not be shaded; in the center of the line were rumpled Sam Reed and erect, high-collared Colonel Seymour with his well-trimmed salt-and-pepper beard, and, next to him, the dusty, slightly stooped Dodge, who had yet refrained from handing a discharge to the consulting engineer.

Brigham Young of course had been invited to ride with them and would have stood with them in the picture, but he was said to have more pressing business in the southern part of the territory; as his emissaries he had sent Bishop John Sharp, Ogden mayor Lorin Farr, and Chauncey West. These Mormon dignitaries, along with the elderly apostle Ezra Taft Benson, had, as Brigham Young's contractors, completed their work for the two railroad companies, which had by then each paid the Mormons $1 million, less than half of what was owed.

Soon after the completion, when it seemed apparent that the Union Pacific and the Central Pacific would not meet their obligations, the contractors were heavily in debt, their laborers destitute. In the summer most of the Utah territory would be thrown upon the barter system—and Chauncey West would go to California to try to collect from Leland Stanford. "He broke down under the strain," West's son would write, "but died before it was accomplished. . . . He was in the prime of life, but 43 years of age." West left a family of several wives and thirty-six children. Bishop Sharp would eventually settle with Stanford, but in the face of continuing resistance from the Union Pacific, with near panic springing forth in Salt Lake City, an apparently ruined Apostle Benson would also collapse and die. With continued suffering in the community, Bishop Sharp would settle for a token cash payment and excess iron and rolling stock for the Mormon's Utah Central line to Salt Lake City. On the day of "the marriage" of the eastern and western railroads, of course, none of this was known, and the Mormon observers participated in the spirit of thankfulness that the faithful's hard work would soon be rewarded with much-needed cash to aid an economy brought to its knees by two years' drought and locusts.[23]

Another Union Pacific train arrived, carrying members of the Ogden Tenth Ward Band, and dignitaries from that city and from Salt Lake City, and more reporters, and behind their train approached another which contained officers and soldiers of the 21st Infantry heading for San Francisco; several of the officers' wives accompanied them, and the regimental band appeared with their instruments ready. Dignitaries joined dignitaries, soldiers formed into ranks at trackside, parade rest, bayonets glinting in the sunlight, and the bands began to alternate tunes. Somewhere in the growing crowd, listening to the patriotic airs, was the fearsome former general Patrick E. Connor. On a flagpole affixed to the top of a telegraph pole, Old Glory, with its thirty-seven white stars, snapped in the breeze.

Strobridge put a squad of Chinese to work on the blank dirt between the rails, and with picks and shovels they loosened the soil and leveled it for the ties. "On the Central Pacific all the timber for telegraph poles, ties, etc., is sawed," noticed Andrew Russell as he set up for more photographs, "while that of the U. P. R. R. is hewed." Now Stanford's train, behind Central Pacific locomotive No. 60, *Jupiter*, engine and tender decorated with little flags and red, white, and blue ribbons, and branches of evergreen cut in the California mountains, slowly chuffed up to where the Cantonese solemnly worked. From the train men carried one last ceremonial tie; donated by a now well-heeled tie contractor and prepared by a San Francisco billiard-table manufacturer, it was cut from a California laurel on Mount Tamalpais, sawed into an eight-foot length, eight inches by eight around and polished until it reflected light, and it bore a silver plate identifying it as the last tie of the Pacific railroad, followed by the names of the directors and officers of the Central Pacific and the presenter, West Evans.[24]

As the Central Pacific men gravely lowered it to the ground, Russell exposed a picture; behind the crowd, Union Pacific No. 119 with its chimneylike stack shone in the sun. The laurel tie's last spike would be businesslike iron, though fittingly there were several ceremonial spikes—which, because of their delicacy, would not stand up to pounding but would be dropped into the predrilled holes. One, eighteen ounces of pure gold, had been presented to Stanford by David Hewes, formerly of Sacramento and now a contractor in San Francisco, and it was engraved with the legend, "The Last Spike," the originally scheduled date of May 8, 1869, and the names of the Central Pacific officers; there was even room for the sentiment "May God continue the unity of our Country as this Railroad unites the two great Oceans of the world." A rough nugget of gold had been fixed to one end, which later Hewes would have broken off and melted down into finger rings; Hewes presented these to Stanford and Oakes Ames, and to President Grant, Secretary of State Seward, and the Reverend Dr. John Todd, the clergyman from Pittsfield, Massachusetts who was to deliver the invocation that day. The rings would be inscribed, "The Mountain Wedding, May 10, 1869."[25] There would be two other spikes—one, of silver, forged with a hundred men each striking a blow, had been presented by F. A. Tritle of Nevada (currently he was serving as a Central Pacific government commissioner but he would soon be elected governor of Nevada), and the other, which had been fabricated of iron, silver, and gold, had been given by Governor Safford of Arizona. Both were sentimentally inscribed.[26] Leland Stanford carried a silver-plated maul given to Stanford by the Pacific Express Company of San Francisco—whom Stanford and the Associates hoped to soon drive out of business—for the light work ahead of him.

Some distance from the murmuring crowd, the two bands of musicians, and the puffing engines, on the edge of that upland valley, a squad of U.S. infantrymen from Fort Douglas were struggling up a dry gulch completely unaware of what was going on at Promontory Summit. J. W. Malloy, one of the soldiers, recalled many years later that they had been sent out after "the notorious Piute Indian, 'Captain Charley,' who had been terrorizing the people of Ogden and vicinity for some time,

> *for Captain Charley was a bad one when he had too much "firewater" under his belt, which was almost always the case. My squad and I set out on the trail of Charley and we were determined to have his scalp. . . . The trail was hard to follow and as we came to forked roads we separated and continued our manhunt. After stumbling through forests of brambles and thickets, we finally came to the clearing where the two engines were to meet, and as we paused to get our bearings and also our wind, we saw the old Central Pacific*

engine slowly making its way to where the Union Pacific's stood puffing and panting, looking for all the world like a victorious foot racer.

"We stood with mouths agape," Malloy continued, "as we realized that the much talked-of line was completed." The soldiers joined the onlookers to watch what happened next; after it was over, Malloy would say, "we continued our hunt for Cap'n Charley, and—well, we found him, and he never gave anyone any more trouble."[27]

Up between the two locomotives, with all the ties but one in place, the two final rails were brought forward—one by a Union Pacific gang of Irishmen, who placed it on the west side of the ties, and one by a Central Pacific gang of Chinese, who set theirs down on the east side. At that moment, it has been said, someone in the crowd hailed photographer Savage. "Now's the time, Charlie!" he yelled. "Take a shot!" The Cantonese knew only one meaning for the word "shot"—a detonation of black powder or nitroglycerine—and they all dove for cover, to great hilarity in the crowd.[28] The rails and the laurel tie waited for their parts in the ceremony.

It was getting toward noon. Andrew Russell set his tripod facing east at trackside and asked Jack Casement to call for the crowd to move back, and a number of the Union Pacific dignitaries compliantly spread across the track embankment. It was a good, clear shot, nearly a hundred people who would hold still long enough to stay in focus, with Union Pacific No. 119 behind them under a complement of balancing men, and again they all doffed their hats: the first row included Dr. Durant, standing on the western ends of two ties; the Reverend Dr. Todd, with his long gray beard; Samuel Reed, some distance, as usual, from his wife and daughter and visiting sister-in-law; John Duff, still looking shaken after the kidnapping up at Piedmont; sober Sidney Dillon, shouldering in front of cocky-looking Silas Seymour; and Grenville Dodge, a little lower on the embankment, who quite uncharacteristically smiled for Russell's camera as he stood leaning jauntily on his uphill leg. The subjects held their breath. The exposure was made. Then Russell and Sedgwick clambered to the tops of the respective engines, the former on the cab roof of Union Pacific No. 119, and they aimed at each other, capturing the ranks of soldiers, the groups of dignitaries and onlookers, the flat sagebrush plain, the enveloping mountains.

It has been said that the representatives of the two railroads, Dodge and Mills, argued at Promontory over who, Stanford or Durant, would strike which blow to the last iron spike. Rationalizations included which company had built more miles or conquered higher terrain, even which officers had suffered more during the construction. Dodge was said to have threatened that the Union Pacific would simply walk away from the ceremony, although that is improbable. But now at least, it was agreed that Governor Stanford would stand on the east side of the track, striking the first blow, and Dr. Durant, on the west side, would follow.[29]

Now it was time to connect them to the rest of the nation. Two telegraph operators, Watson N. Shilling for the Union Pacific, and Louis Jacobs for the Central Pacific, stood by as the Central Pacific's general foreman of telegraph construction, Amos L. Bowsher, expertly wrapped a wire extension from the eastern overhead cable around the spike, doing likewise with a line from the west which was wrapped around the handle of Stanford's silver maul and connected to a copper plate on the striking face. With the first hammer-tap the connection would be made—and the driving of the last spike would be "heard" all around the country. In every city equipped with fire-alarm telegraphs, the alarm would sound. Moreover, in San Francisco and New York, the telegraph was also attached to cannons on the shores of both oceans—at the signal they would fire, as it were, warning shots out over the Pacific and Atlantic to the rest of the world: the Empire Express was won.

In New York City in the downtown area around City Hall, Trinity Church, Wall Street, and Printing House Square, where most of the newspapers congregated their offices, flags, pennants, and bunting were particularly evident that Monday morning, although characteristically for New York, most citizens went about their normal business. Noontime in New York was 10:00 A.M. in Utah, and at noon a special religious service opened at the crowded Trinity Church. A long line of white-robed choristers and clergymen filed in from the vestry, singing the processional:

O come, loud anthems let us sing,
Loud thanks to our Almighty King,
For we our voices high should raise
When our salvation's rock we praise.

The depths of earth are in His hand;
Her secret wealth at His command;
The strength of hills that reach the skies
Subjected to His empire lies.

After prayers and collects were read, the Reverend Dr. Dix, of Trinity, raised his voice in a special prayer.

O God, the Creator of the ends of the earth, Who upholdest all things by the Word of Thy power, without Whom nothing is strong, nothing is holy; we bless and magnify Thy glorious name that by Thy goodness the great work

> *which we commemorate this day has been accomplished, so that the extreme borders of our land have been joined and brought nigh together, and a pathway opened between remote parts of the earth, both for the commerce of the nations and for a highway and a way whereby Thy Gospel may have free course, and Thy holy name may be glorified. We thank Thee that the wilderness and the solitary place are made glad, and that the desert may rejoice and blossom as the rose.*

He continued in such a vein, being followed by the Te Deum, the lesson, the creed, and the offertory. The Reverend Dr. Vinton, also of Trinity, rose to address at length. "This is indeed a great event of the world," he intoned at one point. "It is one of the victories of peace—a victory grander than those of war, which leave in their track desolation, devastation, misery and woe. It is a triumph of commerce—a triumph indicating free trade as a future law of the nation." The Pacific railroad was also, he said, "a means, under Divine Providence, for propagating the Church and the Gospel from this, the youngest Christian nation, to the oldest land in the Orient, now sunk in Paganism and idolatry." After the good Reverend Dr. Vinton finished, the choir pealed forth the Mozart *Gloria* from the Twelfth Mass, and as the immense congregation filed outside into the sunlight, the carillon master struck up the chimes, ringing the eight bells of the changes and following with national and operatic airs. The grandeur of the chimes' display, so unusual for an ordinary Monday in Manhattan, attracted a large crowd to the head of Wall Street and Broadway, where gradually the presence of bunting and pictorial banners, and word of mouth, worked the news through the increasingly enthusiastic crowd. Upon hearing the news of what the railroads were about to bring forth, merchants and traders flocked into the stock exchange where, in minutes, there was a frenzy of speculation.[30]

In Washington a large gathering crowded into the main office of the Western Union Company, where the manager had connected a magnetic ball to the main line and placed it in a conspicuous place. Over at the War Department, Generals Sherman and Rawlins waited by the telegrapher, hoping that Grant would break away from his meetings and join them. In Chicago, Omaha, Philadelphia, Scranton, Buffalo, Boston, New Orleans, and many other cities, the telegraph was fixed to the central fire-alarm bells and to gongs in the Western Union offices, and to many it seemed like an indrawn breath had been taken, anticipating the next event.

For all intents and purposes the entire nation was now connected in one great electrical circuit. The Western Union manager in Washington tapped out that he was ready. Telegraphers in New York, Boston, and New Orleans instantly answered that they were ready. It was about 2:27 P.M. in the east—12:27 P.M. at

Promontory—and there was a sudden clamor on the lines as many offices sent inquiries to the Omaha office, at the head of the circuit. "To everybody," Omaha replied, "keep quiet. When the last spike is driven at Promontory Point we will say 'Done.' Don't break the circuit, but watch for the signals of the blows of the hammer." There was some trouble in the Chicago office. Somewhere west of Buffalo a circuit closed. Washington readjusted and again was ready.

Watson N. Shilling, sitting at a little table with a telegraph key out under the Utah sun, tapped out: "Almost ready. Hats off. Prayer is being offered." There were twelve or thirteen minutes of silence while the Reverend Dr. Todd stood at the joining place and asked the favor of heaven upon the enterprise.

At 2:40 P.M. in Washington the message bell rang again.

"We have got done praying," Watson tapped again. "The spike is about to be presented." Chicago replied, "We understand. All are ready in the East."

The presenters of the ceremonial spikes stepped forward, W. H. Harkness of the *Sacramento Press* handing Governor Stanford the inscribed golden spike. Tritle of Nevada was next. "To the iron of the East and the gold of the West," he called out over the crowd, "Nevada adds her link of silver to span the Continent and wed the oceans." Governor Safford followed. "Ribbed with iron, clad in silver and crowned with gold," he called, "Arizona presents her offering to the enterprise that has banded the Continent and dictated the pathway to commerce."

Sedgwick exposed a photograph of the ceremony from about thirty yards away across the sagebrush, catching the crowd in a large knot between the two engines. Two of the telegraph men had climbed a ladder leaning against the pole to get a better view from above, and one had now perched on the crosspiece, holding on to the flagpole. The breeze was out of the west and the flag was rampant.

Now Stanford was responding on behalf of the Central Pacific, accepting the gifts and expressing—in spectacularly dull sentiments—that "[t]hese gifts shall receive a fitting place in the superstructure of our road. And, before laying the ties and driving the spikes in completion of the Pacific Railway, allow me to express the hope that the great importance which you are pleased to attach to our undertaking may be in all respects fully realized. This line of rails connecting the Atlantic and Pacific, and affording to commerce a new transit, will prove, we trust, the speedy forerunner of increased facilities." The governor went on and on—and if anyone there began to measure the distance between the iron rails and the nearest tent purveying Red Jacket by the shot, their attention would have finally snapped back to the central scene as Stanford rose to perhaps his greatest oratorical moment. "In conclusion," he announced, raising the spirits of the assembly, "I will add that we hope to do ultimately what is now impossible on long lines: transport coarse, heavy and cheap products, for all distance, at living rates to trade."

It was a hard act for Chief Engineer Dodge to follow; he may have been an active back-room kind of a politician in Des Moines, Council Bluffs, and Washington, but he was a man of action, not of words, and when he responded for the Union Pacific, he kept it short. "The great Benton prophecied that some day a granite statue of Columbus would be erected on the highest peak of the Rocky Mountains," he called out over the heads of the tracklayers and stockjobbers, the soldiers and the engineers, the hopeful Mormons, the scribbling correspondents, the token handful of women, "pointing westward, denoting the great route across the Continent. You have made the prophecy today a fact. This is the way to India." In Washington, the clerk in the attorney general's office, Walt Whitman, would avidly devour all press accounts of the ceremony, later in the year beginning his magisterial poem "Passage to India" with these lines:

Singing my days,
Singing the great achievements of the present,
Singing the strong light works of engineers,
Our modern wonders, (the antique ponderous Seven out-vied,)
In the Old World the east the Suez canal,
The New by its mighty railroad spann'd,
The seas inlaid with eloquent gentle wires;
Yet first to sound, and ever sound, the cry with thee O soul,
The Past! the Past! the Past!

Samuel Reed and James Strobridge lowered the laurel tie, and the squads of Chinese and Irish tracklayers put down the last two rails, their foremen quickly fixing them in place. Stanford and Durant dropped the precious spikes into their slots, and, the laborers having gotten the two last common iron spikes started in the wood, the Governor stood at the south rail and the Doctor at the north—"Half the world at their backs," as Bret Harte would write. Telegrapher Watson Shilling tapped out a message: "All ready now, the spike will soon be driven. The signal will be three dots for the commencement of the blows." All that was required was that each of them make a light tap with their hammers; in many subsequent accounts it was imaginatively (and derisively) alleged that both missed their swings. There is no contemporary evidence of that. The Doctor raised his hammer, the Governor raised his silver maul, they looked at telegrapher Shilling, who tapped three dots and said, simply, "O.K."

And the work was done.[31]

As simultaneously as distance and the speed of electricity permitted, a fifteen-inch gun overlooking Fort Point on the Pacific, connected by underground wire to San Francisco City Hall, thundered with the striking of all the fire bells in that city; in City Hall Park in New York, a hundred cannons rattled all the windows of lower Manhattan while bells at Trinity pealed forth, answered by other churches and by fire gongs which brought throngs out in the streets; from the train yards in Sacramento all the locomotive whistles shrilled while cannons boomed and bells rang; the bells on Independence Hall and in fire stations of Philadelphia sounded at the moment that a host of fire engines parked near City Hall let forth with screeching whistles and clanging bells; in Omaha every Union Pacific engine's steamwhistle opened up, joined by a hundred cannons, bells, gongs, small-arms fire, and full-throated cheers from the fifteen thousand assembled; in Buffalo, a large crowd of citizens heard a gong ring out and after cheers broke into "The Star Spangled Banner," followed by prayer; in Chicago, tens of thousands poured into the streets at the sounding of fire alarms, and with only a minimum of advance preparation there commenced the largest and most enthusiastic parade in that city's history, which would stretch fully seven miles long.[32] "It was free," the *Chicago Tribune* editor would comment, "from the atmosphere of warlike energy and the suggestions of suffering, danger, and death which threw their oppressive shadow over the celebrations of our victories during the war for the Union."[33] Church bells and fire alarms rang all over Iowa City, alerting Peter Dey; at Locust Hill, near the Soldier's Home in Washington, the clamor from the Capitol would have reached Asa Whitney; in Greenfield, Massachusetts, the bells of St. John's church were heard by Anna Judah, who had refused herself to all visitors and stayed home alone to read and reflect on that day, May 10, her wedding anniversary. "It seemed as though the spirit of my brave husband descended upon me," she recalled, "and together we were there unseen, unheard of man," as the assembly at Promontory let forth with cheer after cheer, and Leland Stanford heartily shook the hand of Thomas Clark Durant, and Durant exclaimed, "There is henceforth but one Pacific Railroad of the United States." Cheers followed for the engineers, contractors, and the laborers who had done the work, but of course the names of Dey, Judah, and Whitney went unsaid.[34]

The two locomotives drew together until their pilots were all but touching, and the two engineers mounted the pilots to touch celebratory champagne bottles, while other men climbed all over the locomotives to stand or lean against the warm iron flanks triumphantly. Grenville Dodge and Samuel Montague stood below them at the edge of the track and shook hands. Andrew Russell exposed the most famous photograph of the day.

Officers of the Twenty-first came close and took turns soberly tapping the iron and the golden spike with their sword hilts, a few energetically enough to dent the soft gold. Workmen removed the laurel tie with its precious spikes and took them back to the superintendant's car, and an ordinary tie and iron replaced them. Then the crowd moved in, jackknives brandished and souvenirs on their minds, and excitedly reduced the last tie to splinters. It was replaced and in no less time disappeared into pockets. Amos Bowsher, the Central Pacific telegraph foreman, remembered that "several ties had to be put down after the ceremony, for no sooner was one placed than it was pounced on and cut to pieces by treasure hunters. There were probably enough pieces of wood to make a dozen ties taken away from Promontory as coming from the 'original' last tie." The whiskey tents were, by now, doing a rousing business, and entrepreneurs were busy circulating through the milling crowds. "There was a 'shark' from San Francisco," Bowsher recalled, "who signed up purchasers of watch charms to be cast from the original gold spike. I was one of those that 'bit' at $5 a charm. Of course our charms were knocked out of far different stuff than the real gold spike."[35]

The company officers and dignitaries adjourned to the directors' cars for a fitting repast; the photographers continued to circulate with their tall tripods and cameras, getting knots of celebrants to post in front of the trains; telegraph operators were still industriously translating announcements and dispatches to Morse code:

> *To His Excellency Gen. U. S. Grant, President of the United States, Washington D. C.: Sir: We have the honor to report that the last rail is laid, the last spike is driven, the Pacific Railroad is finished. Leland Stanford, President Central Pacific Railroad Company, Dr. Durant, Vice President, Union Pacific Railroad Company.*

> *To President U. S. Grant: The time to which you have looked interested has today arrived. It is now all rail across the Continent. It gives me great satisfaction that the work was completed during your administration. G. M. Dodge.*

> *To Hon. Schuyler Colfax, Vice President: Sir: The rails were connected today. The prophecy of Benton today is a fact. This is the way to India. G. M. Dodge, John Duff, Sidney Dillon, T. C. Durant.*

> *To General John A. Rawlins, Secretary of War: At 12 o'clock M. today the last rail was laid at this point, 1,086 miles from Missouri River and 690 miles from Sacramento. The great work, commenced during the Administration of Lincoln, in the middle of a great rebellion, is completed under that of Grant, who conquered the peace. G. M. Dodge, Chief Engineer.*[36]

> *To General W. T. Sherman, Washington, D. C.: The tracks of the Union and Central Pacific Railroads were joined today at Promontory Utah. . . . Your continuous active aid, with that of the Army, has made you a part of us and enabled us to complete our work in so short a time. I congratulate you upon it, and thank you for all you have done for us. G. M. Dodge.*

Sherman's reply went on its way to Dodge. "In common with millions," he wrote, "I sat yesterday and heard the mystic taps of the telegraphic battery and heard the nailing of the last spike in the great Pacific road. Indeed, am I its friend? Yes. Yet, am I to be a part of it, for as early as 1854 I was vice president of the effort begun in San Francisco under the contract of Robinson, Seymour & Company." Indeed, much had happened to them all in the years since banker Sherman helped launch Colonel Wilson's (and Theodore Judah's) Sacramento Valley enterprise. Sherman promised Dodge to "go out" as soon as responsibilities would permit. "All honor to you, to Durant, to Jack and Dan Casement, to Reed, and the thousands of brave fellows who have wrought out this glorious problem," he concluded, "spite of changes, storms, and even doubts of the incredulous, and all the obstacles you have now rapidly surmounted."[37]

> *To Oliver Ames, North Easton: The last rail was laid today connecting the Union Pacific with the Central Pacific at Promontory. This act quietly performed 2500 miles west of Boston, 690 miles east of Sacramento will have an influence upon the future and upon the commerce and travel of the world that no one can today estimate. We congratulate you upon the success of the enterprise. John Duff, Sidney Dillon, Thomas Durant, G. M. Dodge.*

"Let us rejoice," Oliver Ames would reply to Dodge,

> *that the last rail is laid and we have a road complete as far as engineering and grading and all the exciting contest for length of line is concerned. It is the great event of the age in railroad construction. The question now with us is how shall we put the road in the best running order and our finances in satisfactory shape? We have used up all our securities except a few that Duff has and these we can't get. We could give $300,000 more if Mr. Duff would let us have the $200,000 of first mortgage bonds he has. I have loaned the Company every first mortgage I have after buying them of the company and putting up my money on them. We are doing everything we can but it is impossible for us to raise money without securities.*[38]

Celebration was in earnest at Promontory Summit, whether out in the open air, beneath tents, in the shade of construction cars, or in the palace coaches. James Strobridge sumptuously entertained the press and officers of the Twenty-first Infantry. "When the other guests arose from the table," the *Chicago Tribune* correspondent noted, "Mr. Strobridge introduced his Chinese foreman and laborers, who had been with him so long, took the head of the table, making some excellent remarks, and inviting them to the banquet. This manly and honorable proceeding was hailed with three rousing cheers by the Caucasian guests, military and civilian, who crowded around Strobridge to congratulate and assure him of their sympathy."[39]

Over at Stanford's car, where the Governor had entertained executives of both companies with California wine and fresh fruit and other victuals, the numerous champagne toasts were getting a little out of hand. There were several impromptu speeches and all went fine until Stanford rose to speak and launched into what Dodge recalled as "a severe attack upon the Government. He went so far as to claim that the subsidy, instead of being a benefit, had rather been a detriment, with the conditions they had placed upon it." Time and time again the companies had been scrutinized not only by its own government commissioners but by one government committee after another—members of which, they all were aware—sometimes needed a little greasing before approving all the hard work. And now there were the "five eminent citizens" sent out by Congress, before the last special commissioners had written their report. Where was it to end? It was the wrong time and place to be complaining, and Dodge saw that everyone was discomfited by the diatribe. Dan Casement was by then feeling very merry. He climbed aboard his brother's shoulders and called out: "Mr. President of the Central Pacific: If this subsidy has been such a detriment to the building of these roads, I move you, sir, that it be returned to the United States Government with our compliments!" This dare "brought a great cheer," Dodge said, "but put a very wet blanket over the rest of the time."[40]

There were other wet blankets thrown—though far away from that upland Utah valley so awash at the moment in rotgut whiskey and champagne. In New York, Collis Huntington sat down as the city bells, alarms, and whistles held forth outside his downtown office on that Monday, May 10, and he scrawled off a biting letter to Charley Crocker, who was enjoying the accolades in Sacramento. "I notice by the papers," Huntington wrote, "that there was ten miles of track laid in one day on the Central Pacific, which was really a great feat, the more particularly when we consider that it was done after the necessity for its being done had passed."[41]

His next morning's city newspapers would carry delirious accounts of the events in Utah and Manhattan and elsewhere, of course, but it would be hard to miss the occasional sour notes in the editorial columns. "The golden spike that closes the final link of the Pacific Railroad is suggestive," the *Herald* would comment. "It is unfortunate for travelling communities that golden spikes, in the shape of fat dividends, have been the ambition of many directors, who manage to secure them at the sacrifice of the lives and limbs of passengers." And the *Tribune* would shoe-horn into the accounts of crowded parades and demonstrations a recollection of the recent "Great Union Pacific Railroad suit" of Jim Fisk. "All these things were unpleasant," editor Horace Greeley said, "not only to the directors and stockholders pecuniarily interested in the well-being and successful completion of the gigantic work, but to the American public as well. The suits will recur; charges and counter-charges will again be made; the characters of directors and managers will be again subjected to the fiery ordeal of private attack and public criticism; but while the wordy war is waging the people will know that the highway which makes the distance between New York and San Francisco only a journey of a few days is an accomplished fact, and will be content to let the litigants tear each other to pieces, if need be, so that they enjoy the facilities afforded by the road."[42]

Back on Promontory, the mixed celebrants in the car of the Central Pacific's James Strobridge were still greatly enjoying themselves, with none of the discordant notes as sounded in Governor Stanford's entertainment of the executives. The men in the latter car were presumably taking their leave of each other while, in Strobridge's, the Central Pacific superintendent of rolling stock, James Campbell, rose to give the last speech. "The work is finished," he said. "Little you realize what you have done. You have this day changed the path of commerce and finance of the whole world." The Iron Horse, he said, would finally replace the slow seagoing trade. "Where we now stand," he continued, "but a few months since could be seen nothing but the path of the red man or the track of the wild deer. Now a thousand wheels revolve and will bear on their axles the wealth of half the world, drawn by the Iron Horse, darkening the landscape with his smoky breath and startling the wild Indian with his piercing scream. Philosophers would dream away a lifetime contemplating this scene, but the officers of the Pacific Railroad would look and exclaim: 'We are a great people and can accomplish great things.'"[43]

And then the party broke up for good. The journalists returned to file and fill. The Chinese and Irish laborers withdrew to their respective camps on the western and eastern slopes of the Promontories, with an enormous amount of corrective work ahead of them and, after that, to take their meager pay and skills to other railroad enterprises, where most of the surveyors and engineers were already bound, to fill in other blank spaces on the map of the West. The infantrymen climbed aboard their San Francisco–bound train. Durant, still tortured by his headache, would go back east with the other executives, who would combine with the Ameses and Hazard to prepare for the Doctor's ouster at the very next directors' meeting in Boston; Durant would, of course, beat them to the punch by resigning first, turning his gaze upon the untrammeled Adirondacks and enlisting Colonel Seymour as his chief engineer for a railroad into its heart. Stanford's special train would turn on the Y-track at Promontory and hook up six beautiful first-class passenger cars just arrived from Huntington in New York, and the Governor would return home to do the minimum necessary as the octopus of the Central and Southern Pacific began to stretch over all of the Pacific states and the southwest. By 5:00 P.M. the upland valley of Promontory was deserted except for the few canvas tents flapping in the wind. Elsewhere, in all the great cities in the East and far West, and many of the smaller towns, the skies that evening were bright with fireworks and the air was filled with the sounds of happy anticipation, but at Promontory there would have been only the voice of wind through sagebrush and rabbit grass, vibrating telegraph wires, and perhaps the call of coyotes, while the rails went west, went east, into the future.

Part X

1872–73
Scandals, Scapegoats, and Dodgers

Epilogue

"Trial of the Innocents"

"As Maine goes, so goes the Nation." So went the old political proverb. With but two months to go in the presidential campaign of 1872, the lead-off electoral contest in Maine was to take place on September 5. It always drew enormous attention—those early local returns were an accurate forecast of who would prevail in November.

Anticipation was especially high this splintered, contentious season. In May, liberal Republicans had met in Cincinnati under the leadership of Carl Schurz, Gideon Welles, Charles Sumner, E. L. Godkin, and Charles Francis Adams to show their opposition to "Grantism," as they bitterly called it—"bayonet rule" in the South, the civil service corrupted, graft and plunder widespread, especially at the higher altitudes of federal office, and nepotism and favoritism on the part of the president himself. The liberals wanted no part of Grant's reelection, widely thought a "sure thing," and had chosen the *New York Tribune* founder and editor, Horace Greeley, as their presidential candidate. He was a romantic but singularly weak choice: probably the best-known editor in the country, and one of the earliest abolitionists, he was a man of enthusiams which put him far ahead of the mainstream—he had advocated temperance, women's rights, and the radical economic and social experiments of Fourierism; such free-thinking would be looked on with suspicion by the conservative multitude. Moreover, at sixty-one, Greeley was not in the greatest of health. But he was still full of pluck, and game for a war against the nation's most popular warrior, and he even accepted with good humor the support of the Democratic Party (which also chose him as its candidate in July) despite the fact that he was "the most vituperative persecutor of the Democrats in the whole country," as one commentator wrote.[1] "Straight" Democrats (as they called themselves) who could not countenance their old enemy would ultimately bolt from the bolters, meeting to nominate their own candidates and fracturing the opposition to the stalwart Republicans of Grant. In June, in Philadelphia, President Grant was renominated on the first ballot just as Vice President Schuyler Colfax was cast aside on his first ballot—penalized for appearing to flirt with the liberals in Cincinnati. Colfax was replaced with Massachusetts senator Henry Wilson. And thus the memorably bitter contest had begun, with stump speakers and newspaper editorials predicting that if this true war of good versus evil did not end with the virtuous as victors, disaster—and the end of democracy—would certainly follow.

Nominated in Grant's home state of Ohio, in his punishing regimen of public addresses Greeley headed, naturally, to other loci of the administration's power, taking in several western states but more pointedly going to Senator Henry Wilson's Massachusetts and to Maine, home state of House Speaker James G. Blaine. Blaine, forty-two, was a onetime newspaperman and state legislator who had first been elected to the House in 1863 and had thereupon forged strong political bonds with Westerners like Ohio's James Garfield and Iowa's William Boyd Allison; now they controlled the Republican Party in the House, and were obviously presidential timber for 1876 and 1880; in the 1872 electoral contest between Grant and Greeley, James Gillespie Blaine was drawing as much fire in the "Greeley press" as was General Grant. There was not much to pin on Speaker Blaine—he was one of the most well-liked politicians in Washington—but he had become wealthy without visible means of income and had recently purchased a large house, and oppositionists were certain that such a skillful swimmer in the polluted waters of the Grant administration had probably picked up a taint himself.

Then, bare days before the Maine election, on Thursday, September 4, after Speaker Blaine had spent several assiduous months working to deliver his state for his president, the liberal *New York Sun*—the fiercely independent Charles A. Dana's gadfly daily which had seized on the Greeley candidacy as the country's only salvation from corrupt Grantism—dropped a bombshell, right on page one.

> THE KING OF FRAUDS.
>
> How the Crédit Mobilier Bought its Way Through Congress.
>
> COLOSSAL BRIBERY.
>
> Congressmen who Have Robbed the People, and who now Support the National Robber.
>
> HOW SOME MEN GET FORTUNES.
>
> Princely Gifts to the Chairmen of Committees in Congress. From 2,000 to 3,000 Shares Each to Henry Wilson, Schuyler Colfax, George S. Boutwell, John A. Bingham, James A. Garfield, the Pattersons, Eliot, Brooks, Dawes and James G. Blaine.

The news was datelined Philadelphia, and trumpeted as "the most damaging exhibition of official and private villainy and corruption ever laid bare to the gaze of the world." It involved the vice president, the Speaker of the House of Representatives, the new candidate for vice president, and the chairmen of almost every important committee in the House—"all of them are proven, by irrefutable evidence, to have been bribed." The correspondent reminded his readers that "the public has long known, in a vague sort of way, that the Union Pacific Railroad was a gigantic steal": the federal bond subsidies had been enough to build and equip the road, he said, but the public had been bilked when the Union Pacific was allowed to issue first-mortgage bonds ahead of the government's, which had been the treasury's only security. On top of that were the tens of millions of acres of the public domain. Finally, a handful of cagey financiers had bought a Pennsylvania corporation, renamed it the Crédit Mobilier of America, and made "millions upon millions of dollars in building and equipping" the Union Pacific. Only when the robbers had fallen out with one another was the scandal of congressional influence for sale laid bare. One trustee and stockholder, Henry S. McComb of Delaware, claimed he was owed a large amount of Crédit Mobilier stock—and sued the corporation and its officers—including Sidney Dillon, Rowland G. Hazard, Thomas C. Durant, John Duff, former representative John B. Alley, and sitting congressman Oakes Ames of Massachusetts. The suit had dragged through the Pennsylvania Supreme Court for years, and even had some unnoticed repercussions in the Fisk suit in New York, but recently testimony taken some fifteen months earlier had been wormed out of a lawyer's office by the newspaperman; much later he said the leak had cost him $500.[2]

The testimony—of McComb himself—fanned out across the full seven columns of the *Sun*'s first page, took up all of the second page, and draped onto the third. It spread out for all to see the dark disclosures of Doctor Durant's "Inside Ring"; the founding concept of stockholders' "limited liability"; the Hoxie building contract; the Williams and the Oakes Ames contracts; the diversion of profits from the Union Pacific to the "Inside Ring" of the Crédit Mobilier through princely dividends of cash and stock; the quarrels, cabals, and conspiracies between principals, especially between the Durant and Ames factions; and the smaller history of complainant Henry McComb's bitter suit to obtain more of the magic shares than he already held. And not until the last two columns of that long, densely printed exposé did the *Sun* correspondent disclose the dramatic revelations at the root of the headlines about corruption in the Capitol: Oakes Ames had written McComb three letters from the Capitol in the winter of 1867–68 in which he disclosed the selling of deeply discounted Crédit Mobilier shares to congressmen, the discreet withholding of most of the shares "in trust," to keep the politicians' names off the books, the divisions and bonuses, the quick payment of dividends, and the rationale for such generosity: current unfriendly legislation, and the desire, Ames had written, to place stock "where it will produce most good to us."

Ames had consulted with McComb about political recipients: "You say I must not put too much in one locality," he wrote McComb. "I have assigned as far as I have given to—four from Massachusetts; one from New Hampshire; one, Delaware; one, Tennessee; one-half, Ohio; two, Pennsylvania; one, Indiana; one, Maine; and I have three to place, which I shall put where they will do most good to us. I am here on the spot, and can better judge where they should go." In pencil on the bottom of one of Ames's letters, the transcript indicated, was this notation from McComb:

Oakes Ames's list of names, as shown to-day to me for Crédit Mobilier, is:

BLAINE of Maine	3,000
PATTERSON of New Hampshire	3,000
WILSON, Massachusetts	2,000
PAINTER (Rep.) for Quigley	3,000
S. COLFAX, Speaker	2,000
SCOFIELD and KELLEY, Pa.	2,000 each
ELIOT, Massachusetts	3,000
DAWES, Massachusetts	2,000
FOWLER, Tennessee	2,000
BOUTWELL, Massachusetts	2,000
BINGHAM and GARFIELD, O.	2,000

Endorsed: Oakes Ames, Jan. 30, 1868.

More than two columns of the *Sun*'s editorial page focused on the scandal, reiterating the choicer statements of Ames and reprinting McComb's penciled, annotated list. Almost every day in the coming two months that list was reprinted—the original boldface printer's type wearing down with overuse—and the list, and the letters, and the damning statements, and deeper, darker speculations about all of the named and unnamed stalwarts who "had been seen" by Oakes Ames, sprang up in "the Greeley Press" like the mythical soldiers from serpents' teeth. The last question in the September 5 *Sun*'s "THE KING OF FRAUDS" editorial was perhaps the last temperate statement for the rest of the campaign: "The pertinent question," wrote editor Dana, "which these facts must suggest to all honest and intelligent men is, Shall these men be continued in office?"

Of course the *New York Tribune* picked up the news immediately—with Horace Greeley out on the hustings, the paper was edited by Whitelaw Reid. Many others climbed over the ropes into the ring and joined the fray, including the Democratic *New York World*, the liberal *Springfield Republican* and *Boston Herald* of Massachusetts, the *Cincinnati Commercial*, and numerous others, each quoting from the other, each racheting up the indignation a few more notches. "If this exposure is true, and the truth or falsehood of it must soon be demonstrated," said the *New York Tribune*, "the duty of Congress and of the constituencies of these unfaithful stewards is very plain. They must instantly be expelled by their colleagues and repudiated by their fellow-citizens, or any pretense of honest legislation becomes a farce."[3] The reaction in the pro-administration press, such as the *New York Times*, was to ignore what the Greeley papers were calling "the scandal" as long as possible, and when it was finally unavoidable the matter was deemed "The Crédit Mobilier Slander."

With the Maine contest fairly hours away, Speaker Blaine denounced the "utterly baseless and groundless" charges. "I never owned a dollar in the company in my life," he told the *Kennebec Journal*, "either directly or indirectly." Vice presidential candidate Wilson exclaimed that he had "never owned a bond or a share of stock in the Union Pacific Railroad Company, or in any other railroad or corporation that has been before Congress." Representative Garfield said that "he never subscribed for a single share of that stock, and that he never received or saw a share of it." Representative Dawes said, "Neither Oakes Ames nor any other man, dead or alive, ever gave me, directly or indirectly, a penny of the stock of the Crédit Mobilier." From the other named politicians—most notably Oakes Ames—there was only haughty silence. To the antiadministration press, the silence was a confession, the Dawes denial a crafty evasion (since it was said he had paid for the stock), the Garfield denial also an evasion (since his had been reportedly held in trust), the Wilson denial as suspicious as his slippery campaign claims that he had never belonged to the nativist, anti-Catholic Know-Nothings, and the Blaine denial distinctly hard to swallow: "Mr. Blaine," the *Sun* commented astringently, "was a poor man when he became a member of Congress in 1864. He is now a millionaire."[4]

Of course truth blended with inaccuracy, distortion, and leaps of logic in the original *Sun* exposé and in subsequent accounts; only one side of the Philadelphia case, complainant McComb's, was the source of the anti-Grant press campaign, although it was swiftly announced that Oakes Ames and the other officers had evaded the state supreme court's summonses for years and had no comparable transcripts to release as rebuttal. Ironically enough, in the *Sun*, the *Tribune*, and others, Henry McComb and Thomas Durant were portrayed as innocent victims instead of scoundrels of a like stripe to the Washington influence peddlers and Wall Street buyers. Although it would remain unclear for some time, Oakes Ames's "fatal list" of congressmen enumerated dollar figures of stock at par, not individual shares, as advertised by the papers; Vice President Colfax's number, for instance, was $2,000, not two thousand shares, which in the court transcript and subsequent reports represented a hefty inflation of the value of a politician's corruption, especially with McComb's estimations that the shares were worth between $600 and $1,000 each.

And the leaked "fatal list" of congressmen represented salesman Ames's list of customer contacts, not clients, as of January 28, 1868. There were twelve penciled congressional names; the thirteenth person, Uriah H. Painter, was the Washington correspondent of the *Philadelphia Inquirer*—but also a Union Pacific lobbyist working through Grenville Dodge and the attorney William E. Chandler. Of the first twelve congressmen, three had subsequently declined Ames's offer—Blaine, Boutwell, and Eliot—on the grounds that it seemed a bad investment. (Another named by McComb in testimony, Bayard of Delaware, had also declined.) Ames's money-back guarantee option had, moreover, been exercised in 1868 or 1869 by buyers Allison, Henry Wilson, and J. F. Wilson because of possible suits against the Crédit Mobilier or even ethical considerations; Ames had bought their shares back for cost and interest. A more accurate talley, including those remaining on the penciled list, others named in McComb's leaked testimony, or others subsequently named (who had obtained shares from Ames, Durant, or other officers), would look like this:

Garfield, of Ohio
Colfax, of Indiana
Allison, of Iowa
Kelley and Scofield, of Pennsylvania
Bingham, of Ohio
Logan, of Illinois
Patterson, of New Hampshire
Dawes, of Massachusetts
Brooks, of New York
Hooper and Alley, of Massachusetts
Grimes, of Indiana
Boyer, of Pennsylvania

Several—Hooper and Alley—were now retired, with John B. Alley now so energetically linked to the Union Pacific and Crédit Mobilier that memory of his congressional service was all but erased. And there were quite probably more beneficiaries in the Capitol. Ames's own clientele had bought, or borrowed for, or accepted, some 160 shares and collected the dividends, bonuses, and distributions. Another 183 shares—as Union Pacific historian Maury Klein discovered—were held by Ames but never accounted for, and may have made their way undetected into politicians' portfolios. Little of this would be clear for some time. Suffice it to say that the level of discomfort in Washington only went upward as the weeks of the 1872 political campaigns moved forward.[5]

Even in the face of the mounting bribery accusations, James G. Blaine handily won another congressional term. The *New York Tribune* was quick to comment. "Now that he is safe in his place, he will have time to throw some light on this shady matter of the Crédit Mobilier," ran Whitelaw Reid's editorial on September 11. "Mr. Blaine is one of the names so inscribed. This is one of the things that cannot be disposed of by silence or mere denial. It must be investigated, and the sooner the better for all parties." The *Sun* was tougher. "This officeholders' Republican party, with its noisy claims to principle and to conscience," Dana wrote on September 14, "which stands by and witnesses such sickening corruption and thinks that it may all be ignored and lightly passed over, and that a little whitewash will hide the wickedness—verily the day of its peril is at hand, for the just anger of the people waxes hot against it." With Maine returns, prognosticators happily foresaw a victory over the liberals in November, and though one would see few hints of concession in the Greeley press, it was plain to those closest to Greeley himself that he had not only given up but lost nearly all hopes for any future; his wife was now on her deathbed, and Greeley at last abandoned the campaign, sitting up nights with his wife and getting virtually no sleep for weeks, eating little, taking less and less interest in even mundane daily affairs.[6]

Though they did not get a rise out of the bereft, exhausted Greeley, the daily reports of outrages committed by the "universally corrupt" Grant administration certainly gained in coloration as the increasingly lopsided presidential campaign wobbled on toward November: there were reported frauds in the Postmaster General's office and the Indian Affairs Agency; there was the barefaced sale of consular offices by Grant's own brothers-in-law, the Dents; there was even the disclosure that President Grant himself had surrendered strategic land overlooking a military reservation and worth, in practical terms, millions of dollars, to a ring of real estate speculators for the bargain price to the treasury of $3,000.

The Crédit Mobilier, though, stood at the pinnacle, and there were no end of angles to it. The *Sun* reported that on Friday, September 14, a meeting of prominent Grant managers, including Senator Henry Wilson, took place at the Fifth Avenue Hotel. Oakes Ames, "bribery agent of the Crédit Mobilier" was summoned to meet them—and asked to sign their draft of a "broad, square, and comprehensive denial of all the facts stated in his letters" to McComb. The statement was to be published in the Saturday journals and telegraphed all over the Union. "But when it came to the point of signing and sending out this enormous lie," said the *Sun*, "Mr. Ames did not dare to do it." To statements like this, pro-administration papers had one answer: the slander was all the invention of editor Dana, "the most malignant and prolific libeler ever connected with the American Press."[7]

On September 17 in Washington, Asa Whitney died of typhoid fever at the age of seventy-five, his death attracting scant notice in the reckless modern era which had replaced his own.

There was little room in the newspapers for much beyond scandal, recriminations, and politics. An enterprising reporter at the *New York World* had scrutinized the *Congressional Record*, turning up ammunition against future denials that any pertinent legislation involving the railroads or the Crédit Mobilier had appeared in Congress after the winter of 1867–68, when the Ames distribution was made. In the spring of 1869, in fact, legislation to move the Union Pacific headquarters from New York to Boston in response to the Fisk lawsuit had sped through the House under a suspension of the rules, with Representative Blaine in the chair, and all the other House names in Ames's list voting to approve; in the Senate, Henry Wilson had spoken eloquently about Oakes Ames's high character and the merit of the bill, and resisted his opponents' attempts to attach riders opening the railroad to greater accountability or scrutiny. Moreover, in February 1871, there had been a move to revoke the original understanding in the Pacific Railroad Act of 1862 that in return for federal munificence, the railroads would transport the mails, federal freight, and the military for free. This was at the end of the Forty-first Congress, as legislators were rushing to finish and leave, and an amendment directing the secretary of the treasury to pay the Union Pacific and others "one-half of the compensation for all services heretofore or hereafter rendered" was attached to the army appropriations bill. The Senate passed it with little debate on the day it was introduced; the House received the army bill without the amendment being read, and though suspicious members attempted to alert the House and check the gift to the railroads, tenacious work by the Railroad Congressmen and Speaker Blaine pushed the bill through as the session squeaked to a close. "Thus the Union Pacific Company was saved" from the bankruptcy which had threatened it, the *Sun* commented, "and all danger of an exposure of the Crédit Mobilier frauds was for a time removed. Mr. Ames was clearly right when he said: 'I have used this stock where it will produce most good to us I think.'"[8]

On September 18, Oakes Ames broke his silence of two weeks to issue a widely published denial. "I never gave a share of stock of [Crédit Mobilier] or any other Company, directly or indirectly, to any member of Congress," he wrote, offering to "place the sworn declarations of any one of these individuals, of the highest character and reputation, against the affidavits of Mr. McComb." In reply the *Sun* and the *New York Tribune* fastened on the word "gave" as not being the same as "sold" or "lent," and the evasion of the matter of dividends paid, and the newspapers brandished the court evidence of Ames's letters being in his own handwriting. And the *Chicago Tribune* weighed in with its exclusive interview with Henry McComb, in which he elaborated his charges and averred that his suit was not politically motivated: he was emphatically not a Greeley supporter, and though Grant was not his first choice he was going to vote for him. During the interview, conducted at New York's Fifth Avenue Hotel, several of McComb's associates, prominent railroad men, listened in. "What surprises me most," commented H. D. Newcomb, president of the Louisville and Nashville Railroad, "is that the newspaper profession, with all its acuteness, did not discover this matter long ago—four years ago—it being an old subject of conversation among railway men and operators." Punctuating this was the statement from none other than George Francis Train, who dispatched a messenger from his Europe-bound ocean steamer as it was casting off. "The Pacific Railway bill of 1864 I machined through Congress," he boasted to the *Sun*, "saving it with forty-four Democratic votes, who never asked or received a cent. The Radicals were all on the make. Thad. Stevens's share was seventy thousand."[9]

Buffoon as he might have been—Train had announced he was running for president and then hopped aboard the steamer to spend the campaign "among the effete monarchies of the Old World," as a disparager commented—he was still credited as being enough of an intimate with people like Ames and Durant to know whereof he spoke. A little more biting comic relief was the response of the liberal *Cincinnati Commercial*, with a bit of doggerel, widely reprinted:

THE CRÉDIT MOBILIER FRAUDS

The good Mr. Blaine
Came up out of Maine
As poor as the proverb's church mouse is,
But when he returned
His pockets all burned
With riches in gold and in houses.

How came Mr. Blaine,
So poor down in Maine,
In Congress to fatten so quickly?
Why, Mr. Oakes Ames
Had a long list of names
With shares written down to them thickly.

And poor Mr. Blaine,
Well knowing that Maine
Could seldom her church mice enrichen,
Made a trade of his wares
For some sheaves of those shares,
And fed on the Oakes like a lichen.[10]

The poem could have been addressed to dozens in Congress as effectively. But it was no laughing matter in many quarters. Speaker Blaine suffered new charges when the *Tribune* publicized old documents reaching back to the machinations behind passage of the Pacific Railroad Act of 1862, in which a lobbyist drew up a list of recipients of shares in the Leavenworth, Pawnee, and Western Railroad (later renamed the Union Pacific, Eastern Division, and subsequently the Kansas Pacific). Congressmen, senators, and even a cabinet secretary were on the list, and there was an entry of eight thousand shares for "Blaine, of Maine." Blaine called Horace Greeley "a slanderer" for allowing the charges to be published, not knowing that the candidate was despondent after his wife's death and now wholly uninvolved in the affairs of the day. And soon it became clear that the stock entry referred to James G. Blaine's brother, a Kansas resident; the former was then serving in the Maine legislature, presumably far from the Capitol influences of western railroads.[11] Schuyler Colfax had no such convenient excuse, but after a month of what the *Sun* called the "most suggestive silence," the vice president stated, in a long, self-righteous South Bend speech, that no one had given him shares in anything, nor had he received any dividends. About all of this Oakes Ames remained "mute as an oyster," said the *Tribune.*[12]

But soon it would be time for the electorate to speak. Despite all the editorial ink employed over the ceaseless Washington scandals, the public seemed almost unswayed, politically, regardless of what they may have personally felt about fraud and corruption. Local contests in significant states like Pennsylvania, Ohio, and Indiana came and went in October, giving liberal Republicans and Democrats minimal hopes; the emotions of the late Civil War—especially about the distant, taciturn, insulated war hero in the White House, whose durable reputation as a binge drinker also seemed to have no effect on voters—were helping to carry incumbents through. A brawny business climate, however illusory, was also buoying the status quo. And on November 5, it was Grant over Greeley by a majority of 763,000 votes, and 286 to 66 in the electoral college. "Four Years More of Fraud and Corruption," shouted the *Sun*. "Death of the Democratic Party." The defeated Horace Greeley would not even live long enough for the electoral votes to be officially tallied. Broken by his wife's death; crushed by a delusion that his Quixotic campaign had ruined him, the *Tribune,* and his partner-publisher; agitated by a squabble with his editor, Whitelaw Reid; Greeley slid from despondency into something his friends feared was mania; they had him taken to a private asylum, where he died on November 29.[13]

The following Monday, December 2, was a balmy Indian summer day in Washington and the Capitol building was crowded with legislators returning to work, with family members and other spectators, and with lobbyists, lawyers, and other dignitaries. Satisfaction over the past national and local elections warmed the mood, as did the prospect for many—an unusually high number in Congress—of retirement from public life at the end of the session in three months. Members of the pro-administration Republican majority caucuses of both houses were quietly moving toward the ouster of all liberals from positions of any significance in committees, though they remained outwardly friendly. "There was a hearty exhibition of good-nature on the floor" of the House of Representatives, a *New York Times* correspondent noted, "prior to the fall of the Speaker's gavel, with many warm greetings and still warmer congratulations, and when the House came to order it seemed as if the genial weather had bred a magnanimous and forgiving spirit among the members." One of the first orders of business was the resolution of Henry Dawes, of Massachusetts, to mark the House's respect to the memory of Horace Greeley, which was passed after a brief but touching speech. Vice President Colfax was to attend the funeral as the administration's representative, and several legislators planned to go.

After small items of business were disposed, Speaker Blaine did something unusual: he descended from his chair, appointing Ohio's Samuel "Sunset" Cox as Speaker pro tempore, and from the floor, to the chamber's surprise, Blaine began to address the Crédit Mobilier "slanders," recapitulating the serious matters lately trumpeted in the liberal press. The floor and the galleries were in complete silence as Blaine moved the creation of a special committee of five to investigate the grave charges. After short discussion, the resolution passed with only a single dissenting vote—that of the Democrat Archer of Maryland, who explained that "the people had passed a verdict of condemnation upon these slanders at the polls" and an official investigation was redundant.[14]

Temporary Speaker Cox, a supporter of the Union Pacific since its earliest lobbying days in Washington for the 1864 Pacific Railroad Act, quickly named the House Crédit Mobilier Committee, comprised of five lawyers: Judge Luke Poland of Vermont, who formerly served on the state supreme court, as chair; Major General Nathaniel Banks of Massachusetts, a political appointee to the general's rank during the war who had been removed by Lincoln after numerous ineptitudes; George McCrary of Iowa; William Niblack of Indiana; and J. B. Beck of Kentucky. The last soon begged to be excused and was replaced by William Merrick of Maryland. Merrick and Niblack were Democrats, the others Republicans. The committee would, the *New York Times* commented, "be recognized as composed of men abundantly able and willing to probe the matter to the bottom."

Predictably, the *Sun* and other antiadministration papers disagreed, finding the Poland committee "very favorably composed for whitewashing purposes. Mr. Poland of Vermont is a namby-pamby sort of a man, and Mr. McCreary of Iowa is directly interested in more than one railroad job." The three others were "good men" but already extremely overworked with other important committee duties, and were anything but enthusiastic about taking on more hearings. With the Poland Committee established and hearings announced to commence as soon as practicable, the House would have turned to finish its day's business before the president's message to Congress was to be read, but for some moments of visible consternation when Benjamin Franklin Butler, Republican of Massachusetts, rose to propose a change in the Poland Committee authorization. Butler, another political general during the war who may have been Lincoln's most incompetent officer, was still nicknamed "Spoons" Butler by detractors in honor of his service as military governor of Louisiana, where he was accused of corruption—even down to the theft of household silverware from his official residence. A criminal lawyer and, in the House, a relentless political operator who had played a major role in the Johnson impeachment efforts, Butler was trusted by few, even among those close to him. Now, squinting at his colleagues, he proposed blandly that the scope of the Crédit Mobilier inquiry into the names revealed in the published McComb testimony be widened to "any other persons or corporations." A palpable shudder moved across the floor of the House. The resolution found no supporters.[15]

The Poland Committee commenced hearings on Thursday, December 13, in a first-floor committee room in the Capitol, and before they began Judge Poland announced that the committee had decided to sit with closed doors. Protesting, correspondents and private citizens were ushered out. A number of congressmen in the corridors objected. "The accused gentlemen are doing themselves great injustice by allowing the committee to exclude the public," complained one prominent Republican to a *Sun* reporter. "The very fact that the proceedings are secret places the accused at the mercy of the committee. If I was accused, I would refuse to make any statement or answer any questions until the doors were thrown open and I was allowed to make my defense to the people." The next day, and for the week following, until the committee recessed for Christmas and New Year's, newspaper editorials excoriated the closed-door policy while news columns brimmed with full accounts of the secret hearings: how, on the first day, Speaker Blaine was invited to make a statement without being sworn or submitting to questions; how Henry McComb appeared to testify, and beyond repeating the names he claimed were read to him by Oakes Ames five years before, he named under oath one prominent congressman whom he personally knew to have bought the railroad securities; how John B. Alley appeared to impugn McComb's character and veracity, calling him a war-profiteering swindler; and how Oakes Ames was preparing a written statement to be read before he was questioned. Soon the *Sun* had bestowed a supremely ironic title on the Crédit Mobilier hearings: it was the "Trial of the Innocents."[16]

"They are, of course, all innocent," explained the *Sun*'s chief Washington correspondent, A. M. Gibson, "for did not the people so pronounce them in November last? and haven't they demanded an investigation?" Nevertheless, he had noticed "unusual agitation . . . troubled expression of countenance . . . distracted manners" and an inordinate number of conferences with each other. "I have been a pretty close observer of the House of Representatives for several sessions," Gibson said,

> *and I never remember to have seen certain members so careless of what was under consideration by that body as they were last Thursday and Friday, nor do I remember having seen Messrs. Garfield, Bingham, Dawes, Scofield, and Kelley so frequently dodging in and out of committee rooms, and hobnobbing with Oakes Ames and John B. Alley in the coat rooms and lobbies. Can it be possible that these pure and unspotted statesmen are afraid of this investigation, which they so defiantly demanded? or are they afraid of Ames and Alley?*

He had even seen Vice President–elect Henry Wilson, "nervous and fidgety," slipping into the lobby of the Arlington House to ask the rooms of Ames and Alley. Wilson "walked uneasily through the lobby into the reading room and back, casting a furtive look at every guest he met," and after a short delay he was sent upstairs. "The great teetotaller's visit may have been perfectly honest and honorable," Gibson wrote. "I don't say it wasn't, but there was certainly a sheep-stealing air about it."[17]

Mysterious meetings continued. The *Tribune* correspondent saw Ames, Alley, and three or four of the suspected members gather in a Capitol committee room on Saturday, December 14, and wrote that it created "much talk. These gentlemen were closeted together for an hour as if arranging some plan of action or harmonizing their testimony in advance." Another reporter commented that McComb and his attorney, Judge Black (who, saying little, playfully encouraged the newspapermen to keep digging) wore increasingly satisfied expressions in the Capitol corridors as the days progressed. This seemed especially evident on December 17, when Oakes Ames finally appeared to have his prepared statement read to the committee and begin to answer questions. Meanwhile, on the floor of the House there were increasing sentiments to open the doors of the inquiry to the public. A move by a Georgia representative on December 16 to open and broaden the hearings had been killed, but he persisted the next day. Judge Poland announced that his five-man committee was unanimous in desiring secrecy until their job was done—he was, he said, as distressed by the journalists' inaccuracies as he was by the actual leaks—and in response, calls for obtaining a "sense of the House" resolution as to open hearings began to gain ground.

Back in the committee room, Ames had reportedly admitted that he had sold shares to all on the published list but Blaine and a few others in a "strictly honest and honorable" transaction, a friendly gesture to friends and not an attempt to influence colleagues to all their gain. He said, however, that a number of these friends had, after the scandal broke, asked him to buy back their stock, and that he had done so in a few cases, and that his friends had begged him to shield them before the committee. He maintained that since there was in his eyes no illegality there was no reason to withhold any names. Outside the committee room that day, McComb's attorney said he planned to offer in evidence "the votes of these gentlemen on measures passed by Congress which were of vast importance to the Union Pacific Railroad Company as proof that something influenced them to do wrong." These fragmentary bits of news only excited more speculation and curiosity.

The tantalizing drama behind closed doors that day was dwarfed by the commotion when New York representative James Brooks, the lone Democrat to be named in the inquiry, obtained the floor on a question of personal privilege to deny McComb's leaked statements in the *New York Herald* and *World* that he had obtained stock to influence himself and other Democratic members in legislation affecting the Union Pacific Railroad. "If this charge is true," he declared, "I am unfit to be a member of the House and ought to be expelled—not only from the House, but from all association with decent men." He spoke for almost an hour. "The whole house listened throughout with the closest attention," the *Tribune* commented, "the occupants of seats distant from Mr. Brooks's desk, standing in a dense group in front of him." Toward the end of his speech he raised his hand "in the manner of a witness taking an oath," saying that he had never tried to influence a colleague's vote to his financial advantage, that he had always refrained from voting on such legislation himself, and that he had never owned, "directly or indirectly," any Crédit Mobilier stock. His son-in-law, a Wall Street investor, in fact owned the fifty shares in question, and he had the signed receipt to prove it. After his angry speech, it was apparent that all fellow Democrats would shun him—that Brooks must, several said, take the consequences. He was in the same boat with Garfield, Dawes, Bingham, and the rest of the Republicans.[18]

Brooks was one of the principal subjects during the ensuing two days of testimony before the holiday adjournment—following Ames was John B. Alley and then McComb, on reexamination—and, in answer to the growing clamor, rebellious House members vowed to offer a resolution on the first Monday after recess to suspend the rules and compel majority leaders to open the hearings to the public—and to print the evidence already taken.[19] "A secret investigation will not do," the pro-Grant *Milwaukee Sentinel* averred. "Let us have the whole matter opened to the daylight. A ghost has been raised before the eyes of the people that cannot be laid with the waving of a whitewash brush. The Opposition began this business. Let the Administration party end it."[20]

Leaked, pieced-together testimony had a kind of mysterious allure to it, but it had nothing against the authority of actual transcripts, as with the endlessly recycled McComb avowals. The public demand for open hearings made itself known to congressmen home for the holidays, and by the first week of January a *New York Tribune* correspondent could find no man in Washington in favor of continued secrecy. "No Star Chamber inquiry, conducted by friend or foe and ending with a report without the full evidence, is satisfactory," he declared. One Republican congressman talked at length about his colleagues now in the harsh spotlight. "Oh, you know all about these sons-in-law and brothers-in-law," he scoffed, "that don't help it in the least." If the transactions were all right, he thought, why should Vice President–elect Wilson register the stock in his wife's name, or Representatives Eliot and Brooks have drafted their sons-in-law? He was sure that sooner or later there would be a resolution to expel Oakes Ames as a briber. "The worst of this whole business is that the men who are most corrupt will escape," he said, "while those who are little to blame will be made scape-goats. It is always so." The lucky invisible ones were the congressmen and senators who were concurrently paid attorneys and lobbyists of the corporations, many of whom had been elected with railroad money. The Crédit Mobilier scandal, he said dejectedly, was but a small part of a long-term corruption of far greater magnitude.[21]

Only sixteen members of the House were in favor of continued secrecy when the vote was called on January 6, though this number was reduced to seven when the division was to be put on record. Transcripts of the December hearings were delayed while Judge Poland allowed the witnesses, including Speaker Blaine, to correct such statements as they wished, but when excerpts began to leak out, the chairman ordered the transcripts released to the press. The morning editions of January 7 brimmed over with the testimony of Ames, Alley, and McComb. "I never dreamed" of corrupting colleagues, Ames was quoted as saying. "I did not know that they required it, because they were all friends of the road and my friends; if you want to bribe a man you want to bribe one who is opposed to you, not to bribe one who is your friend."

Despite the oft-stated fears of Poland that his little committee room would suffocate under the crush of outsiders, only four or five correspondents were in attendance in the first open hearing on January 7. Vice President Colfax appeared to elaborate, with a written statement, upon denials made during the fall campaign. He painted a confusing picture. He had never been offered stock in the Crédit Mobilier or the Union Pacific, nor collected any dividends, Colfax said, although then he admitted that Ames had approached him and he had agreed to buy twenty or thirty shares and had made a down payment of $500 against a promise to pay the rest later. Some time later, he claimed, he had decided to get out because of the McComb and Fisk litigation, but Ames had never repayed his $500. When he had finished his curious statement there was almost no reaction from the committee.

There was, however, much drama to see after John Alley was sworn in again, especially in the personal emnity between McComb and Alley. This came to a head during a break; McComb called Alley a "dirty scoundrel" and a liar for making the incredible statement that the Crédit Mobilier "had never made a dollar out of the Union Pacific Railroad" and that it "had never paid more than seven per cent dividends." Alley ignored him as if he were litter which had not yet fallen to the floor. "Oh," McComb exclaimed, "I wish you were about 25 years younger, Mr. Alley. I would comb you down in more ways than one!" But later, outside in the corridor, Alley reserved his own scorn for Oakes Ames's congressional colleagues, who had been ceaselessly badgering Ames to recall transactions as they claimed to remember them. Ames was an honest man with good intentions, he told a *Tribune* reporter, and he was giving his old friends too much latitude. "If any action is taken toward making him carry the sins of the members who were supplied with stock, the country will find out that other people know something which it would not be healthy for some other people to have known. . . . [H]e had more offers for Crédit Mobilier stock from members of Congress than he could supply."[22]

Down the hall on the House floor, a reporter found the mood to be "unusually dull and languid"—members spent the morning at their desks ignoring the proceedings and working through newspaper accounts of the Crédit Mobilier hearings, dejectedly complaining to one another that their credibility with the public was shot. Most of the ordinary business of the day was dispatched with little enthusiasm. Not even at the end of the day, when Indiana representative Jeremiah Wilson introduced a resolution to create a new five-man select committee to inquire if the Crédit Mobilier or the Union Pacific had defrauded the federal government, did House members rise to any animation. The second committee was authorized without opposition, though its broader scope of investigation into their own body caused any number of pained expressions during passage. With Wilson as head, the committee was named by the Speaker pro tempore Cox as Samuel Shellaburger of Ohio, George Hoar of Massachusetts, Thomas Swann of Maryland, and Henry Slocum of New York, a West Point–trained officer who had commanded well at Antietam, Gettysburg, and Chickamauga. The last two were Democrats; all were lawyers. The Wilson Committee would commence hearings on January 10.[23]

With daylight flooding the two committee rooms, there was no end of interesting, if not always illuminating, testimony. It filled the newspapers daily. The Poland Committee examined thirty-nine witnesses; its printed report, generated after the close of the investigation on February 19, ran some 523 pages. The Wilson Committee heard thirty-three people in depth; its report would be 831 pages. Verily, it would seem like old-home month in the Capitol building. Beyond Oakes Ames, Alley, McComb, Colfax, and the named, sitting congressmen, there were numerous Union Pacific officials: Cornelius Bushnell, Sidney Dillon, John Duff, John M. S. Williams, and of course Oliver Ames, who had remained a company director after his presidency ended in 1871, and who would stay on the board until death unseated him in 1877.

Dr. Durant, a little grayer and more stooped when he testified than when he commanded an empire, had officially severed ties with the Union Pacific shortly after the golden-spike ceremony. But his connections through speculations as well as legalities were ongoing, especially in the drawn-out and bitter suit brought against him by Rowland Hazard, and in a case Durant pressed against the Union Pacific through his Omaha intermediary, James Davis, the former tie contractor. Most of Durant's energies had been devoted, though, to building a railroad into the heart of the Adirondacks, aiming to open the great unspoiled center of New York state to tourism, settlement, timber interests, and mining; his land company had bought up some seven hundred thousand acres of thickly forested mountains and exquisite, isolated lakes—he owned more land in the mountains for a time than did the state of New York—but, fortunately for that wilderness region, adequate financing had eluded the Doctor and he was able to build only sixty miles from Saratoga Springs to the hamlet of North Creek. Slow business, and the national financial panic in the fall of 1873, would plunge the company into insolvency; though after a slight recovery he would try to guide his son, William West Durant, at the helm, the Doctor would find little but failure, even past the grave. Three years after Durant's death in his elaborate mansion at North Creek in 1885, Hazard and associates would win a $20 million judgment against Dr. Durant's estate.[24]

But for the fact that they testified at different times, it was nearly a reunion for Durant and other onetime Union Pacific figures such as former operations superintendent Webster Snyder (whose brief, subdued testimony only recounted one bribe-demanding government commissioner, and nothing more of the catalog of interior corruption he might have revealed) and former chief engineer Peter Dey. The latter's contributions to the Wilson Committee inquiries into overruns and possible fraud against the government shed light on Durant's earliest construction cost inflations through the Hoxie contract, which had precipitated Dey's resignation in 1864 and set a model for a great deal of subsequent billing.

Notably absent from the hearings, however, was Peter Dey's successor as chief engineer of the Union Pacific: former U.S. congressman Grenville M. Dodge, who had been a busy man since the golden-spike ceremony. Seriously considered for the post of war secretary after Grant's adviser John Rawlins died in the summer of 1869, Dodge had been passed over by Grant and Sherman because of his overwhelming conflicts of interest. (Grant's new secretary of war, William W. Belknap, would be impeached by the House in 1876 for selling Indian trading posts, and resigned before his Senate trial; Dodge was a principal beneficiary of Belknap's patronage, winning numerous contracts to supply food and materials to his former mortal enemies in the Plains reservations beginning in 1871, being paid well over $1 million for his services. Unsavory business practices and associates would earn Dodge the stern disapproval of the congressional committee indicting Belknap.)

Though Dodge had joined the Union Pacific board of directors, and though after the golden-spike ceremony there were sundry demands as chief engineer in the struggle to bring the company line into compliance with government orders, he confessed to Sherman that he had "no taste" for administrative duties. He thereupon resigned his post as chief engineer in January 1870.

Now one of the richest and most politically powerful men in Iowa, Dodge constructed a lavish, fourteen-room, three-story Victorian mansion overlooking Council Bluffs and the Missouri Valley, and became even more deeply involved in local and federal politics, being instrumental in boosting his friend and onetime Crédit Mobilier congressman William B. Allison into the U.S. Senate. Dodge had not stayed out of railroading for very long. In March 1871 he joined Thomas A. Scott (vice president of the Pennsylvania Railroad and just named Union Pacific president) and a handful of southern politicians to incorporate the Texas and Pacific Railway Company, which was planned to build along the 32nd parallel from Marshall, Texas, to El Paso and San Diego. Dodge was named the well-paid chief engineer in February 1872, and helped set up a separate construction company modeled on the Crédit Mobilier, enabling them to let business to themselves at high profits.[25]

Dodge was busy at work in Texas in the fall and early winter of 1872–73 as the Crédit Mobilier scandal blew into Washington. He was greatly worried that it would touch him. Writing to Union Pacific lobbyist friends in Washington on January 7, he begged them not to "let that Committee do me any dirt for my skirts are clean no matter what others may have done. All I want is to have plainly shown who I was and what I was to the U. P. if they try to connect me with bribery." The next day he wrote Iowa representative George McCrary, of the Poland Committee. Only "Durant and the other thieves that I drove out of the company" would dare attack him in the hearings, he said. He did admit, though, that his wife had purchased one hundred shares in the Crédit Mobilier while he was still in Congress—though "she paid for [it] with her own money and the dividends of which she has today." In the spring of 1866, he claimed, the shares had been worth little, but they "became valuable under my surveys and reports. Then the company put me off and although I done the hard work I got none of the profits."[26]

As a former congressman, Dodge, like a number of others, did not suffer the attention of the Poland Committee, which had absolutely no jurisdiction over him or any interest in pursuing the lead. But his name swiftly came before the Wilson Committee when Cornelius Bushnell and a Union Pacific accountant disclosed his role during a Washington lobbying effort in March 1871 over a federal mortgage on the proposed Missouri River railroad bridge between Council Bluffs and Omaha. The Capitol paymasters in that lobbying were Dodge and his former congressional colleague James Falconer Wilson—

also his partner in Indian trading, bridge approach real estate, and many other ventures. The Union Pacific had disbursed $126,000, with the lobbying partners getting nearly a third and the rest presumably distributed among congressmen. The bill had passed.[27]

Wanting details, the committee dispatched a process server to locate Dodge in Texas. But the engineer had skipped—and though the server pursued him to St. Louis, where Dodge hid for a time under an assumed name, the quarry escaped on night train to the West. When informed, the committee chairman, Jeremiah Wilson, vowed that federal agents would track Dodge down. Peter Dey told him he was dreaming: Dodge was "a man of wonderful resources, and can live in Texas all winter, out of doors, if he wants to, and where none of your marshals can go, and if he don't want to come he will not come." If cornered, Dodge would offer a $10,000 bribe, "and you have not got a marshal in the state of Texas that will stand that sort of treatment. . . . Money is no object to him." The Wilson Committee's term would run out without ever seeing or hearing from Chief Engineer Dodge.[28] Railroad matters—and the immense wealth and power which flowed from them—would consume him for the rest of his long, eventful life, as he built lines across the south and the southwest, up into the mountain states, and down into Mexico. When he died in 1916 at his home in Council Bluffs, he had long outlived all the other principal players in the transcontinental drama. His great admirer, former president Theodore Roosevelt, would call him "essentially American."[29]

Waiting like a forlorn suitor for Dodge to appear before the Wilson Committee in those winter months of 1873, attempting to follow the labyrinthine explanations, with all of their evasions and half-truths, of the good Dr. Durant, trying to draw a bird's eye view of the Union Pacific and Crédit Mobilier in Washington with the grudging testimony of the summoned lawyers, accountants, and politicians, the five inquiring congressmen easily and repeatedly lost their way. The path of the Poland Committee investigators, though simpler, was still by appearances uncharted. But in the face of Oakes Ames's matter-of-fact recollections of giving certain friends an insider's price on Crédit Mobilier stock, with no intent to bribe, the members strove to put daylight between them and the Massachusetts congressman. Taking advantage of holes in Ames's memory and his continuing feelings of amity toward them, their statements in and out of the hearing room alleged that Ames was at best in error or at worst a liar. By the third week in January, Ames was reported in the newspapers to be "growing uneasy under the load of opprobrium and guilt which his victims have tried to slip off their own shoulders on to his." His friend John B. Alley continued to threaten that if they persisted "in making Mr. Ames a liar," the whole truth would be told. "Now, it may be averred that Ames is a liar and a scoundrel," the *New York Sun* commented, "but, pray, what are they? The main difference between him and them seems to be that he has some of that honor which is said to obtain among thieves, and they have none."[30]

By the first week in February, the *Sun* was commenting that "[p]erhaps there never was a more frightened set of lobbyists than those who now frequent the Capitol and throng the rooms of the investigating committees." There were more cases in point than the Union Pacific—there were, for instance, the "rascalities" of the Kansas Pacific Railroad and its predecessors, and the Burlington and Missouri River Railroad, which had contaminated nearly every department of the government. The reporter followed leads through the Senate and House, often in the footsteps of the unnamed lobbyist Thomas Ewing, and he mentioned a number of commissioners and cabinet secretaries, including Orville Hickman Browning; more often than not, he said, the respondents were afraid to go on the record, even anonymously. Unless people began to come forward, "the plundering of the public" would continue.

That night, on Wednesday, February 5, a sellout crowd gathered in New York City's Steinway Hall to hear Samuel L. Clemens, Mr. Mark Twain, lecture on the Sandwich Islands. He was by now an immensely popular humorist, known for his western sketches, for his comical travel narrative, *The Innocents Abroad* (1869), and for his funny Nevada memoir, *Roughing It* (1872), all of which had done well. Now married and living in Hartford, Connecticut, Clemens had been at dinner a few weeks before with his wife and two friends, the *Hartford Courant* editor Charles Dudley Warner and his wife. Kidding the women about the current novels they enjoyed reading, Clemens and Warner were challenged to write something better. Clemens had never written a novel, but Warner, who had reviewed many, said they could do it on the dare, so the two began to write alternating chapters. Over the next several months the novel evolved with Clemens concocting the main plot and Warner doing the subplot. The latter was a dull and conventional love story, but Clemens drew his inspiration from the daily newspapers and his natural inclination as a social critic—and so the novel, *The Gilded Age,* was born to give a name to the era.

While the revelations of the Wilson and Poland Committee dominated the newspapers, with names like Ames, McComb, Durant, Garfield, Colfax, and Wilson making daily headlines, and while further scandals within the Grant administration began to unfold, Clemens spun out the characters of Colonel Eschol Sellers, the rascally railroad and land speculator; Laura Hawkins, the coquettish Washington lobbyist; Senator Abner Dilworthy, the elected swindler; and many others whose antecedents were pulled out of unfolding events in early 1873. The central swindle of *The Gilded Age* was a real estate bubble in Missouri, connected to a railroad project, with assorted shenanigans in Washington. But there was a minor character named Fairoaks (Oakes

Ames) and a congressional bill called the National Internal Improvement Directors' Relief Measure, whose lobbying transactions came right out of the Crédit Mobilier. Early drafts of Clemens's manuscript contained even more pointed references to the real-life scandal and characters; Colfax, Wilson, and Patterson were easily identifiable.[31] Announced as early as April, *The Gilded Age* would be published in December 1873 and would sell forty thousand copies within two months despite its many flaws as a collaborative novel; though reviews would be mixed, *Appleton's Journal* was most accurate when it said that Twain did not allow his humor to blind him to existing evils in society, and that his views of the falsities of society were more penetrating than the "Olympian bolts of the thunderers in [the] daily press or [the] Sunday pulpit."[32]

As a novel, *The Gilded Age* was a mewling infant when Clemens interrupted his writing that February 5 to earn some cash at New York's Steinway Hall talking about the curious and amusing practices of the innocent, picturesque Hawaiians. "In the Sandwich Islands," he told the chuckling audience, "everything is done in an 'upsidedown' manner. Among other foolish things that they do is to elect the most incorruptible men to Congress."[33]

The longer the Crédit Mobilier hearings went on in Washington, the greater chance there was that fresh revelations about the Union Pacific, or any other railroad, or public works project, or government contract, would send the simmering pot to boiling over, scalding who knew how many public figures. The antiadministration papers were drawing lines between the elected officials named by McComb and diffidently defended by Oakes Ames and their positions of influence in transportation matters before the government. Only weeks remained in the session, after which so many would retire. Even the blameless Speaker Blaine, who still entertained hopes of someday running for president, had been revealed to be a speculator in a $32,000 Sioux City investment, also brokered by Oakes Ames and complete with suspicious dividends. Where did it all end? And there was the dramatic development involving Vice President Colfax. Investigators looking into Colfax's accounts had found a $1,200 depost in June 1868, which coincided by amount and timing with Ames's testimony; after tripping through a tortuous explanation, Colfax would hire a New York criminal attorney and aggressively let it be known that if he were permitted a face-to-face showdown with Ames in front of the Poland Committee, he would dispense with his former friend and unmask him as a liar of the worst sort. For his part, Oakes Ames had watched old friends—courtly colleagues of a thousand battles and deals over many years—turn their backs. He had grown tired of evasions, half-truths, obfuscations, and character assassination through leaks and rumors. It seemed clear that he was being set up as the wicked tempter who had taken advantage of his purer colleagues' credulity. Ames thereupon told the Poland Committee the hitherto unrevealed, explosive news that he had kept records of all transactions with his colleagues—in a little memorandum book. He promised to reveal it to the committee on February 11—also the day when Colfax would confront him.

An extraordinary number of spectators crowded into the Poland Committee hearing room early that morning. People filled every inch of standing space from the walls to the edge of the committee table, literally breathing down the necks of the examiners, and Colfax, and counsel. Thirty or forty newspapermen shoved inside. Judge Poland called in two Capitol policemen to clear breathing space between the table and the open window, but even this was filled when a number of women pushed and elbowed their way through the densely packed room to take advantage of the old judge's courtliness; Poland let them remain, and the policemen left in disgust. "As it was impossible for all to elbow their way to the table," the *Sun* correspondent noted, "the rear ranks mounted every piece of furniture in the room. At one time I noticed three women perched upon the top of a high desk, and at another time two or three correspondents were seated on the arms of chairs occupied by members of the committee."

Fifty minutes late, Oakes Ames finally pressed his way into the center of the room. He had left the book with his attorney, Horace F. Clark, at the Arlington Hotel, to force the committee to order its submission, which was swiftly done, and a messenger went off to fetch Clark. Some people in the room thought that Ames looked better physically than he had in weeks, but they were only comparing two extremely low points. "Ames had been bullied and badgered till his patience and good-nature were exhausted," one correspondent who knew Ames better recalled. "Colfax, Brooks, and Garfield persisted in raising the issue of veracity with him, and he was literally driven into a corner. It was a sight to see the rugged, stalwart old man as he took his seat at the right hand of Judge Poland. . . . Sorrow and determination were written in every line of his strong face. He looked broken. It was manifest that he had grown older within a few days. The struggle evidently had been an awful one. He realized fully that he was to some extent declaring his own infamy, but his character for truth and veracity had been wantonly assailed. Cost what it might he was determined to vindicate himself."[34]

"You could have heard a pin drop," said a *Times* man, "when Mr. Ames produced his red morocco covered memorandum book." And then the names, dates, and amounts tumbled out, with each new name causing a murmured exclamation in the crowd: Senator Patterson, Vice President-elect Wilson, Vice President Colfax, Congressmen Dawes, Scofield, Bingham, Garfield, Kelley, Logan, Allison, J. F. Wilson, and Eliot (or, rather, "Stetson, son-in-law of Eliot"). (Brooks, having gone through the New York officers to get his shares, had already been named by the New Yorkers.)

After Ames had read the entries into the record, Colfax's attorney cross-examined him mercilessly, trying to point out contradictions in his previous testimony about the vice president—in December, Ames had stated that he could not recall paying Colfax any dividends. "When I made that statement I had no memoranda here, and had not examined them to refresh my memory," Ames replied. "These men had all been running after me, one wanting me to make this statement and another that, and I really wanted to make it as favorable as I could. They wanted to be got out of a bad scrape which they thought would damage them, and I probably erred on their side to save them." In the fashion of criminal practice, the lawyer pressed him closer, browbeating him with personal taunts in hopes of breaking him down, but Ames remained stoic, and finally Judge Poland rebuked Colfax's counsel and halted the inquiry. He permitted the vice president to reply, and Colfax—whose earlier sickly smiles had, by this time, been replaced with a "death-like pallor," said one observer—began to defend himself. The rambling statement touched on a wealthy, conveniently deceased New York stationer and contributor, mysterious and improbable cash contributions arriving in the morning's mail, and even something about trying to buy a Steinway piano at a celebrity's discount; it provoked nothing but stony faces from the congressional investigators and a few snickers from the audience. It was, the *Sun* correspondent wrote, "decidedly too thin to stand before the well-corroborated testimony of Ames." And with that, Judge Poland said the committee was finished examining witnesses, and would retire to deliberate and make its report.[35]

A short distance from the Capitol at the attorney general's office, the clerk Walt Whitman had followed the breaking newspaper stories all during the fall campaign. Troubled by the slurs against the beloved Grant, disturbed by the charges and innuendos against solid-seeming citizens and public servants, Whitman was most strongly upset by the virtual indictment of the modern miracle in America, as he had celebrated it. Technology, industry, capital, individualism, and the vital and muscular spirit of the Republic, had, he thought, survived the awful test of war; the nation was rising to the beginning of a triumphal new stage in civilization: "Toil on heroes! toil well!" he wrote of the men wielding the new "weapons" replacing the late war's, the "human-divine inventions, the labor-saving implements," the engines, machines, the locomotives. There was now a betrayal at work—surely it could not be the system, but rather something dark and personal—and though he had struggled to will it away, the disillusionment which had been growing in him for several years could not be stilled. He had voiced it in his prose pamphlet, *Democratic Vistas,* published a little over a year before, in 1871. "The spectacle is appalling," he wrote. "The depravity of the business classes of our country is not less than it has been supposed, but infinitely greater. The official services of

America, national, state, and municipal, in all their branches, except the judiciary, are saturated in corruption, bribery, falsehood, mal-administration; and the judiciary is tainted. The great cities reek with respectable as much as non-respectable robbery and scoundrelism." In his poetry, too, were there new notes of desperation and disillusion, as in the 1871 revision of "Respondez!":

Stifled, O days! O lands! In every public and private corruption!
Smother'd in thievery, impotence, shamelessness, mountain-high;
Brazen effrontery, scheming, rolling like ocean's waves around and upon you, O my days! my lands!
For not even those thunderstorms, our fiercest lightnings of the war, have purified the atmosphere.

Recently transferred to a desk in the Treasury Department, that winter of 1872–73, Whitman watched the spreading stains of scandal touching the blue hem of Grant's trousers. It moved him to compose a new poem:

Nay, tell me not to-day the publish'd shame,
Read not to-day the journal's crowded page,
The merciless reports still branding forehead after forehead,
The guilty column following guilty column,
To-day to me the tale refusing,
Turning from it—from the white capitol turning.

He sent it to the *New York Daily Graphic,* where it would appear in March. There was much to trouble Whitman in that month of scandal and inquiry—the most recent edition of *Leaves of Grass* was tied up and difficult to find after a series of bankruptcies had beset his publishers and their successors; his mother had entered what was clearly her final illness; and Whitman himself awoke on the morning of January 23 to find his left side paralyzed from a stroke. In a few months, his mother dead and himself recovered only a little, the poet would leave Washington for a room in his brother's house in Camden, New Jersey, the drab railway hub and factory town across the Delaware from Philadelphia. Though his work would increasingly turn toward the subject of old age and death, there would still be room for a transcendent hope that his nation and its people would rise through the cloud layer of greed and corruption, into true and lasting greatness. Watching a Camden and Amboy railway locomotive chugging up from the riverside terminal one day in 1876, he would compose "To a Locomotive in Winter," celebrating the "Fierce-throated beauty," which was, he saw, the "Type of the modern—emblem of motion and power—pulse of the continent." In some ways the leviathan merged with his own personality, with its "lawless music," its "madly-whistled laughter," but it was also, of course, symbolic of the nation's spirit: "Law of thyself complete, thine own track firmly holding."

Thy trills of shrieks by rocks and hills return'd,
Launch'd o'er the prairies wide, across the lakes,
To the free skies unpent and glad and strong.[36]

The Poland Committee expended five days in deliberating and writing its conclusions. Oakes Ames spent much of the tense, interim period with James Blaine at the Speaker's house; he had been a frequent visitor in the winter months, the vaporous charges against Blaine notwithstanding, for the two had remained close even as Blaine directed the tactical defense of their colleagues. Mary Abigail Dodge, the Massachusetts antisuffrage essayist and kinswoman to Blaine, who wrote under the pseudonym Gail Hamilton, looked in often. "I have seldom seen a more pathetic sight than that of Oakes Ames," she would recall in her adulatory biography of Blaine,

a man of honored ancestry and stainless name, the modest hero of the great Pacific Railroad, the man whose energy had wrenched it from failure when to a less patriotic insight the nation itself seemed a failure, and had made its final link a guaranty of national peace and union, sitting silent, stunned into immobility before Mr. Blaine's library fire with his head bowed on his breast, while the younger man, alert and intent, applied himself indefatigably in and out of the house, arranging for his defense and for that of the other men who were implicated with him and who were equally guiltless of bribery.[37]

Guilt of one or another was certainly debatable, but that the defense interests of one versus the others could be uttered in the same breath is stunning. Later, Blaine would write to Ames's sons to commend the older man's "solidity and uprightness of character," his "sterling sense," "sound judgment," and "manly courage," his generous guidance when the young and eager Blaine first took his House seat.[38] But even as he worked as Ames's defender, the shovel-maker seemingly was being prepared for ritual sacrifice, that their House might be spared further opprobrium. On Tuesday, February 18, in the packed House of Representatives, with Sunset Cox temporarily inhabiting the Speaker's chair and Blaine down among his allies, Judge Poland handed the report to the clerk for reading. Ames was seated as if at the gallows, between the chaplains of the House and Senate, and it was noted that through the reading he smiled ironically, derisively, while most of his named colleagues seemed smug and sanctimonious. Only James Brooks seemed mortally shaken, as indeed he was, weeks away from his deathbed; he was "ghastly pale, his hands of bloodless hue," betraying "his mental and physical suffering."[39]

The report exonerated Blaine (as, in this case, it should have), and found that nearly all the others had been indiscreet but not guilty of accepting bribes. Kelley and Garfield, the latter almost exactly eight years away from his short residence in the White House, were slightly more difficult to handle, having permitted Ames to pay for their Crédit Mobilier stock himself and then graciously accepted their large dividends. But they were not only declared innocent of bribery but Garfield was deemed blameless in perjuring himself in his sworn testimony before the committee. Colfax, having risen from the House to the vice presidency, was unnamed in the report. Brooks, the lone Democrat, was a cut-and-dried case: as a government director he had been forbidden by law to own the stock, and, after assiduously badgering Ames he had gone around his colleague to his friends Durant and Dillon, and he had clearly tried to hide the connection by listing his son-in-law as owner; Brooks himself almost immediately betrayed himself, according to testimony, by tending the investment and dividends personally, as if they were his own: no number of sworn denials could hide that. For this, the committee recommended his expulsion.

Oakes Ames, though, had brought the House into "contempt and disgrace," and was guilty of offenses of "dangerous character" by selling stock to colleagues at insider's prices with the intent of influencing their votes and decisions. He was a briber—but, with the exception of James Brooks, the others were not bribed. At the conclusion of the report, Brooks rose to stammer out another protest of innocence, but Ames was silent, though sardonic, as he walked to the block.[40] That night he talked to a reporter—that the man was from the *World,* so despised by the Republican establishment, was an indication that Ames had already left his own world though his body remained. "It's like the man in Massachusetts who committed adultery," he said, "and the jury brought in a verdict that he was guilty as the devil, but that the woman was as innocent as an angel. These fellows are like that woman."[41]

"Whitewash!" bellowed the *Sun,* the *Tribune,* the *World,* and dozens of other antiadministration organs. By recommending the expulsion of Ames and Brooks and then "whitewashing their accomplices," the Poland five "have committed a grave offence against official purity, orderly government, public morals, and common decency," the *Sun* charged, "which every member of the committee will live long enough to bitterly regret. All right-minded persons, of whatever political party, will indignantly resent the partial, unwise, unjust, and illogical conclusion to which the committee have arrived." Even the pro-Grant papers seemed surprised, almost squeamish at the results, though their mild disclaimers were drowned out in the roar of the reformists. "Hardly worth discussing," said E. L. Godkin in *The Nation.* Unfair, partisan, mediocre melodrama, Whitelaw Reid said in the *Tribune:* "The Republicans were given minor roles and allowed to slip off the stage unnoticed, but when Brooks came on, the sheet thunder was sounded, the calcium lights were burned, and he was shown up in the most gorgeous colors as the chief villain of the plot." That is, after the Prince of Darkness, Oakes Ames. As the *Tribune* and other commentators noted, the vice president, the vice president–elect, the chairmen of the Ways and Means Committee, the Appropriation Committee, the Judiciary Committee, the Naval Committee, the Banking and Currency Committee, and several stalwart party leaders, had slipped out of the noose. No one bothered to mention the important party members who had since escaped across the jurisdictional borderline into private life. Out in the Capitol corridors, at least, Representative Fernando Wood, Democrat from New York, began talking about offering an impeachment resolution for the Judiciary Committee against Vice President Colfax, especially with the fresh revelation of other mysterious deposits into his personal bank account. "Of course," commented the *New York Tribune* correspondent, "no such resolution will be allowed to come to a vote."[42]

At the suggestion of Spoons Butler of Massachusetts, deliberation on the Poland report and the fate of Ames and Brooks was set for Tuesday, February 25, but in the interim the House was presented with the Wilson Report, on February 20. As the Union Pacific historian Maury Klein has written, the report was potentially more important than the narrowly limited bribery inquiry of the Poland Committee, but it made little impression on the public, and despite its great length and complexity the report was flawed and incomplete. "No one could have sorted out in a short time the convolutions within the management during the construction years," wrote Klein, "especially when bitterness and hatred still galled some of the participants," with many answers incomplete and misleading.

For their part, the five Wilson Committee members had little interest in acknowledging the already-old conflicts of power and personality or the extraordinary difficulties faced by the builders; even recent history was too much for the congressmen to learn or learn from. The urgencies and stringencies of war were somehow entirely forgotten. The huge costs of mismanagement and experimentation were overlooked in favor of the easily seen villainy of a relative few. The committee made an assumption that the government had provided more than enough support to build the railway—which it had not. The congressmen got closer to the truth when they fastened on the use of the Crédit Mobilier construction company. It had been employed, after all, to enrich the directors at the expense of the Union Pacific and the public; in that sense the treasury had been cheated of millions in unpaid loans; in that sense the directors had bequeathed to the nation a crippled and vulnerable railway. But the Wilson Committee, by highlighting the stock dividends awarded to the directors by themselves, failed to venture into all of the directors' enriching sidelines—their loans to the Union Pacific at high interest, their sales and commissions of supplies, their land sales, their hidden interests in subcontracts, their inside tracks into mining, lumbering, and water diversion projects. Not only was it impossible for the committeemen to absorb it all in five weeks—not even after 130 years is it possible to estimate what treasures were conveyed by the Empire Express. When the Wilson Committee submitted its report, and its recommendation that the government sue to recover the Crédit Mobilier's illegal profits, it had glimpsed only a glimmer of reality.[43]

With the Poland Committee resolution on Ames and Brooks scheduled for action in a few days, it was said around Washington that Speaker Blaine had little appetite for extreme punishment. This provoked more outraged or derisive accusations of a whitewash in the critical press, which had other occasions that week to criticize elected officials' lack of curiosity or enthusiasm. On February 20, there was a House vote on the Democratic resolution to send the question of impeaching Vice President Colfax to the Judiciary Committee. The resolution was voted down by the Republican majority; the vote was so close, however, with nine Republicans defecting, that a counterresolution was immediately offered by the majority to save them of being accused of defending Colfax. It was identical to the earlier one except that Schuyler Colfax's name and office were replaced by the phrase, "any officer of the United States not a member of this House." It passed without opposition, but with eleven days remaining in the vice president's term there was limited eagerness for proceeding; on February 24 the Judiciary Committee, chaired by the Crédit Mobilier congressman Bingham, reported there were no grounds for impeaching either Colfax or Brooks. Charges published in the *New York World* by a former diplomat in the Johnson administration that the late Thaddeus Stevens had received $80,000 for services during the passage of the Pacific railroad bills in 1862 and 1864 attracted but a little attention in the Capitol; cynics predicted that issue was as dead as the onetime "scourge of the South."

Over in the Senate, a special committee chaired by Republican Lot Morrill had heard testimony that Secretary James Harlan had been given $10,000 by Dr. Durant to help finance his successful senatorial campaign. The committee seemed to be interested in pursuing the issue in the short tenure left, but it was divided on tactics; there was the clearer Crédit Mobilier evidence collected against Senator James Patterson, but the old man was retiring on March 4; members would decide to let the matters lie without action. There were other likely candidates for probes, commented the *Sun*, such as Senator John Sherman, the general's brother—fifteen or sixteen years before, commencing his first term, "supposed to be worth $20,000 at the utmost" but "today estimated to be a millionaire four times over." Indeed, the paper continued, "It is easy to count twenty or twenty-five Senators who came here poor, and who are now revelling in wealth. Everybody knows it. The surprise is that honest men countenance such unblushing rogues, and have not demanded investigation. Tolerance has encouraged crime, until seats in the Senate are sold at legislative auctions." Apparently, the status quo simply paid too well.[44]

Not even during the impeachment hearings against President Andrew Johnson was the Capitol as crowded as it was on Tuesday, February 25, when congressmen found it nearly impossible to shove their way through the corridors into their chamber. When the doors to the spectators' gallery were opened, an unruly crowd of elegantly dressed Washingtonians, mostly women, plunged down the aisles and scrambled over the backs of seats "utterly regardless of good breeding or decency," sniffed a correspondent. So many were denied entrance that dispensation was given for "wives and ladies of members' families" to come onto the floor, but a motley throng pushed past attendants—"bummers and lobbyists were allowed to elbow their way in," an observer complained, so that members were crowded from their seats, forced to stand in the aisles: "not unfrequently a bummer of the lowest order was to be seen occupying a seat well advanced toward the chair, while the member to whom it belonged was forced to stand or perch himself on a neighbor's desk." And so, in what seemed like a beer-garden atmosphere, the Poland report was considered by the assembly.[45]

Acting as prosecutor, Judge Poland spoke first and presented the case against Ames and Brooks over the course of an hour and a half. Then Oakes Ames's long defense statement was read—from the invocation of Lincoln's urging him to "take hold" of the railway project through a history of his involvement, admitting no consciousness of guilt but pointedly asking if he were a briber where were the indictments for the bribed? As it was read Ames slumped in his seat, for the first time openly emotional—to the great discomfort of his fellows—weeping at several points, and at one time lowering his head to his desk and crying like a child. Then it was time for the debate; the arguments for and against the defendants went until late in the evening, across the entire day of February 26 until nearly midnight, and into the twenty-seventh, with no flagging of interest in the galleries.

Accounts in the press crowded other news to the margins, and even the usually pro-administration papers were disturbed by the spectacle of selectivity. "To refuse to censure the holders of that stock is to say that the Congressional standard of morals is not high enough to condemn it," said the *New York Times* editor. "There is a quiet feeling in the public mind that the Crédit Mobilier is a bad business, and that a House which will not treat it as such is certainly cowardly, and possibly corrupt." The gadfly press had no doubts of the state of affairs, however: all of Congress was on trial, not just the two uninfluential scapegoats—one chosen from each party by a hoodwinking committee in, as the *Sun* scoffed, "an apparent display of impartiality."[46]

On the last day, Thursday, February 28, an even greater feeling of urgency and pressure seemed to fill the chamber almost to bursting: managers limited speeches to ten minutes while an extraordinary amount of lobbying went on in the aisles—Blaine had temporarily vacated his Speaker's chair, and seemed to be everywhere, huddling with leaders all around the chamber; California congressman Aaron Sargent was also extremely mobile. Then, as Judge Poland summed up his case, Sargent rose to introduce a substitute resolution—Ames and Brooks to be censured, not expelled. When the Sargent substitution came up for a vote, it seemed at first as if it had been defeated, but then magically, as correspondents watched "one of the largest lobbies ever seen on the floor," the tide reversed itself, and the votes went for the substitution, 115 to 110. Then, quickly, like a locomotive gaining speed, came the two separate votes on censure; they were passed, by a margin of about six to one.

But the day was far from over; it was only early in the afternoon: Democrats sat up as if recovering from a deep sleep, realizing that political gains might be made now that the lesser penalty of censure was in the air, and what followed was a veritable fusillade of efforts against the ruling party; a call for censuring other Crédit Mobilier congressmen—the powerful Garfield, Kelley, Bingham, Dawes, and Scofield—was thwarted by Speaker Blaine only through the parliamentary tactic that a resolution could not include more than one name. Democrats then rose to begin naming them one by one, causing an uproar on the other side of the aisle as Representative William "Pig Iron" Kelley of Pennsylvania was the first to be cited. Within minutes the floor of the House had become as partisan and warlike as any could remember, with fusillades of parliamentary maneuvers and countermaneuvers flying over the heads of members until the wearied congressmen lost track of what they were voting on and who they were attacking or defending in speeches. But the move to censure Kelley was tabled, and Blaine parried all subsequent attacks. When Aaron Sargent rose again to resolve that the Poland Committee be discharged and the House move on to other matters, the exhausted and battered representatives passed it.

With but four days remaining in the session, censured members Oakes Ames and James Brooks went home. The day before the Forty-second Congress expired on March 4, members passed a resolution to raise members' salaries by 50 percent (from $5,000 to $7,500) and to award each a bonus for the session just ending, of $5,000; the president's salary was doubled to $50,000, and other government officials were raised.[47] Brooks, all too evidently in his final decline, died within weeks. Oakes Ames, back in Massachusetts to take up again his many business interests, surrounded by well-wishers in his old district, remained defiant in his defense that he had done nothing criminal or unethical, that the "notorious" Poland Committee and his craven former colleagues had wasted six months of the nation's time. "I have got home among my friends," he wrote a friend in Boston. "They know me, and I mean to stay." The wealthy merchants and businessmen of Boston were readying plans for a sumptuous banquet to welcome him back to productive private life when, on May 5, Ames suffered a stroke, soon complicated by pneumonia, and died three days later. He was sixty-nine.[48] Later, at Sherman Summit in Wyoming Territory, at the highest elevation of a railroad, the Union Pacific erected a large, pyramidal stone monument dedicated to the memory of Oakes Ames and Oliver Ames: marking, it was expected, the two men's extraordinary efforts on behalf of the Empire Express, and crushing, it was hoped, though unsuccessfully, the memory of the scandals which had briefly stopped the nation in its tracks and came to symbolize the dark underlay of the Gilded Age.

"Why is it," wrote Collis Potter Huntington to Mark Hopkins in February 1873, "that we have so many enemies out there?" He was referring to the consistently hostile press on the Pacific Slope, for the Associates had continued to expand across the western coast, engulfing all the weaker lines, and had begun to move westward from the town of Los Angeles toward an eventual meeting point with the Texas and Pacific, showing an insatiable appetite for land grants and subsidy bonds all over the slope. The control of virtually all of the western freight business had hardly endeared the Central and Southern Pacific, and their many subsidiaries, to the hearts and minds of those who had to pay the rates, nor had the railroads' power in the statehouse pleased their adversaries.

But Huntington was also referring to the inquisitors in Washington: the Poland Committee, with its fixation on keeping the inquiry as near to Oakes Ames as possible, had nothing on Huntington, but the Wilson Committee had its broader mandate for poking into possible frauds and influence-buying; nearly all of its energy in the first two months of the year was expended on the closer target of the Union Pacific, and that would be the focus of its first report, dated February 20, but Wilson did summon Huntington to a close questioning. The congressman received aid and encouragement from two colleagues in the other end of the Capitol building—California senators Cornelius Cole and Eugene Casserly. Both were enemies of the Central Pacific—but Cole was especially bitter. Having been credited with defeating Huntington in the latter's monumental efforts to obtain craggy Goat Island for a port terminal in 1870, Cole brought a hailstorm of political trouble upon his own head. Huntington backed Congressman Aaron Sargent's senatorial bid, and together they virtually bought a majority of California legislators, and by this maneuver Sargent thereupon trounced Cole in the 1872 election. As a lame-duck congressman, Sargent had, of course, helped to solve the Oakes Ames problem in February 1873; as a lame-duck senator, finding an eager ally in

Casserly, Cole did what he could in his last month of service to wound Huntington and his railroads. "These hellhounds," Huntington groused to Hopkins.[49]

The Poland Committee expired, but Representative Wilson's would live on into the Forty-third Congress, and the spectacle of the Crédit Mobilier scandals haunted Huntington and the Associates. It took no leap of the imagination to see that the Central Pacific would be next.

Huntington had still not learned to relax—indeed, would he ever?—and this was no time for letting up, with the Central Pacific still in relatively weak financial shape despite its many gains and successes, particularly in a recent, caustic political fight in Los Angeles with the Texas and Pacific's Tom Scott, who had designs on the West Coast. In contrast to the relentless Huntington there was Charley Crocker, who, after his frantic years at the head of Central Pacific construction, bossing men like animals through brawn and fearless willpower, was unsuited and indifferent to a sedentary management job in Sacramento. With the specter of his paralyzed brother, Ed, who had sacrificed his mind to the corporation, Charley had announced in 1870 that the brothers wanted to be bought out. Huntington, Hopkins, and Stanford could either do so, or the Crockers would look beyond the Associates.

Against that horror of admitting outsiders into their circle, the partners had no choice but to agree. After nearly seven months of bargaining and bickering, Huntington, Hopkins, and Stanford contracted to pay each Crocker $900,000 for his one-fifth interest—assets included the Contract & Finance Company, Central Pacific land-grant bonds, railroad bonds of the California and Oregon, San Joaquin, and other lines, and a number of other franchises and charters. They were each to be paid in $300,000 installments over three years—but the brothers did not part with their portion of Central Pacific stock, itself a large block of potential wealth. Soon thereafter Charley had taken his family on a long tour of Europe, leaving behind him a glow of resentment from his former partners for deserting them at a difficult time.

Later in the scandal-tinged year of 1873—the year in which the Central Pacific would move its general offices from Sacramento to San Francisco—Charley Crocker would return to the fold. In the midst of the great financial panic rocking New York and about to devastate the nation's economy, he came back from Europe to New York, demanding his second $300,000 installment and being told by Huntington that there was nothing to give him. Crocker helpfully returned his first installment and added many other assets and rejoined the company as director and second vice president.[50]

He would happily serve in a largely ceremonial position, which involved a good deal of world travel, building a huge, ostentatious mansion on San Francisco's Nob Hill and filling it with as many nearly priceless European works of art as were left by other acquisitive moguls of the era. His many personal business interests expanded in inverse proportion to the the yardage of his waistcoat—at five feet, ten inches, though he still weighed over 235 pounds, he was down from a high of 274 during the building of the Central Pacific—and Crocker came to own large blocks of real estate in San Francisco, the San Joaquin Valley, and Nevada, ranching, banking and insurance concerns, and farming irrigation projects, setting up several financial dynasties for his children. His final enthusiasm was a sprawling resort hotel beside Monterey Bay, the Del Monte, which is where he would die, weeks shy of his sixty-sixth birthday in 1888, of a diabetic coma, the end result of decades of the excessive culinary habits of the Gilded Age. His silent, immobile older brother, Edwin Bryant Crocker, had preceded him by two months and ten years.[51]

Nearly that same period separated Charley Crocker's death from that of the least extravagant Associate, Mark Hopkins, who died in 1878 at the age of sixty-five, after long service as treasurer to the Central Pacific and its associated interests. The abstemious vegetarian had lived like one of his lowest clerks, even after his status as multimillionaire was a fact; after the corporate move to San Francisco, Hopkins and his wife had lived in a rented cottage. At his wife's insistence, and with the example of his partners staring him in the face, Hopkins had agreed to buy half a block on the weedy, windy Nob Hill, next to Stanford's. The treasurer turned his personal treasury over to his wife, telling her to handle the details of design and building; released from her many years of quiet privation, Mary Hopkins indulged herself in a fabulous, turreted castle which would take so long to construct that Mark Hopkins would die before it was finished. His end came in a director's car, in his sleep, on a railroad siding near Yuma, Arizona, where Hopkins had traveled via the newly opened Sunset Route of the Southern Pacific, hoping that the desert heat would give him relief from a crippling spell of sciatica and rheumatism. Mary Hopkins would inter him in a palatial mausoleum in Sacramento, constructed of highly polished blocks of red and black granite—the former extracted from Sherman Summit in the Union Pacific's quarry in the Rocky Mountains, and the latter from a Central Pacific cut in the Sierra Nevada.[52]

The Stanford mansion in Sacramento, where Leland and Jane Stanford hovered anxiously over their precocious, five-year-old son, Leland Jr., had ceased to give the couple a reassurance of sufficient grandeur in 1873, and when the corporate move was effected, the Stanfords moved to San Francisco also. They leased a luxurious house while the Governor squinted about town, finally focusing on Nob Hill for their edifice. The always-simmering resentment between Stanford and Charley Crocker and Huntington rose that year of 1873, which had begun in scandal and would end in panic, when Stanford prematurely withdrew a significant portion of his interest in the Contract & Finance Company to underwrite his real estate investments in San Francisco. After the Stanford mansion was completed, and it was surpassed by their neighbors the Hopkinses and the Crockers, and then the nouveaux riches from the Nevada mining camps, the Govenor lost his enthusiasm for San Francisco, focusing on a nine-thousand-acre farm in Palo Alto (and yet another mansion), on breeding race horses to run on his two private racetracks, on developing what was called "the largest vineyard in the world" in the northern Sacramento Valley, and, most important, on training his young son to be a cultured and educated leader of the future. Virtually none of the Governor's time was devoted to management; Huntington, nominally vice president, saw to it all. The Stanfords spent two years in Europe, affording Leland Jr. every opportunity for enculturation, and there were plans to enroll him in Harvard. The doting parents leased a house on New York's Fifth Avenue so as to be nearer to him when he matriculated. But the boy's persistent sickliness sent the family back to Europe again, on doctors' orders. He was but fifteen when he died, in Florence, in early 1884, of typhoid.[53]

Hopelessly bereft, the parents decided to endow a center of learning in California in their son's memory; originally the governor fastened on the idea of a technical school but over the next several years the notion blossomed into a full-fledged private university, the Leland Stanford Junior University, on the rolling, oak-studded Palo Alto acreage. Even in mourning the Stanfords were extravagant. What impatience Collis Huntington may have felt as Stanford continued to dip into the shared partner pool (Collis cynically called the university "the circus") was replaced in 1885 by outright fury when Stanford contested—and won—the senatorial seat of Huntington's friend Aaron Sargent. Huntington seethed about that for several years until he could make his move for a fitting revenge against "the Senator," as the long-ago one-term governor now preeningly called himself, when Stanford was ready to seek another term.

Huntington's battleground this time was one in which he could exert more political authority—the presidency of the Southern Pacific Railroad, the Central Pacific's holding-company successor. If the coup in 1890 was bloodless it was not without numerous bruises, but Huntington at last took away from Stanford the titular leadership of the company he had coveted since the Central Pacific's inception. There would never be peace between them, and a number of Huntington's dire predictions about his partner's lush life would come true. Although Senator Stanford began to enjoy the fruits of his second term, even as the university he created opened with the promise of continuing, the years of lavish spending caught up with him shortly before his sudden death, in 1893 at age sixty-nine. His estate in tatters, only Jane Stanford's sharp retrenchment saved both it and the Palo Alto memorial they had created together. There was enough left for the widow to live and travel like a duchess, and for "the circus" to survive, and finally prosper.[54]

Collis Huntington was left as the lone survivor of the partnership which had begun in a hotel meeting room and taken shape above a hardware store—and eventually draped itself across a good portion of the continent. Like a proud old lion, Huntington relished all the scars and satisfactions of a lifetime of battles, savoring all those victories.

His solution to the panic-stricken, scandal-tinged difficulties exactly two decades before, in the spring and summer of 1873, had been characteristically quick, simple, and to the point. With the Wilson Committee still breathing its menace in Washington, with some financial difficulties in the Central Pacific, Huntington was confronted with more problems than he needed at that time. He had hoped to have Central Pacific stock listed on the New York Stock Exchange that year as a way to raise money through sales and collateral loans, but listing required disclosing the number of shareholders, the amount of issued stock and various debts, and the cost of the Central Pacific road.

There was simply too much to hide. Within six months of the golden-spike ceremony, Central Pacific stockholders had begun to complain of no dividends and no annual reports. Huntington had tried to put them off, but several groups had banded together and sued for a full accounting. The financier brothers Lambard had persuaded Theodore Judah's widow, Anna, who owned a nominal twenty-five shares, to join them in one action for the bad publicity, and another stockholder, the busted railroad promoter Sam Brannan, joined what quickly became a siege—with the principal weapon being daylight. Their lawyers concocted a wildly excessive estimate of Central Pacific costs and profits, which was enough to wear down the Associates into a settlement lest they be forced to disclose the true figures. Granted, the settlement was greatly reduced from what the plaintiffs sought. But putting out subsequent brushfires of small stockholder suits became almost a full-time job, and the original court complaints—and cost and profit estimates—of the Lambards and Sam Brannan were to smolder along undetected until 1873, the year of the Crédit Mobilier publicity. Just as the McComb suit against the Crédit Mobilier and Oakes Ames had ignited the firestorm of scandal when a leak had led right to the *New York Sun,* the Associates feared that their old lawsuit would disclose details about their construction company. New legal actions might affect their control of the corporation. Huntington had cause to worry, also, that this would lead to inquiries about his role in legislation stretching back for years.

Needing to have some kind of convincing accounting for the stock exchange, Huntington arrived at a highly inflated figure for all of the Central Pacific's construction, one which hid between $100 million and $125 million over estimates named in the old lawsuit and, most recently, by accountants working for the Wilson Committee. Then, having cooked the books, it was time to incinerate the originals. Fifteen volumes of the Contract & Finance Company were burned in the basement furnace of the Central Pacific office in Sacramento.[55]

As Huntington's biographer, David Lavender, has commented, "They did it just in time." On Saturday, July 19, 1873, the *New York Sun* led with a shattering new front-page story—once again based on leaked, yellowing court documents. "THE ACME OF FRAUD," the headline trumpeted. "$211,299,328.17—GOBBLED!" As with McComb and Ames nearly a year before, the account with its laborious court depositions sprawled over columns, and continued over subsequent days in the editorials. Congressman Wilson of course reacted, and Huntington was summoned to the Capitol on July 28 for more grilling. The committee demanded to see the Contract & Finance Company records. But Huntington told them apologetically that alas, they were no more. They were dissolving the corporation—and Treasurer Hopkins had burned the books, to save space in their new offices.[56]

Over the course of his long and eventful life, the onetime pushcart peddler from Oneonta did rather well, building railroads all over the ancient territories of Oregon, the Southwest, and Texas, even as far as New Orleans—and, radiating beyond, up the old trading rivers of the Mississippi and the Ohio and continuing a big loop back—the Chesapeake and Ohio—toward the bay tidewater of the Atlantic, where even shipbuilding became a going concern, and Washington. And Washington—where generations of congressmen and senators came to depend on the inexhaustible Huntington largesse.

Aging, like a barnacle-encrusted old tortoise, Huntington vowed he would see one hundred. Now easily one of the wealthiest men in the country, he lived nearly as frugally as his long-dead partner Hopkins. After his long-suffering wife, Elizabeth, died in 1883, he married a woman friend who may have been his mistress, Arabella Yarrington Worsham, about thirty years his junior. As a grand gesture to Arabella, Collis adopted her grown son, Archer, as his own. He also brought up into the ranks of the Southern Pacific his nephew, Henry E. Huntington. Then, at his new wife's direction, Collis Huntington began finally to live like a grand duke. There was a succession of grand and grander houses and estates in New York and elsewhere. Fittingly, Clara Huntington, the daughter Collis and Elizabeth had adopted back in California and raised in New York as their own, married a German prince, albeit a wastrel seeking a Huntington dowry. Now living conspicuously, Collis even dispensed some of his money to charities, notably to African American educational institutes in the South, hearkening all the way back to his idealistic young abolitionist days.

One of his last enthusiasms was at the growing vacation colony for the extremely wealthy in New York's Adirondack Mountains—Raquette Lake, developed by the son of his old archenemy, Dr. Durant. Huntington bought a rustic-facaded "camp" and land (as had Morgan and Alfred Vanderbilt) from William West Durant, not satisfied until he had spent a quarter million on buying and refurbishing it. He even joined young Durant in a short-line railroad project from the lake down to the Delaware and Hudson (in the opposite direction, ironically, from the Doctor's original railhead at North Creek), to make the lake resort more convenient for moguls and their palace cars. There, at Raquette Lake in the summer of 1900, after an exceptionally companionable day, Huntington retired and died, of a heart attack, aged 78.[57] His generous estate was divided into thirds, two shares going to Arabella and her son, Archer, and one to his nephew, Henry. Poetically, some of the money flowed back together when Henry E. Huntington, who had moved back to California to found a ranch at San Marino, married his uncle's widow. By then the younger Huntington had built, with family money in San Marino, what would become one of the country's greatest private libraries and art collections.

Perhaps, as was the case with so many other moguls of the Gilded Age—those once-ragtag peddlers and shopkeepers and apprentices with names like Vanderbilt, Morgan, Carnegie, Frick, and Rockefeller, who came to be the heedless royalty of the developing republic, crushing enemies, exploiting the powerless, building empires—the solid new walls of culture at San Marino began to polish off the taint from the millions hauled in by the Empire Express, just as would the university in Palo Alto and the art gallery in Sacramento, and more modest library counterparts in places like North Easton, Massachusetts, Des Moines, and elsewhere. The mansions, or at least many of them across the nation, would burn or be pulled down, the yachts would founder and sink, most of the palace cars would be scrapped, the edifices and monuments would erode—even, in generations, many of the rails would, sadly, be pried up. The libraries, lecture halls, and galleries would remain, perhaps even lowering the temperature being applied to the souls of the departed empire-builders. Indelibly, though, they left their footprints and the paths and networks they set out, creating—like them or not, like it or not, the American nation we know today.

Notes

Part I—1845–57: A Procession of Dreamers

Chapter 1—"For All the Human Family"

1. Asa Whitney, *Diary* (1842–44), as transcribed by Margaret L. Brown [hereafter *Diary*], 4. This chapter would not have been possible without the prior scholarship of Dr. Margaret L. Brown, who was rightly chosen to write the entry on Asa Whitney in the *Dictionary of American Biography* following her seminal work on him. Primarily, see her *Asa Whitney: Projector of the Pacific Railroad*, Ph.D. Thesis, University of Michigan, 1930; also, "Asa Whitney and His Pacific Railroad Publicity Campaign," *Mississippi Valley Historical Review* XX, 2 (September 1933), and her introduction to Whitney's *Diary*, now held with miscellaneous Whitney papers at the University of Michigan. Note that Brown's scholarship refutes many misconceptions promulgated by Edwin Sabin (see below) and Nelson H. Loomis in the latter's unavailing "Asa Whitney: Father of Pacific Railroads," *Proceedings, Mississippi Valley Historical Association*, VI (1912–13), 166–75. Brown's otherwise trustworthy thesis errs when discussing Whitney's China trip; postdissertation discovery of his diary allowed her to later correct the mistakes.
2. *Diary*, 7–8; 12, 16.
3. Ibid., 15.
4. M. F. Maury, *The Physical Geography of the Sea;* Carl C. Cutler, *Greyhounds of the Sea*, 76–86, 106–17.
5. *Diary*, 27.
6. Ibid., 27–32.
7. Ibid., 13.
8. Ibid., 4.
9. Ibid., 62.
10. Ibid., 64.
11. Ibid., 71–72.
12. Whitney, Senate Memorial (1846), 29 Cong., 1 Sess., S. Doc. 161, p. 2.
13. Ibid., 2.
14. Whitney, *A Project for a Railroad to the Pacific* (1849), iii, 14–16 [hereafter *A Project*].
15. Ralph C. Morris, "The Notion of a Great American Desert East of the Rockies," *Mississippi Valley Historical Review* XIII (1926), 190–200.
16. Ibid.
17. Whitney, *Railroad from Lake Michigan to the Pacific: Memorial of Asa Whitney, of New York City, relative to The construction of a railroad from lake Michigan to the Pacific ocean* (1845), 28 Cong., 2 Sess., H. Doc. 72, pp. 1–4 [hereafter referred to as *Memorial* (1845)].

Chapter 2—"Who Can Oppose Such a Work?"

1. Albright, *Official Explorations for Pacific Railroads, 1853–1855*, 25; Margaret L. Brown, *Asa Whitney: Projector of the Pacific Railroad* (1930), 69–73 [hereafter cited as *Thesis*]; Sabin, *Building the Pacific Railway*, 14–22, 23.
2. William C. Redfield, *Sketch of the Geographical Rout of a Great Railway . . .* 2d ed. (New York: G & C & H Carvill, 1830), 2423.
3. *American Railroad Journal* I (February 4, 1832), 83.
4. Bailey, *The First Transcontinental Railroad*, 11–12; Brown, *Thesis*, 70–72; Carter, *When Railroads Were New*, 227.
5. Also see Riegel, *The Story of the Western Railroads*, 2–3, for interesting stories.
6. Jack T. Johnson, "Plumbe's Railroad to the Moon," *The Palimpsest* (1938), 92.
7. John Plumbe, *Sketches of Iowa and Wisconsin* (reprint, Iowa City: Athens Press, 1948).
8. George Wilkes, *The History of Oregon, Geographical and Political* (New York: William H. Colyer, 1845), 48–61, 128.
9. Ibid., 4.
10. *Cong. Globe.*
11. Undated *New York Tribune* article, cited in Brown, *Thesis*, 21.
12. *New York Commercial Advertiser*, October 6, 1845.
13. *Daily National Intelligencer*, August 15, 1845.
14. *Cleveland Leader*, June 1868.
15. *New York Commercial Advertiser*, op. cit.
16. Ibid.
17. *United States Magazine and Democratic Review*, August 1845 (NYPL).
18. *New York Commercial Advertiser*, op. cit.
19. *New York Evening Post*, October 6, 1845.
20. *New York Tribune*, dispatch dated November 10, 1845, cited in Brown, *Thesis*, 29.
21. Brown, *Thesis*, 38.
22. Asa Whitney, "A Reply to Stephen A. Douglas" (1845), 2.

Chapter 3—"I Must Walk Toward Oregon"

1. 29 Cong., 1 Sess.
2. I have borrowed this Thoreau quotation from the epigraph of DeVoto, *The Year of Decision, 1846*; it is from *Walden*.
3. Whitney, *Memorial* (1846), 8. The tribal elders he had met who indicated such willingness were outnumbered by young warriors willing to fight for their land. Whitney naively assumed one side would win the other over.
4. Ibid., 10.
5. *Cong. Globe*, 29 Cong., 1 Sess., 1171.
6. Whitney's recollection, in *New York News*, June 28, 1868.
7. Ibid.
8. Henderson to Breese, September 19, 1846, reprinted in *Washington Daily Union*, November 5, 1846.
9. *Philadelphia Public Ledger*, September 8, 1846.
10. *Philadelphia Public Ledger*, December 22, 1846.
11. "Address of Mr. Asa Whitney Before the Legislature of Pennsylvania" (Harrisburg: McKinley & Lescure, 1848).
12. *Nashville Triweekly Union*, December 3, 1846; *American Railroad Journal*, December 19, 1846.
13. *Cincinnati Daily Commercial*, November 4, 1846.
14. *New York Herald*, January 5, 1847; see also *New York Tribune*, September 29, 1849.
15. Carver, "A Proposal for a Charter to Build a Railroad from Lake Michigan to the Pacific Ocean" (Washington: J. & G. S. Gideon, 1847). Interestingly, a letter (1868) from the then-eighty-year-old Carver, to Union Pacific contractor J. S. Casement, continues his claim of originator. "I am the first man who ever thought of the thing and have got the printed documents to prove the facts."

Carver to Casement, Pittsford, NY, July 31, 1868, Casement Papers, American Heritage Center, University of Wyoming Library.

16. Wilkes, "An Inquiry Into the Practicability and Benefits of a National Railroad to the Pacific," Senate Memorial (1846).
17. Whitman, *The Gathering of the Forces*, eds. Cleveland Rodgers and John Black (New York: Putnam, 1920).
18. Plumbe, "Memorial Against Mr. Asa Whitney's Railroad Scheme" (Washington, 1851), 16–17; William J. Peters, "Historical Introduction," in Plumbe, *Sketches of Iowa and Wisconsin.*
19. Albright, 11–12.
20. See his letter in *DeBow's Review* IV (October 1847), 165–94; see also the interesting account, *Memphis Daily Enquirer,* November 15, 1849.
21. Brown, "Publicity Campaign," op. cit., 216; Albright, 16; H. Rept. 140, 31 Cong., 1 Sess., 6.
22. DeVoto, *The Year of Decision, 1846*, 496.

Chapter 4—"The Great Object for Which We Were Created"

1. *New York Tribune*, May 3, 1847.
2. Whitney, *Memorial* (1847).
3. *Cong. Globe*, 30 Cong., 1 Sess., 1011.
4. Whitney, "Reply to Senator Benton," *Hunt's Merchants' Magazine* XIX (November 1848), 527–31.
5. Whitney, *A Project*, 42.
6. *Cong. Globe*, December 5, 1848.
7. Sabin, 20.
8. *Cong. Globe*, 30 Cong., 2 Sess., 25, 388, 472; Benton, *Highway to the Pacific*, 1850.
9. Albright 10–13; *Cong. Globe*, 30 Cong. 2 Sess., 550–51.
10. Whitney, *A Project*, iv.
11. Niles to Whitney, May 13, 1850, reprinted in *New York Evening Post*, May 30, 1850.
12. H. Rept. 140, 31 Cong., 1 Sess., 47.
13. *New York Tribune*, March 18, 1850.
14. H.R. Rept. 140, 31 Cong., 1 Sess., 6
15. *New York Tribune*, January 21, 1851.
16. *New York Tribune*, June 2, 1851.
17. *Cong. Globe*, 37 Cong., 2 Sess., 1727.
18. *New York Sunday News*, June 28, 1868.
19. Noted in *New York Evening Post*, October 6, 1852.
20. Asa Whitney to Lucy Whitney Wright, November 6, 1865, cited in Brown, *Thesis*, 143.
21. *New York Tribune*, September 18, 1872.
22. Richard Smith Elliot, *Notes Taken in Sixty Years* (St. Louis, 1883), 213. Sabin, and others, promulgated a myth that Whitney passed his declining years on a Washington milk route, an unhelpful error. In truth, he lived very comfortably as a gentleman farmer in Maryland; one of his hobbies there included looking after his farm's dairy business.

Chapter 5—"An Uninhabited and Dreary Waste"

1. *New York Evening News*, June 28, 1868.
2. For a fuller discussion of this subject, see Allan Nevins, *Frémont*, 377–78; Russel, 168, 184.
3. Allan Nevins, *Ordeal of the Union* II, 86–87, 201.
4. Goetzmann, *Army Exploration in the American West, 1803–1863*, 262–67; Russel, 100–107.
5. Goetzmann, *Exploration and Empire*, 266–70; Nevins, *Frémont*, 408–20.
6. Goetzmann, *Exploration and Empire*; Stansbury, *Exploration and Survey of the Valley of the Great Salt Lake of Utah*, 32 Cong., Spec. Sess., Sen. Exec. Doc. 3 (1851).
7. Goetzmann, *Army Exploration*, 305; see also his *Exploration and Empire*, 265–93.
8. U.S. War Department, *Reports of Explorations and Surveys . . .* (1855–60). See Library of Congress catalogue cards for the best contents list. Also see Goetzmann, *Army Exploration*, 262–337.
9. Theodore D. Judah, "A Practical Plan for Building The Pacific Railroad" (1857), 5–6.

Part II—1860–61: Union, Disunion, Incorporation

Chapter 6—"Raise the Money and I Will Build Your Road"

1. Junior League of Washington, *The City of Washington* (New York: Knopf, 1977), 182–201; Leech, 1–13.
2. Howard, 29–30, in drawing on the Anna Judah ms. and on Carl Wheat, exercised a romantic imagination involving conjunctions in the Albany-Troy area, as he confessed to the author in 1987.
3. Anna Judah ms., Bancroft Library, UCB, 1–2.
4. Author's site visit; Robert C. Jones, *Railroads of Vermont* II (1993), 153, provides many interesting facts about regional railroads in Judah's era.
5. Edward T. Williams, *History of Niagara County, New York* (1921), 267–68; see also *Burke's Guide to Niagara Falls* (1856); E. T. Williams, *Commercial Development at Niagara, 1805–1913.*
6. Anna Judah ms., op. cit., 2–3.
7. Robert O. Briggs, *The Sacramento Valley Railroad* (unpublished M.A. thesis, Sacramento State College, 1954), 14.
8. It has been alleged that Horatio Seymour was the brother of Silas Seymour, later of the Union Pacific; they were in fact cousins: Alexander J. Wall, "A Sketch of the Life of Horatio Seymour, 1810–1886," privately printed (1929).
9. See Anna F. Judah, in "Early Pacific Railroad History," *Themis*, December 14, 1889, 3; also, Anna Judah ms., op. cit., 3.
10. Ibid., 4–5.
11. Works Progress Administration, California: *A Guide to the Golden State* (1939), 251–57.
12. *Territorial Enterprise,* February 25, 1866.
13. *Report of the Chief Engineer, Sacramento Valley Railroad,* 1854.
14. *Sacramento Union,* June 5, 1854.
15. *Report of Committee of Board of Directors of the Sacramento Valley Railroad Company,* August 7, 1855.
16. *San Francisco Alta California,* February 17, 1855.
17. *Sacramento Union,* January 1, 1856.
18. *Alta California,* July 30, 1855.
19. *Sacramento Union,* January 1, 1856.
20. *San Francisco Herald,* February 24, 1856; *Sacramento Union,* February 23, 1856.
21. See L. L. Robinson, "Evidence Concerning . . ." cited in Wheat, 224n.

Chapter 7—"There Comes Crazy Judah"

1. Anna Judah ms., Bancroft Library, UCB, 6; Wheat, 227 (n. 45).
2. *Report of the Chief Engineer upon the Preliminary Survey, San Francisco and Sacramento Railroad,* February 1856, Railroad Pamphlets VII:11, Bancroft Library, UCB..
3. Ibid.
4. *Sacramento Union,* June 15, 1855. Other Judah dispatches in *Sacramento Union,* April 19, June 30, July 18, 1856.
5. Judah, "A Practical Plan . . . ," Library of Congress.
6. Ibid., 13, 16.
7. Ibid., 13–14.
8. Ibid., 27–28.
9. Judah, "A Practical Plan," 9.
10. *Nevada Journal,* October 5, 1860; *Sacramento Union,* undated clipping (summer 1860); T. D. Judah notebook, Bancroft Library, UCB.
11. *Nevada Journal,* November 9, 1860.
12. Anna Judah ms., op. cit.
13. John McIntyre to Carl I. Wheat, May 9, 1925, quoted in Wheat, 227.
14. *Sacramento Union,* September 24, 1859; see also *Themis* (Sacramento), August 31, 1889, wherein Catlin claims credit for himself, Judah, and three others for the first transcontinental. They "actu-

ally surveyed," he said, "and laid out the trans-continental railroad as far as Roseville," a distance of some fifteen miles. See also Campilo, 9.

15. *Sacramento Union*, April 5, 1859.
16. *Sacramento Union*, September 24, 1859.
17. See Hittell, *History of California*, IV, 453–54; Bancroft, *History of California*, VII, 452.
18. *Sacramento Union*, September 26, 1859.
19. *San Francisco Alta California*, October 20, 1859.
20. *Sacramento Union*, July 25, 1860.
21. John C. Burch, address, Territorial Pioneers of California, April 13, 1875, *First Annual of the Territorial Pioneers* (San Francisco, 1877), 30–43.
22. *Sacramento Union*, July 25, 1860.
23. Ibid.
24. Ibid.
25. Burch, op. cit.
26. *Sacramento Union*, July 25, 1860.

Chapter 8—"The Marks Left by the Donner Party"

1. Daniel Strong testimony, USPRRC 5–6, 2838.
2. The subscription list: USPRRC 5–6, 2959. See Elliott's report, *Dutch Flat Enquirer*, November 7, 1860, in T. D. Judah notebook, Bancroft Library, UCB. A month after Judah began announcing the route's discovery, the quiescent Elliott was goaded by backers to issue the report. He announced he was forming a railroad company—but nothing came of it.
3. Anna Judah ms., 6, Bancroft Library, UCB.
4. "California Central Railroad," "Eastern Extension," *Sacramento Union*, two undated clippings, T. D. Judah notebook, Bancroft Library, UCB. See also *Nevada City Morning Transcript*, October 4, 1860; *Nevada City Journal*, October 5, 1860.
5. Anna Judah ms., 6, Bancroft Library, UCB.
6. *Nevada City Morning Transcript*, October 4, 1860; *Nevada City Journal*, October 5, 1860. Judah's relationship with the Robinsons receives fuller treatment below.
7. Daniel Strong testimony, USPRRC 5–6, 2839.
8. T. D. Judah, "A Practical Plan for Building the Pacific Railroad," Washington, 1857, 14–15.
9. Daniel Strong testimony, USPRRC 5–6, 2839.
10. Ibid., 2840.
11. USPRRC 5–6, 2960; *Sacramento Union*, November 10, 1860.
12. USPRRC 5–6, 2961.
13. Anna Judah ms., 7, Bancroft Library, UCB.
14. His presentation would likely have been drawn from his public letter and pamphlet. See *Sacramento Union*, November 10, 1860.
15. Anna Judah ms., 8, Bancroft Library, UCB.

Chapter 9—"The Most Difficult Country Ever Conceived"

1. Accounts of this meeting—its place, time, and composition—have varied: Anna Judah ms., 8–9, Bancroft Library, UCB; Cornelius Cole, *Memoirs*, 148–49 (1908); Robert L. Fulton, *Epic of the Overland* (1925); C. C. Goodwin, *As I Remember Them*, Salt Lake City (1913) 33; Daniel Strong testimony, USPRRC V, 2841; Stanford testimony, USPRRC V, 2617–18; Hopkins ms., 11–12, Bancroft Library; Crocker ms., 14–15, Bancroft Library; H. H. Bancroft, *Chronicles of the Builders of the Commonwealth* V, 48. Text scenario seems the most likely, and most previous scholars have more or less agreed.
2. This matter is highly controversial. Curiously, all previous scholars have chosen to take the bitter Lester Robinson at his word. See L. L. Robinson to Charles A. Sumner and Henry Epstein, Nevada Legislature, February 3, 1865, Bancroft Library, UCB; SacVRR, "Report of the President, Trustee, and Superintendant," December 21, 1860, California State Library. For a convincing refutation of Robinson's slander: Leland Stanford, "Reply to the Letter of L. L. Robinson," February 14, 1865, Syracuse University Library; D. H. Haskell, "Speech Before Nevada Legislature," February 21, 1865, Carson City (1865), *Pamphlets on California Railroads* IV, 9, Bancroft Library. Judah's notebook is replete

with newspaper clippings from 1860 which clearly give priorities to the engineer's various callings: "freight agent" for the SacVRR is always last and least. But the most telling statement is Charles Crocker's, printed in Stanford's "Reply" (p. 10):

> *Mr. Judah, in his lifetime, exhibited to me a letter from L. L. Robinson to him, in which he, Robinson, stated that unless the Central Pacific Railroad Company purchased his interest in the Sacramento Valley Railroad upon his own terms, which he fixed at an extravagant price, that he would throw every obstacle in our way that he could; that he, Judah, was well aware of the difficulties in the way of building railroads in California, with no opposition, and all interests favorable to it, but with the active opposition of his Company, wielding a money influence of $30,000 per month, we could not hope to succeed; and that he, Robinson, would wield that influence with all his power and energy against the Company, both here and at the East, unless they complied with his terms. The Company did not purchase his interest, and he has been fulfilling his threat ever since, and has done his utmost, hesitating at no means which he thought would accomplish his object.*

3. Another, lesser, controversy. Most historians have chosen to believe Anna Judah's interpretation of her husband's proposals to the Sacramento merchants, in the Anna Judah ms., Bancroft Library: "The wire Mr. Judah could pull on these 'far-seeing wise men' was this, it was purely local." Her ironic charge, written many years later, was that the Big Four (Huntington, Hopkins, Stanford, and Crocker) had been too parochial or dimwitted to see the Dutch Flat–Donner Pass route as more than a purely local benefit. But her ms. was written in December 1889, long after her settlements and investments had dwindled, and she was struggling to make ends meet in Greenfield, Massachusetts. When the Judahs' old friend and former attorney, Amos P. Catlin, wrote to her (June 9, 1889) at the behest of H. H. Bancroft's researcher, David R. Sessions, it unleashed a flood of letters filled with valuable information but also tempered by bereavement and bitter memories. Particularly hard for the widow was the fact that by 1889 her husband's name and achievements were forgotten by all but their surviving friends, while the Big Four were moguls—their wives living in queenly splendor—and were claiming credit for the genesis of the Central Pacific Railroad. The Anna Judah ms. has been the most popularly consulted source, but also see the A. P. Catlin papers, California State Library, for many useful letters written by Theodore and Anna Judah, Amos Catlin, Daniel Strong, and D. R. Sessions. With this in mind I have chosen to give the Sacramento merchants the benefit of the doubt: they would have instantly and clearly seen what was being dangled in front of their noses.
4. Huntington ms., Bancroft Library, UCB.
5. Biographical details derived from Huntington ms., Bancroft Library; Cerinda W. Evans, *Collis Potter Huntington* (1954); David Lavender, *The Great Persuader* (1969); Oscar Lewis, *The Big Four* (1938).
6. Oscar Lewis, ibid., 229; David R. Sessions, "Huntington Notes," 128, Bancroft Library, UCB.
7. David Lavender, *The Great Persuader,* contains a useful description of early California politics, the first to really do so in transcontinental railroad literature.
8. Arthur McEwen, quoted in Oscar Lewis, ibid., 211.
9. It seems most likely that Judah's argument would have taken this tenor, given his experiences, writings, and public statements, and taking into account the contemporary context of politics which he surely gleaned in Washington.
10. Daniel Strong testimony, USPRRC, 2840.
11. Huntington ms., Bancroft Library, UCB. The dynamics of this are a surmise, but totally in keeping with Huntington's personality.
12. Biographical details taken from Hopkins ms. (typescript draft of biography assembled by H. H. Bancroft staff), Bancroft Library, UCB; Ralph Cioffi, "Mark Hopkins, Inside Man of the Big Four" (unpublished M.A. thesis, University of California, Berkeley, 1950); Oscar Lewis, op. cit.
13. Hopkins ms., 23, Bancroft Library, UCB.
14. Huntington Autobiographical Notes, 57, Bancroft Library, UCB.
15. Huntington ms., Bancroft Library, UCB.
16. Biographical details collected from Stanford testimony, USPRRC V, 2618-19; George T. Clark, *Leland Stanford* (1931); Oscar Lewis, *The Big Four* (1938).
17. Biographical details principally taken from Crocker ms. (dictated autobiography), Bancroft Library, UCB. Also: Oscar Lewis, ibid.; H. H. Bancroft, *Chronicles of the Builders* VI, 33–47; Bunyan

H. Andrew, "Charles Crocker" (unpublished M.A. thesis, University of California, Berkeley, 1931). Quotation from Lewis, 54.

18. T. D. Judah account book, Bancroft Library, UCB.
19. Anna Judah ms., 9, Bancroft Library, UCB.
20. Charles T. Clark, Ibid., 98–99.
21. Daniel Strong testimony, USPRRC 5–6, 2838.
22. Details of their expedition taken from a series of letters found in the A. P. Catlin Papers, California State Library: from Theodore to Anna Judah (May–June 1861); from Anna Judah to A. P. Catlin (1865) and from Daniel Strong to Anna Judah (1863–66). Also: T. D. Judah, "Memorial of the Central Pacific Railroad Company of California," 37 Cong., 2 Sess., H.D. 12 (December 9, 1861), later published as "Report of the Chief Engineer," CPRR, Sacramento, October 22, 1862, NYPL.
23. Theodore to Anna Judah, May 18, 1861, Catlin Papers, California State Library.
24. Theodore to Anna Judah, undated letter ("3 miles above Dutch Flat"—early June 1861), Catlin Papers, California State Library.
25. Theodore to Anna Judah, June 9, 1861, Catlin Papers, California State Library.
26. Theodore to Anna Judah, June 13, 1861, Catlin Papers, California State Library.
27. T. D. Judah, "Memorial," op. cit., 8–9.
28. Daniel Strong testimony, USPRRC 5–6, 2841–43. Note, though, that Strong had become hazy about dates.
29. T. D. Judah, "Memorial," op. cit., 23–24. Of course he underestimated what a really harsh Sierra winter could deposit on the slopes. The cold seasons would be exceptionally mild, in fact, until 1865.
30. Anna Judah ms., 4, 12, Bancroft Library, UCB; Theodore to Anna Judah, June 13, 1861, and Daniel Strong to Anna Judah, December 16, 1865, Catlin Papers, California State Library.
31. T. D. Judah, "Memorial," op. cit., 15–16.
32. Shelby Foote, *The Civil War, A Narrative: Fort Sumter to Perryville* (1958), 85.
33. These are listed at the end of Judah's "Memorial," op. cit., 31–32.
34. USPRRC 5–6, 2964.
35. Carl I. Wheat, "Sketch of the Life of Theodore D. Judah," op. cit., 251; Aaron Sargent to Leland Stanford, in *Sacramento Union*, February 21, 1865.

Chapter 10—"We Have Drawn the Elephant"

1. The Junior League of Washington, *The City of Washington: An Illustrated History* (1977), 201–31; Glenn Brown, *History of the United States Capitol* (1970), 136–37. Margaret Leech's incomparable *Reveille in Washington 1860–65* (1941) was also of great help in understanding the atmosphere of wartime Washington.
2. This is a good point to acknowledge Wallace D. Farnham's important paper, "The Pacific Railroad Act of 1862," *Nebraska History* 43 (September 1962), 141–67, and his "The Weakened Spring of Government," *American Historical Review* LXVII (April 1963), 662–80. Farnham provided the regional political context to this national legislation, particularly the influence of the Leavenworth, Pawnee, and Western Railroad. However, he did not go as far as he could in examining the countervailing influence of the Mississippi and Missouri Railroad, the officers of which got on top of the Railroad Act (and its successors) and rode it for all it was worth.
3. See Stanley P. Hirshson, *Grenville M. Dodge* (1967), 12–14, 24–36, 60, for many valuable and unvarnished insights into the Mississippi and Missouri Railroad and its personnel. However, he resisted going into detail about the Pacific Railroad Act of 1862 and the political machinations allowing the M & M to prevail over the LP&W.
4. Wallace D. Farnham, "The Pacific Railroad Act of 1862," op. cit., 146–47. See also Eugene Glaab, *Kansas City and the Railroads*, 104ff.
5. Lavender, 107.
6. James C. Stone testimony, USPRRC, 1595–1622; John P. Usher testimony, USPRRC, 1672–1716; Thomas Ewing, Jr. testimony, USPRRC, 3849–52; also see documents in Thomas C. Durant papers in the Levi O. Leonard Collection, University of Iowa Library, particularly in folder V:1:11, for other particulars about what changed hands.
7. T. D. Judah account book, Bancroft Library, UCB; Leland Stanford testimony, USPRRC 5–6, 2619.
8. James C. Stone testimony, USPRRC, 1609.

9. T. D. Judah, "Report to the President and Directors of the Central Pacific Railroad Company of California," September 1, 1862, 1–4.
10. Ibid., 4–5.
11. "Speech of Hon. Aaron A. Sargent," Washington, January 31, 1862, NYPL.
12. Anna Judah ms., 11–12.
13. Collis P. Huntington to T. D. Judah, December 26, 1861, Catlin Papers, California State Library.
14. Robert L. Fulton, op. cit., 30.
15. Daniel Strong testimony, USPRRC 5–6, 2845.
16. H. H. Bancroft, *Chronicles of the Builders* VI, 217.
17. Cornelius Cole, op. cit., 181.
18. T. D. Judah, "Report," op. cit., 6–7, 9. Judah's report contains a valuable digest of the bill's progress, one that will aid the reader when turning to the *Cong. Globe* (37 Cong., 2 Sess.) for actual proceedings.
19. *Cong. Globe,* 37 Cong., 2 Sess., April 17, 1862, 1704.
20. *Cong. Globe,* May 1, 1862, 1909.
21. T. D. Judah, "Report," op. cit., 22.
22. *Statutes at Large,* XII, 489, contains the full text of the act.
23. Theodore H. Hittell, *History of California* IV, 461; Carl I. Wheat, op. cit., 255 n. 114.
24. T. D. Judah, "Report," op. cit., 28–29.
25. Anna Judah ms., 9, 26–28, Bancroft Library, UCB.

Part III—1863: Last of the Dreamers

Chapter 11—"Speculation Is as Fatal to It as Secession"

1. *Sacramento Union,* January 8, January 9, January 30, 1863; *Railroad Pamphlets* VII, 18, California State Library; *San Francisco Bulletin,* January 9, 1863; *San Francisco Call,* January 9, 1863. *Annual Report, 1863,* Central Pacific Railroad Company; Crocker testimony, USPRRC, 3113; *Sacramento Union,* November 13, 1863; Lavender, 142.
2. *Chicago Times,* September 2–5, 1862.
3. Union Pacific Railroad, *Report of the Organization and Proceedings of the Union Pacific Railroad Co.,* 1864; Griswold, 45.
4. *Sacramento Union,* October 16, October 18, October 22, 1862.
5. Crocker ms., 30, 32, Bancroft Library, UCB.
6. T. D. Judah, *Report of the Chief Engineer,* October 22, 1862, presents the data from his preliminary survey as published in 1861, with the addition of nearly eighteen pages outlining benefits of the line such as timber yields and freight revenues; Huntington to Julius Speyer, December 6, 1899, Huntington Papers, Syracuse University Library, 8.
7. *Sacramento Union,* December 20 and 22, 1862.
8. Crocker ms., Bancroft Library, UCB, 12, 14; Crocker testimony, USPRRC, 3651.
9. Daniel Strong testimony, USPRRC 2846–47.
10. *San Francisco Alta California,* January 29, 1863, prints an anonymous letter of accusation. *Sacramento Union,* February 6, 1863, prints the "refutation." The discount he supposedly secured was 30 percent for the bonds and 50 percent for the stock. Williams, 61, misattributes Lavender, who seems to have said nothing on the subject.
11. *Sacramento Union,* December 25, December 29. 1862, January 3, January 5, 1863; Crocker ms., Bancroft Library, UCB.
12. *Sacramento Union,* January 9, 1863.
13. *Sacramento Union,* January 5, 1863.
14. *Sacramento Union,* December 10, 1862; Bancroft, *Chronicles of the Builders,* VI, 214.

Chapter 12—"I Have Had a Big Row and Fight"

1. See *Diary of Gideon Welles* (ed. Howard K. Beale), New York, 1960, 228–29.
2. *New York Tribune,* February 26, 27, 28, 1863.
3. Huntington ms., 75, Bancroft Library, UCB

4. Huntington ms., 37–41, Bancroft Library, UCB; see also H. H. Bancroft, *Chronicle of the Builders of the Commonwealth,* V, 58.
5. Huntington interview with H. T. Holmes, Bancroft Library, UCB.
6. Huntington Papers, C–D, 773, 2/12, Bancroft Library, UCB.
7. In answer, beginning in August 1862 the Central Pacific had sponsored well-advertised barometrical examinations of the Yuba and Beckwourth routes, including several approaches, rejecting them either because they were impassable for a railroad, were too expensive, or too long. See *Sacramento Union,* August 17, November 11, 1862, January 1, 1863, and also Judah's reports of October 22, 1862 and July 1863.
8. *Sacramento Union,* April 23, April 25, 1863; *San Francisco Alta California,* April 25, 1863. The $15 million figure had been originally designated during the pioneer Pacific Railroad Convention held in Sacramento in 1859.
9. *Sacramento Union,* May 19, 1863.
10. Stuart Daggett, *Chapters on the History of the Southern Pacific,* New York, 1922, 32–33, 38; "The Great Dutch Flat Swindle! An address to the Board of Supervisors, Officers and People of San Francisco," 1864, California State Library.
11. *Sacramento Union,* May 21, 1863, June 8, 1863; *San Francisco Alta California,* May 25, 1863.
12. Daggett, 33–38; "Old Block," *The Central Pacific Railroad; or, '49 and '69,* San Francisco, 1868, 11.
13. He withdrew from the 1863 race when the party faithful in Sacramento turned to U.S. Rep. Frederick Low: *Sacramento Union,* June 4, 1863.
14. Charles Crocker testimony, USPRRC, 3680–81; Edward H. Miller, Jr. testimony, USPRRC, 3568–70; USPRRC, 3643; Stanford testimony, USPRRC, 2764–65. These transcripts do not include derisive snorts, but certainly some were heard when Crocker declared, "[Sacramento] is where the true base ought to have been, but we were a little modest and did not ask for it." They would not have dared to go that far.
15. Daniel Strong testimony, USPRRC, 2860–62; T. D. Judah, Report, October 22, 1862, 11.
16. *Sacramento Union,* February 24, 1864.
17. Anna Judah ms., 9, Bancroft Library, UCB.
18. T. D. Judah to D. W. Strong, July 10, 1863, USPRRC, 2966–67.
19. *Sacramento Union,* April 27, 1863.
20. T. D. Judah to D. W. Strong, May 13, 1863, USPRRC, 2965.
21. Anna Judah ms., 10, Bancroft Library, UCB.
22. T. D. Judah to D. W. Strong, USPRRC, 2965.
23. T. D. Judah to D. W. Strong, May 13, 1863, USPRRC, 2965.
24. *Sacramento Union,* May 21, 1863, advertises the solicitation for sealed bids.
25. Geoffrey Ward, Ken Burns, and Ric Burns, *The Civil War: An Illustrated History,* 215.
26. *Sacramento Union,* July 15, 1863; *Annual Report,* CPRR, 1863.
27. *Sacramento Union,* November 11, 1862, January 1, 1863.
28. C. P. Huntington to E. B. Crocker, Jr., May 1868, *Letters from Collis P. Huntington to Mark Hopkins, Leland Stanford, Charles Crocker, E. B. Crocker, and D. D. Colton, August 20, 1867–August 5, 1869,* New York, 1892.
29. Charles Crocker testimony, USPRRC, 3642.
30. T. D. Judah to D. W. Strong, July 10, 1863, USPRRC, 2966.
31. USPRRC 7, 4527.
32. T. D. Judah to D. W. Strong, July 10, 1863, USPRRC, 2966; Daniel Strong testimony, USPRRC, 2868–69.
33. *Sacramento Union,* July 16, 1863; Huntington ms., 79, Bancroft Library, UCB.
34. Huntington ms., 58, Bancroft Library, UCB; H. H. Bancroft, *Chronicles of the Builders of the Commonwealth* V, 62–64; C. P. Huntington to Julius Speyer, December 6, 1899, Huntington Papers, Syracuse.
35. *History of Sacramento County, California,* Oakland, CA: Thompson and West, 1880.
36. Huntington ms., 59, Bancroft Library, UCB.
37. Document of Conveyance, Hopkins papers, Stanford University. James Bailey sold his interests in the Nevada Central to Philip Stanford.
38. T. D. Judah to S. S. Montague, Sepember 30, 1863, Catlin Papers, California State Library.

39. T. D. Judah to D. W. Strong, September 9, 1863 (an error; it must have been October 9, 1863 as it was written on the steamer St. Louis, which left on October 3), USPRRC, 2966–67. Regarding the verbal agreement, no documents have been uncovered to refute this, nor have any surfaced to support the contention of some (such as Lester L. Robinson; see L. L. Robinson to Joint Committee on Railroads of the Nevada Legislature, February 3 and 23, 1865, Railroad Pamphlets, California State Library) that when he left for New York, Judah had completely sold out for $100,000 (Robinson charged that it was a "bribe" to keep silent over the impossibility of the Donner Pass). Anna Judah's ms. is quite specific on her husband's intent to buy (and the Associates' acquiescence).
40. Anna Judah ms., 10, Bancroft Library, UCB.
41. Carl I. Wheat, "Sketch of the Life of Theodore D. Judah," *California Historical Quarterly* IV-3, 263, asserts that Anna Judah told her nephew, H. R. Judah, that it was the Vanderbilt group; Wheat interviewed the nephew in 1924.
42. John C. Burch, "Theodore D. Judah," *First Annual of the Territorial Pioneers of California* (San Francisco, 1877).
43. T. D. Judah to D. W. Strong, September 9, 1863 [sic], USPRRC, 2967.
44. *Sacramento Union*, September 30, 1863.
45. *Sacramento Union*, September 4, 1863, September 30, 1863.
46. *Sacramento Union*, October 16, 1863.
47. *Sacramento Union*, October 20, 1863.
48. Anna Judah ms., 15, Bancroft Library, UCB.
49. Burch, op. cit.
50. Leland Stanford to D. W. Strong, October 3, 1864, USPRRC, 2967.
51. Anna Judah ms., 10, 15–16, Bancroft Library, UCB.
52. *Huntington Manuscript*, 82, Bancroft Library, UCB.
53. Anna Judah ms., 10, 16–17, Bancroft Library, UCB; John C. Burch, "Theodore D. Judah," op. cit.
54. *Sacramento Union, San Francisco Evening Bulletin, San Francisco Call*, November 5, 1863; Daniel Strong testimony, USPRRC, 2847. It is interesting to note that during the latter testimony Strong claimed to have been done out of his investments in the Central Pacific; subsequent testimony (3552) showed this to be in error, although the committee neglected to recall Strong himself to allow him to correct the record.
55. *San Francisco Daily Bulletin*, December 15, 1863.
56. Burch, op. cit., also Daniel Strong testimony, USPRRC, 2847, and A. P. Catlin to Anna Judah, June 9, 1889 (which repeats statements Theodore Judah made in 1863 to Colonel Charles Wilson and Judah's assistant engineer, Watson) A. P. Catlin Papers, California State Library.
57. Again, see L. L. Robinson to Nevada Legislature, February 1865, op. cit., along with Stanford's and Crocker's refutation; Stanford repeated the official word that Judah was on the payroll until his death. Judah's appointment letter to S. S. Montague (September 30, 1863, Catlin Papers, California State Library) should kill any lingering doubts.
58. *Sacramento Union*, November 7, 10, and 11, 1863.
59. *Sacramento Union*, November 10, 1863.

Part IV—1864: Struggle for Momentum

Chapter 13—"First Dictator of the Railroad World"

1. Levi O. Leonard, "The Founders and Builders of the Rock Island: II: Thomas Clark Durant," *Rock Island Magazine*, October 1926, 17.
2. Albert D. Richardson, *Garnered Sheaves* (Hartford, 1871), 266.
3. George L. Miller (a rare supporter), in J. Sterling Morton, *Illustrated History of Nebraska* II (Lincoln, 1907), 86n.
4. *New York Herald*, October 6, 1885; Albert Richardson, "Through to the Pacific," *New York Herald*, May 29, 1869; "Thomas Clark Durant," *Dictionary of American Biography*; Leonard, op. cit.
5. Charles Edgar Ames, *Pioneering the Union Pacific*, 21.
6. Dwight L. Agnew, "Iowa's First Railroad," *Iowa Journal of History*, 48 (January 1950); J. R. Perkins, *Trails, Rails and War: The Life of General G. M. Dodge*, 38–41; Stanley P. Hirshson, *Grenville M. Dodge*, 25–26.

7. Henry Walcott Farnam, *Henry Farnam* (privately printed, New Haven, Connecticut, 1889), 65–67.
8. Farnam, 51, 55–56.
9. William E. Hayes, *Iron Road to Empire: The History of the Rock Island Lines,* 39–53; Grenville M. Dodge, *How We Built the Union Pacific,* 60.
10. Hirshson, 27.
11. Farnam, 126.
12. 42 Cong., 3 Sess. (1873), H.R. 78, 515.
13. Dey to G. M. Dodge, quoted in Perkins, 123–24; Johnson, *Peter Anthony Dey,* is the standard biography.
14. Dey to G. M. Dodge, Omaha, January 21, 1864, Dodge Papers, Iowa State Department of History and Archives, Des Moines.
15. Hirshson (1967) is the most dependable biographical source—far superior to Perkins's (1929) or Dodge's own revisionist memoirs.
16. G. M. Dodge, *How We Built the Union Pacific Railway* (1910), 6.
17. John W. Starr, *Lincoln and the Railroads* (1927).
18. G. M. Dodge, *How We Built the Union Pacific,* 10; G. M. Dodge, "What I Saw of Lincoln," *Appleton's Magazine* XIII (February, 1909), 134; Perkins, 46–53.
19. G. M. Dodge, Address at Omaha, November 25, 1901, in J. Sterling Morton, *Illustrated History of Nebraska* II (1907), 117.
20. G. M. Dodge to Anne Dodge, Washington, March 17, 1861, Dodge Papers, Iowa State Department of History and Archives, Des Moines.
21. G. M. Dodge to Julia Phillips Dodge, Washington, June 21, 1861, Dodge Papers, Iowa State Department of History and Archives, Des Moines.
22. Hirshson, 50–53.
23. Hirshson (1967) was apparently the first out of many chroniclers of Grenville Dodge and the western railroads to actually check the records as opposed to merely repeating the web of self-aggrandizing myths Dodge had spun. For this Hirshson deserves enormous credit for piecing together the true facts of Dodge's life and his contributions to the Union Pacific. Wounds: Hirshson, 57–58.
24. Perkins, 107.
25. Grenville M. Dodge, *How We Built the Union Pacific Railway,* Washington, Government Printing Office, 1910, 10; *How We Built the Union Pacific Railway,* Council Bluffs, Monarch, n.d., 61–63; *Personal Recollections of President Abraham Lincoln, General Ulysses S. Grant and General William T. Sherman,* Council Bluffs, Monarch, 1914.
26. Dodge, *How We Built the Union Pacific,* Washington, 11.
27. Grenville Dodge, *Autobiography* (unfinished ms.), Dodge papers, Iowa State Department of Archives and History, Des Moines (also at Council Bluffs, Iowa, Free Public Library), 498–99.
28. Reed to Dodge, New York, August 20, 1863, Dodge Papers, Box 5, Iowa State Archives; also see Reed to Dodge, Washington, May 10, 1863, Dodge Papers, Box 3.
29. *Council Bluffs Nonpareil* (September 5, 1863) reports on his triumphal return.
30. G. M. Dodge to Nathan P. Dodge, New York, September 25, 1863, Dodge Papers, III, 624–25.
31. Dodge to Durant, October 11, 1863, L. O. Leonard Collection, Iowa.
32. Augustus Kountze, a lawyer, pillar of the Lutheran Church in Omaha, and a director of the Union Pacific, was the agent; he had been suggested by both Dodge and Dey. See: Dey to Durant, October 19, October 21, and October 26, 1863, and Durant to Dey, November 23 and November 30, 1863, L. O. Leonard Collection, Iowa.
33. Usher testimony, Sen. Ex. Doc. 51, 1675.
34. Perkins, 286–92; Lincoln's order is reproduced in full on p. 289; it was also reproduced in the *New York Times,* March 12, 1864.

Chapter 14—"Dancing with a Whirlwind"

1. *Report of the Organization and Proceedings of the Union Pacific Company,* 1864, Appendix; Dey's report is fully reprinted in Johnson, *Peter Anthony Dey: Integrity in Public Service* (1939), 91–106; see also Dey testimony, Wilson Report, HR 78, 42 Cong., 3 Sess. (1873), January 10, 1873, 238–68.

2. Dey to Durant, January 4, 1863, Leonard Collection, Iowa.
3. Durant to Dey, November 30, 1863, Leonard Collection, Iowa.
4. See Dey to Dodge, December 15, 1863, in Dodge, *Autobiography*, 500.
5. Dey to Durant, December 4, 1863, Leonard Collection, Iowa.
6. *Report of the Organization and Proceedings of the Union Pacific Railroad Co.*, Appendix 1-a, 2.
7. Dey to Durant, December 24, 1863, Durant to Dey, January 2, 1864, Leonard Collection, Iowa; U.P. Exec. Comm. Minutes, December 30, 1863, UPRRC, NSHS.
8. Dey to Dodge, January 21, 1864, Dodge Papers, Iowa State Archives.
9. Dey to Durant, January 27, January 28, 1864, UPRRC, NSHS.
10. Kountze to Dodge, February 4, 1864, Dodge Records, Des Moines.
11. Perkins, 137; Dodge to Durant, February 4, 1864, UPRRC, NSHS.
12. Nathan Dodge to G. M. Dodge, March 25, 1864, Dodge Papers, Box 6, Iowa State Archives; see also Perkins, 138: "It is now history," he writes, "that Rock Island stock went down to one hundred and ten and Galena and Chicago rose to one hundred forty-nine, and some one must have cleaned up a quarter-million dollars." Stanley P. Hirshson (1967) makes even more of the matter courtesy of Nathan Dodge. The Doctor was certainly mercurial—but he was never far off the scent of money.
13. Glaab, *Kansas and the Railroads*, contains the most complete modern account of the controversies, and should be considered the most reliable source for names, events, and dates.
14. Wilson Report, 103–13, 137, 173–79, 388–94, 514–21; Hazard, 11–12.
15. Train, *My Life in Many States and in Foreign Lands* (New York, 1902); *Dictionary of American Biography*; *National Cyclopedia.*
16. Train to Durant, March 4, June 18, 1863, UPRRC, NSHS; Wilson Report, op. cit., 599–600; Train to Durant, April 11, June 11, June 13, 1864, UPRRC; Richardson, *Beyond the Mississippi* (New York, 1867), 565; Nichols, *Condensed History of the Construction of the Union Pacific Railway* (Omaha, 1892), 51–58.
17. Morton, 89, 90.
18. Richardson, 565.
19. Johnson, *Peter Anthony Dey*, 120.
20. Train, 284–86; Hazard, *The Crédit Mobilier of America* (Providence, 1881), 15–17; documents of the Pennsylvania Fiscal Agency and the Crédit Mobilier's infancy may be found in the Union Pacific papers (ms 3761), NSHS, SG 23, 1:1.
21. Johnson, 122–23; Dey to Dodge, Omaha, April 11, 1864, Dodge Records, Iowa State Archives, IV, 504–505.
22. *Sacramento Union*, February 18, 1864; *Report of the Chief Engineer*, December 1864, 9–13; George Kraus, *High Road to Promontory* (1969), 87; Neill Wilson and Frank Taylor, *Southern Pacific* (1952), 19–20; *WPA Guide to California* (1939), 569–70.
23. *Report of the Chief Engineer, CPRR*, December 1864, 9; USPRRC, 3769 (C. P. Huntington); USPRRC, 2621 (Leland Stanford).
24. *Sacramento Union*, March 21, 1864.
25. *Sacramento Union*, March 23, 1864; see also Best, *Iron Horses to Promontory*, for a correction.
26. *Sacramento Union*, March 26, April 26, 1864; *Report of the Chief Engineer* (December 1864); *Sacramento Union*, May 3, 1864.
27. Durant testimony, Wilson Report, op. cit., 516, 520–21.
28. Charles Edgar Ames, *Pioneering the Union Pacific* (1969), 79–81; *The Palimpsest*, December 1962, 550.
29. Russel, *Improvements of Communication*, 319–20. For an interesting contrast, compare Lavender, 152–53, with Griswold, 66–67, and bet on Lavender's interpretation of events.
30. Huntington typescript, 78–79, UCB.
31. *Cong. Globe*, May 23, 1864.
32. *Cong. Globe*, 40 Cong., 2 Sess., March 26, 1868, 2135–37.
33. Ibid.; John P. Davis, *The Union Pacific Railway: A Study in Railway Politics, History, and Economics* (1894), 120–26.
34. *Cong. Globe*, June 21, 1864.
35. Klein, *Union Pacific*, 32–33; Wilson Report, op. cit., 103–13, 175, 457–67; correspondence in UPRRC, NSHS., especially Stewart to W. D. Bain, January 5, 1864; Hoxie to Durant, April 21, 1864; Sherman to Durant, July 25, 1864; *New York Times*, July 20, 1865.

Chapter 15—"Trustees of the Bounty of Congress"

1. Dey to Durant, Omaha, September 14, 1864, Leonard Collection, Iowa.
2. Dey to Durant, Omaha, May 14, 1864, Leonard Collection.
3. Dey to Durant, Omaha, June 20, 1864, Leonard Collection.
4. Dey to Durant, Omaha, April 9, 1864, Leonard Collection.
5. Dey to Durant, Omaha, April 28, 1864, Leonard Collection.
6. H. Bissell, "Samuel B. Reed," *Proceedings of American Society of Civil Engineers,* v. 19 (July 1892), 184–85; L. O Leonard, "Builders of the Union Pacific: Samuel B. Reed," *Union Pacific Magazine,* October 1922, 18–19.
7. S. B. Reed to Jane (Jenny) Reed, June 12, 1864, Reed Papers, Leonard Collection; L. O. Leonard, "Building the Union Pacific from the Utah End," *Union Pacific Magazine,* May 1922, 10, based on personal correspondence with Reed's family.
8. S. B. Reed to Jenny Reed, June 12, 1864, op. cit.
9. S. B. Reed to Jenny Reed, June 18, 1864.
10. Peter Dey, in his Fall 1862 letter to Grenville Dodge ("Durant has eternally damned the Mississippi and Missouri"), reports seeing Gov. Evans at the convention. Leonard Collection.
11. Ralph K. Andrist, *The Long Death* (Macmillan, 1964), 69–75; Robert M. Utley, *The Indian Frontier of the American West 1846–1890* (New Mexico, 1984), 86; Dee Brown, *Bury My Heart at Wounded Knee* (1970), 68–70. Alvin M. Josephy, Jr., in *The Civil War in the American West* (Knopf, 1991), 294–316, provides the indispensible key—politics and greed—by illuminating Governor Evans's role.
12. Ibid.
13. Ibid., 297–98.
14. Ibid., 300.
15. Andrist, op. cit., 76–77.
16. Andrist, 77–78; Brown, 70–72; Josephy, 300.
17. Quoted in Josephy, 300. General Curtis, of course, was the original sponsor of the 1862 Railroad Act during his service in Congress.
18. Josephy, 301–2; Andrist, 80–82.
19. Josephy, 302–3; Andrist, 82–83.
20. Brown, 75.
21. As quoted in Josephy, 303.
22. Brown, 75–79.
23. Ibid., 79–83.
24. Ibid., 83; Alvin Josephy (op. cit., 307) cites rumors that Chivington slipped word of Wynkoop's humane policy to Gen. Curtis, who moved quickly to quash it.
25. Andrist, 89.
26. Josephy, 307–12; Brown, 83–94; also see Sol Lewis, *The Sand Creek Massacre: A Documentary History* (1973).
27. Josephy, 312–16; Brown, 92–99.
28. A highly detailed participant's account of the Evans expedition may be found in the narrative of Andrew Rosewater, "Finding a Path Across the Rocky Mountain Range," *Union Pacific Magazine,* December 1922, 6–8, and January 1923, 6–8.
29. *Report of Peter A. Dey,* December 30, 1864, 2–3.
30. Dey to Durant, Omaha, November 17, 1864, Leonard Collection, Iowa.
31. Alan W. Farley, "Samuel Hallett and the Union Pacific Railway Company in Kansas," *Kansas Historical Quarterly* 25: 1–16 (Spring 1959); Charles N. Glaab, *Kansas City and the Railroads* (Madison: State Historical Society of Wisconsin, 1962), 118–21, 231–32.
32. Seymour to Durant, Washington, December 5, 1864, UPRRC (ms 3761), SG 4, S3, R4, NSHS.
33. *Sacramento Union,* June 1, 1864.
34. *Sacramento Union,* July 2, 1864.
35. Crocker ms., 19, Bancroft Library, UCB.
36. *Sacramento Union,* September 19, 1864.
37. Crocker ms., 16, Bancroft Library, UCB.

38. H. C. Crane to T. C. Durant, New York, July 30, 1864, UPRRC (ms 3761), SG 4, S3, R4, NSHS; Seymour biography: Silas Morrow Seymour file, Adirondac Railroad file, UPRRC.
39. Seymour to Durant, Washington, August 6, 1864, August 15, 1864, UPRRC (ms 3761), SG 4, S3, R16, NSHS.
40. Seymour to Durant, Washington, November 14, 1864, UPRRC (ms 3761), SG4, S3, R16, NSHS.
41. Hirshson, 102–9.
42. Hoxie to Dodge, New York, November 21 and December 8, 1864, Box 8, Dodge Papers, Iowa State Archives, Des Moines.
43. Hirshson, 34–35, 43–45.
44. Dey to Dodge, Omaha, December 8, 1864, Dodge Records, Dodge Papers, Iowa State Archives, Des Moines.
45. UPRRC (ms 3761), SG4, S3, R12, NSHS.
46. There has been some controversy and confusion over this episode: after earlier mistakes and hazy dating, the prevailing interpretation is found in Maury Klein's *Union Pacific* (1987), that when Dey resigned he did so over the Hoxie contract only. However, there is room to believe that when Dey composed his resignation letter to J. A. Dix (on December 7, 1864) he had already received Seymour's oxbow proposal. Seymour to Durant, Washington, December 12, 1864 (UPRRC, ms 3761, SG4, S1, R16, NSHS) reports: "On receiving an answer to my dispatch of yesterday to Mr. Dey I shall make you a written report on the location of the 1st 22 miles west of Omaha." Dey to Durant (see note 112), December 12, 1864, begins, "I have a letter from Mr. Seymour criticising our location from Omaha to the Elkhorn river, and making suggestions at great length. His earnestness is further evinced by a telegram sent a few days after his letter was mailed, urging an immediate and full answer from me." Plainly, Seymour's follow-up telegram arrived on December 11, and his original letter some days earlier—possibly before Dey wrote his resignation on the 7th, and his complaint to Dodge on the 8th.
47. Dey to Durant, Omaha, December 12, 1864, supplied (with other material) by Dey to J. Sterling Morton, ed., *Illustrated History of Nebraska* II (Lincoln: North & Co., 1907), 94–95.
48. Seymour to Simpson, New York, August 29, 1865, Morton, 96.
49. Dey to Dix, December 7, 1864, UPRRC (ms 3761), SG4, S3, R6, NSHS.
50. Ibid (it is a separate letter).
51. General Dix did not bother to reply to Peter Dey for three months: see Dix to Dey, February 27, 1865, Wilson Report, 670, in which he defended the Hoxie contract as the only expedient route. But see Dix to Crane (UP Secretary), Albany, March 23, 1874, UPRRC, Special Collections, Serial AP2.C4 7/8, NSHS, for a more telling view of Dey's "hero" and his nonintervention. It was written in response to an anonymous report in *Scribner's Monthly* about the Union Pacific and its current troubles before congressional investigators. (Dey's resignation letter to Dix was quoted in the article, which went on to characterize the letter and resignation as "noble.") Dix's letter betrays his ignorance of events in 1864 and his patrician scorn for the Irish engineer. He was, moreover, well rewarded as officer and stockholder for looking the other way in 1864.

Part V—1865: The Losses Mount

Chapter 16—"The Great Cloud Darkening the Land"

1. The following were most useful for study of the Chinese in California: Gunther Barth, *Bitter Strength: A History of the Chinese in the United States, 1850–1870* (Cambridge: Harvard Univ. Press, 1964); Stuart Cretin Miller, *The Unwelcome Immigrant: The American Image of the Chinese, 1785–1882* (Berkeley: Univ. of California Press, 1969); William L. Tung, *The Chinese in America, 1820–1973* (Dobbs Ferry, NY: Oceana, 1974); Sung, Betty Lee, *Mountain of Gold: The Story of the Chinese in America* (NY: Macmillan, 1967). George Krause, in "The Chinese Laborers and the Central Pacific," *Utah Historical Quarterly* (Winter 1969), 37, 1, has performed a useful service in identifying primary sources.
2. Tung, 7–11.
3. *Sacramento Union*, January 11, 1862; Griswold, 110.

4. Mark Twain, *Roughing It* (New York, 1872), 395.
5. *Hutchings' Magazine* 4: 238 (November 1859); Ping Chiu, *Chinese Labor in California: An Economic Study* (Madison, Wis., 1963), 41.
6. Sen. Rept. 689, 44 Cong., 2 Sess., 667.
7. R. L. Fulton, a telegraph operator for the Union Pacific in the 1860s, compiled a lively account of the first transcontinental in 1924, *Epic of the Overland*, drawing on his and other reminiscences; he talked and corresponded extensively with J. H. Strobridge when the latter was near death. Biographical details drawn from Fulton, 32–33, including the fact that Strobridge lost his eye to a powder blast at Bloomer Cut, not later to an unspent nitroglycerin accident as has been erroneously reported by third- and fourth-hand historians, often without citing Fulton for the details they copied correctly. See also Sen. Rept. 689, 44 Cong., 2 Sess., 725.
8. Crocker ms., 14, 49, 51, UCB.
9. Sen. Rept. 689, 44 Cong., 2 Sess., 78 (Frederick F. Low test.), 666–69 (Crocker test.), 723–24, 727 (Strobridge test.); USPRRC (Sen. Exec. Doc. 51, 50 Cong. 1 Sess., 1887), 3659–60 (Crocker test.).
10. Mark Twain, *Roughing It* (1872), 42.
11. Josephy, 312–13; Brown, 94–96; Andrist, 92; George Bird Grinnell, *The Fighting Cheyennes* (1915), 181–87; George E. Hyde, *Spotted Tail's Folk* (1961) are also helpful.
12. Josephy, 313.
13. Hirshson, 110–12, 115–17; Perkins, 157, 171–75.
14. Perkins, 173.
15. Hoxie to Durant, St. Louis, May 15, 1865, Leonard Collection, Iowa, quoted in Hirshson, 117.
16. Ames was to testify that his first Crédit Mobilier subscription was in August 1865, a date often cited by rail historians. However, he actually first invested on March 15, 1865, with Oliver Ames—his brother's name serving as the owner of record. Poland Report, 19; Oliver Ames Test., Wilson Report, 160, 266. Moreover, Oakes Ames's letters show he was actively canvassing for new investors during the spring and summer, 1865: Ames to Durant correspondence, Durant Papers, Leonard Collection, Iowa.
17. Wilson Report, 160; Charles Edgar Ames, *Pioneering the Union Pacific* (1969), 58, 107, 109–10, 111, 128, 148–49.
18. Ames, Ibid., 79–86.
19. Ames, 89; Winthrop Ames, *The Ames Family of Easton, Massachusetts*, 145.
20. Huntington Manuscript, 76–77, Bancroft Library, UCB.
21. Ward, Burns, and Burns, *The Civil War: An Illustrated History*, 368.
22. John W. Starr, *Lincoln and the Railroads* (New York: Dodd, Mead, 1927: reprinted by Arno Press, 1981), 244.
23. Surely the present author is not the first to construct a narrative from the famous theater poster, but if he has seen such a narrative before he cannot remember it.
24. Walt Whitman, "When Lilacs Last in the Dooryard Bloom'd," *Leaves of Grass* (1892).
25. Starr, 277.
26. Ibid., 274.
27. Ibid., 269, 274.
28. Ibid., 270–71.
29. Ibid., 272, 274.
30. Ibid., 273.
31. Dodge, *Autobiography*, 343–45 (ms.), Council Bluffs Public Library; his "Personal Recollections of Lincoln" (1914), 28, differs slightly in language.
32. Starr, 275.

Chapter 17—"If We Can Save Our Scalps"

1. Seymour to Durant, Washington, April 18, 1865, Seymour Papers, UPRRC (ms 3761), SG4, S3, R16, NSHS.
2. Crocker to Cole, Sacramento, April 12, 1865, in Cole Papers, University of California, Los Angeles; Cole, *Memoirs*, is replete with revisionist memories.

3. Hopkins to Huntington, Sacramento, May 31, 1865, Huntington Papers, Syracuse University Library.
4. *Sacramento Union*, April 13, 1865; Daggett, 21.
5. Huntington to Stanford, May 13, 1865, Timothy Hopkins Papers, Stanford University Library; Daggett, op. cit.
6. *Annual Report 1865*, CPRR Co. of California; *Reports of the President and Chief Engineer*, CPRR Co. of California, December 1865; Hopkins to Huntington, Sacramento, July 19, 1865, Huntington Papers, Syracuse.
7. Sen. Rept. 689, 667; Sen. Ex. Doc. 51, 3660.
8. Krause, "The Chinese Laborers and the Central Pacific," *Utah Historical Quarterly* 37, 1 (Winter 1969) contains much of this information, as does Corinne K. Hoexter, *From Canton to California: The Epic of Chinese Immigration* (New York: Four Winds, 1976), both drawing from contemporary reports.
9. A. W. Loomis, "How Our Chinamen Are Employed," *Overland Monthly*, March 1869.
10. Sen. Rep. 689, 668.
11. Griswold, *A Work of Giants* (New York: McGraw-Hill, 1964), 117. The book is a splendid, brisk job of writing—it lacks only a more consistent and thorough documentation.
12. Griswold, ibid., imagined the trestles as "transfixed centipedes." See also *Sacramento Union*, May 1, June 16, September 1, October 14, 1865.
13. Samuel S. Montague, *Reports of the President and Chief Engineer*, December 1865.
14. Dodge to Durant, St. Louis, May1, 1865, UPRRC (ms 3761), SG4, S3, R6, NSHS.
15. Again, see J. Sterling Morton, *Illustrated History of Nebraska* II (1907), for details, letters, and for Dey's interesting perspective more than 35 years later.
16. Hoxie to Dodge, June 4, 1865, Dodge, *Autobiography*, Council Bluffs Public Library, 510; N. Dodge to G. M. Dodge, Council Bluffs, May 26, 1865, Dodge Papers, Iowa State Archives, Des Moines.
17. Durant to Edward Creighton, New York, June 9, 1865; Creighton to Durant, Omaha, June 10, 1865; Morton, 100
18. Kansas Senator James H. Lane would write President Johnson in October to complain that Simpson's committee, also charged to examine the competing Kansas line, could never be impartial with Silas Seymour "in the employ of person deadly hostile to all the Rail Road interests of Kansas," meaning Dr. Durant. Johnson apparently forwarded the letter to Interior Secretary Harlan, Durant's man, who naturally rejected Lane's request. J. H. Lane to President, Wyandotte, October 3, 1865, Tel, DNA-RG107, Tels Recd., President, Vol. 4 (1865–66), in *The Papers of Andrew Johnson*, IX (Univ. Tennessee Press, 1991).
19. *Chicago Tribune*, August 14, 1865; Hirshson, 121–22.
20. Charles T. Sherman test., Wilson Report, 664; Oakes Ames test., Wilson Report, 252.
21. Lambard to Durant, July 11, 1865, Leonard Collection, Iowa.
22. *The War of the Rebellion: Official Records*, Ser. I, v. 50, pt. 2, p. 144; also cited in Josephy, 253.
23. Josephy, 257–59. Bear Creek was, in Josephy's words, "one of the largest, most brutal, and, because of its eclipse by other Civil War news, least-known massacres of Indians in American history."
24. Strobridge interview in Fulton, *Epic of the Overland*, 39.
25. *War of the Rebellion*, Ser. I, vol. 48, pt. 2, 1048–49.
26. H. E. Palmer, "History of the Powder River Indian Expedition of 1865," *Transactions and Reports*, II, Nebraska State Historical Society; Hirshson, 124–25; Dee Brown, 104–19; Josephy, 314–15.
27. Utley, 95. The Indian Affairs Commission would, some four years later, call the summer campaign "useless and expensive. Fifteen or twenty Indians had been killed at an expense of more than a million dollars apiece, while hundreds of our soldiers had lost their lives. Many of our border settlers had been butchered and much property destroyed." (Andrist, 95) One's breath is taken away, however, by this "official" casualty rate of "fifteen or twenty"—the opposite of a latter era's "body count," deflation instead of inflation.
28. Hirshson, 125.
29. DAB; Connor contract, Salt Lake, UPRRC, ms 3761, SG4, S3, R3, F199, NSHS.

30. Reed to Durant, Joliet, November 1, 1865, SG4, #15, NSHS; "Report of Samuel B. Reed . . . 1865," April 1866, NSHS.
31. Evans, *Report*, contained in an appendix to *Report to the Commissioners*, April 10, 1866, NSHS; Klein, 62.
32. Dodge, "How We Built the Union Pacific Railway," S.D. 447, 61 Cong., 2 Sess. (March 22, 1910), 17; also see a book of the same title published in Council Bluffs, and compare to Dodge's unfinished *Autobiography, 1831–71* (408–10), Council Bluffs Free Public Library. Jacob Perkins, in *Trails, Rails and War* (186–90), provides a vivid, somewhat embellished account of the episode, while interestingly, Stanley Hirshson skips it entirely in *Grenville M. Dodge*.
33. Ibid.

Chapter 18—"I Hardly Expect to Live to See It Completed"

1. Richardson, *Beyond the Mississippi*, 438–40.
2. Bowles, *Our New West*, 397, 410–13.
3. Ibid., 32, 34, 36.
4. Ibid., 38.
5. Ibid., 151.
6. Richardson, op. cit., 331–32; 338–40.
7. Bowles, op. cit., 313–14.
8. Hopkins to Huntington, Sacramento, September 29, 1865, Huntington Papers, Syracuse University Library.
9. *W.P.A. Guide to California* (1939), 568–69; *Sacramento Union*, September 4, 1865.
10. Richardson, op. cit., 461–62; 389.
11. Ibid., 463.
12. Ibid., 463–65.
13. *Sacramento Union*, October 14, 1865; Hoexter, 74–75; Thomas W. Chinn (ed.), *A History of the Chinese in California: A Syllabus* (San Francisco: Chinese Historical Society, 1969), 44–45; Samuel S. Montague, *Report of the Chief Engineer*, CPRR, Sacramento, November 25, 1865 (17), states that "The work at Cape Horn has proved less difficult and expensive than was first anticipated," presumably thanks to the Cantonese basket-drillers.
14. *Sacramento Union*, June 16, 1865; Leland Stanford, "Statement Made to the President of the United States, and Secretary of the Interior, of the Progress of the Work," October 10, 1865, Sacramento, 1865.
15. USPRRC (Strobridge test.), 3146.
16. Montague, *Report of the Chief Engineer*, 13–18.
17. Lynne R. Mayer and Kenneth E. Vose, *Makin' Tracks* (1975), 40; Kraus, 141.
18. W. T. Sherman, *Memoirs* (New York: Webster, 1891), v. 2, 411.
19. Hoxie to Dodge, Omaha, October 18, 1865, in Dodge, *Autobiography*, 420.
20. Morton, 107.
21. Williams to Durant, Boston, November 6, 1865, NSHS; Williams had bought 500 shares of Crédit Mobilier stock at $100 par: Wilson Report, 160.
22. Dey to Morton, July 9, 1903, in Morton, 93n. Efforts to verify this in published stockholder lists have been unavailing, but Dey's investment may have been made as part of a group, or in another's name. There is no apparent motivation for his making up such a story, thirty-eight years later.
23. Dey to Train, Omaha, July 5, 1865, Train papers, UPRRC, ms 3761 SG4, S3, R18, NSHS.
24. *Annual Report*, December 4, 1865, 39 Cong., 1 Sess., in *The Cong. Globe.*
25. Sherman, op. cit.
26. USPRRC (Strobridge test.), 3147, 3150.
27. USPRRC (Stanford test.), 2762–63.
28. USPRRC (Crocker test.), 3661. Snow removal came under "extra work," and was billed above Crocker's ordinary remuneration.
29. USPRRC (Stanford test.), 2762.
30. Montague, op. cit., 19; Stanford, op. cit., 9–10, 11–12; *Annual Report*, CPRR, 1865.
31. Lavender, 169.

Part VI—1866: Eyeing the Main Chance

Chapter 19—"Vexation, Trouble, and Continual Hindrance"

1. S.B.R. to Jennie Reed, Omaha, January 14 , 17, 21, 1866, UPRRC, ms. 3761, SG4, R15, NSHS.
2. Wilson Report, 738; Ames, 126–27.
3. *American Biography: A New Cyclopedia* , American Historical Society (1922), 3–5; "Mrs. J. S. Casement," *Painesville Telegraph*, August 17, 1928, courtesy Western Heritage Center, Univ. Wyoming; Works Progress Administration, *West Virginia: A Guide to the Mountain State* (NY, 1941, 1943), 50; Rice, Otis K. and Stephen W. Brown, *West Virginia: A History* (University Press of Kentucky, 1993), 127–29; *War of the Rebellion*, S1, V5, 118–19.
4. See NSHS, SG4, S3, Roll 4., for Durant correspondence regarding "Cotton Expedition." See Casement Papers, Kansas State University, for 1865 "Cotton Speculation."
5. J. S. Casement to Frances Casement, Charlotte, NC, August 2, 1865, Casement Papers, Kansas State University.
6. J. S. Casement to Frances Casement, Des Moines, December 21, 1865, Casement Papers, Univ. Wyoming (duplicate at Kansas State University).
7. Casements to Durant, New York, February 6, 1866 (3 drafts), Casement to Durant, February 27, 1866, Durant Papers, ms. 3761, SG4, S3, R2, NSHS; Durant to Reed, February 24, 1866, Durant Letterpress Book, ms. 3761, SG4, S3, R7, NSHS.
8. Dodge to Nathan Dodge, Ft. Leavenworth, January 26, 1866, Dodge Letter Books, Council Bluffs Public Library; Durant to Dodge, New York, March 1, 1866, Durant Letterpress Book, NSHS.
9. Dodge, *Autobiography*, 554–55, Council Bluffs Public Library; see chapter 20, n.5 below as to correcting the dates traditionally given.
10. Hopkins to Huntington, Sacramento, February 20, 1866, Huntington Papers, Ser.IV, Reel 1, Syracuse University Library. Unless otherwise noted, all subsequent Huntington office correspondence is from Syracuse.
11. *Sacramento Union*, January 8, 1866.
12. Hopkins to Huntington, February 10 and 16, 1866; *Sacramento Union*, January 1, 1866.
13. *Sacramento Union*, January 5, 1866.
14. Ibid.; also, Hopkins to Huntington, February 20, 1866.
15. Hopkins to Huntington, March 12 and 17, May 5, 1866.
16. Hopkins to Huntington, March 23, May 5, 1866.
17. Hopkins to Huntington, February 16, 1866.

Chapter 20—"The Napoleon of Railways"

1. J. S. Casement to Frances Casement, Omaha, March 8, 14, 18, April 6, 1866, Casement Papers, Univ. of Wyoming (some duplicates at Kansas State Univ.); E. C. Lockwood, "With the Casement Brothers While Building the Union Pacific," *Union Pacific Magazine*, February 1931.
2. All telegrams are at Univ. Iowa and NSHS, including Durant to Reed, New York, April 14, 1866, Durant Corres., UPRRC, ms 3761, SG4, S3, R7, NSHS.
3. Durant to Harbaugh, April 17, 1866, Durant Corres., op. cit., NSHS.
4. Durant to Williams, Omaha, May 2, 1866, Durant Corres., op. cit., NSHS.
5. Dodge himself, writing decades after the fact, said he arrived in Omaha on May 6th, a mistake repeated by most rail historians and biographers. He was still at Fort Leavenworth on May 8th when he telegraphed Durant at Omaha to say he would leave in a few days. Dodge to Durant, May 8, 1866, Dodge Corres., NSHS ms. 3761, SG4, S3, R6, NSHS.
6. It is intriguing how many chroniclers of the railroad have taken Dodge at his word about this when ample evidence to the contrary can be found nearly everywhere. Stanley Hirshson (page 43) cogently sets the record straight, finally, in 1967.
7. Lloyd Lewis, *Sherman: Fighting Prophet* (1932), 595.
8. Sherman to Grant, May 14, 1866, cited in Athearn, *William Tecumseh Sherman and the Settlement of the West* (1956), 47–48.

9. Durant General Order, Omaha, May 15, 1866, Dodge Corres., op. cit., NSHS.
10. Dodge to Evans, Omaha, May 29, 1866, Dodge Corres., op. cit., NSHS; *Report of the Chief Engineer* (1866), June 19, 1867, Ibid.; Dodge, *How We Built the Union Pacific Railway* (1910), 18–19, *Autobiography*, 563.
11. Beadle, 81.
12. W. A. Bell, "Pacific Railroads," *Fortnightly Review*, May 1869, 572–73; also found in Glenn C. Quiett, *They Built the West* (New York: Appleton-Century, 1934), 38–39.
13. E. C. Lockwood, "With the Casement Brothers While Building the Union Pacific," *Union Pacific Magazine*, February 1931; Ames, op. cit., 124.
14. Durant to Reed, New York, June 1, 1866, Durant Corres., op. cit., NSHS.
15. W. T. Sherman to John T. Sherman, January 6, 1857, as quoted in Lewis, 109.
16. W. B. Doddridge, "Then and Now: Pioneer Days on the Union Pacific," *Union Pacific Magazine*, May 1923, 10.
17. E. C. Lockwood, op. cit.; George E. Hyde, *Spotted Tail's Folk* (Norman, OK, 1961), 119.
18. I am indebted to many Nebraska histories, particularly Morton's, but most fondly to the cornhusker guide of the W.P.A. (1939).
19. Durant to Reed, June 6, 7, 8, 11, 1866, Durant Corres., op. cit., NSHS.
20. Andrew Rosewater, "Finding a Path Across the Rocky Mountain Range," *Union Pacific Magazine*, January 1923, 6.
21. *Omaha Herald* articles quoted in Morton, 108–9.
22. Durant to Bushnell, June 13, 1866, Durant Corres., op. cit., NSHS.
23. *Cong. Globe*, June 13, 1866, 3261; David Lavender has a very good account in *The Great Persuader*, 173–74.
24. Oakes Ames's speech before Congress, *Cong. Globe*, February 25, 1873 (also quoted in Ames, 140).
25. Huntington Manuscript, 79, UCB.
26. Richardson, *Beyond the Mississippi*, 542.
27. *Panama Star*, April 5, 1866 (quoted in *San Francisco Alta California*, April 23, 1866); *Alta California*, April 21, 1866.
28. *Alta California*, 16, 18, 21, 22 1866; *Sacramento Union*, April 20, 1866.
29. *Alta California*, April 17, 18, 1866.
30. *Sacramento Union*, April 18, 1866.
31. *Sacramento Union*, April 20, 1866; *Alta California*, April 19, 21, 22, 1866.
32. *Alta California*, April 18, 1866; John R. Gilliss, "Tunnels of the Pacific Railroad," *Transactions*, American Society of Civil Engineers, I, 1872; Crocker ms., 58, UCB.
33. Hopkins to Huntington, May 5, 1866, Huntington Papers, Syracuse.
34. Mark Twain, *Roughing It* (1872), 144–45.
35. Butler Ives to William Ives, April 2, 1866; Samuel S. Montague to Butler Ives, Sacramento, April 10, June 2, August 11, 1866, (M97-6:2) Stanford University Special Collections.
36. Crocker ms., 52, UCB.
37. J. O. Wilder, quoted in Kraus, 134.
38. *Sacramento Union*, July 6, 1866.
39. Huntington Manuscript, Bancroft Library, UCB.
40. Griswold's assessment is in Griswold, op. cit., 153; Nathan Dodge to Dodge, April 11, 1864, Dodge Records, IV, 504–5.
41. Huntington Manuscript, 17, Bancroft Library, UCB. David Lavender, Huntington's indefatigable official biographer, looked at Collis's figures and detected an amusing exaggeration: Huntington would not have bought 60,000 tons in 1866 and could not have used them, nor would 23 ships have been able to carry it all. Lavender says it is more like 6,000. "But what's an extra zero in the spinning of a good yarn?" he says (p. 399), and one cannot disagree.

Chapter 21—"We Swarmed the Mountains With Men"

1. Richardson, *Beyond the Mississippi*, 565–66; S. B. Reed to Jennie Reed, July 18, 1866, Reed Papers, UPRRC, ms. 3761, SG4 R15, NSHS.

2. Hirshson, 132–38.
3. Ibid., 138–39.
4. *Dictionary of American Biography,*175–76; Glaab, 146–47, 165. There is an interesting contemporary profile of Browning reprinted in the *San Francisco Alta,* September 3, 1866.
5. Hyde, *Spotted Tail's Folk,* 113–17, including a paraphrase of George Hyde's poignant analysis; Andrist, 99–104; Utley, *The Indian Frontier of the American West,* 99–101.
6. Dodge to Grant, Omaha, July 16, 1866, Dodge Papers, Iowa State Department of History and Archives, Des Moines; Dodge, *How We Built the Union Pacific,* 16, 36.
7. Reed to Durant, Omaha, September 17, 1866, NSHS.
8. Sherman to Rawlins, Omaha, August 17, 1866, HR Exec. Doc. 23, 39 Cong., 2 Sess.
9. Sherman to Rawlins, Fort Kearny, August 21, 1866; Sherman to Rawlins, Fort Laramie, August 31, 1866; Sherman to Rawlins, September 21, 1866; HR Exec. Doc. 23, 39 Cong., 2 Sess.
10. Dodge, *Autobiography,* 575, 604, Council Bluffs Public Library.
11. Richardson, *Beyond the Mississippi,* 562–67, *Garnered Sheaves,* 265.
12. Seymour, *Incidents of a Trip Through the Great Platte Valley, to the Rocky Mountains and Laramie Plains* (New York, Van Nostrand, 1867), 38–40; Seymour to Durant, Ft. John Beauford, September 27, 1866, Seymour Papers, UPRRC, ms 3761, SG4, S3, R16, NSHS.
13. Seymour, ibid., 40–45.
14. Dodge to Durant, Omaha, September 8, 1866, Dodge Papers, UPRRC, ms. 3761, SG4, S3, R6, NSHS.
15. Dodge, *Autobiography,* 580, 593.
16. Durant to Reed, August 29, 1866, Durant Papers, UPRRC, ms 3761, SG4, S3, R7, NSHS.
17. Seymour, *Incidents,* 81, 94–97; Hoxie to Durant, St. Joseph, October 15, 1866, Durant Corres., NSHS; Reed to Jennie Reed, Omaha, September 23, 1866, Reed Papers, op. cit., NSHS.
18. Silas Seymour, *Incidents of a Trip,* 82–109; Lockwood, "With the Casement Brothers."
19. *Sacramento Union,* July 6, 1866.
20. *San Francisco Alta California,* July 15, 1866. It was reprinted in *New York Herald,* September 25, 1866, and *Sacramento Union,* October 18, 1866.
21. *Sacramento Union,* July 21, 1866.
22. *Sacramento Union,* August 14, 1866.
23. As quoted in *Sacramento Union,* August 20, 1866.
24. Recollections of J. O. Wilder as reprinted in Kraus, *High Road to Promontory,* 134; Gilliss, "Tunnels of the Pacific Railroad," *Transactions,* American Society of Civil Engineers, I: xiii, 154.
25. Gilliss, 161.
26. Reminiscences of J. O. Wilder, Kraus, 134, 152–53.
27. Gilliss, 161–62; USPRRC (Stanford test.), 2762.
28. As quoted in *Sacramento Union,* September 24, 1866.

Chapter 22—"Until They Are Severely Punished"

1. Reed to Durant, Omaha, November 20, 1866.
2. Unsigned letter to Durant (handwriting is Train's), September 27, 1866, Train folder, Durant Papers, UPRRC, ms 3761, SG4, S3, R18, NSHS.
3. Maury Klein's account of board goings-on is, thankfully, the clearest and most accurate in a previously obfuscated field: Klein, 80–81. He also has credit for uncovering Williams's first proposal in the Union Pacific board minutes, which eluded earlier researchers.
4. Dodge, "Report of the Chief Engineer, 1866," 20; *Autobiography,* 589–95.
5. Durant to Pullman, New York, December 25, 1866, Durant Papers, NSHS.
6. Dodge, *How We Built the Union Pacific,* 110; Hirshson, 141–42; Dodge to Rawlings, New York, December 10, 1866, Dodge to Durant, December 12, 1866, Durant Papers, NSHS; Ames to Browning, North Easton, MA, December 10, 1866, Secretary of the Interior, RG 48 (Ames folder), NARA.
7. Wilder interview, quoted in Kraus, 136; *Sacramento Union,* October 24, 1866.
8. *Sacramento Union,* October 8, 1867; Mark Twain, *Roughing It,* 582–86; *Territorial Enterprise,* February 25, 1866; Partridge, *Reminiscences,* SPRR, California State Railroad Museum, Sacramento, and cited in Kraus, 141; E. B. Crocker to C. P. Huntington, Sacramento, December 22, 1866, Huntington Papers, Syracuse University Library.

9. *Sacramento Union*, November 11 and 16, 1866.
10. *Sacramento Union*, November 5, 1866; Gilliss, "Tunnels of the Pacific Railroad," 154–55; *Sacramento Union* , November 11 and 12, December 31, 1866.
11. E. B. Crocker to C. P. Huntington, Sacramento, December 28, 1866, Huntington Papers, Syracuse Univ. Library.
12. J. S. Casement to Frances Casement, December 2, December 7, December 23, 1866, Casement Papers, Western History Collection, Wyoming.
13. *Nebraska: A Guide to the Cornhusker State* (1939), 211–17.
14. Frances Casement to J. S. Casement, Painesville, OH, November 25, December 11,1866; J. S. Casement to Frances Casement, North Platte, NE, December 23, 1866, Casement Papers, American Heritage Center, University of Wyoming.
15. Patricia L. Faust (ed.), *Historical Times Illustrated Encyclopedia of the Civil War* (1986), 115–16; Andrist, 108; Utley, *The Indian Frontier of the American West*, 103–5; Grinnell, *The Fighting Cheyennes*, 230–44.
16. Andrist, 117.
17. Andrist, 109–20; Grinnell, *The Fighting Cheyennes*, 234–44; Elbert D. Belish, "American Horse: The Man Who Killed Fetterman," *Annals of Wyoming* 63 (Spring 1991), 54–67; Robert W. Larson, *Red Cloud: Warrior-Statesman of the Lakota Sioux* (Norman, OK: University of Oklahoma Press), 1997, 97–104.
18. Sen. Exec. Doc. 13, 40 Cong., 1 Sess.
19. Dodge to Sherman, Council Bluffs, January 14, 1867, Dodge Records I, 643–48, Iowa State Dept. of History and Archives, Des Moines. See also Dodge to Sherman, New York, December 10, 1866, Sherman Papers (v. 20), Library of Congress.
20. Sherman to Dodge, St. Louis, January 5, 18, 22, 1867, Dodge Papers, Iowa State Archives.

Part VII—1867: Hell on Wheels

Chapter 23—"Nitroglycerine Tells"

1. E. B. Crocker to C. P. Huntington, Sacramento, January 7, 9, and 10, July 6, 1867; Hopkins to Huntington, Sacramento, January 2, 1867, Huntington Papers, Series IV, Reel 1, Syracuse University Library. Note: hereafter in Chapter 23, it is to be assumed that all correspondence to Huntington was written in Sacramento, and is from the Syracuse University collection, unless otherwise noted. It may be useful to researchers to know that Maury Klein in his fine *Union Pacific* (1987) erroneously cites this collection as belonging to the University of Iowa; in fact, Iowa has the Syracuse Microfilm Edition.
2. C. Crocker to Huntington, February 4, 1867; E. B. Crocker to Huntington, January 31, 1867.
3. Lewis Clement statement, USPRRC V, 2577; Gilliss, 155–59.
4. Ibid., 156–57.
5. E. B. Crocker to Huntington, April 23, 1867. Previous histories seem to have erred in extreme, imaginative detail about the ox-team transport of rails and rolling stock over the summit at this time. Hopkins to Huntington, February 15, 1867, may have misled them with its stated firm intention. But Judge Crocker, who would have known, said it plainly two months later that their hope of doing it had "failed." Not until July 1867 was the transport to take place: E. B. Crocker to Huntington, Sacramento, July 23, 1867.
6. Hopkins to Huntington, March 30, 1867.
7. Clark, *Leland Stanford*, 224; Stanford test., USPRRC, V, 2581; Kraus, 159; E. B. Crocker to Huntington, January 31, 1867.
8. E. B. Crocker to Huntington, February 12, March 9 and 20, 1867; Gilliss, 162.
9. E. B. Crocker to Huntington, March 8 and 9, 1867.
10. Hopkins to Huntington, February 15, 1867, Huntington Papers, Syracuse; Huntington to Hopkins, March 12 and 29, 1867, Hopkins Papers M97, Box 20, Stanford.
11. Wilson Report, 66–69; Poland Report, 54. Durant later claimed to congressional interrogators that he was away during the January 5 meeting and that the Boomer contract was negated behind his back; taken on faith by most railway historians, this was false, as the reliable Maury Klein demonstrated in 1987 by simply referring to the minutes on file in the Union Pacific archives.

They can be found on microfilm in Nebraska and on paper in Iowa: Durant was present, but keeping to himself. See Klein, 90–91 and also the Klein notes (published separately), and compare Ames,163–64.

12. Reed to Jenny Reed, New York, January 16 and 22, 1867, Reed Papers, UPRRC, ms 3761, SG4, S3, R15, NSHS; Casement to Frances Casement, New York, January 13, 1867, Casement Papers, American Heritage Center (AHC), Univ. Wyoming.
13. Klein, 91–92; Ames, 163–64; Reed to Jenny Reed, January 24, 27, and 31, 1867, NSHS.
14. Reed to Jenny Reed, Omaha, February 13, 1867, NSHS.
15. Dodge to Durant, Omaha, January 29, 1867, Dodge Papers, UPRRC, ms. 3761, SG4, S3, R6, NSHS.
16. Sherman to Dodge, St. Louis, January 18, 1867, Dodge Papers, Iowa State Archives; Dodge, *Report of the Chief Engineer, with Accompanying Reports of Division Engineers, for 1866; The Union Pacific Railroad . . . Its Construction, Resources, Earnings, and Prospects.* N.Y.: Wynkoop & Hallenbeck, 1867.
17. Reed to Jenny Reed, Omaha, March 27, 1867, NSHS.
18. Richardson, *Garnered Sheaves,* 265; Klein, 92–93; Poland Report, 171–72; Ames, 165–69; Alley testimony, Wilson Report, 4.
19. Klein, 93; Wilson Report, 70–71; Ames, 169.
20. Reed to Jenny Reed, Omaha, March 27, April 18 ,19, and 27, 1867; Reed to Durant, April 5 and 6, 1867, Reed Papers, NSHS; Casement to Frances Casement, Omaha, April 13, 1867, AHC.

Chapter 24—"Our Future Power and Influence"

1. *Sacramento Union,* April 8, 1867.
2. C. Crocker to Huntington, April 25, 1867; E. B. Crocker to Huntington, April 23 and 27, 1867, Huntington Papers, Series IV, Reel 1, Syracuse University Library. Note: hereafter in Chapter 24, it is to be assumed that all correspondence to Huntington was written in Sacramento, and is from the Syracuse University collection, unless otherwise noted.
3. Ibid.; E. B. Crocker to Huntington, May 16, 1867; *Sacramento Union,* April 8, 1867.
4. C. Crocker to Huntington, April 25, 1867.
5. E. B. Crocker to Huntington, January 31, 1867; *Sacramento Union,* April 11, 1867.
6. Hopkins to Stanford, April 1, 1867; E. B. Crocker to Hopkins, April 1 and 15, 1867; Stanford to Hopkins, April 16, 1867; all M97, Box 20, SUL. Also cited in Kraus, 153, 157.
7. E. B. Crocker to Huntington, April 17 and May 16, 1867; Stanford's letters are in Clark, *Leland Stanford,* 224–25.
8. Lavender, 181–82, 400; Huntington to E. B. Crocker, September 27, 1867, Huntington to Stanford October 26, 1867, *Collected Letters,* I.
9. *Sacramento Union,* April 1, 1867.
10. E. B. Crocker to Huntington, April 15, 1867.
11. *Sacramento Union,* March 22 and 27, 1867.
12. E. B. Crocker to Huntington, January 2 and 7, 1867; Lavender, 183–89; Hopkins to Huntington, February 15, April 23, 1867. Note: for the sake of clarity I have employed the later town name of Niles instead of the original, Vallejo Mills.
13. Lavender, 178–79, 184–86, was invaluable here; also, "Map of the Sacramento Valley, California," 1870's, in Ward McAfee, *California's Railroad Era 1850–1911* (San Marino: Golden West), 1973, 139.
14. E. B. Crocker to Hopkins, San Francisco, April 5, 1867, cited in Clark, *Leland Stanford,* 235.
15. E. B. Crocker to Huntington, April 23, 1867; Hopkins to Huntington, April 10, April 20, 1867.
16. Hopkins to Huntington, April 23, 1867.
17. E. B. Crocker to Huntington, April 23, 1867. Not until June 25 would the Judge learn why Huntington was silent—he had not read, or claimed to have missed, Crocker's "many letters" about the construction costs "and other matters" pertaining to the Western Pacific. This left Crocker almost as much in the dark as previously. "Probably," he wrote his friend, "it was overlooked, & I will remedy it." E. B. Crocker to Huntington, June 25, 1867.
18. Clark, 238, 241.
19. Hopkins to Huntington, April 24, 1867.

20. Hopkins to Huntington, May 11, 1867; E. B. Crocker to Huntington, May 3, 8 and 17, 1867.
21. Huntington to Hopkins, March 15, and December 8, 1867, M97, Box 20, SUL.
22. Browning, *The Diary of Orville Hickman Browning*, II, 144–45; E. B. Crocker to Huntington, April 27, 1867; Kraus, 160.
23. Dodge to Anne Dodge, Washington, March 12, 1867, Dodge to Lettie and Ella Dodge, Washington, March 10, 1867, Dodge Papers, Iowa State Archives, cited in Hirshson, 148; *Washington Star*, February 21, 1867.
24. Dodge to Caleb Baldwin, Ft. Leavenworth, April 7, 1866, Dodge Papers, Council Bluffs Public Library; Dodge to Anne Dodge, Washington, March 5, 1867, Dodge Papers, cited in Hirshson, 147–48; Claude G. Bowers, *The Tragic Era* (Boston: Houghton Mifflin, 1929), 164–65. Bowers's book, narrating the "tragedy" of Reconstruction and the pillorying of Andrew Johnson, may be excessively argued and worse, racist, but it nonetheless opens the door on many valuable primary sources from the Gilded Age.
25. Hirshson, 147, 149–51; Dodge to Durant, Washington, April 6, 1867, NSHS.
26. Dodge to Durant, Council Bluffs, April 19, 1867, NSHS; Reed to Jennie Reed, Omaha, April 27, 1867, NSHS.
27. Reed to Jenny Reed, May 4, 1867, NSHS.
28. Reed to Jenny Reed, Omaha, May 6 and 7, 1867, NSHS; Oliver Ames Diary, May 4 and 5, 1867, Stonehill College Library, North Easton, MA (cited in Klein, 95).
29. Reed to Jenny Reed, Omaha, April 2, 1867, NSHS; Henry C. Parry, "Observations on the Prairies," *Montana* (Autumn 1959), 24; also in Parry, "Letters from the Frontier," *Annals of Wyoming* (October 1958), 30:128. Hereafter only the former will be cited.
30. Stanley, *My Early Travels and Adventures in America* (1895) (Lincoln: University of Nebraska Press, 1982), 107. The biographical foreword, by Dee Brown, is a small masterpiece.
31. D. S. Tuttle, "Tales From Old Timers—No. 4," *Union Pacific Magazine*, July 1923, 35.
32. Grinnell, *The Fighting Cheyennes*, 247–58, is a meticulous account; Brown, 150–57; Andrist, 137–40; Utley, 106–8. Stanley, who was present during the first fatal weeks of the campaign, witnessed the soldiers' looting of the village, typically finding nothing objectionable in any of Hancock's actions against the "savages": Stanley, 26–40. Custer's court-martial, the boy general felt, was "an attempt by Hancock to cover up the failure of the Indian expedition": Paul Andrew Hutton, *Phil Sheridan and His Army* (Lincoln: Univ. Neb. Press, 1985), 33; Custer, *My Life on the Plains* (Norman: Univ. Okla. Press, 1962), 53–123.
33. Oliver Ames diary, April 6, 1867, cited in Klein, 94–95, a very well-researched account; Ames, 168–69, is also useful.
34. Klein, 96–97; Ames, 170–71; Oliver Ames to Dodge, N. Easton, MA, May 25, 1867, Dodge Papers, Iowa State Archives.
35. Oliver Ames to Dodge, New York, June 6, 1867, Dodge Papers, Iowa State Archives.
36. Sherman to Dodge, May 7, 1867, Dodge to Sherman, Council Bluffs, May 20, 1867, Dodge Papers, Iowa State Archives (also cited in its entirety in Morton, 121); C. H. Sharman, "Tales From Old Timers—No. 24," *Union Pacific Magazine*, October 1925, 6. As Sharman notes, his original place in Browne's team was taken by Clarke when Browne was transferred to a construction site; thus, Sharman felt, sparing him from an Indian's arrow in favor of Clarke. See Andrew Rosewater, "Finding a Path Across the Rocky Mountain Range," *Union Pacific Magazine*, January 1923, 6–7, for a full account of the attack on Clarke.
37. Stanley, 108.
38. Dodge to Durant, Council Bluffs, May 20, 1867, Dodge, *Autobiography*, 621, 626; Durant to Grant, New York, May 23, 1867, NSHS.
39. Dodge, *How We Built the Union Pacific Railroad*, 18; Dodge, *Autobiography*, 628–29.
40. Sherman to Dodge, St. Louis, May 27, 1867, Dodge Papers; Sherman to Grant, St. Louis, May 27 and 28, 1867, quoted in Athearn, 105–51, 152.
41. Reed to Jenny Reed, May 27, 1867, NSHS.
42. Ferguson Journal, June 2 and 12, 1867, NSHS.
43. *Cheyenne Daily Leader*, September 8, 1880, American Heritage Center, Laramie, and cited in Athearn, 201. "Sherman's pride in his namesake would be somewhat wounded if he were to see the place today that lost its name of Altimont," writes Athearn. "It is little more than a signpost."

44. *New York Times*, June 7 and 11, 1867; *Omaha Weekly Herald*, June 7 and 14, 1867; both are cited in Athearn, 152–55, which contains an imaginative account of the excursion; Reed to Jenny Reed,undated June 1867 (first page missing), NSHS.
45. Sherman to A. C. Hunt, Fort Sedgwick, CO, June 6, 1867; *Rocky Mountain News*, June 7, 1867; Parry, 26.
46. *Diary of Orville Hickman Browning*, II, 147; Utley, 108–9.
47. "Pawnee," *Handbook of American Indians* II (Washington, D.C.: Bureau of Ethnology, 1910), 10, also in condensed form in Bruce, 7–9; U.S. Army Topographical Engineers, "Military Map of Nebraska and Dakota, 1855–6–7," reprinted in Bruce, 4.
48. Bruce, 69. A note about Robert Bruce, *The Fighting Norths and Pawnee Scouts* (N.Y. and Lincoln, Neb., NSHS). This enthusiast's scrapbook was based on extensive research and correspondence, notably with Frank North's brother, Luther H. North, who was also connected with the Pawnee Battalion. Bruce's narrative almost defeats the logical reader; it is a mishmash of undigested facts, requiring patient study. See notes following. But if a reader will only index it and arrange chronologically, it yields an immense amount of material.
49. Bruce, 25, 2, 12–14, 70, 11.
50. Grinnell, *Two Great Scouts and Their Pawnee Battalion* (Cleveland: Arthur H. Clark), 1928, 20.
51. Bruce, 32, 33–34.
52. Ibid., 3.
53. Bruce, 26; Grinnell, *Two Great Scouts*, 139–40.
54. Dodge, *Autobiography*, 635–37; Dodge, *How We Built the Union Pacific*, 24; Dodge, *Reports of the Chief Engineer*, 1867, 3–4, 30; Ferguson Diary, June 17, 1867, January 7, 1868; Dodge to Durant, Council Bluffs, May 20, 1867, NSHS; Andrew Rosewater, "Finding a Path Across the Rocky Mountain Range," *Union Pacific Magazine*, January 1923, 8.
55. Reed to Jenny Reed, June 1867 (first page missing), NSHS.

Chapter 25—"They All Died in Their Boots"

1. E. B. Crocker to Huntington, June 4, 1867, Huntington Papers, Series IV, Reel 1, Syracuse University Library. Note: hereafter in Chapter 25, it is to be assumed that all correspondence to Huntington was written in Sacramento, and is from the Syracuse University collection, unless otherwise noted.
2. Hopkins to Huntington, June 26, 1867; E. B. Crocker to Huntington, June 27 and 28, 1867.
3. E. B. Crocker to Huntington, June 27, 1867; Hopkins to Huntington, Ibid.
4. Hopkins to Huntington, June 28, July 1, 1867; Crocker ms., 53, UCB.
5. E. B. Crocker to Huntington, July 2, 1867.
6. E. B. Crocker to Huntington, July 10, 1867.
7. E. B. Crocker to Huntington, July 23, 1867 (There are two letters of this date, one meant for Huntington's eyes only and one for public consumption: "I have exaggerated nothing in it," Crocker assured him); Partridge, *Reminiscences*, SPRR, California State Railroad Museum, Sacramento, and cited in Kraus, 163.
8. Butler Ives to William Ives, June 8, and November 17, 1867, Syracuse Univ. Library.
9. E. B. Crocker to Huntington, July 27, 1867.
10. E. B. Crocker to Huntington, June 27, July 11, 23, and 30, August 2, 1867.
11. E. B. Crocker to Huntington, July 6, 1867, July 11, 1867 (letter 1—there are two written on this date, the second after receiving Huntington's letter).
12. E. B. Crocker to Huntington, July 11, 1867 (letter 2); see also E. B. Crocker to Huntington, August 1, 1867.
13. He must have peered into it more than once. "I have an idea," he confided to Huntington a few weeks later, "that in 6 mos. or a year from the time the roads are completed, the 2 companies will be consolidated." He would be right—though he would be off by merely a century and twenty-eight years. E. B. Crocker to Huntington, July 23, 1867.
14. E. B. Crocker to Huntington, July 14, 1867.
15. E. B. Crocker to Huntington, July 23, 1867 (letter 2).
16. E. B. Crocker to Huntington, June 8, 1867; Crocker ms., 35, UCB. Huntington, too, thought the

job printing contract was at the base, as if the growing monopoly were not enough: Huntington to J. Speyer & Co., December 6, 1899, Huntington Papers, IV, 1, Syracuse.

17. *Sacramento Union,* March 7, 1868; Hopkins to Huntington, June 26, July 10, 1867.
18. E. B. Crocker to Huntington, August 10, 1867.
19. E. B. Crocker to Huntington, June 4, July 2, 6, and 23, August 16, 1867; Stanford to Browning, San Francisco, August 9, 1867; *Poor's Manual,* 1867.
20. E. B. Crocker to Huntington, July 23, 1867.
21. E. B. Crocker to Huntington, August 2, 10, and 22, 1867.
22. E. B. Crocker to Huntington, August 19 and 29, 1867.
23. Parry, 29.
24. *Cheyenne Leader,* March 20, 1914, Western Heritage Center, Laramie; Dodge, *How We Built the Union Pacific,* 116; Dodge, *Autobiography,* 637–41, Council Bluffs Public Library; Hirshson, 153.
25. Hirshson, 152; *Iowa State Register,* July 18, 1867; Klein (p. 103 and notes, p. 21) was aware of Hirshson's assertion that Dodge was faking, as he concedes, but he was inclined to give Dodge the benefit of the doubt, which may have been overly generous, as a complete reading of Hirshson would convey.
26. Oliver Ames to Dodge, June 13, 1867, Dodge Papers, Iowa.
27. Parry, 27; Stanley, 125–26; the *WPA Guide to Colorado* locates and carefully distinguishes between the four Julesburgs, three of which had been covered over by irrigated farms.
28. Parry, 28.
29. Reed to Jenny Reed, Julesburg, July 15, 1867, NSHS.
30. Dodge, *Autobiography,* 639.
31. Dodge, *Autobiography,* 636–41; Seymour to Dodge, July 17 and 20, 1867, Dodge Papers.
32. Dodge, *Autobiography,* 643–44; (unsigned, probably Appleton) to F. S. Hodges, August 21, 1867, ms. 3761, SG14, Ser. 4, NSHS; Andrew Rosewater, "Finding a Path Across the Rocky Mountain Range," *Union Pacific Magazine,* January 1923, 7–8, offers a heretofore overlooked participant's account.
33. Dodge, *Autobiography,* 644–45; he was never able to spell Browne's name correctly, nor, in fact, was Rosewater (above note).
34. Dodge, Ibid., 642.
35. Dodge, Ibid., 645; Parry, 29.
36. Dodge, Ibid., 642, 645.
37. Wilson Report, 40–41.
38. Klein, 110–12, is invaluable at making sense of the summer 1867 machinations; his is a well-written, eminently reasonable, and highly detailed account; see also Ames, 177–82.
39. Hazard to Oliver Ames, July 3, 1867, Hazard Papers, Rhode Island Historical Society, and cited in Klein, 112.
40. Oliver Ames Diary, July 14 and 15, 1867; Klein, 113.
41. Klein, 113; Ames, 177; Oliver Ames Diary, July 16 and 17, 1867.
42. Klein, 113–15; Ames, 182–85; Charles Sherman to "Genl.," August 17, 1867, ms. 3761, SG4, Ser. 3, folder 819. While there is no concrete indication that this letter was in fact addressed to W. T. Sherman, examination of other correspondence between the brothers lends support.
43. Reed to Jenny Reed, Julesburg, July 30, 1867, in Dodge, *Autobiography,* 646.
44. Stanley, 165–67.
45. Dodge, Ibid., 646–47.
46. Hirshson, 154, as Dodge to Anne Dodge, Fort Sanders, July 28, 1867, Dodge Records VI, 636–37, Dodge Papers, Des Moines. Note: virtually identical letter cited in Dodge, *Autobiography* , 641n, as Dodge to Dillon, Ft. Sanders, July 2, 1867. Weighing the number of Dodge's errors and misrepresentations against Hirshson, one is inclined to follow the latter.
47. Dodge to Anne Dodge, North Platte, August 3, 1867, Dodge Records, VI, 651; Dodge, *Autobiography,* 700.
48. Dodge, Ibid., 700.
49. Dodge, Ibid., 648–49.
50. Dodge, Ibid., 648–57, 698–703; Dodge, *How We Built the Union Pacific,* 20–21; Dodge to Anne Dodge, September 14, 1867, Dodge Records VI, 713–15; Works Progress Administration, *Wyoming: A Guide to Its History, Highways, and People,* 1941, 234–47.

51. Robert W. Larson, *Red Cloud: Warrior-Statesman of the Lakota Sioux* (Univ. Okla. Press, 1997), 110.
52. Andrist,126
53. Larson, 110–11; Andrist, 127.
54. Andrist, 128–31, offers a gripping account of Wagon Box; Larson,112–13, is useful for its reliance on Red Cloud's several subsequent accounts.
55. Grinnell, *The Fighting Cheyennes*, 263–68; Stanley, 154–61. Another participants' account is in John Stands in Timber and Margot Liberty, *Cheyenne Memories* (Yale Univ. Press, 1967), 173–76. Former bridge carpenter H. W. Guy, in "Tales From Old Timers—No. 20," *Union Pacific Magazine*, February 1925, mentions the wreck and the death of the engineer, Billy Brooks. He also knew H. M. Stanley, calling him "a jolly good fellow."
56. Grinnell, *Two Great Scouts*, 145–47; Grinnell, *The Fighting Cheyennes*, 268; Bruce, 26.
57. Grinnell, *The Fighting Cheyennes*, 268–69; Utley, 109; Stanley, 202, 204–5.

Chapter 26—"There Are Only Five of Us"

1. Huntington Typescript, 17, Bancroft Library, UCB; Huntington to Hopkins, New York, November 7, 1867, Huntington Papers, Syracuse University Library (unless otherwise noted, all correspondence between Huntington and his partners is from the Syracuse Library, with all Huntington letters written from New York and his partners' from Sacramento).
2. *New York Times*, September 7, 1867.
3. *New York Times*, September 9, 1867; Huntington to Hopkins, October 17, 1867.
4. Huntington to E. B. Crocker, September 6, 1867.
5. Huntington to E. B. Crocker, October 9, December 7, 1867, April 21, 1868. Note: Lavender (402) asserts these were written to Charles Crocker, using the transcriptions in *Collected Letters*, I, but internal evidence in the original handwritten letters at Syracuse—references to earlier messages from E. B. C.—disputes this. Granted, it is confusing since C. P. H. used the greeting "Friend Crocker" regardless of which brother he addressed.
6. E. B. Crocker to Huntington, September 17, 1867.
7. E. B. Crocker to Huntington, September 12, 1867.
8. Ibid.
9. E. B. Crocker to Huntington, September 12, 1867.
10. Stanford to Huntington, September 9, 1867; Huntington to E. B. Crocker, October 3, 1867.
11. E. B. Crocker to Huntington, September 12 and 17, 1867; Stanford to Huntington, September 9, 1867; Hopkins to Huntington, September 12, 1867.
12. E. B. Crocker to Huntington, September 26, 1867; Huntington to E. B. Crocker, December 28 and 29, 1867, February 25, 1868.
13. E. B. Crocker to Huntington, September 12 and 26, 1867; see also August 22, 1867; Huntington to E. B. Crocker, September 9, 18, and 23, 1867, overruled the Judge's pleas. Stanford consulted with Tevis, Ralston, and others in August, and they were eager to join with the Associates in improving Goat Island: E. B. Crocker to Huntington, August 29, 1867.
14. Montague to Ives, September 13, 1867, Stanford University Library.
15. Dodge *Diary*, August 28, 1867, *Dodge Records* VI, 881–82; Dodge to Anne Dodge, Salt Lake City, September 3, 1867, *Dodge Records* VI, 705–6; Dodge, *Autobiography*, 920; Dodge notes on Miss Stanley's painting of Dodge and Rawlins in 1867, Dodge Papers, Council Bluffs Public Library; Hirshson, 155–57.
16. Works Progress Administration, *Wyoming*, 137–38, 185.
17. Stanley, 191–96.
18. Maj. Gen. C. C. Augur to War Department, Omaha, September 30, 1867, NARS; also quoted, though abridged, in Bruce, 10–11.
19. Reed to Jenny Reed, Julesburg, September 19, 1867, NSHS.
20. Oliver Ames to Dodge, August 5 and 10, 1867, Dodge Papers, Iowa State Archives.
21. Oliver Ames to Dodge, August 21 and 28, September 3, 1867, Dodge Papers; Dodge, *Autobiography*, 683.
22. Dodge, *Autobiography*, 682–87; Dodge to Oliver Ames, October 7, 1867, Dodge Papers; Oliver Ames to Dodge, October 7, 1867, Dodge Papers; Seymour to Oliver Ames, October 10, 1867, Dodge Papers.

23. Dodge, *Autobiography*, 682.
24. Oliver Ames to H. S. McComb, September 17, 1867, NSHS; Klein, 122–28, is invaluable in understanding the complicated machinations of September-October 1867; he has the care of a scholar yet the eye for dramatic detail of a prizefight announcer. Also see: Ames Diary, September 26 and 27, 1867, Oliver Ames Papers, Stonehill College; Hazard Diary, September 17 and 18, 1867, Rhode Island Historical Society.
25. Allen to Oliver Ames, September 19, 1867, Oliver Ames Papers; Ames Diary, September 21 and 23, 1867, Oliver Ames Papers; Hazard Diary, September 18, 22, and 23, 1867; Klein, 123; Huntington to E. B. Crocker, September 27, 1867.
26. Huntington to E. B. Crocker, September 27, 1867.
27. Klein, 124–25; Ames Diary, September 26 and 27, 1867; Hazard Diary, September 24-27 and 30, 1867; Huntington to E. B. Crocker, October 3, 1867. As a matter of interest, see Ames, *Pioneering the Union Pacific*, 187, for a milder family interpretation of events. The directors' minutes are at NSHS and in the Leonard Collection, University of Iowa.
28. Klein, 126; Ames Diary, October 3, 1867; Hazard Diary, October 3, 1867.
29. Huntington to E. B. Crocker, October 5 and 9, 1867; Huntington to Hopkins, October 11, 1867. See also Klein, 126–27; Union Pacific Directors' Minutes, October 4 and 9, 1867; Ames Diary, October 8–10, 15, and 16, 1867; Hazard Diary, October 4, 1867.
30. Klein, 126–27; Union Pacific Directors' Minutes, October 4, 1867; Huntington to E. B. Crocker, October 9, 1867.
31. Klein, 127–28; Hazard Diary, October 1, 1867; Ames to Dodge, October 7, 1867, Dodge Papers.
32. E. B. Crocker to Huntington, October 14, 1867. And same on October 25, 1867: "What a loving crowd [these] Union Pacific men must be."
33. E. B. Crocker to Huntington, September 12 and 26, 1867; Stanford to Huntington, October 11, 1867; E. B. Crocker to Huntington, October 14, 1867.
34. E. B. Crocker to Huntington, October 10, 1867.
35. E. B. Crocker to Huntington, October 14 , 25, and 29, 1867. The Associates had a Sacramento merchant, Theodore J. Milliken, together with Stanford's former secretary in the statehouse, W. E. Brown, and B. R. Crocker of the Central Pacific, sign as incorporators; several weeks later they all transferred their stock to Charles Crocker, who divided it equally between the five Associates: E. B. Crocker to Huntington, February 25, 1868; Hopkins to Huntington, March 16, 1868. Several years later, when the Associates were under governmental scrutiny in the wake of the Crédit Mobilier scandals, they claimed that the Contract & Finance Company was begun to attract new partners; as the above letters indicate, this is not true. Sen. Exec. Doc., 50 Cong. 1 Sess., No. 51, II, 10–11; IV, 2624, 2637. Harry J. Carman and Charles H. Mueller, "The Contract and Finance Company and the Central Pacific Railroad," *Mississippi Valley Historical Review*, XIV, 3, is based largely on the federal sources but not at all on the Associates' correspondence. The Contract & Finance Company signed its first construction contract on December 3, 1867: USPRRC, 3436–37.
36. E. B. Crocker to Huntington, October 29, 1867.
37. Hopkins to Huntington, October 26, 1867; E. B. Crocker to Huntington, October 29, 1867; C. Crocker to Huntington, October 30, 1867.
38. Reed to Jenny Reed, October 29, 1867, NSHS.
39. Grinnell, *Two Great Scouts*, 148–51; Bruce, 19.
40. Utley, 114–17; Andrist, 132; Larson, 116–17; Hyde, 290; Grinnell, *The Fighting Cheyennes*, 274–75.
41. Ames Diary, October 22 to November 6, 1867.
42. Reed to Jenny Reed, undated Cheyenne letter, November 1867, NSHS.
43. Works Progress Administration, *Wyoming*, 186, 109; *Cheyenne Leader*, November 10, 11, 14, 15, 1867, January 6, 7, 1868.
44. E. B. Crocker to Huntington, December 5, 1867; B. P. Avery to wife, Sacramento, December 1, 1867, ms. 2959, California Historical Society. Crocker's letter notes that the unnamed Associated Press correspondent-on-retainer sent the first dispatch from the summit; Avery tells his wife that he sent the first dispatch. It would seem a safe guess to name Benjamin Parke Avery of the *Bulletin* as the Central Pacific's secret press agent. Of course there was more than one; another was Thomas Magee of San Francisco, apparently a freelancer whose laudatory articles on the Central Pacific found homes locally and on the wire to the East: E. B. Crocker to Huntington, April 15,

1867. Crocker asked that Huntington use his influence to get Magee appointed West Coast correspondent for the *New York Tribune.*

45. Hopkins to Huntington, December 1, 1867.
46. *Sacramento Union,* November 30, 1867; B. P. Avery to wife, Sacramento, December 1, 1867, op. cit.
47. E. B. Crocker to Huntington, November 9 and 18, 1867; Stanford to Huntington, November 23, 1867.
48. E. B. Crocker to Huntington, November 9, 1867.
49. *Virginia Enterprise,* December 6, 1867, reprinted in *Sacramento Union,* December 9, 1867.
50. *Sacramento Union,* December 9, 1867. Also see Hopkins to Huntington, December 6, 1867 (the day before the excursion), in which Hopkins says confidently, "We have never failed on such occasions to make a favorable impression. . . ."
51. E. B. Crocker to Huntington, November 7, 1867; Huntington to E. B. Crocker, November 2, 29 and 30, 1867; Bancroft, *Chronicle* V, 57. Franchot was paid $25,000 per year as the Central Pacific's chief lobbyist.
52. Huntington to Hopkins, November 27, 1867.
53. Huntington to E. B. Crocker, October 29, November 2, 1867, February 21 , March 13, 1868; Huntington to Hopkins, December 17 and 29, 1867.
54. Huntington to Hopkins, November 7, December 21, 1867; Huntington to E. B. Crocker, December 6, 1867; E. B. Crocker to Huntington, December 19, 1867.
55. Huntington to E. B. Crocker, November 1, December 28, 1867; Huntington to Hopkins, December 29, 1867.
56. Klein, 142; Ames, *Pioneering,* 196; Wilson Report, 725–27.
57. Poland Report, iv, 16, 32; Ibid., 4–7; Oakes Ames to Oliver Ames, December 8, 1867, January 30, 1868, Oliver Ames Papers, Stonehill College.
58. Poland Report: Oakes Ames test.: Durant: 57, 18, 335; 174, 241–42, 386; Alley: 406; Ames, 197–98.
59. Poland Report, iii, 57; Ames, 198.
60. Poland Report, 59–60; Wilson Report, 725–27; Ames, ibid.
61. Poland Report, iii, 59–60; Wilson Report, 725–27; Ames, ibid.; Klein, 144–45.
62. Ames, 203, features a handy table of the twenty-three U.S. congressmen called in by the Poland Committee, although the individual testimony is fascinating for personal details and ethical dodges (all selected references here are in Poland Report, which contains a decent name index, though not as good as that in the Wilson Report): Allison: 9, 41, 111, 292, 304–8; Bingham: vi, 21, 34–35, 181, 191–97, 336, 458; Colfax: 81–84, 279, 325–28, 451, 481–92, 501–16; Dawes: v, 6, 20, 32–36, 112–15, 449–50; Garfield: vii, 6, 21, 128–31, 295–98, 355–61, 459–60; Kelley: vii, 6, 21, 197–204, 297–99, 313–18, 330, 451, 466–68; Logan: 335, 346–47, 352–53, 460; Patterson: 184–86, 261–66, 266–73, 270–72, 336–37, 356, 457–58, 480; Scofield: vi, 6, 204–7, 39–40, 181, 299–300, 336, 352, 455–56; H. Wilson: 6, 20, 26, 78–80, 136, 186–90, 288–89, 336, 448–49, 461, 481; J. Wilson: 21, 29–30, 38, 40, 211–20, 212, 333, 336, 408, 457.
63. Poland Report: Hooper: 12, 19, 21, 40, 59–60, 195, 205; Alley: 74–80, 84–109, 311, 322, 405–8, 420–23, 429; Grimes: 10, 19, 21, 220, 306; Boutwell: 6, 21, 28, 32, 80, 92, 103, 186, 303–4, 406; Eliot: 6, 21, 28, 32; Blaine: 1, 2, 6, 10, 31, 181, 293, 323, 339; Bayard: 4, 9, 15, 30, 33, 74, 293, 336, 408; Fowler: 4, 6, 9, 21, 303, 408, 516-19; Boyer: 207-11, 133, 138, 243, 272; Brooks: xii, 243-61, 439-41, 361. Harbaugh and Morehead: Harbaugh to Durant, April 5, 1867, NSHS.
64. Klein, 145; Klein *Notes,* 28: 145–2.
65. Wilson Report, xii. Dodge would testify that his wife bought the shares with her own money and collected the dividends herself, a statement that may have elicited derisive snorts in the committeeroom, as the Dodges, like most people, supported their household and affairs as a team despite his long absences.
66. Huntington to E. B. Crocker, December 6, 1867.
67. Snyder to Dodge, Omaha, December 13, 1867, Dodge Papers, Iowa State Archives; Dodge to Crane, Washington, January 23, 1868, NSHS; Durant to Snyder, New York, December 5, 1867, NSHS; Dodge to Reed, Washington, December 27, 1867, NSHS.

68. Sheridan Report, September 26, 1868, Box 83, Sheridan Papers, Library of Congress; also cited in Paul Andrew Hutton, *Phil Sheridan and His Army* (Lincoln: Univ. Neb. Press), 1985, 38.
69. Dodge, *Report of the Chief Engineer,* 1867.
70. Casement to Crane, December 20, 1867, NSHS.
71. Hopkins to Huntington, December 13, 1867; E. B. Crocker to Huntington, December 23, 1867. The train accident is also mentioned in Huntington to Hopkins, January 6, 1868.
72. E. B. Crocker to Huntington, December 19 and 23, 1867; John R. Brown (Cisco freight clerk), *Diary,* California State Library; Huntington to E. B. Crocker, December 29, 1867.
73. Partridge, *Reminiscences,* SPRR, California State Railroad Museum, Sacramento, and cited in Kraus, 182.
74. *Sacramento Union,* December 21, 1867; E. B. Crocker to Huntington, December 19, 1867.
75. E. B. Crocker to Huntington, December 12, 1867 (two letters).
76. E. B. Crocker to Huntington, December 5, 1867.
77. Huntington to E. B. Crocker, December 27, 1867.

Part VIII—1868: Going for Broke

Chapter 27—"More Hungry Men in Congress"

1. Dodge, *Autobiography,* 705, Council Bluffs Public Library.
2. Grant to Sherman, Washington, February 10, 1868, Sherman to Grant, St. Louis, February 14, 1868, *The Papers of Ulysses S. Grant* XVIII (Carbondale: So. Illinois Univ. Press, 1995), 138–39.
3. Lloyd Paul Stryker, *Andrew Johnson: A Study in Courage* (NY: Macmillan, 1930), 518.
4. Hirshson, 160; Benjamin Thomas and Harold Hyman, *Stanton, The Life and Times of Lincoln's Secretary of War* (NY: Knopf, 1962); Frank A. Flower, *Edwin McMasters Stanton* (Akron: Saalfield Pub. Co., 1905), 336–38; the quotations are from notes taken by Maj. A. E. H. Johnson. Eric L. McKitrick, *Andrew Johnson and Reconstruction* (Chicago: Univ. Chicago, 1960), a temperate and carefully reasoned work, was also invaluable in sorting through these events.
5. The temperature of the body of work on Reconstruction and the Johnson impeachment is feverish but certainly makes for interesting reading (some titles noted above, and also Bowers, *The Tragic Era*). Most make a direct connection between congressional corruption and radical Republicanism: "But it was by no means to prophesy alone that these Representatives of a corrupt Congress confined themselves," writes Stryker. "A Congress in the pockets of many of whose members the gold of Crédit Mobilier was jingling! A House of Representatives whose Speaker was a bribe taker! They knew how trivial a thing was the removal of Edwin Stanton." Most books will devote a chapter to the Oakes Ames efforts, overlooking Huntington's. But as any student of Congress knows, corruption has had no limitation as to party, background, or beliefs, over two centuries, though admittedly this time—at the dawn of the Gilded Age—was particularly florid.
6. Dodge to John Duff, Washington, January 4, 1868, in Dodge, *Autobiography,* 706–7.
7. Oliver Ames to Dodge, North Easton, January 8, 1868, Dodge Papers, Iowa State Archives; Dodge to Reed, Washington, January 4, 1868, in Dodge, op. cit., 708.
8. Dodge to Reed, Washington, January 4, 1868, in Dodge, op. cit., 707–8.
9. Reed to Dodge, Cheyenne, early January 1868, in Dodge, op. cit., 709.
10. Reed to Jenny Reed, Cheyenne, January 18, 1868, NSHS.
11. Reed to Crane, Cheyenne, January 20, 1868, NSHS.
12. Snyder to Dodge, Cheyenne, dated January 12 and 13, 1868, in Dodge, op. cit., 713–15, but as January 22, in Dodge Papers.
13. Snyder to Dodge, February 7, 1868, Dodge Papers, and in Dodge, op. cit., 724–25.
14. Dodge, op. cit., 726.
15. Dodge, op. cit., 569–70. See also Brigham D. Madsen, *Glory Hunter: A Biography of Patrick Edward Connor* (Salt Lake City: Univ. of Utah, 1990), 163-65.
16. Dodge, op. cit., 717.
17. Klein, 142. "From such sources," Klein writes, "came large returns that never found their way into the tortuous calculations by later generations of the profits realized from construction of the Union Pacific."

18. Wilson Rept., 88 (Durant test.), 542, 544 (Bushnell), 679–80 (Oakes Ames).
19. Poland Rept., 4–7, reprints the three Ames letters. Oakes later disputed that he had dictated recipients' names to McComb (see Chapters 26 and 31), and after the letters were publicized, some of the named parties also denied signing on. As Maury Klein notes (143), the incriminating statements to McComb "were essentially rewrites of text from letters written to Oliver, which tended to be more candid." These confidences to his brother did not surface during the ensuing investigations, tending to dig Oakes a little deeper in a hole despite his protests of innocence.
20. *Sacramento Union,* February 3, 1868. Train's 8-day imprisonment and eventual release was covered in the *Union* well into February, in great detail, including reprints from the *Cork Examiner: Union,* February 26 and 27, 1868.
21. Huntington to E. B. Crocker, January 1, 1868, Huntington Papers, Syracuse University Library (unless otherwise noted, all correspondence between Huntington and his partners is from the Syracuse Library, with all Huntington letters written from New York and his partners' from Sacramento).
22. Huntington to E. B. Crocker, January 3, 1868; Huntington to Hopkins, January 6, 1868.
23. C. Crocker to Huntington, January 3 and 16, 1868; E. B. Crocker to Huntington, January 16, 1868.
24. Huntington to E. B. Crocker, January 21, February 3, 1868; Hopkins to Huntington, January 27, 1868.
25. E. B. Crocker to Huntington, January 16, 1868.
26. Huntington to E. B. Crocker, January 13, February 22, 1868; Huntington to Hopkins, February 5, 1868.
27. Hopkins to Huntington, February 12, 1868; E. B. Crocker to Huntington, February 11, 1868.
28. Huntington to E. B. Crocker, February 21, 1868; Huntington to C. Crocker, February 22, 1868.
29. Huntington to E. B. Crocker, January 20, 1868.
30. Huntington to E. B. Crocker, March 14 and 21, 1868; Huntington to Stanford, February 7, 1868; Huntington to Hopkins, March 7 and 31, 1868; *Dictionary of American Biography* (Conkling), 346–47.
31. Huntington to E. B. Crocker, February 3, 1868; Huntington to Hopkins, March 31, 1868.
32. Huntington to E. B. Crocker, March 12, 1868.
33. Huntington to C. Crocker, January 24, 1868; Huntington to E. B. Crocker, March 28, 1868.
34. E. B. Crocker to Huntington, February 6 and 13, 1868.
35. Hopkins to Huntington, January 30, 1868.
36. Huntington to E. B. Crocker, January 20, March 12, 1868; Huntington to Hopkins, March 31, 1868.
37. Stanford testimony, USPRRC (S. Exec. Doc. 51, 50 Cong., 1 Sess.), vol. V, 2807.
38. E. B. Crocker to Huntington, January 8, 1868; Hopkins to Huntington, January 8, 1868.
39. *San Francisco Bulletin,* March 14, 1868.
40. E. B. Crocker to Huntington, January 20, 1868.
41. Hopkins to Huntington, March 21, 1868.
42. E. B. Crocker to Huntington, March 28, 1868.
43. Hopkins to Huntington, March 15, 1868; *Sacramento Union,* February 7, March 7, 1868; *Auburn Stars and Stripes,* January 9, 1868, quoted in *Sacramento Union,* January 11, 1868.
44. E. B. Crocker to Huntington, March 9 and 20, 1868.
45. Huntington to E. B. Crocker, March 28, 1868.
46. Hopkins to Huntington, March 31, 1868; E. B. Crocker to Huntington, March 31, 1868.
47. Huntington to Hopkins, February 5, 1868; *Virginia Trespass,* February 17, 1868, quoted in *San Francisco Alta California,* February 25, 1868.
48. Hopkins to Huntington, March 16, 1868 (two letters).
49. C. Crocker to Huntington, March 29, 1868.
50. *Wyoming: A Guide to Its History, Highways, and People* (NY: Oxford), 1941.
51. Sherman to Dodge, St. Louis, March 11, 1868, cited in Dodge, *Autobiography,* 739.
52. Gibbon to Dodge, Ft. Sanders, early March 1868, cited in Dodge, op. cit., 730.
53. Reed to Crane, March 13, 1868, NSHS; Reed to Dodge, Cheyenne, March 29, 1868, cited in Dodge, op. cit., 745.

54. Christen Nelson, "Tales From Old-Timers—No. 32," *Union Pacific Magazine,* June 1926, 16.
55. Reed to Crane, March 28, 1868, NSHS.
56. Reed to Crane, March 12, 1868, NSHS; Snyder to Dodge, Omaha, March 26, 1868, cited in Dodge, op. cit.
57. Dodge, op. cit., 733, 743.
58. Dodge, op. cit., 729; Blickensderfer to Dodge, Salt Lake City, March 7, 1868, Dodge to Blickensderfer, Washington, March 20, 1868, cited in Dodge, op. cit., 737, 740. 743.
59. Blickensderfer to Dodge, Salt Lake City, March 24, 1868, Dodge to Blickensderfer, Washington, March 30, 1868, cited in Dodge, op. cit., 744, 746.
60. Dodge to W. B. Shattuck, Washington, March 6, 1868, cited in Dodge, op. cit., 741–42.
61. Sherman, *Recollections,* I, 425; *Letters,* 315.
62. Browning, *Diary,* 188–89; Dodge, op. cit., 745.
63. Dodge, op. cit., 746–47; Hirshson, 160–61.
64. Dodge, op. cit., 739–40.
65. Dodge, op. cit., 742–44.
66. Dodge, op. cit., 746; Dodge to Durant, Washington, March 19, 1868, Dodge Papers, (ms. 3761, SG4, S3, R6), NSHS.

Chapter 28—"Bring On Your Eight Thousand Men"

1. C. Crocker to Huntington, April 4, 1868; E. B. Crocker to Huntington, April 13 and 23, 1868, Huntington Papers, Syracuse University Library (unless otherwise noted, all correspondence between Huntington and his partners is from the Syracuse Library, with all Huntington letters written from New York and his partners' from Sacramento);*Virginia City* (Nev.) *Enterprise,* April 11 and 12, 1868, also quoted in *Sacramento Union,* April 14, 1868.
2. *Sacramento Union,* February 3, 1868.
3. Hopkins to Huntington, April 23, 1868; Works Project Administration, *Nevada: A Guide to the Silver State* (Portland, OR: Binfords & Mort, 1940), 138–40.
4. E. B. Crocker to Huntington, April 21, 1868; Huntington ms., 65–66, UCB.
5. E. B. Crocker to Huntington, April 1 , 7, 13 and 21, 1868. The Cold Stream gap was variously given as six and seven miles; I have sided with seven after counting and judging the sources. The discrepancy might be explained by a plan for a siding there.
6. E. B. Crocker to Huntington, April 27, 1868.
7. Huntington to E. B. Crocker, April 6 and 7, 1868; Huntington to Hopkins, April 13, 1868; Huntington to C. Crocker, April 15, 1868.
8. Huntington to E. B. Crocker, April 2 and 6, 1868; E. B. Crocker to Huntington, April 13, 1868.
9. Huntington to Hopkins, April 13, 1868; Huntington to E. B. Crocker, April 24 and 28, 1868.
10. Huntington to E. B. Crocker, May 4 and 9, 1868; E. B. Crocker to Huntington, April 21, 1868.
11. Huntington to E. B. Crocker, April 17, 1868. According to the original handwritten letter, this is dated erroneously and misquoted in Lavender (203), who rather artfully depicts Huntington in the act of drawing a fraudulent red line on the location map; he also misattributes the letter's recipient as Charles Crocker, a frequent mistake; Klein (149) picks up the image of the red line to dramatize the dramatization.
12. Ibid.; Huntington to C. Crocker, April 24, 1868.
13. Huntington to E. B. Crocker, May 7, 1868. I elected not to continue to quote into Huntington's unfortunate racist "in the woodpile" expression—not to insulate him from embarrassment but to avoid a distractive effect right at the end of this section.
14. Dodge to O. H. Browning, Sherman Summit, April 16, 1868, Sec. Int. Files (RG48), National Archives; *New York Tribune,* April 18, 1868; *American Railroad Journal* XLI (April 25, 1868), 408.
15. Durant to Stanford, Sherman Summit, April 16, 1868 (as transmitted through Cheyenne, April 17), and Stanford to Durant, Sacramento, April 17, 1868, Huntington Papers, Syracuse University Library. The exact altitude at Sherman Summit was 8,242 feet.
16. Reed to Jenny Reed, April 23, 1868, Reed Papers (ms. 3761, SG4, S3, R15), NSHS.
17. Dodge, *Autobiography,* 751, Council Bluffs Public Library; Works Progress Administration, *Wyoming,* 197.
18. Ibid.

19. C. C. Cope, "Tales From Old-Timers—No. 11," *Union Pacific Magazine*, February 1924, 10.
20. Ibid.; L. O. Leonard, "Lived After Being Scalped," *Union Pacific Magazine*, April 1925, 9. Note: Cope's date for this episode, fogged by some 56 years, is mistaken; Leonard corrected it for his article, having examined Union Pacific telegrams in the mean time. Cahoon Street in Ogden is named after the Union Pacific veteran.
21. Dodge, op. cit., 748–49; Grinnell, *Two Great Scouts*, 153; Bruce, 34; Oliver Ames to Dodge, April 1868, cited in Dodge, ibid., 748.
22. Casement to F. Casement, April 21, 26, and 28, May 8, 1868, Casement Papers, Univ. Wyoming.
23. Evans to Dodge, Green River, May 7, 1868, Dodge Papers, Iowa State Archives; Dodge, op. cit., 748.
24. Hirshson, 160–61; David M. DeWitt, *The Impeachment and Trial of Andrew Johnson*, 515–96.
25. Huntington to E. B. Crocker, May 5, 9, 1868.
26. Oakes Ames to Oliver Ames, Washington, May 12, 1868; Oliver Ames Papers, Stonehill College; Oliver Ames to Dodge, New York, May 15, 1868; Dodge, *Autobiography*, 755.
27. Huntington to E. B. Crocker, May 13, 1868; Huntington to Hopkins, May 16, 1868.
28. Huntington to Hopkins, May 26, 1868.
29. Huntington to E. B. Crocker, May 13, 1868; Huntington to Hopkins, May 30, 1868.
30. Dodge to H. C. Crane, Washington, May 11, 1868, Dodge Papers (ms. 3761, SG4, S3, R6), NSHS.
31. Dodge, *Autobiography*, 750.
32. Ibid., 748.
33. Dodge Papers, NSHS.
34. Evans to Dodge, Ft. Sanders, May 11, 1868; ibid., 754; Evans to Durant, Ft. Sanders, May 9, 1868, Evans Papers (ms. 3761, SG4, S3, R8), NSHS.
35. Ibid., 755.
36. Dodge to Oliver Ames, Washington, May 14, 1868, Oliver Ames Papers, Stonehill College.
37. Hopkins to Huntington, May 14 and 27, 1868; Stanford testimony, USPRRC V, 2523; E. B. Crocker to Huntington, June 6, 1868.
38. C. Crocker to Huntington, May 20, 1868; Hopkins to Huntington, May 26, 1868; E. B. Crocker to Huntington, May 21, 1868.
39. Huntington to E. B. Crocker, May 13 and 21, 1868; Hopkins to Huntington, May 20 and 26, 1868; Stanford to Huntington, May 20, 1868; E. B. Crocker to Huntington, June 6, 1868.
40. Huntington to Hopkins, June 1, 1868; Huntington to C. Crocker, June 8, 1868.
41. E. B. Crocker to Huntington, May 21 and 23, June 6, 1868.
42. Hopkins to Huntington, May 6, 1868 (he had meant to write "organism" but instead tellingly wrote "organization"); Huntington to E. B. Crocker, June 9, 1868.
43. E. B. Crocker to Huntington, May 23, 1868.
44. Huntington to Hopkins, May 23, 1868; Stanford to Jane Stanford, May 30, 1868, Stanford Univ. Library; Hopkins to Huntington, June 2, 1868; E. B. Crocker to Huntington, April 23, 1868.
45. E. B. Crocker to Huntington, May 28, June 4, 1868; Huntington to Stanford, May 29, 1868; Hopkins to Huntington, May 14, 1868; Huntington to Hopkins, June 1, 1868.
46. Huntington to E. B. Crocker, April 2 and 25, May 2, 8, 15, 1868; Huntington to Hopkins, May 14 and 27, 1868.
47. Huntington to E. B. Crocker, May 28, 1868; Huntington to Hopkins, May 26, June 6, 1868.
48. Huntington to Hopkins, June 6, 1868; Browning, *Diary*, II, 201.
49. Huntington to Hopkins, June 10, 1868; Huntington to E. B. Crocker, June 12 and 13, 1868.
50. Huntington to Hopkins, June 11 and 19,1868.
51. E. B. Crocker to Huntington, May 27, 1868.
52. Hopkins to Huntington, May 19, 1868.
53. E. B. Crocker to Huntington, June 9, 1868; C. Crocker to Huntington, May 20, 1868.
54. Oliver Ames to Dodge, North Easton, May 16, 1868; Dodge to Blickensderfer, Washington, May 24, 1868, both cited in Dodge, op. cit., 456–57.
55. Evans to Dodge, Fort Sanders, in ibid., 757A.
56. Leonard J. Arrington, *Great Basin Kingdom*, 261–63, 269–70; David J. Croft, "The Private Business Activities of Brigham Young," *Journal of the West* XVI (October 1977).
57. Reed to Jenny Reed, May 18 and 20, June 7, 1868, Reed Papers (ms. 3761, SG4, S3, R15) NSHS; Dodge, ibid., 763; USPRRC, 2154, 2173; Joseph A. West, "Construction of the Union and Central

Pacific Railroads, Across Utah, 55 Years Ago," *Union Pacific Magazine,* October 1922; Durant to Snyder, undated draft (June 1868), Snyder Papers (ms. 3761, SG4, S3, R17), NSHS; Huntington to Stanford, June 15, 1868.

58. Evans to Dodge, Fort Sanders, in Dodge, op. cit., 757A.
59. Blickensderfer to Dodge, late May 1868, cited in Dodge, op. cit.; Works Progress Administration, *Utah,* 354–57; Twain, *Roughing It,* 90, 91.
60. Blickensderfer to Dodge, in Dodge, op. cit.
61. Dodge, ibid., 759.
62. Dodge to Blickensderfer, Washington, June 4, 1868, Dodge to Oliver Ames, Washington, June 8, 1868, both cited in Dodge, ibid., 760–61.
63. Dodge, ibid., 763.
64. Works Progress Administration, *Utah,* 356–61; Dodge to Blickensderfer, ibid.
65. Durant to Dodge, New York, June 4, 1868, Dodge Papers (ms. 3761, SG4, S3, R6), NSHS. There are two copies, one in Durant's hand, one in Dodge's scrawl, the latter possibly copied off to show to Ames or even U. S. Grant.
66. Dodge to House, Washington, June 6, 1868, in Dodge, op. cit., 761; Dodge to Oliver Ames, Washington, June 8 and 10, 1868, in Dodge, ibid., 761–62, 766A; Oliver Ames to Dodge, New York, June 11, in Dodge, ibid., 762.
67. Dodge, ibid., 762.
68. Hopkins to Huntington, June 16, 1868; E. B. Crocker to Huntington, Sacramento, June 16, 1868, Telegram Book, Stanford Papers (SC33A, B5, F44), Stanford Univ. Archives.
69. Hopkins to Huntington, June 16, 1868. Note: Gov. Bigler wired his own resignation as commissioner while in transit to the convention in New York City: E. B. Crocker to Huntington, June 23, 1868.
70. Huntington to Hopkins, June 11, 1868 (Hopkins Corres. File), Stanford Univ. Archives.
71. Huntington to E. B. Crocker, June 9, 1868.
72. E. B. Crocker to Huntington, June 16, 1868.
73. Stanford to Hopkins, June 9, 1868 (Hopkins Corres. File), Stanford Univ. Archives.
74. E. B. Crocker to Huntington, June 16 and 23, 1868; Huntington to Stanford, June 26, 1868; Stanford to Huntington, July 18, 1868.
75. *San Francisco Alta California,* June 20, 1868; Montague to Ives, Sacramento, June 20, 1868, Stanford Papers, Stanford Univ. Archives.
76. *Sacramento Union,* June 19, 1868 (editorial and news item).
77. Huntington to E. B. Crocker, July 2 and 28, 1868; Huntington to Hopkins, June 19, 1868.
78. Huntington to Hopkins, June 19, 1868; E. B. Crocker to Huntington, July 2, 1868.
79. Montague to Ives, Sacramento, June 16, 1868, Ives Papers, Stanford Univ. Archives.
80. E. B. Crocker to Huntington, June 23, 1868.

Chapter 29—"We Are in a Terrible Sweat"

1. Benedict to Reed, Laramie, June 7, 1868, Reed Papers (ms. 3761, SG4, S3, R15), NSHS.
2. Dodge, *Autobiography,* 763, Council Bluffs Public Library.
3. Reed to Jenny Reed, June 12, 1868, Dodge Papers, Iowa State Archives; Dodge, op. cit., 763–64; Reed to Jenny Reed, June 23, 1868, Reed Papers, NSHS; Jenny Reed to Dodge, June 15 and 22, 1868, Dodge Papers, Iowa; Dodge to Jenny Reed, Council Bluffs, June 25, 1868, Reed Papers, NSHS.
4. Snyder to Dodge, Omaha, June 4, 1868, Dodge Papers, Iowa State Archives; Dillon to Dodge, June 16, 1868, Dodge Papers, Iowa; Dodge, op. cit., 765–66.
5. Dodge, op. cit., 774.
6. Dodge, ibid., 775, 776.
7. Dodge, ibid., 778; Snyder to Dodge, Omaha, July 1, 1868, Dodge Papers, Iowa State Archives.
8. Seymour to Durant, Salt Lake City, June 23, 1868, Seymour Papers (ms. 3761, SG4, S3, R16), NSHS.
9. Seymour to Durant, Weber, June 28 and 29, 1868, Seymour to Crane, Weber, July 2, 1868, Seymour Papers, NSHS.
10. Dodge, op. cit., 778–82.

11. Dodge, ibid., 782, 784; Beadle, 87–90. He devotes nearly a chapter to observations there.
12. Tuttle to Dodge, New York, July 17, 1868, Dodge Papers (ms. 3761, SG4, S3, R6), NSHS; Dodge, op. cit., 784.
13. See note 19.
14. Dodge, ibid., 785–86.
15. Evans to Durant, Laramie, June 3, 18, and 19, 1868, Durant to Evans, New York, undated c. June 15, June 19, 1868, Evans Papers (ms. 3761, SG4, S3, R8), NSHS.
16. Durant to Evans, New York, June 19 and c. 25, 1868, Evans to Durant, Laramie, June 21, 1868, Evans Papers, NSHS.
17. Evans to Durant, Laramie, June 27, 1868, Evans Papers, NSHS; *Frontier Index*, July 1, 1868, reprinted in *Sacramento Union*, July 17, 1868.
18. Klein, 162–6; Ames, 280–82.
19. Dodge, op. cit., 785–86. His publications, *Personal Recollections of Lincoln, Grant, and Sherman* (1914), draw partly from the unpublished memoirs but differ in interesting ways regarding the Fort Sanders meeting. In the Grant paper (102–3), Dodge says he sent his resignation to Durant from Salt Lake City (he did not actually reach that city for some weeks), precipitating the Laramie showdown; the memoirs manuscript does not note this, but the former telling makes sense if one overlooks the small mistake about Salt Lake City. In the Grant article (102), Dodge contends he threatened to resign and "The Government heard of this action of the Company, and [the generals] came . . . to visit me." In the Sherman article (187–89), Dodge says that Sherman read of his threat to resign in the newspapers—and, getting Grant, hastened "thousands of miles" to his side to sustain him. The latter assertion was obviously not true, as the following notes show. The same article claims that Durant met Dodge's stage at Benton and capitulated: "General, I want you to withdraw your dispatch; the lines you want you may have. I am convinced that you are right." This somewhat self-serving anecdote does not jibe with other evidence. But what nags—what remains to be proved—is whether it was Dodge who invited his high-powered sponsors to Fort Sanders, instead of Durant (as Dodge would claim, much later, in his unpublished memoirs). After all, Durant would have realized he had everything to lose by inviting them to a private company meeting involving Dodge. I suspect that Dodge wired Rawlins or Sherman from Utah or Green River, summoning Grant to a prebattle conference at Benton.
20. Grant to C. R. Morehead, Jr., St. Louis, July 14, 1868, Grant to Julia D. Grant, Ft. Leavenworth, July 17, 1868, *The Papers of Ulysses S. Grant* XIX, 8–9; William McFeely, *Grant* (New York: Norton, 1981), 280.
21. Grant to Julia D. Grant, ibid.; Athearn, *William Tecumseh Sherman and the Settlement of the West*, 214; *Cheyenne Leader*, July 25, 1868.
22. See note 18 above. Dodge, op. cit., 787–89, contains the bare bones of the story. His first biographer, the Reverend Jacob Perkins of Council Bluffs (1929) constantly invented dialogue and "scenes" in his work, which was rather naively picked up and repeated as gospel by Charles Edgar Ames (*Pioneering the Union Pacific*) and Robert Athearn in his work on Sherman.
23. E. B. Crocker to Huntington, July 2, 8, 10 and 11, 1868, Huntington Papers, Syracuse University Library (unless otherwise noted, all correspondence between Huntington and his partners is from the Syracuse Library, with all Huntington letters written from New York and his partners' from Sacramento.); C. Crocker to Huntington, July 11, 1868, Telegraph Book, Stanford Papers, Stanford University Library.
24. C. Crocker to Huntington, July 15, 1868; E. B. Crocker to Huntington, July 17, 1868.
25. E. B. Crocker to Huntington, July 28, 1868.
26. Clement statement, USPRRC V, 2577–78; Strobridge recollections, Fulton, *Epic of the Overland*, 38.
27. *Sacramento Union*, July 17, 1868; Works Progress Administration, *Nevada*, 138–39; Twain, *Roughing It*, 193.
28. E. B. Crocker to Huntington, July 31, 1868; Clement statement, USPRRC V, 2577–78.
29. Works Progress Administration, *Nevada*, 137.
30. C. Crocker to Huntington, August 3, 1868.
31. E. B. Crocker to Huntington, July 2, 15, and 17, 1868; Stanford to Huntington, July 15, 1868.
32. E. B. Crocker to Huntington, July 17, 1868; Huntington to Crocker, July 15, 1868, Telegram

Book, Stanford Papers, Stanford Univ. Archives. See also Lavender, 222, and Stewart Daggett, *History of the Southern Pacific* (1922), 120–23.

33. E. B. Crocker to Huntington, July 17, 1868; Stanford to Huntington, July 18, 1868.
34. Bruce, 32, 33; Grinnell, *Two Great Scouts*, 153–57; Hyde, *The Pawnee Indians*, 293–95.
35. Hyde, ibid., 125–29; Sherman to Ellen Sherman, July 15, 1868, cited in Paul A. Hutton, *Phil Sheridan and His Army* (Lincoln: Univ. Neb. Press, 1985), 32–33.
36. Hyde, 125–27; Arthur Ferguson, *Diary* (ms. 3761, SG14, S1, Box 3), NSHS.
37. Beadle, 87–88, 90–92; Emmett D. Chisum, "Boom Towns on the Union Pacific," *Annals of Wyoming* 53 (Spring 1981), 2–13; E. C. Lockwood, "With the Casement Brothers," *Union Pacific Magazine*, February 1931.
38. Casement to Frances Casement, August 1, 13, 17, and 19, September 19, 1868, Casement Papers, Univ. Wyoming; Casement telegrams and dispatches (ms. 3761, SG20, S1), NSHS; Casement Papers (Box 1, f2), Kansas State Univ. Library; Snyder to Dodge, Omaha, August 26, 1868, Dodge, op. cit., 806; Beadle, 104.
39. Carver to Casement, Pittsford, NY, July 31, 1868, Casement Papers (ms. 3761, SG20, S1), NSHS.
40. John H. Gilliss, "Tunnels of the Pacific Railroad," *Transactions* XIII (1870), American Society of Civil Engineers, 163–66; Reed to Durant, Weber, August 7, 11 and 12, 1868, Dodge Papers, NSHS.
41. Oliver Ames to Dodge, North Easton, July 26 and 27, 1868; Dodge, op. cit., 790–91.
42. Ames to Dodge, North Easton, August 23, 1868, in Dodge, op. cit., 805–06.
43. Dodge, op. cit., 800–1; Perkins, 229–30.
44. Young to F. D. Richards, August 4, 1868, cited in Arrington, *The Kingdom Threatened*, 263.
45. Dodge, op. cit., 801–2.
46. *Deseret News*, August 17, 1868. Courtesy Ronald G. Watt and staff, L. D. S. Historical Department.
47. Dodge, op. cit., 802.
48. Young to F. D. Richards, May 23, 1868, cited in Arrington, 262.
49. *New York Herald*, July 30, 1868; *New York Tribune*, July 30, 1868; Klein, 163; J. L. Williams to Dodge, New York, September 1868, in Dodge, op. cit., 818–19.
50. Klein, 167.
51. UP Exec. Comm. Minutes, September 4 , 9, and 11, 1868, 53, 58–61; Oliver Ames to Dodge, August 23, 1868, Dodge Papers, Iowa State Archives.
52. E. B. Crocker to Huntington, August 10, 1868; Stanford to Huntington, August 28 and 30, 1868; Joseph West, "Construction of the Union and Central Pacific Railroads," *Union Pacific Magazine*, October 1922; Arrington, 262–63.
53. E. B. Crocker to Huntington, September 10, 1868.
54. C. Crocker to Huntington, August 22, 1868; E. B. Crocker to Huntington, August 24, 1868. An account of the 6-mile day is in the *Reno Crescent*, August 22, 1868, and in the *Sacramento Union*, August 20, 1868.
55. E. B. Crocker to Huntington, September 13, 1868; C. Crocker to Huntington, August 31, 1868.
56. C. Crocker to Huntington, August 31, 1868.
57. E. B. Crocker to Huntington, August 10, September 10, 1868.
58. Ibid.; E. B. Crocker to Huntington, August 13, September 16, 1868.
59. E. B. Crocker to Huntington, September 7, 1868; E. H. Miller, Jr. to Huntington, September 5, 1868.
60. E. B. Crocker to Huntington, September 10 and 26, 1868; Works Progress Administration, *Nevada*, 129–30, 139–40.
61. Stanford to Huntington, September 15 and 25, 1868; E. B. Crocker to Huntington, September 28 and 29, 1868; Gray to Stanford, September 25, 1868.
62. Gray to Stanford, September 22 and 25, 1868; Gray to Huntington, September 25, 1868.
63. E. B. Crocker to Huntington, August 6, 10, and 24, September 29, 1868; Crocker ms., 38–40, Bancroft Library, UCB.
64. C. Crocker to Huntington, August 31, 1868; E. B. Crocker to Huntington, September 28 and 29, 1868.
65. E. B. Crocker to Huntington, September 28, 1868.

66. Anne Dodge to Dodge, Council Bluffs, August 30, 1868, in Dodge, op. cit., 809.
67. Dodge to Durant, August 27, 1868, in Dodge, op. cit., 807; Williams to Dodge, August 31, 1868, in Dodge, ibid., 809–10.
68. Dodge to Oliver Ames, September 4, 1868, Dodge, op. cit., 812A; Dodge to Durant, Red Dome Pass, August 27 and 28, 1868, Dodge Papers, NSHS.
69. Hirshson, 162–63, 176–77.
70. Anne Dodge to Dodge, September 1868, Dodge, op. cit., 814.
71. C. C. Cope, "Tales from Old Timers—No. 11," *Union Pacific Magazine*, February 1924. Cope, who died at age seventy-nine after writing this story, was mistaken in his dates—he said this transpired in the spring of 1867, when no such episode had occurred.
72. Reed to Dodge, September 1868, in Dodge, op. cit., 818; Durant to Reed, Chicago, September 25, 1868, Durant Outgoing Travelling Copybook, Durant Papers, NSHS.
73. Williams to Dodge, September 5, 1868, in Dodge, op. cit., 813. Interestingly, Robert L. Fulton, *Epic of the Overland* (43), says that the waters of Salt Lake were at their highest in 1868, and over the succeeding thirty years they fell seventeen feet. This cycle has been reversed for quite some time.

Chapter 30—"A Man for Breakfast Every Morning"

1. Browning, *Diary*, II, 212–19.
2. Huntington to Hopkins, September 14, 1868, printed in *Collected Letters* I, Huntington Library. Around this time in late 1868, Huntington misfiled or neglected to keep copies of many of his own letters to Hopkins, Stanford, and the Crockers, therefore leaving significant holes in his New York office files now in the possession of Syracuse University Library; most of his Associates' letters, though, are there. Late in his life his own letters from 1867 on were lightly edited, transcribed, and printed for his private use; copies of this four-volume work may be found at the Huntington Library in San Marino, California, and the Mariners' Museum in Newport News, Virginia. Some letters here were wrongly said to be directed to Charles Crocker when they were actually responses to E. B. Crocker's letters. Because of the convenience of this printed collection, most previous chroniclers—including Huntington's official biographer, David Lavender—have heavily relied on these edited and transcribed letters, though some originals do reside in the Stanford University collection. I have preferred to use Huntington's original, handwritten copies in the Syracuse collection whenever possible, though there have been occasions when only the later transcripts seem to have survived.
3. Huntington's "Go and see him" scheme dribbles across a number of letters and telegrams in the correspondence, most clearly in Huntington to E. B. Crocker, October 22, 1868, in *Collected Letters*, I.
4. Stanford to Huntington, October 1, 1868 (two letters), Huntington Papers, Syracuse University Library (as noted in previous chapters, unless otherwise noted, all correspondence between Huntington and his partners is from the Syracuse Library, with all Huntington letters written from New York and his partners' from Sacramento.)
5. Gray to Huntington, Salt Lake City, October 10, 1868, Huntington Papers, Syracuse. All subsequent Gray letters are from Syracuse. It is impossible to substantiate these particular rumors about inducements to Brigham Young, so tangled were his personal finances in those of the church; certainly, as Leonard Arrington has demonstrated, it is just as hard to read Young's intentions.
6. Gray to Stanford, October 9,1868, copy sent to Huntington October 10, by mail and telegraph.
7. Durant to Stanford, transcribed in Gray to Huntington, Salt Lake City, October 11, 1868.
8. E. B. Crocker to Huntington, October 12, 1868.
9. See Note 6.
10. Huntington to Stanford, September 29, 1868, *Collected Letters*, I; Hopkins to Huntington, October 20, 1868; E. B. Crocker to Huntington, October 26, 1868; Huntington to Browning, *Papers Submitted to the House Committee on the Pacific Railroad. . . . Relative to the Issue of Bonds to the Central Pacific Railroad Company*, Washington, 1869.
11. Huntington to E. B. Crocker, October 14, 17, 21, 1868, *Collected Letters*, I; Lavender, 230, de-

serves credit for stitching this episode together from sundry fragments. As is often the case there is little to document any illegalities; Huntington was becoming more careful even in writing his partners, allowing them, as must we, to read between the lines.

12. Browning, op. cit., 221.
13. Ames to Browning, October 15, 1868, Sec. Int. Correspondence, RG48, NARS.
14. E. B. Crocker to Huntington, October 19, 1868, Central Pacific Telegram book, Stanford Univ. Library; Hopkins to Huntington, Reno, October 19, 1868.
15. Browning, op. cit., 222.
16. Browning did not record appointment of the Central Pacific commission; his first mention was December 4, when its report was cabled to him. The appointment date has been variously given, with this date seeming the likeliest, given the tenor of the cabinet discussion on October 16.
17. Huntington to Hopkins, October 23, 1868, *Collected Letters,* I.
18. Huntington to C. Crocker, October 22 and 29, 1868, *Collected Letters,* I.
19. E. B. Crocker to Huntington, October 27, 1868; Stanford to Huntington, November 2, 1868. Huntington's "Go and see him" wire to Hopkins was dated October 21, 1868 and received the next day: Hopkins to Huntington, October 23, November 19, 1868.
20. Gray to Huntington, Salt Lake City, October 23, 1868.
21. Dodge, *Autobiography,* 824–29, Council Bluffs Public Library.
22. Oliver Ames to Dodge, October 19 and 20, 1868, in Dodge, ibid., 829–30; Oliver Ames to E. H. Rollins, New York, October 10, 1868, Ames to Durant, New York, October 10, 1868, Durant Papers (ms. 3761, SG4, S3), NSHS.
23. Williams to Dodge, Dodge, op. cit., 830.
24. Sen. Ex. Doc. 51, 50 Cong., 1 Sess., 2968.
25. Wilcox to Carlisle, October 26, 1868, Casement Papers (ms. 3761, SG20, S1), NSHS; Oliver Ames to Durant, October 28, 1868, Durant Papers, NSHS.
26. W. C. A. Smoot, "Tales From Old Timers—No. 9," *Union Pacific Magazine,* December 1923.
27. H. W. Guy, "Tales From Old-Timers—No. 20," *Union Pacific Magazine,* February 1925; Works Progress Administration, *Wyoming,* 197.
28. H. Clark Brown, "Tales From Old Timers—No. 31," *Union Pacific Magazine,* May 1926.
29. Fulton, 69–70, 73–74.
30. Theodore Haswell, "Driving Golden Spike, May 10, 1869," *Union Pacific Magazine,* May 1925; Fulton, 78. Haswell mistakenly said that the troops were from Fort Douglas down in Salt Lake Valley. The *Salt Lake City Daily Telegraph,* November 20, 1868, reported that three men, not one, had been lynched in the prompting incident, and that the riot resulted in twenty-five killed and sixty wounded; Leigh Freeman's *Frontier Index* is preserved, though not microfilmed, at the Bancroft Library, UCB.
31. Dodge, op. cit., 833–38; Oliver Ames to Dodge, October 24, 1868, in ibid., 830.
32. Hopkins to Huntington, October 23 and 26, 1868; E. B. Crocker to Huntington, October 27, 1868.
33. C. Crocker to Huntington, October 29, 1868.
34. C. Crocker to Huntington, November 2, 1868; Stanford to Huntington, November 10, 1868.
35. Hopkins to Huntington, November 2, 11, 12, 13, 14, 16, 1868.
36. Hopkins to Huntington, November 16, 1868.
37. Stanford to Huntington, November 4, 1868.
38. Stanford to Huntington, November 10, 1868 (2 letters).
39. C. Crocker to Huntington, November 13, 1868; Hopkins to Huntington, November 16, 1868.
40. E. H. Miller to Huntington, November 13, 1868; Hopkins to Huntington, November 16 and 20, 1868.
41. Stanford to Huntington, Salt Lake City, November 21, 1868, Ogden, November 24,1868.
42. Ibid.; Huntington to Stanford, November 13, 1868, *Collected Letters* I.
43. Hopkins to Huntington, November 25, 1868; Huntington to Hopkins, November 18, 1868, *Collected Letters* I.
44. J. S. Casement to Frances Casement, October 31, November 6 and 15, 1868, Casement Papers, Univ. Wyoming; Hutton, 52–55; Dodge, op. cit., 834.
45. Evans to Durant, Bear River, November 19, 1868, Evans Papers (ms. 3761, SG4, S3, R8), NSHS.

46. Dodge, op. cit., 829–30, 842.
47. Ibid., 861.
48. Ames to Dodge, October 24, 1868, Dodge Papers, Iowa State Archives; Bushnell to Durant, October 28, 1868, Bushnell Papers (ms. 3761, SG4, S3, R2), NSHS.
49. Wilson Report, 765-66; Davis & Assoc. contract, November 1, 1868, Davis Papers (ms. 3761, SG4, S3, R6), NSHS.
50. Poland Report, 4, 414 (McComb test.); Oliver Ames to Hazard, November 17, 1868 (2 letters), Hazard to Oliver Ames, November 30, 1868, Hazard Papers, Rhode Island Hist. Soc.; Klein, 176–77.
51. House Exec. Doc. *15*, 40 Cong., 3 Sess., 1–16.
52. Dodge, op. cit., 851; Williams report, House Exec. Doc. *15*, 27–31.
53. Blickensderfer to Dodge, Omaha, late November 1868, in Dodge, op. cit., 857.
54. Durant to Johnson, October 29, 1868, Durant Outgoing Travelling Copy Book UPRC (ms. 3761, SG4, S3, R7), NSHS; Williams to Dodge, New York, September 5, 1868, in Dodge, op. cit., 813; Browning, *Diary*, 228–29; Secretary of the Interior, Annual Report, November 30, 1868, R. G. 48, National Archives.
55. Oliver Ames Diary, December 1, 1868, Ames Papers, Stonehill College; Klein, 175; Dodge, op. cit., 857.
56. C. Crocker to Huntington, November 21, 1868; Lavender, 234; E. B. Crocker to Huntington, November 24, 1868.
57. Crocker ms., 36–37, Bancroft Library, UCB.
58. Browning, *Diary*, 229; *Report of the Special Commissioners*, Secretary of the Interior, R. G. 48, National Archives; *Papers Submitted to the House Committee on the Pacific Railroad . . . Relative to the Issue of Bonds to the Central Pacific Railroad Company*, Washington, 1869 (the report was also published in pamphlet form separately in San Francisco, 1869); Huntington to C. Crocker, December 16, 1868, *Collected Letters* I: "There is," he added, "nothing like sleeping with men, or women either for that matter." Contemporaries said that in conversation Huntington had a ribald sense of humor, but it rarely worked its way into his business correspondence.
59. C. Crocker to Huntington, December 1, 1868; Hopkins to Huntington, December 2, 1868.
60. Stanford to Huntington, December 4, 1868; Stanford to Hopkins, December 10, 1868, disclosed that he had already commissioned Chauncey West to purchase the rights-of-way. See also Stanford to Hopkins, December 4 and 13, 1868, and Stanford to E. B. Crocker, December 8, 1868, Stanford University Library.
61. E. B. Crocker to Huntington, December 9, 1868.
62. Hopkins to Huntington, December 10, 1868; E. B. Crocker to Huntington, December 8, 1868.
63. Stanford to Huntington, December 13, 1868. See also Stanford to Hopkins, December 13, 1868, Stanford University Library.
64. For such a dramatic homecoming, very little survives beyond telegrams of the journey's arrangements and progress (Central Pacific Telegram Book, Correspondence, Stanford University Library); in the Syracuse letters as well as the Huntington transcripts there are hardly any references to specific conversations held during the trip. David Lavender (235–36) supposes that Huntington "railed" at his friends face-to-face, basing this on letters written by Huntington after he returned to New York. Though Huntington and Charley Crocker had reputations as blunt talkers, it's more likely that they, like the others, reserved spleen for the written page. There are almost no references to angry verbal confrontations, even with people like Cole and the larcenous government director Denver, in all their many hundreds of pages of letters. I assume a tighter rein in their meetings.
65. Oliver Ames, *Diary*, December 3–5, 1868, Oliver Ames Papers, Stonehill College; Browning, *Diary*, 229–30; Oliver Ames to Durant, December 3, 1868, Ames Papers, UPRC (ms. 3761, SG4, S3, R1), NSHS.
66. Dodge, op. cit., 851.
67. Ibid., 861–62; Oliver Ames to Browning, New York, December 30, 1868, cited in ibid., 854.
68. Ibid., 862–63.
69. Browning, *Diary*, 231–33; B. F. Ham to Oliver Ames, December 29, 1868, Oliver Ames to Oakes Ames, December 29, 1868, Ames Papers, Stonehill College; Klein, 178.

70. Oliver Ames to Durant, Washington, December 23, 1868, Ames Papers, NSHS; Durant to Snyder, December 9, 1868, Durant to Reed, December 17, 1868, Durant Outgoing Travelling Copy Book, NSHS.
71. Browning, *Diary,* 232; Oliver Ames to Durant, New York, December 24, 1868, Ames Papers, NSHS.
72. Oliver Ames to Durant, New York, January 1, 1869, Ames Papers, NSHS.
73. Apparently complaints reached Secretary Browning, who conferred with President Johnson on December 26, 1868, resulting in Wendle's discharge. John F. Coyle of the *National Intelligencer* seems to have carried not only the complaint but the nomination for a replacement, Chauncy Snow, also of that newspaper. See Browning, *Diary,* 232.
74. Reed to Jenny Reed, December 16 and 28, 1868, Reed Papers, (ms. 3761, SG4, S3, R15), NSHS; Dodge, *How We Built the Union Pacific,* 117.
75. Dodge, *Autobiography,* 859, 863, 866.
76. Works Progress Administration, *Nevada,* 120; Beadle, 156–58; C. Crocker to Huntington, December 15, 1868.
77. Hopkins to Huntington, December 15, 1868.
78. Hopkins to Huntington, December 21, 1868.
79. Hopkins to Huntington, November 25, 1868, December 2, 15, and 21, 1868; E. B. Crocker to Huntington, November 26, 1868, December 5 and 8, 1868; Huntington to Hopkins, December 2, 1868, *Collected Letters* I.
80. *Annual Report,* Central Pacific Railroad Company of California, 1868.
81. *Reno Crescent,* November 28, 1868.
82. Trustees Minutes, December 28, 1868, 55, Board/Exec. Papers (ms. 3761, SG1, S2);Wilson Report, 552 (Bushnell test.), 736, 760–61.
83. Browning, *Diary,* 231–33.

Part IX—1869: Battleground and Meeting Ground

Chapter 31—"A Resistless Power"

1. Huntington Typescript, 12–13, UCB. The date of his trip west is erroneously given there as December 1867—a mistake of Bancroft's biographer, David Sessions, which may be found added to his handwritten first draft (Huntington Manuscript), inked in. As has been seen, Huntington did not go to California in 1867.
2. Stanford to Hopkins, Ogden, January 15, 1869, Huntington-Hopkins Correspondence, Stanford; Stanford to Huntington, Ogden, January 18, 1869, Huntington Papers, Syracuse Univ. Library. (Unless otherwise noted, all correspondence between Huntington and his partners is from the Syracuse Library, with all Huntington letters written from New York and his partners' from Sacramento.)
3. Stanford to Huntington, Salt Lake City, January 22, 1869.
4. C. Crocker to Huntington, January 20, 1868.
5. Oliver Ames to Browning, New York, December 30, 1868, cited in Dodge, *Autobiography,* 851–54.
6. Dodge, ibid., 854.
7. Oliver Ames to Dodge, New York, January 8, 1869 (two letters), in Dodge, ibid., 876–77.
8. Dodge, ibid., 854, 877.
9. Blickensderfer to Dodge, January 2, 5 and 6, 1869, in Dodge, ibid., 866–78; the last-referred letter is erroneously said to have been received in Washington on January 6, an impossibility given its content.
10. Snyder to Dodge, January 1869, in Dodge, ibid., 869.
11. *New York Tribune,* January 15, 1869.
12. Charles Francis Adams, Jr., "Railroad Inflation," *North American Review,* January 1869, 144–45, 147–48.
13. Ibid., 164. It is a delicious irony that Adams would one day come to preside over the Union Pacific.
14. *New York Tribune,* January 15, 1869.
15. Dodge to Browning, January 11, 1869, in Dodge, op. cit., 844, 845–48. Apparently the first news

of Blickensderfer's appointment came to him from Dodge: Blickensderfer, January 16, 1869, Dodge, op. cit., 886.

16. Oliver Ames Diary, January 14, 1869, Ames Papers, Stonehill College; Oliver Ames to Dodge, January 15, 1869, in Dodge, ibid., 883.
17. *New York Tribune*, January 15, 1869.
18. Huntington to Stanford, January 25, 1869, Huntington to Hopkins, January 26, 1869, *Collected Letters*, I.
19. Huntington to Johnson, Andrew Johnson Papers, Library of Congress; partially transcribed in Dodge, op. cit., 855.
20. Huntington to Hopkins, January 20 and 26, 1869, Huntington to C. Crocker, February 11, 1869, *Collected Letters*, I.
21. Stanford to Huntington, January 24, 1869; Stanford to Hopkins, January 25, 1868, Huntington-Hopkins Correspondence, Stanford University.
22. Huntington to Stanford, January 30, 1869, *Collected Letters*, I.
23. E. B. Crocker to Huntington, January 28, 1869.
24. Warren to Stanford, Stanford to Warren, January 29, 1869; Stanford to Huntington, January 30, 1869.
25. Hopkins to Huntington, January 31, 1869; *Sacramento Union*, January 19, 1869.
26. Stanford to Huntington, February 8, 1869; Hopkins to Huntington, February 8, 1869.
27. Reed to Jenny Reed, January 12, 1869, Reed Papers, UPRC (ms. 3761, SG4, S3, R15), NSHS.
28. Joseph A. West, "Construction of the Union and Central Pacific Railroads Across Utah, Fifty-five Years Ago," *Union Pacific Magazine*, October 6, 1922.
29. Klein, 192; Oliver Ames to Durant, January 1 and 16, 1869, Ames Papers, UPRC (ms. 3761, SG4, S3, R1), NSHS.
30. Young to Albert Carrington, January 5 and February 4, 1869, in Arrington, *The Kingdom Threatened*, 264.
31. Gilliss, 163–66.
32. Beadle, 138–39.
33. Works Progress Administration, *Utah*, 357.
34. Ibid., 361–62; West, 8; Arrington, 264–65; Beadle, 120–25; Bernice Gibbs Anderson, "The Gentile City of Corinne," *Utah Historical Quarterly* IV (1941), 141–54; B. D. and B. M. Madsen, "Corinne, The Fair," *Utah Historical Quarterly*, XXXVII (1969)102–23.
35. West, 8; Beadle, 140, 154.
36. Hoxie to Dodge, January 17, 1869, in Dodge, op. cit., 886.
37. Lockwood, 36; Griswold, 301–2.
38. Lockwood, 35–36: "Dan Casement told him that if he ever needed help to come to him," Lockwood continues. "Years afterward, the captain needing some help, went to Painesville to see Dan Casement and found he was dead. The captain then went to General Casement and told him the story and promise. General Casement said he knew nothing about it and told him to see Mr. Lockwood. He hunted me up, so I introduced him to my sister, Mrs. Dan Casement, and told her how he had saved Dan's life. She helped him generously as she knew her husband would have done had he been alive." It is interesting that Dan Casement was said to have nearly perished walking seventy-seven miles to Laramie in the subsequent blizzard (J. Casement to Frances Casement, March 12, 1869, Casement Papers, Univ. Wyoming). It is difficult to judge which account is true: Lockwood was closer to the action in Echo Canyon at the time, but his memory may be fogged by fifty years; Jack Casement, for that matter, although writing only a few weeks later, may have confused events during the back-to-back snowstorms of February 1869.
39. Casement to Frances Casement, February 8 and 11, 1869, Casement Papers, Univ. Wyoming; Reed to Jenny Reed, February 10, 1869, Reed Papers, NSHS.
40. J. Casement to Dodge, February 13, 1869, Dodge, op. cit., 893; D. Casement to Snyder, February 27, 1869, Dodge, ibid., 897; Evans to Dodge, February 18, 1869, Dodge, ibid., 1172.
41. *Sacramento Union*, March 6, 1869; Lambard et. al. to Huntington, Rawlins, February 22, 1869, Hopkins to Huntington, March 17, 1869.
42. Reed to Jenny Reed, February 27, 1869, Reed Papers, NSHS; Seymour to Durant, February 28, 1869, Seymour Papers (ms. 3761, SG4, S3, R17), NSHS.
43. *Sacramento Union*, February 20 and 22, 1869.

44. E. B. Crocker to Huntington, February 23, 1869; Stanford to Huntington, February 14, 1869.
45. *Reno Crescent,* January 23, 1869; Griswold, 298.
46. Stanford to Huntington, January 22 and 24, 1869; Lavender, 238; Dodge, op. cit., 855, 917–18; Browning, *Diary,* 242.
47. Axtell to Huntington, February 22, 1869; Browning, *Diary,* 238–39, 242.
48. Huntington Typescript, 92–93, UCB; Lavender, 239; Huntington to Hopkins, March 5, 1869, *Collected Letters,* I; Dodge, op. cit., 856; CPRR, *Papers Submitted to the House Committee on the Pacific Railroad . . . Relative to the Issue of Bonds to the Central Pacific Railroad Company.* Note that nearly every account above quotes a different sum received by Huntington; the last-cited pamphlet seems to be the correct one. Also see Klein, *Source Notes: Union Pacific* (1987), note 196–1, 35.
49. Browning, *Diary,* 244; *Sacramento Union,* January 8, 1869; Dodge, op. cit., 898.
50. Dodge, op. cit., 867.
51. Hoxie to Dodge, January 17, 1886; Dodge, op. cit., 886.
52. Dodge, op. cit., 870, 886, 889; Durant to Seymour, February 15, 1869, Seymour Papers, UPRC, NSHS; Durant to Dodge, February 13, 1869, Dodge Papers, UPRC, NSHS.
53. Durant to Dodge, March 8, 1869, Dodge Papers, UPRC, NSHS.
54. Dodge, op. cit., 889–90.
55. Snyder to Dodge, February 4, 1869, Dodge, op. cit., 890.
56. Klein, 198; Wilson Report, 272, 765–66.
57. Hopkins to Huntington, March 9, 1869; Stanford to Huntington, March 16, 1869.
58. Stanford to Huntington, March 13 and 16, 1869; *Sacramento Union,* April 21, 1869.
59. Arrington, 265; Utah Historical Records Survey (WPA), "A History of Ogden" (1940), 45–47, 48–49.
60. Ibid., 49.
61. Works Progress Adminstration, *Utah,* 205.
62. Casement to Frances Casement, March 3, 8, and 12, 1869, Casement Papers, American Heritage Center, University of Wyoming Library; also in Casement Papers (ms. 3761, SG20, S1, F20), NSHS.
63. Stanford to Huntington, March 16, 1869.
64. *New York Times,* March 23, 1869.
65. Ibid.; Klein, 202; Adams, op. cit.
66. Klein, 202.
67. *Sacramento Union,* January 8, 1869; *New York Tribune,* May 11, 1869.
68. Klein, 203; *New York Sun,* March 11, 1869; *New York Times,* March 11 and 12, 1869.
69. Klein, 203; Dillon to Dodge, March 14, 1869, Dodge Papers, Iowa State Archives.
70. *New York Herald,* March 13, 1869.
71. All quoted in *New York World,* March 6, 1869.
72. Bushnell to Durant, March 20, 1869, Bushnell Papers, (ms. 3761, SG4, S3, F137), NSHS.
73. Dodge to Grant, March 26, 1869, Railroad File 254, Secretary of Interior (RG48), NARA; Dodge to Anne Dodge, March 26 and 29, 1869, Dodge Papers, Iowa State Archives.
74. Dodge, op. cit., 915–16.
75. *New York Times,* March 21, 1869; Durant to Oliver Ames, March 19, 1869, NSHS.
76. *New York Times,* March 24 and 31, 1869.
77. *New York Herald,* March 31, 1869; *New York Times,* April 1, 1869.
78. *New York Herald,* April 3, 1869.
79. Ibid.
80. *New York Times,* March 24, 1869.
81. *Sacramento Union* (report datelined April 8), April 22, 1869; *New York Times,* March 31, 1869; Hopkins to Huntington, March 17, 1869; Chittenden to Huntington, March 18, 1869.
82. Stanford to Huntington, March 21, 1869; Hopkins to Huntington, March 22 and 31, 1869.
83. "*Statement of the Central Pacific Railroad Comp'y of California to the Committee of the Senate of the United States on the Pacific Railroad, March 25, 1869* (Washington: Gibson Bros., 1869).
84. Stanford to Huntington, April 5, 1869.
85. Dodge, op. cit., 906–19, 921, 923.
86. Ibid., 919, 920, 930, 932, 933–34; Union Pacific Directors' Minutes, April 9, 1869, NSHS; Klein, 207; Lavender, 241.

87. Stanford to Huntington, April 22, 1869; C. Crocker to Huntington, April 10, 1869; Crocker ms., 58, Bancroft Lib., UCB.
88. Ames to Dodge, April 12, 1869 (it is marked erroneously as March 12), Dodge, op. cit., 920. Charles Edgar Ames (316) was led astray by this misdate, despite ample internal evidence to the contrary.
89. Ibid.; Ames to Dodge, April 12, 1869, Dodge, op. cit., 932. Probably this was written the night before, on the eleventh, as the above-mentioned letter refers to it exactly. This April 11 letter also states, interestingly, based on vague newspaper reports Ames had just read, that "Any settlement is better than a constant fight," but worries about the Central Pacific "only paying us for our road to Promontory Point and probably not paying near as much as it has cost us." Dodge's lengthy retort is in Dodge, op. cit., 933–34.
90. Dodge, ibid., 931.
91. *New York Times,* April 4, 11, and 15, 1869; *New York Herald,* March 31, April 3, 4 , and 9, 1869; Klein, 204.
92. *Sacramento Union,* April 10, 1869.
93. C. Crocker to Huntington, April 10 and 20, 1869; Stanford to Huntington April 22, 1869 (two telegrams, one letter), April 23, 1869.
94. *Daily Colorado Tribune,* April 6, 1869, quoted in *Sacramento Union,* April 21, 1869.
95. "To the Stockholders of the Union Pacific Railroad Company," April 10, 1869, NSHS; Klein, 204.
96. Casement to Frances Casement, April 6 and 9, 1869, Casement Papers, American Heritage Center, University of Wyoming Library; Dodge to Oliver Ames, March 23, 1869, Dodge Papers, NSHS.
97. Snyder to Dodge, Dodge, op. cit., 937.
98. *New York Herald,* April 23, 1869; *New York Times,* April 23, 1869; Dodge, op. cit., 937, 939.
99. Crocker ms., 55–58, Bancroft Library, UCB; Stanford to Huntington April 22, 1869; *San Francisco Bulletin,* April 29, 1869. J. D. B. Stillman, "The Last Tie," *Overland Monthly,* July 1869, 81, names the Irishmen who each bore some 74 tons in 11 hours: Michael Shay, Patrick Joyce, Thomas Dailey, Michael Kennedy, Frederick McNamara, Edward Killeen, Michael Sullivan, and George Wyatt; he does not trouble to single out any Cantonese.
100. Dodge to Anne Dodge, May 2, 1869, Dodge Record, Iowa State Archives.
101. Stanford to Huntington, April 28, 1869; E. B. Crocker to Huntington, May 4, 1869; W. E. Chandler to Dodge, April 25, 1869; Dodge, op. cit., 939.
102. E. B. Crocker to Huntington, April 16, 23, 28, and 29, 1869; Hopkins to Huntington, April 29, 1869; Stanford to Huntington, April 23, 1869.
103. C. Crocker to Huntington, May 1, 1869.
104. Dodge to Oliver Ames, May 2, 1869, Dodge to Evans, May 3, 1869, Dodge Papers, Iowa State Archives; Dodge to Oliver Ames, April 24, 1869, Dodge, op.cit., 938; Dodge to Dillon, May 4, 1869, Dodge, ibid., 943.
105. A. P. Wood, "Tales From Old-Timers—No. 19," *Union Pacific Magazine,* January 1925.
106. S. Schimonsky to Snyder, April 4 and 15, 1869, Snyder to Dodge, May 4, 1869, Dodge Papers, Iowa State Archives. There is an amusing anecdote about Schimonsky in the article by A. P. Wood; see Note 104.
107. *New York Herald,* April 1, 1869; Ham to Oliver Ames, May 1, 3, and 8, 1869 (ms. 3761, SG4, S3, R10), NSHS; Oakes Ames to Dodge, May 2, 1869, Dodge, op. cit., 943; Klein, 216.
108. Oliver Ames to Duff and Dillon, May 1, 1869; Dillon to Glidden, May 1, 1869; Dodge, ibid., 943; Oakes Ames to Glidden and Williams, May 4,1869 (ms. 3761, SG4, S3, R1, R20), NSHS.

Chapter 32—"We Have Got Done Praying"

1. J. D. B. Stillman, "The Last Tie," *Overland Monthly,* July 1869, 78–79.
2. Ibid., 80.
3. There is much misinformation about the episode to follow, even to the extent of which direction Durant was traveling when his train was halted; A. P. Wood ("Tales From Old Timers—No. 19," *Union Pacific Magazine,* January 1925) was a construction engineer at Echo City in May 1869, and asserted that Durant headed east from Echo City; because his dates and several other facts were wrong he may safely be discounted. Wesley Griswold (1962), often a careful writer despite

a deplorable lack of source notes, seems to agree with the eastbound thesis and follows an account in the *San Francisco Alta California*, May 10, 1869, which is also weakened by errors. The always resourceful Maury Klein (1987) is, here, carefully noncommittal. Union Pacific cable traffic in the Leonard Collection (Univ. Iowa or UPRC, NSHS) does not help much. But Dodge's account in his unpublished memoirs, and the contemporary letters in his papers in the Iowa State Archives, seem to indicate a westbound thesis. I followed this, especially after finding an account of the episode in the *New York Herald*, May 11, 1869, datelined Omaha, May 3, 1869. (I cannot resist noting that the inventive John Hoyt Williams places the episode two months earlier, despite overwhelming evidence to the contrary.)

4. Richardson, *Garnered Sheaves*, 277, 262–67.
5. Dodge to Anne Dodge, May 6, 1869, Dodge Record, VIII, 233–34, Iowa State Archives.
6. Dodge to Oliver Ames, May 7, 1869, Dodge Papers, Iowa State Archives.
7. Dodge to Oliver Ames, May 8, 1869, *Memoirs*, 845, Council Bluffs Public Library; Dodge Papers, Iowa State Archives.
8. Dodge, op. cit., 944–45.
9. Oliver Ames to Dodge, May 12, 1869, Dodge Papers, Iowa State Archives.
10. Stillman, 81; Works Progress Administration, *Nevada*, 252–53.
11. Stillman, 81; Richardson, *Garnered Sheaves*, 288–89. Richardson would die in New York in December, shot in the *Tribune* offices by the divorced husband of Richardson's fiancée.
12. Ibid.
13. Stillman, 82.
14. *Sacramento Union*, May 10, 1869, including all details of the city ceremony following these passages.
15. *San Francisco Evening Bulletin*, May 8, 1869.
16. *Sacramento Union*, June 25, 1878; H. H. Bancroft, Bancroft Reference Notes, California Biography, UCB; Bancroft, "Crocker Biographical Sketch," UCB; *Phelps' Contemporary Biography of California's Representative Men* (1881–82), II, 139; Henry B. Nason, *Biographical Record of the Officers and Graduates of the Rensselaer Polytechnic Institute* (1887), 202–8.
17. Stillman, 82.
18. L. H. Eicholtz, Diary, May 7–9, 1869, quoted in Paul Rigdon, *Historical Catalog of the UPRR Museum* (1951), UPRC (ms. 3761, SG197), NSHS.
19. Dodge, *Autobiography*, 958, Council Bluffs Public Library; Rigdon, 863; Lewis, 96.
20. The nearest one can get to this myth is in the *San Francisco Evening Bulletin*, May 8, 1869, which mentions the verbal threats of some Union Pacific graders to "clean out" the Chinese, which is apparently as far as it went. One third-hand-reported event of an unannounced blast occurring too close to a rival gang seems to have involved two Mormon contractors: *Salt Lake City Deseret News*, March 25, 1869.
21. Russell's personal observations of the day are in *Frank Leslie's Illustrated Newspaper*, June 5, 1869; see also Susan E. Williams, "The Great West Illustrated: A Journey Across the Continent with Andrew J. Russell," *The Streamliner* (Union Pacific Historical Society), 1996. The presence of S. J. Sedgwick has usually been overlooked in accounts. It should be noted that Stillman's and Russell's accounts of the ceremony, taken with the other contemporary reports in newspapers, have been relied on here more than Dodge's and Dillon's, which were fogged by the long passage of time and damaged by serious errors of fact, and Todd's, which was very brief. Levi O. Leonard's account in *Union Pacific Magazine*, May 1929, was helpful and relatively free of errors; other helpful but sometimes unreliable accounts of ceremony observers not otherwise cited here are in *Union Pacific Magazine*, May 1922, May and September 1923, and May 1925.
22. Stillman, ibid.
23. Joseph A. West, "Construction of the Union and Central Pacific Railroads," *Union Pacific Magazine*, October 1922; Arrington, *The Kingdom Threatened*, 265–67; Works Progress Administration, *A History of Ogden*, 48, and *Utah*, 130–31; Milton R. Hunter (Ed.), *Beneath Ben Lomond's Peak: A History of Weber County* (1966), 415–16; Robert G. Athearn, "Contracting for the Union Pacific," *Utah Historical Quarterly* XXXVII (Winter 1969).
24. *Sacramento Union*, May 11 and 12, 1869; *San Francisco Alta California*, May 11 and 12, 1869. David Hewes, who donated one golden spike, later claimed that he donated the laurel tie, also, but it was Evans. Late in his life Hewes made elaborate, silver-tipped canes from laurel wood, and pre-

sented them to friends, with engravings purporting them to be "Made from the Tree of the Last Tie" (*Union Pacific Magazine*, May 1926, 5). The bona fide laurel tie burned in the Southern Pacific office after the 1906 San Francisco earthquake. See also J. N. Bowman, "Driving the Last Spike," *California Historical Quarterly* XXXVI (1957), probably the last word on the ceremony.

25. John Todd, *The Sunset Land, or, The Great Pacific Slope* (1870).
26. Golden-spike observers Thomas Rose and Amos Bowsher were interviewed in the *Ogden Standard Examiner*, May 20, 1939.
27. "Veteran Recalls Driving of Golden Spike," *Leavenworth Times*, reprinted in *Union Pacific Magazine*, May 1926.
28. *Ogden Standard Examiner*, May 20, 1939.
29. Throughout the ceremony account I am simplifying directions: since the track ran in a southwest-to-northeast direction in the summit valley, chroniclers have variously described Stanford, for instance, as standing south of the tracks, southeast of the tracks, and east of the tracks. The ceremony participants considered positions in east-west terms for symbolic reasons, as did I.
30. *New York Times, New York Tribune, New York Herald*, May 11, 1869.
31. Ibid., and May 12, 1869. Let us examine the mythology of the missed swings, another durable anecdote. At the crucial moment both Stanford and Durant are said to have missed the swings, causing great hilarity among the workingmen present: what a wonderful send-up of the soft executives, to be so feckless. The sole source seems to be Alexander Toponce, a Utah freight driver said to have a contract to supply beef to the laborers, who set down his recollections long after the golden spike; they were published in 1923. Curiously, none of the other eyewitnesses—the many correspondents, Dr. Stillman, Dodge, Russell, Todd, Dillon, Bowsher, or the Union Pacific veterans collected by Levi O. Leonard (Shilling, Doddridge, Doremus, Hodges, Haswell, Wood, Bissell, O'Donnell, Malloy, and Anna Reed Bennitt, daughter of S. B. Reed), or Lt. J. C. Currier May 10 entry, *Diary*, quoted in J. N. Bowman, recounted such a hilarious moment. Thomas O'Donnell (*Union Pacific Magazine*, May 1922) contends that "the privilege of striking the spike the first blow was given to some lady, whose name I don't know, but she proved a poor 'Spiker' and missed it. Mr. Durant took the maul and started the spike. President Stanford of the Southern Pacific completed the job." He is the sole contender for this. Significantly, Bowsher, who was perched above everyone on the telegraph pole, recalled no problem.
32. Ibid., and *Chicago Tribune*, May 11, 1869.
33. *Chicago Tribune*, May 11, 1869.
34. Anna Judah, ms. 14, Bancroft Library, UCB; *New York Times*, May 12, 1869.
35. Bowman, 99; Erle Heath, "Eye Witness Tells of 'Last Spike' Driving," *Southern Pacific Bulletin*, May 1926; Theodore Haswell, "Driving Golden Spike May 10, 1869," *Union Pacific Magazine*, May 1925; Stillman, 84.
36. *New York Times*, May 12, 1869; *New York Herald*, May 11, 1869; *The Papers of Ulysses S. Grant*, XIX, 468.
37. Dodge, *Personal Recollections of Sherman* (1914), 203. The mistaken date has been corrected here.
38. Oliver Ames to Dodge, May 10, 1869; Dodge, *Autobiography*, 946.
39. *Chicago Tribune*, May 12, 1869.
40. Dodge, op. cit., 953–54.
41. Huntington to C. Crocker, May 10, 1869, *Collected Letters*, I.
42. *New York Herald*, May 11, 1869; *New York Tribune*, May 11, 1869.
43. *Chicago Tribune*, May 12, 1869.

Part X—1872–73: Scandals, Scapegoats, and Dodgers

Epilogue—"Trial of the Innocents"

1. Muzzey, 64–65; *Dictionary of American Biography* (Greeley, Blaine, Garfield, Colfax).
2. *New York Sun*, September 4, 1872; *New York Times*, February 2, 1885.
3. *New York Tribune*, September 7, 1872.
4. *New York Sun*, September 10 and 17, 1872; *New York Times*, September 16, 1872.
5. See Chapters 26 and 27.
6. *New York Tribune*, September 11, 1872; *New York Sun*, September 14, November 30, 1872.

7. *New York Sun,* September 17, 1872; *Worcester* (Massachusetts) *Spy,* September 14, 1872, reprinted in *New York Times,* September 16, 1872.
8. *New York Sun,* September 17, 1872.
9. *New York Times,* September 18, 1872; *Chicago Tribune,* September 11, 1872; *New York Sun,* September 20, 1872.
10. *New York Sun,* September 23 and 26, 1872.
11. *New York Tribune,* September 28, 1872.
12. *New York Tribune,* September 24, 1872; *New York Sun,* September 28, 1872.
13. *New York Sun,* November 6, 7, 29, 30, 1872, January 31, 1873.
14. *New York Times,* December 3, 1872; *New York Sun,* December 3, 1872.
15. Klein, 30; *New York Times,* ibid.; *New York Sun,* ibid.
16. *New York Sun,* December 14 and 16, 1872; *New York Tribune,* December 16, 1872. The extra congressman named by McComb was not revealed in the press of the day, though it is presumed to be James Brooks, as widely rumored in Washington at the time.
17. *New York Sun,* December 16, 1872; Frank M. O'Brien, *The Story of the Sun* (New York: Doran, 1918), 312.
18. *New York Tribune,* December 17 and 18, 1872; *New York Sun,* December 18 and 19, 1872.
19. *New York Sun,* December 20, 1872; *New York Tribune,* December 20, 1872.
20. Excerpted in *New York Tribune,* December 26, 1872.
21. *New York Tribune,* January 6, 1873.
22. *New York Tribune,* January 7 and 8, 1873; *New York Times,* January 7, 1873.
23. *New York Tribune,* January 8, 1873.
24. Hochschild, 2–13; Gilborn, 1–14.
25. Hirshson, 174–88.
26. Ibid., 191–92; Dodge to Uriah H. Painter, January 7, 1873, Dodge to George W. McCrary, January 8, 1873, Dodge Papers, Iowa State Archives.
27. Hirshson, 192.
28. Ibid., 192–93.
29. Ibid., 262.
30. *New York Sun,* January 22 and 20, 1873; *New York Tribune,* January 22, 1873.
31. Twain and Warner, *The Gilded Age* (1873); Albert Bigelow Paine, *Mark Twain: A Biography* (1912), 476–77; Bryant Morey French, *Mark Twain and The Gilded Age* (Dallas: Southern Methodist, 1965), 121–24; Philip S. Foner, *Mark Twain: Social Critic* (New York: International, 1958), 69–80.
32. Foner, ibid., 81.
33. *New York Sun,* February 6, 1873.
34. *New York Sun,* February 6 and 12, 1873; *New York Times,* February 2, 1885.
35. *New York Sun,* February 12, 1873; *New York Times,* February 2, 1885.
36. Whitman, "The Return of the Heroes," "To a Locomotive in Winter," *Leaves of Grass* (New York: Modern Library Edition, 1993); *Collect and Other Prose* (ed. Floyd Stovall, New York, 1964), 369–70; *Leaves of Grass, A Textual Variorum of the Printed Poems* (ed. Bradley, Blodgett, Golden, and White, New York, 1980), 261. I am grateful to Justin Kaplan, *Walt Whitman: A Life* (New York, Simon & Schuster, 1980), and David S. Reynolds, *Walt Whitman's America* (New York, Knopf, 1995), for many helpful signposts.
37. Gail Hamilton, *Biography of James G. Blaine* (Norwich: Henry Bill, 1895), 274.
38. "Tributes to Oakes Ames: Extracts from Letters Received," *Oakes Ames: A Memoir* (Cambridge: Riverside, 1883), 80.
39. *New York World,* February 19, 1873; *New York Sun,* February 19 and 20, 1873; *New York Tribune,* February 19, 1873; Bowers, 401.
40. The full report is in the *New York Sun,* February 19, 1873 (with additions on February 20); the report and proceedings of the inquiry are in House Report 77, 42 Cong., 3 Sess. There are useful overviews of the report in most of the newspapers, but the *Sun* (February 22, 1873) is the harshest—and the brightest of the contemporary press.
41. *New York World,* February 19, 1873.
42. *New York Sun,* February 19, 1873; *The Nation,* February 27, 1873; *New York Tribune,* February 19 and 20, 1873; Bowers, 401–2.
43. Klein, 294–95; Prof. Klein has a useful short discussion of the committee reports and the finan-

cial issues involved (293–303), although a careless reader might think that Klein intended to buff the widespread fraud and greed with the value-free sandpaper of directorial "mismanagement." Similarly, an inattentive reader might think that Klein's dismissal of the Gilded Age and the robber barons as mere "historical clichés" means he finds no truths in critical summaries of the era, which surely cannot be the case. Before Prof. Klein's corporate history (1987) there was, of course, Robert Fogel's book *The Union Pacific Railroad: A Case in Premature Enterprise* (1960), which was the first critical discussion of costs and profits of the Union Pacific, although he does not address the sidelines issue.

44. *New York Tribune,* February 21 and 22, 1873; *New York World,* February 19, 1873; *New York Sun,* February 21, 24 and 25, 1873.
45. *New York Sun,* February 26, 1873. Contains, in addition, the full Associated Press report.
46. *New York Times,* February 27, 1873; *New York Sun,* February 27, 1873.
47. The "Salary Grab Act" was repealed after a public outcry by the subsequent Congress (1874), except in the cases of the president and the Supreme Court; Speaker Blaine had, to his credit, excepted himself from the congressional raise and bonus.
48. Sons of Oakes Ames, *Oakes Ames: A Memoir* (1883), 46–47.
49. Huntington to Hopkins, February 20 and 27, March 3, 1873, *Collected Letters* II; Lavender, 287, 291.
50. Lavender, 284, 294; Lewis, 110–11.
51. Crocker ms., 63, Bancroft Library, UCB; Lewis, 111–23.
52. Hopkins Biographical ms., 21, 23, Bancroft Library, UCB; Lewis, 130–40.
53. Lavender, 325; Lewis, 180–81.
54. Lavender, 359–61; Lewis, 188–89.
55. Lavender, 277, 280, 291–93.
56. Ibid., 292–93; *New York Sun,* July 19, 20, 21, 29, 30, 1873; *New York Tribune,* July 29, 30, 1873.
57. Lavender, 375; Gilborn, 20–23, 60, 94–104, 140–43.

Bibliography

Abdill, George B. *Pacific Slope Railroads from 1854 to 1900.* Seattle: Superior, 1959.

Adams, B. B., Jr. "The Everyday Life of the Railroad Men." In *The American Railway: Its Construction, Development, Management, and Appliance.* New York: Scribner's, 1889.

Adams, Charles Francis, Jr. *An Autobiography, 1835–1915.* Boston: Houghton Mifflin, 1916.

———. "The Case of the Union Pacific Railway Company." Statements made before the Committee of the Pacific Railroads, February 24, 1886, and February 11, 1888. Boston: R. A. Supply Co., 1888.

———. *Chapters of Erie and Other Essays.* Boston: Osgood, 1871.

———. *A Cycle of Adams Letters, 1861–1865.* Edited by W. C. Ford. Boston: Houghton Mifflin, 1920.

———. "The Pacific Railroad Ring." *North American Review,* January 1869.

———. *The Railroad Problem: A Lecture Delivered at the Lowell Institute, Boston, February 26, 1875.* New York: Railroad Gazette, 1875.

———. *Railroads: Their Origins and Problems.* New York: Putnam's, 1878.

Adams, Henry. *The Education of Henry Adams.* Boston: Houghton Mifflin, 1918.

Ainsworth, D. H. *Recollections of a Civil Engineer.* Newton, Iowa: Privately printed, 1948.

Albright, George Leslie. *Official Explorations for Pacific Railroads, 1853–1855.* Berkeley: University of California, 1921.

Aldrich, Charles. "Anecdotes of General Grenville M. Dodge." *Magazine of American History* XXIV (October 1890).

———. "An Iowa Emancipator." *Annals of Iowa* XXX (April 1950), 302–305.

Alexander, Edwin P. *American Locomotives, 1829–1900.* New York: Bonanza, 1941.

American Geographical Society of New York. *The Golden Spike: A Centennial Remembrance.* New York, 1969.

Ames, Charles Edgar. *Pioneering the Union Pacific: A Reappraisal of the Builders of the Railroad.* New York: Appleton-Century-Crofts, 1969.

Ames, Oakes. *A Memoir.* Cambridge: Riverside, 1883.

———. The Defense of Oakes Ames. Printed text of statement given before the U.S.H.R., 25 February, 1873.

Ames, Winthrop. *The Ames Family of Easton, Massachusetts.* Privately printed, 1948.

Anderson, Bernice. "The Driving of the Golden Spike." *Utah Historical Quarterly* XXIV (April 1956).

Andrist, R. K. *The Long Death: The Last Days of the Plains Indians.* New York: Macmillan, 1964.

Arrington, Leonard J. *Brigham Young: American Moses.* New York: Knopf, 1985.

Ashby, George F. *Major General Grenville M. Dodge (1831–1916), Maker of History in the Great West.* New York: Newcomen Society of England, 1947.

Association of American Railroads. *A Bibliography of Railroad Literature,* 10th ed. Washington, 1976.

———. *Railroad History and Sources of Historical Information about Railroads.* Washington, 1940.

Athern, Robert G. "General Sherman and the Western Railroads." *Pacific Historical Review* XXIV, 1 (February 1955).

———. *Rebels of the Rockies.* New Haven: Yale, 1962.

———. *Union Pacific Country.* Chicago: Rand McNally, 1971.

———. *William Tecumseh Sherman and the Settlement of the West.* Norman: University of Oklahoma, 1956.

Avery, B. P. "The Building of the Iron Road." *Overland Monthly,* May 1869.

Bailey, Edd H. *The Century of Progress: A Heritage of Service, Union Pacific, 1869–1969.* New York: Newcomen Society, 1969.

Bailey, W. F. *The Story of the First Transcontinental Railroad.* Pittsburgh: Pittsburgh Printing Co., 1906.

Bancroft, Hubert Howe. *Bancroft's Guide for Travelers by Railway, Stage, and Steam Navigation in the Pacific States.* San Francisco, 1869.

———. *Chronicles of the Builders of the Commonwealth,* vols. VI, VII. San Francisco: The History Company, 1890.

———. *History of California,* vol. VII, *1860–90.* San Francisco: The History Company, 1890.

———. *History of the Life of Leland Stanford, A Character Study.* Oakland: Biobooks, 1952. (Omitted from HHB's Chronicles.)

———. *History of Nevada, Colorado, and Wyoming.* San Francisco: The History Company, 1890.

———. *History of Utah, 1540–1887.* San Francisco: The History Company, 1891.

———. *History of the Pacific States of North America.* New York, 1902.

Barnes, Demas. *From the Atlantic to the Pacific, Overland: A Series of Letters.* New York: Van Nostrand, 1886.

Baster, Maurice G. *Orville H. Browning: Lincoln's Colleague and Critic.* Bloomington: Indiana University, 1957.

Beadle, J. H. *The Undeveloped West, or, Five Years in the Territories.* Philadelphia: National, 1873.

Bean, Walton. *California: An Interpretive History.* New York: McGraw Hill, 1968.

Bearss, Edwin C. "The Battle of Pea Ridge," *Arkansas Historical Quarterly* XX (Spring 1961), 74–94.

———. "The First Day at Pea Ridge, March 7, 1862." *Arkansas Historical Quarterly* SVII (Summer 1958), 132–154.

Bell, William A. *New Tracks in North America: A Journal of Travel and Adventures Whilst Engaged in the Survey for a Southern Railroad to the Pacific Ocean During 1867–68.* London: Chapman & Hall, 1869.

Benton, Thomas Hart. "Highway to the Pacific: Grand National Central High-way." Senate speech on introducing a bill for the construction of a grand national highway from St. Louis to San Francisco, December 16, 1850. Washington: Towers, 1850.

———. *Letter from Col. Benton to the People of Missouri.* Washington: Privately printed, 1853.

Berkhofer, Robert F., Jr. *The White Man's Indian: Images of the American Indian from Columbus to the Present.* New York: Knopf, 1978.

Berner, Bertha. *Mrs. Leland Stanford: An Intimate Account.* Stanford: Stanford University Press, 1935.

Berthrong, Donald J. *The Southern Cheyenne.* Norman: University of Oklahoma, 1963.

Best, Gerald M. *Iron Horses to Promontory.* San Marino, Calif.:Golden West, 1969.

———. "Rendezvous at Promontory: The 'Jupiter' and No. 119." *Utah Historical Quarterly,* XXXVII, 1 (Winter 1969).

———. *Snowplow: Clearing Mountain Rails.* Berkeley: Howell-North, 1966.

Bigelow, John. *Memoir of the Life and Public Services of John Charles Fremont.* Cincinnati: Derby & Jackson, 1856.

Biographical Directory of the American Congress, 1774–1949. Washington: U.S.G.P.O., 1950.

Biographical History of Eminent and Self-Made Men of the State of Indiana, II. Cincinnati: Western Biographical Publishing Company, 1880.

Botkin, B. A., and Harlow, Alvin F., eds. *A Treasury of Railroad Folklore.* New York: Crown, 1953.

Bowers, Claude G. *The Tragic Era: The Revolution After Lincoln.* Cambridge: Houghton Mifflin, 1929.

Bowles, Samuel. *Across the Continent.* New York: Hurd & Houghton,1866.

———. *Our New West: Records of Travel Between the Mississippi River and the Pacific Ocean.* Hartford: Hartford Publishing Co., 1869.

———. *The Pacific Railroad—Open; How to Go; What to See.* Boston: Fields, Osgood & Co., 1869.

Bowman, J. N. "Driving the Last Spike at Promontory, 1869." *Utah Historical Quarterly* XXXVII, 1 (Winter 1969).

Brace, Charles Loring. *The New West, or, California in 1867–1868.* New York: Putnam & Sons, 1869.

Braly, Mary Gramling. "If I Had a Thousand Lives." *Tennessee Historical Magazine,* Series II, I (July 1931), 261–269.

Bringhurst, Newell G. *Brigham Young and the Expanding American Frontier.* Boston: Little, Brown, 1986.
Brodie, Fawn M. *Thaddeus Stevens: Scourge of the South.* New York: Norton, 1959.
Brown, Dee. *Bury My Heart at Wounded Knee.* New York: Holt, 1970.
———. *Fort Phil Kearny: An American Saga.* New York: 1962.
———. *The Galvanized Yankees.* Urbana: University of Illinois, 1963.
———. *Hear That Lonesome Whistle Blow.* New York: Holt, 1977.
Brown, Margaret L. "Asa Whitney and His Pacific Railroad Publicity Campaign." *Mississippi Valley Historical Review* XX, September 1933, pp. 209–224.
———. "Asa Whitney: Projector of the Pacific Railroad." Ph.D. thesis, University of Michigan, 1930.
Browning, Orville H. *The Diary of Orville Hickman Browning,* vol. II, *1865–81.* In Collections of the Illinois State Historical Library, XXII, 1933.
Brownlee, Richard Smith. "Guerrilla Warfare in Missouri, 1861–1865." Ph.D. dissertation, University of Missouri, 1955.
Bruce, Robert. *The Fighting Norths and Pawnee Scouts.* Lincoln: Nebraska State Historical Society, 1932.
Bruchey, Stuart, ed. *Memoirs of Three Railroad Pioneers.* New York: Ayer Co., 1981.
Burch, John C. "Theodore D. Judah." *First Annual of the Territorial Pioneers of California.* San Francisco: Territorial Pioneers, 1877.
Burton, Sir Richard Francis. *The City of the Saints, and Across the Rocky Mountains to California.* London: Longman, Green, Longman, and Roberts, 186l.
Butler, Anne M. *Daughters of Joy, Sisters of Misery: Prostitutes in the American West, 1865–90.* Urbana: University of Illinois, 1985.
———. "The Press and Prostitution: A Study of San Antonio, Tombstone, and Cheyenne, 1865–1890." Master's thesis, University of Maryland, 1975.
——— and Ona Siporin. *Uncommon Common Women: Ordinary Lives of the West.* Logan: Utah State University, 1996.
"California Credit Mobilier, Alias Contract & Finance Co. of the CPRR." *San Francisco Evening Bulletin,* January 4, 1873.
Carman, Harry J., and C. H. Mueller. "The Contract and Finance Co. and the Central Pacific Railroad." *Mississippi Valley Historical Review* 14 (1927), 326–341.
Carpenter, Francis B. *The Inner Life of Abraham Lincoln.* New York: Hurd & Houghton, 1868.
Carr, Sarah Pratt. *The Iron Way: A Tale of the Building of the West.* Chicago: McClurg, 1907.
Central Pacific Railroad Co., *The Central Pacific Railroad Company of California: Character of the Work, Its Progress, Resources, Earnings, and Future Prospects, With the Foundations and Advantages of Its First Mortgage Bonds* New York: Brown & Hewitt, February 1868.
———. *A Trip . . . from Ogden to San Francisco.* New York: Nelson, 1870.
Chambers, William N. *Old Bullion Benton: Senator from the New West.* Boston: Little, Brown, 1956.
Chandler, Alfred Dupont, ed. *The Railroads, the Nation's First Big Business: Sources and Readings.* New York: Harcourt, 1965.
Chiang, Monlin. *Tide From the West: A Chinese Autobiography.* New Haven: Yale, 1947.
Chinn, Thomas W., ed. *History of the Chinese in California—A Syllabus.* San Francisco, 1969.
Clampitt, John W. *Echoes from the Rocky Mountains.* Chicago: Bedford, Clark & Co., 1889.
Clark, George T. "Leland Stanford and H. H. Bancroft's 'History': A Bibliographical Curiosity." *Bibliographic Society of America Papers* 27, part 1. Chicago, 1933, pp. 12–23.
Clarke, T. C., et. al. *American Railroad: Its Construction, Development, Management, and Appliances.* N.P., 1889.
Cleland, Robert G. *A History of California: The American Period.* New York: Macmillan, 1923.
Clews, Henry. *Twenty-Eight Years in Wall Street.* New York: Irving, 1888.
Cochran, Thomas C. *Railroad Leaders, 1845–90: The Business Mind in Action.* Cambridge: Harvard, 1953.
——— and William Miller. *The Age of Enterprise.* NY: Macmillan, 1942.
Coffin, Morse H. *The Battle of Sand Creek.* Waco, Tx.: Morrison, 1965.
Cohen, Norm. *Long Steel Rail: The Railroad in American Folksong.* Urbana: University of Illinois, 198l.
Cole, Cornelius. *Memoirs.* New York: McLoughlin, 1908.
———. "A Guide to the Papers of Cornelius Cole and the Cole Family, Collection 217." In Occasional Papers, no. 4., University of California at Los Angeles.
Collins, Frederick L. *Money Town: The Story of Manhattan's Toe.* New York: Putnam's, 1946.

"The Colton Letters Declaration of Huntington that Congress Men are For Sale," n.p.

Colton, Ray Charles. *The Civil War in the Western Territories: Arizona, Colorado, New Mexico and Utah.* Norman: University of Oklahoma, 1959.

Combs, Barry B. "Union Pacific and the Early Settlement of Nebraska, 1868–1880." *Nebraska History* 50, no. 1 (Spring 1969), pp. 1–26.

———. *Westward to Promontory: Building the Union Pacific Across the Plains and Mountains. A Pictorial Documentary with Text.* Palo Alto: American West, with cooperation of American Geographical Society, 1969.

Conwell, Russell H. *Why and How the Chinese Emigrate.* Boston: Lee, Shepard, 1871.

Cook, the Reverend Joseph W. *Diary and Letters.* Laramie, Wyo.: Laramie Republican Co., 1919.

Coolidge, Mary Roberts. *Chinese Immigration.* New York: Holt, 1909.

Coolidge, Susan. "A Few Hints on the California Journey." *The Century Magazine* VI, no. 5 (May 1873).

Coutant, C. G. *The History of Wyoming.* Laramie, Wyo.: Chaplin, Spafford & Mathison, 1899.

Coy, Owen C. *The Great Trek: Collected Diary Entries of the '49 Emigration to California.* Los Angeles: Powell, 1931.

———. *The Humboldt Bay Region, 1850–1875: A Study in the American Colonization of California.* Los Angeles: California State Historical Association, 1929.

Crawford, J. B. *The Credit Mobilier of America.* Boston: Calkins, 1880.

Crofutt, George A. *Crofutt's Trans-Continental Tourist's Guide,* New York: Crofutt, 1871.

Cruise, John D. "Early Days on the Union Pacific." *Collections of the Kansas State Historical Society* XI, 1909–1910.

Curtis, Samuel P. "The Army of the South-West and the First Campaign in Arkansas." *Annals of Iowa,* 1st Series, V (October 1867), 930–933.

Cutler, Carl C. *Greyhounds of the Sea: The Story of the American Clipper Ships.* Annapolis, MD: U.S. Naval Institute, 1930.

Daggett, Stuart. *Chapters on the History of the Southern Pacific* (New York, 1922; reprint Fairfield, N.J.: Augustus M. Kelley Co., 1966).

Dana, Charles A. *Recollections of the Civil War.* New York: Appleton, 1898.

Daniels, William. *American Railroads: Four Phases of Their History.* Princeton, N.J.: Princeton University, 1932.

Davies, R. *Forgotten Railways: Chilterns and Cotswolds—Railway History in Pictures.* North Pomfret, Vt.: David & Charles, 1977.

Davis, Elmer O. *The First Five Years of the Railroad Era in Colorado.* Privately printed, 1948.

Davis, George T. M. *Autobiography of the Late Col. George T. Davis.* New York: "Published by His Legal Representatives," 1891.

Davis, John P. *The Union Pacific Railway: A Study in Railroad Politics, History, and Economics.* Chicago: S. C. Griggs, 1894.

Davis, Richard Harding. *The West from a Car Window.* New York: Harper, 1904.

Debo, Angie. *A History of the Indians of the United States.* Norman: University of Oklahoma, 1970.

De Golyer, Everett L., Jr. *The Track Going Back: A Century of Transcontinental Railroading, 1869–1969.* Amon Carter, 1979.

D'Elia, Donald J. "The Argument Over Civilian or Military Control, 1865–1880." *Historian* XXIV (February 1962).

Derby, E. H. *The Overland Route to the Pacific: A Report on the Condition, Capacity, and Resources of the Union Pacific and Central Pacific Railways.* Boston: Lee & Shepard, October 1865.

Devens, R. M. *American Progress, or, The Great Events of the Greatest Century.* Chicago: Heron, 1883.

Dictionary of American Biography. New York: Scribner's, 1928.

Dillon, Sidney. "Historic Moments: Driving the Last Spike of the Union Pacific." *Scribner's Magazine,* August 1892.

———. "The West and the Railroads." *North American Review,* April 1891.

Dix, Morgan, compiler. *The Memoirs of John Adams Dix.* New York: Harper, 1883.

Dodge, Major General Grenville M. *Biographical Sketch of James Bridger.* New York: Unz and Co., 1905.

———. "General James A. Williamson." *Annals of Iowa,* 3rd Series, IV (January 1901), 577–594.

———. "Historical Address at the Camp fire of the Crocker Brigade." *Annals of Iowa,* 3rd Series, XVIII (January 1932), 163–179.

———. *How We Built the Union Pacific Railway, and Other Railway Papers Addresses.* Washington: U.S. Government Printing Office, 1910; Ann Arbor: University Microfilms, 1966.

———. "The Indian Campaign of Winter of 1864–65" (1877; reprint Denver, 1907).

———. "Paper Read Before the Society of the Army of the Tennessee, Sept. 15, 1888," on the transcontinental railroad.

———. "Personal Recollections of General Grant and His Campaigns in the West." *Journal of the Military Service Institution of the United States* XXXVI (January 1905), 39–6l.

———. *Personal Recollections of President Abraham Lincoln, General Ulysses S. Grant and General William T. Sherman.* Council Bluffs: Monarch Printing Co., 1914.

———. "Reminiscences of Engineering Work on the Pacific Railways and in the Civil War," *Engineering News* 62 (October 28, 1909), 456–458.

———. *Romantic Realities, The Story of the Building of the Pacific Roads.* Omaha: Union Pacific Railway, 1889.

———. "Surveying the M. and M." *Palimpsest* XVIII (September 1937), 301–3ll.

———. "What I Saw of Lincoln." *Appleton's Magazine* XII (February1909), 134–140.

Dodge, Joseph Thompson. *Genealogy of the Dodge Family of Essex County, Massachusetts.* 2 vols. Madison, WI: Democratic Printing Co., 1894.

Dodge, Nathan P. "Early Emigration through and To Council Bluffs." *Annals of Iowa,* 3rd Series, XVIII (January 1932), 163–169.

Dodge, Colonel Richard I. *The Black Hills: A Minute Description of the Routes, Scenery, Soil, Climate, Timber, Gold, Geology, Zoology, etc.* New York: J. Miller, 1876.

———. *Our Wild Indians.* Hartford: A. D. Worthington & Co., 1882.

Donaldson, Emma Brace. *The Life of Charles Loring Brace, Author of "Hungary in 1851."* New York: Hungarian Reference Library of America, 1941.

Donovan, Frank P. "The Race to Council Bluffs." *Palimpsest* XLIII (December 1962), 545–556.

Downey, F., and J. N. Jacobsen. *The Red Bluecoats: The Indian Scouts.* Old Army Press, 1973.

Drinnon, Richard. *Facing West: The Metaphysics of Indian-Hating and Empire-Building.* Minneapolis, 1980.

Dufwa, Thamar E. *Transcontinental Railroad Legislation, 1835–1862.* New York: Ayer, 198l.

Dugan, Ruth A. "Grenville Mellen Dodge." *Palimpsest* 11 (1930), 160–171.

Dulles, Foster Rhea. *The United States Since 1865.* Ann Arbor: Michigan, 1959.

Dunbar, Seymour. *A History of Travel in America.* New York: Tudor, 1915, 1937.

Dunn, J. P. *Massacres of the Mountains: A History of the Indian Wars of the Far West, 1815–1875* (1886). New York: Archer House, 1958.

Ehrenberger, James L. *Smoke along the Columbia: Union Pacific, Oregon Division.* Callaway, Neb.: E. & G. Publications, 1968.

———. *Smoke Over the Divide: Union Pacific, Wyoming Division.* Callaway, Neb.: E. & G. Publications, 1965.

———. *Smoke Above the Plains: Union Pacific, Kansas Division.* Callaway, Neb.: E. & G. Publications, 1965.

———. *Smoke Across the Prairie: Union Pacific, Nebraska Division.* Golden, Col.: Intermountain Chapter, National Railway Historical Soc., 1964.

Elegant, Robert S. *The Dragon's Seed.* New York: St. Martin's, 1959.

Ellis, Richard N. "Civilians, the Army, and the Indian Problem on the Northern Plains, 1862–66," *North Dakota History* XXXVII (Winter 1970).

———. *General Pope and United States Indian Policy.* Albuquerque: University of New Mexico, 1970.

Ellsworth, George. "H. H. Bancroft and the History of Utah." *Utah Historical Quarterly* 22, no. 2 (April 1954).

Emmons, David M. *Garden in the Grasslands: Boomer Literature of the Central Great Plains.* Omaha: University of Nebraska, 1971.

Evans, Cerinda W. *Collis Potter Huntington.* 2 vols. Newport News: Mariners' Museum, 1954.

Evans, John H. *One Hundred Years of Mormonism: A History of the Church of Jesus Christ of Latter-Day Saints From 1805–1905.* Salt Lake City: Deseret, 1905.

———. *The Story of Utah, The Beehive State.* New York: Macmillan, 1933.

——— and Minnie E. Anderson. *Ezra Taft Benson, Pioneer-Statesman-Saint.* Salt Lake City: Deseret News Press, 1947.

Ewers, John C. "Intertribal Warfare as the Precursor of Indian-White Warfare on the Northern Great Plains." *Western Historical Quarterly* VI (October 1975).

Farmer, Hallie. "The Railroads and Frontier Populism." *Mississippi Valley Historical Review,* XIII (1926–27).

Farnham, Henry W. *Memoir of Henry Farnham.* New Haven: Privately printed, 1889.

Farnham, Wallace D. "Grenville Dodge and the Union Pacific: A Study of Historical Legends." *Journal of American History* LI (March 1965).

———. "The Pacific Railroad Act of 1862." *Nebraska History* 43 (September 1962).

———. "The Weakened Spring of Government." *American Historical Review* LXVII (April 1963).

Farquhar, Francis P. "Exploration of the Sierra Nevada," *California Historical Society Quarterly,* IV (March 1925).

———. *History of the Sierra Nevada.* Berkeley: University of California, 1965.

Farrell, Dennis. "Adventures on the Plains, 186567." *Collections of the Nebraska State Historical Society* XVII (1913).

Field, Homer H., and J. R. Reed. "Short Sketch of the Services of Maj. Gen. Grenville Mellen Dodge . . ." New York, 1926. Reprint from *A History of Pottawattamie County, Iowa, 1907.*

Fischer, Christiane, ed. *Let Them Speak for Themselves: Women in the American West, 1849–1900.* New York: Dutton, 1978.

Fisk, Harvey E. "Fisk & Hatch, Bankers and Dealers in Government Securities, 1862–1885." *Journal of Economic and Business History* II (August 1930).

Flint, Henry M. *The Railroads of the United States* (1868; reprint, New York: Ayer, 1976).

Flower, Benjamin. "Twenty-five Years of Bribery and Corrupt Practices: The Railroads, the Lawmakers, the People." Boston: Arena, 1904.

Fogel, Robert W. *Railroads and American Economic Growth: Essays.* Baltimore: Johns Hopkins, 1964.

———. *The Union Pacific Railroad: A Case in Premature Enterprise.* Baltimore: Johns Hopkins, 1960.

Fogelson, Raymond D. *The Cherokees.* Norman: University of Oklahoma, 1978.

Fritz, Henry E. *The Movement for Indian Assimilation, 1860–1890.* Philadelphia: Lippincott, 1963.

Fuller, Robert H. *Jubilee Jim: Life of Col. James Fisk, Jr.* New York: Macmillan, 1928.

Fulton, Robert Lardin. *The Epic of the Overland.* San Francisco: A. M. Robertson, 1925.

Galloway, John D. *The First Transcontinental Railroad.* New York: Simmons-Boardman, 1950.

Ghent, W.J. *The Road to Oregon.* New York: Longmans, Green, 1929.

Gibson, Otis. *The Chinese in America.* Cincinnati: Hitchcock & Walden, 1877.

Gilborn, Craig. *Durant: The Fortunes and Woodland Camps of a Family in the Adirondacks.* Blue Mountain Lake, New York: The Adirondack Museum, 1981.

Gillette, Edward. *Locating the Iron Trail.* The Christopher Publishing House, 1925.

Gilliss, John R. "Tunnels of the Pacific Railroad." *Transactions* (American Society of Civil Engineers) I, 1872.

Glaab, Charles N. *Kansas City and the Railroads.* Madison: State Historical Society of Wisconsin, 1972.

Glad, Paul. "Frederick West Lander and the Pacific Railroad Movement." *Nebraska History* XXXV, no. 3 (September 1954).

Goetzmann, William H. *Army Exploration in the American West, 1803–1863.* New Haven: Yale, 1965.

———. *Exploration and Empire.* New York: Knopf, 197l.

Golden Spike Symposium, University of Utah, 1969, *The Golden Spike.* Salt Lake City, 1973.

Goodwin, C. C. *As I Remember Them.* Salt Lake City: Salt Lake Commercial Club, 1913.

Granger, John T. *A Brief Biographical Sketch of the Life of Major-General Grenville M. Dodge.* New York: Cash & Styles, 1893.

Grant, Ulysses S. *Personal Memoirs.* 2 vols. New York: Charles L.Webster, 1885–1886.

Gray, Carl Raymond. "The Significance of the Pacific Railroads." Cyrus Fogg Brackett Lectures, Princeton University, 1935.

Greeley, Horace. *An Overland Journey from New York to San Francisco in the Summer of 1859.* New York: Sexton, Barker, 1860.

Greenberg, Dolores. *Financiers and Railroads, 1869–1889.* Newark, DE: University of Delaware Press, 1980.

Grinnell, George Bird. *The Cheyenne Indians: Their History and Way of Life.* 2 vols. New Haven: Yale, 1923.

———. *The Fighting Cheyennes.* Norman: University of Oklahoma, 1956.

———. *Two Great Scouts and Their Pawnee Battalion.* Cleveland: Arthur H. Clark, 1928.

Griswold, Wesley S. *A Work of Giants: Building the First Transcontinental Railroad.* New York: McGraw-Hill, 1962.

Grodinsky, Julius. *Transcontinental Railway Strategy, 1869–1893.* Philadelphia, 1962.

Gross, H. H. "The Land Grant Legend." *Railroad Magazine,* August–December, 1951.

Gudde, Erwin G. *California Place Names: A Geographical Dictionary.* Berkeley: University of California, 1949.

Hafen, LeRoy R., and Carl Coke Rister. *Western America.* New York: Prentice Hall, 1941.

Haney, Lewis Henry. *A Congressional History of Railways in the United States, 1850–1887.* University of Wisconsin Bulletin 342, Economic and Political Series, VI. New York, 1968.

Hart, Alfred A. *The Traveler's Own Book.* Chicago: Norton & Leonard, 1870.

Harte, Bret. *San Francisco in 1866.* San Francisco, 195l.

Hassrick, R. B. *The Sioux: Life and Customs of a Warrior Society.* Norman: University of Oklahoma, 1964.

Hayes, William Edward. *Iron Road to Empire: The History of the Hundred Years of Progress and Achievements of the Rock Island Lines.* New York: Simmons-Boardman, 1953.

Haymond, Creed. *The Central Pacific Railroad Company: Its Relations to the Government.* San Francisco: H. S. Crocker, 1888.

Hazard, Rowland. *The Credit Mobilier of America.* Providence: Sidney S. Rider, 188l.

Heath, Erle. *Trail to Rail.* Photostatic copies of installments in Southern Pacific Bulletin, Southern Pacific Railroad. California State Railway Museum, Sacramento.

Hedges, James Blaine. *Henry Villard and the Railways of the Northwest.* New Haven: Yale, 1930.

Herriott, Frank I. "Iowa and the First Nomination of Abraham Lincoln." *Annals of Iowa,* 3rd Series, VIII (October 1907).

———. "Memories of the Chicago Convention of 1860." *Annals of Iowa,* 3rd Series, VIII (October 1907).

Hewes, David. "An Autobiography." In *Lieutenant Joshua Hewes, a New England Pioneer,* a genealogical history. Privately printed, 1913.

Higham, John. *Strangers in the Land: Patterns of American Nativism, 1860–1925.* New Brunswick, N.J.: Rutgers, 1955.

Hill, Thomas. *The Last Spike, a Painting by Thomas Hill.* San Francisco: E. Bosqui & Co., 1881.

Hinckley, Helen. *Rails From the West: A Biography of Theodore D. Judah.* San Marino, Calif.: Golden Spike, 1970.

Hirshson, Stanley P. *Grenville M. Dodge.* Champaign-Urbana: University of Illinois, 1967.

History of the State of Nebraska. Chicago: Western Historical, 1882.

History of the State of Nevada. Oakland, Calif.: Thompson & West, 1881.

History of Sacramento County, California. Oakland, Calif.: Thompson & West, 1880. Reprinted 1960, University of California, Berkeley.

Hittell, Theodore H. *History of California,* IV. San Francisco: N. J. Stone & Co., 1897.

Hoar, George F. *Autobiography of Seventy Years.* New York: Scribner's, 1905.

Hochschild, Harold K. *Doctor Durant and his Iron Horse.* Blue Mountain Lake, New York: Adirondack Museum, 1961.

Hoebel, E. Adamson. *The Plains Indians: A Critical Bibliography.* Bloomington: Indiana University, 1977.

Hoexter, Corinne K. *From Canton to California: The Epic of Chinese Immigration.* New York: Four Winds Press, 1976.

Hofstadter, Richard, William Miller, and Daniel Aaron. *The United States: the History of a Republic.* Englewood Cliffs, NJ: Prentice-Hall, 1976.

Hogg, Garry. *Union Pacific: The Building of the First Transcontinental Railroad.* New York: Walker, 1967.

Hoig, Stan. *The Sand Creek Massacre.* Norman: University of Oklahoma, 196l.

Holbrook, Stewart H. *The Story of American Railroads.* New York: Crown, 1947.

Holland, Rupert S. *Historic Railroads.* Philadelphia: Macrae-Smith, 1927.

Hoover, Herbert T. *The Sioux: A Critical Bibliography.* Bloomington: Indiana University, 1979.

Howard, Robert West. *The Great Iron Trail: The Story of the First Transcontinental Railroad.* New York: Putnam's, 1962.

Hoyt, A. W. "Over the Plains to Colorado." *Harper's New Monthly Magazine,* June 1867.

Hulse, James W. *The Nevada Adventure: A History.* Reno: University of Nevada, 1965.

Humason, W. J. *From the Atlantic Surf to the Golden Gate.* Hartford: Press of William C. Hutchings, 1869.

Hungerford, Edward. *Men and Iron:The History of the New York Central.* New York: Crowell, 1938.

Hunt, Rockwell D. "Cornelius Cole: A California Pioneer." *Overland Monthly* 68 (1916), 255–261.

Hunter, Milton R. *Beneath Ben Lomond's Peak; A History of Weber County, 1824–1900.* Salt Lake City: Deseret News Press, 1944.

Hutton, Paul A. *Phil Sheridan and His Army.* Lincoln: University of Nebraska Press, 1985.

Hyde, George E. *The Pawnee Indians.* Denver: University of Denver, 195l.

———. *Red Cloud's Folk.* Norman: University of Oklahoma, 1937.

———. *Spotted Tail's Folk: A History of the Brule Sioux.* Norman: University of Oklahoma, 1961.

Inman, Col. Henry, and Colonel William F. Cody. *The Great Salt Lake Trail.* Topeka, Kansas, 1899.

"In Memoriam: Asa Whitney" (pamphlet). Washington, D.C.: September 1872 (Reprints of Obituaries in *Washington Patriot* and *Washington Globe Leader* .)

Jackson, Helen Hunt. *Bits of Travel at Home.* Boston: Roberts Brothers, 1878.

Jackson, William Henry. "The Diaries of William Henry Jackson." In *The Far West and the Rockies History Series, 1820–1875*, vol. 10. edited by LeRoy R. Hafen and Ann W. Hafen. Glendale, Calif.: A. H. Clarke, 1959.

———. *Time Exposure.* New York: Putnam's, 1940.

——— and Howard R. Driggs. *The Pioneer Photographer.* Yonkers-on-Hudson, New York: World, 1929.

Jackson, W. Turrentine. *Wagon Roads West.* Berkeley: University of California, 1952.

Jacobs, Wilbur R. "The Indian and the Frontier in American History—A Need for Revision." *Western Historical Quarterly* IV (January1973).

James, Marquis. *They Had Their Hour.* Indianapolis: Bobbs-Merrill, 1934.

Jeffrey, Julie Roy. *Frontier Women: The Trans-Mississippi West, 1840–1880.* New York: Hill & Wang, 1979.

Jensen, Larry. *The Sierra Railroad: An Adventure in Yesteryear.* Darwin, n.d.

Jensen, Oliver O. *The American Heritage History of Railroads in America.* New York: American Heritage, 1975.

Johnson, Arthur Menzies. "Boston Capitalists and Western Railroads." In *Harvard Studies in Business History*, vol. 23. Cambridge: Harvard, 1967.

Johnson, Enid. *Rails Across the Continent: The Story of the First Transcontinental Railroad.* New York: Messner, 1965.

Johnson, Jack T. *Peter Anthony Dey, Integrity in Public Service.* Iowa City: State Historical Society of Iowa, 1939.

Johnson, Robert Underwood, and Clarence C. Buel, eds. *Battles and Leaders of the Civil War.* 4 vols. New York: Century, 1884–1887.

Jones, Peter, ed. *The Robber Barons Revisited.* Boston: Heath, 1968.

Josephson, Matthew. *The Politicos, 1865–1896.* New York: Harcourt, 1938.

———. *The Robber Barons: The Great American Capitalists, 1861–190l.* New York: Harcourt, 1934.

Josephy, Alvin M., Jr. *The Civil War in the American West.* New York: Knopf, 1991.

———. *Now That the Buffalo's Gone.* New York: Knopf, 1982.

Joslyn, David L. "The Romance of the Railroads Entering Sacramento." *Bulletin of the Railway and Locomotive Society* 48 (March 1939).

Judah, Theodore Dehone. *The Central Pacific Railroad of California.* San Francisco, Calif.: November 1, 1860.

———. *Report of the Chief Engineer on the Preliminary Survey, Cost of Construction, And Estimated Revenue of the Central Pacific Railroad of California Across the Sierra Nevada Mountains from Sacramento to the Eastern Boundary of California.* Sacramento: H. S. Crocker, 1862.

Kenna, Edward D. *Railway Misrule.* New York: Duffield, 1914.

Kennan, George. *Misrepresentation in Railroad Affairs.* Garden City: Country Life Press, 1916.

Kirkland, Edward C. *Charles Francis Adams: Patrician at Bay.* Cambridge: Harvard, 1965.

———. "Divide and Ruin." *Mississippi Valley Historical Review* XLIII (June 1956).

Klein, Maury. *Union Pacific: The Birth of a Railroad 1862–1893.* New York: Doubleday, 1987.

———. *Source Notes: Union Pacific: The Birth of a Railroad.* Omaha: Union Pacific Museum/Doubleday, 1987.

Kline, Allen. "The Attitude of Congress Toward the Pacific Railway, 1856–1862." *Annual Report of the American Historical Association, 1910.* Washington, 1912.

Knight, Oliver. *Following the Indian Wars; The Story of the Newspaper Correspondents Among the Indian Campaigners.* Norman: University of Oklahoma Press, 1960.

Kraus, George. *High Road to Promontory: Building the Central Pacific Across the High Sierra.* Tucson, Ariz.: American West, 1969.

Kraus, Michael. *The United States to 1865.* Ann Arbor, Mich.: University of Michigan, 1959.

Kung, S. W. *Chinese in American Life.* Seattle: University of Washington, 1962.

Kyner, James H. (with Hawthorne Daniel). *End of Track.* Caldwell, Iowa: Caxton, 1937; Lincoln, Neb.: 1960.

Larson, Henrietta M. *Jay Cooke, Private Banker.* Cambridge: Harvard, 1936.

Larson, John Lauritz. *Bonds of Enterprise: John Murray Forbes and Western Development in America's Railway Age.* Cambridge: Harvard, 1984.

Larson, Robert W. *Red Cloud: Warrior-Statesman of the Lakota Sioux.* Norman: University of Oklahoma Press, 1997.

Latta, Estelle C., and Mary L. Allison. *Controversial Mark Hopkins,* 2nd ed. Sacramento: Cothran Foundation, 1963.

Lavender, David. *The Great Persuader: Collis P. Huntington.* New York: Doubleday, 1970.

Leech, Margaret. *Reveille in Washington, 1860–1865.* New York: Harper, 1941.

LeMassona, R. A. *Union Pacific in Colorado, 1867–1967.* Denver: Hotchkiss & Nelson, 1967.

Leonard, Levi O., and Johnson, Jack T. *A Railroad to the Sea.* Iowa City: Midland House, 1939.

Letters from Collis P. Huntington to Mark Hopkins, Leland Stanford, Charles Crocker, E. B. Crocker, and D. D. Cotton [sic], *August 20, 1867–August 5, 1869.* New York: Privately printed, 1892.

Lewis, Lloyd. *Sherman, Fighting Prophet.* New York: Harcourt, 1932.

Lewis, Oscar. *The Autobiography of the West: Personal Narratives of the Discovery and Settlement of the American West.* New York: Holt, 1958.

———. *The Big Four: The Story of Huntington, Stanford, Hopkins and Crocker, and of the Building of the Central Pacific.* New York: Knopf, 1938.

———. *High Sierra Country.* New York: Duell, Sloane & Pierce, 1955.

———. *The War in the Far West.* New York: Doubleday, 196l.

Licht, Walter. *Working for the Railroad: The Organization of Work in the Nineteenth Century.* Princeton: Princeton, 1983.

"The Life, Influence, and the Role of the Chinese in the United States, 1776–1960." *Proceedings/papers of the National Conference held at the University of San Francisco, July 10–12, 1975.* Sponsored by the Chinese Historical Society of America. San Francisco, 1976.

Limerick, Patricia Nelson. *Desert Passages: Encounters with the American Deserts.* Albuquerque: University of New Mexico Press, 1985.

———. *The Legacy of Conquest: The Unbroken Past of the American West.* New York: Norton, 1987.

Loomis, the Reverend A. W. "Chinese Women in California." *Overland Monthly,* April 1869.

———. "Holiday in the Chinese Quarter." *Overland Monthly,* March 1869.

———. "How Our Chinamen Are Employed." *Overland Monthly,* March 1869.

Loomis, Nelson. "Asa Whitney: Father of Pacific Railroads." *Mississippi Valley Historical Association Procedures* 16 (1913).

Low, Frederick F. "Reflections of a California Governor." Edited, with preface and notes, by Robert H. Becker. In *Sacramento Book Collectors Club Publication,* no. 7. Sacramento: Sacramento Book Collectors Club, 1959.

Lowie, Robert H. *Indians of the Plains.* Lincoln, NE: University of Nebraska, 1954.

Lyman, Stanford M. *The Asian in the West.* Reno, 1970.

———. *The Structure of Chinese Society in Nineteenth Century America.* Ph.D. thesis. Berkeley: University of California, 1961.

McCague, James. *Moguls and Iron Men: The Story of the First Transcontinental Railroad.* New York: Harper, 1964.

McCartney, Clarence E. *Grant and His Generals.* New York: McBride, 1953.

McClure, Alexander K. *Three Thousand Miles through the Rocky Mountains.* Philadelphia: Lippincott, 1869.

McHugh, Tom. *The Time of the Buffalo.* New York: Knopf, 1972.
McKitrick, Eric L. *Andrew Johnson and Reconstruction.* Chicago: University of Chicago, 1960.
McLeod, Alexander. *Pigtails and Gold Dust.* Caldwell, Idaho: Caxton, 1947.
Majors, Alexander. *Seventy Years on the Frontier.* Chicago: Rand, McNally, 1893.
Mann, David H. "The Undriving of the Golden Spike." *Utah Historical Quarterly* XXXVII, no. 1 (Winter 1969).
Marryat, Frank. *Mountains and Molehills* (1855; reprint, Philadelphia: Lippincott, 1962).
Marshall, Samuel L. A. *Crimsoned Prairie: The Wars Between the United States and the Plains Indians during the Winning of the West.* New York: Scribner's, 1972.
Marshall, Walter G. *Through America; or, Nine Months in the United States.* London: Sampson Low, Marston, Searle, & Rivington, 1882.
Mauck, Genevieve P. "Council Bluffs Emerges." *Palimpsest* XLII (September 196l).
Mayer, L. R., and K. E. Vose, eds. *Makin' Tracks: A History of the Transcontinental Railroad in the Pictures and Words of the Men Who Were There.* New York: Praeger, 1975.
Merriam, George S. *The Life and Times of Samuel Bowles.* 2 vols. New York: Century, 1885.
Meyer, R. W. *A History of the Santee Sioux.* Omaha: University of Nebraska, 1967.
Miller, Marilyn. *The Transcontinental Railroad.* N.P.: Silver, n.d.
Miller, Stuart Creighton. *The Unwelcome Immigrant: The American Image of the Chinese, 1785–1882.* Berkeley: University of California, 1969.
Miller, William, ed. *Men in Business: Essays in the History of Entrepreneurship.* Cambridge: Harvard, 1952.
Millsap, Kenneth F. "The Election of 1860 in Iowa." *Iowa Journal of History* 48 (April 1950).
Milner, Clyde A., II. *A New Significance: Re-envisioning the History of the American West.* New York: Oxford, 1996.
Miner, H. Craig. *The St. Louis-San Francisco Transcontinental Railroad: The 35th Parallel Project, 1853–1890.* Lawrence: Kansas University Press, 1972.
Mitchell, Thomas W. "The Growth of the Union Pacific and its Financial Operations." *Quarterly Journal of Economics* 21 (1907).
Modelski, Andrew M. *Railroad Maps of North America: The First Hundred Years.* Washington: Library of Congress, 1984.
Monaghan, James. *The Overland Trail.* The American Trails Series. Indianapolis: Bobbs-Merrill, 1947.
Monaghan, Jay. *Civil War on the Western Border, 1854–1865.* Boston: Little, Brown, 1955.
Moody, John. *The Railroad Builders: A Chronicle of the Welding of the States.* New Haven: Yale, 1919.
Morgan, Dale L. *The Great Salt Lake.* The American Lakes Series. Indianapolis: Bobbs-Merrill, 1947.
———. *The Humboldt: Highroad of the West.* New York: Farrar & Rinehart, 1943.
———. *Overland in 1846: Diaries and Letters of the California-Oregon Trail.* Georgetown, Calif.: Talisman Press, 1963.
———. ed. "Diary of William Henry Ashley (1778–1838), March 25–June 27, 1825, A Record of Exploration West Across the Continental Divide, Down the Green River and Into the Great Basin." *Missouri Historical Society Bulletin* 11, no. 1 (October 1954).
Morse, Frank P. *Cavalcade of the Rails.* New York: Dutton, 1940.
Morse, John Frederick. *The First History of Sacramento City,* Sacramento: Colville, 1853.
Morton, J. Sterling. *History of Nebraska* (1881; reprint Lincoln: Western, 1918).
———. *Illustrated History of Nebraska,* vol. II. Lincoln: Jacob North & Co., 1907.
Mountfield, David. *The Railway Barons.* New York: Norton, 1979.
Muzzey, David S. *James G. Blaine: A Political Idol of Other Days.* New York: Dodd, Mead, 1935.
Myrick, David F. *History of Nevada* (1881; reprint, Berkeley: Howell-North, 1958).
National Railway Bulletin, vols. 41- (1976-). Allentown, PA.: National Railway Historical Society.
"Nebraska History and Record of Pioneer Days." *Collections of the Nebraska State Historical Society* VII (April–June, 1924).
Nelson, Thomas, and Sons, Ltd. *The Central Pacific Railroad: A Trip Across the North American Continent from Ogden to San Francisco.* New York: Nelson, 1870.
Nevada Railroads Committee. *Evidence Concerning Projected Railways Across the Sierra Nevada Mountains, from Pacific Tide Waters in California. . . . Procured by the Committee on Rail Roads of the First Nevada Legislature.* Carson City: J. Church, 1865.

New York Public Library. *Index to Uncatalogued U.S. Railroad Pamphlets.* New York, 1936.

Nichols, David A. "Civilization over Savage: Frederick Jackson Turner and the Indian." *South Dakota History* II (Fall 1972).

———. *Lincoln and the Indians: Civil War Policy and Politics.* Columbia: University of Missouri, 1978.

Nichols, Joseph. *Condensed History of the Construction of the Union Pacific Railway.* Omaha: Klopp, Bartlett & Co., 1892.

Nordhoff, Charles. *California: For Health, Pleasure, and Residence.* New York: Harper, 1873.

North, Edward P. "Blasting with Nitro-Glycerine." *Transactions* (American Society of Civil Engineers) I, 1872.

North, Major Frank J. "The Journal of an Indian Fighter" (1869 diary of Major North). *Nebraska History,* June, 1958.

Oberholtzer, Ellis P. *A History of the United States Since the Civil War.* 5 vols. New York: Macmillan, 1917–1937.

O'Connor, Richard. *Iron Wheels and Broken Men: The Railroad Barons and the Plunder of the West.* New York: Putnam's, 1973.

———. *Sheridan the Inevitable.* New York: Bobbs-Merrill, 1953.

Ogburn, Charlton. *Railroads: The Great American Adventure.* Washington, D.C.: The Society, 1977.

Ogden Standard Examiner, Golden Spike Centennial Edition. Ogden, Utah, 1969.

"Old Block." *The Central Pacific Railroad, or, '49 and '69.* San Francisco: White & Bauer, 1868.

Olson, James C. *History of Nebraska.* Lincoln: University of Nebraska, 1955.

———. *Red Cloud and the Sioux Problem.* Lincoln: University of Nebraska, 1965.

Olson, Sherry H. *The Depletion Myth: A History of Railroad Use of Timber.* Cambridge: Harvard, 1971.

Ostrander, Alson B. *An Army Boy of the Sixties: A Story of the Plains.* Yonkers-on-Hudson, New York: World, 1924.

"The Pacific Railroad, a Defense Against Its Enemies." CPRR pamphlets, vol. V (December 1864).

"The Pacific Railroads, the Rights of the People Ignored." N.p., 1874.

"The Pacific Railway." *Engineering* IV (November 8, 1867).

Park, William Lee. *Pioneer Pathways to the Pacific.* Clare, Mich.: Privately printed, 1938.

Parks, Albert S. "James G. Blaine and the Mulligan Letters." *Florida Agricultural and Mechanical College Quarterly Journal,* 1933.

Patterson, James W. "Observations on the Report of the Committee of the Senate of the U.S. Respecting the Credit Mobilier of America." Washington: McGill & Witherow, 1873.

Paxson, Frederic L. *The Last American Frontier.* New York: Macmillan, 1928.

Pease, Theodore C. "The Diary of Orville H. Browning, A New Source for Lincoln's Presidency." Lecture before Chicago Historical Society, March 29, 1923. University of Chicago, 1924.

Peattie, Donald Culross. "Tracks West." *The Saturday Evening Post,* November 5, 1949.

Pence, Mary Lou, and Lola M. Homsher. *The Ghost Towns of Wyoming.* New York: Hastings, 1956.

Perkins, Jacob R. *Trails, Rails and War, The Life of General G. M. Dodge.* Indianapolis, Ind.: Bobbs-Merrill, 1929.

Phelps, Alonzo. *Contemporary Biography of California's Representative Men.* San Francisco: A. L. Bancroft, 188l.

Philipps, Mrs. Catherine. *Cornelius Cole, California Pioneer and U.S. Senator: A Study in Personality and Achievements Bearing Upon the Growth of a Commonwealth.* San Francisco: Nash, 1929.

Poppleton, Andrew J. *Reminiscences.* Omaha: Privately printed, 1915.

Powell, Peter J. *The Cheyennes: A Critical Bibliography.* Bloomington: Indiana University, 1980.

———. *People of the Sacred Mountain: A History of the Northern Cheyenne Chiefs and Warrior Societies, 1830–1879, with an Epilogue, 1969–1974.* N.p.

Priest, Loring B. *Uncle Sam's Stepchildren: The Reformation of U.S. Indian Policy, 1865–1887.* New Brunswick, N.J.: Rutgers University Press, 1942.

Prouty, Annie Estelle. "The Development of Reno." *Nevada Historical Society Papers* IV (1923–1924).

Prucha, Francis Paul, ed. *A Bibliographical Guide to the History of Indian-White Relations in the United States.* Chicago: University of Chicago Press, 1977.

———. *Indian Policy in the United States: Historical Essays.* Lincoln, Neb.: Univ. Neb., 1981.

———. *Indian-White Relations in the United States: A Bibliography of Works Published 1975–1980.* Lincoln: University of Nebraska, 1982.

———. "New Approaches to the Study of the Administration of Indian Policy." *Prologue: The Journal of the National Archives* III (Spring 1971).

———. *United States Indian Policy: A Critical Bibliography.* Bloomington: University of Indiana, 1977.

Quiett, Glenn Chesney. *They Built the West: An Epic of Rails and Cities.* New York: Appleton-Century, 1934.

Rae, W. F. *Westward by Rail: The New Route to the Far East.* New York: Appleton-Century, 1871.

"The Railroad in the American Landscape, 1850–1950." Wellesley College Museum, April 15–June 18, 1981. Susan Danly Walther, guest curator. Wellesley, Mass.: the Museum, 1981.

Railway History Monograph, The, vol. 1- (1972-).

Ralston, Leonard F. "Railroads and the Government of Iowa, 1850–1872," Ph.D. dissertation, State University of Iowa, 1960.

Rand McNally's Pioneer Atlas of the American West (facsimile maps from 1876 ed.). Chicago: Rand McNally, 1969.

Randall, James G., and David Donald. *The Civil War and Reconstruction.* Boston: Heath, 1961.

Rasmussen, Louis J. *Railway Passenger Lists of Overland Trains to San Francisco and the West.* 2 vols. San Francisco Historical Records, 1966, 1969.

Redding, B. B. *A Sketch in the Life of Mark Hopkins, of California.* San Francisco: A. L. Bancroft, 1881.

Reed, J. Harvey. *Forty Years an Engineer: Thrilling Tales of the Rail.* Prescott, WA: Charles H. O'Neil, 1915.

Reinhardt, Richard, ed. *Workin' on the Railroad: Reminiscences from the Age of Steam.* American West, 1970.

Rhodes, James Ford. *History of the United States from the Compromise of 1850.* 7 vols. New York: Macmillan, 1893–1906.

Rice, Harvey. *Letters From the Pacific Slope, or, First Impressions (1855–70).* New York: D. Appleton, 1870.

———. "A Trip to California in 1869." *Magazine of Western History* VII (November 1887–April 1888).

Richardson, Albert D. *Beyond the Mississippi: From the Great Rim to the Great Ocean.* Hartford, Conn.: American Pub. Co., 1867.

———. *Garnered Sheaves from the Writings of Albert D. Richardson,* collected and arranged by his wife. Hartford, Conn.: Columbian, 1880.

———. *Our New States and Territories, Being Notes of a Recent Tour of Observation Through Colorado, Utah, Idaho, Nevada, Oregon, Montana, Washington Territory, and California.* New York: Beadle, 1866.

Richardson, H. R. *Public Relations of the Railroad Industry in the U.S., a Bibliography, 1808–1955.* Washington: Bureau of Railway Economics, 1956.

Riegel, Robert Edgar. *America Moves West.* New York: Holt, 1930.

———. *The Story of the Western Railroads, from 1852 through the Reign of the Giants.* Lincoln: University of Nebraska Press, 1926, 1964.

Ripley, W. Z. *Railroads: Finance and Organization,* 1915. New York: Longmans, 1915.

Robertson, William T., and W.F. *Our American Tour: Being a Run of Ten Thousand Miles from the Atlantic to the Golden Gate, in the Autumn of 1869.* Edinburgh: privately printed, 1871.

Robinson, Doane. *History of the Dakota or Sioux Indians.* Minneapolis: Ross & Haines, 1956, 1967.

Robinson, John R. *The Octopus: A History of the Construction, Conspiracies, Extortions, Robberies & Villainous Acts of Subsidized Railroads* (1894; reprint, New York: Ayer, 1981).

Rusling, Brigadier General James F. *Across America, or, The Great West and the Pacific Coast.* New York: Sheldon, 1874.

Russell, A. J. *The Great West Illustrated in a Series of Photographic Views Across the Continent; taken along the line of the Union Pacific Railroad west from Omaha, Nebraska.* New York: Union Pacific Railroad Company, 1869.

Russell, Robert A. *Improvements of Communication with the Pacific Coast as an Issue in American Politics, 1783–1864.* Cedar Rapids, Iowa, 1948.

Sabin, Edwin L. *Building the Pacific Railway.* Philadelphia: Lippincott, 1919.

Sacramento Union, The Golden Spike Centennial. (22 pp. vintage newspapers on microfilm.) Sacramento,California State Library, 1969.

Salt Lake Tribune. Feature story on Bear River City riot, published in June 1961. Courtesy Union Pacific Railroad.

Sanborn, John B. *Congressional Grants of Land in Aid of Railways.* (1869; reprint, New York: Ayer, 1981).

Sandburg, Carl. *Abraham Lincoln: The War Years,* vol. II. New York: Harcourt, 1936.
Sanders, Donald G. *The Brasspounder.* New York, 1978.
Sandmeyer, Elmer C. *The Anti-Chinese Movement in California.* Urbana: University of Illinois, 1939.
Sandoz, Mari. *The Buffalo Hunters.* New York: Hastings, 1954.
Savage, James W., and J. T. Bell. *History of the City of Omaha, Nebraska.* New York: Munsell, 1894.
Scott, Hugh L. *Some Memories of a Soldier.* New York: Century, 1928.
Sexton, Alexander. "The Army of Canton in the High Sierra." *Pacific Historical Review,* May 1966.
Seymour, Silas S. *Incidents of a Trip Through the Great Platte Valley, to the Rocky Mountains and Laramie Plains in the Fall of 1866, with a synoptical statement of the various Pacific Railroads, and an account of the Great Union Pacific Railroad Excursion to the One Hundredth Meridian of Longitude.* New York: Van Nostrand, 1867.
Sharp, Mildred J. "The M. and M. Railroad." *Palimpsest* III (January 1922).
Shelton, William A. *Atlas of Railway Traffic Maps.* Chicago: La Salle, 1913.
Sherman, John. *Recollections of Forty Years in the House, Senate, and Cabinet.* Chicago: Werner, 1895.
Sherman, General William T. *Memoirs,* 2nd ed. New York: Appleton, 1887.
Shuck, Oscar T. *Representative Men of the Pacific.* San Francisco: Bacon & Co., 1887.
"Sidney Dillon." *Transactions* (American Society of Civil Engineers) XXXVI, December 1896.
Sienkiewicz, Henryk. "The Chinese in California." *California Historical Society Quarterly,* December 1955.
Sillcox, L.K. *Safety in Early American Railway Operation, 1853–1871.* Princeton, N.J.: Princeton University, 1936.
Smith, Philip R. *Improved Surface Transportation and Nebraska's Population Distribution, 1860–1960.* New York: Ayer, 1981.
Smith, W. P. *Book of the Great Railway Celebrations of 1857, 1858.* N.P.
Sobel, Robert. *The Fallen Colossus.* New York, 1976.
———. *Panic on Wall Street: A History of America's Financial Disasters.* New York: Macmillan, 1969.
Sparks, David S. "Iowa Republicans and the Railroads, 1856–1860." *Iowa Journal of History* 53 (July 1955).
Stanley, Henry M. *Autobiography of Sir Henry Morton Stanley, G.C.B.* Boston: Houghton Mifflin, 1909.
———. *My Early Travels and Adventures in America and Asia.* New York: Scribner's, 1895.
Starr, John W. *Lincoln and the Railroads.* New York: Dodd, Mead, 1927.
———. *One Hundred Years of American Railroading.* New York: Dodd, Mead, 1928.
Starr, Kevin. *Americans and the California Dream, 1850–1915.* New York: Oxford, 1973.
Steele, W. R., ed. *The Trans-Continental: Published Daily on the Pullman Hotel Express Between Boston and San Francisco, May 23–July 4, 1870.* Boston: Board of Trade, 1870.
Sterne, Simon. *Railways in the United States.* New York: G. P. Putnam, 1912.
Stevenson, Robert Louis. "Across the Plains." In *From Scotland to Silverado.* Edited James D. Hart. Cambridge: Belknap Press of Harvard University Press, 1966.
Stewart, Joseph B. "The Bastile Testimony of Joseph B. Stewart Before a 'Select Committee,' Given on the 18th and 29th of January, 1873, With Remarks About His 'Contempt.'" Washington: Gibson Bros., 1873.
Stiles, Edward H. *Recollections and Sketches of Notable Lawyers and Public Men of Early Iowa.* Des Moines: Homestead, 1916.
Stilgoe, John R. *Metropolitan Corridor: Railroads and the American Scene.* New Haven: Yale, 1983.
Stillman, Dr. J. D. B. "The Last Tie." *Overland Monthly,* July 1869.
Stoddard, William O. *Lincoln's Third Secretary: The Memoirs of William O. Stoddard.* New York: Exposition Press, 1955.
Stokes, Anson Phelps. *Stokes Records.* New York: Privately printed, 1910.
Stone, Irving. *Men to Match My Mountains.* Garden City, New York: Doubleday, 1956.
Stover, John F. *American Railroads.* Chicago: University of Chicago, 1961.
———. *Iron Road to the West.* New York: Columbia University Press, 1978.
———. *Life and Decline of the American Railroad.* New York: Oxford University Press, 1970.
Summers, Mark W. *Railroads, Reconstruction, and the Gospel of Prosperity: Railroad Aid Under the Republicans, 1865–1877.* Princeton, N.J.: Princeton University, 1984.
Swanberg, W. A. *Jim Fisk: the Career of an Improbable Rascal.* New York: Scribner's, 1959.
Swasey, W. F. *The Early Days and Men of California.* Oakland: Pacific Press, 1891.

Taft, Robert. *Photography and the American Scene.* New York: Macmillan, 1938.

Taylor, George R., and Irene D. Neu. *The American Railroad Network, 1861–1890.* Cambridge: Harvard, 1956.

"Thomas C. Durant." *Transactions* (American Society of Civil Engineers) XXXVI, December1896.

Thompson and West. *History of Sacramento County.* Oakland, 1880.

———. *History of the State of Nevada.* Facsimile ed. Berkeley: Howell-North, 1958.

Thorndike, Rachel Sherman, ed. *The Sherman Letters: Correspondence Between General and Senator Sherman from 1837 to 1891.* London: Samson Low, Marston & Co., 1894.

Thornton, Willis. *The Nine Lives of Citizen Train.* New York: Greenburg, 1948.

Todd, John, D. D. *The Sunset Land; or, The Great Pacific Slope.* Boston: Lee & Shepard, 1870.

Toponce, Mrs. Kate, *Reminiscences of Alexander Toponce, Pioneer.* Ogden, Utah: Privately printed, 1923.

Townsend, George Alfred. *Washington, Outside and Inside.* Hartford, Conn.: Betts, 1874.

Travelers' Official Railway Guide of the United States and Canada. National Railway Publication Co., June 1869.

Trottman, Nelson. *History of the Union Pacific.* New York: Ronald Press Co., 1923.

Tung, William L. *The Chinese in America, 1820–1973.* Dobbs Ferry, New York: Oceana, 1974.

Turner, George E. *Victory Rode the Rails: The Strategic Place of the Railroads in the Civil War.* Indianapolis: Bobbs-Merrill, 1953.

Tutorow, Norman E. *Leland Stanford: Man of Many Careers.* Menlo Park, Calif.: Pacific Coast Publishers, 1971.

Union Pacific Historical Museum. *Early Western Photographs, 1862–1897.* Omaha, Neb.: Union Pacific Railroad, 1930.

———. *Photographic Views of Construction and Early History of the Union Pacific Railroad.* Omaha, NE: Union Pacific Railroad, 1874.

Union Pacific Magazine. Omaha: Union Pacific Railroad Co., 1922–33.

Union Pacific Railroad Company. *The Great Union Pacific Railroad Excursion to the Hundredth Meridian: From New York to Platte City. Incidents of the Excursion—Character of the Country—Statistics of the Road—Its Progress and Trade.* Chicago, 1867.

———. *Guide to the Union Pacific Railroad Lands.* Omaha, 1870.

———. *The Illustrated Railroad Guide, Giving a Description of the Union and Central Pacific Railroads, with Illustrations Along the Line and Branches.* Chicago: Adams, 1879.

———. *Progress of the Road West from Omaha, Nebraska, Across the Continent.* New York, 1870.

———. *Progress of the Road West . . . Five Hundred Miles Completed October 25, 1867.* New York: C. Alvord, 1868.

———. *Progress of the Union Pacific Railroad West . . . Making, With Its Connection, An Unbroken Line from the Atlantic to the Pacific Ocean. 820 Miles Completed Sept. 20, 1868.* New York: The Company, 1868.

———. *Report of the Chief Engineer on Bridging the Missouri River.* New York: D. Van Nostrand, 1867.

———. *Report of the Consulting Engineer on the Location Between Omaha City and the Platte Valley, dated Dec. 21, 1864.* New York: W. C. Bryant, 1865.

———. *Report of Thomas C. Durant, Vice-President and General Manager to the Board of Directors, in Relation to the Operations of the Engineer Department, and the Construction of the Road, Up to the Close of the Year 1865.* New York: W. C. Bryant, 1866.

———. *Report of Thomas C. Durant . . . In relation to the Surveys Made Up to the Close of the Year 1864.* New York: W. C. Bryant, 1866.

———. *A Trip Across the North American Continent from Omaha to Ogden.* New York: Nelson, 1873.

———. *The Union Pacific Railroad: The Great National Highway Between the Missouri River and California.* Chicago: Horton & Leonard, 1868.

———. *Union Pacific Railroad: The Great National Highway Between the Mountains and the East. The Direct Route to Colorado, Utah, Idaho, Montana, Nevada and California. Open from Omaha to North Platte.* Omaha: Omaha Herald, 1867.

———. *The Union Pacific Railroad . . . Its Construction, Resources, Earnings, and Prospects.* New York: Wynkoop & Hallenbeck, 1867.

United States Library of Congress, Geography and Map Div., *Railroad Maps of the United States: A Selective Annotated Bibliography of Original 19th Century Maps.* Washington: U.S. Government Printing Office, 1975.

U.S. Senate Committee on the Pacific Railroad, *The Policy of Extending Government Aid to Additional Railroads to the Pacific, by Guaranteeing Interest on their Bonds, February 19, 1869.* Washington: U.S. Government Printing Office, 1869.

U.S. Works Progress Administration. *California: A Guide to the Golden State.* New York: Hastings House, 1939.

———. "Historical Sketch of Weber County." Utah Historical Records Survey, 1940.

———. "History of Ogden." Utah Historical Records Survey, 1940.

———. *Nebraska: A Guide to the Cornhusker State.* New York: Viking, 1939.

———. *Nevada: A Guide to the Silver State.* Portland, Oreg.: Binfords & Mort, 1940.

———. *A Guide to Utah.* New York: Hastings House, 1941.

———. *Wyoming—A Guide to its History, Highways and People.* New York: Oxford, 1941.

Utley, Robert M. *Frontier Regulars: the United States Army and the Indian, 1866–1891.* New York: Macmillan, 1973.

———. *The Indian Frontier of the American West, 1846–1890.* Albuquerque: University of New Mexico, 1984.

———. *Last Days of the Sioux Nation.* New Haven: Yale, 1963.

Vestal, S. *New Sources of Indian History, 1850–1891.* Norman: University of Oklahoma, 1934.

Ware, Eugene Fitch. *The Indian War of 1864.* (1911; reprint, New York: St. Martin's, 1960).

Warman, Cy. *The Last Spike and Other Railroad Stories.* New York: Scribner's, 1906.

———. *The Story of the Railroad.* New York: Appleton, 1898.

———. *Tales of an Engineer.* New York: Scribner's, 1895.

———. "A Wild Night at Woodriver," "The Battle of the Snow-Plows," *McClure's Magazine,* VIII (November 1896–April 1897).

Washburn, Wilcomb E. *The American Indian and the United States: A Documentary History.* New York: Random House, 1973.

———. "A Moral History of Indian-White Relations: Needs and Opportunities for Study," *Ethnohistory* IV (Winter 1957), 47–61.

———. *Red Man's Land, White Man's Law: A Study of the Past and Present Status of the American Indian.* New York: Scribner's, 1971.

Weber, Thomas. *The Northern Railroads in the Civil War, 1861–1865.* New York: Columbia University Press, 1952.

Welles, Gideon. *Diary of Gideon Welles,* 3 vols. Boston: Houghton Mifflin, 1911.

Wellman, Paul I. *Death on Horseback: Seventy Years of War for the American West.* Philadelphia: Lippincott, 1947.

Wells, the Reverend Charles W. *A Frontier Life.* Cincinnati: Jennings & Pye, 1902.

Wenzel, Caroline. "Finding Facts About the Stanfords in the California State Library." *California Historical Society Quarterly* XIX (September 1940).

Wheat, Carl I. "A Sketch in the Life of Theodore D. Judah." *California Historical Society Quarterly* IV (September 1925).

White, Henry Kirke. *History of the Union Pacific Railway.* Chicago: University of Chicago, 1895.

White, John H., Jr. *American Locomotives: An Engineering History, 1830–1880.* Baltimore: Johns Hopkins, 1968.

White, Richard. "The Winning of the West: The Expansion of the Western Sioux in the Eighteenth and Nineteenth Centuries." *Journal of American History* LXV (September 1978).

Whitney, Asa (1797–1872). *A Project for a Railroad to the Pacific.* New York: G. W. Wood, 1849.

Wibberly, Leonard. "The Coming of the Green." *American Heritage,* August 1958.

Williams, H. J. *The Pacific Tourist: Williams' Illustrated Trans-Continental Guide of Travel, from the Atlantic to the Pacific Ocean . . . Being a Complete Traveler's Guide of the Union and Central Pacific Railroads.* New York: H. J. Williams, 1876.

Williams, Kenneth P. *Lincoln Finds a General: Military Study of the Civil War,* 5 vols. New York: Macmillan, 1949–1959.

Williams, John Hoyt. *A Great and Shining Road: The Epic Story of the Transcontinental Railroad.* New York: Times Books, 1988.

Wilson, James Harrison. *The Life of John A. Rawlins.* New York: Neale, 1916.

Wilson, Neill C. and Frank J. Taylor. *Southern Pacific: The Roaring Story of a Fighting Railroad.* New York: McGraw-Hill, 1952.

Winther, Oscar Osburn. *The Transportation Frontier: Trans-Missisippi West, 1865–1890.* New York: Holt, 1964.

Wissler, Clark. *North American Indians of the Plains,* 3rd ed. New York, 1927.

Wu Cheng-Tsu. *"Chink!" A Documentary History of Anti-Chinese Prejudice in America.* New York: World, 1972.

Young, Otis E. *The West of Philip St. George Cooke, 1809–1895.* Glendale, Calif.: Arthur H. Clark, 1955.

Younger, Edward A. *John A. Kasson: Politics and Diplomacy from Lincoln to McKinley.* Iowa City: State Historical Society of Iowa, 1955.

Manuscripts and Letters

Adjutant General's Files, Record Groups 94 and 98, National Archives and Records Service, Washington.

Andrew, Bunyan Hadley. "Charles Crocker," unpublished M.A. thesis, University of California, Berkley, 1931.

Avery, Benjamin Parke. Papers, California Historical Society.

Bissell, Hezekiah. Recollections, Wyoming State Archives and Historical Department, Cheyenne.

Briggs, Robert O. "The Sacramento Valley Railroad," unpublished M.A. thesis, Sacramento State College, Sacramento, 1954.

Browne, John Ross. Diary, Calfornia Section, California State Library, Sacramento.

Cary, Thomas G. "Pacific Rail Road," Manuscript Division, Library of Congress, Washington.

Casement, Brigadier General John S. Papers, Kansas State University Library (Easterling Collection), Manhattan, Kansas.

Casement, John S. Papers, Western Heritage Collection, University of Wyoming, Laramie.

Catlin, A. P. Papers, California State Library.

Central Pacific Collection. California State Railroad Museum, Sacramento.

Central Pacific Railroad. Telegram Copies, 1868–69, Special Collections, The Stanford University Libraries.

Cioffi, Ralph. "Mark Hopkins, Inside Man of the Big Four," unpublished M.A. thesis, University of California, Berkeley, 1950.

Cole, Cornelius. Papers, Department of Special Collections, University of California Library, Los Angeles.

Conness, Senator John. Letter, to James McClatchy, December 18, 1863, California Section, California State Library, Sacramento.

Crocker, Charles. Reminiscences, Bancroft Library, University of California, Berkeley.

Dodge, Major General Grenville M. *Autobiography, 1831–71,* Council Bluffs, Iowa: Free Public Library.

———. Transcribed Letter Books, Free Public Library, Council Bluffs, Iowa.

———. Papers, Iowa State Department of History and Archives, Des Moines.

Dodge, Nathan P. Scrapbooks, Council Bluffs Public Library.

Dodge Family Papers. Denver Public Library.

Ferguson, Arthur N. Journals, Union Pacific Historical Museum, Omaha.

Hopkins, Mark. Biographical MS., Bancroft Library, University of California, Berkeley.

Hopkins-Huntington Letters. Special Collections, The Stanford University Libraries, Stanford.

Hopkins, Timothy. Random notes on Central Pacific History, Special Collections, The Stanford University Libraries, Stanford.

Huntington, Collis P. Papers, Syracuse University Library.

Huntington, Collis P. Reminiscenses, Bancroft Library, University of California, Berkeley.

Records of the Bureau of Indian Affairs, Record Group 75, National Archives, Washington.

Secretary of the Interior's Files, Record Group 48, National Archives, Washington.

Jackson, William Henry. Diary, Manuscript Division, New York Public Library.

Judah, Anna Ferona. Reminiscences, Bancroft Library, University of California, Berkeley.

Judah, Theodore Dehone. Papers, Bancroft Library, University of California, Berkeley.

Kasson, John A. Papers, Iowa State Department of History and Archives, Des Moines.

Leonard, Levi O. Papers, State University of Iowa, Iowa City.

Miller, E. H., Jr. Ledger, 1862–71, Special Collections, The Stanford University Libraries, Stanford.

Montague, Samuel S. Letters, Special Collections, The Stanford University Libraries, Stanford.

Partridge, A. A. Reminiscences, The Southern Pacific Company, San Francisco.

Root, Henry. Papers, California Historical Society.
Sherman, William T. Papers, Hayes Library, Fremont, Ohio.
———. Papers, Library of Congress, Washington.
Stanford, Leland. Letters, Stanford Collection, The Stanford University Libraries, Stanford.
Towne, A. N. Letter, to H. H. Bancroft, April 16, 1889, Bancroft Library, University of California, Berkeley.
Union Pacific Railroad Collection. Nebraska State Historical Society, Lincoln.

Government Documents

Act to Aid in the Construction of a Railroad and Telegraph Line from the Missouri River to the Pacific Ocean. Approved July 1, 1862. Chicago: Tribune Book and Job Steam Prtg. Off., 1862.
Acts of Congress Relating to the Union Pacific Railroad Passed Between the Years 1862 and 1873. Washington: U.S. Government Printing Office, 1873.
*Acts of Congress Relating to the Union Pacific Railroad and Branches.*Compiled December, 1877. Washington: J. L. Pearson, 1877.
Affairs of the Union Pacific Railroad Company, 42 Cong., 3 Sess. (1873), H. Rept. 78.
Annual Reports of the Secretary of the Interior, 1863–69.
Annual Reports of the Secretary of War, 1863–69.
Crédit Mobilier Investigation, 42 Cong., 3 Sess. (1873), H. Rept. 77. Known as Wilson Report.
Investigation on the Conduct of Indian Affairs, 43rd Cong., 1 Sess., H. Rept. 778.
(N.Y. State Supreme Court), James Fisk, Jr., *Against the Union Pacific Railroad Company, the Credit Mobilier of America, Oliver Ames, and Others . . . Proceedings for the commencement of the action, July 3, 1868 to the end of March, 1869.* New York: J. Polhemus, 1872.
Protection Across the Continent, 39 Cong., 2 Sess. (1867), H. Exec. Rept. 23.
Report of the Commission and of the Minority Commissioner of the United States Pacific Railway Commission, 5 vols., 1887.
Reports of the Government Directors of the Union Pacific Railroad Company, 1864–1884, 49 Cong., 1 Sess. (1886), S. Exec. Rept. 69.
Report of the Joint Special Committee to Investigate Chinese Immigration, 44 Cong., 2 Sess. (1887), Sen. Rept. 689.
Report of the Select Committee to Investigate the Alleged Credit Mobilier Bribery Made to the House of Representatives . . . , 1873. Known as Poland Report.
Report of Lieut. Col. James H. Simpson, Corps of Engineers, U.S.A., on the Change of Route West from Omaha . . . together with the President's Decision. Dept. Interior, 1865.
Report of U.S. Special Commissioners to Hon. Orville H. Browning, Secretary of the Interior, on the Central Pacific and Western Pacific Railroads, San Francisco.
Report on Transcontinental Railways, 1883, by Col. O.M. Poe, U.S. Engineers, 48 Cong., 1 Sess. (1884), H. Exec. Rept. 1, Part 2.
Reports of the Goverment Directors of the UPRR Made to the Secretary of Interior, From 1864 to 1885.
Testimony Taken by the United States Pacific Railway Commission. 8 vols. 50 Cong., 1 Sess., S. Exec. Doc. 51. Known as USPRRC.

Railroad Documents

Bishop, Francis A. *A Report of the Chief Engineer on the Survey and Cost of Construction of the San Francisco and Washoe Railroad of California: Crossing the Sierra Madre Mountains from Placerville to the Eastern Boundary of California, on the Line of Business from San Francisco to the Silver Mines of Nevada,* San Francisco, 1865.
Central Pacific Railroad. *Letter to Hon. C. A. Sumner and H. Epstein,Chairmen of R.R. Committees, Nevada State Legislature, from Leland Stanford, Pres., Central Pacific R.R.,* February 14, 1865.
———. *Railroad Communication with the Pacific, with an Account of the Central Pacific Railroad of California: The Character of the Work, Its Progress, Resources, Earnings and Future Prospect, and the Advantages of Its First Mortgage Bonds.* Fisk & Hatch, 1867.
———. *Report of the Chief Engineer of the Central Pacific Railroad Company of California, on his Operations in the Atlantic States,* 1862.

———. *Report of the Chief Engineer to the Board of Directors and President, Central Pacific Railroad Co.*, July 1, 1863.

———. *Report of the Chief Engineer upon Recent Surveys, Progress of Construction, and Estimate Revenue of the Central Pacific Railroad of California*, December, 1864.

———. *Report of the Chief Engineer upon Recent Surveys and Progress of Construction of the Central Pacific Railroad of California*, December, 1865.

———. *Report of the Acting Chief Engineer to the Board of Directors of the Central Pacific Railroad Co. of California*, January 5, 1867.

———. *Rules and Regulations for Employees* (c. 1868)

———. *Statement Made to the President of the United States, and Secretary of the Interior, on the Progress of the Work*, October 10, 1865.

———. *Statement Made to Senate Committee of the Nevada Legislature*, January 14, 1865.

Pacific Railroad Convention: Memorial to the President of the United States, Heads of Departments, Senate and House of Representatives, 1859.

———. *Report of Theodore D. Judah, accredited agent Pacific Railroad Convention, Upon His Operations in the Atlantic States*, August 1860.

Union Pacific Railroad: Report of the Chief Engineer, with Accompanying Reports of Division Engineers, for 1866.

———. *Report of the Chief Engineer on Bridging the Missouri River, 1867.* New York: Van Nostrand, 1867.

———. *Report of the Consulting Engineer (S. Seymour) on the Location Between Omaha City and the Platte Valley*, Dec. 21, 1864. New York: Bryant, 1865.

———. *Report of the Consulting Engineer, With Reference to the Operation of the Engineer Department*, New York, 1865.

———. *Report of T. M. Case, of Surveys of Cache La Poudre and South Platte Routes, and Other Mountain Passes in Colorado.*

———. *Report of G. M. Dodge, Chief Engineer, with accompanying Reports of Chiefs of Parties for 1868–69.*

———. *Report of G. M. Dodge on A Railroad Line from the Union Pacific Railroad to Idaho, Montana, Oregon, and Puget's Sound.* Washington: Philip & Solomons, 1868.

———. *Report of James R. Evans of Exploration from Camp Walbach to Green River*, 1864.

———. *Report of James T. Hodge on the Coal Deposits in the Rocky Mountains*, 1864.

———. *Report of Samuel B. Reed of Surveys and Explorations from Green River to Great Salt Lake City*, 1864.

———. *Report of the Organization and Proceedings of the Union Pacific Railroad Co.*, 1864.

———. *Union Pacific Railroad Company, Chartered by the United States; Progress of their Road West from Omaha, Nebraksa, Across the Continent*, June 18, 1868.

Periodicals

American Railroad Journal
Cheyenne Daily Leader
Chicago Tribune
Congressional Globe
Corinne (Utah) *Weekly Reporter*
Council Bluffs Bugle
Council Bluffs Nonpareil
Denver Rocky Mountain News
Des Moines Iowa State Ledger
The Friend (Honolulu)
The Frontier Index
Frank Leslie's Illustrated Newspaper
Nevada City (California) *Journal*
New York Herald
New York Post
New York Sun

New York Times
New York Tribune
New York World
North American Review
Omaha Weekly Herald
Painesville (Ohio) *Telegraph*
Placerville (California) *Mountain Democrat*
Railway Age
Railway Gazette
Rocky Mountain News (Denver)
Reno Crescent
Sacramento Bee
Sacramento Union
Salt Lake City Deseret News
Salt Lake City Tribune
San Francisco Alta California
San Francisco Chronicle
San Francisco Newsletter
Southern Pacific Magazine
Union Pacific Magazine

Index

A. A. Sargent (locomotive), 258–59
Adams, Charles Francis, Jr., 599, 621, 624, 638, 675
Adams, John Quincy, 599
Alabama, 445–46, 582
Alameda Railroad, 540
Albany and Vermont Railroad, 217
Allen, W.F., 403
Alley, John B., 227, 270, 271, 291, 328, 348, 378–79, 405, 406, 419, 420, 423, 589, 616, 620, 677, 680, 686, 689–90, 693
Allison, William B., 422, 676, 680, 692, 696
Ames, Oakes, 133, 143, 177, 218, 224, 226–27, 242, 243, 253, 270–71, 302, 322, 347, 349, 374, 378, 379–80, 403–5, 435, 481, 485, 535, 536–37, 572–73, 581, 583, 601–602, 613, 614, 616, 620–22, 624, 625, 628, 630, 637, 644, 661, 671, 690, 694–95, 709, 727
  congressional censure of, 704–5
  Crédit Mobilier scandal and, 681–84
  death of, 705
  Dix buyout and, 444
  "fatal list" of, 677–80
  legislators bribed by, 419–424, 444–45, 483, 626, 677
  Lincoln's meeting with, 211–12
  McComb letters of, 444–45, 677–78
  mystery shares and, 421–22
  Poland Committee testimony of, 686–89, 693, 695–97, 699–701, 702, 703
Ames, Oliver, 133–34, 143, 211, 270, 302, 322, 324, 329, 372, 376–77, 401, 409, 411, 435, 438, 439, 441–42, 444, 448, 453, 454, 472, 479, 481, 485, 494, 500, 504, 508–10, 515, 529–33, 535, 545, 555, 557, 559, 563, 569, 572–75, 581–84, 589, 594–97, 601–3, 605, 612, 614–15, 616, 620–623, 624, 632, 637, 638
  death of, 690
  Durant's leadership conflict with, 343–44, 347–49, 377–80, 402–7, 466–468
  elected president of Union Pacific Board, 304–5
  Huntington and, 393, 403, 407, 419
  Piedmont kidnapping episode and, 649–51
  Promontory Summit agreement opposed by, 634
Andrist, Ralph K., 310
Anthony, James, 366–67, 413, 459, 504
Anthony, Scott J., 190
Antietam, Battle of, 126
Apache, 227, 410
Appleton, Frank, 383–84
Arapaho, 185–88, 190, 227–228, 230–31, 260, 268, 284, 294, 311, 410
Army Topographical Corps, U.S., 48–49, 557, 575
Arthur, Chester A., 452
Associated Press, 366, 428
Astor, John Jacob, 268
Atlantic and Great Western Railroad, 161, 170
Augur, Christopher Columbus, 326, 349–50, 351, 354–355, 356, 374, 380, 389, 399–400, 426, 461, 516, 517
Avery, Benjamin Parke, 412–413, 740
Axtell, Samuel B., 418, 452, 505, 540

Bad Wound, 188
Bailey, James, 85, 92, 96, 99, 103, 127, 137, 141–42, 144
Baker, Edward Dickinson, 105
Baldwin, Matthew, 16

Ball's Bluff, Battle of, 105
Baltimore and Ohio Railroad, 17, 118, 132, 216
Bancroft, H. H., 279
Bank of California, 565–66
Banks, Nathaniel, 685
Barker, Fred, 548
Barlow, Samuel, 18
Barnard, George, 621–23, 625–27, 632, 635, 643
Barnes, James, 557, 563, 573
Bates, Benjamin E., 211, 406, 589
Bates, Thomas H., 263, 303, 321, 358, 383, 497, 614, 637
Bayard, Richard, 422, 423, 680
Beadle, Edward Fitzgerald, 50, 137
Beadle, John H., 512, 526, 527–28, 606–7
Bean, A.A., 441, 442
Bear Creek Massacre, 229, 244
Bear Hunter, 229
Bear River City, Wyoming Territory, 562–63
Beck, J.B., 685
Beckwith, E.O., 50, 51
Beecher, Henry Ward, 169
Behale, Baptiste, 357
Belknap, William W., 691
Bell, Clark, 405
Bell, G.W., 273
Bell, John, 38
Benedict, S.S., 506–7
Bennett, Henry, 108
Bennett, James Gordon, 170, 345, 459
Benson, Ezra Taft, 537–38, 541, 607, 659–60
Bent, William, 187–88
Benton, J.A., 655
Benton, Thomas Hart, 27–28, 37–38, 41, 42, 47, 48, 50, 161, 218
Benton, Wyoming Territory, 561–63
Bidwell-Bartleson party, 13
Bingham, John A., 422, 625, 630, 676, 678, 680, 687, 688, 696, 702, 704, 758
Bissell, Hezekiah, 476
Black Bear, 231
Blackfoot, 400
Black Kettle, 187, 189–90
Blaine, James G., 423, 452, 676, 678–87, 689, 695, 699, 702, 704, 758
Blair, Francis P., 351
Blair, John, 324, 405–6
Blickensderfer, Jacob, 372, 374, 464–65, 480, 494, 496, 497–99, 500, 506, 509–10, 513, 516, 529–530, 548, 557, 563, 570, 574, 596, 597–98, 601, 603, 604, 611, 613, 614, 617, 638, 641
Bloomer Cut, 174, 176, 194
Blue Jay (locomotive), 611
Boomer, L.B., 302, 304, 476, 546, 641–42
Booth, John Wilkes, 213–14, 217
Booth, Junius Brutus, 23
Booth, Lucius A., 85, 96, 127, 129, 141
Booth, Newton, 72
Borland, Solon, 41
Boston and Albany Railroad, 18
Boston and Maine Railroad, 315
Boston and Worcester Railroad, 17
Boutwell, George S., 423, 624, 625, 633, 676, 678, 680
Bowers, Brooks, 548
Bowles, Samuel, 234, 235–36, 308, 527, 622, 624
Bowsher, Amos L., 663, 668
Boyd, John, 394
Boyer, Benjamin M., 423, 680
Boynton, Henry Van Ness, 601
Bozeman, John M., 284, 385
Bozeman Trail, 244, 284, 309, 354, 384, 386, 389
Brady, James, 418
Brannan, Sam, 337
Breese, Sidney, 26, 28
Brooks, James, 423, 515, 575, 581, 620, 637, 676, 680, 688, 689, 696, 699–701, 702, 703, 704, 757
Brooks, Noah, 137
Brooks, Preston, 68
Bross, William, 234, 235–36
Brown, Arthur, 320
Brown, H. Clark, 561
Brown, John, 74
Browne, Percy, 263, 303, 349, 350, 358, 374–75, 377, 382–83
Browning, Orville Hickman, 283, 304, 340–41, 346, 354, 364, 372, 435, 465, 466, 475, 481–82, 483, 489, 491–92, 501, 523, 524, 535–36, 543, 545, 550–51, 554–55, 556, 557, 569, 574–75, 576, 581–82, 584–85, 589, 595–98, 601, 602, 603, 611, 612, 613, 625
Brulé Sioux, 188, 209, 266–268, 285, 410, 524–525
Buchanan, James, 57, 69, 71, 74–75, 153, 491
Buffalo and New York Railroad, 60
Bull, Alpheus, 417
Bull Run, battles of, 101, 126, 151, 228
Burch, John C., 74, 75, 76, 84
Bureau of Indian Affairs, 284
Burke, John, 307
Burlington and Missouri River Railroad, 694
Burnside, Ambrose, 126
Bushnell, Cornelius S., 171–172, 199, 211, 270, 302, 348, 377–78, 379, 394, 402–3, 407, 409, 422, 444, 515, 572, 589, 620–21, 625, 629, 637, 643, 690, 692
Butler, Andrew P., 68
Butler, Benjamin F., 342, 686, 701

Cahoon, Tom, 478–79
California, 12, 23, 43, 48, 61, 71, 493
  anti-Chinese legislation of, 206

Chinese immigration to, 205–7
elections in, *see* elections, California
first railroad ride in, 65
California and Oregon Railroad, 336, 337, 408, 414, 417, 490, 504–5, 587, 641, 706
California Central Railroad, 71, 79–80, 176, 336–37, 395–96, 490
California Crédit Mobilier, 408
California Northern Railroad, 336–37
California Pacific Railroad, 397
California Trail, 341
Cameron, Simon, 159
Carmichael, Lewis, 439, 441, 462, 463, 476, 479–80, 485, 511, 514
Carpentier, Horace W., 336, 456–57, 473–74, 523, 565
Carr, William B., 474, 490
Carrington, Henry B., 284, 309–11
Carter, T.J., 542–43
Carver, Hartwell, 31–32, 33, 123, 528–29, 714
Case, Frances M., 183, 192, 570
Casement, Dan, 253, 254, 265, 291, 293, 301, 312, 323, 328, 344, 380, 426, 441, 476, 485, 527, 570, 607–8, 659, 669, 670
Casement, Frances Jennings, 254, 261, 308, 328, 528, 609, 619
Casement, John Frank, 308
Casement, John Stephen "General Jack," 253, 261, 263, 269, 291, 293, 301, 307, 309, 312, 323, 328, 344, 351, 371, 372, 374, 381, 399, 412, 424–27, 441, 448, 475, 476, 479–80, 485, 486, 522, 527–29, 535, 544, 570–71, 607–608, 610, 619–20, 624, 637–38, 639, 652–53, 658, 659, 662, 669
background of, 254–55
Cass, Lewis, 40
Casserly, Eugene, 705–6
Catlin, Amos P., 72, 85
Cedar Rapids and Missouri Railroad, 167–68, 177, 211, 279, 324, 393, 409
Central Georgia Railroad, 18
Central Pacific Railroad Company, 86, 101–2, 103, 115, 117–18, 127, 131, 132, 169, 177, 194, 196, 212, 219, 229, 259–60, 270, 279, 290, 308, 333, 366–67, 404, 435, 464, 470, 474, 476, 481–83, 513, 532, 533, 542, 545–46, 555, 577, 582, 583, 585–86, 594, 599, 602, 625, 630, 671
accidents and mishaps of, 427–28, 608, 611–12, 635–36
Ames's deal with, 133–34
Bailey's resignation from, 144
Bank of California episode and, 565–66
board of directors named for, 96
Boomer Cut of, 174, 176, 194
Carter visit to, 542–43
Chinese laborers of, 207–9, 219–23, 235, 239, 245, 278, 296, 297, 331–32, 360–62, 369, 395, 407, 428–29, 447, 470, 493–94, 669, 671
Colfax Excursion and, 234–238
congressional lobbying by, 106, 108–10
corporate officers elected to, 99
county support controversy and, 134–35, 143
Denver's shakedown attempt on, 501–2
directors' financing of, 127–128
in drive to Promontory Summit, 636–38, 641, 644
Dutch Flat wagon road of, 258–59
earnings of, 220–21, 258, 367, 392, 431, 568, 588
1864 amendment to Railroad Bill and, 178–79
1866 amendment to Railroad Act and, 271
first rail laid by, 145
first revenue trains of, 175–176
first through service to Reno of, 503–4
Forty-Mile Desert survey for, 276–77
Goat Island terminal sought by, 334–36, 340–41, 397, 416–19, 451–54, 456, 491, 504, 565, 588, 641, 705
groundbreaking ceremonies of, 122–24, 129–30
"high ground" scheme of, 492–93
investors sought by, 125–27
iron shortage of, 471, 487, 522
Judah's conflict with board of, 137–42, 143
Judah's route survey for, 94, 96–101
labor problems of, 277–78, 360–62, 605–6
last spike ceremony and, *see* Promontory Summit last spike ceremony
lobbyists paid off by, 451–52
meeting place debate and, 601–2, 632–34
Mormon contractors of, 364–65, 537–38, 541–42, 551–52, 554, 556, 564, 605–7, 631–32, 659–60
Nevada link of, 111–12
new special commissioners and, 601–4, 611–12, 617–18

Central Pacific Railroad Company (*cont.*)
nitroglycerine experiments by, 272–75, 315–16, 320–21
onset of construction by, 128
onset of regular passenger service of, 176
original backers of, 91–94
parallel grading issue and, 553, 658
rift within, 99, 128–30
Robinson brothers and, 194–95
Rockies' base question and, 136–37
Sacramento Union's rift with, 366–67, 413
San Francisco and Oregon acquired by, 539–40, 543
senators bribed by, 418–19
small shareholders' suit against, 709
snow as obstacle to, 245–246, 305–7, 317–20, 330, 360, 407, 428–29, 447, 460, 469–70, 485–86, 609–10
Southern Pacific and, 456–458, 473–74, 490–91, 523–24, 540, 587–88, 671
special commissioners and, 575–76
Strong's resignation from, 146
summit excursion of, 412–416
ten-mile feat of, 638–40
through rate dispute and, 640–41
tunnel projects of, 240, 246, 298–300, 306–7, 315–18, 320–21, 332, 367–70, 407, 605–6
Vallejo ring and, 417
Wadsworth supply base of, 520–21
Western Pacific acquired by, 335–39, 368
Central Transportation Company, 258
Chancellorsville, Battle of, 139
Chandler, William E., 624, 680
Chapman, Oliver S., 211
Charles Crocker & Company, 129, 138, 408
Chase, Salmon, 226, 437
Cheyenne, 185–90, 227, 228, 231, 266, 268, 311, 346–47, 349–50, 353–354, 384–86, 389, 410
Cheyenne, Wyoming Territory, 370–71, 462–63, 478, 483, 648
growth of, 399, 411–12
Vigilance Committee of, 411
Cheyenne Leader, 399, 411, 517
Chicago, Iowa and Nebraska Railroad, 177, 211
Chicago and Alton Railroad, 218
Chicago and Northwestern Railroad, 76, 441
Chicago and Rock Island Railroad, 34, 152, 153, 157–158, 161, 196
Chicago Tribune, 225, 226, 234, 291, 574, 667, 682
Chickamauga, Battle of, 142
China:
laborers and immigrants from, 205–7
Whitney's sojourn in, 3–8
Chittenden, L.E., 630
Chivington, John M., 186–91, 227, 230
Cincinnati Commercial, 29–30, 526–27, 622, 678, 683
Cisco, John J., 304, 392, 403, 405, 537, 627
Civil War, U.S., onset of, 95–96
Clark, Horace F., 696
Clark, Lewis Gaylord, 18
Clark, William, 11
Clark Dodge and Company, 378–79, 392
Clarke, Stephen, 349
Clay, Henry, 12, 39, 201
Clemens, John, 34
Clemens, Samuel (Mark Twain), 62, 209, 276, 305, 496, 521
on Chinese immigrants, 206–7
Crédit Mobilier scandal satirized by, 694–95
Clement, Lewis M., 223, 238, 297, 319, 369, 505, 522, 532, 552, 604, 611, 617, 641
Clews, Henry, 392
Clinton, DeWitt, 59
Cody, William Frederick ("Buffalo Bill"), 294, 355
Cold Harbor, Battle of, 178, 179
Cole, Cornelius, 85, 177, 180, 212, 219, 260, 418, 588, 705–6
Cole, George, 418
Cole, Nelson, 228, 231
Colfax, Schuyler, 234, 235–237, 422, 437, 466, 481, 512, 528, 557, 622, 668, 676, 678–80, 684, 685, 689, 694–96, 697, 700, 701, 702
Comanche, 188, 227, 410
Compromise of 1850, 43
Confederate States of America, 95, 104
Congress, U.S., 8, 19, 23, 26, 28, 31, 34, 37, 40–43, 47–49, 51, 83, 84, 101–102, 108, 124–25, 126, 131, 159, 168, 184, 185, 196, 212, 224, 244, 246–47, 253, 281, 335, 341, 365, 389, 391, 402, 416–17, 429, 438, 453, 473, 489, 504, 525, 535, 540, 573, 597, 599, 600, 602, 615, 624, 628, 670, 684–85, 686, 688, 700
bribery and, 418–24, 444–445, 483, 626, 677
1866 amendment to Railroad Act in, 270–71
1871 amendment to Railroad Act in, 682
Judah as lobbyist in, 74–76
Judah as observer of, 67, 68–69
Practical Plan and, 70–71

Reconstruction and, 332–333
Whitney's railroad proposal to, 13–15
see also House of Representatives, U.S.; Senate, U.S.
Congressional Pacific Railroad Committee, 610
Congressional Record, 681
Conkling, Roscoe, 452, 528, 588
Conness, John, 180, 256, 270, 367, 419, 430, 474, 491, 505, 587
Connor, Patrick Edward, 228–31, 235–36, 244, 277, 284, 310, 356, 443, 606, 607, 660
Contract and Finance Company, 408, 489, 568, 706, 707, 710, 739
Cooper, Peter, 16
Cope, C.C., 478, 547–48
Corbett, Henry Winslow, 418, 419
Corinne, Utah Territory, 607, 618
Corning, Erastus, 132, 539, 586
Cox, Jacob, 254, 624, 625, 630, 633, 638
Cox, Samuel "Sunset," 685, 690, 699
C. P. Huntington (locomotive), 175
Cragin, Aaron H., 451
Crane, A.M., 123
Crane, Henry C., 199, 211, 378–79, 425, 440, 462, 463, 483, 510, 643–44
Crazy Horse, 385
Crédit Foncier, 282
Crédit Mobilier de France, 172
Crédit Mobilier Frauds, The, 683
Crédit Mobilier of America, 227, 247, 253, 256, 262, 270, 271, 302, 322–24, 327–28, 343, 347–48, 377, 378, 379, 409, 420–23, 435, 438, 444, 453, 536, 568, 572–73, 575, 601, 602, 623–27, 632, 635, 643
Adams on, 599–600
December 1868 dividend of, 589
establishment of, 172, 199
major stockholders of, 211
obscure accounting practices of, 242–43
Train and, 172, 181, 199, 515
see also Crédit Mobilier scandal; Durant, Thomas Clark; Union Pacific Railroad
Crédit Mobilier scandal, 675–710
Ames's congressional testimony in, 686–90, 693, 695–97
Ames's "fatal list" and, 677–680
Clemens's send-up of, 694–695
McComb's revelations in, 677–80, 686–89, 695
N. Y. Sun's exposure of, 676–679
poem on, 683
Poland Committee and, *see* Poland Committee
Whitman's view of, 697–99
Wilson Committee and, 690, 692–94, 701–2, 705–10
Crocker, Charles, 92, 93–94, 96, 101, 111, 123, 125–129, 136, 137, 139, 142, 147, 174, 194, 195, 213, 220–22, 239, 245–46, 260, 274–75, 278, 295, 307, 316, 317, 320, 330–32, 360–62, 367, 368, 370, 393, 394–97, 407–9, 412, 416, 428–429, 431, 446–47, 450–451, 453, 460, 469, 472, 486, 489, 492, 505, 518–20, 522–23, 538–539, 551, 553–56, 564–565, 567–68, 577, 578, 580, 586–88, 604, 617, 621, 631, 633, 636
Carter visit and, 542–43
death of, 707
fate of, 706–7
Goat Island terminus and, 336, 397
labor problems and, 276–277
at last spike ceremony, 645–646, 654–56
special commissioners and, 575–76
Strobridge and, 207–8
ten-mile feat of, 638–40, 670
Western Pacific acquisition and, 337–39
Crocker, Edwin Bryant, 93, 94, 129, 136, 219, 260, 275, 278, 297, 306, 307, 315–16, 317, 320–21, 331, 332–33, 335, 341, 360–69, 404, 407, 409, 412–14, 416–19, 424, 427–31, 446–47, 450–453, 455–59, 470–75, 481–83, 487–91, 501–502, 504, 505, 518–522, 538, 541–43, 564, 565, 567, 576–78, 580, 587, 603, 611, 621, 706
Contract & Finance Company formed by, 408
death of, 657, 707
"high ground" scheme and, 492–93
at last spike ceremony, 646, 654–57
rail laying scheme of, 448
San Francisco and Oakland Railroad and, 539–540
Southern Pacific and, 523–524
strokes suffered by, 471–472, 488, 539, 656–657
through rate dispute and, 640–41
Western Pacific acquisition and, 336–40
Crow, 309, 385, 411
Curtis, Samuel R., 76, 90, 101, 107, 112, 113, 158, 187, 188–89, 191
Cushing, Caleb, 624

Custer, George Armstrong, 310, 346, 347, 353, 386, 570

Daggett, Stuart, 135
Dana, Charles A., 528, 676, 678, 681
Dana, James Dwight, 136
Dana, Richard Henry, 599
Dana, Richard Henry, Sr., 599
Dane, Timothy, 127
Davis, George T. M., 132, 199, 279, 572
Davis, James W., 572, 614, 616–17, 691
Davis, Jefferson, 48, 49, 50–51, 95, 212, 214, 392
Davis, T.T., 393
Dawes, Henry Laurens, 422, 676, 678, 679, 680, 685, 687, 688, 696, 704
Day, Sherman, 575
Dehone, Theodore, 58
Democratic Party, U.S., 13, 37, 69, 77, 281, 367, 416, 429, 572, 675, 684, 702, 704
  1844 election and, 11–12, 201
  1868 National Convention of, 502
Democratic Vistas (Whitman), 697–98
Dent, Frederick T., 517
Denver, Frank, 502
Depew, Chauncey M., 217
Des Moines Register, 282, 283
DeVoto, Bernard, vii, 26, 35
Dexter, F.G., 406
Dexter, Samuel, 17, 19, 53
Dey, Anthony, 181
Dey, Kate, 181
Dey, Peter Anthony, 125, 156–157, 160, 162, 163, 171, 172, 181–83, 192, 196, 213, 225, 288, 371–72, 667, 691, 693
  Durant and, 156–57, 165–167, 173
  Hoxie contract opposed by, 197–201, 726
  oxbow route opposed by, 199–200, 726
  resignation of, 201–2
  Union Pacific stock purchased by, 243
Dillon, Sidney, 211, 243, 291, 302, 343, 348, 349, 352–53, 374, 375, 409–10, 412, 421–24, 468, 476, 480, 508, 515–18, 529–30, 535, 589, 615, 616, 620–24, 627, 630, 635, 636–637, 638, 639, 642, 644, 650, 659, 662, 668, 669, 677, 690, 700
Disraeli, Benjamin, 170
Dix, John Adams, 152–53, 155–56, 199, 201, 270, 303, 406, 444
Doddridge, W.B., 266
Dodge, Anne, 343, 382, 384, 398, 534, 543–44, 546–547, 625, 640
Dodge, Grenville Mellen, 157–163, 166, 173, 176, 191, 197–99, 213, 229, 230, 241, 244, 254, 262–63, 269, 279, 287, 307, 309, 312, 323, 324–26, 348–351, 355, 357, 364, 407, 411, 426, 427, 435–39, 441, 443, 463, 464, 475, 476, 479–82, 495, 504, 507–8, 511, 528, 537, 543–45, 548–49, 552–553, 558, 597, 603, 607–9, 612, 616, 623–625, 635–44, 680, 691
  attempted shakedown of, 571–72
  Browne incident and, 374–375
  Browning and, 582–83, 595–96
  Cheyenne established by, 371–74
  as congressman, 341–42, 371, 465–66
  Crédit Mobilier scandal and, 692–93
  death of, 693
  Durant correspondence of, 224–25
  Durant's clashes with, 461–462, 466–68, 483–85, 494, 509–10, 512–15, 529–32, 551, 556–57, 614–15, 746
  Durant's hiring of, 255–56
  Echo Canyon debate and, 486–501
  1868 final report of, 585
  elected to Congress, 282–283, 304
  at Fort Sanders meeting, 516–18
  "General Order No. 1" and, 483–85, 494, 509–10
  "General Order No. 7" and, 529–32
  Grant's meeting with, 613
  Hundredth Meridian Excursion and, 290–92
  Huntington's agreement with, 632–34
  Indian campaigns of, 209–210, 227–28, 231
  Johnson-Stanton imbroglio and, 437
  at last spike ceremony, 658–659, 662, 666–70
  Lincoln's conversations with, 158–61
  in Lincoln's funeral procession, 218, 224
  Lone Tree route found by, 232–33, 288, 290, 303, 304–5, 374
  meeting point issue and, 632–34
  Missouri River bridge issue and, 546–47, 692
  new special inspectors and, 601–2
  at Pea Ridge, 159, 371
  Piedmont kidnapping episode and, 649–51
  in Plum Creek expedition, 285–86
  reconnaissance expedition of, *see* Dodge expedition
  Seymour's "reconnaissance" and, 288–90

special commissioners and, 563–64, 570, 582–83
tariff speech of, 465–66
Union Pacific resignation of, 692
Union Pacific shares bought by, 424–25
at White House dinner, 342–43
Williams's final report and, 573–74
Wilson Committee's pursuit of, 692–93
at Young's sermon, 533–35
Dodge, Mary Abigail, 699
Dodge, Nathan, 167, 225, 279
Dodge, William E., 134
Dodge expedition, 370–76, 380–84, 394, 398–402
Browne incident and, 374–375
Cheyenne city and, 370–71, 373–74, 398–99
Mormons and, 398–99
Rawlins in, 371, 382–84, 398, 436, 438
Seymour in, 371–74, 377, 382, 400–402
Dole, William P., 186
Donahue, Peter, 127
Donnelly, Ignatius, 418, 451, 473
Donner, Jacob, 81
Donner party, 31, 33, 81, 99, 496, 568, 646, 657
Doolittle, James R., 227
Douglas, Stephen A., 23, 47, 48, 68, 283
Downing, Jacob, 186
Drew, Daniel, 403, 621, 624
Duff, John R., Sr., 211, 243, 291, 302, 343, 373, 422, 515, 623, 638, 639, 644, 649, 650, 659, 662, 668, 669, 677, 690
Dull Knife, 230, 385
Durant, Thomas Clark, 107, 114, 148, 151–57, 159, 161, 162, 177, 181–82, 192, 193, 210, 211, 213, 218, 219, 232, 234, 240, 247, 252–54, 262–63, 265, 269, 281–82, 286–290, 321, 326, 329, 332, 340, 351, 373, 374, 376–77, 393–94, 409–410, 419, 425, 430, 439, 441, 443–44, 446, 451, 466–68, 472, 475, 489, 507–8, 535, 541, 544, 546–49, 552–56, 563, 570–71, 577, 581, 584, 594, 601–2, 604–5, 609–610, 613, 620, 625, 641, 644, 658, 677, 679, 680, 683, 694, 700, 702, 711
Ames construction contract and, 572–73
Ames's leadership conflict with, 343–44, 347–49, 377–80, 402–7, 466–468
background of, 151–52
Blickensderfer fired by, 597–598, 614
board of directors' clashes with, 302–4, 322–24, 327–28, 347–49, 352–53, 377–80, 402–7, 615–16, 623, 638, 691
Boomer contract and, 302–304, 322–24, 616
Davis contract and, 616–17, 623
death of, 691
deranged behavior of, 614–615
Dey and, 156–57, 165–67, 173
Dodge hired by, 255–56
Dodge's clashes with, 461–462, 466–68, 483–85, 494, 509–10, 512–15, 529–32, 551, 556–57, 614–15
Dodge's correspondence with, 224–25
Echo Canyon debate and, 497–500
Farnam's partnership with, 152–55
Fisk suit and, 621, 626–29, 643
at Fort Sanders meeting, 517–18
"General Order No. 1" of, 483–85, 494, 509–10
"General Order No. 7" of, 529–32
Hoxie contract and, 198–200
Hundredth Meridian Excursion and, 290–94
Huntington-iron deal and, 278–79
Huntington's confrontation with, 179
at last spike ceremony, 659, 662–63, 666–69, 671
Lincoln's terminus decision and, 163–64, 165
M&M-Cedar Rapids scheme and, 167–68
mystery shares of, 421–22, 423
as "Napoleon of Railways," 270
oxbow controversy and, 225–26
partners' unease with, 242–243
Piedmont kidnapping incident and, 649–51
Poland Committee and, 691
Railroad Act Amendment and, 176, 178–79, 180
revenue bill and, 168–69
Richardson's description of, 648–49
Seymour and, 195–97
Stanford's meeting with, 567
Strong's labor contract with, 558–59
"suspense account" of, 169, 327, 380, 536–37, 573–74
train and, 170–71, 172
Wilson Committee and, 693
Durant, William Frank, 247, 252
Durant, William West, 691, 711
Durkee, Charles, 364, 489
Dutch Flat and Donner Lake Wagon Road Company, 112, 127

Dutch Flat Enquirer, 79, 297, 306

Eayre, George S., 187
Echo Canyon, 496–97
Echo City, Utah Territory, 606–607
Edmundson, Wilkes, 478–79
Edwards, Ogden, 183, 232
Eicholtz, Leonard, 651
elections, California:
of 1855, 94
of 1857, 94
of 1861, 96, 101
elections, U.S.:
of 1844, 11–12, 13, 30, 201
of 1846, 30–31
of 1848, 37, 39–40
of 1856, 69, 247
of 1860, 76, 77, 85, 157–158, 198
of 1863, 676
of 1864, 180, 181, 185, 191, 193
of 1866, 281–83, 304
of 1867, 416
of 1868, 436, 451, 466, 480–81, 482, 502, 516, 554, 557, 572, 624
of 1872, 675–77, 679, 680–681, 683, 684, 705
of 1876, 676
of 1880, 676
Eliot, Thomas D., 423, 676, 678, 680, 689, 696
Eliot, William Greenleaf, 283
Elkhorn, 262, 343, 344
Elliott, Samuel G., 79
Emancipation Proclamation, 118, 126
Emigrant's Trail, 49, 184
Emory, W.H., 49
Erie and Kalamazoo Railroad, 18
Erie Railroad, 403, 599, 620–22
Evans, James A., 183, 184, 191–92, 232, 262, 263, 269, 288–90, 358–59, 371, 374, 376, 401, 402, 411, 464, 480, 483–84, 485, 494–95, 496, 498, 506, 508–9, 514, 515, 544, 571, 596, 609, 642, 643
Evans, John, 184–91, 227, 263
Evarts, William M., 550, 554, 583, 584, 612, 624
Ewing, Thomas, Jr., 107, 110, 112, 159, 283, 475, 481–82, 483, 491–92, 555, 584, 694

Farnam, Henry, 107, 152–55, 156, 157, 158, 159, 165, 176, 177, 210
Farr, Lorin, 537–38, 541, 607, 619, 659–60
Felch, Alpheus, 37
Ferguson, Arthur N., 352, 358, 526
Fetterman, William J., 310–11
Fetterman Massacre, 309–11, 346, 354, 384, 385
Fillmore, Millard, 69
Fisk, James "Jubilee Jim," 403, 515, 535, 536, 572, 573, 599, 620–29, 630, 635, 643–44, 670–71, 682
Fitch, Thomas, 587–88
Five Forks, Battle of, 214
Flint, Edward P., 417, 419, 451
Flint, James P., 417
Folsom, Joseph L., 65
Ford, James A., 227
Ford, John T., 214
Foreign Miners' License Tax Law (California), 206
Forrest, Nathan Bedford, 160
Forshey, C.J., 33
Forty-Mile Desert, 276–77, 341, 486, 505, 522, 575, 647–48
Foster, Lafayette, 270
Fourteenth Amendment, 281, 557
Fowler, Joseph S., 422, 423, 678
Franchot, Richard, 112, 333, 417, 453, 473, 505
Freedmen's Bureau, 281
Freeman, Leigh, 562–63
Frémont, Jessie Benton, 48
Frémont, John Charles, 13, 22, 26, 27, 34, 37–38, 39, 42, 48, 49, 50, 69, 94, 161, 168, 193, 247, 268, 382, 384, 657
Fulton, Robert Lardin, 561, 563

Gadsden, James, 33
Galena and Chicago Union Railroad, 167, 168
Garfield, James A., 422–23, 452, 676, 678, 679, 680, 687, 688, 694, 696, 700, 704
General Sherman (locomotive), 241
Gessner, N.A., 302
Gettysburg, Battle of, 139, 346, 557
Gibbon, John, 461–62, 479, 517
Gibson, A.M., 686–87
Gilded Age, The (Twain), 694–695
Gilliss, John R., 298, 306, 318–20
Glidden, William T., 211, 406, 409, 637, 638, 644
Goat Island terminal, 334–36, 340–41, 397, 416–19, 431, 451–54, 456, 491, 504, 565, 588, 641, 675
Gobright, Horace, 217, 218
Godkin, E.L., 675, 700
Goetzmann, William H., 49
gold rush, 40–41, 48, 60–63
Gorham, George B., 474
Gould, Jay, 403, 599, 620–621
Governor Stanford (locomotive), 143, 145, 146–47, 173, 174–75, 176
Grant, Ulysses S., 139, 160, 161, 178, 179, 197, 210, 214, 227, 256, 263, 285, 286, 304, 311, 326, 342, 351, 371, 416, 426, 436, 461, 466, 516, 525, 529,

612, 615, 618, 624, 625, 633, 634, 638, 661, 664, 668, 681, 682, 691, 697, 698
Dodge's meeting with, 613
1872 election and, 675–76, 681, 684
at Fort Sanders meeting, 516–18
inauguration of, 603, 613
Johnson-Stanton clash and, 436–37
nomination of, 480–81
Gray, George M., 258, 471, 489, 502–3, 539, 541–542, 552–53, 556, 564, 568–69, 617
Gray, H.W., 199, 211
Great American Desert, 11, 12, 18, 26
Great Britain, 7, 8, 9, 12, 19, 26, 27, 35, 170
Alabama claims and, 445–446, 582
Confederacy recognized by, 104
Train incident and, 445–46
Greeley, Horace, 234, 599, 601, 622, 638, 670–71, 675–76, 678, 682, 685
Green, Duff, 515, 573
Gregg, Josiah, 11
Grimes, James W., 114–15, 423, 680
Grinnell, George Bird, 386, 409–10
Griswold, Wesley S., 223, 279
Gunnison, John W., 50
Guy, H.W., 560–61
Gwin, William M., 74, 95

Haight, Henry H., 655–56
Hale, John P., 38
Hall, S.W., 59
Halleck, Henry W., 116, 453, 586
Hallett, Samuel, 161, 168, 169, 176, 180, 193, 195
Hallett, Tom, 193
Halstead, Murat, 622
Ham, Benjamin, 627, 643–44
Hancock, Winfield Scott, 346–347, 350, 354, 386, 426
Hannibal and St. Joseph Railroad, 34, 106, 112, 241, 283
Harbaugh, Springer, 226, 262, 343, 379, 423, 443
Harding, William Henry, 172
Hardy, John, 411
Harkness, W.H., 647, 665
Harlan, James, 114–15, 180, 226, 239, 244, 245, 258, 283, 451–52, 491, 702
Harney, General, 389, 517
Hart, Albert, 655
Hart, Alfred, 646, 653–54, 657, 659
Harte, Bret, 666
Hay, John, 216
Hayes, Rutherford B., 291, 452
Hayfield Fight, 385, 386
Hazard, Isaac, 211
Hazard, Rowland G., 211, 347–48, 349, 378, 379, 403, 405, 406, 420, 536, 598, 615–16, 620, 621, 677, 691
Henry, Joseph E., 198
Hewes, David, 661, 756
High Backbone, 385
Hills, L.L., 263, 357–58, 370–371, 373, 384
Hirshson, Stanley, 282–83
Hoar, George, 690
Hodges, Fred, 321, 464, 513
Holladay, Ben, 224, 237, 587
Holman, William, 180
Homestead Act, 244
Hooker, Joseph, 139
Hooper, Samuel, 211, 227, 423, 632, 680
Hopkins, Mark, 89, 91–94, 96, 99, 101, 117, 127–128, 138–41, 144, 147, 173, 174, 194, 213, 220–21, 236, 245, 256–260, 278, 295, 302, 316, 317, 320–22, 332, 333, 340, 360–61, 369, 391, 393, 396, 406, 417–18, 427–28, 431, 446–50, 452, 453, 455, 459–60, 472, 473, 485–91, 505, 523, 540, 551, 555, 564, 567, 568, 570, 577–79, 580, 586–88, 602–3, 604, 617–18, 621, 630–631, 641, 645, 654, 655, 705, 706, 710
Bank of California episode and, 565–66
death of, 707
Denver's shakedown attempt and, 501–2
Glidden and, 408–9
Goat Island terminus and, 334–35, 336, 419
"High Ground" scheme and, 492–93
Summit Tunnel visit described by, 412–13
Western Pacific acquisition and, 337–39
western railroad purchases opposed by, 456–58
Hopkins, Mary, 707
House, Jacob, 225, 464, 499–500
House, J.E., 263
House of Representatives, California, 457
House of Representatives, U.S., 13, 18, 75, 96, 132, 342, 343, 491, 504, 685, 694, 699, 700, 704
Ames's payoffs in, 419–24, 444–45, 483, 626, 677
Appropriation Committee of, 701
Banking and Currency Committee of, 701
Committee of the Whole on the State of the Union of, 110
Crédit Mobilier Committee of, *see* Poland Committee
Curtis Bill passed in, 95
Dodge's tariff speech in, 465–66
1862 railroad bill passed by, 114
1864 amendment to Railroad Act in, 176–80

House of Representatives, U.S. (*cont.*)
1866 amendment to Railroad Act in, 270–71
1871 amendment to Railroad Act in, 682
Impeachment Committee of, 437
Judiciary Committee of, 435, 701, 702
Naval Committee of, 701
Pacific Railroad Committee of, 109–10, 112, 114, 178, 451, 625
Post Offices and Post Roads Committee of, 76
Public Lands and Railroads Committee of, 417–18, 451
Roads and Canals Committee of, 15, 19, 44
Territories Committee of, 23
Ways and Means Committee of, 176, 701
Whitney's address to, 44
see also Congress, U.S.; Wilson Committee
Houston, Alexander, 127
Houston, Samuel, 33–34, 41–42
Howden, James, 315–16, 321
Hoxie, Herbert M., 107, 162, 197–99, 210–11, 225, 227, 241, 255, 282, 291, 302, 322–23, 353, 441, 443, 513, 571, 572, 596, 607, 610, 613–14
Hudnutt, J.O., 480, 513
Hundredth Meridian Excursion, 290–95
Huntington, Arabella Yarrington Worsham, 710–11
Huntington, Clara, 391, 710
Huntington, Collis Potter, 92–94, 96, 99, 101, 111–12, 117, 127–28, 130, 138–41, 143–47, 155, 159, 168–69, 176–78, 180, 194, 212–13, 219–21, 236, 246, 256, 259, 270, 271, 273, 281, 297, 306, 315–17, 321, 330–31, 333–34, 360–62, 364–366, 368–70, 391–97, 412–14, 420, 427–30, 446, 451, 453, 457–60, 465, 469–75, 481–83, 485–90, 494, 501–3, 505, 518–24, 537–43, 546, 549, 553–56, 564–570, 572, 575–78, 582–584, 586–88, 595–96, 602–4, 610–11, 617–18, 621, 624–25, 630–31, 636–37, 640–41, 657, 670–71, 707–8
Ames brothers deal and, 133–34
background of, 87–89
congressional lobbyists paid off by, 450–51
death of, 711
Dodge's agreement with, 632–34
Durant confronted by, 179
Durant and iron purchasing deal of, 278–80
fate of, 710–11
gauge decision and, 131–32
Goat Island Terminal and, 335–36, 340–41, 397, 416–19, 431, 454–55, 491, 504
"high ground" scheme and, 492–93
Judah's first meeting with, 87, 89–91
Lincoln and, 213
McCulloch and, 612–13
Mormon contractors and, 551–52
Oliver Ames and, 393, 404, 407, 419
Pennsylvania Railroad and, 454–55
Southern Pacific and, 473–475, 491
spy network of, 393–94, 474–75, 495
Stanford's Omaha-Salt Lake conferences with, 579–581, 594
Stanton and, 213
U.P.'s leadership conflict monitored by, 405–7
Western Pacific and, 337, 338–39
Wilson Committee and, 705–6, 709–10
Huntington, Elizabeth Stoddard, 87, 391, 580, 710
Huntington, Henry E., 710–11
Huntington, Solon, 87, 88
Hyde, William B., 274

Illinois Central Railroad, 43, 67, 157
Indians, American, 235–36, 256, 265, 276, 326, 344, 371, 373, 384, 410, 411, 464, 470–71, 478–79, 510, 541, 549, 550, 570
Bear Creek Massacre and, 229–31
Browne killed by, 374–75
Connor's campaign against, 229–31
Dodge's campaigns against, 209–10, 227–28, 231
Evans's war against, 185–191
Fetterman massacre and, 309–11
in Fort Laramie peace talks, 284–85
genocide policy considered for, 244–45
Hancock's Kansas campaign against, 346–47
in Hayfield Fight, 385, 386
Hundredth Meridian Excursion and, 292–94, 295
Julesburg raids by, 209, 266, 353–54
Laramie Treaty signed by, 525–26
Medicine Lodge Treaty signed by, 525–26
in Mud Creek attack, 524–525
North Platte prisoner exchange and, 389–90
North Platte Treaty signed by, 400
Plum Creek attacks by, 285–286, 386–89, 399, 547–548

Sand Creek Massacre and, 190–91
Sherman's perception of, 286–87, 311–12
Union Pacific harassed by, 349–52
Union Pacific's Grand Island encounter with, 266–68
in Wagon Box Fight, 385–386
Whitney's encounters with, 21, 22
see also specific tribes
Indian Territory, 11, 41–42
Innocents Abroad, The (Twain), 694
Interior Department, U.S., 109, 131, 137, 192, 195–96, 243, 247, 284, 341, 347, 365, 366, 461, 475, 481, 493, 494, 542, 548–49, 555, 568, 572, 573–74, 596, 597–98, 600, 602, 612, 613, 625
Iowa Land Company, 211
Isaacs, A.J., 107
Ives, Butler, 276–77, 341, 363–64, 394, 398, 464, 475, 481–82, 503, 505, 553–54
Ives, William, 363

Jackson, Andrew, 12
Jackson, Thomas J. "Stonewall," 101, 126, 139
Jacobs, Louis, 663
Jay, John, 4, 5
Jefferson, Thomas, 11, 16
Johnson, Andrew, 218, 219, 226, 227, 239, 246, 258, 271, 281, 283, 284, 291, 303, 304, 312, 324, 346, 354, 392, 416, 424, 441, 445, 446, 545, 554, 574–575, 585, 596, 601, 602, 613, 624
impeachment movement against, 341–42, 435–436, 437, 686
impeachment trial of, 465, 480
Stanton suspended by, 416, 424, 436–37
Jones, George Wallace, 18
Joy, James F., 283
Judah, Anna Ferona Pierce, 59, 60, 61, 66, 67, 72, 74, 80, 83, 84, 95, 96, 97, 100, 103, 109, 110–111, 118, 130, 142, 143, 144–45, 298, 667, 709
Judah, Edward, 96
Judah, Henry Moses, 59
Judah, Henry R., 58
Judah, Theodore Dehone, 52–53, 116, 117–18, 127, 129, 130, 132, 137, 155, 156, 173, 174, 195, 240, 258, 297, 298, 334, 337, 366, 435, 448, 482, 558, 646, 657, 669, 709
background of, 58–59
Congress observed by, 68–69
CP's board of directors' conflict with, 137–42, 143
death of, 144–46
Donner Pass route found by, 81–84
as engineer, 59–60
government surveys criticized by, 52–53
Huntington's first meeting with, 87, 89–91
as lobbyist, 74–76, 79, 108–9
mysterious backers of, 143, 145
Pacific Central route surveyed by, 96–101, 223–224
at Pacific Railroad Convention, 73–74
Practical Plan of, 69–71, 82
railroad bill and, 109–10, 112–13, 114
Robinson brothers and, 80–81, 85–86, 146
Sacramento Valley Railroad and, 61–66
Jupiter (locomotive), 660
Justice Department, U.S., 466
Kansas-Nebraska Act, 68
Kansas Pacific Railroad, 346, 347, 684, 694
Kasson, Jack, 282–83
Kearny, Stephen Watts, 27, 29, 37, 38
Keene, Laura, 214
Kelley, William D., 422–23, 678, 680, 687, 696, 700, 704
Kennedy, William, 584
King, Clarence, 430
Kinnie, Billy, 548
Kiowa, 188, 227, 410
Kirkwood, Samuel J., 107
Klein, Maury, 423–24, 680, 701, 732, 742, 758
Know-Nothing Party (American Party), U.S., 69, 94
Kountze, Augustus, 167

Lafox, Eli, 304
Lake, M.C., 368
Lakota Sioux, 384–86, 410
Lambard, Charles A., 199, 211, 227, 242, 291, 304, 396, 406, 515, 621, 627, 709
Lambard, O.D., 558–59
Lambard, Orville, 133
Lane, Frederick, 620
Lane, James H., 114
Lane, Joseph, 74
La Ramie, Jacques, 461
Laramie, Wyoming Territory, 461, 476, 483, 609–10, 637, 648
blizzard of 1869 and, 609–610
lawlessness in, 477–78, 560–61
Laramie *Frontier Index*, 515, 562
"Last Spike, The," 661
see also Promontory Summit last spike ceremony
Latham, Milton, 131, 333
Latrobe, Benjamin, 16
Lavender, David, 333–34, 554, 710, 732
Lawrence, R.J., 614

Lay, George Tradescant, 5
Lean Bear, 187
Leavenworth, Pawnee, and Western (LP&W) Railroad, 107–8, 109, 110, 112–15, 117, 118, 131, 155, 161–62, 186, 684
Leaves of Grass (Whitman), 114, 698
Lee, Robert E., 126, 139, 178, 212, 214, 346, 550
Leete, Benjamin Franklin, 85, 86
Left Hand, 188, 190
Leutze, Emmanuel, 105, 108
Lewis, Fred, 548
Lewis, Lloyd, 263
Lewis, Meriwether, 11
Lexington and Ohio Railroad, 17
Lincoln, Abraham, 76, 85, 95–96, 101, 105, 107, 108, 109, 110–11, 113, 115, 117, 123, 131, 140, 147, 176, 180, 184, 185, 187, 191, 193, 197, 198, 199, 209, 226, 283, 394, 437, 685, 686, 703
  Ames's meeting with, 211–212
  assassination of, 214–15, 326, 342, 436
  Dodge's conversations with, 158–61
  Emancipation Proclamation issued by, 118, 126
  funeral train of, 215–18
  gauge decision of, 132–33
  Huntington and, 213
  terminus decision of, 160–161, 163–64, 165, 167
Lincoln, Mary Todd, 215–16
Lincoln, Robert, 216
Lincoln, Tad, 216
Lincoln, Willie, 215–16
Lincoln and the Railroads (Starr), 215
Liverpool and Manchester Railroad, 9
Livingstone, David, 345
Lockwood, Erastus, 265, 293, 527, 608–9, 752
  Grand Island encounter described by, 266–68
Logan, John A., 75, 422–23, 680, 696
Long, Stephen H., 11
Long Death, The (Andrist), 310
Loomis, Augustus, 222
Louisiana Territory, 11
Louisville, Cincinnati and Charleston Railroad, 18
Louisville and Nashville Railroad, 683
Lyons and Iowa Central Railroad, 76

McCallum, D.C., 215
McClellan, George B., 104, 113, 116, 126, 191
McComb, Henry S., 199, 211, 378, 379, 403, 421–23, 443, 536, 572–73, 575, 589, 598, 627, 681, 682, 694, 695, 709
  Alley and, 689–90
  Oakes Ames's correspondence with, 444–45, 677–78
  Poland Committee and testimony of, 686–88
McCormick, Cyrus, 406
McCrary, George, 685, 692
McCulloch, Hugh, 554, 583, 612–13
McDougall, James A., 109, 112, 114, 115, 124, 131, 132, 179
McDowell, James H., 107
McIntire, John, 72
McLane, Charles E., 195
McLane, Joseph, 237
McLaughlin, Charles, 127, 142, 336, 337, 339, 362, 368
McLean, Wilmer, 214
Malloy, J.W., 661–62
"Manifest Destiny," 22, 23
Manual of Geology (Dana), 136
Marsh, Charles, 96, 140
Marshall, James W., 63, 64, 369
Maxwell, James, 497, 513, 548
Meade, George Gordon, 139
Medill, Joseph, 291
Memphis and Charleston Railroad, 17
Memphis Railroad Convention, 23, 24
Merrick, William, 685
Mexican-American War, 22, 26–28, 34–35, 40
Mexico, 10, 12
Michigan Central and Southern Railroad, 18
Military Railroad, U.S., 215
Miller, Andrew J., 307–8
Miller, E.H., Jr., 91–92, 144, 540
Mills, Darius Ogden, 133, 273, 577–78, 641
Mills, Edgar, 659, 662
Mills, Robert, 16, 53
Mississippi and Missouri (M&M) Railroad, 106–7, 113, 114, 115, 125, 131, 153, 154–57, 161, 162, 167–68, 177, 199, 201, 224, 225, 262, 279, 324, 355, 435
Missouri Pacific Railroad, 43
Mitchell, Robert B., 188
Mobile and Ohio Railroad, 34, 159–60
Mohawk and Hudson Railroad, 645
Monroe Doctrine, 26
Montague, Samuel Skerry, 127, 142, 146, 223, 238, 274, 275, 276, 321, 398, 407, 472, 487, 503–5, 555, 567, 576, 577, 617, 631, 667
Morehead, J.K., 423
Mormons, 34, 183–84, 228–229, 230, 236, 394, 509, 511, 559
  Central Pacific and, 364–65, 537–38, 541–42, 551–52, 554, 556, 564, 605–7, 631–32, 659–60
  Dodge expedition and, 398–399
  in Echo Canyon, 496–97

last spike ceremony and, 659–60
Union Pacific and, 364–365, 489, 494–97, 502, 509–10, 532–533, 619–20, 659–660
Morrill, Lot, 702
Morris, Theodore M., 613–14
Morse, John, 141
Morton, J.Sterling, 242
Moss, J. Mora, 65, 417
Mullaley, Pat, 266–67

Napoleon I, Emperor of France, 3, 5, 8, 29
Nelson, Christen, 462–63
Nevada Central Railroad, 111, 112
Nevada Rail Road Company, 142
Newcomb, H.D., 682–83
New Haven, Hartford and Springfield Railroad, 59
New Mexico, 30, 35, 43
New York Central Railroad, 60, 132, 217, 635
New York and Erie Railroad, 156
New York and Harlem Railroad, 17
New York Herald, 45, 170, 171, 287, 316, 345, 459, 623–24, 627–28, 629, 630, 635, 670, 688
New York Stock Exchange, 709
New York Sun, 528, 676–79, 681–84, 685, 686–87, 693–94, 696, 697, 700, 702–3, 709, 710
New York Times, 392, 621, 622, 628, 630, 679, 685, 696, 703
New York Tribune, 20, 23, 234, 512, 528, 599, 601, 602, 670–71, 675, 678–84, 687, 688, 689, 690, 700
New York World, 624, 678, 682, 688, 700, 702
Niagara Falls and Lake Ontario Railroad, 59–60
Nicaragua, 42, 68
Nicolay, John, 216
Niles, John, 38, 43
nitroglycerine, 272–75, 315–316, 320–21, 605, 606
North, Frank, 355–57, 359, 387, 389, 399–400, 409, 479, 524–25
North, Luther, 355, 356–57, 409–10
Northern Central Railway, 216
Northern Pacific Railroad, 474
North Platte, 344–46, 353, 373
Indians-U.S. prisoner exchange at, 389–90
Norvell, Caleb, 622
Nounan, Joseph F., 495
Nye, James Warren, 418–19, 630

Oakland Waterfront Company, 565
O'Connor, Michael, 635–36
Ogden, Peter Skene, 521
Ogden, Utah Territory, 618–619
Ogden, William B., 125, 126
Ogdensburg (locomotive), 520
Oglala Sioux, 285, 385
Olcutt, Thomas, 125
Old Ironsides (locomotive), 16
Opium Wars, 7
Oregon Central Railroad, 587
Oregon Territory, 10, 12, 19, 23
Oregon Trail, 13, 73, 209, 230, 526
Oscar, 3, 5–6, 8
O'Sullivan, John Louis, 22
Otis, F.N., 145
Our American Cousin (Taylor), 214
Overland Trail, 209, 230
Owen, Robert Dale, 19

Pacific (locomotive), 175–76
Pacific Express Company, 641, 661
Pacific Postal Telegraph Company, 124
Pacific Railroad Act of 1862, 106–18, 136, 147, 159, 259, 335, 365, 430, 453, 578, 612, 632, 684
contending railroads and, 106–7
1864 amendment of, 176–180, 212, 219, 270, 271, 290, 581, 683, 685, 702
1866 amendment of, 270–271
1871 amendment of, 682
House passage of, 114
Judah and, 109, 110, 112–113, 114
loopholes and controls in, 116–17
LP&W scheme and, 107–8
need for amendment of, 124–27, 132, 161, 168–69
opposition manipulation of, 112–14
Sargent and, 109–10
Senate passage of, 114–15
Pacific Railroad Convention of 1859, 72–74, 79, 83, 85
Pacific Railroad Museum, 58, 75, 77, 90, 110–11, 435
Painter, Uriah H., 678, 680
Paiute, 326, 470–71, 541
Pallady, Leon, 233
Palmer, William Jackson, 457, 473
Panama, 42, 68
Panama Star, 272
Panic of 1837, 4, 18, 23
Panic of 1857, 154
Parke, John G., 50, 51
Parker, Samuel, 18
Parkman, Francis, 11
Parrot, Thomas, 397
Parry, Henry C., 344, 353–354, 372, 373, 376–77, 399
Partridge, A.P., 306, 363, 428
"Passage to India" (Whitman), 666
Patterson, James W., 291, 422, 451, 676, 678, 680, 695, 696, 702

Pawnee, 230–31, 265–66, 286, 292–93, 294, 354–358, 399–400, 524–25
Pawnee Battalion, 357, 399–400
Pawnee Killer, 389
Pea Ridge, Battle of, 113, 159, 371
Peel, James, 92
Peña y Peña, Manuel de la, 35
Pennsylvania Fiscal Agency, 172, 199
Pennsylvania Railroad, 34, 216, 454, 692
Perham Peoples' Railroad, 106
Perpetual Emigrating Fund, 537
Perry, John D., 193, 454, 457, 473
Petersburg, siege of, 178, 179, 214
Philadelphia and Columbia Railroad, 17
Philadelphia Inquirer, 172, 445, 680
Philadelphia and Reading Railroad, 18
Pierce, Franklin S., 59, 68
Pike, Zeb, 11
Pioneer Band, 34
Pioneer Stage Lines, 124
Placerville Railroad, 237
Platte River, 263–64, 268
Plumbe, John ("Iowaian"), 18–19
Poland, Luke, 685–86, 689, 696, 697, 699, 703, 704
Poland Committee, 685–705, 706
  Ames's testimony to, 686–90, 693, 695–97, 699–701
  establishment of, 685–86
  final report of, 699–701, 702, 703
Polk, James K., 12, 13, 19, 22, 26, 30–31, 35, 38, 39–40, 201
Pondir, John, 226
Poor, Henry Varnum, 125
Pope, John, 50, 51, 126, 227–228, 230, 244
Porcupine (warrior), 386–88
Porter, Noah, 155
Port Royal, Battle of, 104–5
Powell, James W., 385–86
Practical Plan for Building the Pacific Railroad, A (Judah), 68–71
Pratt, Orson, 496
Pratt, Zadock, 13
Price, Hiram, 176, 180, 623
Project for a Railroad to the Pacific, A (Whitney), 42
Promontory Summit last spike ceremony, 645–72
  aftermath of, 667–68
  celebrations of, 654–56, 663–64, 669–70
  Central Pacific's excursion to, 645–47, 657
  last spike ceremony in, 661–666
  last tie ceremony in, 660–661
  Mormons and, 659–60
  Union Pacific's excursion to, 648–53
Pullman, George M., 291, 304

railroads:
  first passenger service of, 17
  growth of, 34–35
  impact of, 9
  T-shaped rail innovation and, 17–18, 58
Railway Construction Company, 368
Railway Journal, 125
Railway Pioneer, 293
Ralston, W.C., 397, 523
Ralston, W.H., 565–66
Randall, Alexander, 554
Rawlins, John Aaron, 160, 286, 624, 625, 664, 668, 691
  in Dodge expedition, 371, 382–84, 398, 436, 438
  Raymond, Henry J., 622
  Reconstruction, 219, 234, 270, 281, 283, 342–343, 426
  Red Cloud, 230, 266, 269, 284–85, 309, 310–11, 385–86, 410, 525
Redfield, William C., 16
Red Wolf, 387
Reed, James A., 568
Reed, Jenny, 183, 253, 328, 329, 345, 373, 400, 409, 507, 528, 585, 643
Reed, Peter R., 107, 161
Reed, Samuel Benedict, 183–84, 232, 252–53, 254, 255, 262, 263, 265, 269, 282, 286, 290, 291, 301, 323–324, 327, 328, 343, 344, 345, 349, 352, 353, 358–59, 371, 372, 373, 374, 380, 398, 400, 402, 409, 411, 425, 438–40, 443, 446, 451, 462–463, 468, 476, 484, 494–98, 500–502, 506, 507–16, 529–31, 533, 541, 553, 571, 584, 585, 596, 598, 604–5, 609, 610, 614, 638, 639, 642, 643, 650, 659, 662, 666
Reid, Whitelaw, 678, 681, 684, 701
"Reign of King Cotton, The" (Dana), 599
Reno, Jesse Lee, 368
Reno, Nevada Territory, 470
Reno Crescent, 588, 611
Rensselaer and Saratoga Railroad, 217
Republican Party, U.S., 69, 76, 77, 85, 89, 90, 93, 94, 107, 131, 157, 177, 180, 198, 281, 283, 304, 342, 367, 416, 436, 572, 587, 597, 675, 681, 684, 685, 701
  1868 National Convention of, 480–81, 482
Richardson, Albert D., 171, 234–35, 236, 237–38, 272, 282, 287–88, 327, 512, 528, 648–49, 652–653, 755
Richly, John, 465–66

Richmond and Petersburg Railroad, 18
Robinson, John P., 64, 65, 80–81, 85, 124, 128, 134, 135, 146, 337
Robinson, Lester L., 64, 65, 80–81, 85–86, 128, 134, 135, 146, 195, 337, 397
Rocket (locomotive), 9
Rock Island Line, 106, 115, 125, 131
Roebling, John A., 60
Rollins, James, 108, 112
Roosevelt, Theodore, 693
Rosecrans, William S., 142, 253
Rosewater, Andrew, 269–70, 374–75
Roughing It (Twain), 305, 694
Russell, Andrew J., 518, 659, 660–61, 662
Rutland & Burlington Railroad, 59

Sabin, Edwin, 40
Sacramento (locomotive), 65, 298–99
Sacramento Pioneers, 414
Sacramento, Placer and Nevada Railroad, 86
Sacramento and San Jose Railroad, 601
Sacramento Union, 64, 69, 72–73, 80, 85, 86, 97, 101, 122, 123, 125, 129, 130, 134, 137, 139, 145, 173, 175, 207, 239, 258, 273, 297, 306, 307, 330, 331, 334, 412–16, 446, 457, 458–459, 504, 520, 576, 630, 636, 654, 656
Sacramento Valley Railroad, 61–66, 71, 72, 73, 80–81, 85, 86, 124, 128, 146, 240, 298, 337
Safford, Governor, 646, 661, 665
Sand Creek Massacre, 190–91, 209, 210, 227, 230, 244, 266, 311, 346, 347
San Francisco Alta California, 134, 195, 272–73, 274, 295, 297, 503
San Francisco Bulletin, 300, 367, 412, 656
San Francisco and Marysville Railroad, 207
San Francisco and Oakland Railroad Company, 539–540, 543
San Francisco and Sacramento Railroad, 67
San Francisco and San Jose Railroad, 106, 127, 207, 336, 456–58
San Joaquin Valley Railroad, 490, 491, 504, 706
Santa Fe Trail, 188, 346
Sargent, Aaron Augustus, 101, 103, 109–10, 112–113, 117, 118, 140, 258–59, 260, 367, 704, 705, 708
Satana, Chief, 188
Saunders, Alvin, 147, 243
Savage, Charles R., 659, 662
Schenectady and Troy Railroad, 59
Schofield, John M., 582–83, 595
Schurz, Carl, 675
Scofield, G.W., 422, 678, 680, 687, 696, 704
Scott, Thomas A., 454–55, 457, 473, 692, 706
Scott, Winfield, 35, 39, 104
Sedgwick, S.J., 659, 662, 665
Senate, California, 457
Senate, U.S., 13, 43, 95, 96, 107–8, 117, 132, 167, 168, 342, 436, 453, 490–91, 504, 625, 631, 692, 694, 702–3
Committee on Indian Affairs of, 227, 389
Committee on Military Affairs of, 436
Committee on Pacific Railroads of, 109, 110, 112, 115, 417, 630
Committee on Public Lands of, 26–27, 37, 38, 114–15
1864 amendment to Railroad Act in, 178–80
1871 amendment to Railroad Act in, 682
Johnson's impeachment trial in, 437, 465, 480
railroad bill in, 114–15
Select Committee of Five of, 38
Sumner attacked in, 68
Whitney's proposal in, 27–28
see also Congress, U.S.
Seward, William Henry, 177, 445, 582, 661
Seymour, Horatio, 61, 516
Seymour, Silas Morrow, 195–197, 199–200, 201, 212, 219, 225–26, 247, 252, 348, 425, 438, 439, 441, 446, 467, 476, 480, 483–84, 494, 500–502, 506, 508, 509–10, 513, 514, 530–32, 541, 544, 546, 552, 557, 563, 573, 574, 597, 598, 610, 614, 616, 621, 634, 637, 638, 639, 642, 643, 659, 662, 671
in Dodge expedition, 371–374, 377, 382, 400–402
Echo Canyon debate and, 496–99
Hundredth Meridian Excursion and, 291–94
"reconnaissance" trip of, 288–90, 303
Sharman, C.H., 349, 735
Sharp, John, 495, 605–6, 659–60
Shaughnessy, John, 411
Shellaburger, Samuel, 690
Sheridan, Philip H., 214, 254, 342–43, 426, 516, 517–518, 526, 550, 570, 618
Sherman, Charles, 240, 254, 379, 380, 492
Sherman, John, 291, 465, 702
Sherman, William T., 61, 64, 101, 107, 111, 160, 191, 212, 233, 244, 245, 253, 256, 265, 324–26, 346, 353, 379, 380, 389, 390,

Sherman, William T. (*cont.*) 399, 426, 461, 516, 525, 529, 550, 582, 624, 646, 664, 668–69, 691, 692
at Fort Sanders meeting, 517–18
Indians as perceived by, 286–87, 311–12
Indian uprising and, 345–352, 354
Johnson-Stanton clash and, 436
Union Pacific and, 240–41, 245, 263, 286–87
Shilling, Watson N., 663, 665, 666
Shiloh, Battle of, 113
Shoshoni, 229
Sickles, Daniel, 139
Simpson, James H., 226, 252, 351
Sioux, 21, 22, 186, 188, 228, 230, 266–68, 284–85, 286, 294, 309–11, 346, 350, 371, 375, 385–86, 389, 478–79, 525–26, 547–48
Sioux City and Pacific Railroad, 106, 211
Sketches of Iowa and Wisconsin (Plumbe), 19
slavery issue, 30, 37, 43–44, 68, 107, 110, 115
Slocum, Henry, 690
Smith, Charles, 384–85
Smoot, W.C.A., 559–60
Snyder, Webster, 269, 344, 349, 424–25, 438, 439–443, 463, 465–66, 476, 495, 498, 506–9, 514, 535, 571, 579, 584, 596, 598, 609, 611, 613–14, 616, 638, 640–41, 643, 650, 691
Southern Pacific Railroad, 456–57, 459, 473–74, 490–91, 523–24, 540, 587–88, 641, 671, 705, 707, 708, 711
Spotted Bear, 389
Spotted Tail, 188, 209, 266, 267–68, 269, 285, 357, 389, 410, 524, 525
Spotted Wolf, 386–88
Sprague, M.B., 182
Springfield Republican, 234, 527, 622, 678
Standing Elk, 284, 309, 389
Stanford, Jane Lathrop, 92–93, 489, 707–8
Stanford, Leland, 92, 94, 96, 99, 108–9, 111, 123, 127, 128, 134–36, 137, 139, 140, 144, 146, 147, 173, 174, 194, 195, 206, 228, 239, 240, 245, 246, 260, 275, 278, 296, 299–300, 316, 317, 320, 332–33, 361, 364, 368, 369, 407, 412, 429, 431, 452, 455–58, 473, 476, 482, 486–88, 523, 532, 541–42, 545, 546, 551–556, 568–69, 570, 577, 603, 604, 611, 612, 617–18, 631–33, 636–637, 640, 641, 657, 706
background of, 92–93
Bank of California episode and, 565–66
California Central and, 395–396
Colfax Excursion and, 236–238
as CP's Salt Lake representative, 537–38, 566–67, 578–79
death of, 708
Durant's meeting with, 567
elected governor, 101
fate of, 707–8
Huntington's Omaha-Salt Lake conferences with, 579–81, 594
at last spike ceremony, 646–47, 651–53, 658–63, 665, 667–68, 670–71
Salt Lake mission of, 489–491, 495, 502–3
Southern Pacific and, 523–24
university founded by, 708
Western Pacific acquisition and, 336–39
Stanford, Leland, Jr., 489, 707–8
Stanford, Philip, 135, 141
Stanley, Henry Morton, 345–346, 350, 388, 390
New Julesburg described by, 372–73, 380–81, 399–400
Stansbury, Howard, 49, 657
Stanton, Edwin M., 159, 189, 215, 341, 346, 354, 416, 424, 461
Huntington and, 213
Johnson's suspension of, 416, 424, 436–37
Thomas and, 437
Starr, John William, 215, 216, 217
Steele, Frederick, 382
Stephenson, George, 9
Stevens, Isaac I., 49, 51
Stevens, Robert Livingston, 17, 58
Stevens, Thaddeus, 114, 176–177, 179–80, 281, 342, 416, 451, 452, 550, 683, 702
Stewart, A.T., 624
Stewart, Joseph P., 168–69, 176, 180
Stewart, William Morris, 418–419, 452, 491, 505, 540, 624–25, 630
Stillman, James D. B., 645–48, 651–53, 657, 659
Stone, James C., 107, 109
Stow, W.W., 397
Strobridge, James Harvey, 207–9, 221, 229, 239, 245, 259, 277–78, 306, 331, 332, 360, 362, 367, 409, 429, 450, 469, 471, 486, 487, 492, 493, 501, 505, 520–21, 552, 568, 594, 617, 636, 639, 646, 658, 660, 666, 669–70, 671, 727
Strong, Daniel W., 78–79, 81–82, 85, 86, 111, 129, 137, 138, 140, 141, 142–43, 144, 146, 223, 646
Durant's labor contract with, 558–59
in Pacific Central route survey, 96–101

Sully, Alfred, 354
Summit Tunnel, 298, 315, 317–18, 320, 321, 331, 332, 333, 340, 367, 407, 412–15, 470, 485
  breakthrough in, 369–70, 394
Sumner, Charles, 68, 452, 675
Supreme Court, California, 135–36, 206, 207, 220
Supreme Court, Pennsylvania, 627, 677
Supreme Court, U.S., 437, 465
Sutter, Johannes, 62–63, 64, 369
Sutton, E.B., 279–80
Sutton, Robert, 32
Swann, Thomas, 690
Sweeny, Peter B., 621
Swift Bear, 410, 525

Talcott, Orlando A., 193
Tarbell, Ida, 215, 216, 217
Taylor, E.B., 284–85
Taylor, Nathaniel G., 389–90
Taylor, Tom, 214
Taylor, Zachary, 22, 26, 28, 30, 39–40, 107
T. D. Judah (locomotive), 175
Tenure in Office Act, 416, 436–37, 480
Terminal Central Pacific Railroad, 397, 417
Territorial Enterprise, 62, 305
Tevis, Lloyd, 397, 456–57, 473–74, 490, 523, 565, 575, 641
Texas, Republic of, 10, 12, 13, 22, 26
Texas and Pacific Railway Company, 692, 705, 706
Thomas, Lorenzo, 437
Thompson, William, 387–89
Thoreau, Henry David, 26
Tilden, Samuel J., 348
Todd, John, 661, 662, 665
"To a Locomotive in Winter" (Whitman), 698–99
Tom Thumb (locomotive), 16
Tour Beyond the Rocky Mountains (Parker), 18
Tracy, Charles, 628–29, 635
Tracy, John F., 195–96
T-rail, 17–18, 58
Train, Enoch, 170
Train, George Francis, 169–170, 180, 241, 243, 252, 255, 291, 293, 301, 324, 352, 353, 409, 572, 616
  British arrest of, 445–46
  Crédit Mobilier and, 172, 181, 199, 515
  Durant and, 170–71, 172
  1866 election and, 281–82
  1872 election and, 683
Train, Oliver, 170
transcontinental railroad:
  eastern terminus and, 153, 160, 162–64
  gauge issue and, 118, 124, 131–33
  gold rush and, 40–41, 48, 60–63
  official surveys for, 48–52
  opposition to, 27–28, 37–39
  proposals for, 13–19, 31–33, 41–42
  settlement affected by, 549
  Whitney's idea for, 7–10
Traveling Bear, 410
Treasury Department, U.S., 613
Trist, Nicholas, 35
Tritle, F.A., 646, 661, 665
Troy Union Railroad, 217
Truckee, Chief, 541
Tuttle Charles, 406, 475, 512–13, 629–30
Twain, Mark, *see* Clemens, Samuel
Tweed, William Marcy, 621, 624
Tweed, William Marcy, Jr., 626–30, 635
Twenty Thousand Leagues Under the Sea (Verne), 448
Two Moons, 385
Two Years Before the Mast (Dana), 599
Tyler, Erastus, 253

Union Pacific Museum, 389
Union Pacific Railroad, 107, 112, 113, 115, 117, 126, 155, 160, 166, 168, 170, 172, 177, 184, 191–92, 194, 196, 211–12, 262, 265, 270–71, 283–84, 316, 332–33, 351, 361, 370, 373, 392, 393, 398, 408, 410–11, 418, 419, 437–40, 446, 451, 466, 472, 474–75, 481–82, 492–93, 520, 529, 537, 540–42, 545, 551–52, 566–67, 569, 570, 578, 580, 581, 587, 594, 595–97, 612–13, 618, 637, 652, 657, 688, 693, 695, 710
  Alley-Crane contract and, 379, 407
  Ames contract and, 407, 409, 572–73, 616–17, 677
  Benton station of, 526–27
  Boomer contract and, 302–304, 322–23, 616–17
  Clark Dodge loan and, 378–379
  company safe episode and, 628–30
  CP's Salt Lake route and, 364–66
  Davis contract and, 616–17, 623
  December 1868 dividend of, 589
  Dey's purchase of stock of, 243
  Dey's resignation from, 201–2
  in Dodge-Durant correspondence, 224–25
  Dodge hired by, 255–56
  Dodge's 1868 final report to, 585
  Dodge's resignation from, 692
  in drive to Promontory Summit, 636, 638–41, 644
  Durant's clashes with board of, 302–4, 322–24, 327–28, 347–49, 352–53, 377–80, 402–7, 615–16, 623, 638, 691

Union Pacific Railroad (*cont.*)
earnings of, 426–27
Echo Canyon debate and, 486–501
1864 amendment to railroad bill and, 178–79, 180
final construction commissioners and, 634–35
first nails laid by, 227
Fisk suit against, 515, 535–536, 572–73, 620–29, 630, 635, 637, 643–644, 670–71, 682, 689
"General Order No. 1" and, 483–85, 494, 509–10
"General Order No. 7" and, 529–32
Grand Excursion of, 240–241, 245
Grand Island encounter and, 266–69
groundbreaking by, 147–148, 165–66
Hazard suit against, 573, 591
Hoxie contract and, 198–201, 210–11, 302, 322–323, 327–28, 407, 677, 691
Hundredth Meridian Excursion of, 290–95
labor problems of, 181–82, 439, 514–15, 559–60
last spike ceremony, *see* Promontory Summit last spike ceremony
legislators bribed by, 418–423
location engineers deployed by, 183–84, 191–92, 231–32
Lone Tree route of, 232–33, 288, 290, 303, 304–5, 374
McComb suit against, 677, 682, 689, 709
media criticism of, 598–601
meeting point issue and, 632–33, 634
Missouri River bridge project and, 546–47
Mormon contractors of, 364–65, 489, 494–97, 502, 509–11, 532–33, 619–20, 659–60
mystery shares of, 421–422
new special commissioners and, 601–4, 641–42
organizing convention of, 124–25, 185
oxbow controversy and, 199–200, 225–26, 241, 243, 288
parallel grading issue and, 553, 564, 658
Platte River controversy and, 263–64
in receivership, 626–30
reserve fund debate of, 535–536
Sherman and, 240–41, 245, 263, 286–87
Sherman Summit ceremony of, 476
snow and, 438–39
Snyder-Reed conflict and, 507–9, 514
special commissioners and, 555–58, 563–64, 570, 573–75, 582–83
terminus decision and, 160–164, 167
through rate dispute and, 640–41
ties shortage of, 241–42
Wendle's shakedown of, 571–72
Williams's report on, 573–74
see also Crédit Mobilier of America; Crédit Mobilier scandal
Union Pacific Railway, Eastern Division (UPED), 161–62, 168, 192–94, 263, 454, 457, 473, 504, 684
Upson, Lauren, 64, 69, 72, 85, 134, 145, 173–75, 297, 366
Usher, John Palmer, 108, 131–132, 163, 168, 186, 193, 197, 219
Utah Central Railroad, 660
Van Buren, Martin, 12, 39–40
Vanderbilt, Alfred, 711
Vanderbilt, Cornelius, 143, 599, 621, 624
Vanderbilt, Morgan, 711
Van Dorn, Earl, 113
Vermont Central Railway, 207
Verne, Jules, 170, 448
Vicksburg campaign, 139, 160, 161
Virginia City Enterprise, 414, 610–11

Wade, Benjamin Franklin, 112, 291
Wagon Box Fight, 385–86
Walker, Samuel, 228, 231
Walker, William, 35, 68
War Department, U.S., 159, 182, 186, 213, 218, 354, 400, 436
War of 1812, 16, 104
Warner, Charles Dudley, 694
Warren, Gouverneur Kemble, 557, 563, 601, 603–4, 611, 617–18, 641
Washburn, Cadwalader, 424, 435, 445, 453, 465
Washburne, E.B., 179–80
Weber, Charles, 407
Weber, John G., 499
Welles, Gideon B., 132, 675
Wells, Fargo and Company, 273, 340
Wendle, Cornelius, 571–72, 585
West, Chauncey W., 537–38, 541–42, 566, 569, 605, 607, 659–60
West, Joseph A., 605
Western Pacific Railroad, 127–128, 134, 142, 362, 407–8, 429, 431, 446, 455, 456, 459, 471, 641
Central Pacific's acquisition of, 335–39, 368
"Westward the Course of Empire Takes Its Way" (Leutze), 105–6
Whig party, U.S., 12, 30
Whipple, Amiel W., 50, 51
Whirlwind (locomotive), 658

White, W.M., 351
Whitman, Walt, 33, 114, 215, 217, 218, 283, 666, 697–99
Whitmore, Adin H., 626, 627
Whitney, Asa, vii, 47, 53, 60, 70, 96, 123, 152, 218, 528, 657, 667
  background of, 4–5
  Benton's personal attack on, 38–39
  British support sought by, 45
  China sojourn of, 3–8
  congressional railroad proposal of, 13–15, 26–27, 41, 42
  death of, 46, 681
  Exploring Expedition of, 19–22
  House of Representatives addressed by, 44
  publicity campaigning by, 19–20, 28–30, 34–35, 37, 42–43
  in return to private life, 46
  transcontinental railroad idea of, 7–10
Whitney, Catherine Campbell, 46
Whitney, Herminie Antoinette Pillet, 4
Whitney, Josiah, 136–37
Whitney, Sarah Jay Munro, 4–5
Whitney, Sarah Mitchell, 4
Whitney, Shubael, 4
Wilberforce, William, 5
Wilder, J.O., 278, 298–99, 305
Wilderness, Battle of the, 178
Wilkes, George, 19, 32–33
Williams, George H., 491
Williams, Jesse L., 288–89, 500, 510, 535–36, 544–545, 550–51, 557–58, 573–74, 598, 614, 620, 624, 638, 644
  Interior Department report of, 548–49
Williams, John M. S., 211, 242–43, 262, 302–3, 323, 327–28, 405–6, 409, 638, 690
Williamson, R.S., 50, 575, 601, 602, 604, 611, 641
Wilmot Proviso, 30
Wilson, Charles Lincoln, 60–61, 64, 65, 71, 79–80, 128, 337, 395–96, 397, 416, 490, 669
Wilson, Henry, 422, 676, 678–82, 687, 694–96
Wilson, James F., 113–14, 422–23, 424, 435–36, 692–93, 696, 710
Wilson, Jeremiah, 690, 693, 705
Wilson Committee, 691–94
  Dodge sought by, 692–93
  establishment of, 690
  final report of, 701–2
  Huntington and, 705–6, 709–10
Wood, A.P., 642
Wood, Fernando, 701
Woodhull, Victoria, 169
Works Progress Administration, 619
Wynkoop, Edward W., 189–90

Yates, Richard, 147
Young, Brigham, 34, 125, 147–48, 183–84, 224, 228, 232, 263, 364–65, 398, 465, 489, 494–95, 502–3, 508, 510, 532–535, 537, 541, 552–53, 564, 567, 569, 578, 605–6, 618–20, 632, 633, 659
Young, Joseph A., 495, 605–606
Yuba Railroad, 336, 641